The Most Successful Way of JEE Preparation

DPP

Daily Practice Problems

JEE Main & Advanced

Ray Optics & Wave Optics

with Electromagnetic Waves

PHYSICS Vol-8

DB Singh

arihant

ARIHANT PRAKASHAN (SERIES), MEERUT

DPP
Daily Practice Problems

arihant

ARIHANT PRAKASHAN (SERIES), MEERUT

Administrative & Production Offices

Corporate Office 'Ramchhaya' 4577/15, Agarwal Road, Darya Ganj, New Delhi -110002
Tele: 011- 47630600, 43518550; Fax: 011- 23280316

Head Office Kalindi, TP Nagar, Meerut (UP) - 250002
Tel: 0121-2401479, 2512970, 4004199; Fax: 0121-2401648

Sales & Support Offices

Agra, Ahmedabad, Bengaluru, Bareilly, Chennai, Delhi, Guwahati, Hyderabad, Jaipur, Jhansi, Kolkata, Lucknow, Meerut, Nagpur & Pune

ISBN : 978-93-13193-38-8

Price : ₹260

Published by Arihant Publications (I) Ltd.

Production Team

Publishing Managers	Mahendra Singh Rawat & Keshav Mohan	*Inner Designer*	Mazher Chaudhary
Project Head	Sorabh Chaudhary	*Page Layouting*	Sunil Sharma, Hemant
Cover Designer	Shanu Mahad	*Proof Reading*	Ashish

For further information about the products from Arihant, log on to www.arihantbooks.com or email to info@arihantbooks.com

/arihantpub /@arihantpub Arihant Publications /arihantpub

ABOUT DPP
Daily Practice Problems

Indian Institutes of Technology better known as IITs have always been the dream destination of each science group students who want to pursue their career as Engineers. These are the institutes which have got very high value not even in India but in the whole world. To get the admission in such institutes one need to face entrance Examinations; JEE Main & JEE Advanced, out of which JEE Main is considered as the preliminary exam & JEE Advanced is the final exam to get into the IITs. Approximately per year 12 lakh students enroll themselves for JEE Main and only the top 2.25 lakh in JEE Main merit can attempt for JEE Advanced, the fact clearly express the degree of toughness of JEE Main & JEE Advanced.

To have success in such an exam one needs to be very planned & systematic in its preparation. A very popular wise saying 'Rome was not built in a day' suggests that to achieve a big target like IITs we need to work continuously. For continuous work, one need to have proper planning that how much work he/she has to do on regular basis, i.e., on daily basis. Each student preparing for JEE has many books but unfortunately none of them gives the students a plan for daily studies, which is the most effective success key for JEE.

Keeping the daily practice needs of the students in mind we have come up with DPP Series (Daily Practice Problems) having approximately 26 books. The primary focus of this series is on achieving success through practice & proper planning. For that we have divided the whole syllabus of JEE Main & JEE Advanced; Physics, Chemistry & Mathematics, into different volumes (around 8 volumes of Mathematics & 9 volumes each of Physics & Chemistry), each volume has Daily Practice Problems Sheets having specific questions on various Topics of the individual chapter of the volume ensuring the complete Practice & Assessment of the topic.

DPP

Daily Practice Problems

In this way, we have tried to give you an effective daily planning for JEE keeping in mind the daily hours that a student can devote to his practice, you are just required to practice one sheet of each subject on daily basis. This habit of daily practice will keep you away from exam stress and thoughts like; how much I have prepared/am I preparing properly or not/ do I need to study more on daily basis or what I am doing is ok, etc., which regularly tease your mind and can give you panic at any stage of your preparation.

Some special features that we have tried to incorporate in our Daily Practice Problems (DPP) series are discussed below

1. Micro Level coverage of all the topics of a Chapter/ Unit given in JEE Main & Advanced Syllabus.
2. Each DPP has various types of questions; Subjective Questions, Single/ Multiple Correct Options, Statements type, Comprehension based, Integer Value Answer type, Matching, etc., making you practice & ready for all the question formats of JEE.
3. New & innovative questions in each DPP challenging & improving your problem solving skills.
4. Along with Topical Coverage, Revisal DPPs for JEE Main & JEE Advanced with each Chapter, covering multidisciplinary questions proving comprehensive practice.
5. JEE Main & JEE Advanced Archive (collection of last 13 years' exams questions) with each chapter.
6. Complete Solutions for each DPP.

We hope that DPP series will prove to be very beneficial for the students preparing for JEE, by giving them daily practice in a systematic & planned manner. The suggestions/criticism from the students, teachers & parents about DPP series will be welcomed whole heartedly and we promise to update this series accordingly from time to time.

DPP
Daily Practice Problems

DAILY TASKS...

1 GEOMETRICAL OPTICS

DPP
Daily Practice Problems

2 WAVE OPTICS

1. Geometrical Optics

DPP
DAILY PRACTICE PROBLEMS

DPP-1 Plane Mirror-I

Conceptual Questions

Working Space

Direction *(Q. Nos. 1-13) These questions are conceptual in nature, need to be solved completely on notebook.*

1 Mark the correct option(s).

i. According to electrodynamic theory of light
   a. when a charged particle is accelerated energy is radiated
   b. light is a form of radiant energy of visible region (400 nm to 700 nm)
   c. when a charged particle is accelerated light is not produced
   d. None of the above

ii. Mark the correct options.
   a. White light is mixture of Violet, Indigo, Blue, Green, Yellow, Orange and Red
   b. In visible region, red has greatest wavelength
   c. In visible region, violet has least wavelength
   d. In a homogeneous medium, light propagates on straight line

iii. Mark the correct options.
   a. The interaction of light with retina is discussed in photobiology
   b. In homogeneous medium, the momentum of corpuscles remains conserved
   c. The motion of corpuscle is just like elastic ball
   d. None of the above

iv. When light incident on a plane mirror, then
   a. the momentum of corpuscles remains conserved
   b. the component of momentum of each corpuscle along the surface of the mirror remains conserved
   c. incident ray, reflected ray and normal on incidence point are coplaner
   d. the angle of incidence is equal to angle of reflection

v. Newton's corpuscular model of light,
   a. successfully explain the rectilinear propagation of light, reflection of light and refraction of light
   b. does not explain the phenomena of interference of light, diffraction of light and polarisation of light
   c. is based upon particle theory of light
   d. None of the above

vi. According to Huygen's theory of light
   a. light is stream of photon
   b. light is wave
   c. light behaves as particle
   d. None of the above

vii. Huygen's theory of light
   a. explain reflection and refraction of light
   b. explain scattering of light
   c. explain interference of light, diffraction of light
   d. None of the above

viii. The path of light is
   a. ray b. image
   c. wave d. None of these

Working Space

ix. Mark the correct option(s).
- a. Energy of photon = hc/λ, where c is speed of light, λ is wavelength and h is Planck's constant
- b. Energy of photon = $h\nu$, where ν is frequency of light
- c. Momentum of photon = $h/\lambda = h\nu/c$
- d. $c = \nu\lambda$

x. Which of the following is convergent beam?
- a. 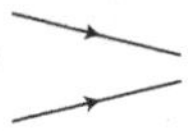
- b. 
- c. 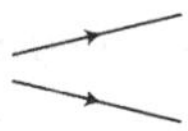
- d. None of these

xi. In the case of convergent beam,
- a. intensity increases upto converging point
- b. intensity decreases upto converging point
- c. intensity remains constant
- d. None of the above

xii. Which of the following are parallel beam?
- a. 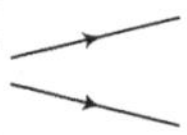
- b. 
- c. 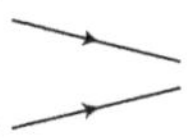
- d. None of these

xiii. In the case of parallel beam,
- a. intensity gradually increases
- b. intensity gradually decreases
- c. intensity remains constant
- d. None of these

xiv. Parallel beam is
- a. produced by a source at infinity
- b. produced by a source at finite distance
- c. produced by a virtual source
- d. None of these

xv. Sun rays on the surface of the earth
- a. form convergent beam
- b. form divergent beam
- c. form parallel beam
- d. None of these

xvi. The law of reflection is applicable for
- a. plane mirror as well as curved mirror
- b. specular reflection as well as diffused reflection
- c. paraxial rays as well as marginal rays
- d. None of the above

xvii. Due to reflection of light,
- a. frequency does not change
- b. wavelength of light changes
- c. speed of light does not change
- d. intensity of light may be decrease

xviii. If reflection takes place from optically denser medium, the change in phase is
- a. zero
- b. π
- c. $\pi/2$
- d. $\pi/3$

xix. If reflection takes place from optically rarer medium the change in phase is
- a. zero
- b. π
- c. $\pi/2$
- d. $\pi/3$

xx. If you read the words of your book, the reflection is
- a. specular reflection
- b. diffused reflection
- c. may be specular may be diffused
- d. None of these

xxi. Diffused reflection
- a. may produce image
- b. never produce image
- c. must produce image
- d. None of these

xxii. Image formation takes place by
- a. specular reflection
- b. diffused reflection
- c. Both (a) and (b)
- d. None of these

2 **Situation** A ray of light incident on a plane mirror in *XY*-plane.

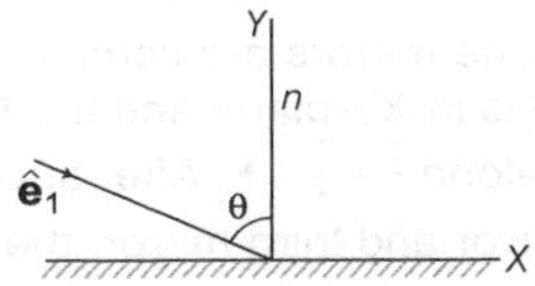

i. The unit vector along in incidence ray is $\hat{\mathbf{e}}_1$. Then, $\hat{\mathbf{e}}_1$ is

a. $\hat{\mathbf{e}}_1 = \cos\theta\,\hat{\mathbf{j}} + \sin\theta\,\hat{\mathbf{i}}$

b. $\hat{\mathbf{e}}_1 = -\sin\theta\,\hat{\mathbf{i}} + \cos\theta\,\hat{\mathbf{j}}$

c. $\hat{\mathbf{e}}_1 = \sin\theta\,\hat{\mathbf{i}} - \cos\theta\,\hat{\mathbf{j}}$

d. None of the above

ii. In previous problem, the unit vector along the normal on incidence point is

a. $\hat{\mathbf{n}} = \hat{\mathbf{j}}$ b. $\hat{\mathbf{n}} = \hat{\mathbf{k}}$ c. $\hat{\mathbf{n}} = \dfrac{\hat{\mathbf{i}} + \hat{\mathbf{j}}}{\sqrt{2}}$ d. $\hat{\mathbf{n}} = \hat{\mathbf{i}}$

iii. The value of $\hat{\mathbf{e}}_1 \cdot \hat{\mathbf{n}}$ is

a. $\sin\theta$ b. $\cos\theta$ c. $-\cos\theta$ d. $-\sin\theta$

iv. If the angle of incidence is θ. Then, the value of $\cos\theta$ is

a. $\cos\theta = \hat{\mathbf{e}}_1 \cdot \hat{\mathbf{n}}$ b. $\cos\theta = |\hat{\mathbf{e}}_1 \cdot \hat{\mathbf{n}}|$

c. $\cos\theta = -|\hat{\mathbf{e}}_1 \cdot \hat{\mathbf{n}}|$ d. None of these

v. The angle of reflection is α. Then

a. $\cos\theta = \hat{\mathbf{e}}_1 \cdot \hat{\mathbf{n}}$ b. $\cos\alpha = |\hat{\mathbf{e}}_1 \cdot \hat{\mathbf{n}}|$

c. $\theta = \alpha$ d. None of these

vi. The unit vector along reflected ray is

a. $\hat{\mathbf{e}}_2 = \cos\alpha\,\hat{\mathbf{i}} + \sin\alpha\,\hat{\mathbf{j}}$ b. $\hat{\mathbf{e}}_2 = \sin\alpha\,\hat{\mathbf{i}} + \cos\alpha\,\hat{\mathbf{j}}$

c. $\hat{\mathbf{e}}_2 = \sin\theta\,\hat{\mathbf{i}} + \cos\theta\,\hat{\mathbf{j}}$ d. $\hat{\mathbf{e}}_2 = \hat{\mathbf{e}}_1 - 2(\hat{\mathbf{e}}_1 \cdot \hat{\mathbf{n}})\hat{\mathbf{n}}$

vii. Mark the correct options.

a. The component of unit vector along incident ray and reflected ray taking along the surface of mirror are same

b. The component of unit vector of incident ray and reflected ray taking along normal on incidence point are opposite to each other

c. The component of unit vector of incident ray and reflected ray are taken along normal on incidence point are same

d. None of the above

3 **Situation** The mirror is in *XY*-plane. The incident ray is along the vector $3\hat{\mathbf{i}} - 4\hat{\mathbf{k}}$.

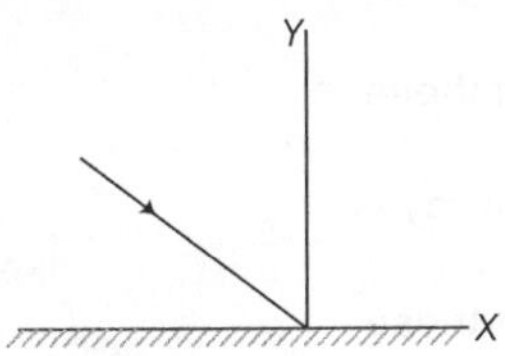

i. The unit vector along incidence ray is

a. $\hat{\mathbf{e}}_1 = \dfrac{3\hat{\mathbf{i}} - 4\hat{\mathbf{k}}}{5}$ b. $\hat{\mathbf{e}}_1 = \dfrac{3}{5}\hat{\mathbf{i}} + \dfrac{4}{5}\hat{\mathbf{k}}$ c. $\hat{\mathbf{e}}_1 = \dfrac{3}{5}\hat{\mathbf{i}}$ d. None of these

ii. The unit vector along normal on incidence point is

a. $\hat{\mathbf{n}} = \hat{\mathbf{i}}$ b. $\hat{\mathbf{n}} = \dfrac{\hat{\mathbf{i}} + \hat{\mathbf{j}}}{\sqrt{2}}$ c. $\hat{\mathbf{n}} = \hat{\mathbf{j}}$ d. $\hat{\mathbf{n}} = \hat{\mathbf{k}}$

iii. The angle of incidence is

a. 37°   b. 53°   c. 90°   d. zero

iv. The unit vector along reflected ray is

a. $\hat{\mathbf{e}}_2 = -\dfrac{3}{5}\hat{\mathbf{i}} - \dfrac{4}{5}\hat{\mathbf{k}}$ b. $\hat{\mathbf{e}}_2 = \dfrac{3}{5}\hat{\mathbf{i}} - \dfrac{4}{5}\hat{\mathbf{k}}$

c. $\hat{\mathbf{e}}_2 = \dfrac{3}{5}\hat{\mathbf{i}} + \dfrac{4}{5}\hat{\mathbf{k}}$ d. $\hat{\mathbf{e}}_2 = -\dfrac{3}{5}\hat{\mathbf{i}} - \dfrac{4}{5}\hat{\mathbf{k}}$

Working Space

4 **Situation** Three plane mirrors are combined such that, first plane mirror is *YZ*-plane, second is in *XZ*-plane and third is in *XY*-plane. A ray of light incident on first plane mirror along $\hat{i} + \hat{j} + \hat{k}$. After successive reflections through the first mirror, second mirror and third mirror, the ray finally emerges out.

i. The unit vector along incidence ray on the first mirror is

a. $\hat{e}_1 = \frac{\hat{i}+\hat{j}+\hat{k}}{\sqrt{3}}$ b. $\hat{e}_1 = \frac{\hat{i}+\hat{j}}{\sqrt{2}}$

c. $\hat{e}_1 = \frac{\hat{i}+\hat{k}}{\sqrt{2}}$ d. None of these

ii. The unit vector along normal on the first plane mirror is

a. $-\hat{i}$ b. $\hat{i}$

c. $\hat{k}$ d. $\hat{j}$

iii. The unit vector along reflected ray on the first plane mirror is

a. $\hat{e}_2 = \frac{\hat{i}+\hat{j}+\hat{k}}{\sqrt{3}}$ b. $\hat{e}_2 = \frac{-\hat{i}+\hat{j}+\hat{k}}{\sqrt{3}}$

c. $\hat{e}_2 = \frac{\hat{i}+\hat{j}-\hat{k}}{\sqrt{3}}$ d. $\hat{e}_2 = \frac{-\hat{i}-\hat{j}-\hat{k}}{\sqrt{3}}$

iv. The unit vector along incident ray on second plane mirror is

a. $\frac{\hat{i}+\hat{j}+\hat{k}}{\sqrt{3}}$ b. $\frac{-\hat{i}+\hat{j}+\hat{k}}{\sqrt{3}}$

c. $\frac{\hat{i}+\hat{j}-\hat{k}}{\sqrt{3}}$ d. None of these

v. The unit vector along reflected ray from second mirror.

a. $\frac{\hat{i}+\hat{j}+\hat{k}}{\sqrt{3}}$ b. $\frac{-\hat{i}-\hat{j}+\hat{k}}{\sqrt{3}}$

c. $\frac{-\hat{i}-\hat{j}-\hat{k}}{\sqrt{3}}$ d. None of these

vi. The unit vector along reflected ray from third mirror.

a. $\frac{\hat{i}+\hat{j}+\hat{k}}{\sqrt{3}}$ b. $\frac{-\hat{i}-\hat{j}+\hat{k}}{\sqrt{3}}$

c. $\frac{\hat{i}-\hat{j}-\hat{k}}{\sqrt{3}}$ d. None of these

vii. The angle between incident ray and final emergent ray is

a. 0° b. 90°

c. 180° d. None of these

viii. A ray of light incident on a mirror making 30° angle with the mirror. Find the deviation in the ray.

a. 30° b. 60°

c. 120° d. None of these

5 **Situation** Two plane mirrors are arranged to such as a ray of light incident on the first mirror is reflected towards second mirror and after being reflected from second mirror, the ray emerges parallel to incoming ray.

i. The deviation in the ray is

a. 0° b. 180°

c. 90° d. None of these

ii. The angle between mirrors is

a. 0° b. 60°

c. 90° d. None of these

Working Space

6 **Situation** A ray of light incident on first plane mirror parallel to second plane mirror. After successive reflections through mirrors, finally ray becomes parallel to first mirror.

i. Draw the ray diagram,

a. 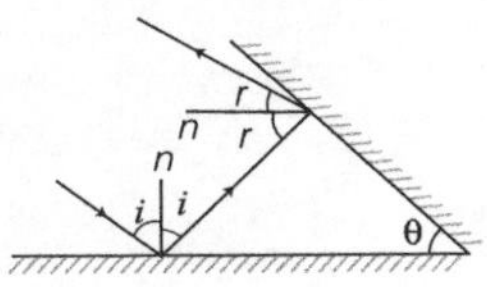

b. 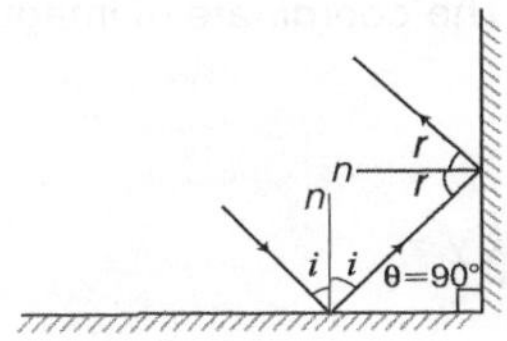

c. 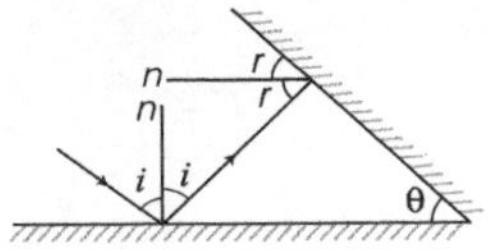

d. None of these

ii. In previous problem,

a. $2i = 90° - r$ b. $2r = 90° - i$ c. $i + r = 0$ d. None of these

iii. The angle between mirrors is

a. $\theta = 30°$ b. $\theta = 90°$ c. $\theta = 60°$ d. None of these

7 **Situation** Consider the figure given below, and mark the correct option(s) from the given questions.

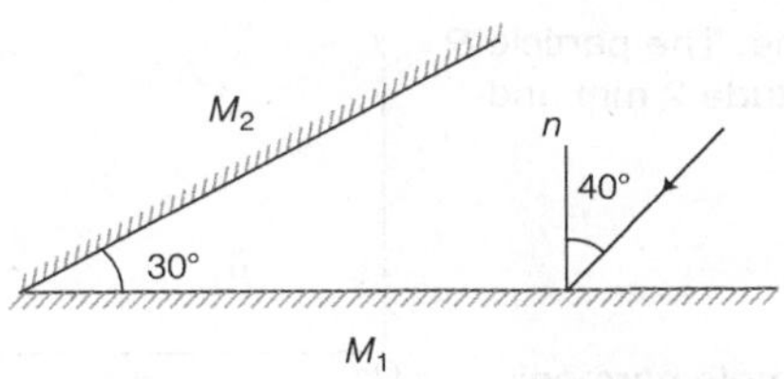

i. Draw ray diagram for three successive reflections.

a. 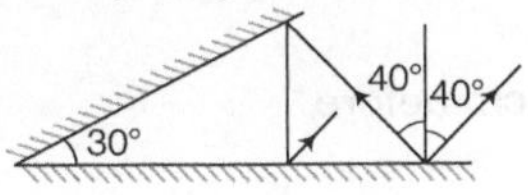

b. 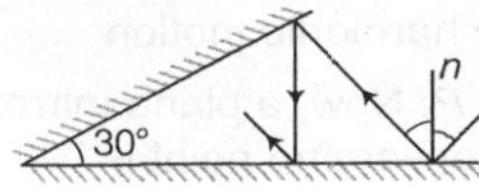

c. 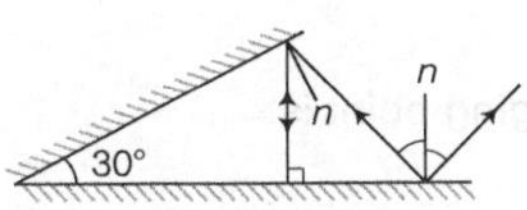

d. None of these

ii. The deviation produced by the first plane mirror during the first reflection is

a. $\delta_1 = 0°$
b. $\delta_1 = 100°$ clockwise
c. $\delta_1 = 100°$ anticlockwise
d. $\delta_2 = 280°$ clockwise

iii. The angle of incidence for second plane mirror is

a. 20° b. 10° c. 30° d. 40°

iv. The deviation produced by second plane mirror is

a. zero
b. 160° clockwise
c. 160° anticlockwise
d. 140° anticlockwise

v. The angle of incidence for third reflection is

a. 10°
b. 20°
c. 30°
d. None of these

vi. The deviation in third reflection is

a. 200° anticlockwise
b. 140° anticlockwise
c. 100° clockwise
d. None of these

vii. The net deviation in ray in three successive reflection is

a. 200° anticlockwise
b. 200° clockwise
c. 160° clockwise
d. None of these

Working Space

viii. The time in a clock is 8 : 40 : 10. The time in the image of the clock formed by plane mirror is

a. 8 : 40 : 10 b. 8 : 50 : 30

c. 3 : 19 : 50 d. None of these

ix. The plane mirror is in YX-plane. The coordinate of image of point P is

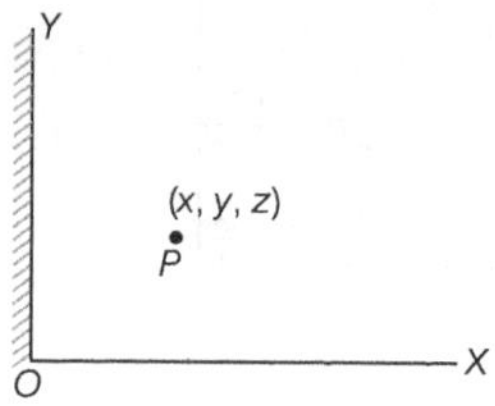

a. $(x, -y, -z)$ b. $(-x, y, -z)$

c. $(-x, y, z)$ d. $(-x, -y, -z)$

x. In previous problem, the equation of a straight line is $y = 2x$. The slope of image of the straight line is

a. +2 b. −2 c. +1 d. −1

xi. In previous problem, if a circle is drawn in XY-plane, taking origin as centre the shape of image of the circle is

a. ellipse b. circle in XY-plane

c. semicircle in XY-plane d. None of these

xii. Assume that, the plane mirror is in YZ-plane. The particle P executes simple harmonic motion of amplitude 2 mm and time period 4s.

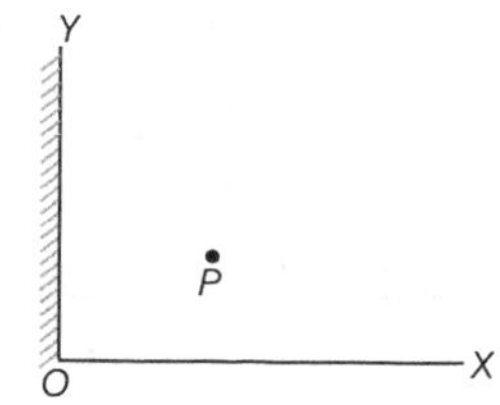

Mark the correct options.

a. The image of particle P is virtual

b. The image of particle P is real

c. The image of the particle P executes simple harmonic motion with amplitude 2 mm and time period 4s

d. Image does not execute simple harmonic motion

xiii. A beam of light converges at point P. Now, a plane mirror is placed 20 cm before the converging point P. The new converging point is

a. 20 cm infront of the mirror

b. 20 cm behind the mirror

c. The separation between new converging point and previous converging point is 40 cm

d. None of the above

xiv. Two plane mirrors are placed as shown in the figure.

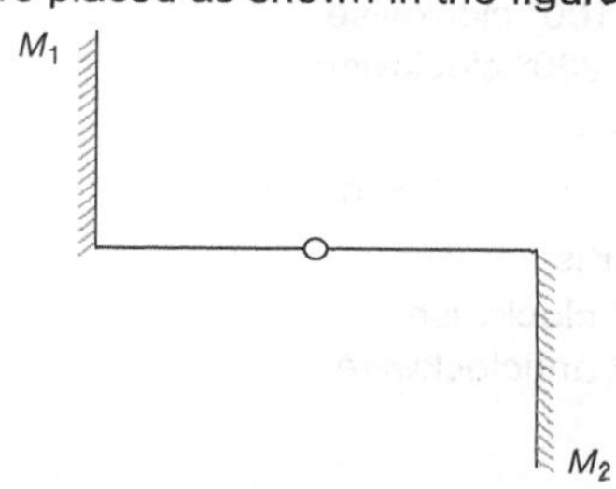

Mark the correct options.

a. A ray incidence from O on mirror M_1 does not go on mirror M_2 after reflection

b. A ray incidence from O on mirror M_2 may go on mirror M_1 after reflection

c. Maximum number of image of point O is 2

d. None of the above

8 Mark the correct option(s).

i. Generally, for calculating number of images of an object by combination of two plane mirrors, we apply the formula $n = 360°/\theta$, where θ is angle between mirrors, then

a. if n is even number, number of image is $n - 1$

b. if n is odd number and object is placed symmetrically, then the number of images is $n - 1$

c. if n is odd number and object is placed symmetrically, then the number of images is n

d. if the plane mirrors are placed parallel to each other the number of images is infinite

ii. A light bulb is placed between two plane mirrors inclined at and angle of 60°. The number of images formed are

a. 6 b. 2 c. 5 d. 4

iii. A thick plane mirror shows a number of images of the filament of an electric bulb. Then, the brightest image is

a. first b. second c. fourth d. last

iv. A person is in a room whose ceiling and two adjacent walls are mirrors. How many images are formed?

a. 5 b. 6 c. 7 d. 8

v. If a point object P is approaching the mirror with velocity $v\hat{i}$. The velocity of image is

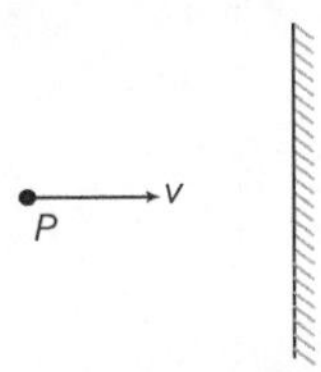

a. $v\hat{i}$

b. $-v\hat{i}$

c. $v\hat{j}$

d. $-v\hat{j}$

vi. The velocity of object is $v\,\hat{j}$. The velocity of image is

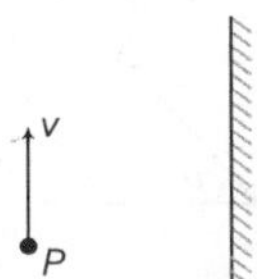

a. $v\hat{j}$

b. $-v\hat{j}$

c. $v\hat{i}$

d. None of the above

vii. The velocity of image is

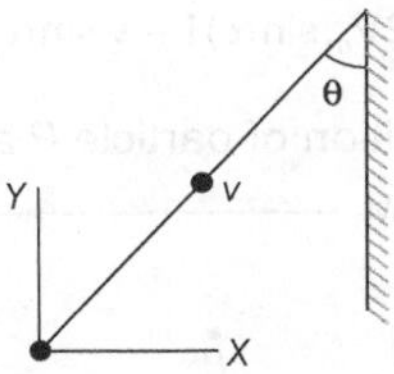

a. $v\sin\theta\hat{i} + v\cos\theta\,\hat{j}$

b. $-v\sin\theta\hat{i} + v\cos\theta\,\hat{j}$

c. $-v\sin\theta\hat{i} - v\cos\theta\,\hat{j}$

d. None of the above

viii. The velocity of mirror is $-v_m\,\hat{i}$.

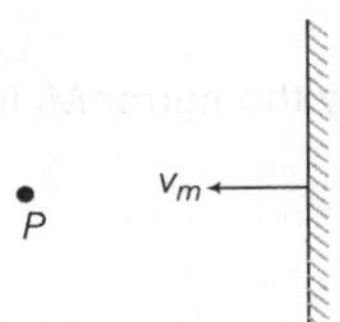

a. $v_m\hat{i}$ b. $2v_m\hat{i}$ c. $-2v_m\hat{i}$ d. None of these

ix. The velocity of mirror is $v_m \hat{\mathbf{j}}$ as shown in the figure. The velocity of image of stationary object P is

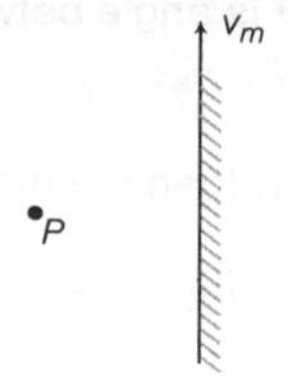

a. $v_m \hat{\mathbf{j}}$ b. $v_m \hat{\mathbf{i}}$ c. $2v_m \hat{\mathbf{i}}$ d. zero

x. The velocity of image is

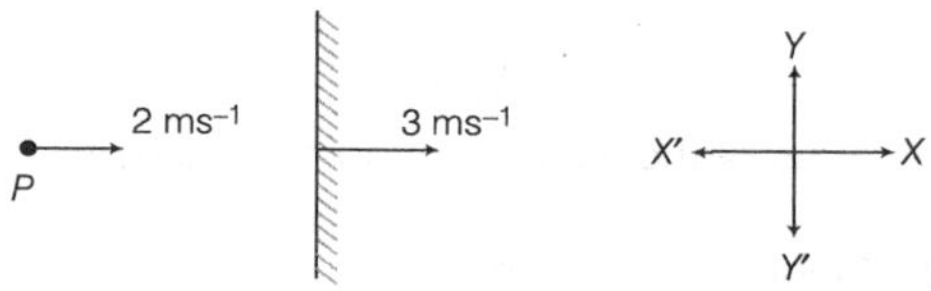

a. $-2\hat{\mathbf{i}}$ b. $4\hat{\mathbf{i}}$
c. $-4\hat{\mathbf{i}}$ d. None of these

xi. The velocity of image of stationary object P is

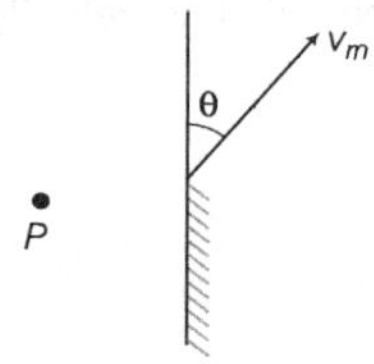

a. $2v_m \cos\theta$ b. $2v_m \sin\theta$
c. $2v_m \tan\theta$ d. $2v_m$

xii. The velocity of image of point object P is

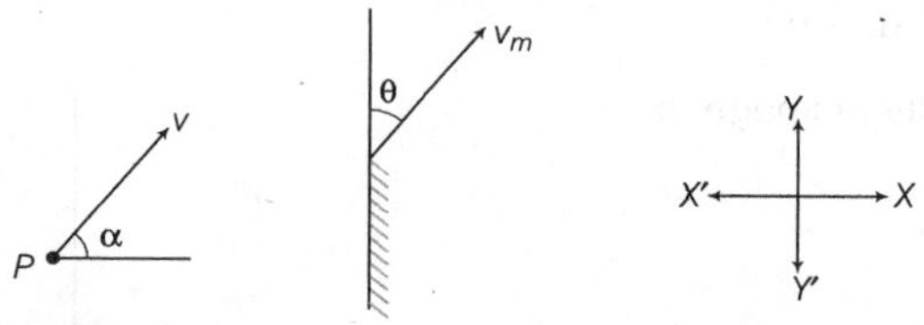

a. $(2v_m \sin\theta - v\cos\alpha)\hat{\mathbf{i}}$ b. $(2v_m \sin\theta - v\cos\alpha)\hat{\mathbf{i}} + v\sin\alpha\,\hat{\mathbf{j}}$
c. $(v_m \cos\theta - 2v_m \sin\alpha)\hat{\mathbf{i}} - v\sin\alpha\,\hat{\mathbf{j}}$ d. None of these

9 **Situation** The position of particle P at $t = 0$ s as shown in the figure.

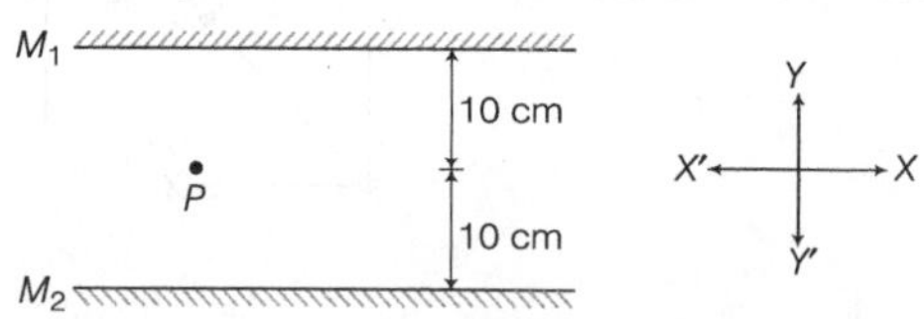

i. The distance of first image of point P at $t = 0$ formed by the mirror M_1 from the point P is

a. 10 cm b. 20 cm
c. 30 cm d. 40 cm

ii. The distance of first image of point P at $t = 0$ formed by the mirror M_2 from the point P is

a. 10 cm b. 20 cm
c. 30 cm d. 40 cm

Working Space

iii. The distance of second image of point object P formed by the plane mirror. M_1 from the point object P is

(Assume that, the point P is moving with velocity $\mathbf{v} = 3\,\text{cms}^{-1}\,\hat{i} + 4\,\text{cms}^{-1}\,\hat{j}$)

a. 20 cm b. 40 cm c. 60 cm d. 80 cm

iv. The velocity of first image of point P formed by mirror M_1 is

a. $3\hat{i} + 4\hat{j}$ b. $-3\hat{i} + 4\hat{j}$ c. $3\hat{i} - 4\hat{j}$ d. $-3\hat{i} - 4\hat{j}$

v. The velocity of first image formed by mirror M_2 is

a. $3\hat{i} + 4\hat{j}$ b. $-3\hat{i} + 4\hat{j}$ c. $3\hat{i} - 4\hat{j}$ d. $-3\hat{i} - 4\hat{j}$

vi. The velocity of second image formed by mirror M_1 is

a. $3\hat{i} + 4\hat{j}$ b. $-3\hat{i} + 4\hat{j}$ c. $3\hat{i} - 4\hat{j}$ d. $-3\hat{i} - 4\hat{j}$

10 **Situation** Two plane mirrors are combined at 90°. A point P is moving with velocity $3\hat{i} + 4\hat{j}$.

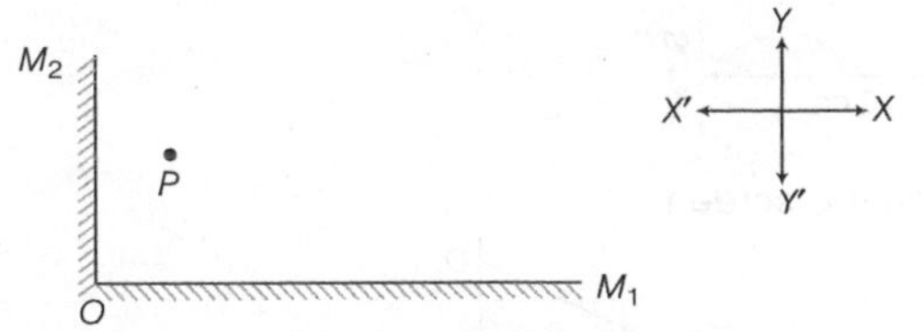

i. The number of images formed by combination is

a. 1 b. 2 c. 3 d. 4

ii. The images formed

a. are always lying on circle, whose centre is at O
b. are lying on a straight line
c. are lying on an ellipse
d. None of the above

iii. The velocity of image formed by only reflection through mirror M_1 is

a. $3\hat{i} + 4\hat{j}$ b. $-3\hat{i} - 4\hat{j}$ c. $-3\hat{i} + 4\hat{j}$ d. $3\hat{i} - 4\hat{j}$

iv. The velocity of image formed by only reflection through mirror M_2 is

a. $3\hat{i} + 4\hat{j}$ b. $-3\hat{i} - 4\hat{j}$ c. $3\hat{i} - 4\hat{j}$ d. $-3\hat{i} + 4\hat{j}$

v. The velocity of image formed by successive reflections through mirror is

a. $3\hat{i} + 4\hat{j}$ b. $-3\hat{i} - 4\hat{j}$ c. $3\hat{i} - 4\hat{j}$ d. $-3\hat{i} + 4\hat{j}$

11 **Situation** Consider the figure given below and mark the correct option(s) from the given questions.

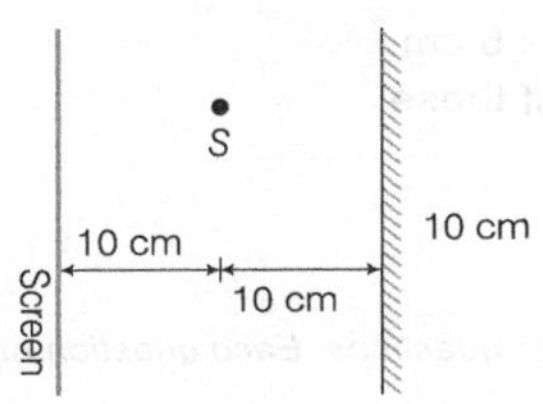

i. Show the image of point source S by taking two rays from source incidence at the ends of mirror.

a. 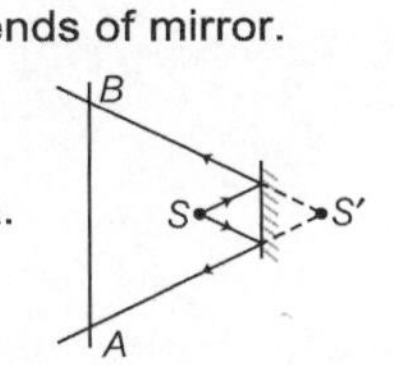

b. 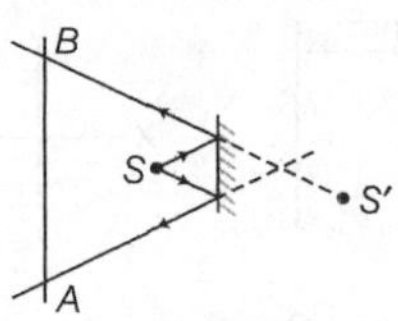

c. Both (a) and (b) d. None of these

ii. Mark the correct options.
   a. If observer is situated at any points between A and B,he can see the image of point source S
   b. If the point source is assumed as observer,he can see the image
   c. If the point source S is assumed as observer, he can not see the image
   d. None of the above

iii. The length of field of view of image of the point source S on the screen is
   a. 10 cm b. 20 cm
   c. 30 cm d. None of these

12 **Situation** Consider the figure given below and mark the correct option(s) for the given questions.

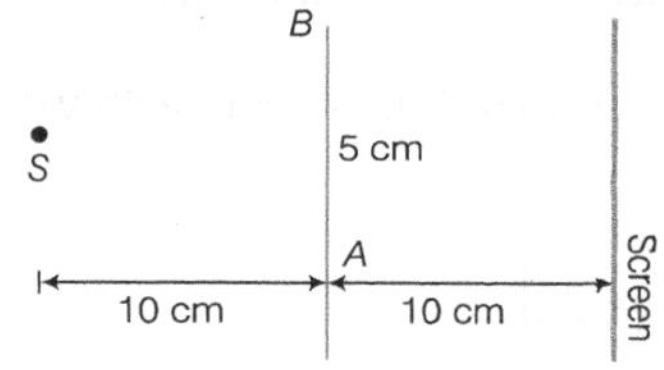

i. Show the shadow of linear object AB on the screen

a. 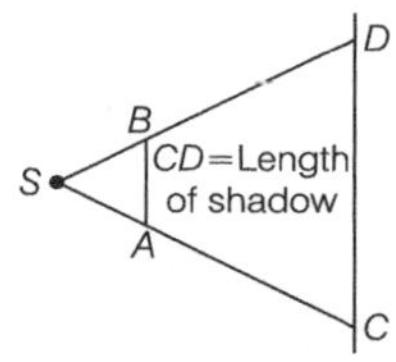

b. 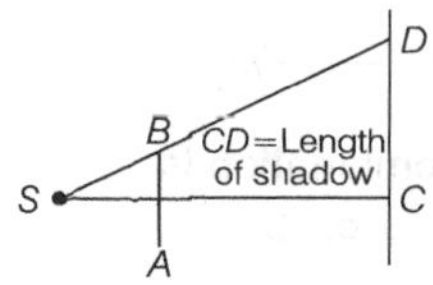

c. no shadow is found on the screen d. None of these

ii. The length of shadow on the screen is
   a. 5 cm b. 10 cm c. 15 cm d. 20 cm

13 **Situation** The size of the face of a dancer is 24 cm × 16 cm.

i. If dancer has only one eye. The minimum size of mirror is
   a. 24 cm × 16 cm b. 12 cm × 8 cm
   c. 12 cm × 6 cm d. None of these

ii. If dancer has two eyes. The distance between eyes is 4 cm. She has desire to see left ear by left eye and right ear by right eye. The size of mirror is
   a. 12 cm × 8 cm b. 12 cm × 6 cm
   c. 12 cm × 10 cm d. None of these

iii. In previous problem, minimum width of mirror is
   a. 12 cm × 8 cm b. 12 cm × 6 cm
   c. 12 cm × 10 cm d. None of these

Only One Option Correct Type

Direction (Q. Nos. 14-37) *This section contains 24 multiple choice questions. Each question has four choices (a), (b), (c) and (d), out of which ONLY ONE is correct.*

14 The velocity of image of P_1 seen by observer P_2 is

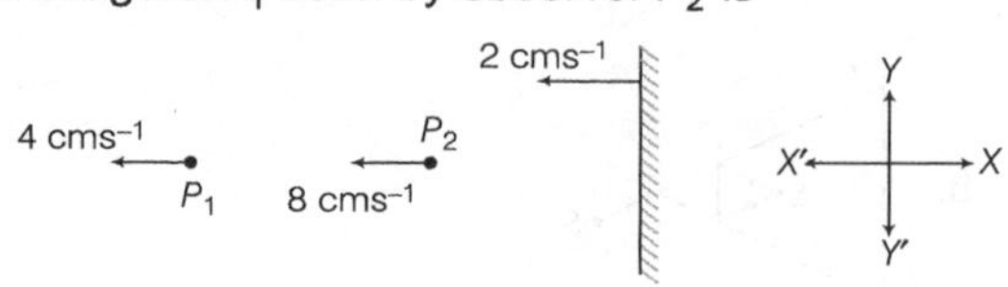

a. $-(8\text{cms}^{-1})\hat{i}$ b. $(8\text{cms}^{-1})\hat{i}$ c. 0 d. $(10\text{cms}^{-1})\hat{i}$

Working Space

15 Find the velocity of image of point object P formed by the plane mirror.

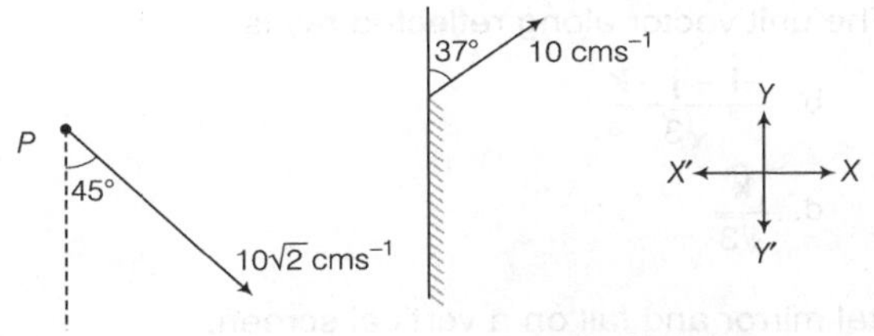

a. $(2\hat{i} - 10\hat{j})\text{cms}^{-1}$ b. $(2\text{cms}^{-1})\hat{i}$

c. $(-4\hat{i} - 10\hat{j})\text{cms}^{-1}$ d. $-(4\text{cms}^{-1})\hat{i}$

16 A reflective surface is given by $y = 2\sin x$. The reflective surface is facing towards positive Y-axis. Find the least values of coordinate of the point, when a ray parallel to positive X-axis becomes parallel to positive Y-axis after reflection?

a. $\left(\sqrt{3}, \frac{\pi}{3}\right)$ b. $\left(\frac{\pi}{3}, \sqrt{3}\right)$

c. $\left(\frac{\pi}{6}, 1\right)$ d. $\left(\frac{\pi}{6}, \sqrt{3}\right)$

17 A plane mirror is placed along X-axis facing towards Y-axis. The one end of mirror is at the origin. The equation of image of an object $y = a\sin\omega x$ is

a. $y = \dfrac{a\sin\omega x + a|\sin\omega x|}{2}$

b. $y = a|\sin\omega x|$

c. $y = -\dfrac{(a\sin\omega x + a|\sin\omega x|}{2}$

d. $y = \dfrac{a\sin\omega - a|\sin\omega x|}{2}$

18 A 50 cm high pole is placed on horizontal surface. Sun rays are incident at an angle 37° with vertical. The length of shadow on the horizontal surface is

a. 37.5 cm b. 66.67 cm

c. 30 cm d. 40 cm

19 A square plane mirror of side length 50 cm is suspended parallel to a vertical wall of a room with its lower edge of 50 cm above the ground. An observer stands infront of the mirror at a distance 200 cm away from the mirror. If his eyes are at a height of 300 cm above the ground, find the length of floor visible to him reflected by the mirror.

a. 10 cm b. 20 cm

c. 40 cm d. 60 cm

20 A plane mirror which rotates 10^4 times per minute reflects light onto a stationary mirror 50 m away. This mirror reflects the light normally, so that it strikes the rotating mirror again. The image observed in the rotating mirror is shifted through θ from the position it occupies when the rotating mirror is stationary. If speed of light is $3\times10^8\ \text{ms}^{-1}$, find the value of θ (in minute).

a. 1.2 b. 0.6

c. 2.4 d. 3.6

21 One end of a plane mirror of length 50 cm is fixed. The other end rests on the top of a vertical rod. A beam of light is reflected from the mirror and forms a spot on a screen 4 m away from the mirror. If the top of the rod is moved upward 0.1 mm. Find the distance through, which spot on screen moved?

a. 1 mm b. 0.8 mm

c. 0.4 mm d. 1.6 mm

22 A ray of light incident on a plane mirror along a vector $\hat{\mathbf{i}} + \hat{\mathbf{j}} - \hat{\mathbf{k}}$. The normal on incidence point is along vector $\hat{\mathbf{i}} + \hat{\mathbf{j}}$. The unit vector along reflected ray is

a. $-\hat{\mathbf{i}} - \hat{\mathbf{j}} - \hat{\mathbf{k}}$
b. $\dfrac{-\hat{\mathbf{i}} - \hat{\mathbf{j}} - \hat{\mathbf{k}}}{\sqrt{3}}$
c. $\dfrac{-\hat{\mathbf{i}} - \hat{\mathbf{j}}}{\sqrt{2}}$
d. $\dfrac{\hat{\mathbf{k}}}{\sqrt{3}}$

23 Sun rays are reflected from a horizontal mirror and fall on a vertical screen.

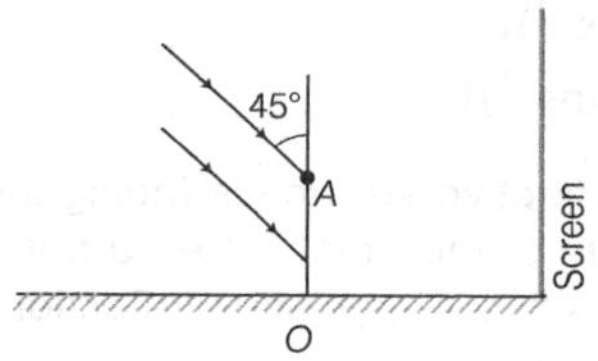

An object *OA* of length 2 m is placed on the mirror as shown in the figure. The distance of screen from object is 4 m. Sun rays make 45° angle with vertical. Find the length of shadow on screen, if only reflected light incident on the screen.

a. 1 m
b. 2 m
c. 3 m
d. 4 m

24 The image of *AB* by the plane mirror is

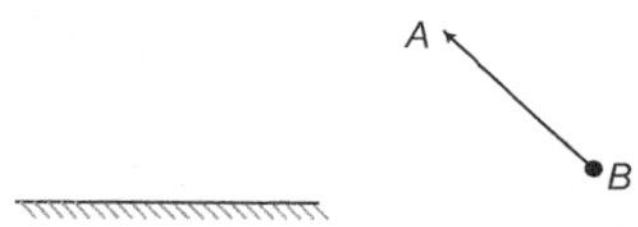

a. 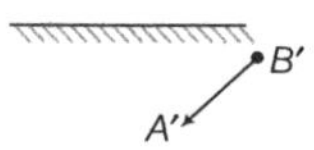

b. 

c. 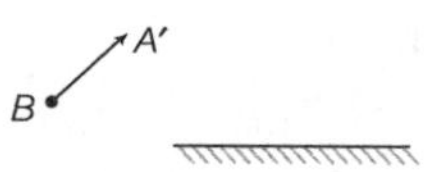

d. 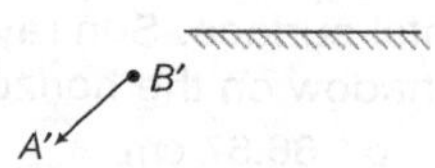

25 A small plane mirror starts from the rest and rotates with a constant angular acceleration. It is found that a light spot is moving on a spherical screen of radius 10 m, whose centre is at the mirror with a speed of 80π ms^{-1} at t = 1s. Find the angular acceleration of the mirror.

a. 4π rads^{-1}
b. 8π rads^{-1}
c. 2π rads^{-1}
d. 80 rads^{-1}

26 A circular plane mirror of area 1m^2 is hung on front wall. A point source of light is at middle of the room, so that centre of mirror and the point source are on a horizontal straight line. Find the area of reflected patch on the opposite wall.

a. 1 m^2
b. 2 m^2
c. 3 m^2
d. 9 m^2

27 On a plane mirror a ray of light is incident at an angle of 60°. To make the reflected ray vertical, at what angle with vertical must a plane mirror be placed?

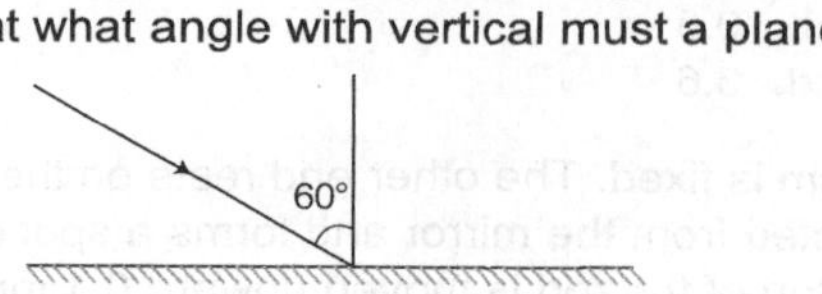

a. 30°
b. 45°
c. 60°
d. 150°

Working Space

28 In the given figure, find the angle between reflected rays.

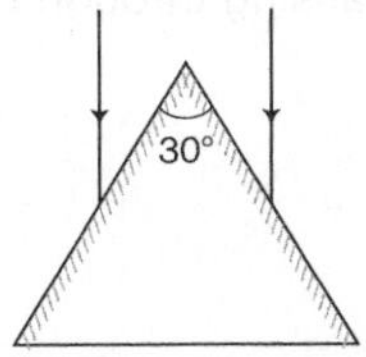

a. 15° b. 30°
c. 45° d. 60°

29 A point object is moving with a speed 2 cms^{-1} infront of a combination of two plane mirrors as shown in the figure. The magnitude of velocity of image in mirror M_1 with respect to image formed by mirror M_2.

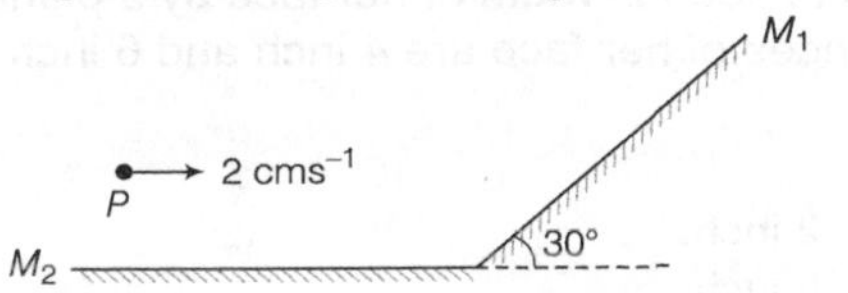

a. 2 cms^{-1} b. $2\sqrt{3}$ cms^{-1}
c. 4 cms^{-1} d. 8 cms^{-1}

30 Four plane mirrors are arranged as shown in the figure. The velocity of image observed by observer is

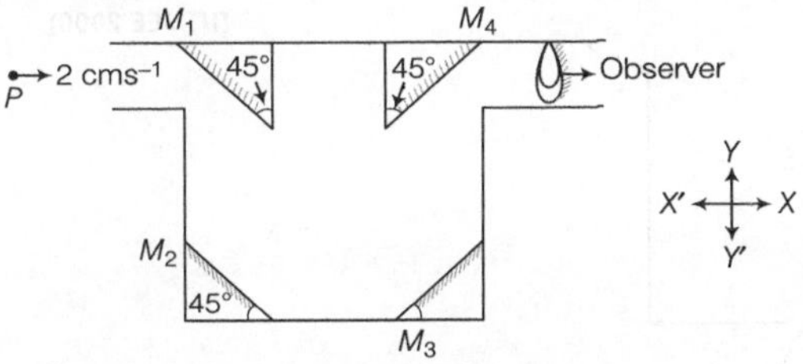

a. $(1\text{ cms}^{-1})\hat{\mathbf{i}}$ b. $(2\text{cms}^{-1})\hat{\mathbf{i}}$
c. $\left(\frac{1}{\sqrt{2}}\text{cms}^{-1}\right)\hat{\mathbf{i}} - \left(\frac{1}{\sqrt{2}}\text{ cms}^{-1}\right)\hat{\mathbf{j}}$ d. $(1\text{ cms}^{-1})\hat{\mathbf{i}} + (1\text{cms}^{-1})\hat{\mathbf{j}}$

31 A point source has been placed as shown in the figure. What is the length on the screen that will receive reflected light from the plane mirror?

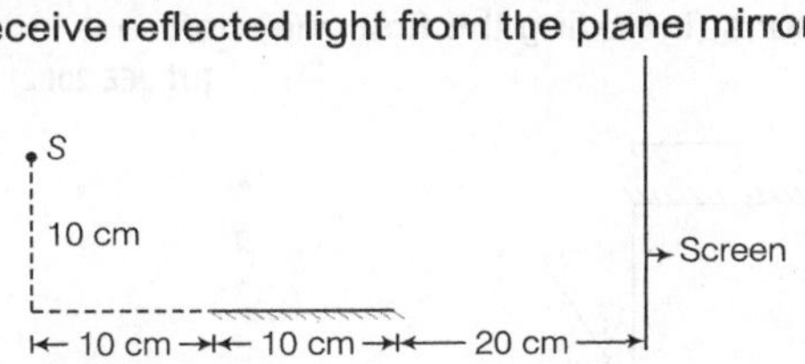

a. 10 cm b. 20 cm c. 30 cm d. 40 cm

32 After successive reflections from plane mirrors M_1 and M_2, the ray is passing through the point C. Find the value of θ.

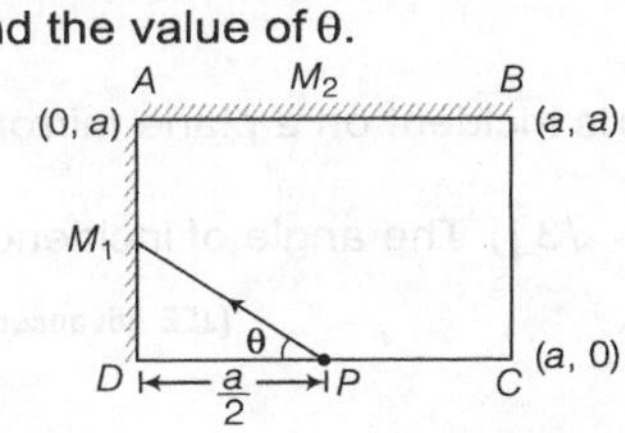

a. 30° b. 37° c. 45° d. 53°

Working Space

33 A point object P is situated in front of a plane mirror as shown in the figure. The width of mirror is 10 cm. The field of view on the line passing through the point P and parallel to mirror is

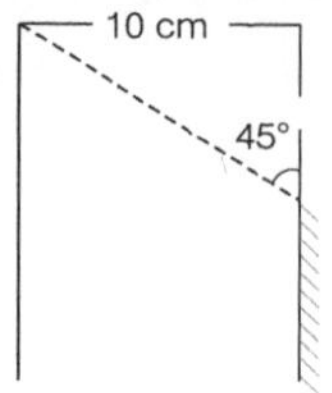

a. 5 cm
b. 10 cm
c. 15 cm
d. 20 cm

34 A beautiful girl with two normal eyes wants to see full width of her face by a plane mirror. The eye to eye and ear to ear distances of her face are 4 inch and 6 inch respectively.

The minimum width of mirror is

a. 1 inch
b. 2 inch
c. 3 inch
d. 5 inch

35 A point source of light S, placed at a distance L infront of the centre of a plane mirror of width d, hangs vertically on a wall. A man walks infront of the mirror along a line parallel to the mirror at a distance $2L$, from it as shown in the figure. The greatest distance over, which he can see the image of the light source in the mirror is **[IIT JEE 2000]**

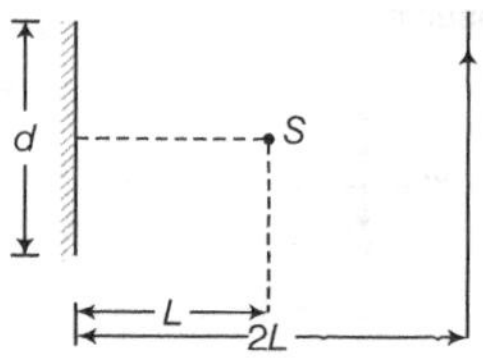

a. $d/2$
b. d
c. $2d$
d. $3d$

36 Two plane mirrors A and B are aligned parallel to each other as shown in the figure. A light ray is incident at an angle 30° at a point just inside one end of A. The plane of incidence coincides with the plane of the figure. The maximum number of times the ray undergoes reflections (including the first one) before it emerges out is **[IIT JEE 2002]**

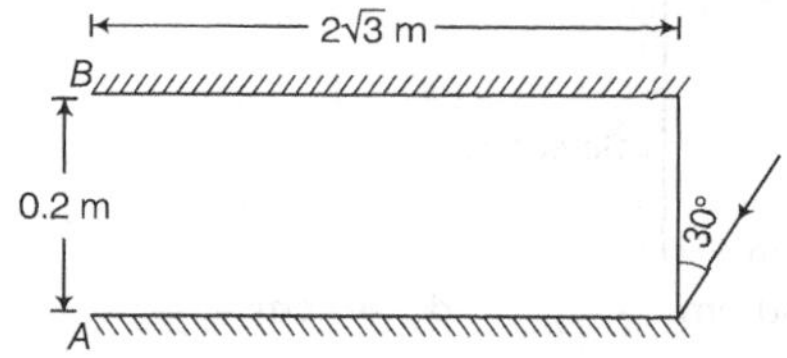

a. 28
b. 30
c. 32
d. 34

37 A ray of light travelling in the direction $\frac{1}{2}(\hat{i} + \sqrt{3}\,\hat{j})$ is incident on a plane mirror. After reflection, it travels along the direction $\frac{1}{2}(\hat{i} - \sqrt{3}\,\hat{j})$. The angle of incidence is **[JEE Advanced 2013]**

a. 30°
b. 45°
c. 60°
d. 75°

Working Space

One or More than One Options Correct Type

Direction *(Q. Nos. 38-41) This section contains 4 multiple choice questions. Each question has four choices (a), (b), (c) and (d), out of which ONE or MORE THAN ONE are correct.*

38 A plane mirror of width L is arranged parallel to wall at distance l from it. The mirror is moving towards wall with a speed v_o. A point source S is situated on the wall as shown in the figure.

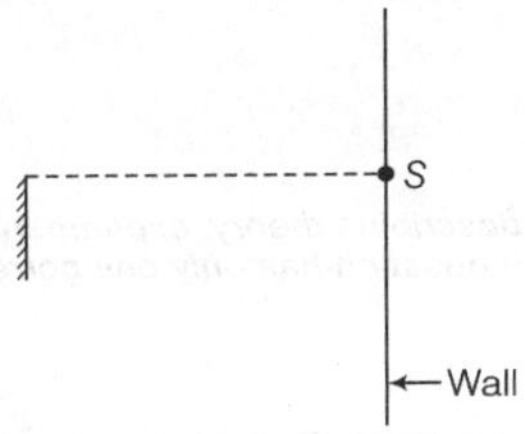

a. The size of spot on the wall is $2L$
b. The size of spot on the wall depends on the position of mirror from wall
c. The spot remains stationary
d. The spot moves on wall with speed v

39 An observer observed light repeatedly reflected from moving plane mirrors as shown in the figure. The mirrors were placed on discs revolving in same direction. The angular speeds of each disc is ω.

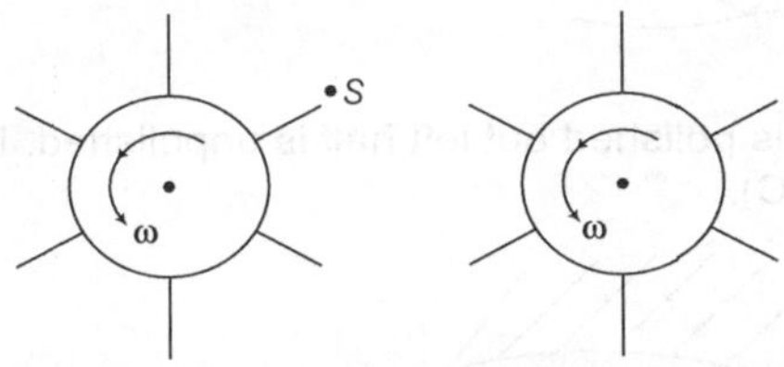

The angular speed of rotation of a beam that is consecutively reflected n times from the mirrors is

a. zero, if n is even
b. 2ω, if n is even
c. zero, if n is odd
d. 2ω, if n is odd

40 In the given figure, two plane mirrors are parallel to each other. They are moving in same direction. The speed of each mirror is v_o. Find the speed of nth image of source S formed by mirror M_1.

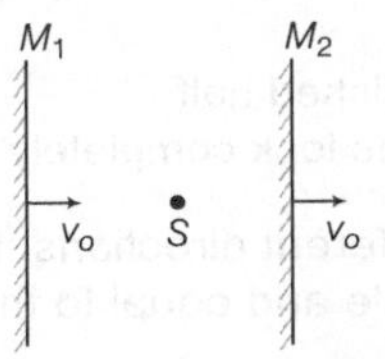

a. zero, if n is even
b. zero, if n is odd
c. $2v_o$, if n is odd
d. $2v_o$, if n is even

41 Figure shows time varry angle $\theta = 2 + 6t^2$. The mirror is rotating about horizontal axis in clockwise direction.

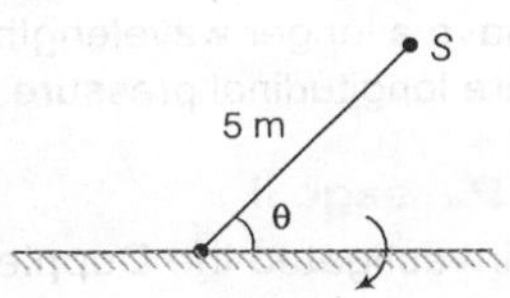

a. The locus of image is circle
b. The angular velocity of image at $t = 1$s is 24 rads^{-1}
c. The speed of image at $t = 1$ s is 120 ms^{-1}
d. The speed of image at $t = 1$s is 60 ms^{-1}

DPP-2 Plane Mirror-II

Comprehension Type

Working Space

Direction (Q. Nos. 1-5) This section contains 2 paragraphs, each describing theory, experiments, data, etc. Five questions related to the paragraphs have been given. Each question has only one correct answer among the four given options (a), (b), (c) and (d).

Passage I

At the microscopic level, the law of reflection holds correct. But at a macroscopic level, light bounces off some surfaces at all different angles. This is known as diffused reflection. It happening because all but the smoothest surfaces contains microscopic nooks and crevices as shown in Fig. (A).

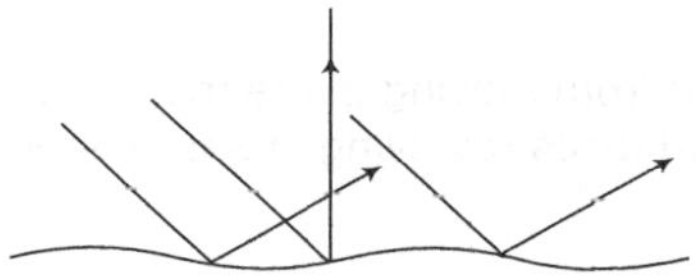

Fig. (A)

Consider a metal coin in Fig. (B), right part is polished but left half is unpolished. It is illuminated by sunlight as shown in Fig. (C).

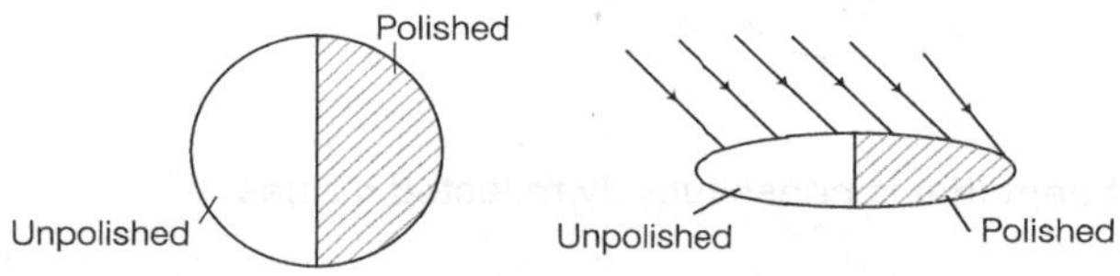

Fig. (B) Fig. (C)

1 According to Dhanad, which half of the coin is brighter?
a. Polished half
b. Unpolished half
c. Both (a) and (b)
d. None of these

2 According to Rishu, which half coin is brighter?
a. The polished half
b. The unpolished half
c. both halves look equally bright
d. both halves look completely dark

3 Light striking a painted concrete wall reflects in all different directions. But a beam of sound hitting the same wall reflects at a single angle and equal to the angle of incidence. Why?
a. Diffuse reflection happens only to light waves, not to other kinds of wave
b. Sound waves are slower than light waves
c. Sound waves have a longer wavelength than light waves
d. Sound waves are longitudinal pressure waves, not transverse wave

Passage II

When a scientist investigated the Doppler optical phenomenon he observed light repeatedly reflected from moving plane mirrors as shown in the figure. The mirrors were fixed on discs revolving in different directions.

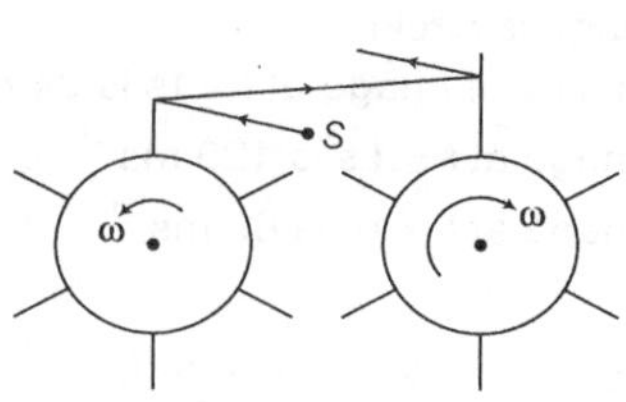

Working Space

4 If the angular speed of rotation of the discs are 1 $rads^{-1}$, find the angular speed of rotation of a beam that it consecutively reflected five times from the mirrors.

a. 1 $rads^{-1}$ b. 2 $rads^{-1}$
c. 10 $rads^{-1}$ d. 18 $rads^{-1}$

5 Determine the linear speed of nth image at the moment, when the mirrors are parallel to each other and their reflecting portions move with a speed v_o in different directions?

a. nv_o b. $2nv_o$
c. $(2n-1)2v_o$ d. v_o

Statement Type

Direction *(Q. Nos. 6-10) This section is based on Statement I and Statement II. Select the correct answer from the codes given below.*

a. Both Statement I and Statement II are correct and Statement II is the correct explanation of Statement I
b. Both Statement I and Statement II are correct and Statement II is not the correct explanation of Statement I
c. Statement I is correct but Statement II is incorrect
d. Statement II is correct but Statement I is incorrect

6 **Statement I** Two plane mirrors are combined at an angle $\frac{\pi}{3}$. A ray of light incident on the mirror as shown in the figure. The deviation in ray during two successive reflection does not when the mirror is rotated through small angle θ about point O in anticlockwise direction.

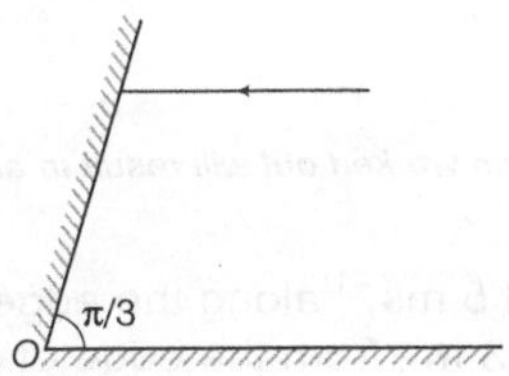

Statement II When a plane mirror is rotated through angle θ about an axis in the plane of mirror, the reflected ray rotated through angle 2θ.

7 **Statement I** The colours of the surface of bodies that can be impregnated with water appear richer after moistening the surface.

Statement II Due to moistening surface irregularities are covered with water films, which reduces diffused reflection of light.

8 **Statement I** When a plane mirror is rotated through the angle θ about an axis perpendicular to plane of mirror, reflected ray does not rotate.

Statement II The law of reflection is applicable for regular reflection as well as irregular reflection.

9 **Statement I** When a ray of light strike with a plane of mirror, the momentum of photon along the surface of mirror remains conserved.

Statement II Angle of incidence is equal to angle of reflection.

10 **Statement I** The shadow will be equally distinct everywhere only, if a point source of light is used.

Statement II The shadow of man's legs on the ground sharp and that of his head diffused.

Working Space

Matching List Type

Direction *(Q. No. 11) Choices for the correct combination of elements from Column I and Column II are given as options (a), (b), (c) and (d), out of which one is correct.*

11 In the given figure, four particles are moving infront of mirrors. Match the Column I with Column II and mark the correct option from the codes given below.

	Column I		Column II
i.	The velocity of particle A is $\mathbf{v}_A = \hat{\mathbf{i}} + (2\hat{\mathbf{i}} + \hat{\mathbf{j}})\ \text{ms}^{-1}$, and mirror is stationary. The speed of image as $t = 1$ s.	p.	$\sqrt{42}\ \text{ms}^{-1}$
ii.	The velocity of particle B is $\mathbf{v}_B = (-\hat{\mathbf{i}} + 3\hat{\mathbf{j}})\ \text{ms}^{-1}$, and mirror is moving with a velocity $\mathbf{v}_m = (3\hat{\mathbf{i}} + 4t\ \hat{\mathbf{j}})\ \text{ms}^{-1}$. The speed of image at $t = 1$ s.	q.	$\sqrt{10}\ \text{ms}^{-1}$
iii.	The velocity of particle C is $\mathbf{v}_c = \{5\hat{\mathbf{i}} + (6 + 2t)\hat{\mathbf{j}}\}\ \text{ms}^{-1}$ and mirror is moving with velocity $\mathbf{v}_m = (3t + \hat{\mathbf{i}})\ \text{ms}^{-1}$. The speed of image at $t = 1$ s.	r.	$\sqrt{58}\ \text{ms}^{-1}$
iv.	The velocity of particle D is $\mathbf{v}_D = (3\hat{\mathbf{i}} + 4\hat{\mathbf{j}} + 5\hat{\mathbf{k}})\ \text{ms}^{-1}$ and mirror is moving with velocity $(\hat{\mathbf{i}} + \hat{\mathbf{j}} + \hat{\mathbf{k}})\ \text{ms}^{-1}$. The image at $t = 1$ s.	s.	$\sqrt{65}\ \text{ms}^{-1}$

Codes

	i	ii	iii	iv		i	ii	iii	iv
a.	p	q	r	s	b.	q	r	s	p
c.	q	p	s	r	d.	p,q	r,s	p	q

One Integer Value Correct Type

Direction *(Q. Nos. 12-25) This section contains 14 questions. When worked out will result in an integer from 0 to 9 (both inclusive).*

12 A man of height 2 m is moving with constant speed 5 ms^{-1} along the edge of straight horizontal side walk at night. At a distance 3 m , from the edge of side walk, there is a lamppost as shown in the figure. The lamp is fixed at a height of 4 cm above the ground. As the man walks, his shadow on the ground shifts continuously as well as changes its size.

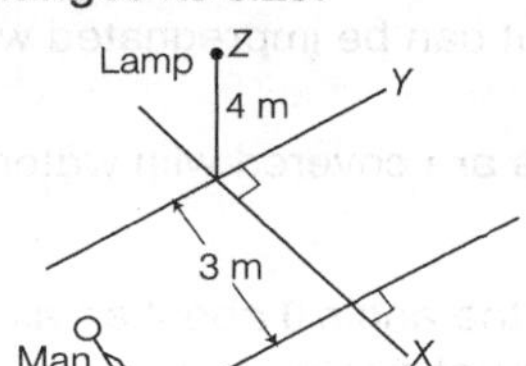

Find the rate of change in length (in ms^{-1}) of shadow at position (3, 4) m.

13 In the given figure, mirror starts to fall under gravity from the position as shown in the figure. The time after which the man sees the reflection of his feet is t_0. If $t_0^2 = 1015 \times 10^{-x}\ \text{s}^2$. Find the value of x.

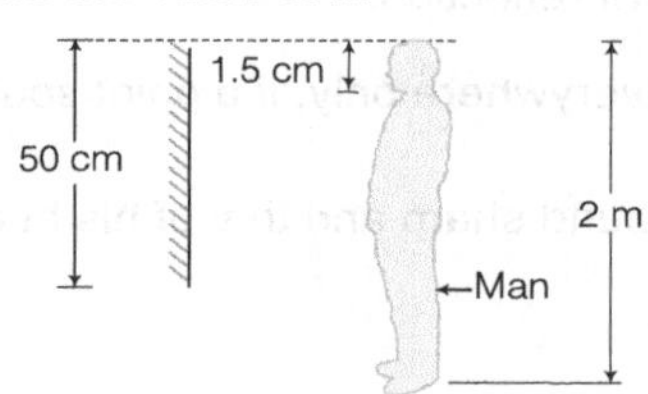

14 A child of height 2.4 ft is standing in front of a plane mirror. His brother of height 4.8 ft is standing behind him as shown in the figure. What is the minimum height (in ft) of the mirror required, so that the child can completely see his own image and his brother's image in the mirror.

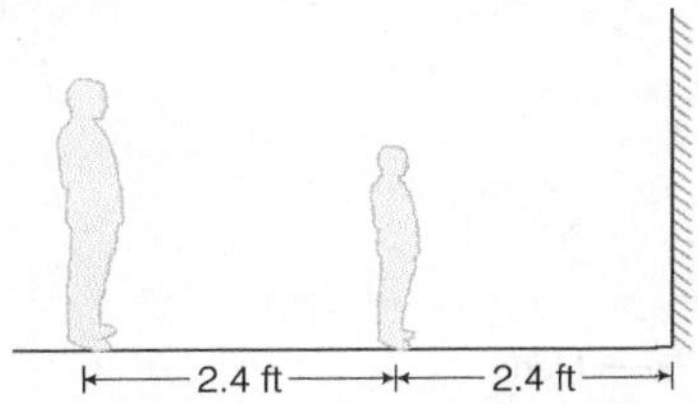

15 Two plane mirrors are combined at 70°. A point object is placed on the bisector of angle between mirrors. Find the number of images formed by combination.

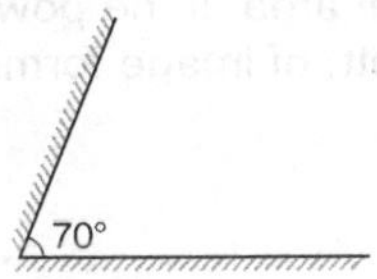

16 Find the number of images of point source S formed by combination of three plane mirrors as shown in the figure.

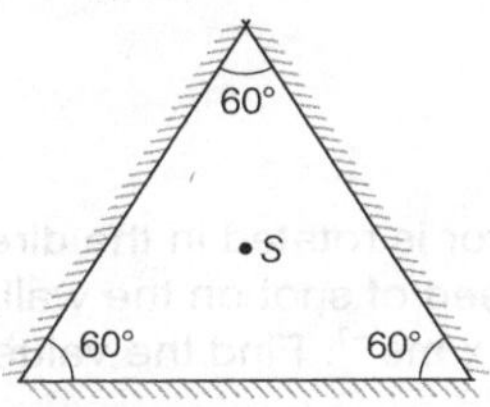

17 A circular plane mirror is placed on a table and a hemisphere of radius 10 cm is placed on it. The centre of hemisphere is at the centre of mirror.
The minimum radius of circular mirror for which image of point object P is formed by the mirror is $\dfrac{25}{n}$ cm. Find the value of n.

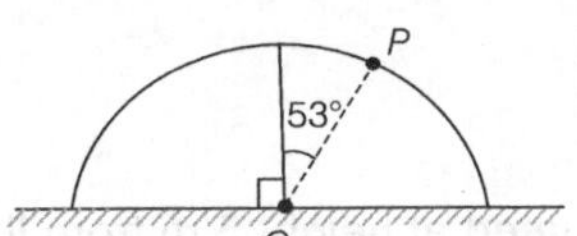

18 The part of his body seen by the man is $2h$ (in cm). Find the value of h. Assume that, eye is 10 cm below.

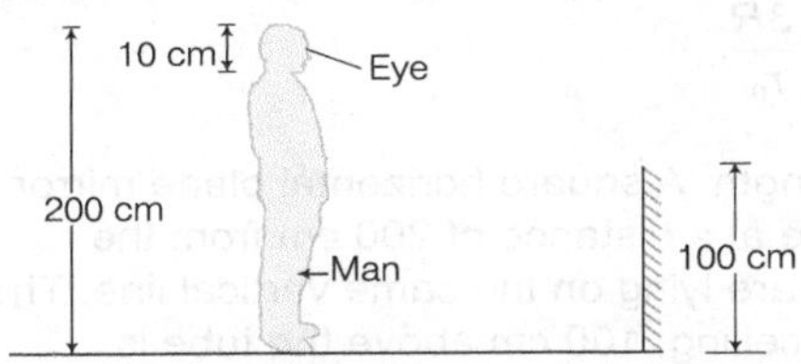

19 One night a man of height $h = 1.8$ m is standing on the bank of a straight pond. The water surface is calm on the opposite bank, a lamp is at height $H = 5.4$ m from surface of water. Light emanated from the lamp, appears as a flare (bright spot) after reflection from the water surface. When the man starts walking along the bank, the flare appears to him moving at constant speed $u = 3\ \text{ms}^{-1}$ relative to ground. Find the speed (in ms^{-1}) of the man.

Working Space

20 A plane mirror with its bottom edge on the floor is tilted at an angle 30° to horizontal. A man is standing before the mirror. The eye of the man is 2 m above the ground. Find the minimum length of the mirror (in metre), so, that he can see the image of the portion of his body below eye.

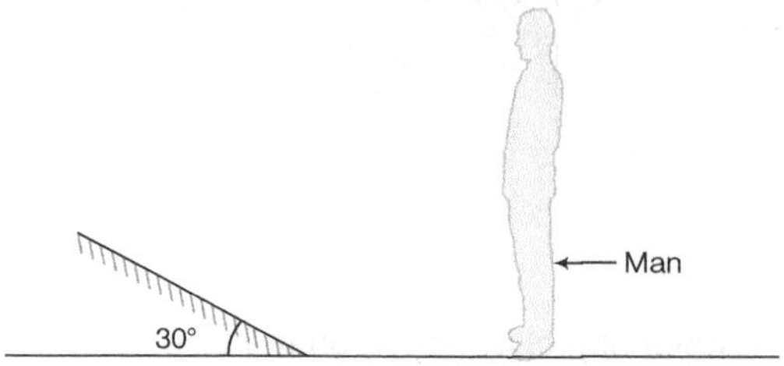

21 Figure shows two small plane mirrors A and B of equal area. If the power of image formed by plane mirror A is 8 W, find the power (in watt) of image formed by the plane mirror B.

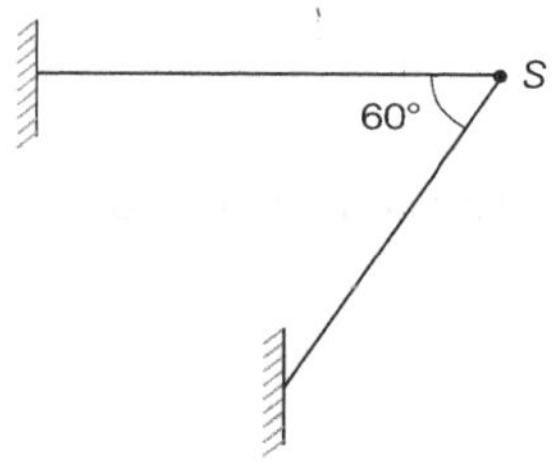

22 A light ray is incident on a plane mirror M. The mirror is rotated in the direction shown with angular velocity 18 rads^{-1}. Find the speed of spot on the wall, when the angle of incidence is 37°. If your answer is 500 x ms^{-1}. Find the value of x.

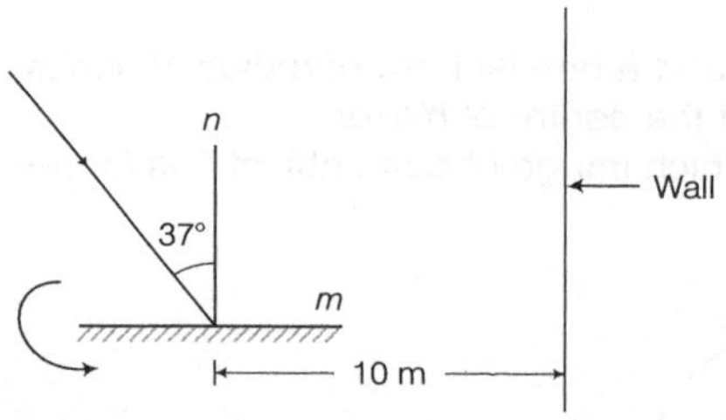

23 A glowing spherical bulb of diameter 10 cm is situated at height 5 m from the horizontal ground. An opaque spherical bulb of diameter 5 cm is situated at a height of 1 m from the horizontal ground. The centres of the bulbs are lying on the same vertical line. If the radius of umbra is r_0 and outer radius of penumbra is R on the horizontal ground. Find the value of $\frac{3R}{r_0}$.

24 A horizontal fluorescent tube is of 80 cm length. A square horizontal plane mirror each of side 20 cm is placed below the tube at a distance of 200 cm from the tube. The centre of the tube and the mirror are lying on the same vertical line. The length of the reflected patch of light on the ceiling, 100 cm above the tube is 34x cm. Find the value of x.

25 When the sun rays pass through a small opening in the foliage at the top of a high tree, they produce an elliptical spot on the ground. The major and minor axes of the ellipses are $a = 12$cm and $b = 10$ cm respectively. If the angular dimensions of the sun's disc is $\theta = \frac{1}{108}$ rad, find the height of tree (in metre).

DPP-3 Spherical Mirror-I

Conceptual Questions

Working Space

Direction *(Q. Nos. 1-6) These questions are conceptual in nature, need to be solved completely on notebook.*

1 **Situation** A linear object of length 2 cm is placed at distance 30 cm from pole of the concave mirror of radius of curvature 40 cm.

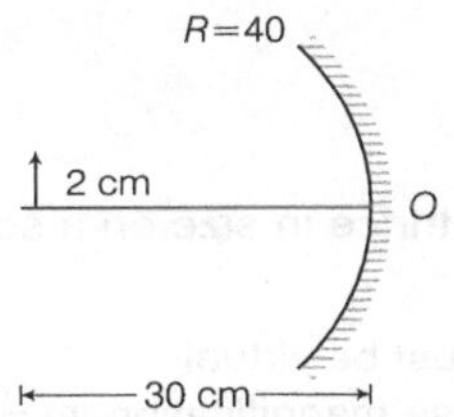

i. The position of image from pole of the mirror is

a. 30 cm infront of mirror
b. 60 cm behind the mirror
c. 60 cm infront of mirror
d. None of these

ii. The magnification is

a. +2
b. −2
c. +(1/2)
d. −1

iii. The length of image is

a. 2 cm
b. 4 cm
c. 6 cm
d. 1 cm

iv. The nature of image is

a. real, diminished and inverted
b. virtual, diminished and errect
c. real, magnified and inverted
d. real, magnified and errect

v. The ray diagram for image formation is

a. 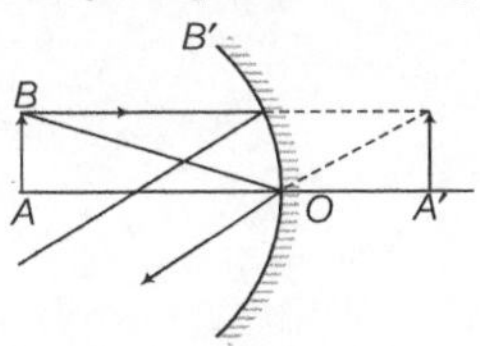

b. 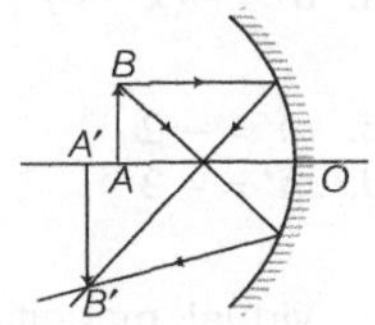

c. 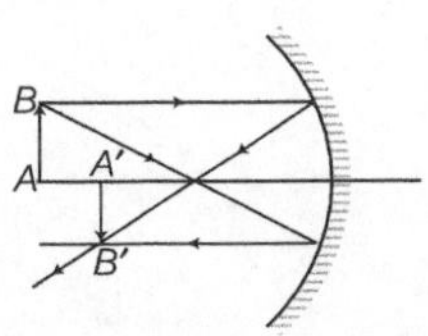

d. None of these

2 **Situation** A converging beam of light converges at the point P. A convex mirror of focal length 5 cm is placed 20 cm before the converging point.

i. Mark the correct option.

a. The converging at point P behaves as real object for the mirror
b. The converging at point P behaves as virtual object for the mirror
c. The converging point does not change due to introduction of the mirror
d. None of the above

Working Space

ii. The new converging point of the beam is

a. (20/3) cm before the mirror
b. (20/3) cm behind the mirror
c. 10 cm behind the mirror
d. None of these

iii. Draw ray diagram

a. 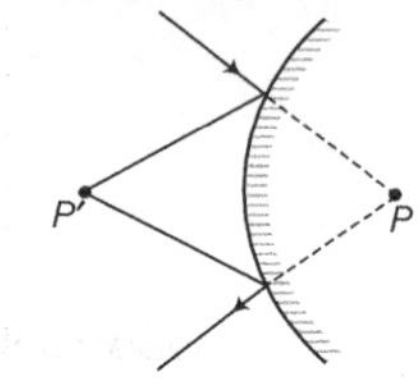

b. 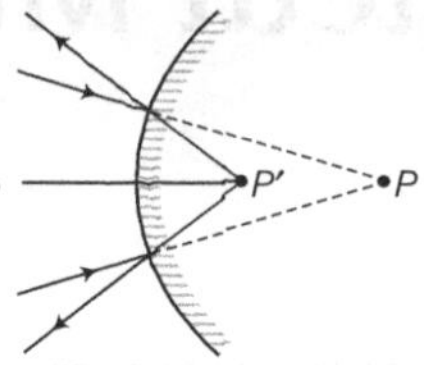

c. 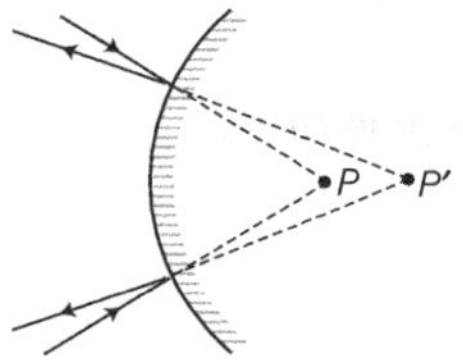

d. None of these

3 **Situation** A concave mirror forms image of an object thrice in size on a screen.

i. Mark the correct options.

a. Image must be real
b. Image must be virtual
c. Image may be virtual
d. Transverse magnification, $m = -3$

ii. If the object is placed x distance from the pole of mirror. Then, the image distance is

a. $v = -2x$
b. $v = 3x$
c. $v = 2x$
d. $v = -3x$

iii. If the radius of curvature of mirror is R. Then, the correct relation is

a. $\frac{1}{v}+\frac{1}{u}=\frac{1}{f}$
b. $\frac{1}{-3x}+\frac{1}{-x}=-\frac{2}{R}$
c. $\frac{1}{3x}+\frac{1}{x}=\frac{2}{R}$
d. None of these

If object and screen are shifted such that size of image on screen becomes twice in size. During this shifting in object is 6 cm.

iv. New object distance with sign convention is

a. $u' = x + 6$
b. $u' = -(x + 6)$
c. $u' = x - 6$
d. $u' = -(x - 6)$

v. The transverse magnification is

a. $m' = 2$
b. $m' = -2$
c. $m' = 3$
d. $m' = -3$

vi. New image is

a. real, magnified and inverted
b. virtual, magnified and errect
c. real, diminished and inverted
d. virtual, diminished and inverted

vii. The correct equation is

a. $\frac{1}{v'}+\frac{1}{u'}=\frac{1}{f}$
b. $\frac{1}{v'}-\frac{1}{(x+6)}=-\frac{2}{R}$
c. $\frac{-1}{2(x+6)}-\frac{1}{(x+6)}=-\frac{2}{R}$
d. $\frac{1}{2(x-6)}+\frac{1}{(x-6)}=\frac{2}{R}$

viii. The radius of curvature of the mirror is

a. 48 cm
b. 72 cm
c. 108 cm
d. None of these

ix. The value of x is

a. 48 cm
b. 72 cm
c. 108 cm
d. None of these

Working Space

x. The distance between images for the two positions of the object is

a. 48 cm
b. 36 cm
c. 108 cm
d. 72 cm

xi. The shifting in the screen is

a. 48 cm
b. 36 cm
c. 108 cm
d. 72 cm

4 **Situation** The successive reflection takes place from concave mirror, then plane mirror. The position of object P is shown at $t = 0$ s. The separation between mirrors is 70 cm.

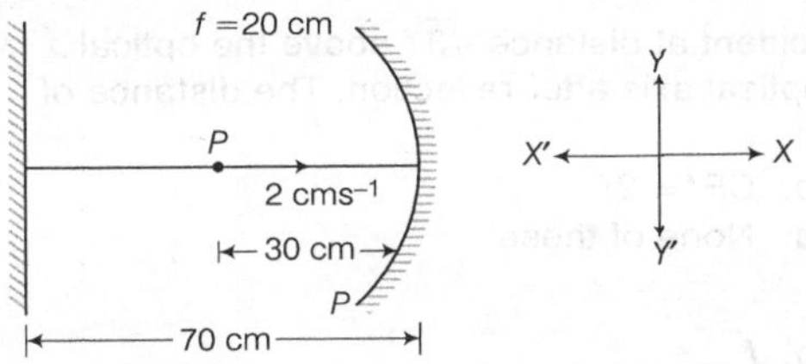

i. The position of image due to reflection from the concave mirror at $t = 0$ is

a. 60 cm infront of mirror
b. 60 cm behind the mirror
c. at the pole of plane mirror
d. None of these

ii. The velocity of image formed by concave mirror at $t = 0$ is

a. 4 cms^{-1} along negative X-axis
b. 8 cms^{-1} along negative X-axis
c. 3 cms^{-1} along positive X-axis
d. 4 cms^{-1} along positive X-axis

iii. The position of image due to successive reflection from the concave mirror, then plane mirror at $t = 0$ is

a. 10 cm behind the plane mirror
b. 10 cm infront of plane mirror
c. 20 cm behind the mirror
d. None of these

iv. In previous problem, at $t = 0$ the image is moving

a. with velocity 8 cms^{-1} along negative x-direction
b. with velocity 8 cms^{-1} along positive x-direction
c. with velocity 2 cms^{-1} in negative x-direction
d. with velocity 2 cms^{-1} in positive x-direction

5 **Situation** The focal length for paraxial ray in the case of concave mirror is f. The mirror is of large aperture. A ray parallel to optical axis is incident at distance f along side the optical axis as shown in the figure. The ray is

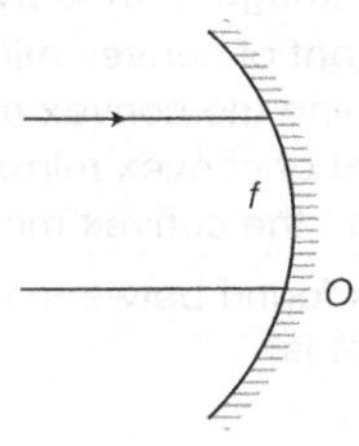

a. paraxial
b. marginal
c. may be paraxial may be marginal
d. None of these

i. Show the reflected ray.

a. 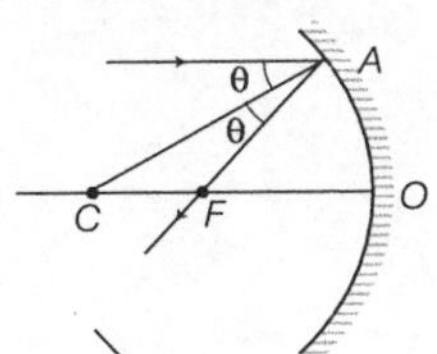

b. 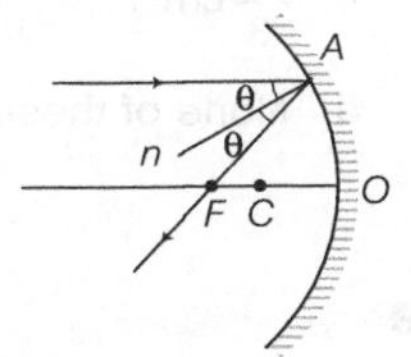

c. 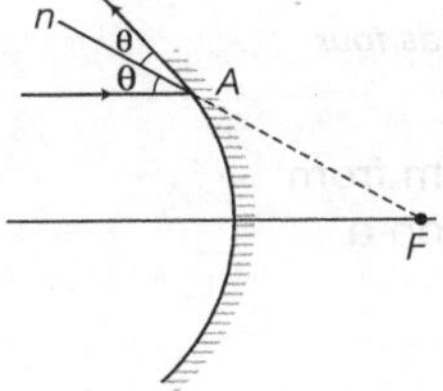

d. None of these

Working Space

ii. The radius of curvature of the mirror is

a. $R = 2f$ b. $R = f$

c. $R < 2f$ d. None of these

iii. If the angle of incidence of ray is θ, then

a. $\sin\theta = f/R$ b. $\sin\theta = 1/2$

c. $\theta = 30^\circ$ d. None of these

iv. In previous problem, the distance of F from centre of curvature of mirror is

a. $CF = (R/2)\sec\theta$ b. $CF = f\sec\theta$

c. $CF = 2f/\sqrt{3}$ d. None of these

v. If another ray parallel to previous ray incident at distance $\sqrt{3}f$ above the optical axis. Assume that, it crosses at F' on optical axis after reflection. The distance of F' from centre of curvature is

a. $CF' = (R/2)\sec 60^\circ$ b. $CF' = 2f$

c. F' is at the pole of mirror d. None of these

vi. The distance between F and F' is

a. $(2 - 2/\sqrt{3})f$ b. f

c. $2f$ d. None of these

6 **Situation** A point object P is placed at distance 50 cm from the convex mirror. The focal length of convex mirror is 25 cm. A plane mirror is placed covering lower half the convex mirror.

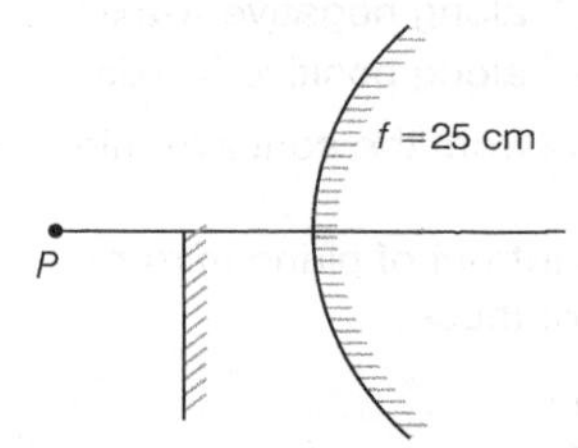

i. Maximum number of images of point object P is

a. 1 b. 2

c. 3 d. more than 3

ii. The position of image formed by convex mirror is

a. 50/3 cm infront of convex mirror

b. 50/3 cm behind the convex mirror

c. 50 cm infront of convex mirror

d. 50 cm behind the convex mirror

iii. If no parallax is found between image formed by the both mirror. The separation between mirrors is

a. $\frac{100}{3}$ cm b. $\frac{50}{3}$ cm

c. $\frac{200}{3}$ cm d. None of these

Only One Option Correct Type

Direction *(Q. Nos. 7-26) This section contains 20 multiple choice questions. Each question has four choices (a), (b), (c) and (d), out of which ONLY ONE is correct.*

7 What is the radius of curvature of a concave mirror at a distance of 200 cm from the face of a man, if he sees in it his image that is 1.5 times greater than on a plane mirror placed at the same distance from the face?

a. 6 m b. 12 m

c. 18 m d. 24 m

Working Space

8 A point source of light is placed 10 cm before a convex mirror of focal length is 10 cm and aperture 1 mm. A screen is at 10 cm from the source. Find the length of field of view of image of point source S on the screen.

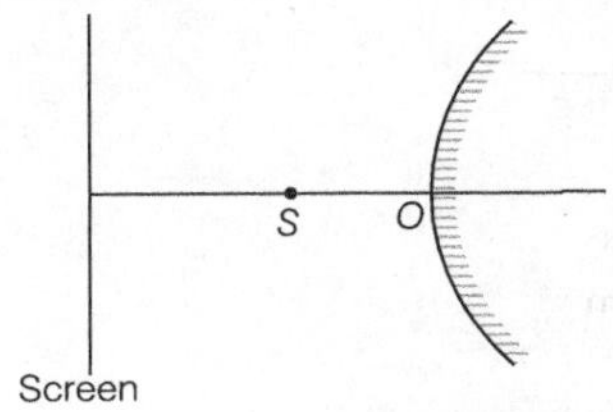

a. 1 mm
b. 0.5 mm
c. 5 mm
d. 10 mm

9 A point object P is placed 30 cm before the concave mirror as shown in the figure. Now, mirror is rotated through 2° in anticlockwise direction. Find the coordinate of image formed by the mirror after rotation.

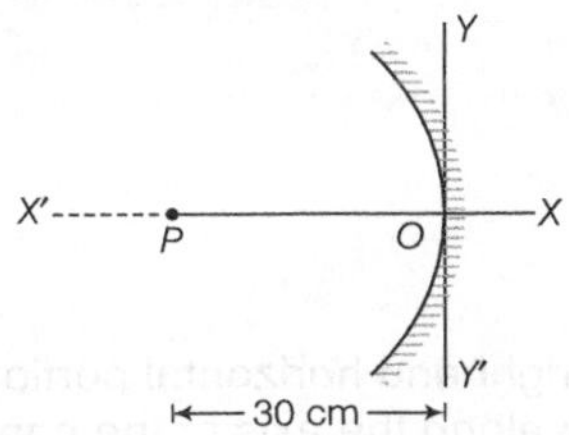

a. $\left(60\text{cm}, \frac{4\pi}{3}\text{cm}\right)$
b. $\left(-60\text{cm}, \frac{-4\pi}{3}\text{cm}\right)$
c. $\left(-90\text{cm}, \frac{-4\pi}{3}\text{cm}\right)$
d. $\left(-60\text{cm}, \frac{-2\pi}{3}\text{cm}\right)$

10 A cube of side length 1 mm is placed at distance 30 cm from the pole of a concave mirror of focal length 20 cm. Its one side is parallel optical axis as shown in the figure.

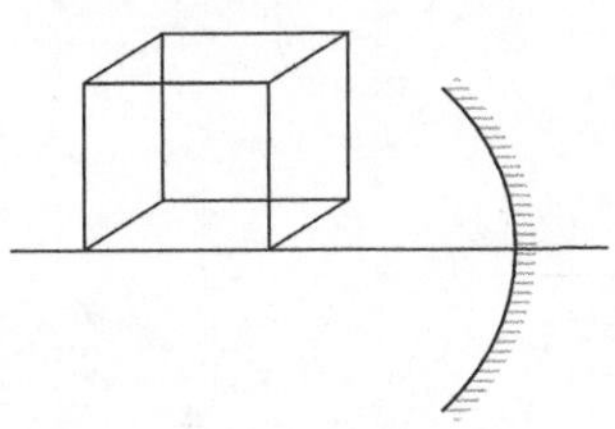

Find the magnification in the volume of cube.

a. 4
b. 2
c. 16
d. 24

11 A convex mirror starts simple harmonic motion along optical axis with amplitude 5 cm. The distance between mean position of mirror object is 20 cm. The focal length of mirror is 10 cm. Find the range of vibration of image.

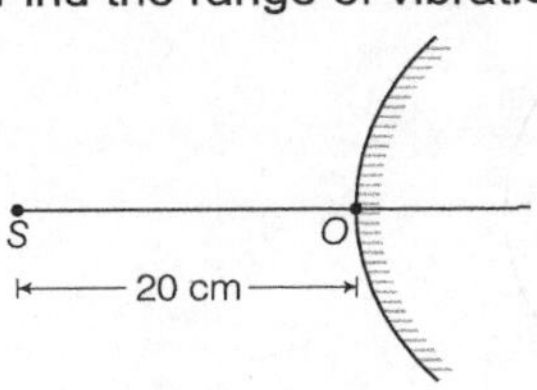

a. 11.14 cm
b. 20 cm
c. 6 cm
d. 16 cm

Working Space

12 Find the length of image of AB.

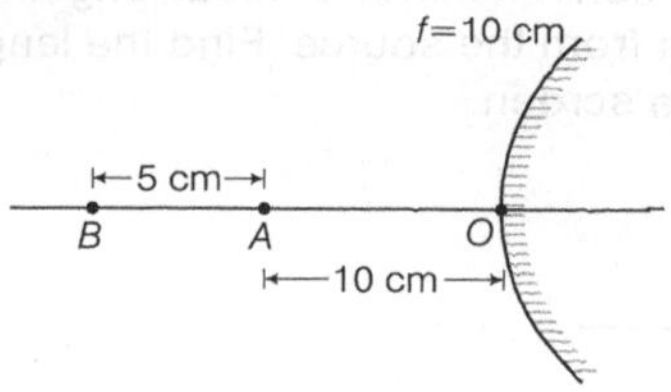

a. 5 cm
b. 2.5 cm
c. 1.25 cm
d. 1 cm

13 Find the angle subtended by image of AB at the eye.

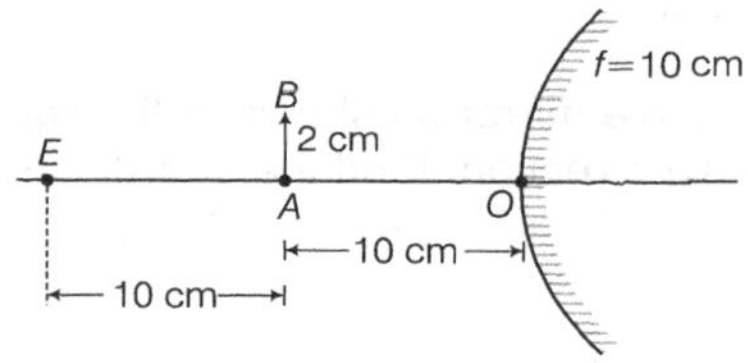

a. 0.04°
b. 2.3°
c. $\left(\frac{1}{25}\right)^{\circ}$
d. 4.6°

14 A small piece of wire bent into an L shape with upright and horizontal portions of equal lengths, is placed with the horizontal portion along the axis of the convex mirror of focal length 10 cm. If the bend is 20 cm from the pole of the mirror, then the ratio of the lengths of the images of the upright and horizontal portions of wire is

a. 1 : 3
b. 3 : 1
c. 2 : 3
d. 3 : 2

15 A convex mirror of focal length 10 cm is cut into two halves and placed before an object at a distance of 10 cm as shown in the figure. The separation between images formed by two halves is

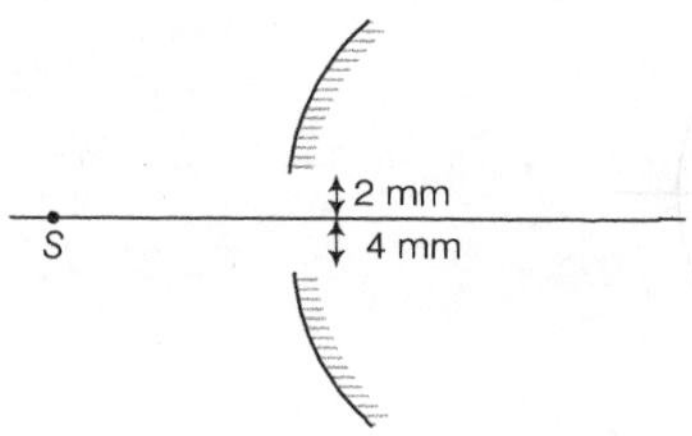

a. 1 mm
b. 2 mm
c. 3 mm
d. 9 mm

16 Moon disc subtends an angle 1° at the pole of a concave mirror. The radius of curvature of concave mirror is 300 cm. Find the diameter of image formed by the concave mirror

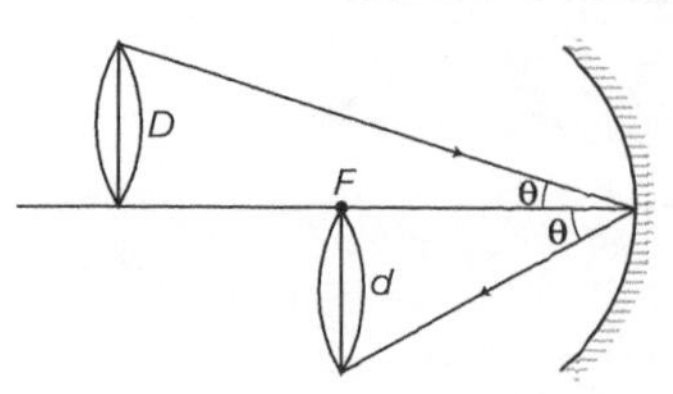

a. 1 cm
b. 2.62 cm
c. 3.24 cm
d. 5 cm

Working Space

17 An astronomer use a parabolic mirror. At what position should a photo film be put to get a clear picture of a distant star?

a. At pole
b. At focus
c. At far distance from mirror
d. Beyond the focus

18 A short linear object of length 2 mm lies on the axis of a convex mirror of focal length 20 cm and distance 20 cm from the mirror. The length of image is

a. 1 mm   b. 2 mm
c. 0.5 mm   d. 0.25 mm

19 A point source of light S is placed between concave mirror. The beam of light incident on plane mirror becomes parallel to optical axis after reflection through the concave mirror. The focal length of concave mirror is 30 cm and the point source is midway between concave mirror and plane mirror. Find the separation between concave mirror and plane mirror.

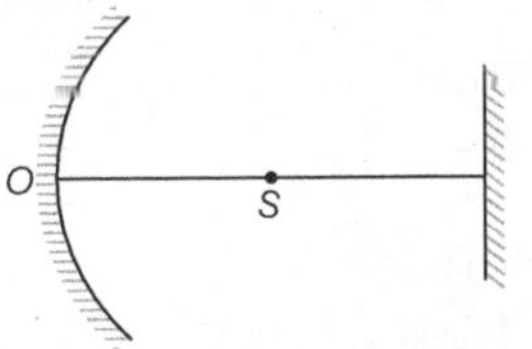

a. 10 cm   b. 20 cm
c. 30 cm   d. 60 cm

20 In the figure, a point source of light is placed 20 cm before a plane mirror. Find the distance between images formed by mirrors.

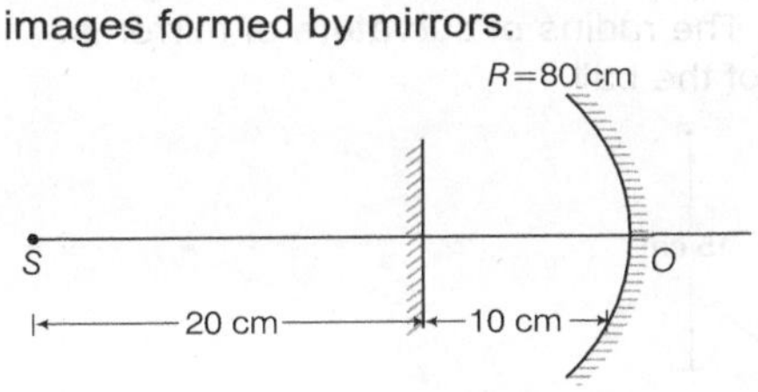

a. 100 cm   b. 120 cm
c. 110 cm   d. 130 cm

21 The coordinates of image of point object S formed after two successive reflections in situation as shown in the figure, considering first reflection at mirror M_1 and then at mirror M_2.

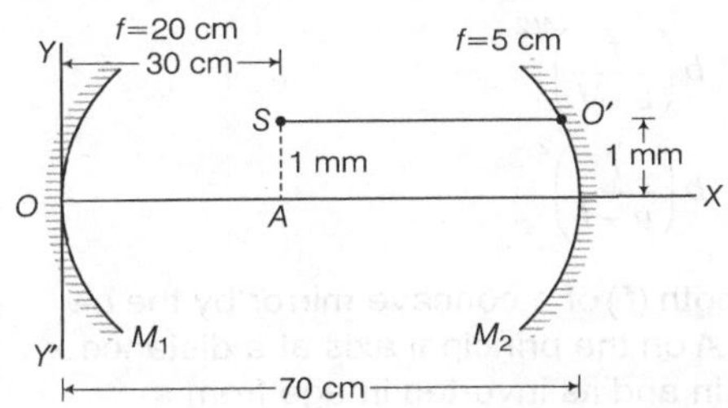

a. (60 cm, 4 mm)   b. (– 60 cm, 3 mm)
c. (–10 cm, 4 mm)   d. (60 cm , – 3 mm)

22 A plane mirror and a convex mirror are arranged as shown in the figure and S is a point object. Find the coordinate of image formed by two reflection first on plane mirror, then on convex mirror.

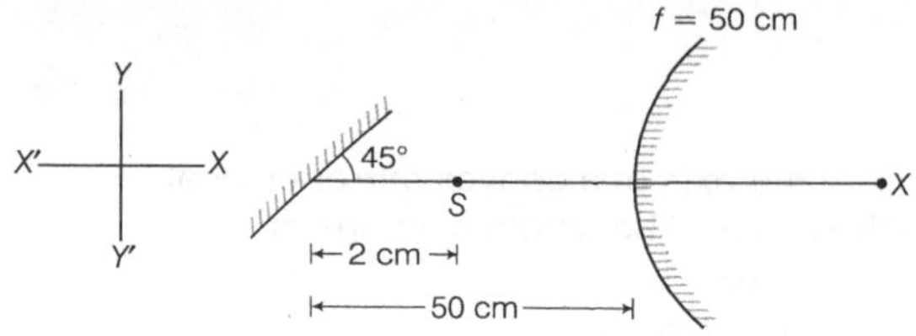

a. 75 cm, – 1 cm
b. 50 cm, 0
c. 75 cm, 1 cm
d. 75 cm, – 2 cm

23 Find the velocity of image at the instant as shown in the figure.

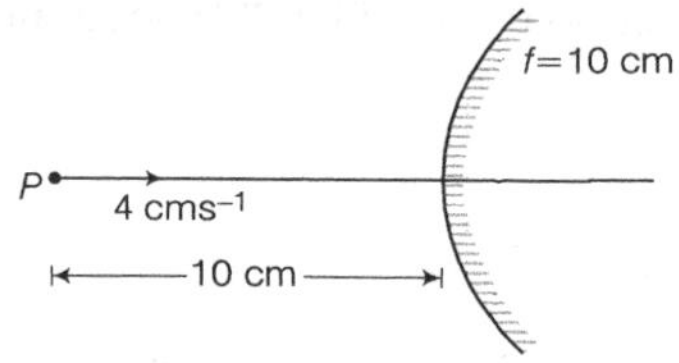

a. 1 cms^{-1} in leftward direction
b. 2 cms^{-1} in rightward direction
c. 1 cms^{-1} in rightward direction
d. 4 cms^{-1} in leftward direction

24 A small ball is dropped along the optical axis of a concave mirror from height 15 cm on the mirror as shown in the figure. The radius of curvature of mirror is 60 cm. Find the maximum speed of image of the ball.

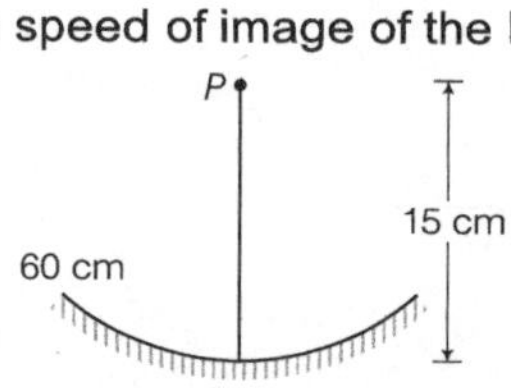

a. 1 ms^{-1}
b. 2 ms^{-1}
c. 2.25 ms^{-1}
d. 4.5 ms^{-1}

25 The short linear object of length b lies along the axis of a concave mirror of focal length f at a distance u from the pole of the mirror. The size of the image is approximately equal to

a. $b\left(\frac{u-f}{f}\right)^{1/2}$
b. $b\left(\frac{f}{u-f}\right)^{1/2}$
c. $b\left(\frac{u-f}{f}\right)$
d. $b\left(\frac{f}{u-f}\right)^{2}$

26 In an experiment of determine the focal length (f) of a concave mirror by the u-v method, a student places at the object pin A on the principal axis at a distance x from the pole P. The student looks at the pin and its inverted image from a distance keeping his/her eye in line with PA. When the student shifts his/her eye towards left, the image appears to the right of the object pin. Then,

a. $x < f$
b. $f < x < 2f$
c. $x = 2f$
d. $x > 2f$

Working Space

One or More than One Options Correct Type

Working Space

Direction *(Q. Nos. 27-32) This section contains 6 multiple choice questions. Each question has four choices (a), (b), (c) and (d), out of which ONE or MORE THAN ONE are correct.*

27 For concave mirror,

a. virtual image of real extended object is always magnified
b. the image of a virtual object is always real
c. image size may be equal to object size
d. the real image may be magnified

28 A linear object AB is placed along optical axis of a concave mirror of focal length 20 cm as shown in the figure.

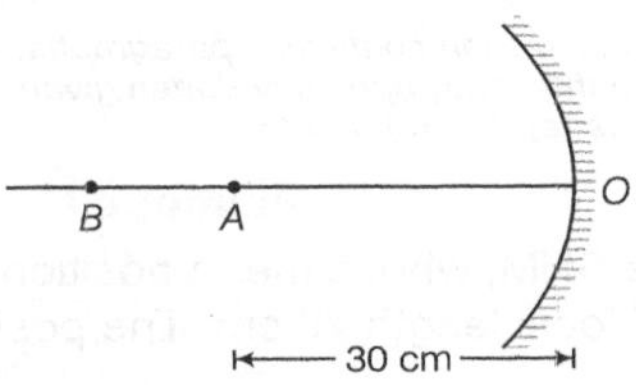

a. If $AB = 5$ cm, the size of image is 13.33 cm
b. If $AB = 5$ mm, the size of image is 10 mm
c. If $AB = 5$ cm, the size of image is 20 cm
d. If $AB = 5$ cm, the size of image is 20 mm

29 A square board of side length 2 mm is placed at distance of 30 cm from a concave mirror of focal length 20 cm. The area of image of the board is

a. $16\ \text{mm}^2$, if optical axis is perpendicular to the plane of board
b. $32\ \text{mm}^2$, if optical axis is perpendicular to the plane of board
c. $32\ \text{mm}^2$, if optical axis is along one of the side
d. $64\ \text{mm}^2$, if optical axis is along one of the side

30 A point source S is placed between a plane mirror and a convex mirror as shown in the figure.

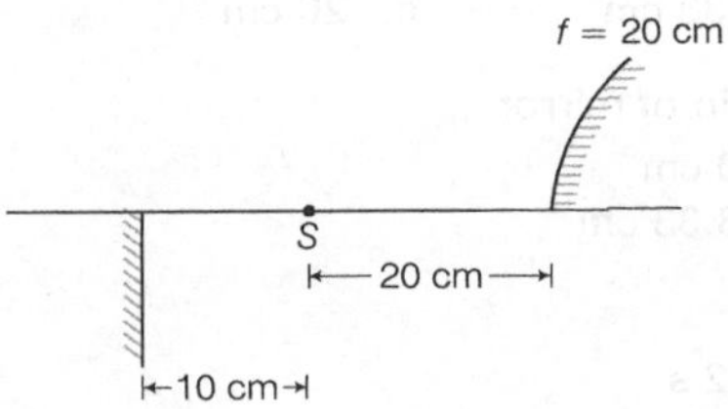

a. The maximum number of images of point source S is two
b. The separation between any two images is 50 cm
c. The maximum number of images of point source S is three
d. Infinite number of images are formed due to combination of these two mirrors

31 A small block of mass 100 g is tied to one end of spring of constant 20 Nm^{-1}. The distance between pole of convex mirror of focal length 10 cm and mean position of the block is 20 cm. The block is released from rest, when spring is in its natural length.

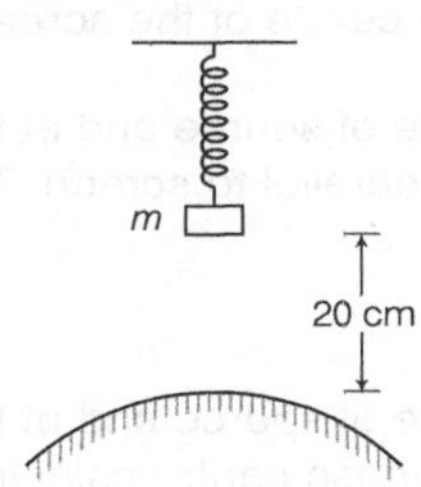

a. The time period of vibration of image is 1s
b. The time period of vibration of image is 0.44 s
c. The range of vibration of image is $\frac{8}{7}$ cm
d. The range of vibration of image is 10 cm

32 A student performed the experiment of determination of focal length of a concave mirror by u-v method using an optical bench of length 1.5 m. The focal length of the mirror used is 24 cm. The maximum error in the location of the image can be 0.2 cm. The 5 sets of (u, v) values recorded by the student (in cm) are (42, 56), (48, 48), (60, 40), (66, 33), (78, 39). The data set(s) that cannot come from the experiment is (are) incorrectly recorded is (are) **[IIT JEE 2009]**

a. (42, 56) b. (48, 48) c. (66, 33) d. (78, 39)

DPP-4 Spherical Mirror-II

Comprehension Type

Working Space

Direction *(Q. Nos. 1-6) This section contains 2 paragraphs, each describing theory, experiments, data, etc. Six questions related to the paragraphs have been given. Each question has only one correct answer among the four given options (a), (b), (c) and (d).*

Passage I

A particle execute SHM, whose mean position is 55 cm from the pole of a concave mirror of focal length 20 cm. The position of particle from mean position is $x = 5\cos 10\pi t$.

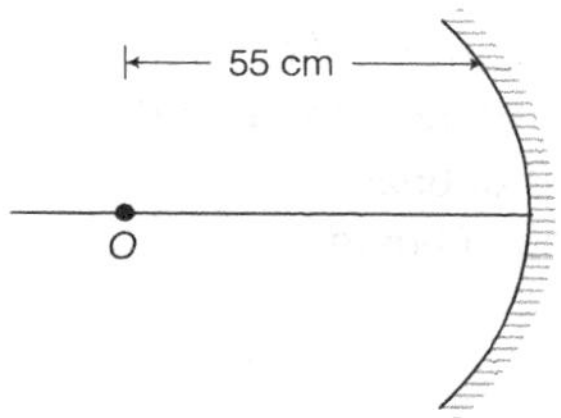

1 Find the range of vibration of image.

a. 10 cm b. 5 cm c. 3.33 cm d. 20 cm

2 Find the maximum distance of image from pole of mirror.

a. 10 cm b. 30 cm
c. 25 cm d. 33.33 cm

3 Find the time period of oscillation of image.

a. 0.1 s b. 0.2 s
c. 0.4 s d. 0.6 s

Passage II

A point source of light placed at a distance from a screen creates intensity of 2.25 Wm^{-2} at the centre of the screen.

4 If on the other side of source and at the same distance from it, we place an infinite size plane mirror parallel to screen. The intensity of centre is

a. 2.5 Wm^{-2} b. 0.25 Wm^{-2}
c. 2.25 Wm^{-2} d. 0.5 Wm^{-2}

5 If on the other side of source and at the same distance from it, we place a concave mirror, whose centre coincides with the centre of screen. The intensity at the centre of screen is

a. 2.25 Wm^{-2} b. 4.5 Wm^{-2}
c. 2.5 Wm^{-2} d. 3 Wm^{-2}

6 Now on the other side of source and at the same distance from it, we place a convex mirror of aperture 1 mm and focal length f. If the distance between source and screen is f. The intensity at the centre of screen is

a. 2.25 Wm^{-2} b. 4.5 Wm^{-2}
c. 2.34 Wm^{-2} d. 2.5 Wm^{-2}

Working Space

Statement Type

Direction (Q. Nos. 7-9) *This section is based on Statement I and Statement II. Select the correct answer from the codes given below.*

a. Both Statement I and Statement II are correct and Statement II is the correct explanation of Statement I
b. Both Statement I and Statement II are correct but Statement II is not the correct explanation of Statement I
c. Statement I is correct but Statement II is incorrect
d. Statement II is correct but Statement I is incorrect

7 **Statement I** In the case of concave mirror, the minimum distance between real object and its real image is zero.

Statement II If concave mirror forms virtual image of real object, the image is magnified.

8 **Statement I** In the given figure, OO' is the principal axis of a spherical mirror. The image of point object S is S' $(h_2 < h_1)$. Then, the mirror is concave mirror placed right side of image S'.

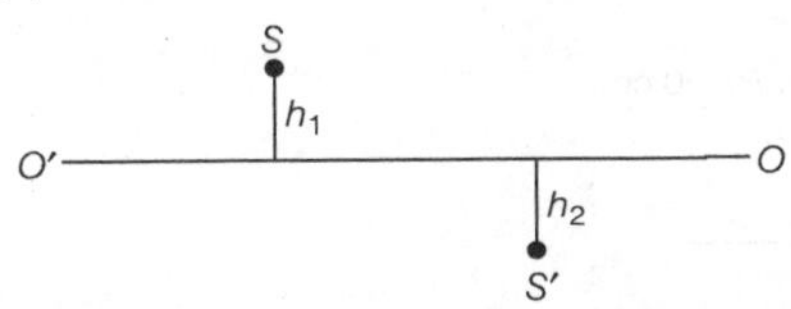

Statement II If object and image both real, transverse magnification is negative.

9 **Statement I** Image formed by a spherical mirror of a point object is always lying on a straight passing through point object and centre of curvature of mirror.

Statement II The ray which is incident on the spherical mirror, which is passing through centre of curvature returns back on the same path.

Matching List Type

Direction (Q. No. 10) *Choices for the correct combination of elements from Column I and Column II are given as options (a), (b), (c) and (d), out of which one is correct.*

10 Match the Column I with Column II and mark the correct option from the codes given below.

	Column I		Column II
i.	For real image of real object	p.	Concave mirror
ii.	For virtual image of real object	q.	Convex mirror
iii.	For virtual image of virtual object	r.	Plane mirror
iv.	For real image of virtual object	s.	If the object is placed at centre of curvature of concave mirror

Codes

	i	ii	iii	iv
a.	p,s	p,q,r	q	p,q,r
b.	p	p,q,r	q	p,q
c.	p	r	q	p
d.	p,s	q,r	p,q	p,r

One Integer Value Correct Type

Working Space

Direction (Q. Nos. 11-23) This section contains 13 questions. When worked out will result in an integer from 0 to 9 (both inclusive).

11 The greatest integer value of speed of image of point object P at the instant shown in the figure, is $17x$ cms^{-1}. Find the value of x.

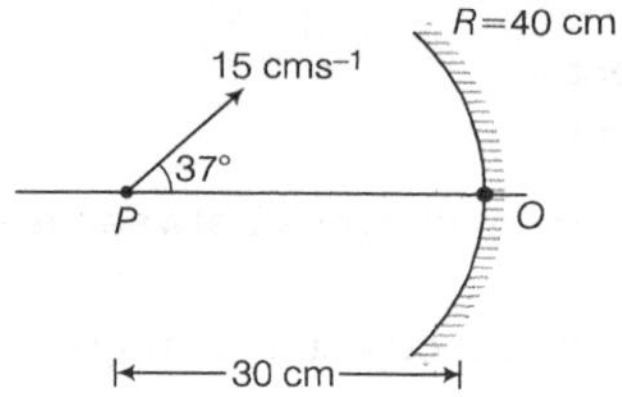

12 If the component of velocity of the image along optical axis and perpendicular to optical axis at the instant as shown in the figure, are v_x and v_y respectively. Find the value of $\frac{10\,|v_y|}{|v_x|}$.

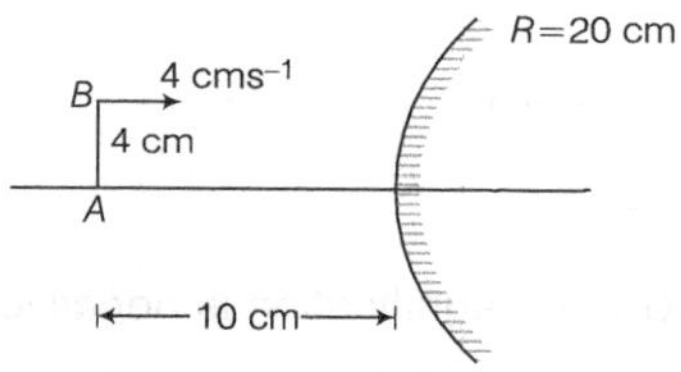

13 A wide parallel beam of light parallel to optical axis incident on concave mirror of radius 40 cm. A small opaque disc of radius 1 mm made of highly insulating material is placed at a distance 10 cm from the pole of the mirror perpendicular to optical axis. In steady state, both surfaces acquire different temperature that are slightly above the room temperature. Find the ratio of temperature $\frac{\Delta\theta_1}{\Delta\theta_2}$, where $\Delta\theta_1$ and $\Delta\theta_2$ are the temperature difference of the left and the right surfaces of the disc and the surroundings.

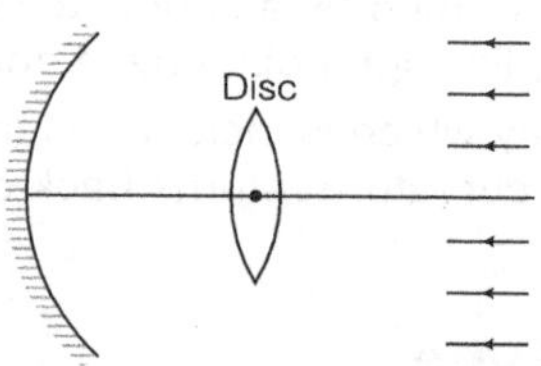

14 The length of image of linear object AB as shown in the figure, is 288×10^{-n} cm. Find the value of n.

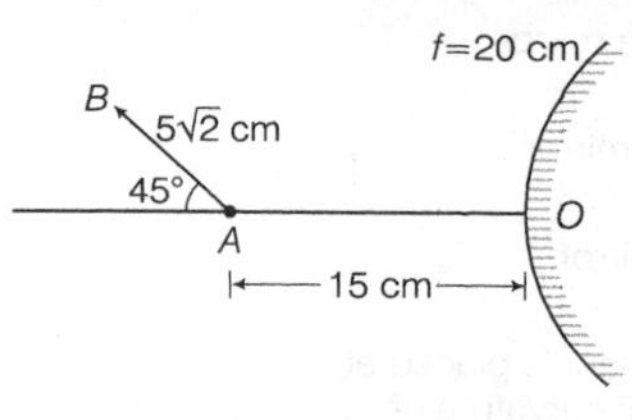

15 Two concave mirrors each of radius of curvature 40 cm are placed at 70 cm apart. A point object S is at distance 30 cm from the concave mirror M_1. The position of image formed due to three successive reflections taking first reflection from the concave mirror M_1 is $\frac{180}{x}$ cm.

Find the value of x.

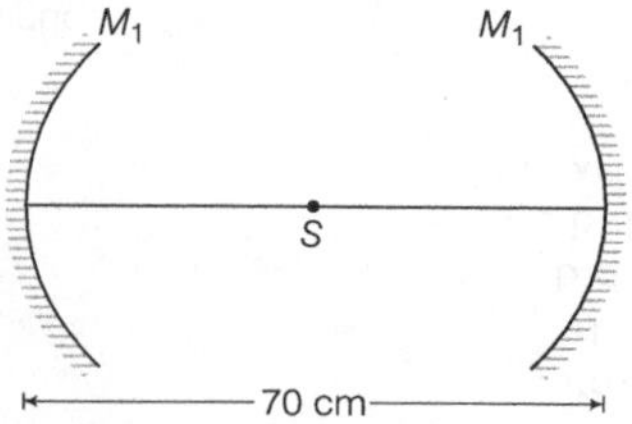

16 Due to successive reflection from concave mirror, then convex mirror final image formed at source itself. The value of $d = (13x + 1)$ cm. Find the value of x.

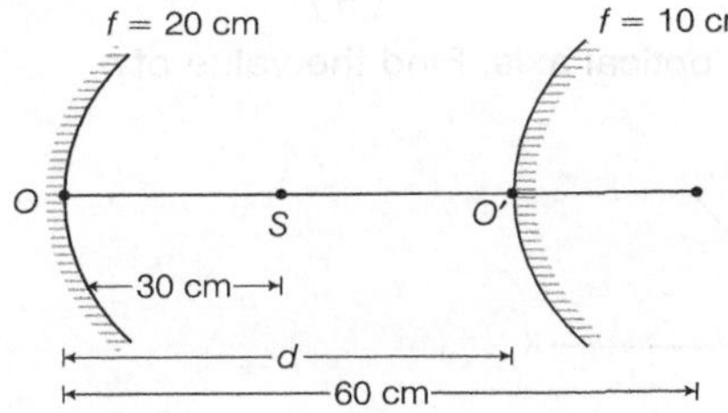

17 The focal length of convex mirror is 100 m and its aperture is 2 mm. Find the maximum length AB (in mm) of object.

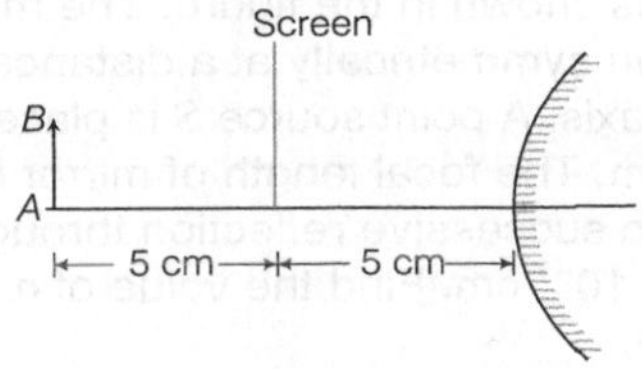

18 Two spherical mirrors, one convex and the other concave, each of focal length 30 cm are arranged co-axially at a distance of 90 cm from each other. A small circle of radius 5 mm is drawn on the convex mirror near the pole as shown in the figure. Find the radius of image (in mm) due to two successive reflection, first from concave mirror, then from convex mirror.

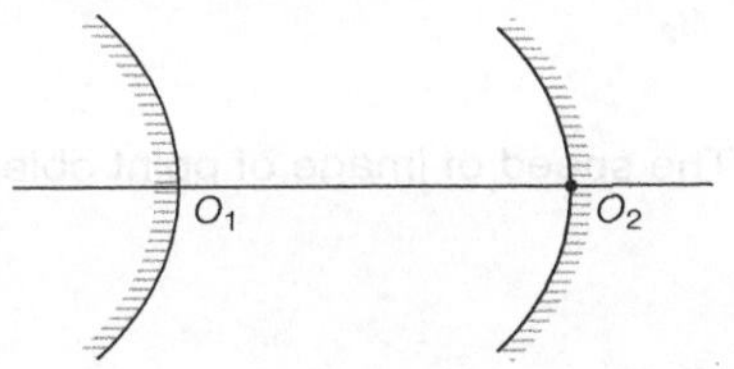

19 The speed of image of point source S is $\frac{29}{2n}$ cms^{-1} at the instant shown in the figure, due to two successive reflections taking first on convex mirror. Find the value of n.

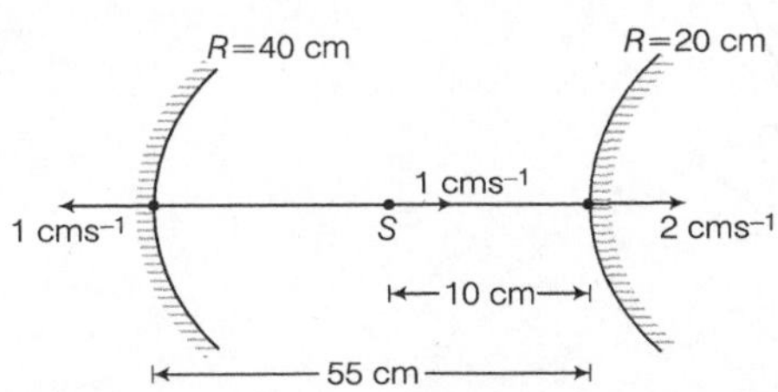

20 A point object S is located at 10 cm from the pole of a convex mirror of focal length 10 cm on its optical axis is moving with velocity as shown in the figure. The mirror velocity is also shown in the figure. The speed of image at the instant shown in the figure, is $\frac{13.9}{x}$ cms^{-1}. Find the value of x.

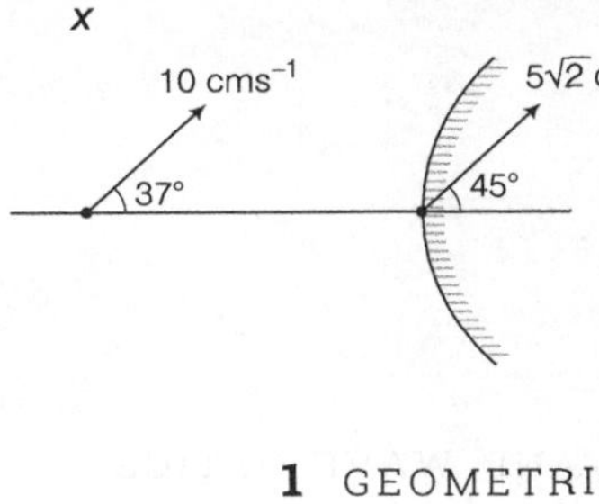

Working Space

Working Space

21 A parallel beam of light is incident on a convex mirror of radius of curvature 30 cm. The y-coordinate of converging point after reflection is $\left(\frac{\pi}{n}\right)$ cm. Assume that, the beam makes an angle of 2° with the optical axis. Find the value of n.

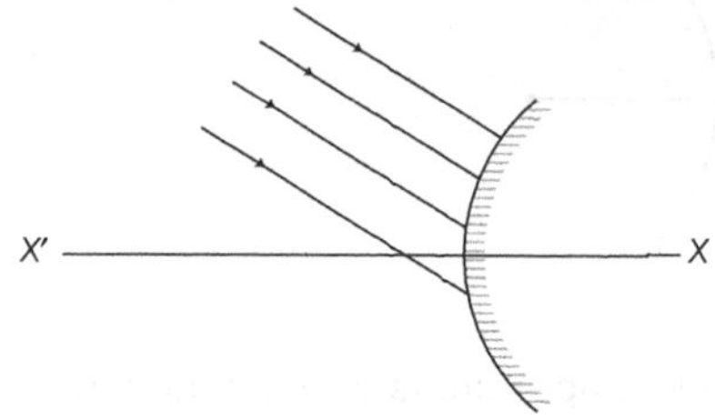

22 Two spherical mirrors M_1 and M_2 are arranged as shown in the figure. The mirror M_2 is cut into two halves and its halves are drawn symmetrically at a distance 1 cm apart in a direction perpendicular to optical axis. A point source S is placed at 30 cm from mirror M_1 whose focal length is 20 cm. The focal length of mirror M_2 is 10 cm. If we consider the image formation due to successive reflection through M_1 and M_2. If the separation between images is $n \times 10^{-1}$ cm. Find the value of n.

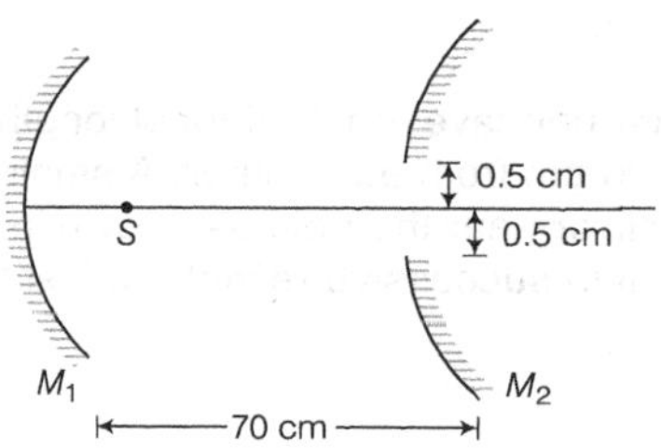

23 In the figure, the situation is shown at $t = 0$ s. The speed of image of point object P at $t = 10$ s is $\frac{n}{2}$ cms^{-1}. Find the value of n.

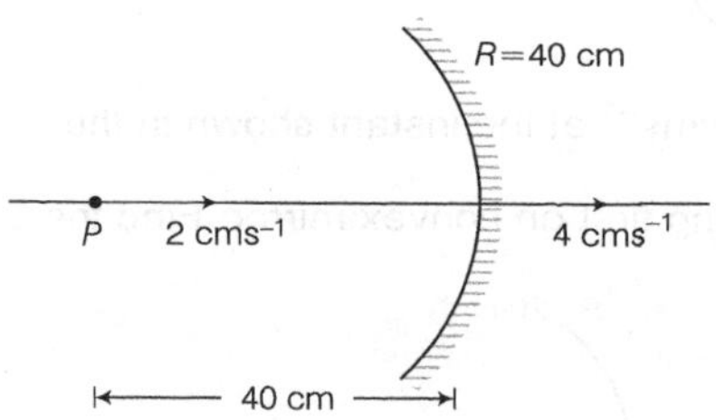

DPP-5 Refraction of Light-I

Conceptual Questions

Working Space

***Direction** (Q. Nos. 1-7) These questions are conceptual in nature, need to be solved completely on notebook.*

1 **Situation** An optical fibre of refractive index μ is placed in air. A ray of light incident at angle θ on one end of the pipe as shown in the figure.

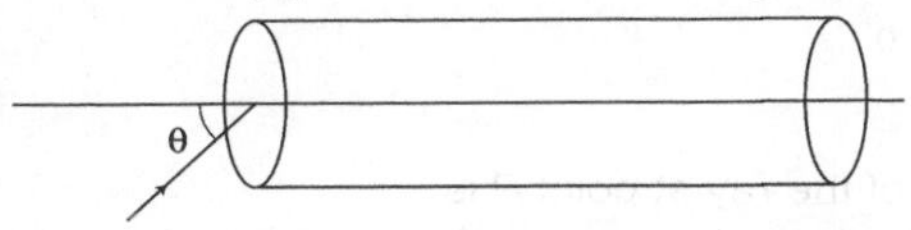

i. Draw ray diagram, if ray totally internally reflected through curve surface.

a. 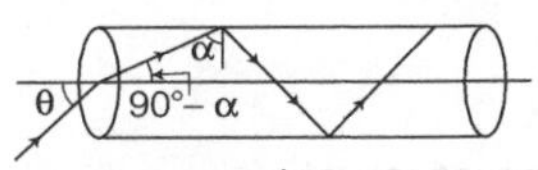

b. 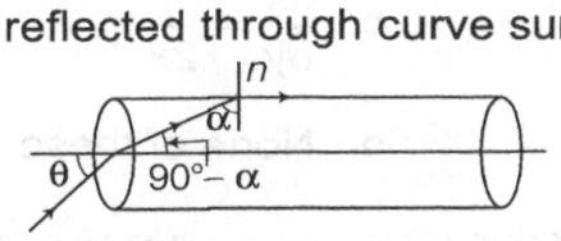

c. 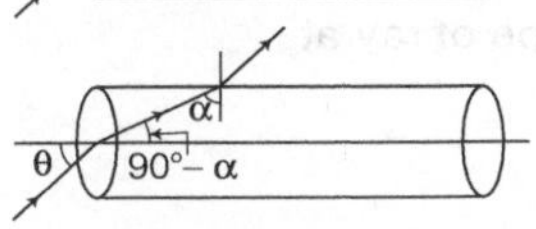

d. None of these

ii. In previous problem, the correct relations are

a. $\sin\theta = \mu \sin(90^\circ - \alpha)$
b. $\sin\alpha > 1/\mu$
c. $\sin\theta = \mu \cos\alpha$
d. None of these

iii. In previous problem, if $\mu = \sqrt{1.25}$. Then,

a. $\sin\alpha > \sqrt{4/5}$
b. $\alpha > 53^\circ$
c. $\sin\alpha > 37^\circ$
d. None of these

iv. In previous problem, the maximum value of θ is θ_0. The value of θ_0 is

a. 37°
b. 53°
c. 30°
d. None of these

v. In previous problem, if $\theta > \theta_0$

a. the ray refracted in air through curved surface
b. the ray emerges out through the other end
c. both (a) and (b)
d. None of the above

vi. Take, $\theta = 45^\circ$ and $\mu = \sqrt{2}$. The length of optical fibre is $10\sqrt{3}$ cm and diameter is 3 mm. The total number of internal reflection is

a. 10
b. 33
c. 40
d. None of these

A cone of angles are called the acceptance cone, exists at the entrance to the core of the optical fibre. Incoming light at an angles within this cone is transmitted through the core and light entering the core from angles outside the cone strikes the core at an angle less than the critical angle (i.e. $\alpha < c$) and goes outside. A light ray entering the fibre within the acceptance cone incident on the curved surface and totally internally reflected and goes out side through other end of the fibre. The angle θ_0 described in previous problems defines the acceptance cone for the fibre. Hence, sine of θ_0 is known as numerical aperture for optical fibre.

vii. If $\mu = \sqrt{1.25}$ and fibre is placed in air. The numerical aperture of the fibre is

a. zero
b. 1/2
c. $1/\sqrt{2}$
d. None of these

Working Space

viii. If optical fibre of refractive index $\mu = \sqrt{3}$ is cladded by a material of refractive index $\mu = \sqrt{2}$. Find the numerical aperture.

a. 2
b. 1
c. $\sqrt{3}$
d. None of these

2 **Situation** A ray travelling in air is incident at grazing incidence on a long rectangular slab, whose refractive index is function of y. The ray follow a parabolic path $y = x^2$ inside the slab.

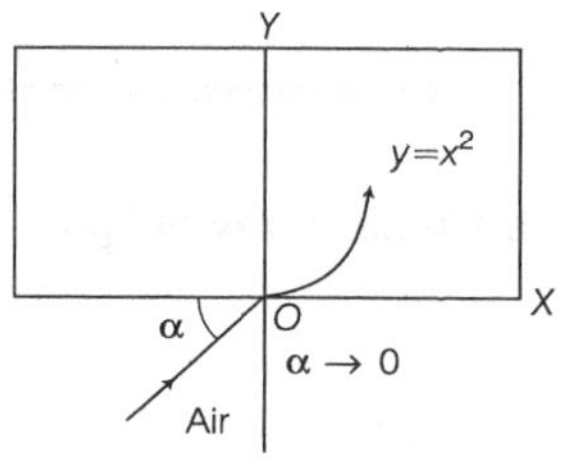

i. If $P(x, y)$ is a point on the ray. The slope of the ray at point P is

a. $\frac{dy}{dx} = 2x$
b. $\frac{dx}{dy} = \frac{1}{2x}$
c. $\frac{dy}{dx} = 0$
d. None of these

ii. In previous problem, if angle of incidence at point P is θ. Then, the slope of ray at the point P is

a. $\frac{dy}{dx} = \tan(90° - \theta)$
b. $\frac{dy}{dx} = \tan\theta$
c. $\frac{dy}{dx} = \cot\theta$
d. None of these

iii. In previous problem, if μ is refractive index at the point P. Then,

a. $\mu \sin\theta$ = constant = C
b. $\mu \sin(90° - \theta)$ = constant = C
c. $\mu \tan\theta$ = constant = C
d. None of these

iv. In previous problem, the value of constant C is

a. 1
b. 2
c. 3
d. $\sqrt{2}$

v. In previous problem, $\tan\theta$ is

a. $\tan\theta = 1$
b. $\tan\theta = \frac{1}{\sqrt{\mu^2 - 1}}$
c. $\tan\theta = \sqrt{\mu^2 - 1}$
d. None of these

vi. In previous problem, the correct relation is

a. $\frac{1}{2x} = \frac{1}{\sqrt{\mu^2 - 1}}$
b. $2x = \sqrt{\mu^2 - 1}$
c. $\mu = \sqrt{4x^2 + 1}$
d. $\mu = \sqrt{4y + 1}$

3 **Situation** A ray of light suffering two internal reflection by water spherical drop of refractive index μ. This is the basic analytical discussion of secondary rainbow as shown in the figure.

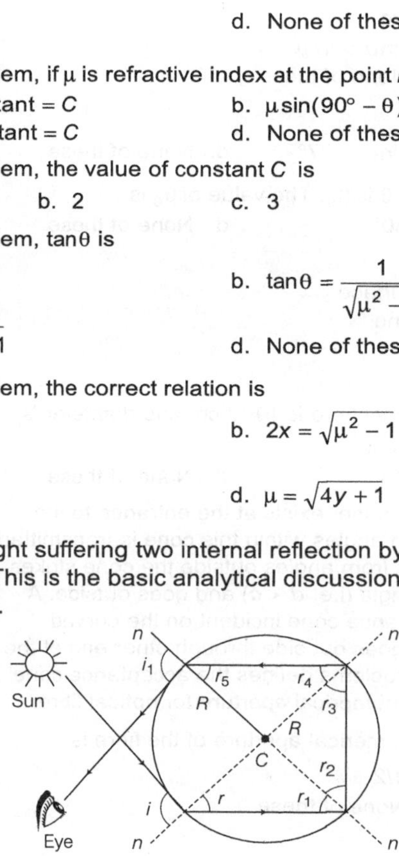

i. Mark the correct option(s).

a. $i = i_1$
b. $r = r_1 = r_2 = r_3 = r_4 = r_5$
c. $i \neq i_1$
d. $r \neq r_2$

ii. In each refraction deviation in ray is

a. $i - r$
b. $180° - 2i$
c. $180° - 2r$
d. None of these

iii. In each reflection deviation in ray is

a. $i - r$
b. $180° - 2i$
c. $180° - 2r$
d. None of these

iv. Net deviation in the rays is

a. $\delta = 2(i - r)$
b. $\delta = 0$
c. $\delta = 2(i - r) + 2(180° - 2r)$
d. None of these

v. $\frac{d\delta}{di}$ is

a. $\frac{d\delta}{di} = 2 - \frac{6dr}{di}$
b. $\frac{d\delta}{di} = 2 - \frac{dr}{di}$
c. $\frac{d\delta}{di} = 2$
d. None of these

vi. For minimum deviation in ray is

a. $\frac{d\delta}{di} = 0$
b. $\frac{di}{dr} = 3$
c. $\frac{di}{dr} = 6$
d. None of these

vii. Mark the correct option(s).

a. $\sin i = \mu \sin r$
b. $\frac{di}{dr} = \frac{\mu \cos r}{\cos i}$
c. $\frac{di}{dr} = 0$
d. None of these

viii. For minimum deviation, the value of $\cos i$ is

a. $\cos i = \sqrt{\frac{\mu^2 - 1}{2}}$
b. $\cos i = \sqrt{\frac{\mu^2 - 1}{8}}$
c. $\cos i = \frac{1}{2}$
d. None of these

4 **Situation** A ray of light travelling in air is incident at grazing angle on a long rectangular slab of a transparent medium of thickness $t = 1\text{m}$.

The refractive index of slab is $\mu = \left[ky^{3/2} + 1\right]^{1/2}$, where $k = 1\text{m}^{-3/2}$. The slab is placed in air. The ray is incidence at the origin.

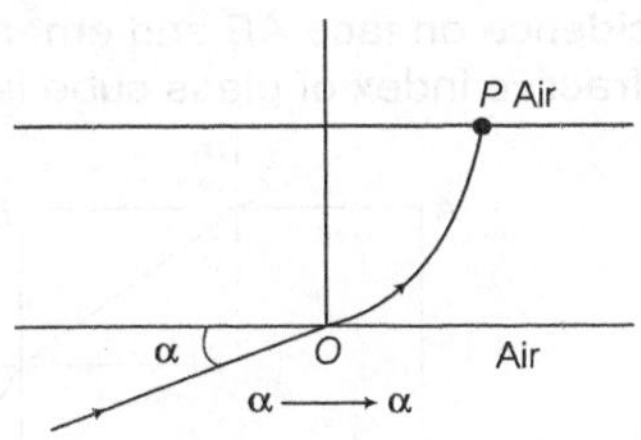

i. The angle of incidence is

a. $\to 0°$
b. $90°$
c. $< 45°$
d. None of these

ii. Consider the ray shown in the figure. The coordinates of point A is (x, y) and is lying on the ray. The ray makes θ angle with Y-axis at point A. Then, according to Snell's law

a. $\mu \sin(90° - \theta) = C = \text{constant}$
b. $\mu \sin\theta = C = \text{constant}$
c. $[k\,y^{3/2} + 1]^{1/2} \sin\theta = C = \text{constant}$
d. None of the above

Working Space

iii. In previous problem, the value of constant C is

a. 1 b. 2 c. 3 d. 4

iv. In previous problem, the value of $\tan\theta$ is

a. $y^{-3/4}$
b. $y^{3/4}$
c. $\sqrt{y}$
d. None of these

v. In previous problem, the correct relation is

a. $\frac{dx}{dy} = y^{-3/4}$
b. $\frac{dy}{dx} = y^{-3/4}$
c. $\frac{dy}{dx} = \sqrt{y}$
d. $\frac{dx}{dy} = \sqrt{y}$

vi. The equation of ray inside the slab is

a. $x = \frac{y^4}{256}$
b. $y = \frac{x^4}{256}$
c. $y = x^{3/4}$
d. None of these

vii. For point P, y-coordinate is

a. 1 m
b. 0.5 m
c. 2 m
d. None of these

viii. The x-coordinate of point P is

a. $x = 1\text{m}$
b. $x = 0.5\text{ m}$
c. $x = 4\text{ m}$
d. None of these

ix. When the ray emerges from slab

a. that must be parallel to initial direction of ray
b. parallel to X-axis
c. parallel to Y-axis
d. None of the above

x. The deviation in ray produced by slab is

a. 0°
b. 180°
c. 90°
d. None of these

xi. The lateral displacement in ray due to slab is

a. 4 m
b. 1 m
c. 0.5 m
d. $\sqrt{17}$ m

5 **Situation** A ray incidence on face AB and emerges out from face BC as shown in the figure. The refractive index of glass cube is μ.

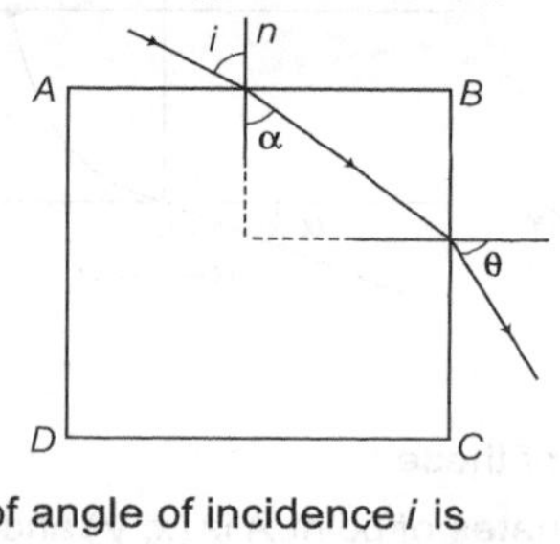

i. The maximum value of angle of incidence i is

a. 0° b. 90° c. 180° d. < 45°

ii. The value of $\sin\alpha$ is

a. $(\sin i)/\mu$
b. $\mu \sin i$
c. $\mu \cos i$
d. None of these

iii. The maximum value of α is

a. $\alpha_{max} = \sin^{-1}(1/\mu)$
b. $\alpha_{max} = \sin^{-1}(\mu)$
c. $\alpha_{max} = \cos^{-1}(1/\mu)$
d. None of these

Working Space

iv. The relation between α_{max} and θ is

a. $\sin\theta = \cot\alpha_{max}$
b. $\sin\theta = \cos\alpha_{max}$
c. $\cot\theta = \sin\alpha_{max}$
d. None of these

v. For emergence of ray from *BC*,

a. $90° - \alpha_{max} \leq \sin^{-1}(1/\mu)$
b. $90° - \alpha_{max} > \sin^{-1}(1/\mu)$
c. $90° - \alpha_{max} = \sin^{-1}(1/\mu)$
d. None of these

vi. The value of μ is

a. $\mu > \sqrt{2}$
b. $\mu \leq \sqrt{2}$
c. $\mu > 1$
d. None of these

6 **Situation** A beam of white light is incident on the curved edge of a semicircular piece of glass. The light enters the curved surface along the normal as shown in the figure.

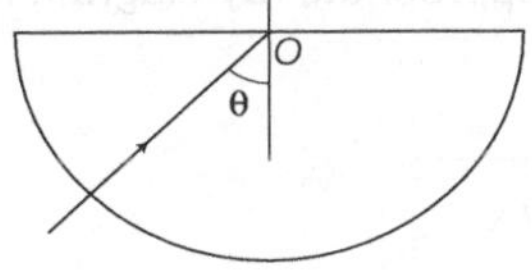

i. At curved edge,

a. refraction does not take place in the ray
b. reflection takes place
c. refraction takes place
d. None of the above

ii. If refractive index of the glass for red and violet light rays are μ_r and μ_v respectively. Then,

a. $\mu_v = \mu_r$
b. $\mu_v > \mu_r$
c. $\mu_v < \mu_r$
d. None of these

iii. Mark the correct option(s).

a. Total internal reflection may take place from flat surface
b. Total internal reflection never takes place from flat surface
c. Total internal reflection takes place for all values of θ
d. None of the above

iv. The critical angle for red light is

a. $C_r = \sin^{-1}(1/\mu_r)$
b. $C_r = \sin^{-1}(\mu_r)$
c. $C_r = \cos^{-1}(1/\mu_r)$
d. None of these

v. If critical angle for red light and violet light are C_r and C_v respectively. Then,

a. $C_r = C_v$
b. $C_r > C_v$
c. $C_r < C_v$
d. None of these

vi. If θ gradually increases from 0°. Then,

a. total internal reflection first takes place for red ray
b. total internal reflection first takes place for violet ray
c. total internal reflection takes place simultaneously for all colours
d. None of the above

vii. In previous problem, just before total internal reflection takes place of all colours, the refracted ray from flat surface is

a. violet
b. red
c. yellow
d. None of these

viii. In previous problem, for what value of θ are refracted beam from flat surface disappears.

a. $\theta > \sin^{-1}(1/\mu_v)$
b. $\theta > \sin^{-1}(1/\mu_r)$
c. $\theta \leq \sin^{-1}(1/\mu_r)$
d. $\theta \leq \sin^{-1}(1/\mu_v)$

Working Space

7 **Situation** A negative refractive index material, can now be artificially prepared. Such type of materials are known as metamaterials. Metamaterials significantly different optical behaviour without violeting any physical laws. Since, refractive index is negative, it results in a change in the direction of propagation of the refracted light. However, the frequency of light remains unchanged upon refraction even in metamaterial.

i. The refractive index of a metamaterial is $\mu = -3/2$. The speed of light in metamaterial is

a. $-2 \times 10^8 \text{ ms}^{-1}$ b. $2 \times 10^8 \text{ ms}^{-1}$

c. $2.25 \times 10^8 \text{ ms}^{-1}$ d. $3 \times 10^8 \text{ ms}^{-1}$

ii. The wavelength of red light in vacuum is 700 nm. A metamaterial has refractive index $\mu = -3.5$. Find the wavelength of red light in metamaterial.

a. 2450 nm b. −2450 nm c. 200 nm d. 700 nm

iii. For light incident from air on a metamaterial, the appropriate ray diagram is

a. 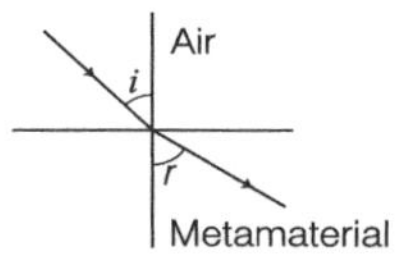

b. 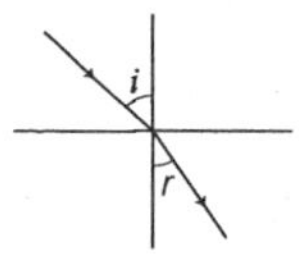

c. 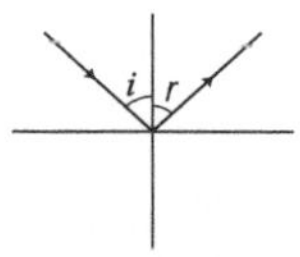

d. 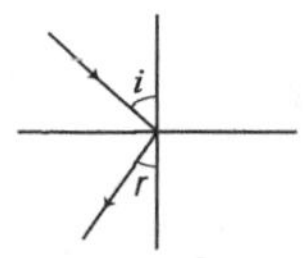

iv. If the angle of incidence $i = 45°$, the refractive index of metamaterial is $\mu = -\sqrt{2}$. Find the value of r in previous problem.

a. 30° b. 45° c. 60° d. None of these

Only One Option Correct Type

Direction (Q. Nos. 8-25) *This section contains 18 multiple choice questions. Each question has four choices (a), (b), (c) and (d), out of which ONLY ONE option is correct.*

8 Find the angle of incidence, θ for which reflected and refracted beams are perpendicular to each other.

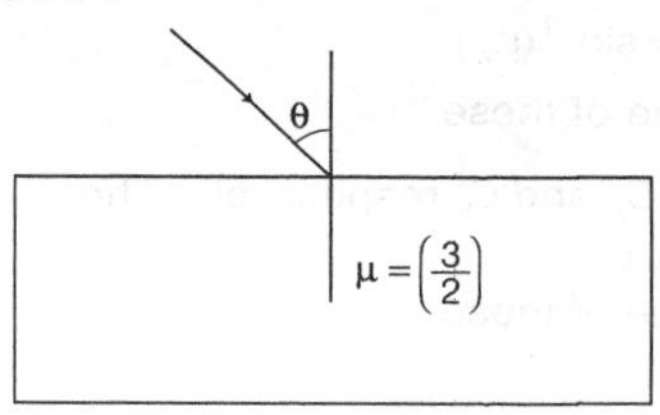

a. $\sin^{-1}\left(\frac{2}{3}\right)$ b. $\cos^{-1}\left(\frac{2}{3}\right)$

c. $\tan^{-1}\left(\frac{2}{3}\right)$ d. $\cot^{-1}\left(\frac{2}{3}\right)$

9 A ray of light incident from vacuum at angle of incidence i and is refracted in a medium of angle of refraction r. The speed of light in the medium is $2 \times 10^8 \text{ ms}^{-1}$. The slope of the graph is

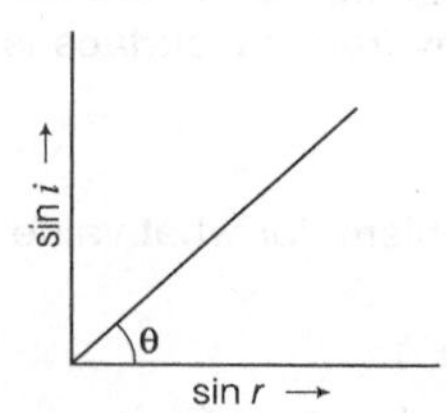

a. $\frac{4}{3}$ b. $\frac{3}{2}$

c. 2 d. $\frac{1}{2}$

Working Space

10 In the given figure, surface *BC* of slab is silvered. A ray of light incident on face *AB* and emerges out from face *CD*. Find the angle of emergence.

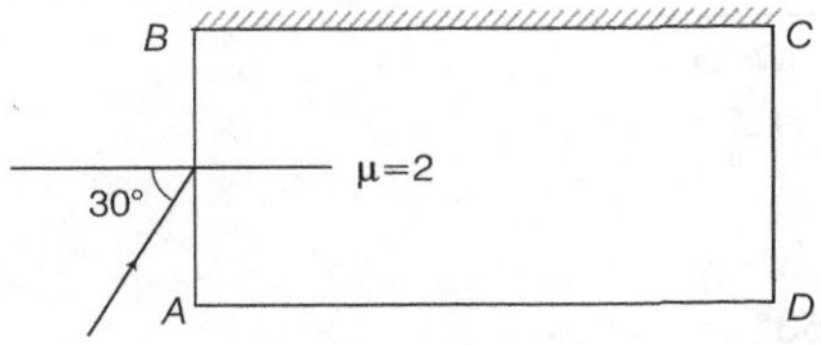

a. 15° b. 30°
c. 45° d. 60°

11 A ray of light propagates from air to glass of as shown in the figure. The ratio of magnitude of momentum of photon in air and glass is

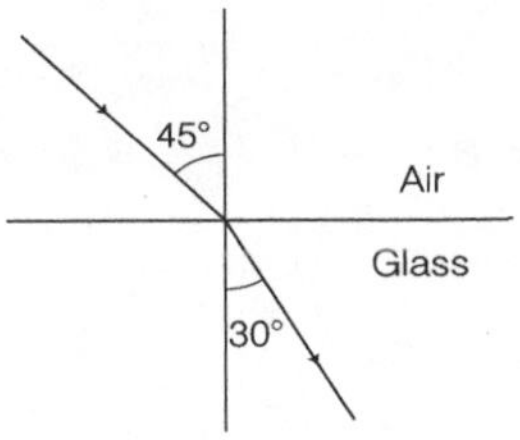

a. $\sqrt{2}$ b. $\sqrt{3}$
c. $\frac{1}{\sqrt{2}}$ d. $\frac{1}{\sqrt{3}}$

12 A plane surface separates two media of refractive indices, $\mu_1 = 2$ and $\mu_2 = 1$. A converging beam of light incident on the surface, whose converging point is 10 cm behind the surface as shown in the figure. The new converging point is

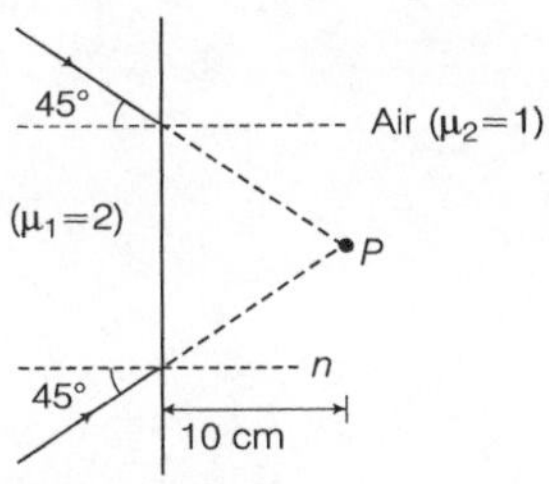

a. before the surface at a distance 10 cm
b. behind the surface at a distance less than 10 cm
c. behind the surface at a distance greater than 10 cm
d. before the surface at a distance greater than 10 cm

13 When ray propagates from denser medium to rarer medium, deviation in ray (δ) *versus* angle of incidence (*i*) graph is

a. 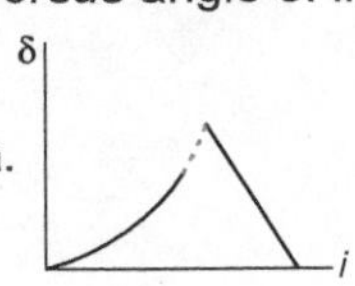

b. 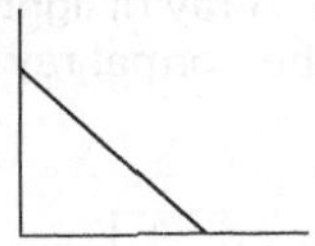

c. 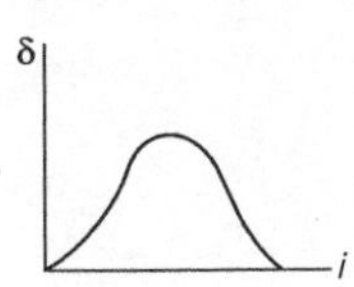

d. 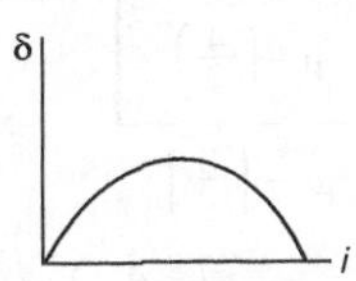

Working Space

14 In the given figure, if $\theta = 53°$ the angle of deviation is

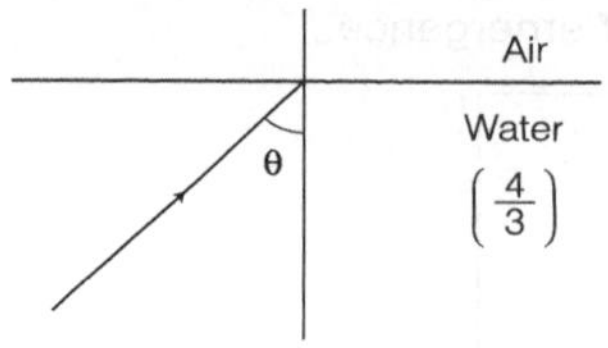

a. 74°
b. 53°
c. 106°
d. 90°

15 In previous problem, a transparent slab of refractive index $\frac{3}{2}$ is placed as shown in the figure, then

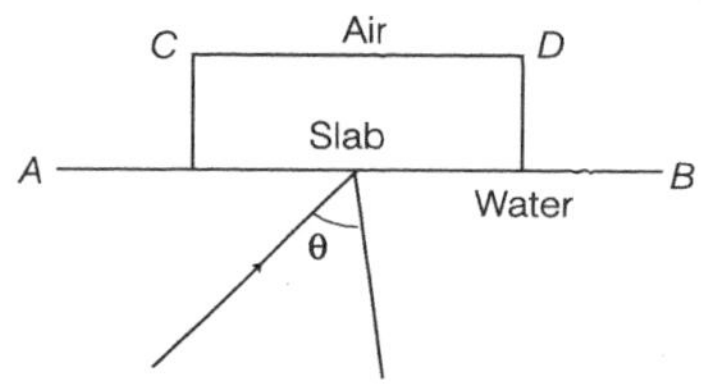

a. the ray will never reach at surface *CD*
b. the ray easily emerges out into air from the surface *CD*
c. the ray returns back into the water
d. total internal reflection does not take place from any surface

16 A ray of light enters into a thick slab of hypothetical medium, whose refractive index varies as $\mu = (\sqrt{3} - y)$. Find the maximum value of *y*-coordinate of the ray.

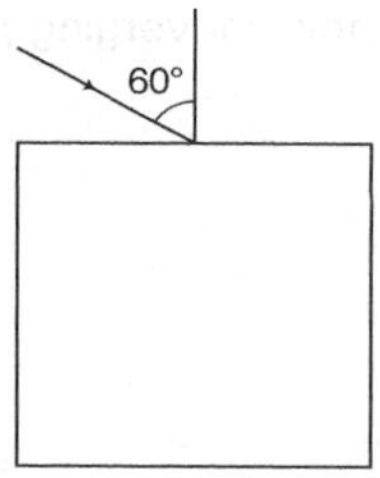

a. $\sqrt{3}$ unit
b. $\frac{\sqrt{3}}{2}$ unit
c. $\frac{\sqrt{3}}{4}$ unit
d. $\frac{3}{2}$ unit

17 A thick slab made of a transparent material of refractive index $\frac{4}{3}$ is placed in a region. The medium above the slab is air and the refractive index of medium below the slab is 3/2 as shown in the figure. A ray of light incident on slab at an angle 60° with vertical. At what angle with the normal ray will leave the plate?

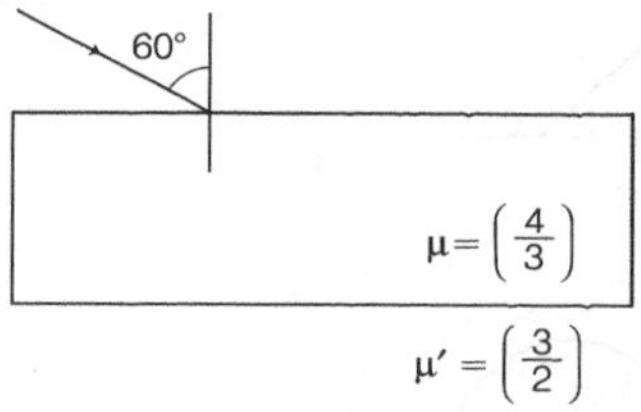

a. 30°
b. 45°
c. $\sin^{-1}\left(\frac{4}{3\sqrt{3}}\right)$
d. $\cos^{-1}\left(\frac{4}{3\sqrt{3}}\right)$

Working Space

18 A slab of variable refractive index is placed in air. The refractive index is $\mu = 1 + \frac{y}{a}$, where $a = 1\,\text{m}^{-1}$. Find the angle of incidence at upper boundary.

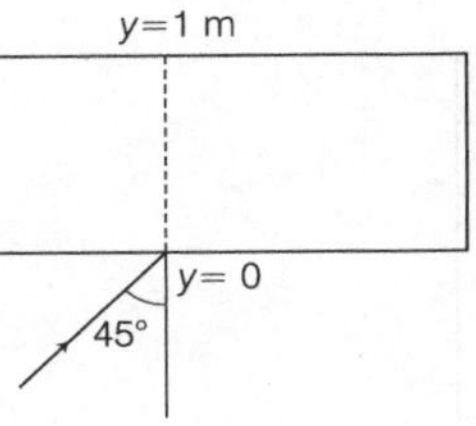

a. $\sin^{-1}\left(\frac{1}{2\sqrt{2}}\right)$

b. $\cos^{-1}\left(\frac{1}{2\sqrt{2}}\right)$

c. $\sin^{-1}\left(\frac{1}{3}\right)$

d. $\sin^{-1}\left(\frac{1}{4}\right)$

19 A ray of light is incident on a slab of refractive index $\mu = 2$ at an angle of incidence 60° as shown in the figure. The thickness of slab is 1 m. The lateral shifting in the ray is

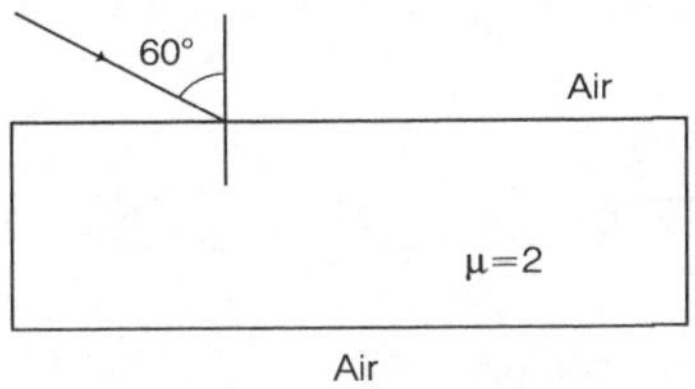

a. 1 m

b. 0.5 m

c. $\left(\frac{\sqrt{39}-\sqrt{3}}{4}\right)$m

d. 2 m

20 A ray of light passes from vacuum into a medium of refractive index n. If the angle of incidence is twice the angle of refraction, then the angle of incidence is

a. $\cos^{-1}\left(\frac{n}{2}\right)$

b. $\sin^{-1}\left(\frac{n}{2}\right)$

c. $2\cos^{-1}\left(\frac{n}{2}\right)$

d. $2\sin^{-1}\left(\frac{n}{2}\right)$

21 A cubic container is filled with a liquid, whose refractive index increases linearly from top to bottom. Which of the following represents the path of a ray of light inside the liquid?

a. 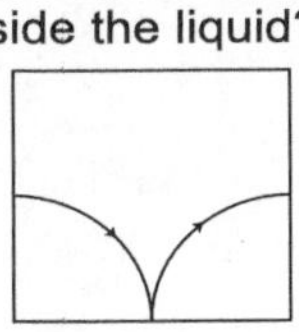

b. 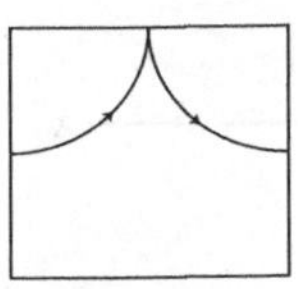

c. 

d. 

Working Space

22 Light is incident on a glass block as shown θ_1 is increased slightly. What happens to θ_2?

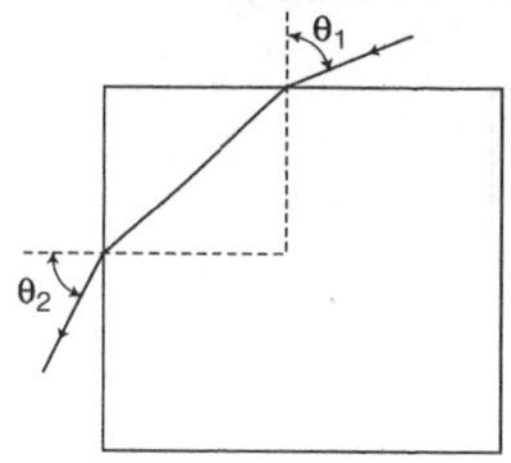

a. θ_2 also increases slightly
b. θ_2 is unchanged
c. θ_2 decreases slightly
d. θ_2 changes abruptly, since the ray experiences total internal reflection

23 A ray incident on the spherical surface separating two media of refractive indices 1 and $\sqrt{3}$ respectively. Find the angle between reflected and refracted ray through the spherical surface.

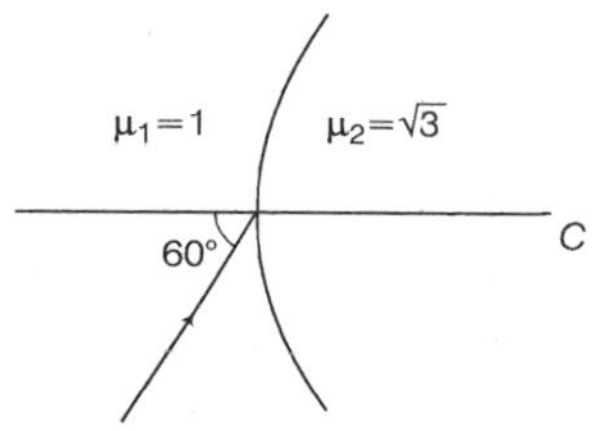

a. 60° b. 30°
c. 90° d. 15°

24 A ray incident at a point at an angle of incidence 60° enters a glass sphere of $\mu = \sqrt{3}$ and is reflected and refracted at the further surface of the sphere. The angle between the reflected and the refracted rays at this surface is

a. 50° b. 90°
c. 60° d. 40°

25 A ray of light travelling in air is incident at angle $\theta = 30°$ on a long rectangular slab of a transparent medium. The point of incidence is origin O of the coordinate system as shown in the figure.

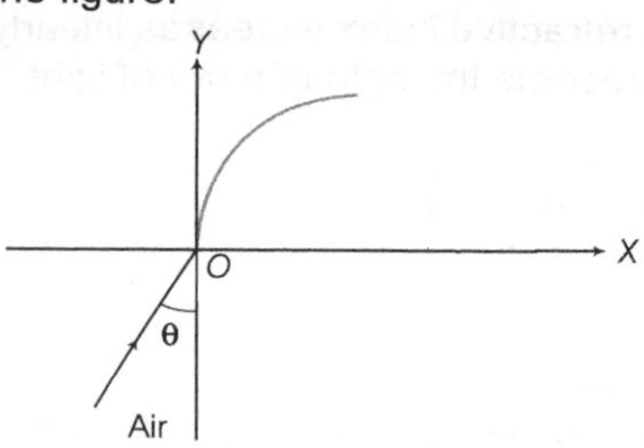

The medium has a variable index of refraction $\mu(y)$ given by $\mu = (0.25 + ky^{-2})^{1/2}$, where $k = 1.0\ m^2$. Calculate the equation for trajectory of the ray in the medium.

a. $x^2 = 4y$ b. $y^2 = 4x$
c. $y^2 = 4x^3$ d. $x^2 + y^2 = 4$

DPP-6 Refraction of Light-II

One or More than one options Correct type

Working Space

Direction (Q. Nos. 1-5) *This section contains 5 multiple choice questions. Each question has four choices (a), (b), (c) and (d), out of which ONLY ONE option is correct.*

1 The speed of light in a medium A is $\sqrt{3} \times 10^8\ \text{ms}^{-1}$. Mark the correct option(s).

a. The refractive index of medium A is $\sqrt{3}$

b. The critical angle for air and medium A boundary is $\sin^{-1}\left(\frac{1}{\sqrt{3}}\right)$

c. The critical angle for air and medium A boundary is $\cos^{-1}\left(\frac{1}{3}\right)$

d. None of the above

2 Two thin slabs of refractive indices $\frac{4}{3}$ and $\frac{3}{2}$ are placed parallel to each other in XY-plane as shown in the figure. A ray of light in XY-plane incident from air at an angle of incidence 45°.

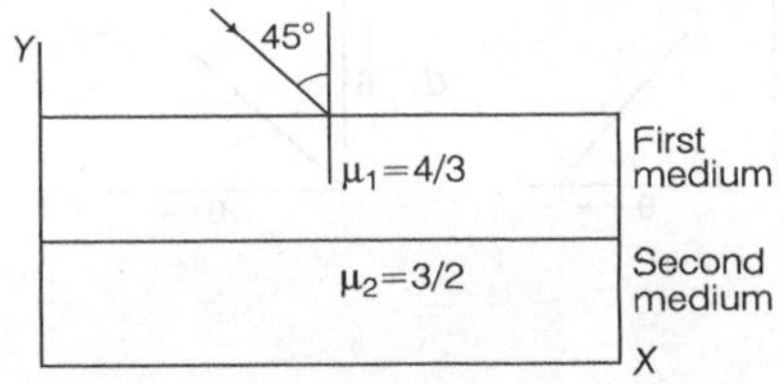

a. The unit vector along ray in first medium $\frac{3}{4\sqrt{2}}\hat{\mathbf{i}} - \frac{\sqrt{23}}{4\sqrt{2}}\hat{\mathbf{j}}$

b. The unit vector along ray in second medium $\frac{2}{3\sqrt{2}}\hat{\mathbf{i}} - \frac{\sqrt{14}}{3\sqrt{2}}\hat{\mathbf{j}}$

c. Unit vector along ray in first medium is $\frac{1}{\sqrt{2}}\hat{\mathbf{i}} - \frac{1}{\sqrt{2}}\hat{\mathbf{j}}$

d. None of the above

3 XY-plane is the boundary between two transparent media. First medium with $z \geq 0$ has refractive index $\sqrt{2}$ and second medium with $z \leq 0$ has a refractive index $\sqrt{3}$. A ray of light in first medium given by vector $\hat{\mathbf{i}} + 2\hat{\mathbf{j}} - 2\hat{\mathbf{k}}$.

a. The angle made by ray with normal in first medium is $\cos^{-1}\left(\frac{2}{3}\right)$

b. The unit vector along ray in second medium is $\frac{\sqrt{2}}{3\sqrt{3}}\hat{\mathbf{i}} + \frac{2\sqrt{2}}{3\sqrt{3}}\hat{\mathbf{j}} - \frac{\sqrt{17}}{3\sqrt{3}}\hat{\mathbf{k}}$

c. The angle made by ray with normal in second medium is $\cos^{-1}\left(\frac{\sqrt{17}}{3\sqrt{3}}\right)$

d. None of the above

4 In the diagram as shown, a light ray is incident on the lower medium boundary at an angle 45° with the normal. Which of the following statements is/are correct?

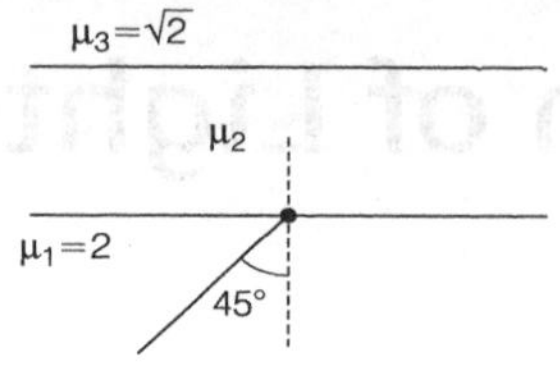

a. If $\mu_2 > \sqrt{2}$ then angle of deviation is 45°
b. If $\mu_2 < \sqrt{2}$ then angle of deviation is 90°
c. If $\mu_2 < \sqrt{2}$ then angle of deviation is 135°
d. If $\mu_2 > \sqrt{2}$ then angle of deviation is 0°

5 A ray of light incident at an angle θ. The angle θ is always lesser than critical angle. Then, which of following graphs represent angular deviation (δ) *versus* angle of incidence (θ) graph?

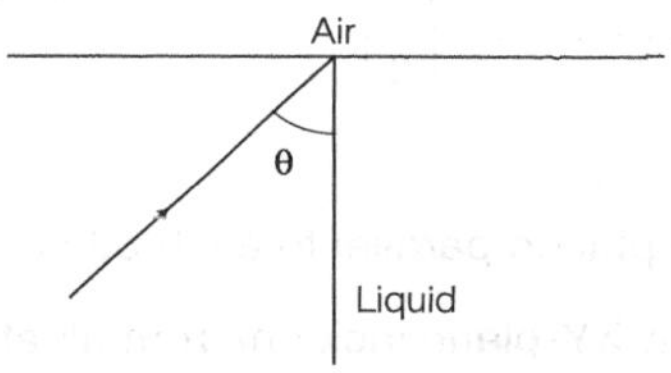

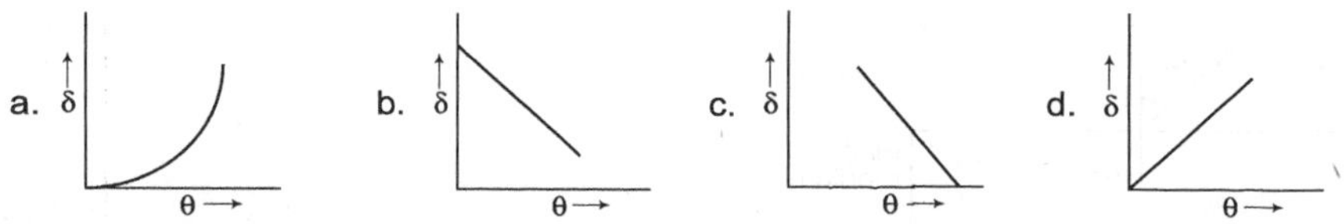

Comprehension Type

***Direction** (Q. Nos. 6-9) This section contains 2 paragraphs, each describing theory, experiments, data, etc. Four questions related to the paragraphs have been given. Each question has only one correct answer among the four given options (a), (b), (c) and (d).*

Passage I

A parallel beam of light falls from vacuum on a surface enclosing a medium of refractive index, $\mu = 2$. The beam of focus at point F (3 cm, 0) after refraction.

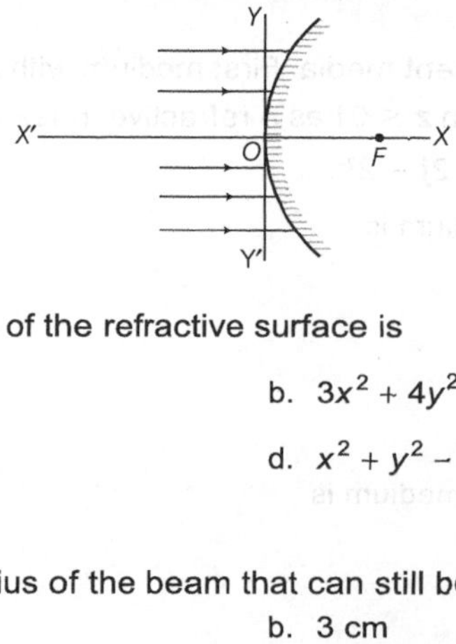

6 The equation of shape of the refractive surface is

a. $\frac{x^2}{3} + \frac{y^2}{4} = 1$
b. $3x^2 + 4y^2 - 12x = 0$
c. $3x^2 - 4y^2 - 12x = 0$
d. $x^2 + y^2 - 12x = 0$
e. $y^2 = 6x$

7 Find the maximum radius of the beam that can still be focussed.

a. 2 cm
b. 3 cm
c. $\sqrt{3}$ cm
d. $3\sqrt{2}$ cm

Working Space

Passage II

The direction of ray is along the Y-axis, when $x < 0$. The refractive index of the rectangular slab varies as, $\mu = \dfrac{\sqrt{1+(1+x)^2}}{1+x}$.

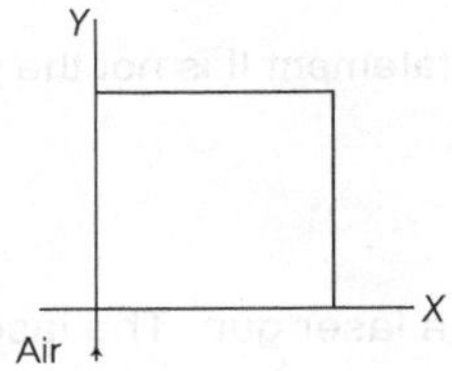

8 Find the angle made by ray with Y-axis, when $x \to 0$.

a. 30° b. 45°
c. 90° d. 0°

9 Find the equation of ray inside the slab.

a. $y = x + 2$ b. $y = x - \dfrac{x^2}{2}$

c. $y = x + \dfrac{x^2}{2}$ d. $2x + x^2$

Matching List Type

***Direction** (Q. No. 10) Choices for the correct combination of elements from Column I and Column II are given as options (a), (b), (c) and (d), out of which one is correct.*

10 Match the Column I with Column II and mark the correct option from the codes given below.

	Column I		Column II
i.	When ray propagates from rarer medium to denser medium deviation *versus* angle of incidence graph is	p.	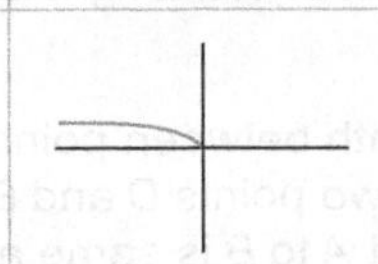
ii.	When ray propagates from denser medium to rarer medium, deviation *versus* angle of incidence graph	q.	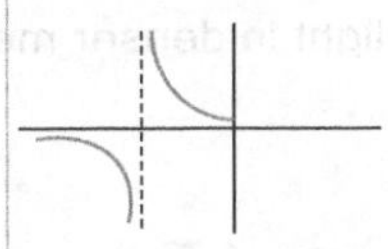
iii.	Image distance *versus* object distance graph for real object in the case of concave mirror	r.	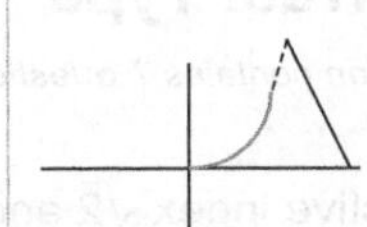
iv.	Image distance *versus* object distance graph for real object in the case of convex mirror	s.	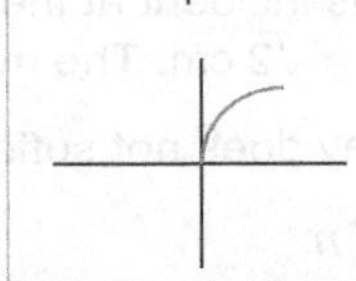

Codes

	i	ii	iii	iv		i	ii	iii	iv
a.	s	r	q	p	b.	p	q	r	s
c.	r	s	p	q	d.	s	r	p,q	p,q

Working Space

Statement Type

Direction (Q. Nos. 11-16) *This section is based on Statement I and Statement II. Select the correct answer from the codes given below.*

a. Both Statement I and Statement II are correct and Statement II is the correct explanation of Statement I
b. Both Statement I and Statement II are correct but Statement II is not the correct explanation of Statement I
c. Statement I is correct but Statement II is incorrect
d. Statement II is correct but Statement I is incorrect

11 **Statement I** If a fish in water is aimed from a duck by a laser gun. The laser gun is exactly aimed at the fish.

Statement II Snell's law is applicable for paraxial rays as well as marginal rays.

12 **Statement I** When a ray of light is refracted from water to glass, its wavelength decreases while frequency does not change.

Statement II When a ray of light is refracted from water to glass, it bends towards normal.

13 **Statement I** When ray propagates from denser medium to rarer medium, for angle of incidence equal to critical angle, refraction takes place.

Statement II When ray propagates from denser medium to rarer medium, for angle of incidence greater than critical angle, total internal reflection takes place.

14 **Statement I** When a ray of light strike with plane water surface, the momentum of photon along the water surface remains conserved.

Statement II Water surface behaves as plane mirror.

15 **Statement I** Critical angle of the light passing from glass to air is minimum for violet colour.

Statement II The wavelength of violet light is greater than the light of the other colours.

16 **Statement I** If the optical path between points A and B in first medium is same as the optical path between two points D and E in second medium. Then, time taken by light to travel from point A to B is same as that of from point D to E.

Statement II The speed of light in denser medium is same as that of in rarer medium.

One Integer Value Correct Type

Direction (Q. Nos. 17-23) *This section contains 7 questions. When worked out will result in an integer from 0 to 9 (both inclusive).*

17 A quarter cylinder of refractive index $\sqrt{2}$ and radius $\sqrt{3}$ cm is placed in air. A ray of light is incident at the point P at an angle of incidence such that, $OP = \sqrt{2}$ cm. The minimum value of θ in radian is $\frac{\pi}{n}$ such that, the ray does not suffer total internal reflection. Find the value of n.

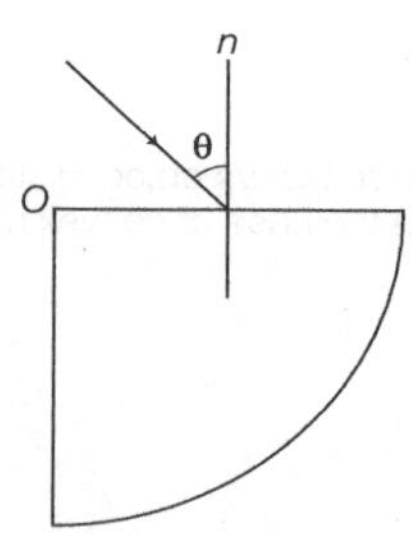

Working Space

18 A parallel beam of light is incident on a quarter cylinder lies on horizontal surface. The refractive index of the quarter cylinder is $\mu = \sqrt{2}$ and radius is $\sqrt{2}$ cm. Find the maximum distance (in cm) from O, where the patch of light be found.

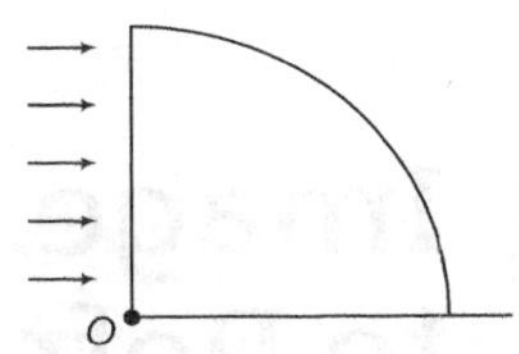

19 A hemispherical glass of radius 20 cm and refractive index $\sqrt{3/2}$ lies on a horizontal plane mirror. A horizontal ray of light is incident on it at height $10\sqrt{3}$ cm above the plane mirror. If the angle of incidence on the plane mirror is $\frac{n\pi}{12}$ rad. Find the value of n.

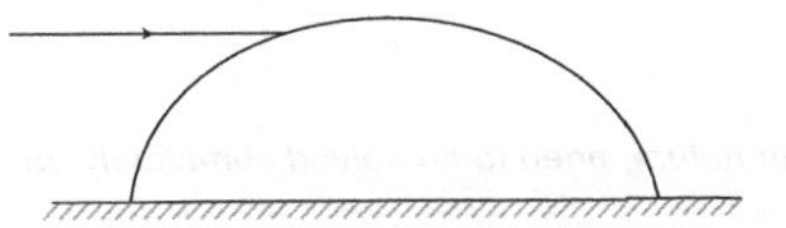

20 A ray of light is refracted through a glass hemisphere of refractive index μ in such a way that, it passes through the extremities of two radii, which make an angle 90° with each other. The deviation in ray is 30°. If refractive index $\mu = \sqrt{\frac{n}{2}}$. Find the value of n.

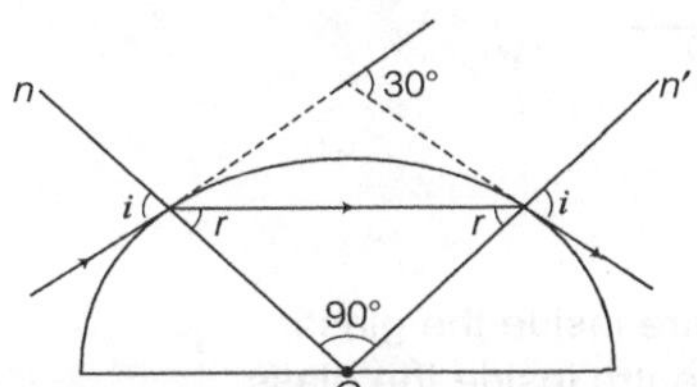

21 A ray of light is incident at origin from air in grazing incidence on a slab of refractive index $\mu = \sqrt{1 + ay}$, where $a = 2 \times 10^{-6} \text{m}^{-1}$. The ray emerges out from point P as shown in the figure. Find x-coordinate (in km) of the point P. The thickness of slab is 2 m.

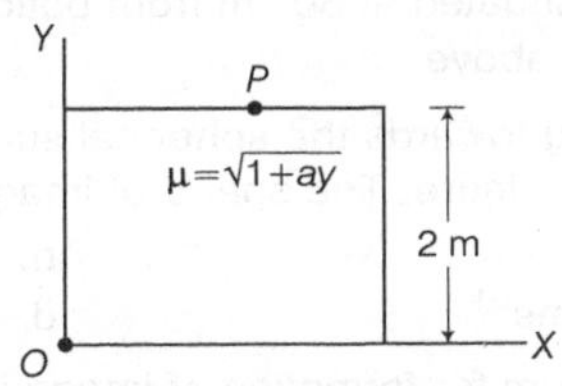

22 A portion of flexible glass rod of diameter 2 cm and refractive index $\frac{3}{2}$ is bent into an arc of circle of radius R. A parallel beam of light is incident on it as shown in the figure. Find smallest value of R (in cm), which permits all the light to pass around the arc.

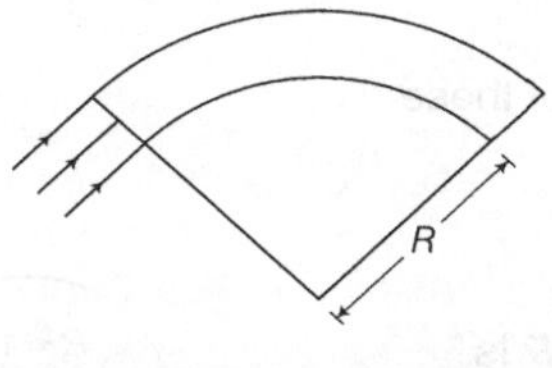

23 A solid hemisphere of refractive index μ is placed in air. The hemisphere is fixed at point O. A horizontal ray is incident at height $(R + h)$ from bottom as shown in the figure. The angular deviation in the ray in degree is

$$\delta = \sin^{-1}\left(\frac{h}{R^2}(\sqrt{\mu^n R^2 - h^2} - \sqrt{R^2 - h^2}\right)$$

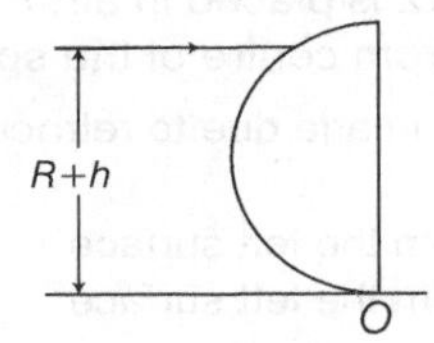

Find the value of n.

DPP-7 Image Formation due to Refraction through Spherical Surface-I

Conceptual Questions

Working Space

Direction *(Q. Nos. 1-3) These questions are conceptual in nature, need to be solved completely on notebook.*

1 **Situation** A spherical surface of radius 10 cm separates two media (water and glass), whose refractive indices are 4/3 and 3/2 respectively. A point object P is at 40 cm from the optical centre of the refractive surface.

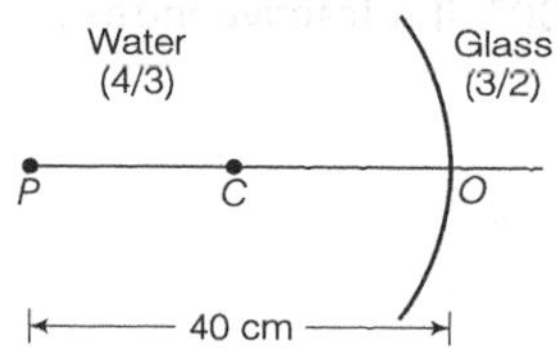

i. The image of point object P is

a. real and situated at 30 cm from optical centre inside the glass

b. virtual and situated at 30 cm from optical centre inside the glass

c. virtual and situated at 30 cm from optical centre inside the water

d. None of the above

ii. If object moving towards the spherical surface with speed 10 cms^{-1} at the position as shown in the figure. The speed of image is

a. 10 cms^{-1} b. 5 cms^{-1}

c. (810/128) cms^{-1} d. None of these

iii. Draw ray diagram for formation of image is

a. 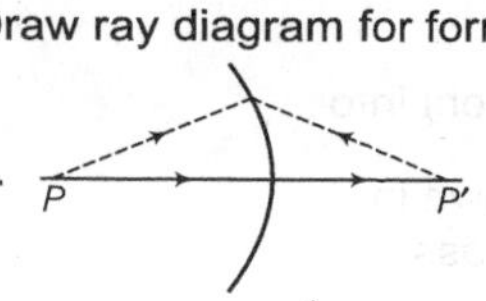

b. 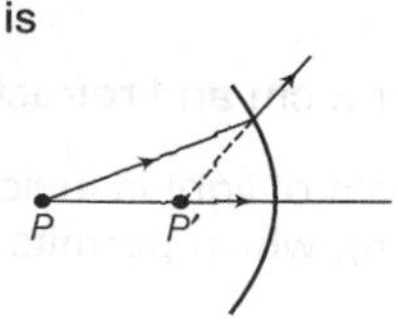

c. 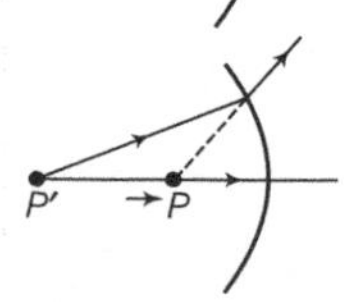

d. None of these

2 **Situation** A solid glass sphere of radius 10 cm and refractive index 3/2 is placed in air. A point object P is placed at 30 cm from centre of the sphere.

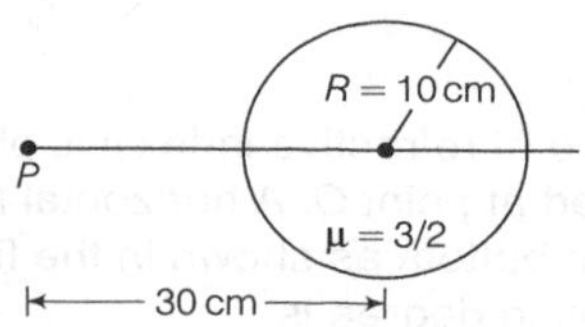

i. The position of image due to refraction through left surface is

a. at 20 cm from the left surface

b. at 30 cm from the left surface

c. at infinity

d. None of the above

Working Space

ii. The final image is

a. real and 20 cm from right surface
b. virtual and 20 cm from right surface
c. at infinity
d. None of these

iii. The ray diagram of the problem is

a. 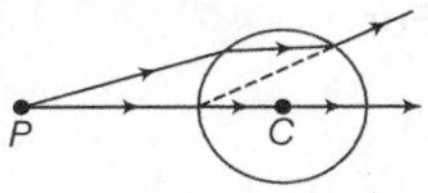

b. 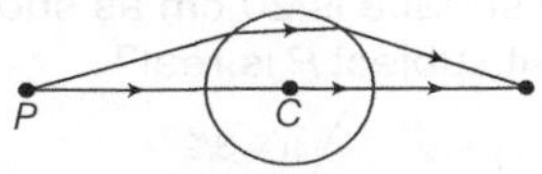

c. 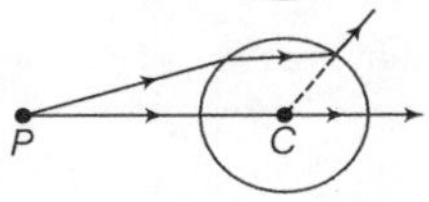

d. 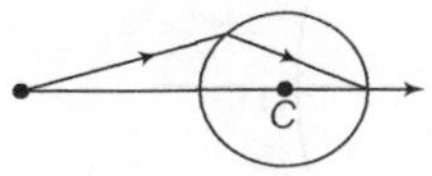

3 **Situation** A cubical block of glass of refractive index 1.5 has a concentric cavity of radius $r = 3$ cm. Each edge of the cube is $l = 12$ cm long. A luminious point object P at a distance, $a = 12$ cm on the left face of the cube as shown in the figure.

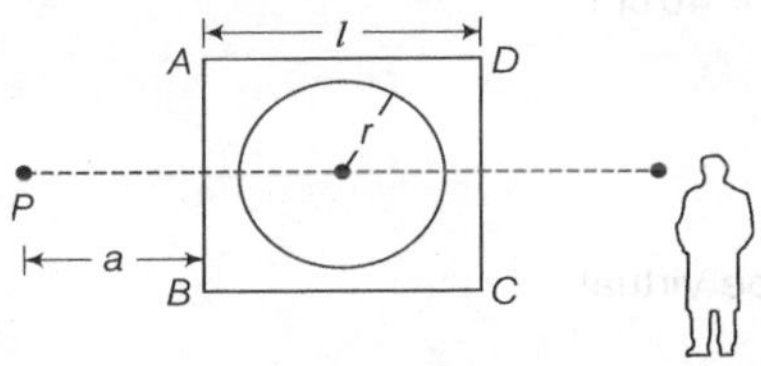

i. The position of image P_1 due to refraction through surface AB is

a. 18 cm right from surface AB
b. 18 cm left from surface AB
c. 21 cm left from surface AB
d. None of these

ii. From previous problem P_1 behaves as

a. object for left surface of cavity
b. image for left surface of cavity
c. the object distance from left surface of cavity is 21 cm in left
d. None of the above

iii. The position of image P_2 due to refraction through left surface of cavity is

a. 21 cm left of left surface of cavity
b. 4.2 cm left from left surface of cavity
c. 17/3 cm left of left surface of cavity
d. None of the above

iv. P_2 behaves as

a. object for right surface of cavity
b. image for right surface of cavity
c. object distance is 10.2 cm left for right surface of cavity
d. None of the above

v. The position of image P_3 formed due to refraction through right surface of cavity is

a. 4.2 cm right of the right surface of cavity
b. 17/3 cm left of the right surface of cavity
c. 17/3 cm right from right surface of cavity
d. None of the above

vi. P_3 behaves as

a. object for right plane surface of the cube
b. image for right plane surface of the cube
c. object distance for right plane surface of cube is 26/3 cm left of the surface
d. None of the above.

vii. Final image is

a. virtual and 52/9 cm left from right surface of the cube
b. real and 52/9 cm right from right surface of cube
c. virtual and 4.2 left of the right surface of cube
d. None of the above

Working Space

Only One Option Correct Type

Direction *(Q. Nos. 4-20) This section contains 17 multiple choice questions. Each question has four choices (a), (b), (c) and (d), out of which ONLY ONE option is correct.*

4 The radius of refractive spherical surface is 20 cm as shown in the figure. For what value of x, the image of point subject P is real?

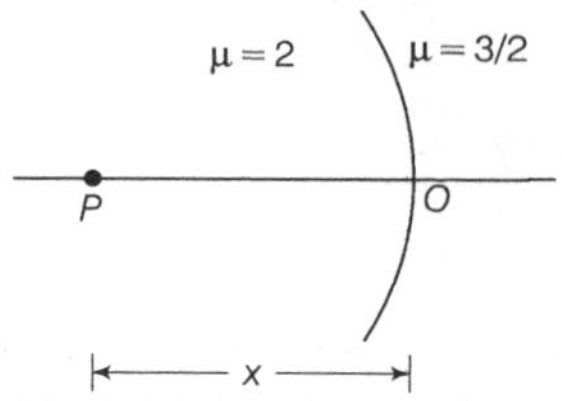

a. $x > 80$ cm
b. $x < 80$ cm
c. $x > 40$ cm
d. $x < 40$ cm

5 In previous problem, for $x = 60$ cm,

a. image of point object P is real
b. image of point object P is virtual
c. image of point object P may be real or may be virtual
d. None of the above

6 A spherical refractive surface of radius 10 cm separates two media of refractive indices $\mu = 1$ and $\mu = 3/2$ respectively. A point object P starts from rest at $t = 0$ with an acceleration 2 cm s^{-2}. Find the speed of image at $t = 1$s.

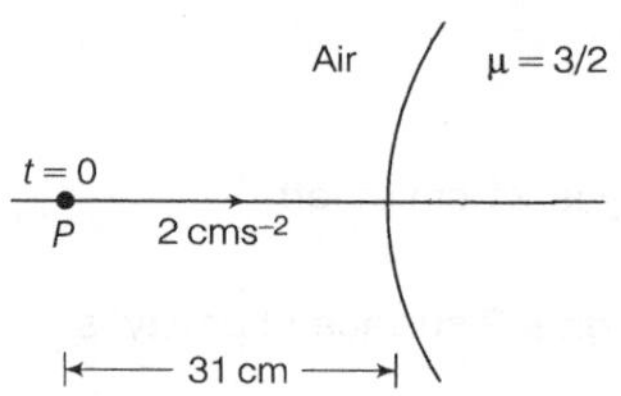

a. 2 cms^{-1}
b. 6 cms^{-1}
c. 12 cms^{-1}
d. None of these

7 A glass spherical paper weight of refractive index $\frac{3}{2}$ and radius 15 cm is placed on a page of a book (choose a letter of the print vertically below the centre of flat base). The position of image of the print from top of sphere is

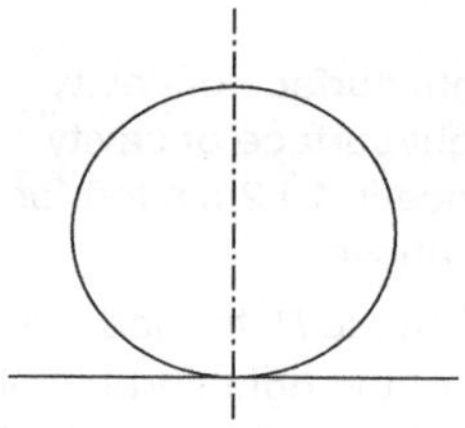

a. 60 cm below
b. 60 cm above
c. 30 cm below
d. 15 cm below

8 A liquid of refractive index $\mu_1 = 2$ is filled in a thin glass sphere of radius 20 cm. A fish is moving towards the centre of sphere with a speed of 2 cms^{-1}. Find speed of image of fish in mirror, when fish is at distance 10 cm from the centre of sphere as shown in the figure.

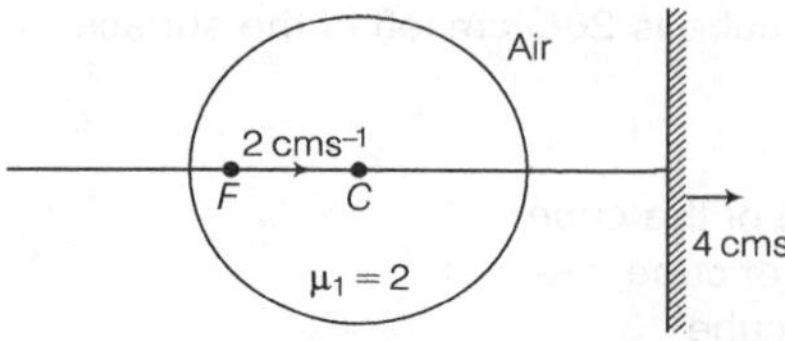

a. 10 cms^{-1}
b. 12 cms^{-1}
c. 24 cms^{-1}
d. 20 cms^{-1}

Working Space

9 A narrow parallel beam of light is incident on a plane mirror as shown in the figure. Find the distance of focus of the beam from the optical centre of spherical refractive surface of radius 10 cm.

a. 10 cm
b. 20 cm
c. 40 cm
d. 60 cm

10 A rectangular slab of refractive $\frac{3}{2}$ has a spherical cavity at its centre. The radius of cavity is 5 cm as shown in the figure. A point mark *P* is at the surface of cavity. Find the distance through which point mark *P* appears to shift for the observer.

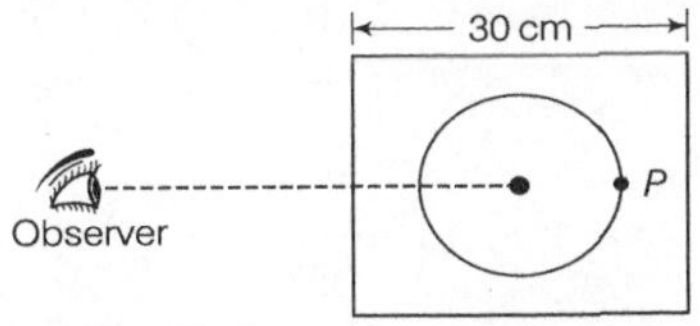

a. 5 cm
b. $\frac{10}{3}$ cm
c. $\frac{25}{3}$ cm
d. 10 cm

11 A spherical air bubble is inside water. A diver sees a linear object *AB* through air bubble as shown in the figure. The image of *AB* seen by diver will be

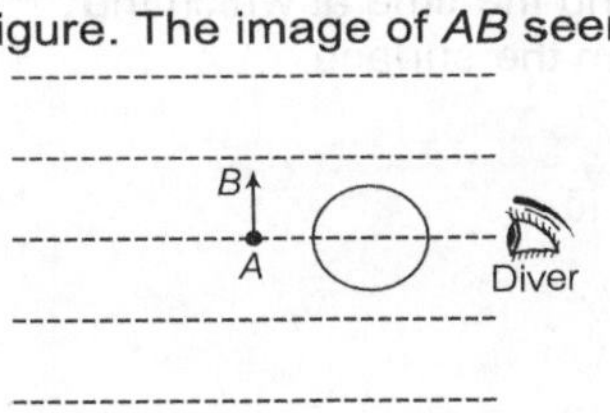

a. smaller and inverted
b. larger and inverted
c. smaller and errect
d. No sufficient information

12 Find the position of focus from plane surface.

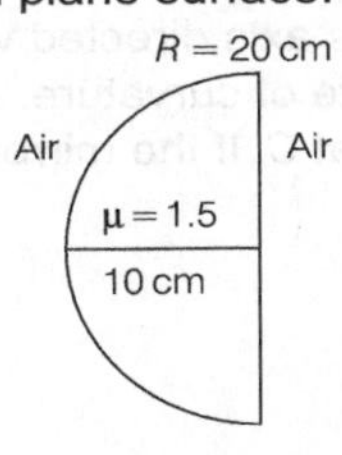

a. 60 cm
b. 50 cm
c. 40 cm
d. 100/3 cm

13 Find the position of an object from spherical surface such that, image coincides with object.

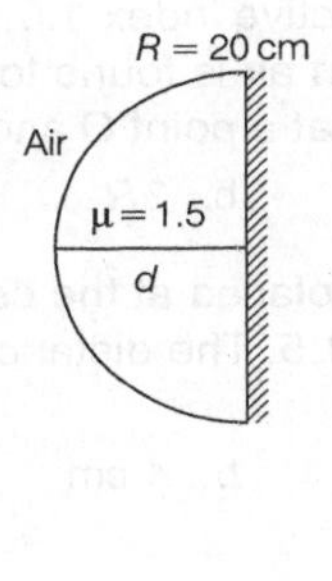

a. 40 cm
b. 20 cm
c. Can not be calculated
d. The position of object come in the form of *d*

Working Space

14 A spherical surface separates two media of refractive indices $\mu_1 = 1$ and $\mu_2 = 3/2$. If the radius of curvatures of the surface is 10 cm, its focal length

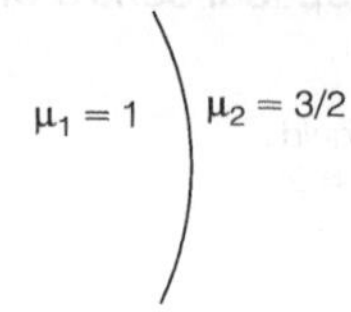

a. + 20 cm b. − 15 cm
c. − 5 cm d. − 30 cm

15 A solid glass sphere of radius 10 cm is placed in air. The refractive index of glass is $\mu = 2$.

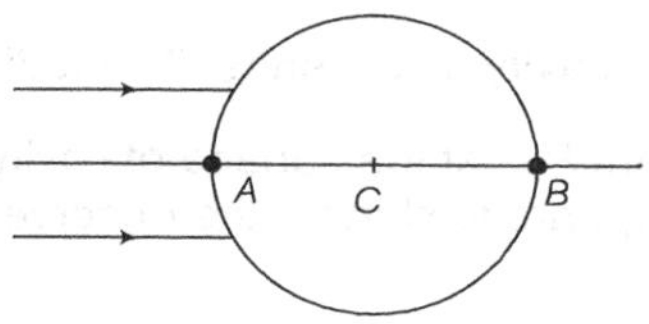

a. The beam converges between points *C* and *B*
b. The beam converges at point *B*
c. The beam converges beyond point *B*
d. The beam converges at point *C*

16 In given figure, spherical refractive surface separates two media. The refractive index of right side of surface is $\mu = 1 + \frac{t}{10}$, where *t* is in second. A paraxial parallel beam of light incident on the surface from air. Find the time at which the converging point of beam is 30 cm rightward from the surface.

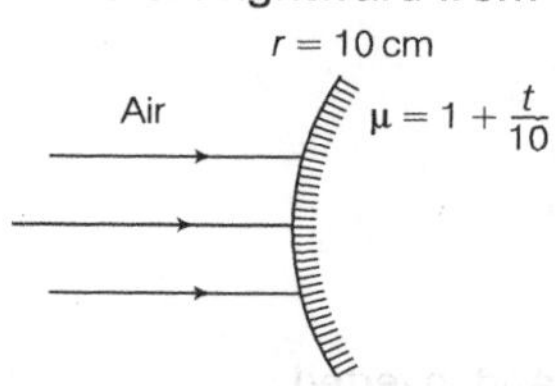

a. 10 s b. 5 s
c. 1 s d. 20 s

17 A concave mirror is placed on a horizontal table with its axis directed vertically upwards. Let *O* be the pole of the mirror and *C* its centre of curvature. A point object is placed at *C*. It has a real image, also located at *C*. If the mirror is now filled with water, the image will be

a. real and will remains at *C*
b. real and located at a point between *C* and ∞
c. virtual and located at a point between *C* and *O*
d. real and located at a print between *C* and *O*

18 A spherical surface of radius of curvature *R*, separates air (refractive index 1.0) from glass (refractive index 1.5). The centre of curvature is in the glass. A point object *P* placed in air is found to have a real image *Q* in the glass. The line *PQ* cuts the surface at a point *O* and *PQ* = *OQ*. The distance is equal to **[IIT JEE 1998]**

a. 5*R* b. 3*R* c. 2*R* d. 1.5 *R*

19 A point object is placed at the centre of a glass sphere of radius 6 cm and refractive index 1.5. The distance of the virtual image from the surface of the sphere is **[IIT JEE 2004]**

a. 2 cm b. 4 cm c. 6 cm d. 12 cm

20 A ball is dropped from a height of 20 m above the surface of water in a lake. The refractive index of water is $\frac{4}{3}$. A fish inside the lake, in the line of fall of the ball, is looking at the ball. At an instant, when the ball is 12.8 m above the water surface, the fish sees the speed of ball as **[IIT JEE 2009]**

a. $9\ ms^{-1}$
b. $12\ ms^{-1}$
c. $16\ ms^{-1}$
d. $21.33\ ms^{-1}$

Working Space

One or More than One Options Correct Type

Direction *(Q. Nos. 21-23) This section contains 3 multiple choice questions. Each question has four choices (a), (b), (c) and (d), out of which ONE or MORE THAN ONE are correct.*

21 A spherical refractive surface 10 cm separates two medium of refractive index $\frac{3}{2}$ and $\frac{4}{3}$ respectively as shown in the figure. A linear object of length 4 cm is placed 20 cm before the refractive surface.

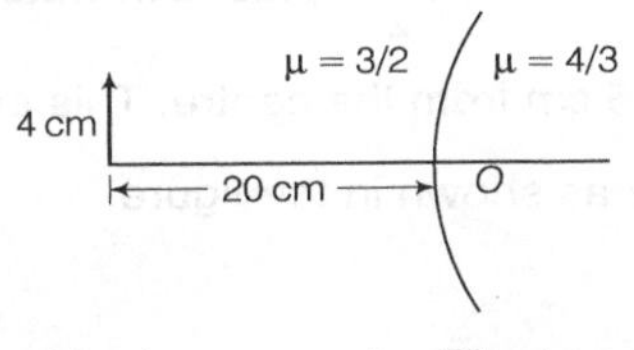

a. The image of object is virtual
b. The image of object is real
c. The size of image is 3.27 cm
d. None of these

22 A spherical refractive surface of radius 20 cm separates two medium of refractive index $\frac{3}{2}$ and $\frac{4}{3}$ respectively as shown in the figure. A linear object of height 2 cm is placed at distance of 30 cm from refractive surface.

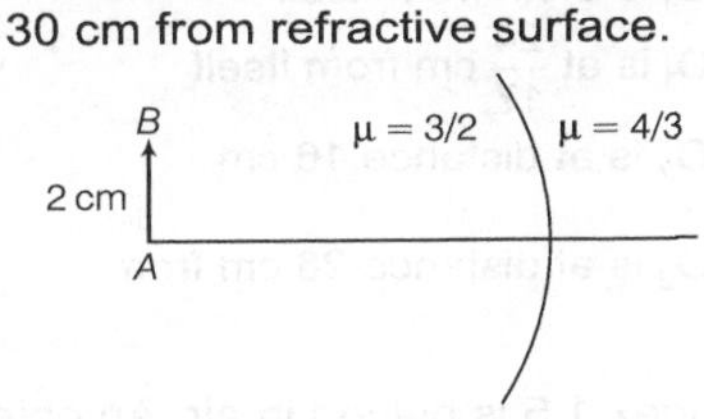

a. The image is virtual
b. The size of image is 2.4 cm
c. The image is 32 cm before the refractive surface
d. The image is 32 cm behind the refractive surface

23 A glass hemisphere of radius 10 cm is placed in air. A point object P is placed at distance 10 cm from curved surface.

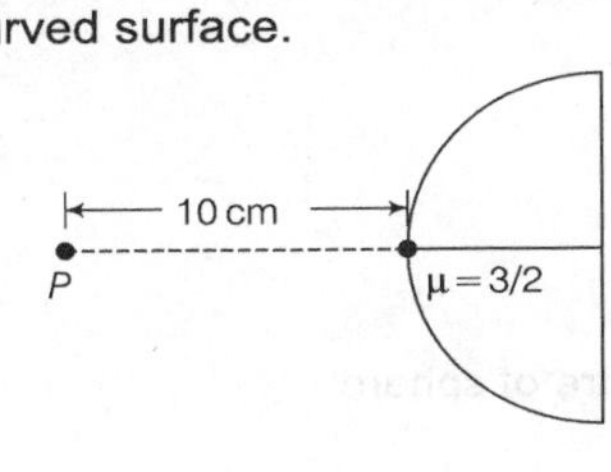

a. Final image is real
b. Final image is virtual
c. Final image is at $\frac{80}{3}$ cm before the plane surface
d. None of the above

DPP-8 Image Formation due to Refraction through Spherical Surface-II

One or More Than One Options Correct Type

Direction (Q. Nos. 1-5) *This section contains 5 multiple choice questions. Each question has four choices (a), (b), (c) and (d), out of which ONLY ONE option is correct.*

Working Space

1 A solid glass sphere of radius 10 cm and refraction index $\frac{3}{2}$ is placed in water of refractive index $\frac{4}{3}$. A point mark P is at distance 5 cm from the centre. This point mark is viewed along the diameter of the sphere as shown in the figure.

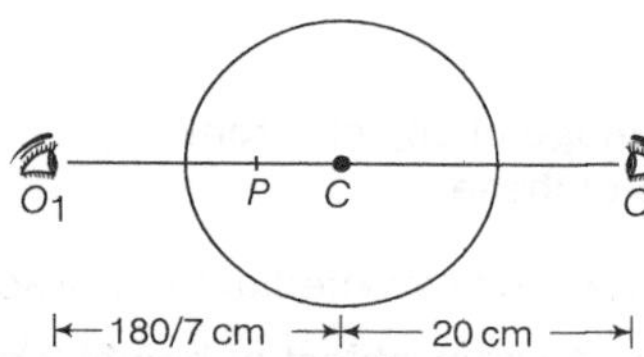

a. The image of point object P seen by observer O_1 is 2 cm from itself

b. The image of point object P seen by observer O_1 is at $\frac{24}{17}$ cm from itself

c. The image of point object P seen by observer O_2 is at distance 16 cm from itself

d. The image of point object P seen by observer O_2 is at distance 26 cm from it self

2 A glass sphere of radius 10 cm and refractive index 1.5 is placed in air. An object with height of 2 cm is placed at distance of 40 cm from the centre of sphere. Then,

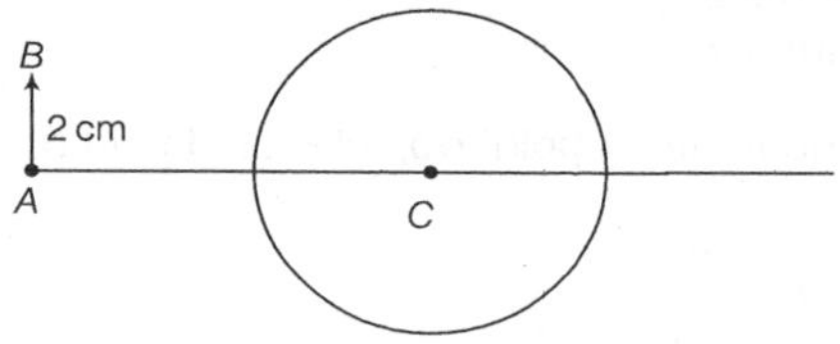

a. the final image is real

b. the final image is virtual

c. the size of final image is 0.6 cm

d. the final image is at distance 24 cm from the centre of sphere

3 In previous problem,

a. no image is seen by an observer which is situated left side of sphere

b. no image is seen by an observer situated inside the sphere

c. an observer situated right side of sphere can see only one image

d. no image is seen by an observer which is situated right side of sphere

Working Space

4 A glass rectangular slab of refractive index $\frac{3}{2}$ and thickness 25 cm is silvered at one face as shown in the figure. A point source S is at distance 15 cm from unsilvered face.

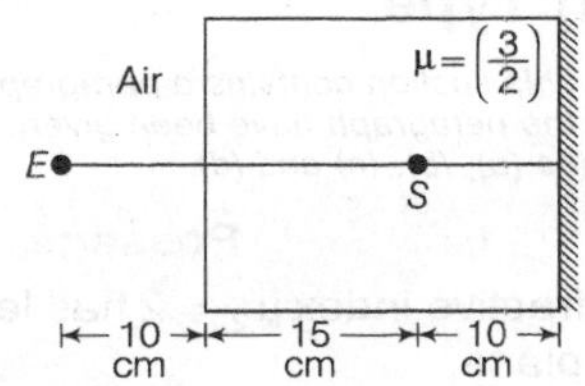

Mark the correct option(s).

a. Two images of point source S are seen by observer E
b. Only one image is seen by observer E
c. One of image seen by observer E is at distance 20 cm from itself
d. One of the image seen by observer E is at distance 33.33 cm from itself

5 Consider a paraxial parallel beam of light incident normally on a solid transparent sphere of radius r and refractive index μ placed in air.

a. If $\mu = 2$, the beam will be focussed rear surface of sphere
b. The beam may focus at the centre of the sphere
c. The beam will focus close to the centre, if the refractive index of the sphere is large
d. The converging of beam at the surface of sphere depends upon radius of sphere only

Matching List Type

Direction (Q. No. 6) *Choices for the correct combination of elements from Column I and Column II are given as options (a), (b), (c) and (d), out of which one is correct.*

6 Match the Column I with Column II and mark the correct option from the codes given below.

	Column I		Column II
i.	$\frac{\mu_2}{v} - \frac{\mu_1}{u} = \frac{\mu_2 - \mu_1}{r}$	p.	When marginal rays are considered
ii.	$\frac{1}{v} + \frac{1}{u} = \frac{1}{f}$	q.	When paraxial rays are considered
iii.	$\frac{\mu_2}{\mu_1} = \frac{\sin i}{\sin r}$	r.	For spherical mirror
iv.	Angle of incidence = angle of reflection	s.	When ray is incident on elliptical refractive surface separating two medium

Codes

	i	ii	iii	iv		i	ii	iii	iv
a.	p	q	r	s	b.	r	q	p	r
c.	r	p	q	s	d.	p	r	s	q

Statement Type

Direction (Q. No. 7) *This section is based on Statement I and Statement II. Select the correct answer from the codes given below.*

a. Both Statement I and Statement II are correct and Statement II is the correct explanation of Statement I
b. Both Statement I and Statement II are correct but Statement II is not the correct explanation of Statement I
c. Statement I is correct but Statement II is incorrect
d. Statement II is correct but Statement I is incorrect

7 **Statement I** The formula $\frac{\mu_2}{v} - \frac{\mu_1}{u} = \frac{\mu_2 - \mu_1}{r}$ is applicable for only paraxial rays.

Statement II The laws of refraction is applicable for plane refractive surface as well as curved refractive surface.

Comprehension Type

Working Space

Direction (Q. Nos. 8-10) *This section contains a paragraph, describing theory, experiments, data, etc. Three questions related to the paragraph have been given. Each question has only one correct answer among the four given options (a), (b), (c) and (d).*

Passage

A glass slab of refractive index $\mu_2 = 2$ has left surface spherical of radius 10 cm and right surface plane.

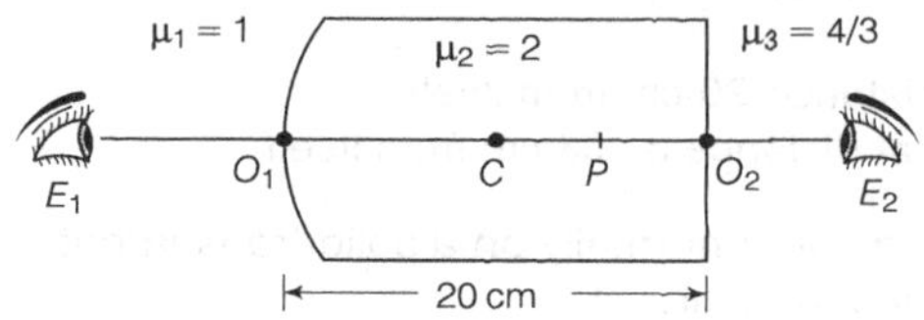

From both surfaces refraction as well as reflection take place. A point object P is placed at distance 15 cm from point O_1.

8 Mark the correct option.

a. Two images are seen by observer E_1
b. Only one image is seen by observer E_2
c. No image is seen by observer E_1
d. No image is seen by observer E_2

9 The distance of final image of point object P from point O_1 seen by observer E_1 will be

a. 30 cm right and 50 cm left
b. 20 cm right and 50 cm left
c. 30 cm right and 20 cm left
d. 20 cm right and 40 cm left

10 The distance of final image of point object P from point O_2 seen by observer E_2 will be

a. 2.5 cm left and 5 cm left
b. 2.5 cm left and 6.25 cm left
c. 2.5 cm left and 6.25 cm right
d. 2.5 cm left and 5 cm right

One Integer Value Correct Type

Direction (Q. Nos. 11-20) *This section contains 10 questions. When worked out will result in an integer from 0 to 9 (both inclusive).*

11 The plane surface of a hemisphere is silvered. The radius and refractive index of hemisphere are 10 cm and 2 respectively. The hemisphere is placed in air. A point object is placed at distance 30 cm from silvered surface. Find the position of final image (in cm) from the curved surface.

12 A sphere of refractive index $\mu = 2$ has inner radius 10 cm and outer radius 20 cm. It is placed in air as shown in the figure. The distance through which point mark P appear to shift is $\frac{10}{3}n$ cm. Find the value of n.

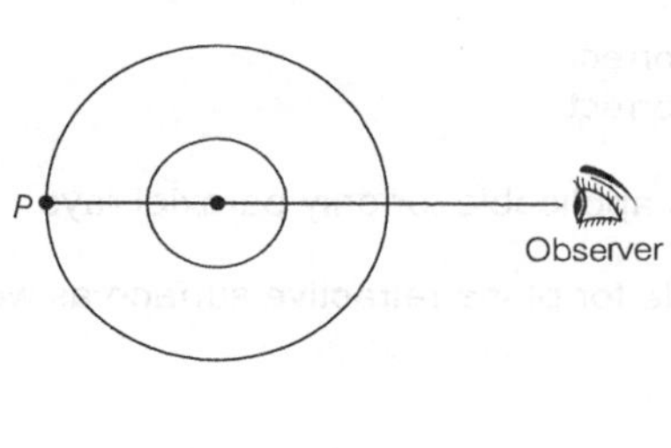

Working Space

13 A needle is embedded in a transparent tube of refractive index 3/2 along its axis. The external radius of tube is much larger than that of the needle. The diameter of the needle appears 3 mm, when looked into the glass tube through its curved surface. What is the actual diameter (in mm) of the needle?

14 A solid glass sphere of radius 10 cm and refractive index $\frac{3}{2}$ is placed in a medium of refractive index $\mu = 2$. The hemispherical portion of the sphere is silvered.

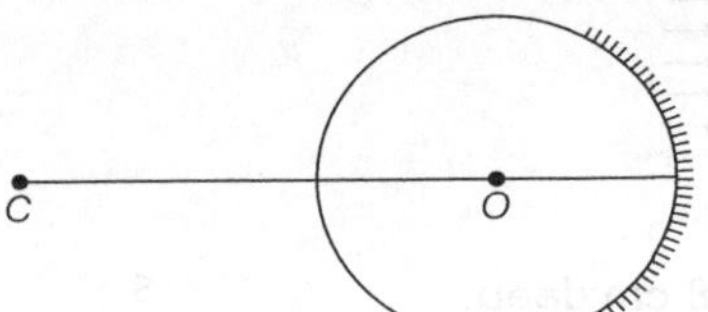

A point object P is placed at distance 30 cm from the centre of sphere. The distance of final image of point object P from centre of sphere is $\frac{50n}{11}$ cm. Find the value of n.

15 The distance between two images seen by observer E as shown in the figure, is $20\,x$ cm. Find the value of x.

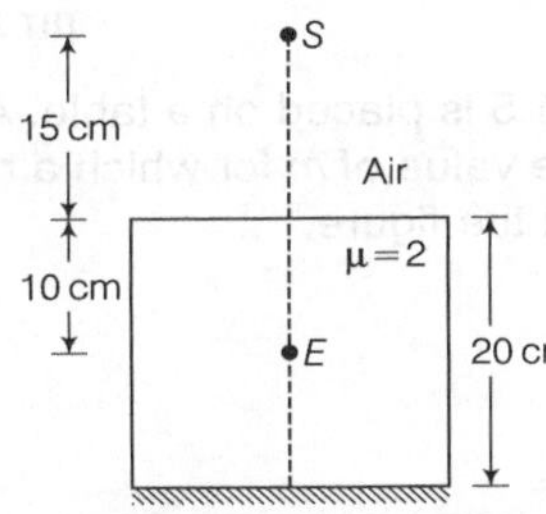

16 A spherical surface of radius 20 cm separates two media of refractive indices $\mu_1 = \frac{4}{3}$ and $\mu_2 = \frac{3}{2}$. A linear object is placed at distance 80 cm from the refractive surface as shown in the figure. What will be magnification of the object?

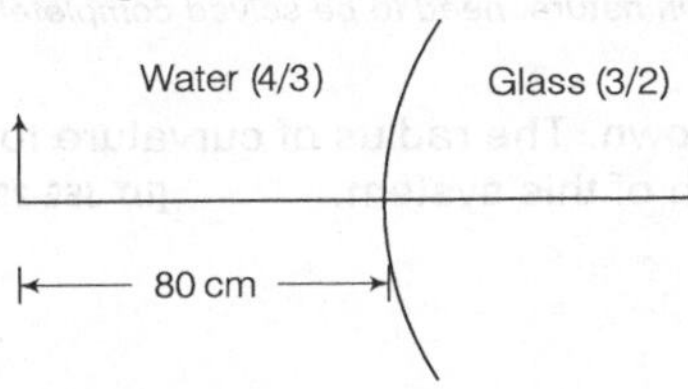

17 If the speed of image at the instant as shown in figure is $11x_0$. Find the value of x_0. (Take, radius of curvature is 10 cm)

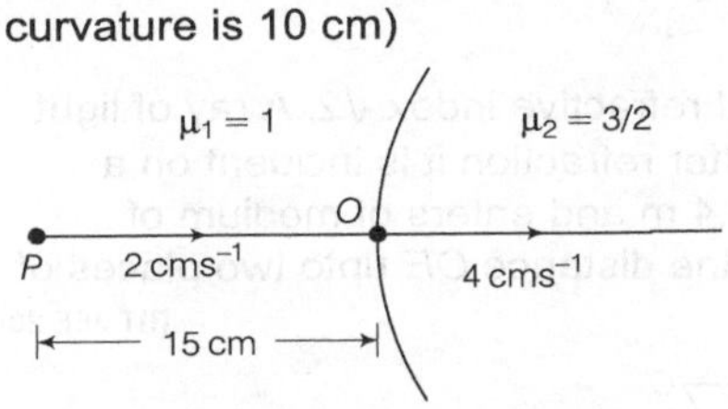

Working Space

18 A paraxial parallel beam of light incident on the refractive surface of small aperture. The intensity of incident beam is 10 Wm^{-2}. If the intensity at the surface of screen is $5n$ Wm^{-2}. Find the value of n. (Absorption and reflection of light from surface may be neglected) (Take, radius of curvature is 30 cm)

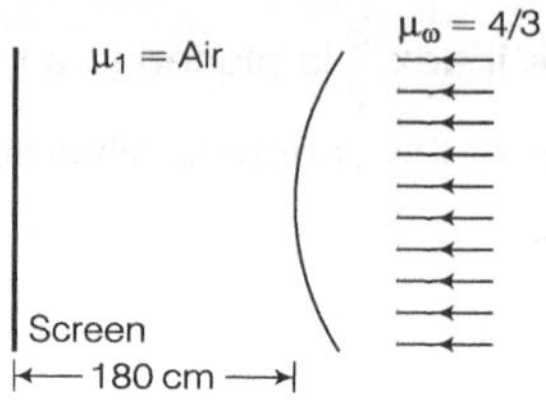

19 Water $\left(\text{with refractive index} = \frac{4}{3}\right)$ in a tank is 18 cm deep. Oil of refractive index $\frac{7}{4}$ lies on water making a convex surface of radius of curvature $R = 6$ cm as shown in the figure. Consider oil to act as a then lens. An object S is placed at 24 cm above water surface. The location of its image is at x cm above the bottom of the tank. Then, x is

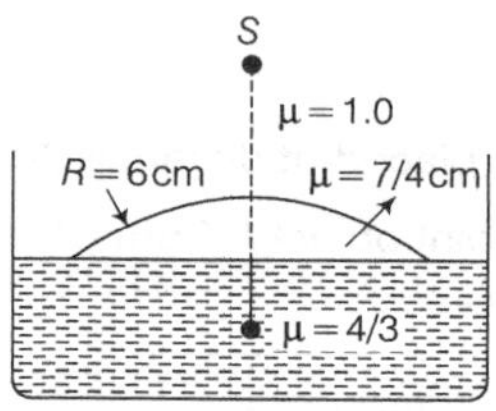

[IIT JEE 2011]

20 A quarter cylinder of radius R and refractive index 1.5 is placed on a table. A point object P is kept at a distance of mR from it. Find the value of m for which a ray from P will emerge parallel to the table as shown in the figure.

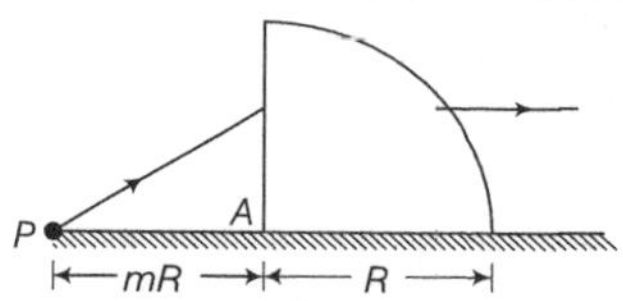

Subjective Questions

Direction *(Q. Nos. 21 and 22) These questions are subjective in nature, need to be solved completely on notebook.*

21 In the figure, light is incident on a thin lens as shown. The radius of curvature for both the surfaces is R. Determine the focal length of this system. **[IIT JEE 2003]**

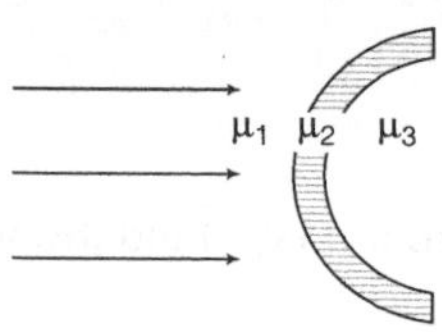

22 Figure shows an irregular block of material of refractive index $\sqrt{2}$. A ray of light strikes the face AB as shown in the figure. After refraction it is incident on a spherical surface CD of radius of curvature 0.4 m and enters of medium of refractive index 1.514 to meet PQ at E. Find the distance OE upto two places of decimal. **[IIT JEE 2004]**

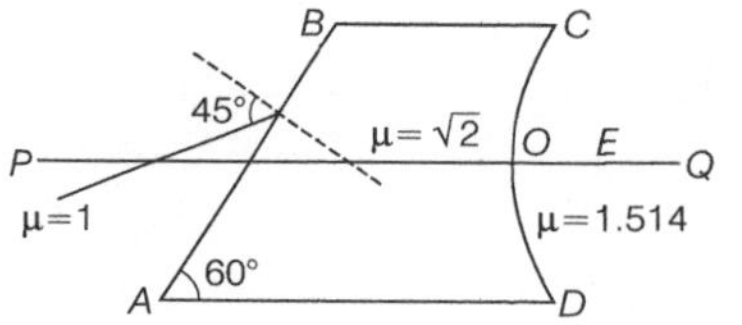

DPP-9 Image Formation due to Refraction through Spherical Surface-III

Conceptual Questions

Working Space

Direction (Q. Nos. 1-2) These questions are conceptual in nature, need to be solved completely on notebook.

1 **Situation** The bird B is 20 cm above the water surface. A fish is 30 cm below the surface of water. The refractive index of water is 4/3.

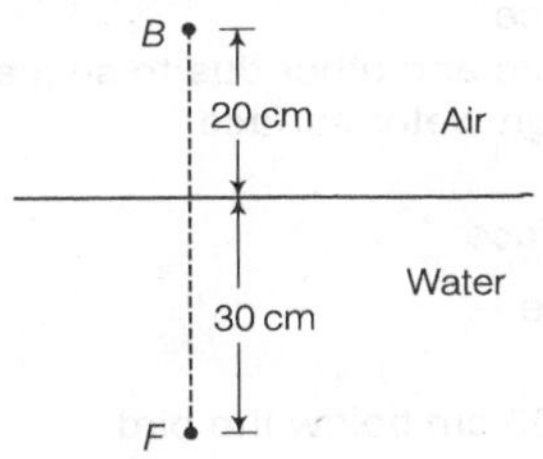

i. The image of fish seen by bird is
   a. 22.5 cm below the surface of water
   b. 40 cm below the surface of water
   c. 42.5 cm below the bird itself
   d. None of the above

ii. The image of bird seen by fish is
   a. (80/3) cm above the water surface
   b. (80/3) cm below the water surface
   c. virtual
   d. None of the above

iii. If fish is moving downward with a speed of 4 cms^{-1}. Find the speed of image of fish observed by the bird.
   a. 1 cms^{-1} in upward direction
   b. (8/3) cms^{-1} is downward direction
   c. 1 cms^{-1} in downward direction
   d. 3 cms^{-1} in downward direction

iv. If bird is moving upward with speed of 6 cms^{-1}. Find the speed of image of the bird.
   a. 4.5 cms^{-1} in upward direction
   b. 8 cms^{-1} in upward direction
   c. 8 cms^{-1} in downward direction
   d. None of these

v. In previous problem, if water surface rises upward with 1 cms^{-1}. Find the speed of image formed by refraction through water surface.
   a. 7 cms^{-1} in upward direction
   b. 6 cms^{-1} in downward direction
   c. (23/3) cms^{-1} in upward direction
   d. None of the above

Working Space

vi. In previous problem, if fish is moving downward with speed of 4 cms^{-1}. The fish observe the speed of image of bird as

a. 35 cms^{-1} in upward direction
b. (28/3) cms^{-1} in upward direction
c. 12 cms^{-1} in upward direction
d. None of the above

2 **Situation** A vessel is filled with water of refractive index 4/3. The height of vessel is 60 cm. A fish *F* is at 40 cm below the surface of water. A bird *B* is 30 cm above the surface of water. The bottom of the vessel behaves as plane mirror.

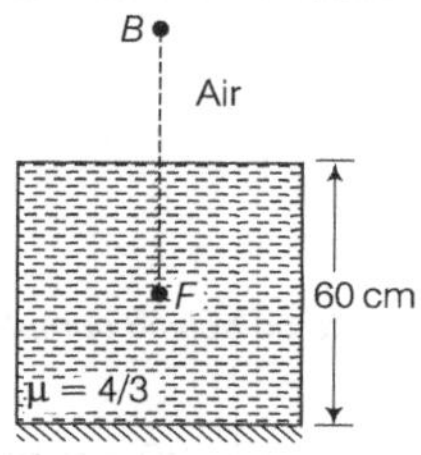

i. The number of image(s) of fish seen by the bird is/are

a. only one due to refraction through upper surface
b. two, one due to refraction through upper surface and other due to successive reflection through mirror, then refraction through water surface
c. two, both due to reflections through mirror
d. two, both due to refractions through water surface

ii. The position(s) image(s) of fish seen by bird is/are

a. only one 60 cm below the bird
b. two, one is 60 cm below the bird and other is 90 cm below the bird
c. two, one is 30 cm below the bird and other is 60 cm below the bird
d. None of the above

iii. If fish is moving upward with a speed of 4 cms^{-1}. Then,

a. only one image is formed and moving upward with a speed of 3 cms^{-1}
b. two images are formed, one is moving upward with velocity 3 cms^{-1} and other is moving downward with speed of 3 cms^{-1}
c. two images are formed and both are moving upward with speeds of 3 cms^{-1}
d. None of the above

iv. The image(s) of bird seen by fish

a. is only one
b. are two, one due to refraction through water surface and other is due to successive refraction through water surface, then reflection through mirror
c. two, both due to only refraction through water surface
d. None of the above

v. The image(s) of bird seen by fish

a. is only one and 80 cm above the fish
b. are two, one 80 cm above fish and other 120 cm below the fish
c. are two, one 80 cm above fish and other 120 cm above the fish
d. None of the above

vi. Assume that fish is in rest. But bird is moving downward with speed of 30 cms^{-1}. Then,

a. only one image of bird is seen by fish moving downward with speed 40 cms^{-1}
b. two images are one is moving downward with speed 40 cms^{-1} and other is moving upward with speed 40 cms^{-1}
c. two images are seen by fish, both are moving upward with speed 40 cms^{-1}
d. None of the above

Working Space

Only One Option Correct Type

Direction (Q. Nos. 3-17) *This section contains 15 multiple choice questions. Each question has four choices (a), (b), (c) and (d), out of which ONLY ONE option is correct.*

3 A bird in air looks at fish vertically below it and inside the water in a tank. If the distance of fish as estimated by the bird is 40 cm and that of bird as estimated by the fish is 50 cm. Then, the refractive index of water is

a. 4/3
b. 1.5
c. 1.25
d. None of these

4 A coin is placed under a glass slab of thickness h, whose refractive index varies as $\mu = (1 + y)\mu_0$, the apparent depth of the coin below the surface of the slab will be

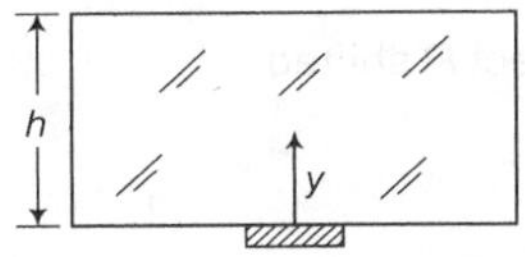

Coin

a. $\frac{\ln(h-1)}{\mu_0}$
b. $\mu_0 \ln(h)$
c. $\frac{\ln(h)}{\mu_0}$
d. $\frac{\ln(h+1)}{\mu_0}$

5 **n** transparent slabs of refractive index 1.5 each having thickness 1 cm, 2 cm, ... to **n** cm are arranged one over another. A point object is seen through this combination with near perpendicular light. If the shift of object by the combination is 1 cm, then the value of **n** is

a. either 2 or 3
b. 2
c. 3
d. 0.3

6 In the given figure, an insect P starts from rest with an acceleration 2cms^{-2} towards an aquarium. Find the time after which it observes that a fish F approaching it with a speed of 9 cms^{-1}.

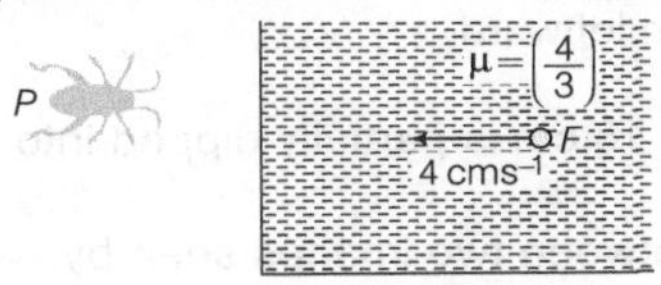

a. 1 s
b. 2 s
c. 3 s
d. 4 s

7 A paraxial beam of light incident on a slab of refractive index $\frac{3}{2}$ and thickness 15 cm. The slab is placed in air as shown in the figure. The converging point of the beam after refraction through slab is

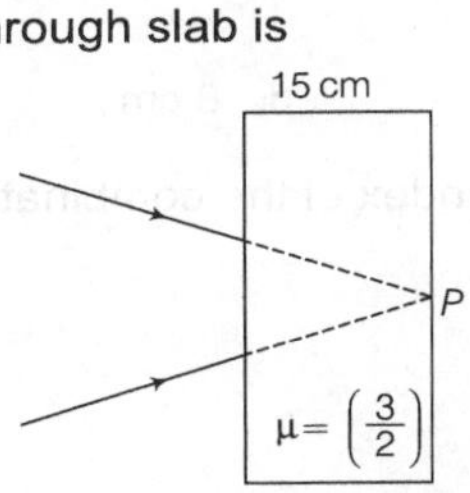

a. at 5 cm rightward from point P
b. at 5 cm leftward from point P
c. at point P
d. None of these

Working Space

8 In the given figure, the distance between two images seen by observer is

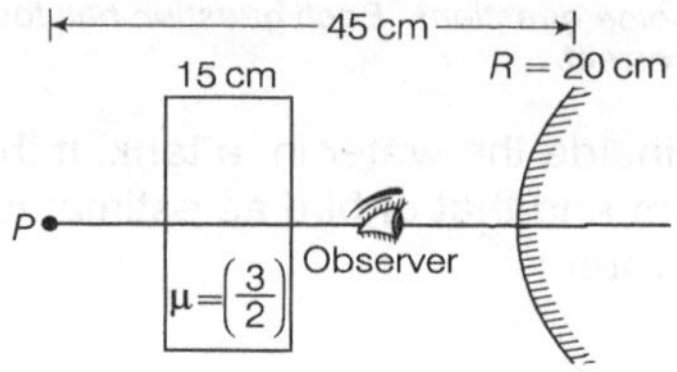

a. 58 cm
b. 48 cm
c. 50 cm
d. 60 cm

9 A slab of refractive index $\frac{3}{2}$ and thickness of 18 cm is placed in water of refractive index $\frac{4}{3}$. Find the distance through which point object P shifted.

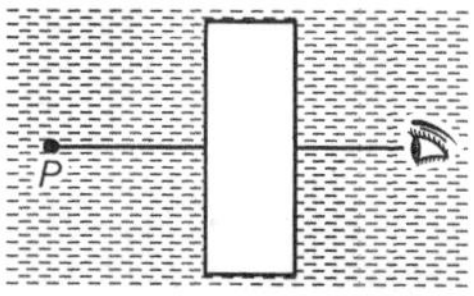

a. 2 cm leftward
b. 2 cm rightward
c. 6 cm rightward
d. 2.25 cm leftward

10 A slab of refractive index $\frac{3}{2}$ and thickness 9 cm placed in a medium of refractive index 2. The point object P as shown in the figure, is shifted through a distance.

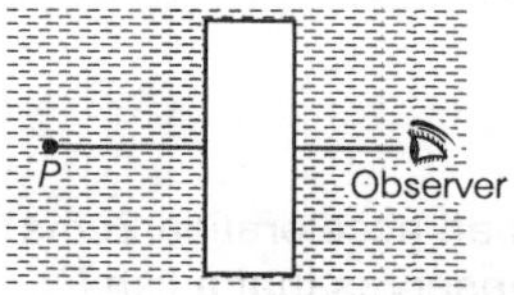

a. 3 cm leftward
b. 3 cm rightward
c. 10 cm rightward
d. 5 cm leftward

11 A glass slab of refractive index $\frac{3}{2}$ and thickness 18 cm is partially dipped into water of refractive index $\frac{4}{3}$. Find the distance between two images seen by observer O at the surface of water as shown in the figure.

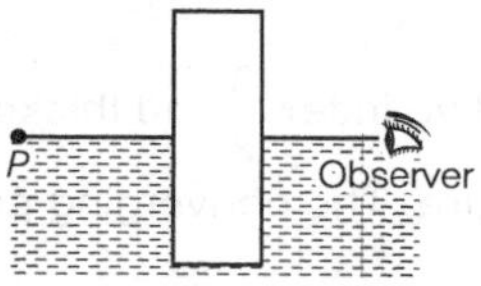

a. 2 cm
b. 4 cm
c. 6 cm
d. 8 cm

12 For normal incidence, find the equivalent refractive index of the combination slabs as shown in the figure.

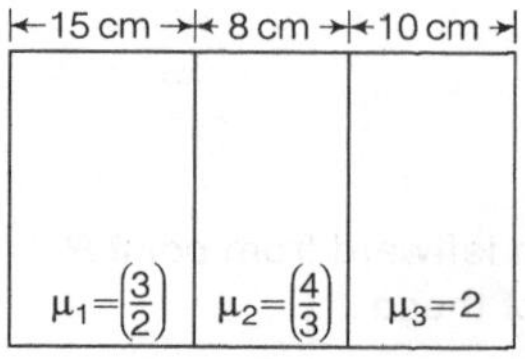

a. 1.7
b. 1.81
c. 11.57
d. 1.9

Working Space

13 Find the distance through which object appears to shift for the observer.

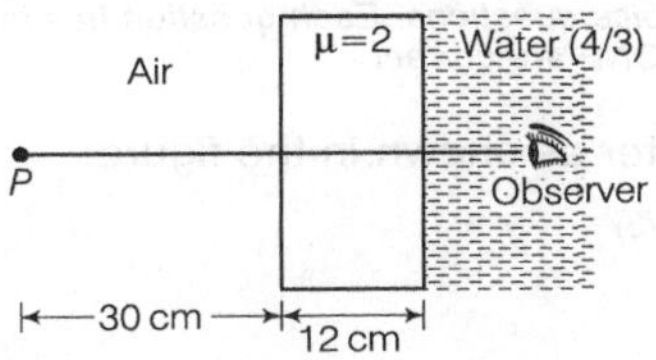

a. 2 cm
b. 3 cm
c. 4 cm
d. 6 cm

14 A cylindrical hollow cylinder has diameter 10 cm and height 8.45 cm. When vessel is empty, observer *E* can see the point mark *A*.

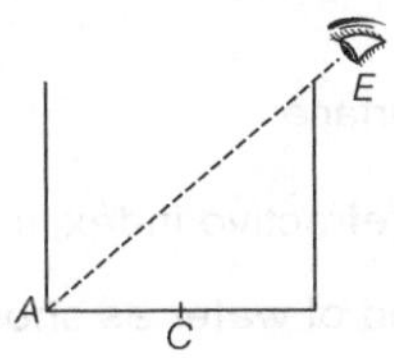

If vessel is filled with a liquid, he can see the point mark at the centre. Find the refractive index of liquid.

a. $\frac{4}{3}$
b. 1.5
c. 1.8
d. 2

15 Find the position of image of point object *P* from silvered surface.

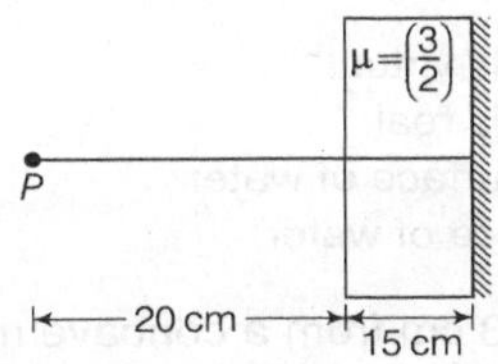

a. 25 cm behind the silvered surface
b. 30 cm behind the silvered surface
c. 20 cm behind the silvered surface
d. None of the above

16 When looking through a glass window to the outdoors at night, you some times see a double image of yourself. The cause behind this is

a. first image is due to reflection and second is due to refraction
b. first image is due to reflection by inner surface of glass and second image is due to reflection by outer surface of glass
c. both images due to refraction
d. None of the above

17 An observer on the west facing of a large lake is watching the begining of Sunset. The water is very smooth except for some areas with small ripples. The observer notices that some areas of the water appear blue and same area pink. Why does the water appear to be different colours in different area?

a. In some area, total internal reflection takes place but in some area refraction as well as reflection take place
b. The smooth areas of the water specularly reflect the light from West, which is the pink light from sunset
c. The areas with small ripples reflect the light diffusely. Thus, light from all parts of the sky is reflected into the observer's eye. Because most of the sky is still blue of the begining of sunset, these area appears to the blue
d. None of the above

One or More than One Options Correct Type

Working Space

Direction *(Q. Nos. 18-21) This section contains 4 multiple choice questions. Each question has four choices (a), (b), (c) and (d), out of which ONE or MORE THAN ONE are correct.*

18 A plane refractive surface separates air and water as shown in the figure.

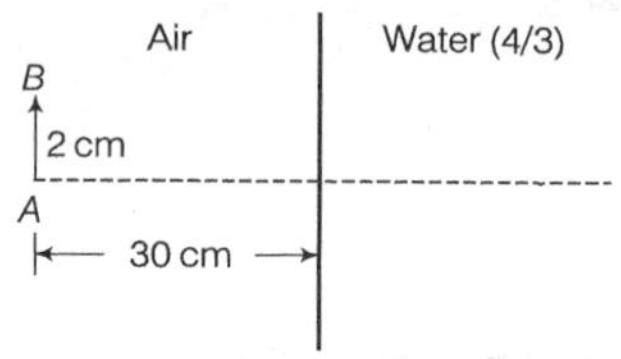

a. The size of image of object *AB* is 2 cm
b. The image is virtual
c. The image of object *AB* is virtual
d. The image of object *AB* is at 40 cm from the plane surface

19 A concave mirror of radius 40 cm is filled with water (refractive index, $\mu = \frac{4}{3}$) upto height 5 cm. A point object *P* is floating on the surfaced of water as shown in the figure.

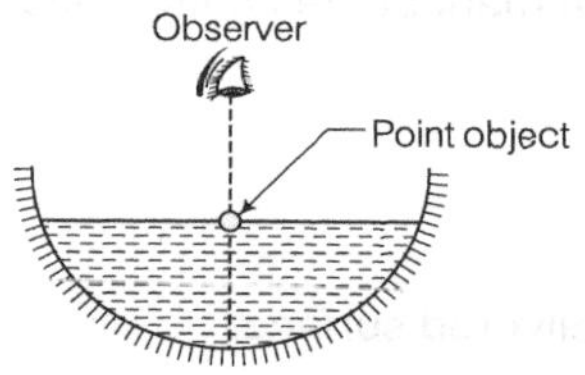

a. The image of point object *P* observed by observer is virtual
b. The image of point object *P* observed by observer is real
c. The image of point object *P* is 8.75 cm below the surface of water
d. The image of point object is 10 cm above the surface of water

20 An object *AB* of height 2 cm is placed at distance 63 cm from a concave mirror of radius of curvature 40 cm. A slab of refractive index $\frac{4}{3}$ and thickness 12 cm is placed between the object and the mirror.

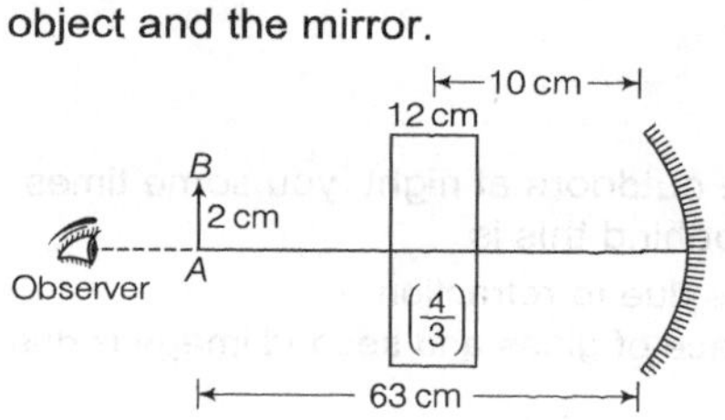

a. Final image is real, inverted and 1 cm high
b. Final image is virtual, inverted and 1 cm
c. Final image is 33 cm from the pole of mirror
d. Final image is 27 cm from the pole of mirror

21 A bird is flying up on an angle 37° with the vertical with a speed of 10 cms^{-1}. A fish in a pond looks at the bird. When it is vertically above the fish? $\left(\text{Take, } \mu_{water} = \frac{4}{3}\right)$

a. The speed of image of bird is 12.24 cms^{-1}
b. The speed of image of bird is 10.67 cms^{-1}
c. The angle at which bird appears to fly to the fish is $\tan^{-1}\left(\frac{9}{16}\right)$ with vertical
d. The angle at which bird appears to fly to the fish to $\tan^{-1}\left(\frac{16}{9}\right)$ with vertical

Working Space

Comprehension Type

Direction (*Q. Nos. 22-25*) *This section contains a paragraph, describing theory, experiments, data, etc. Four questions related to the paragraph have been given. Each question has only one correct answer among the four given options (a), (b), (c) and (d).*

Passage

With a conventional lens, the sharpness of the image is always limited by the wavelength of light, no conventional lens can focus light into an area smaller than square of wavelength (i.e. λ^2). In modern technology an unconventional alternative to a conventional lens i.e. superlens. Superlens is a slab of negative refractive index material. Consider a superlens of refractive index $\mu = -1$ as shown in the figure. (Assume that, superlens obeys Snell's law.)

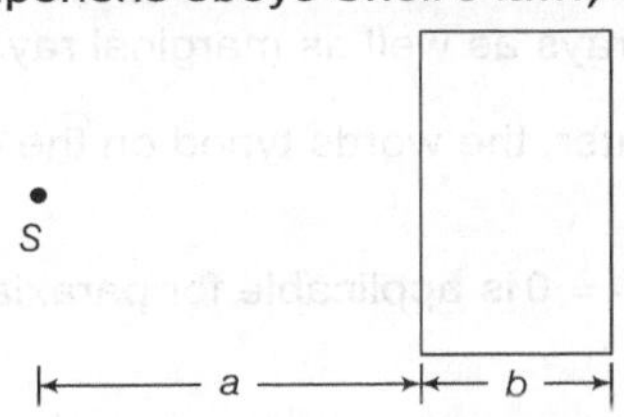

22 Ray diagram for $a = 10$ cm and $b = 20$ cm will be

a. 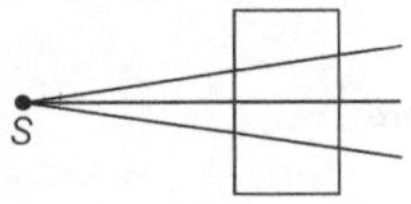

b. 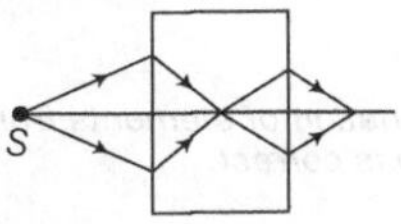

c. 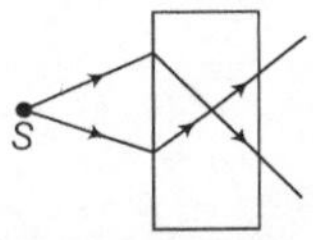

d. None of these

23 In previous problem, the final image of the point source S is

a. real and situated at distance 40 cm from point source S
b. virtual and at distance 20 cm from point source S
c. real and at infinity
d. None of the above

24 In previous problem, the image of the point source S due to refraction, through the first surface is

a. real and distance 20 cm from point source S
b. virtual and at distance 20 cm from point source S
c. image is at infinity
d. None of the above

25 In previous problem, if $a = 10$ cm and $b = 5$ cm. Draw ray diagram.

a. 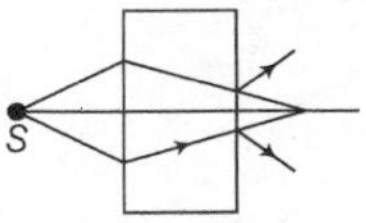

b. 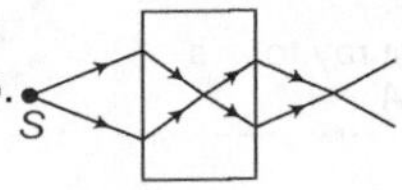

c. 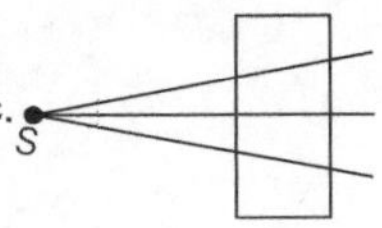

d. None of these

Statement Type

Working Space

Direction (Q. Nos. 26 and 27) *This section is based on Statement I and Statement II. Select the correct answer from the codes given below.*

a. Both Statement I and Statement II are correct and Statement II is the correct explanation of Statement I
b. Both Statement I and Statement II are correct but Statement II is not the correct explanation of Statement I
c. Statement I is correct but Statement II is incorrect
d. Statement II is correct but Statement I is incorrect

26 **Statement I** If you aim the fish through laser gun, you must aim exactly at the fish.

Statement II Snell's law is applicable for paraxial rays as well as marginal rays.

27 **Statement I** If a sheet of paper is covered with water, the words typed on the other side of the sheet can be read.

Statement II For plane refractive surface, $\frac{\mu_2}{v} - \frac{\mu_1}{u} = 0$ is applicable for paraxial rays.

Matching List Type

Direction (Q. No. 28) *Choices for the correct combination of elements from Column I and Column II are given as options (a), (b), (c) and (d), out of which one is correct.*

28 Consider a rectangular slab, whose refractive index varies as $\mu = 1 + \frac{y}{a}$, where $a = 1\ m^{-1}$.

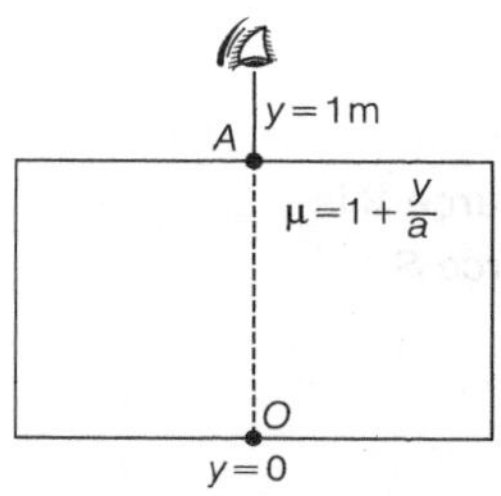

Match the Column I with Column II and mark the correct option from the codes given below.

	Column I		Column II
i.	The equivalent refractive index of slab for normal incidence.	p.	5×10^{-9}
ii.	The apparent depth of point O (in m).	q.	0.693
iii.	The optical path of a ray along OA.	r.	1.5
iv.	The time (in second) taken by light ray to go vertically from point O to point A.	s.	1.44

Codes

	i	ii	iii	iv
a.	p	q	r	s
b.	s	q	r	p
c.	s	r	p	q
d.	s	q	p	r

Working Space

One Integer Value Correct Type

Direction (*Q. Nos. 29-32*) *This section contains 4 questions. When worked out will result in an integer from 0 to 9 (both inclusive).*

29 A cubical block of refractive index $\mu = 2$ and side length 10 cm has a cubical cavity of side length 5 cm. The centre of cavity coincides with the centre of cube as shown in the figure.

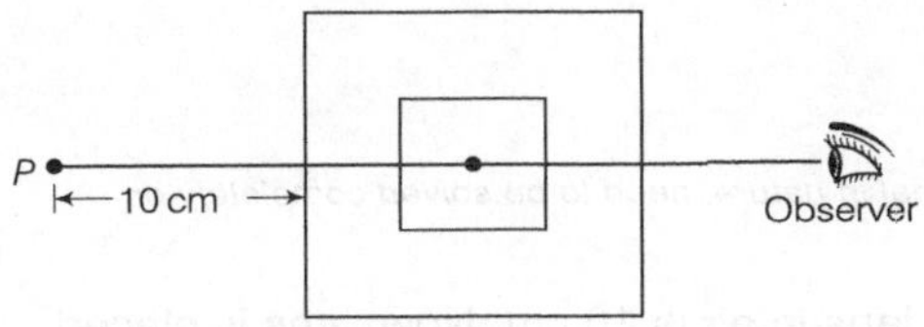

A point object P is placed at distance 10 cm from left face of the cube. Find the distance (in cm) of image seen by observer from the centre of cube.

30 The distance between images of point object formed due to refraction and reflection from water surface is 21cm. The point object P is at distance x cm above the water surface. Find the value of x.

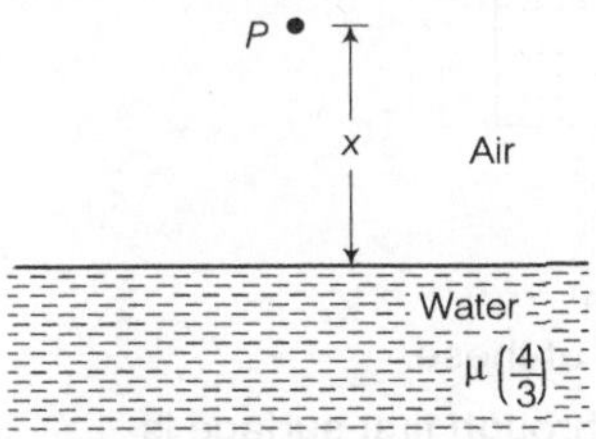

31 A convex mirror and a glass slab of refractive index 1.5 and thickness 15 cm are arranged as shown in the figure. The focal length of convex mirror is 20 cm. A converging paraxial rays is incident on the slab as shown. Find the position of final image (in cm) from the pole of the mirror.

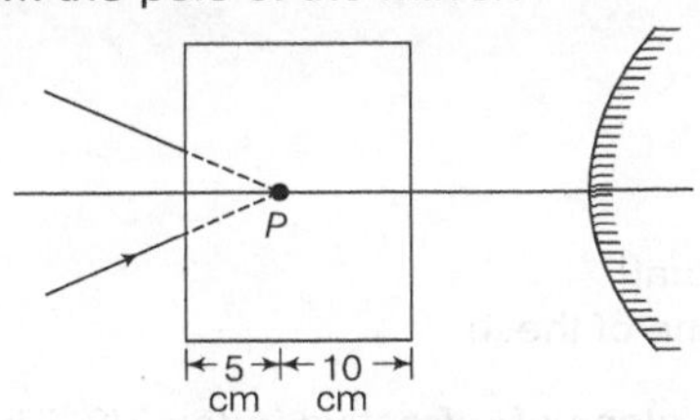

32 A man standing on the edge of a swimming pool looks at a stationary fish lying on the bottom. The depth of the swimming pool is 3 m. The line of vision makes an angle 53° with the normal to the surface 2. The refractive index of water is $\frac{4}{3}$. The distance of image of fish from water surface is $\frac{27n}{256}$ m. Find the value of n.

DPP-10 Lens-I

Conceptual Questions

Working Space

Direction (Q. Nos. 1-4) *These questions are conceptual in nature, need to be solved completely on notebook.*

1 **Situation** The focal length of equiconcave lens in air is 10 cm. Now, lens is placed as shown in the figure. A point object *P* is placed at distance of 20 cm from the lens.

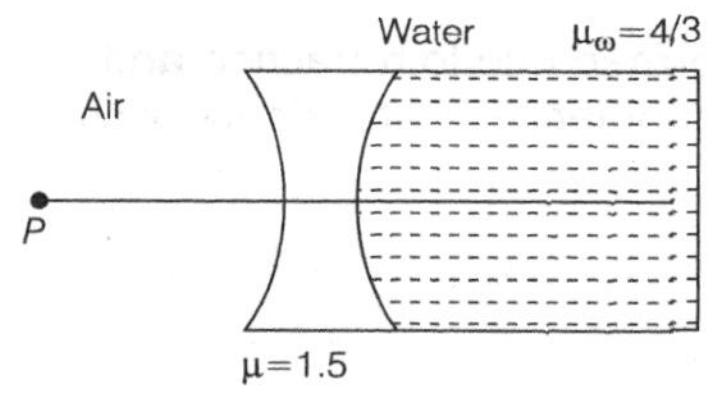

i. The radius of each surface is

a. 10 cm
b. 20 cm
c. 30 cm
d. None of these

ii. The position of image formed due to refraction through first surface is

a. 10 cm left from lens
b. 10 cm right from lens
c. 15 cm left from the lens
d. None of these

iii. The position of final image is

a. 80 cm left from the lens
b. 80 cm right from the lens
c. 16 cm left from the lens
d. None of the above

iv. The final image is

a. real
b. virtual
c. may be real, may be virtuals
d. None of these

2 **Situation** A biconvex thin lens is prepared from glass of refractive index 3/2. The two bounding surfaces have equal radii of 25 cm each. One of the surface is silvered from outside to make it reflecting.

i. The system may be assumed as a combination of

a. a convex lens and concave mirror in contact
b. a convex lens and a convex mirror in contact
c. a convex lens and a concave mirror at a separation
d. None of the above

ii. The focal length of the lens is

a. 12.5 cm
b. 25 cm
c. 20 cm
d. None of these

iii. The power of the silvered lens is

a. $\frac{2}{12.5}+\frac{1}{12.5}$ per cm
b. $\frac{2}{12.5}+\frac{2}{12.5}$ per cm
c. $\frac{2}{25}+\frac{2}{25}$ per cm
d. None of these

iv. If the silvered lens is replaced by equivalent mirror. The equivalent mirror is
   a. convex mirror of focal length 3.75 cm
   b. concave mirror of focal length 6.25 cm
   c. convex mirror of focal length 6.25 cm
   d. None of the above

v. Find the position of an object such that its image coincides with object itself.
   a. 12.5 cm from the silvered lens
   b. 6.25 cm from the silvered lens
   c. 25 cm from the silvered lens
   d. None of the above

vi. A 2 cm tall object is placed at a distance of 18.75 cm from the silvered lens. Find the nature, size and position of image.
   a. Real inverted of size 1 cm at distance 9.375 cm from the lens
   b. Virtual, errect of size 1 cm at distance 9.375 cm
   c. Virtual and magnified
   d. None of the above

3 **Situation** Consider a thin convex lens of focal length 20 cm.

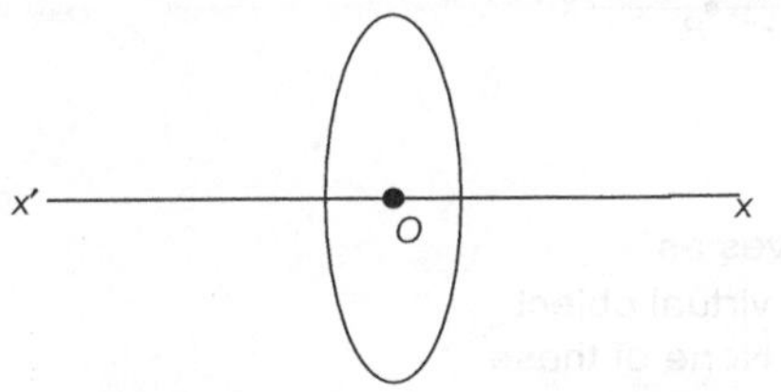

i. If object is placed between $x = 0$ and $x = -20$ cm. The image is
   a. magnified, errect and virtual
   b. diminished, errect and virtual
   c. real and magnified
   d. None of these

ii. In previous problem, *v*-*u* graph is

a. 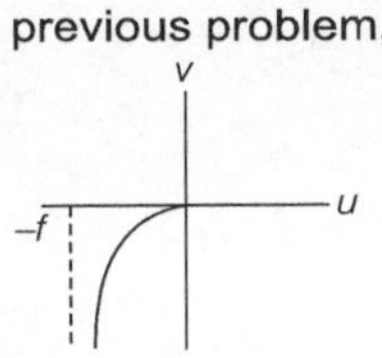

b. 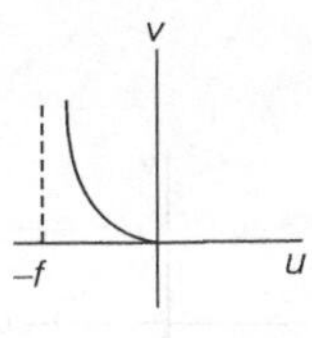

c. 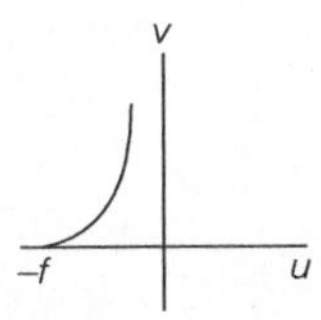

d. None of these

iii. If object is placed between $x = -20$ cm and $x = -40$ cm. Image is
   a. image is magnified and real
   b. image is diminished and real
   c. image is magnified and virtual
   d. None of these

iv. If object is placed at $x = -40$ cm, the image is
   a. real, same size and inverted
   b. real, inverted and magnified
   c. virtual and of same size
   d. None of the above

v. If object is placed between $x = -40$ and $x = -\infty$. The image is
   a. real, inverted and magnified
   b. real, diminished and inverted
   c. virtual, diminished and errect
   d. None of the above

Working Space

vi. If object is placed between $x = 0$ to $x = -\infty$ the v-u graph is

a. 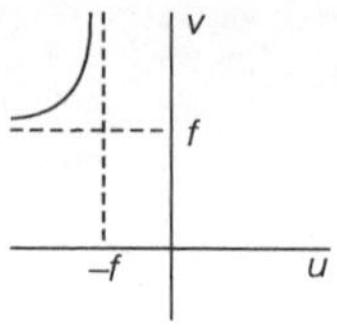

b. 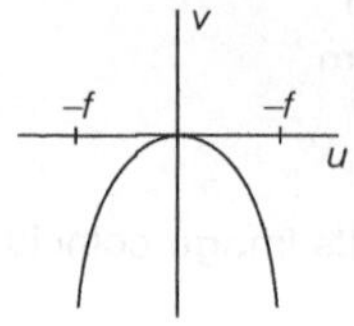

c. 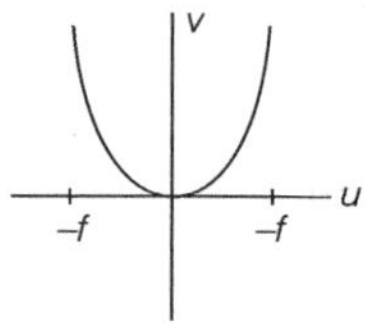

d. None of these

If a converging beam of light incident on the lens.

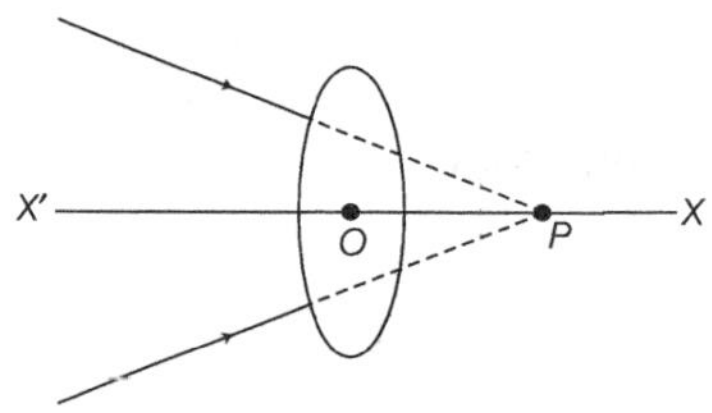

vii. The converging point P of the beam behaves as
   a. real object
   b. virtual object
   c. both (a) and (b)
   d. None of these

viii. The image is
   a. always real, diminished and errect
   b. always virtual, diminished
   c. may be virtual, may be real
   d. None of the above

ix. The v-u graph is

a. 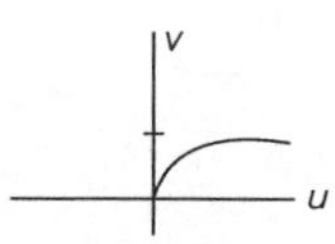

b. 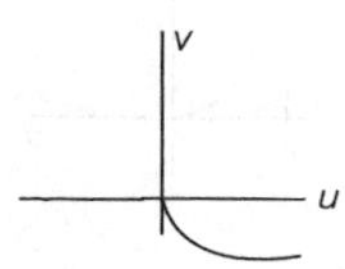

c. 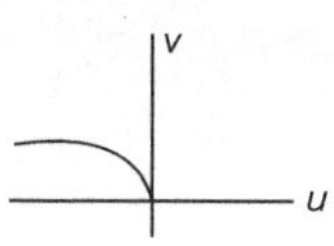

d. None of these

x. In general, the v-u graph for a converging lens

a. 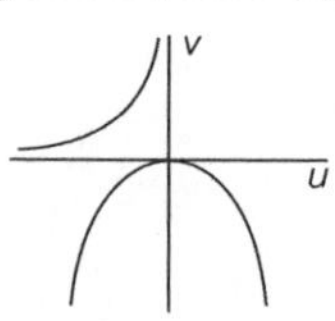

b. 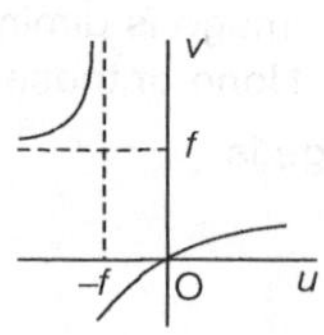

c. 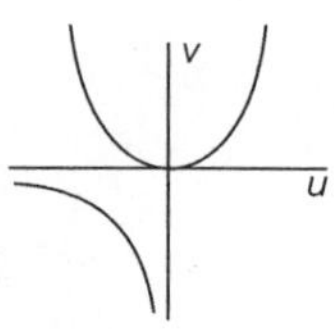

d. None of these

4 **Situation** Consider a thin concave lens of focal length 10 cm.

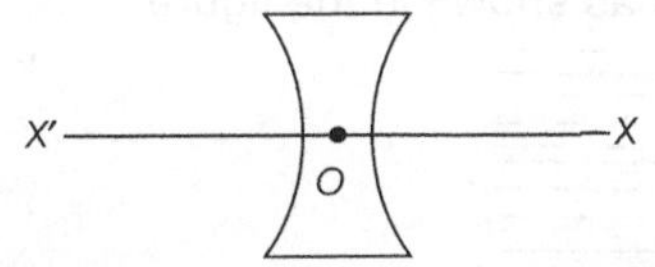

i. If object is placed between $x = 0$ to $x = -\infty$, the image is
  a. virtual, errect and diminished
  b. real, inverted and diminished
  c. may be virtual, may be real
  d. None of the above

ii. In previous problem *v-u* graph is

a. 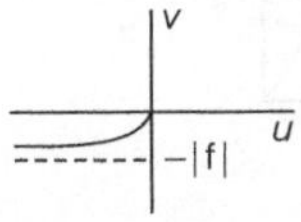

b. 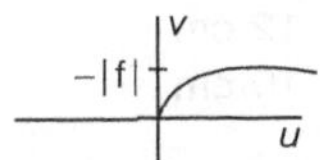

c. 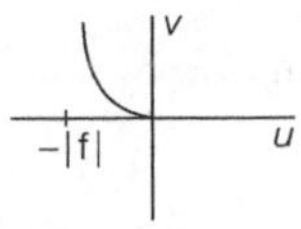

d. None of these

iii. If virtual object is placed between $x = 0$ and $x = 10$ cm. The image is
  a. real, magnified and errect
  b. virtual, inverted and magnified
  c. real, diminished and inverted
  d. None of the above

iv. If virtual object is placed between $x = 10$ cm to $x = \infty$. The image is
  a. virtual
  b. real
  c. may be virtual, may be real
  d. None of the above

v. In general, *v-u* graph for diverging lens is

a. 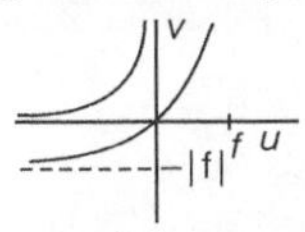

b. 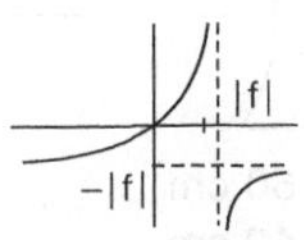

c. 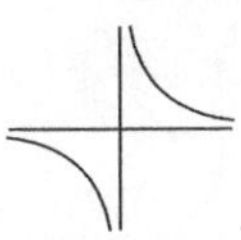

d. None ot these

Only One Option Correct Type

Direction *(Q. Nos. 5-44) This section contains 40 multiple choice questions. Each question has four choices (a), (b), (c) and (d), out of which ONLY ONE option is correct.*

5 The image of point object P is P'. Find the focal length of the lens.

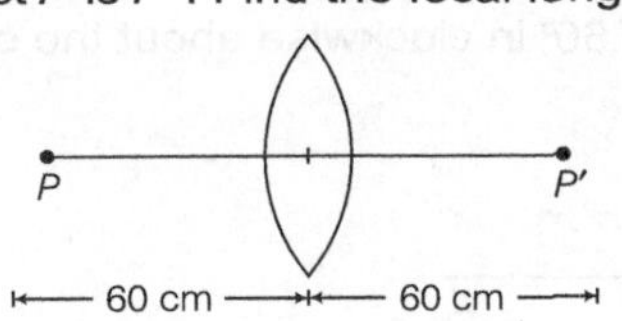

a. 15 cm
b. 30 cm
c. 60 cm
d. 90 cm

Working Space

6 The focal length of equiconvex lens in air is 30 cm. The refractive index of the lens is 1.5. Find the refractive index of the liquid as shown in the figure.

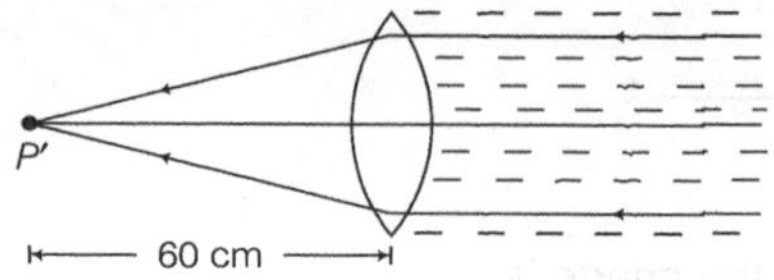

a. 1.2
b. 1.25
c. 1.4
d. 1.8

7 The radius of curvature of the left and right surface of concave lens of refractive index 1.5 are 10 cm and 15 cm respectively. The refractive index of liquid is 2. Find the equivalent of focal length.

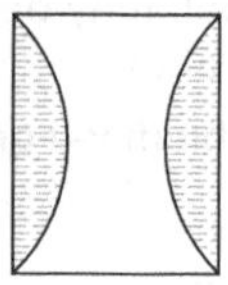

a. −12 cm
b. 12 cm
c. 20 cm
d. 10 cm

8 A lens of focal length 20 cm was cut along the diameter into two identical halves. In the process, the layer of the lens 2 mm in thickness was lost. Then, the halves were put together to form a composite lens. A point object is placed at 30 cm from the lens. Find the separation between images.

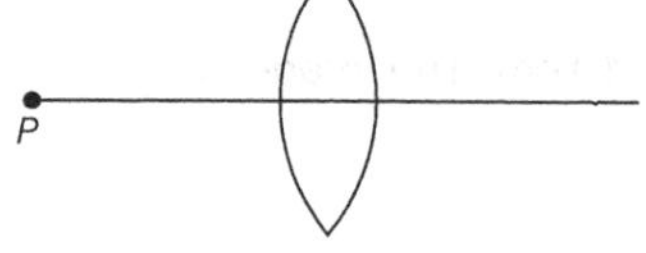

a. 1 mm
b. 2 mm
c. 3 mm
d. 6 mm

9 A parallel beam of light is incident on the system of two lenses of focal length $f_1 = 40$ cm and $f_2 = -10$ cm. What should be the distance between the two lenses, so that rays after refraction from both lenses pass undeviated?

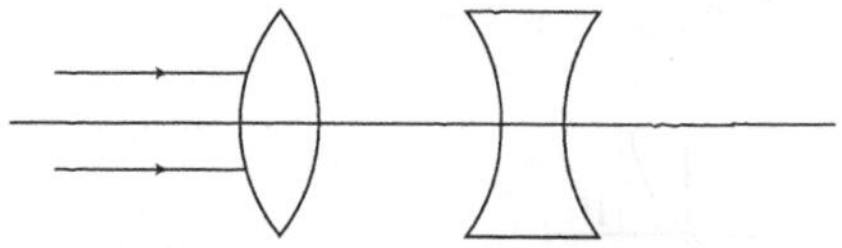

a. 10 cm
b. 50 cm
c. 30 cm
d. 40 cm

10 A converging beam of light converges at point P. A thin convex lens of focal length 20 cm is placed 10 cm before converging point. Find the position of new converging point from the lens.

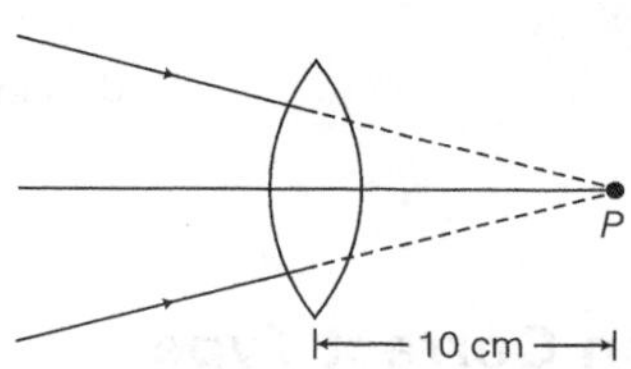

a. 6.67 cm right
b. 6.67 cm left
c. 10 cm left
d. 10 cm right

11 A point object P is placed at distance 20 cm from the optical centre of a lens of focal length 20 cm. Now, lens is rotated through 60° in clockwise about the optical centre. Find the x-coordinate of image.

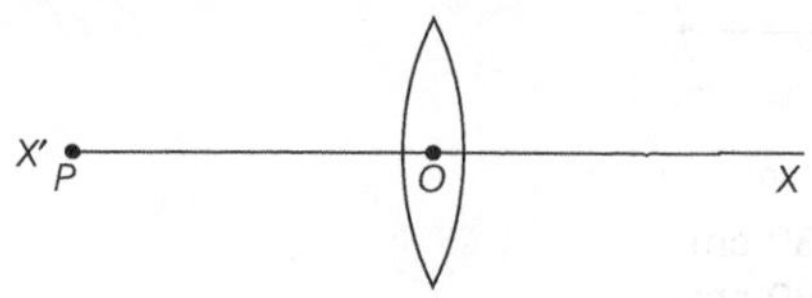

a. 20 cm
b. −20 cm
c. −40 cm
d. +40 cm

Working Space

12 A thin equiconcave lens of focal length 20 cm and refractive index 1.5 is placed in air. A point object P is placed at distance 40 cm from the optical centre of the lens. The surface of lens further away from the object is silvered. The position of image is

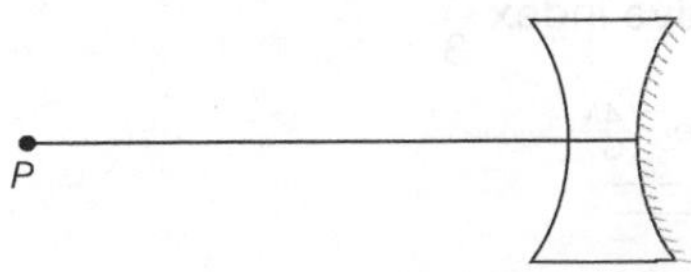

a. 40 cm right of lens

b. $\frac{40}{9}$ cm right of lens

c. 30 cm left of lens

d. $\frac{40}{3}$ cm right of lens

13 A thin equiconcave lens of a material having refractive index 1.5 and radius of each surface 20 cm is placed such that upper half is in a medium of refractive index $\mu_1 = 1.2$ and lower half is in a medium of refractive index $\mu_2 = 2.5$ as shown in the figure. If an object is placed on principal axis, then its two images will form one after refraction from upper part and other after refraction through lower part. If a parallel beam of light parallel to optical axis incident on the lens, find the separation between converging points of the beam.

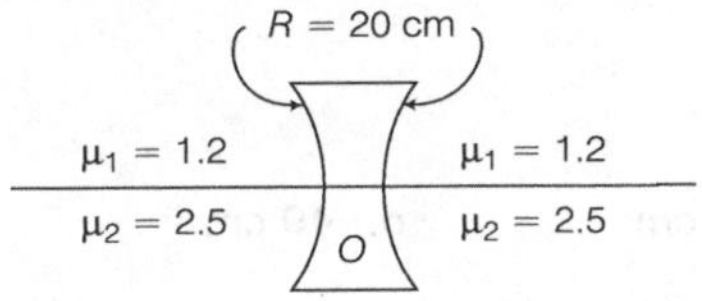

a. 40 cm

b. 25 cm

c. 65 cm

d. 100 cm

14 At $t = 0$, the separation between point object P and the thin convex lens is 40 cm. Find the speed of image at $t = 10$ s.

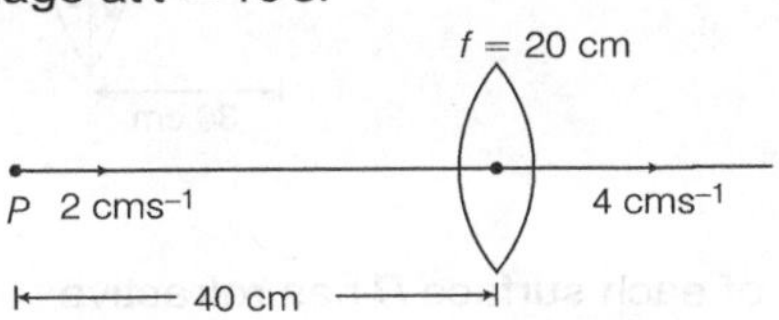

a. $2\ \text{cms}^{-1}$

b. $1\ \text{cms}^{-1}$

c. $4.5\ \text{cms}^{-1}$

d. $3.5\ \text{cms}^{-1}$

15 Find the velocity of image of point object P at the instant as shown in the figure.

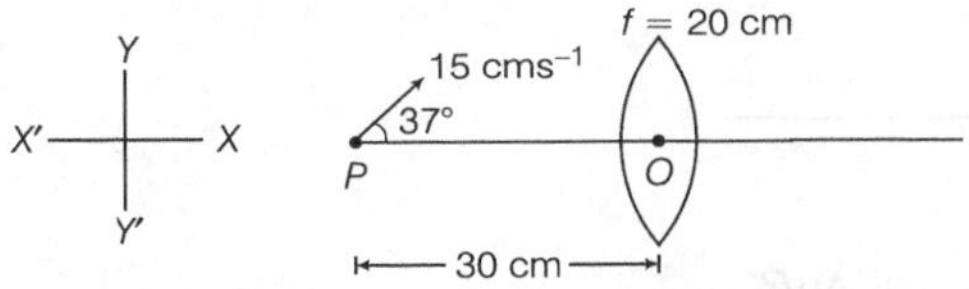

a. $(48\hat{i} - 18\hat{j})\ \text{cms}^{-1}$

b. $(48\hat{i} + 18\hat{j})\ \text{cms}^{-1}$

c. $(18\hat{i} - 48\hat{j})\ \text{cms}^{-1}$

d. $48\ \text{cms}^{-1}$

16 The figure shows an equiconcave lens placed on a plane mirror. The focal length of lens in air is –30 cm. The refractive index of lens is 1.5 and that of liquid is 2. The combination behaves as

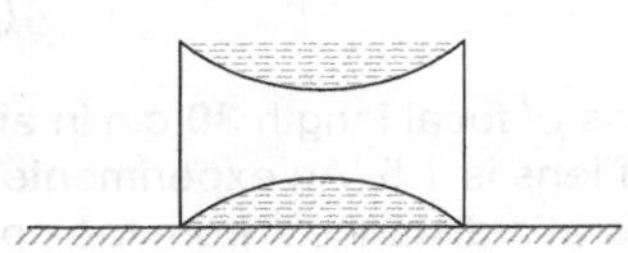

a. a concave mirror of focal length 15 cm

b. a convex mirror of focal length 15 cm

c. a concave mirror of focal length 10 cm

d. a convex mirror of focal length 20 cm

Working Space

17 An object P is placed in air infront of a thin equiconvex lens of focal length 30 cm in air at distance 90 cm from the lens. The refractive index of the lens is $\frac{3}{2}$. The medium towards right of lens is water of refractive index $\frac{4}{3}$.

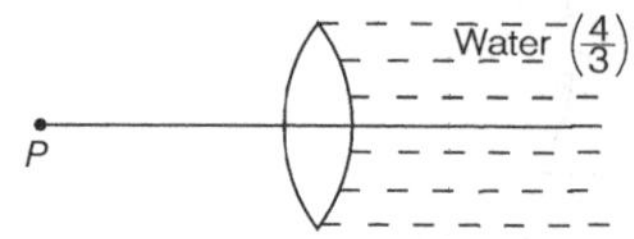

a. The image of point object P is at infinity
b. The image of point object P is at 120 cm right of the lens
c. The image of point object P is
d. None of the above

18 A thick convex lens of thickness 10 cm has radius of each surface 20 cm. Find the position of focus from the rear surface.

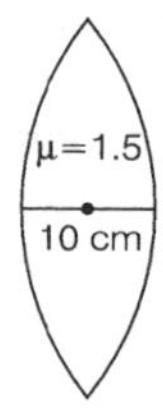

a. 18.18 cm b. 20 cm c. 30 cm d. 40 cm

19 An equiconvex lens is formed by combination of two plano convex lenses of refractive indices 1.5 and 2. The radius of curvature of curved surface is 20 cm. A point object is placed at 30 cm from the lens. The image of the point object is

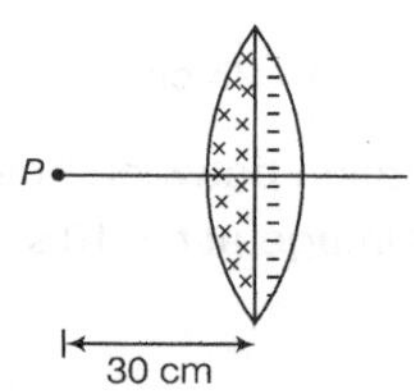

a. 24 cm right from the lens
b. 24 cm left from the lens
c. 12 cm right of the lens
d. None of the above

20 A thin equiconvex lens of radius of curvature of each surface R has refractive index μ in upper half and lower half has refractive index $\mu - \Delta\mu$ ($\Delta\mu << \mu$). A parallel beam of light incident on the lens. Find the distance between converging points of beam by two halves.

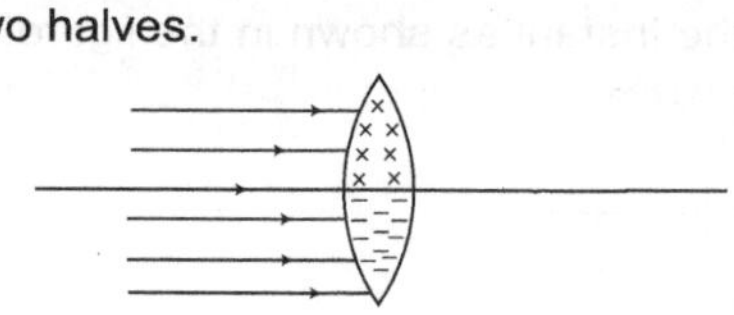

a. $\frac{\Delta\mu R}{2(\mu - 1)}$
b. $\frac{\Delta\mu R}{2(\mu - 1)^2}$
c. $\frac{\Delta\mu R}{2(\mu - 1)^3}$
d. $\frac{\Delta\mu R}{4(\mu - 1)^2}$

21 An equiconvex lens of focal length 30 cm in air is silvered at rear surface. The refractive index of lens is 1.5. An experimenter places the object pin A on the optical axis of lens at a distance x from the optical centre P. The experimenter looks at the pin and its inverted image from a distance keeping his/her eye in line with PA. When the student shifts his/her eye towards right, the image appears to the left of the object pin. Then,

a. $x > 15$ cm
b. $7.5\text{ cm} < x < 15$ cm
c. $x = 15$ cm
d. $x < 7.5$ cm

Working Space

22 You are provided with a convex lens, a concave lens, a screen and light rays, a screen and an optical bench on which all these components can be mounted and arranged in any order. The optical bench has a distance scale, which allows accurate measurement of the distances between the components.
If the concave lens is placed after the convex lens, with both focii at the same points, parallel light entering the convex lens will emerge as

a. parallel light from concave lens
b. converging beam from the concave lens
c. as diverging beam from the concave lens
d. either converging or diverging from the concave lens

23 The figure shows different graphs in case of a thin convex lens for the formation of a real image. u and v are the object and image distances respectively. Find out the correct identification for the different combinations from the following.

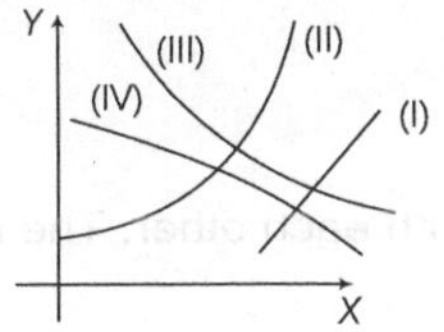

a. graph I, if x is u and y is u/v
b. graph II, if x is u and y is v
c. graph III, if x is v and y is u/v
d. graph IV, if x is $1/u$ and y is $1/v$

24 Mark the correct option.

a. Virtual and real images look the same to the eye and aberrations occur only for real images
b. A diverging lens cannot form a real image from a real object
c. All rays parallel to the axis of a spherical mirror are reflected through a single point
d. A virtual image can be displayed on a screen

25 AB is linear object placed along optical axis as shown in the figure.

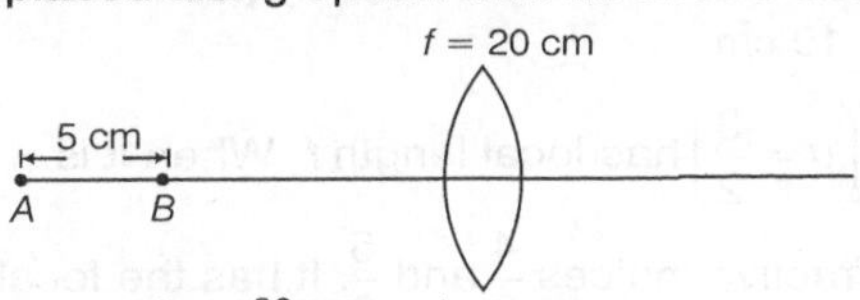

a. The size of image is smaller than 5 cm
b. The size of image is larger than size of object
c. The size of image is equal to the size of object
d. If a transparent rectangular slab of thickness 12 cm and refractive index $\frac{4}{3}$ is placed between object and lens the size of image may increase

26 In the given figure, a thin lens is shown in the figure, PA is a line passing through optical centre of lens. Now, the lens is rotated about optical centre for an angle θ in clockwise direction. The image of point object P is

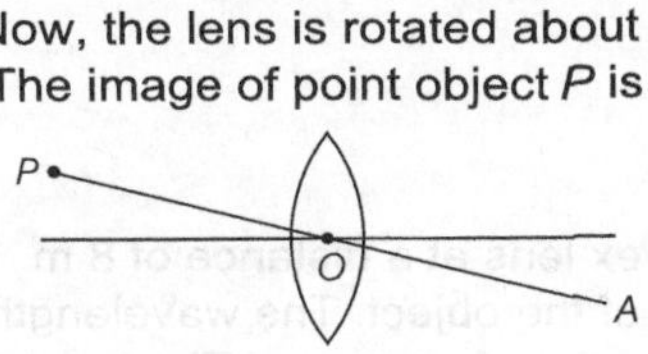

a. rotated in anticlockwise
b. rotated in clockwise direction
c. always lying on line colinear to PA
d. lying anywhere right side of the lens

27 Which of the following is correct ray diagram?

a. 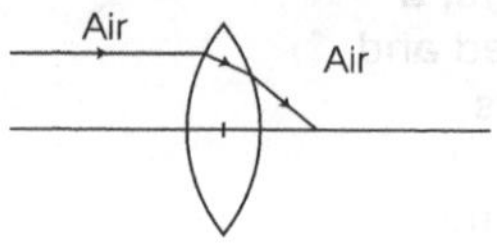

b. 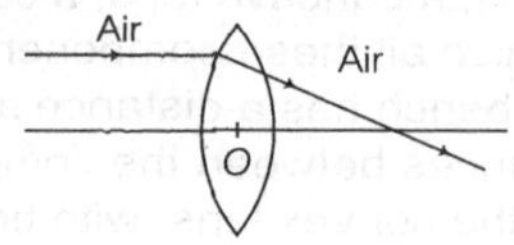

c. 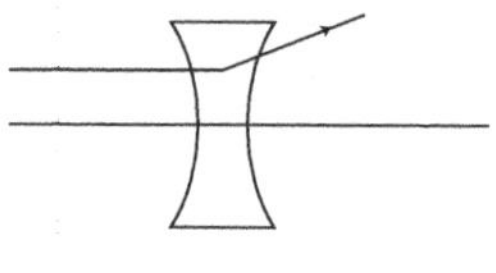

d. 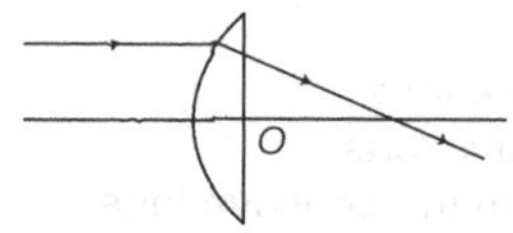

28 A plano-convex lens of refractive index 1.5 and radius of curvature 30 cm is silvered at the curved surface. Now, this lens has been used to form the image of an object. At what distance from this lens, an object should be placed in order to have a real image of the size of the object? **[AIEEE 2004]**

a. 20 cm b. 30 cm
c. 60 cm d. 80 cm

29 Two lenses of power $-15D$ and $+5D$ are in contact with each other. The focal length of the combination is **[AIEEE 2007]**

a. -20 cm b. -10 cm
c. $+20$ cm d. $+10$ cm

30 An object 2.4 m infront of a lens forms a sharp image on a film 12 cm behind the lens. A glass plate 1 cm thick of refractive index 1.50 is interposed between lens and film with its plane faces parallel to film. At what distance from lens should object shifted to be in sharp focus on film? **[AIEEE 2012]**

a. 7.2 m b. 2.4 m
c. 3.2 m d. 5.6 m

31 Diameter of a plano-convex lens is 6 cm and thickness at the centre is 3 mm. If the speed of light in material of lens is $2 \times 10^8 \text{ms}^{-1}$. The focal length of the lens is **[JEE Main 2013]**

a. 15 cm b. 20 cm
c. 30 cm d. 10 cm

32 A thin convex lens made from crown glass $\left(u = \frac{3}{2}\right)$ has local length f. When it is measured in two different liquids having refractive indices $\frac{4}{3}$ and $\frac{5}{3}$. It has the focal lengths f_1 and f_2, respectively. The correct relation between the focal lengths is **[JEE Main 2014]**

a. $f_1 = f_2 < f$ b. $f_2 > f$ and f_2 becomes negative
c. $f_2 > f$ and f_1 becomes negative d. f_1 and f_2 both become negative

33 A converging lens is used to form an image on a screen. When the upper half of the lens is covered by an opaque screen?

a. Half of the image will disappear
b. Complete image will be formed
c. Intensity of the image will increase
d. Intensity of the image will decrease

34 The image of an object, formed by a plano-convex lens at a distance of 8 m behind the lens, is real and is one-third the size of the object. The wavelength of light inside the lens is 2/3 times and the wavelength in free space. The radius of the curved surface of the lens is **[JEE Advanced 2013]**

a. 1 m b. 2 m
c. 3 m d. 6 m

Working Space

35 A biconvex lens is formed with two thin plano-convex lenses as shown in the figure. Refractive index n of the first lens is 1.5 and that of the second lens is 1.2. Both the curved surfaces are of the same radius of curvature $R = 14$ cm. For this biconvex lens, for an object distance of 40cm, the image distance will be **[IIT JEE 2012]**

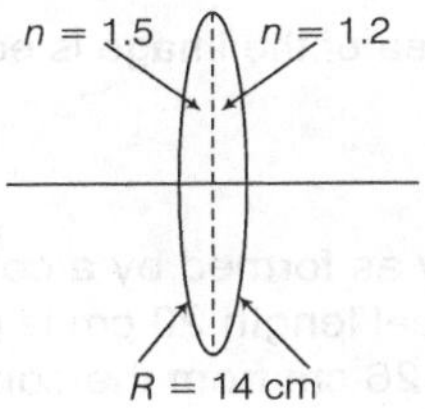

a. −280.0 cm
b. 40.0 cm
c. 21.5 cm
d. 13.3 cm

36 A biconvex lens of focal length 15 cm is infront of a plane mirror. The distance between the lens and the mirror is 10 cm. A small object is kept at a distance of 30 cm from the lens. The final image is

a. virtual and at a distance of 16 cm from the mirror
b. real and at a distance of 16 cm from the mirror
c. virtual and at a distance of 20 cm from the mirror
d. real and at a distance of 20 cm from the mirror

37 A convex lens is in contact with concave lens. The magnitude of the ratio of their focal length is $\frac{3}{2}$. Their equivalent focal length is 30 cm. What are their individual focal lengths? **[IIT JEE 2005]**

a. −75, 50
b. −10, 15
c. 75, 50
d. −15, 10

38 A point object is placed at a distance of 20 cm from a thin plano-convex lens of focal length 15 cm. The plane surface of the lens is now silvered. The image created by the system is at **[IIT JEE 2006]**

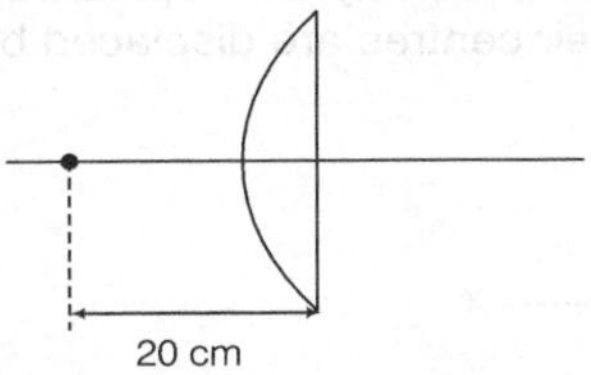

a. 60 cm to the left of the system
b. 60 cm to the right of the system
c. 12 cm to the left of the system
d. 12 cm to the right of the system

39 The graph between object distance u and image distance v for a lens is given below. The focal length of the lens is **[IIT JEE 2006]**

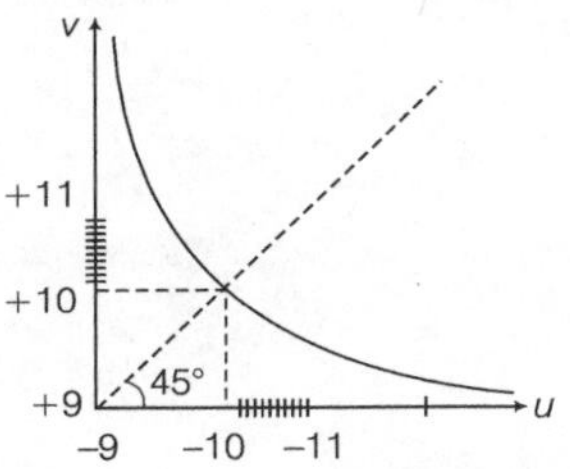

a. 5 ± 0.1
b. 5 ± 0.05
c. 0.5 ± 0.1
d. 0.5 ± 0.05

Working Space

40 A biconvex lens of focal length f forms a circular image of radius r of sun in focal plane. Then, which option is correct? **[IIT JEE 2006]**

a. $\pi r^2 \propto f$

b. $\pi r^2 \propto f^2$

c. If lower half part is convered by black sheet, then area of the image is equal to $\pi r^2/2$

d. If f is doubled, intensity will increase

41 The size of the image of an object, which is at infinity as formed by a convex lens of focal length 30 cm is 2 cm. If a concave lens of focal length 20 cm is placed between convex lens and the image at a distance of 26 cm from the convex lens, calculate the new size of the image. **[IIT JEE 2003]**

a. 1.25 cm
b. 2.5 cm
c. 1.05 cm
d. 2 cm

42 A hollow double concave lens is made of very thin transparent material. It can be filled with air or either of two liquids L_1 or L_2 having refractive indices n_1 and n_2 respectively $(n_2 > n_1 > 1)$. The lens will diverge a parallel beam of light, if it is filled with

a. air and placed in air
b. air and immersed in L_1
c. L_1 and immersed in L_2
d. L_2 and immersed in L_1

43 A concave lens of glass, refractive index 1.5 has both surfaces of same radius of curvature R. On immersion in a medium of refractive index 1.75, it will behave as a

a. convergent lens of focal length 3.5 R
b. convergent lens of focal length 3.0 R
c. divergent lens of focal length 3.5 R
d. divergent lens of focal length 3.0 R

44 Two thin convex lenses of focal lengths f_1 and f_2 respectively are separated by a horizontal distance d (where $d < f_1, d < f_2$) and their centres are displaced by a vertical separation Δ as shown in the figure.

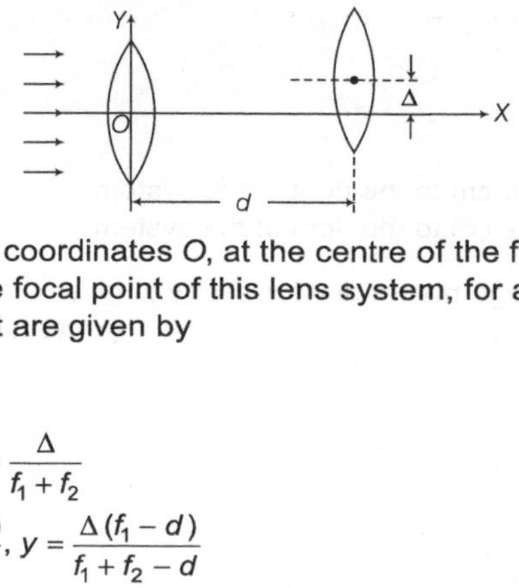

Taking the origin of coordinates O, at the centre of the first lens, the x and y-coordinates of the focal point of this lens system, for a parallel beam of rays coming from the left are given by

a. $x = \dfrac{f_1 f_2}{f_1 + f_2}, y = \Delta$

b. $x = \dfrac{f_1(f_2 + d)}{f_1 + f_2 - d}, y = \dfrac{\Delta}{f_1 + f_2}$

c. $x = \dfrac{f_1 f_2 + d(f_1 - d)}{f_1 + f_2 - d}, y = \dfrac{\Delta(f_1 - d)}{f_1 + f_2 - d}$

d. $x = \dfrac{f_1 f_2 + d(f_1 - d)}{f_1 + f_2 - d}, y = 0$

DPP-11 Lens-II

One or More than One Options Correct Type

Working Space

Direction *(Q. Nos. 1-9) This section contains 9 multiple choice questions. Each question has four choices (a), (b), (c) and (d), out of which ONE or MORE THAN ONE are correct.*

1 A point object is placed at distance 30 cm from a plano-convex lens of focal length 20 cm.

a. The final image is at 60 cm from the lens

b. If the refractive index of lens is $\frac{3}{2}$ and whole system is immersed in a liquid of refractive index $\frac{4}{3}$. In this case, the final image is 48 cm left of the lens

c. If another concave lens of focal length 40 cm is placed in contact to the previous lens, the final image is 120 cm left of the lens

d. None of the above

2 A 2 cm high object is placed at 40 cm from the lens. The focal length of lens is 20 cm and refractive index is $\frac{3}{2}$. The whole system is placed in a medium of refractive index 2.

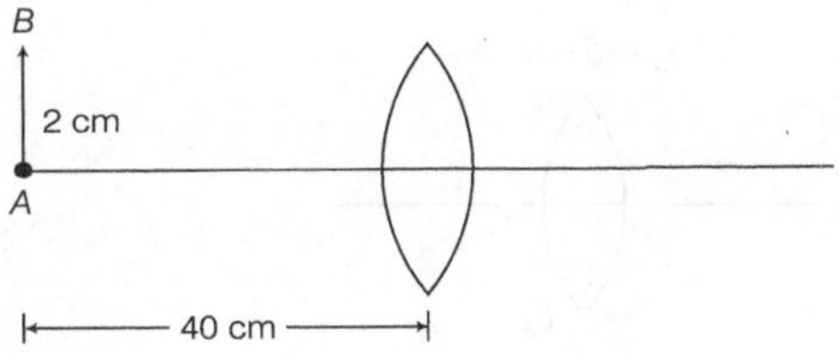

a. The image of *AB* is real, inverted and 1 cm high

b. The image of *AB* is at 20 cm right from the lens

c. The image of *AB* is virtual, errect and 1 cm high

d. The image of *AB* is at $\frac{40}{3}$ cm from the lens

3 A thin convex lens of focal length is able to slide along optical axis. The distance between object plane and screen is 180 cm. The magnification of image produced by initial position of lens is 2.

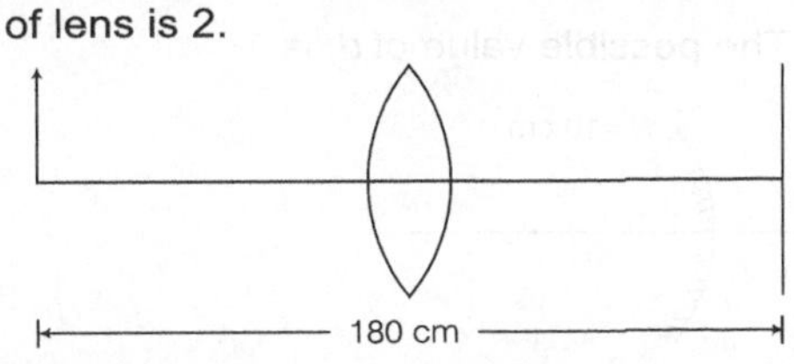

a. The focal length of lens is 40 cm

b. The distance through which lens is shifted to form again real image on the screen is 60 cm

c. The distance through which lens is shifted to form again real image on screen is 30 cm

d. The focal length of lens is 60 cm

Working Space

4 In the given figure, the image of object AB observed by observer is

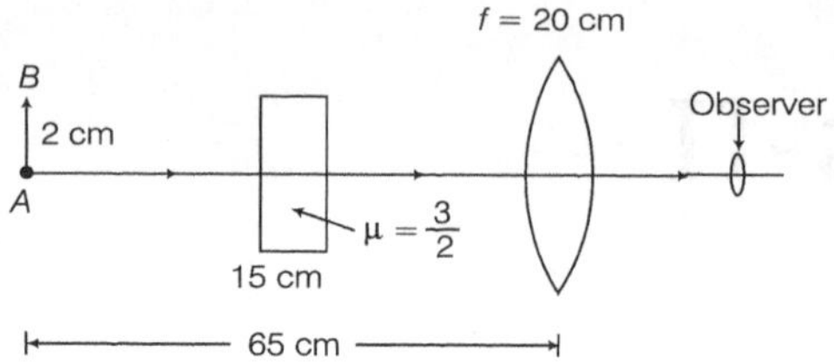

a. real and inverted
b. 1 cm in length
c. 30 cm from the lens
d. 60 cm from the lens

5 A point object P is placed on the common optical axis of lens and mirror as shown in the figure. The possible value of x for which image coincides with object is

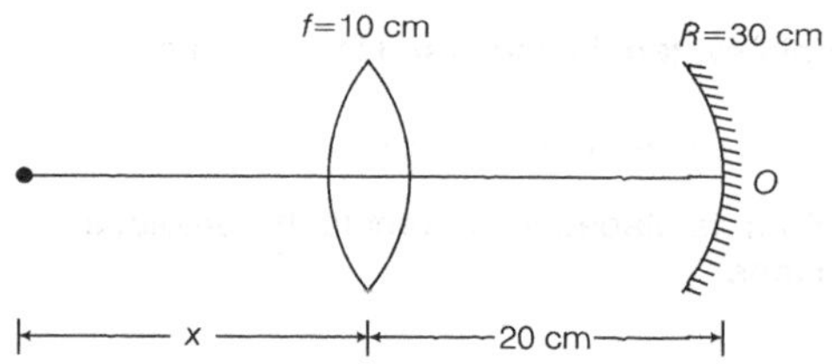

a. 5 cm
b. 10 cm
c. 15 cm
d. 20 cm

6 A parallel beam of light is incident on the co-axial combination of three lenses as shown in the figure.

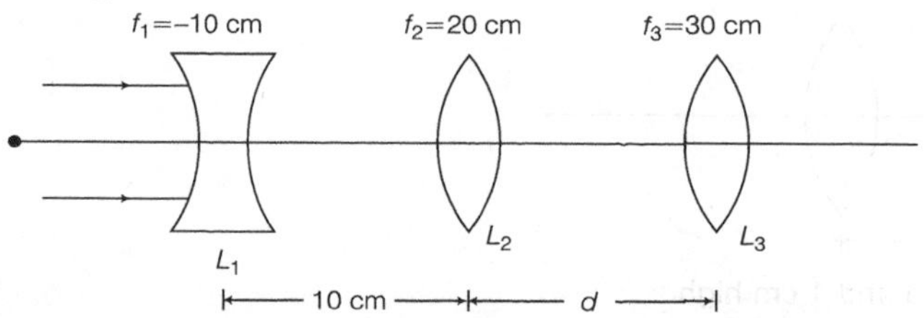

a. For any value of d, the beam converges at, 30 cm right ward from lens L_3
b. For $d = 0$, the beam converges at 30 cm right ward from lens L_3
c. For $d = 0$, the beam converges at 12 cm right ward from lens L_3
d. None of the above

7 If image of point object P coincides with it. The possible value of d, is

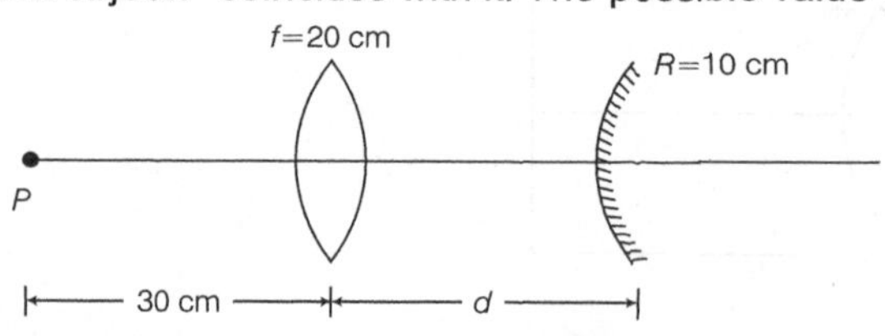

a. 60 cm
b. 50 cm
c. 40 cm
d. 55 cm

Working Space

8 A 2 cm high object is placed 30 cm left of thin lens of focal length 20 cm as shown in the figure.

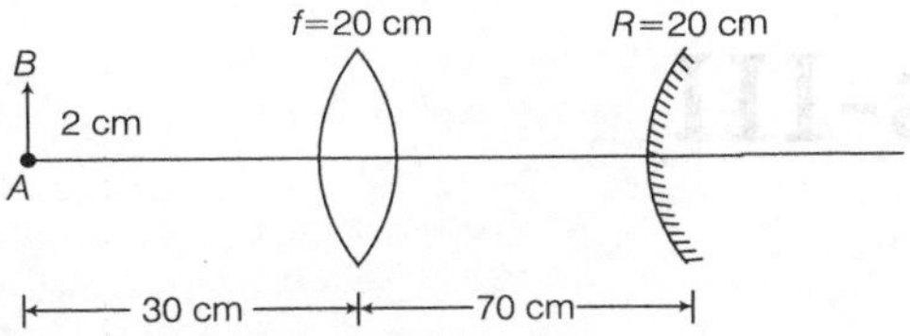

a. The final image is 27.27 cm left of the lens

b. Final image is $\frac{8}{11}$ cm in size

c. Final image is real

d. Final image is virtual

9 Consider a glass thin plano-convex lens-flat on one side and convex on the other side. You project three parallel laser beams through it, as in figure (A) and measure the focal length *f*, that is, the distance from the lens to the point at which the three beams cross. Now, you hold the flat side of the lens against the glass of an aquarium filled with water as shown in figure (B). Laser beams are projected directly on aquarium from right in figure (C).

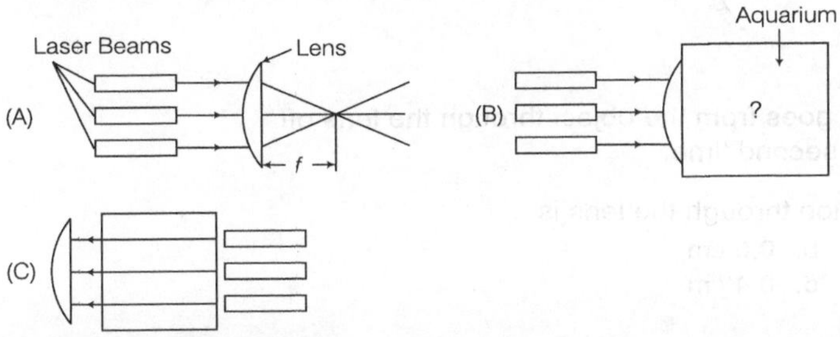

Mark the correct option(s).

a. The laser beam going through the centre of the lens is unaffected by the lens in all three cases

b. All three beams cross at the focal point

c. The laser beams cross at the same distance from the lens in Figure (C) as they did in the Figure (A)

d. The laser beams cross at the larger distance from the lens in Figure (B) as they did in the Figure (A)

DPP-12 Lens-III

Comprehension Type

Working Space

Direction *(Q. Nos. 1-10) This section contains 3 paragraphs, each describing theory, experiments, data, etc. Ten questions related to the paragraphs have been given. Each question has only one correct answer among the four given options (a), (b), (c) and (d).*

Passage I

Shown in the figure below is a system containing an object, a lens and a mirror.

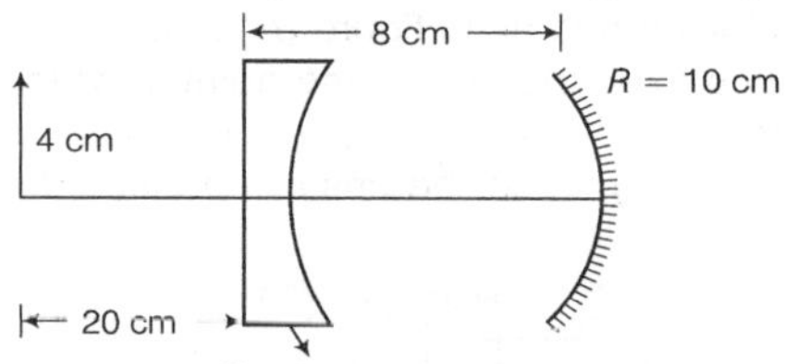

This is a 3-phase system in which light goes from the object through the lens off the mirror and back through the lens a second time.

1 The size of image formed by first refration through the lens is

a. 2 cm
b. 0.8 cm
c. 4 cm
d. 0.4 cm

2 The position of final image is

a. at 8.645 cm from the mirror
b. 8.57 cm from the lens
c. 9.129 cm from the mirror
d. 4 cm from the lens

3 If an observer is situated between lens and mirror on the optical axis. He can see

a. one image
b. two image
c. no image
d. the object

Passage II

A strong source of light when used with convex lens produces a number of images of the source owing to feeble reflections and refraction called flare spots. Assume that the surfaces of a thin equiconvex glass lens has focal length 30 cm in air. The refractive index of lens is 1.5.

4 The radius of each surface of lens is

a. 60 cm
b. 30 cm
c. 15 cm
d. 45 cm

5 The position of image after first refraction is

a. 90 cm
b. 30 cm
c. 60 cm
d. 45 cm

6 The position of final image is

a. $\frac{90}{7}$ cm
b. $\frac{30}{7}$ cm
c. $\frac{90}{13}$ cm
d. 90 cm

Working Space

Passage III

An experimenter wants to determine the refractive index of a liquid. For this, a point source S of monochromatic light is placed at a distance of 80 cm from a thin concave-convex lens made of glass ($\mu = 1.5$) has surfaces of radii 20 cm and 60cm. A flat mirror turned through 45° with respect to the optical axis of the lens, is placed behind the lens at a distance of 160 cm.

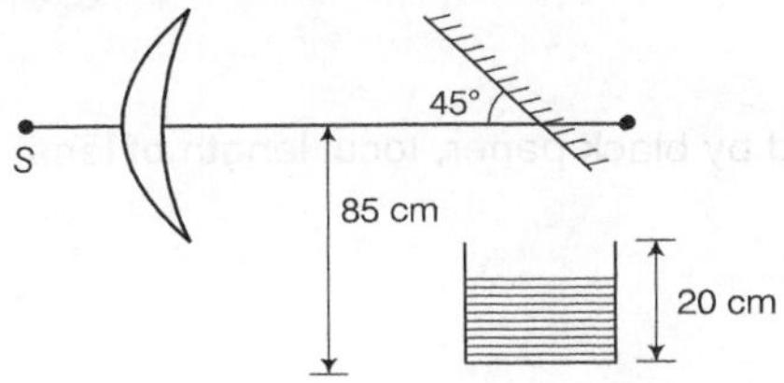

A tray filled with the liquid is placed 85 cm below the optical axis. The liquid in the tray is upto height 20 cm as shown in the figure.

7 The focal length of the lens is

a. 20 cm
b. 10 cm
c. 30 cm
d. 60 cm

8 The image of point source S form by the lens is

a. 80 cm before the mirror
b. 80 cm behind the mirror
c. 80 cm below the optical axis
d. 240 cm behind the mirror

9 The image formed by the plane mirror is

a. must be real
b. must be virtual
c. may be real
d. may be virtual

10 A sharp image of the point source S is observed by the experimenter at the bottom of tray. Then, the refractive index of liquid is

a. $\frac{3}{2}$
b. $\frac{4}{3}$
c. 1.75
d. 1.8

Statement Type

Direction *(Q. Nos. 11-18) This section is based on Statement I and Statement II. Select the correct answer from the codes given below.*

a. Both Statement I and Statement II are correct and Statement II is the correct explanation of Statement I
b. Both Statement I and Statement II are correct but Statement II is not the correct explanation of Statement I
c. Statement I is correct but Statement II is incorrect
d. Statement II is correct but Statement I is incorrect

11 **Statement I** The image of a virtual object formed by a thin converging lens is always real.

Statement II In the case of thin lens, $\frac{1}{v} - \frac{1}{u} = \frac{1}{f}$.

Working Space

12 **Statement I** If a diverging lens is placed in medium, whose refractive index is greater than that of lens, the lens behaves as converging lens.

Statement II According to thin lens maker's formula,

$$\frac{1}{f} = (\mu - 1)\left(\frac{1}{r_1} - \frac{1}{r_2}\right),$$

where, $\mu = \dfrac{\text{Refractive index of lens}}{\text{Refractive index of surrounding}}$.

13 **Statement I** If a portion of thin lens is covered by black paper, focal length of lens does not change.

Statement II $\dfrac{1}{f} = (\mu - 1)\left(\dfrac{1}{r_1} - \dfrac{1}{r_2}\right)$.

14 **Statement I** If a thin convex lens ($\mu = 1.5$) has focal length f_1 in air and f_2 in water $\left(\mu = \dfrac{4}{3}\right)$. Then, f_2 is greater than f_1.

Statement II $\because \dfrac{1}{f} = (\mu - 1)\left(\dfrac{1}{r_1} - \dfrac{1}{r_2}\right)$,

where, $\mu = \dfrac{\text{Refractive index of lens}}{\text{Refractive index of surrounding}}$.

15 **Statement I** If a point object is placed at 10 cm from a concave lens of focal length 10 cm. The image is formed at infinite.

Statement II $\therefore \dfrac{1}{v} - \dfrac{1}{u} = \dfrac{1}{f}$.

16 **Statement I** A thin diverging lens forms virtual as well as real image of a virtual object.

Statement II A diverging lens forms magnified real image of a virtual object.

17 **Statement I** If real object is left ward of a lens, real image of the object is right ward of the lens.

Statement II In the case of real images, refracted rays are converging in nature.

18 **Statement I** Observer O can see two images of point source S. If the plane mirror is rotated through 30° in clockwise direction, the images seen by observer O is only one.

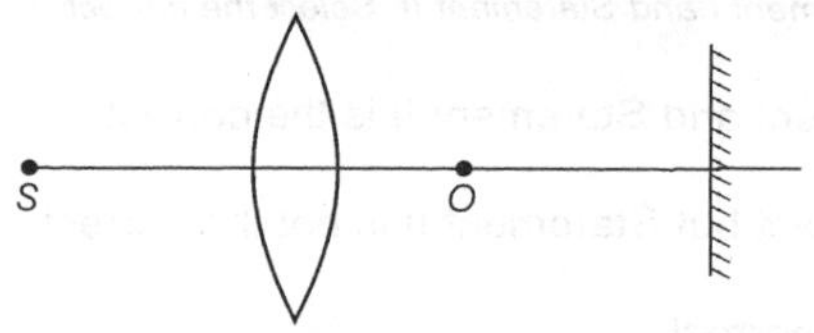

Statement II Due to rotation of mirror, law of reflection becomes failure.

Working Space

Matching List Type

***Direction** (Q. Nos. 19-24) Choices for the correct combination of elements from Column I and Column II are given as options (a), (b), (c) and (d), out of which one is correct.*

19 An equiconcave lens of refractive index μ_2 and focal length – 20 cm (in air) is kept in a medium of refractive index μ_1. Match the Column I with Column II and mark the correct option from the codes given below.

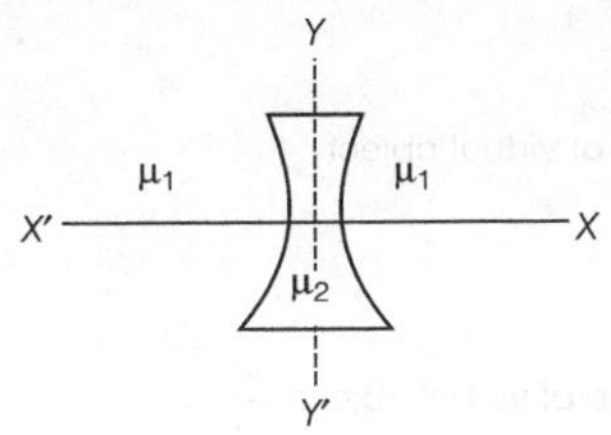

	Column I		Column II
i.	If lens is cut in two equal parts by a plane *yy'* (Take, $\mu_1 = 1$)	p.	Lens will acts as glass slab
ii.	If lens is cut in two equal parts by a plane *xx'*. (Take, $\mu_1 = 1$)	q.	Lens will diverging and focal length will change
iii.	If $\mu_1 = \mu_2$	r.	Lens will be diverging and focal length will remains same
iv.	If $\mu_1 > \mu_2$	s.	Lens will be converging and focal length will be change

Codes

	i	ii	iii	iv
a.	q	r	p	s
b.	p	q	r	s
c.	q	r	s	p
d.	r	q	p	s

20 Consider the situation as shown in the figure.

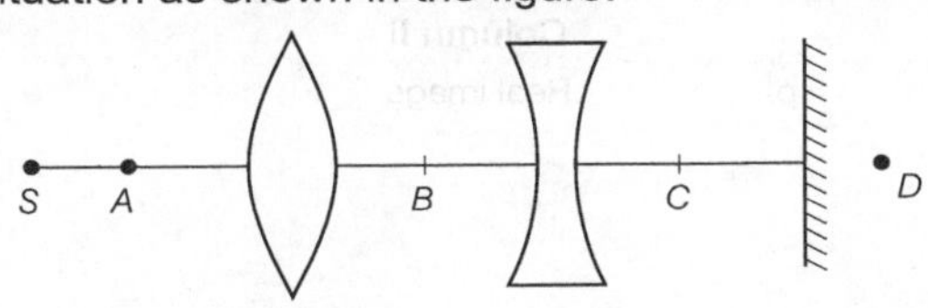

In the situation *S* is a point source and *A*, *B*, *C* and *D* are observers. Match the Column I with Column II and mark the correct option from the codes given below.

	Column I		Column II
i.	Observer *A* will observe	p.	Two images
ii.	Observer *B* will observe	q.	Object
iii.	Observer *C* will observe	r.	No image
iv.	Observer *D* will observe	s.	One image

Codes

	i	ii	iii	iv
a.	q	p	r	s
b.	p	r	s	q
c.	p	q	r	s
d.	q,s	p	p	r

Working Space

21 Match the Column I with Column II and mark the correct option from the codes given below.

	Column I		Column II
i.		p.	Real image of real object
ii.		q.	Real image of virtual object
iii.		r.	Virtual image of virtual object
iv.		s.	Virtual image of real object

Codes

	i	ii	iii	iv		i	ii	iii	iv
a.	p,q,s	r,q,s	s,q,p	p,q,r	b.	s,p,q	r,p,q	q,p,r	r,p,q
c.	r,p,q	s,q,p	s,p,q	p,q,s	d.	p,q,s	p,r,s	p,q,s	p,r,s

22 An optical component and an object S placed along its optic axis are given in Column I. The distance between the object and the component can be varried. The properties of images are given in Column II. Match the Column I with Column II and mark the correct option from the codes given below. **[IIT JEE 2008]**

	Column I		Column II
i.	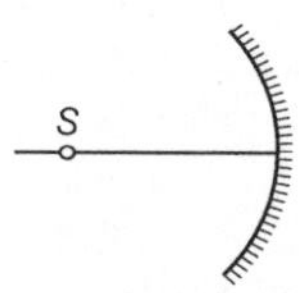	p.	Real image
ii.	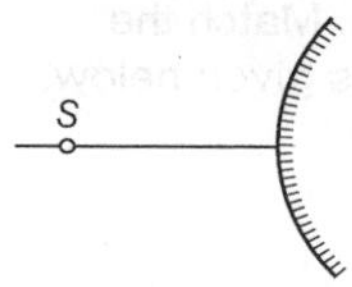	q.	Virtual image
iii.	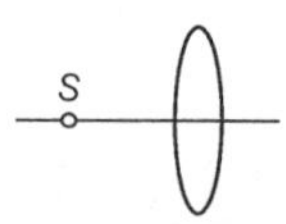	r.	Magnified image
iv.	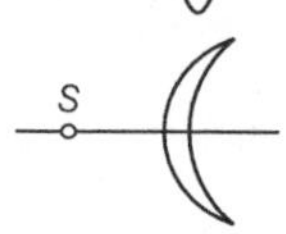	s.	Image at infinity

Codes

	i	ii	iii	iv		i	ii	iii	iv
a.	p,q,r,s	q	p,q,r,s	p,q,r,s	b.	p	q,r,s,p	s	r,p,q,s
c.	s,r,p,q	p	q	r	d.	r	s	p	q

Working Space

23 Two transparent media of refractive indices μ_1 and μ_3 have a solid lens shaped transparent material of refractive index μ_2 between them as shown in figures in Column II. A ray traversing these media is also shown in the figures. In Column I different relations between μ_1, μ_2 and μ_3 are given. Match them to the ray diagram shown in Column II. Match the Column I with Column II and mark the correct option from the codes given below. **[IIT JEE 2010]**

	Column I		Column II
i.	$\mu_1 < \mu_2$	p.	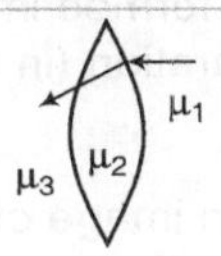
ii.	$\mu_1 > \mu_2$	q.	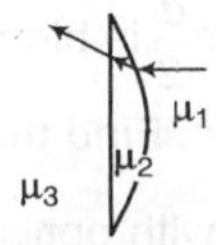
iii.	$\mu_2 = \mu_3$	r.	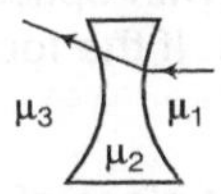
iv.	$\mu_2 > \mu_3$	s.	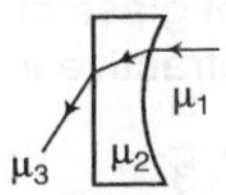
		t.	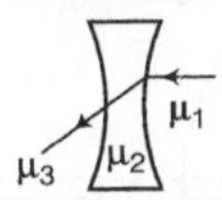

Codes

	i	ii	iii	iv		i	ii	iii	iv
a.	p,r	q,s,t	p,r,t	q,s	b.	p,q	p	q	r,s
c.	p	q,r	p	s	d.	p	s	r	p,s

24 For combinations of two thin lenses are given in Column I. The radius of curvature of all curved surfaces is r and the refractive index of all lenses is 1.5. Match lens combination in Column I with their focal length in Column II and mark the correct option from the codes given below. **[JEE Advanced 2014]**

	Column I		Column II
i.		p.	$2r$
ii.	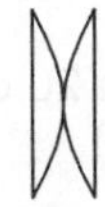	q.	$r/2$
iii.	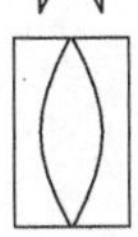	r.	$-r$
iv.		s.	r

Codes

	i	ii	iii	iv		i	ii	iii	iv
a.	p	q	r	s	b.	s	p	r	q
c.	q	s	r	p	d.	q	p	r	s

One Integer Value Correct Type

Direction *(Q. Nos. 25-43) This section contains 19 questions. When worked out will result in an integer from 0 to 9 (both inclusive).*

25 A plano-convex lens is formed by cutting a portion of thickness 2 mm from a solid glass sphere of radius 10 cm. If two such portions are combined to form a convex lens. Find the aperture (in cm) of the convex lens.

26 A thin convex lens of focal length 11 cm formed image of same magnification 4.4 for two positions of object. Find the separation (in cm) between two positions of object.

27 A thin convex lens of aperture d forms an image of intensity 10 Wm^{-2}. Now the central part of the aperture upto diameter $\frac{d}{2}$ is blocked by a black paper. The intensity of image becomes 3.75 x Wm^{-2}. Find the value of x.

28 A parallel beam of light making angle 2° with optical axis converges π cm below the optical axis by a thin converging lens. If the focal length of lens is 10 x cm. Find the value of x.

29 Two identical equiconvex lenses made of glass of refractive index 1.5, placed in contact has power P. When a liquid of refractive index μ is filled in the gap between the lenses, the power becomes $\frac{P}{3}$.

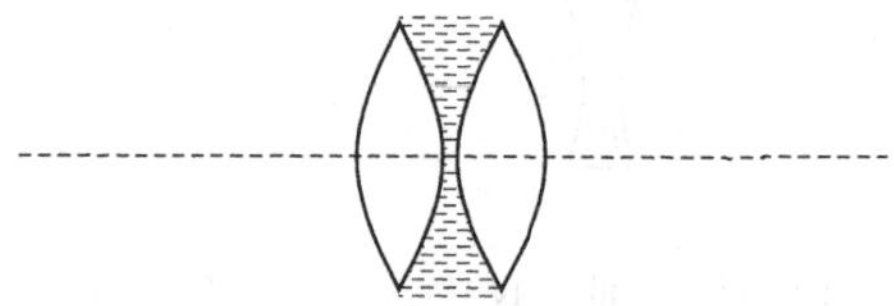

If $\mu = \frac{x}{3}$. Find the value of x.

30 A linear shaped object AB of length to $10\sqrt{2}$ cm is placed before a thin convex lens of focal length 20 cm as shown in the figure. If the length of image of AB is $10\sqrt{n}$ cm. Find the value of n.

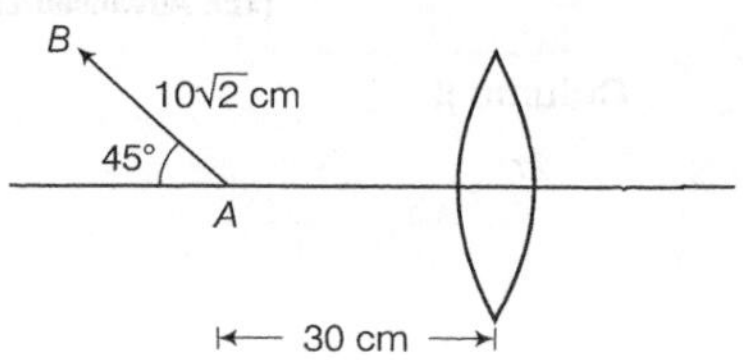

31 A parallel beam of light is incident on a thin convex lens of focal length 20 cm. The beam makes an angle 2° with the optical axis. The y-coordinate of converging point after refraction of $-\left(\frac{2\pi}{n}\right)$ cm. Find the value of n.

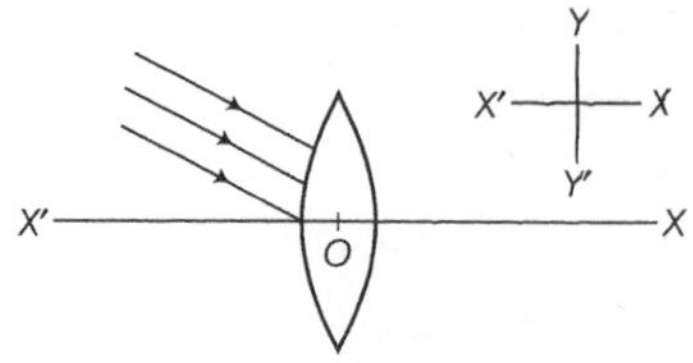

Working Space

32 In figure shown, half above portion of rear surface of the equiconvex lens of refractive index 1.5 is silvered. The radius of each surface is 40 cm. A point object P is placed at 30 cm from the lens. The distance in cm between images formed by silvered part and unsilvered part is $\frac{420}{n}$. Find the value of n.

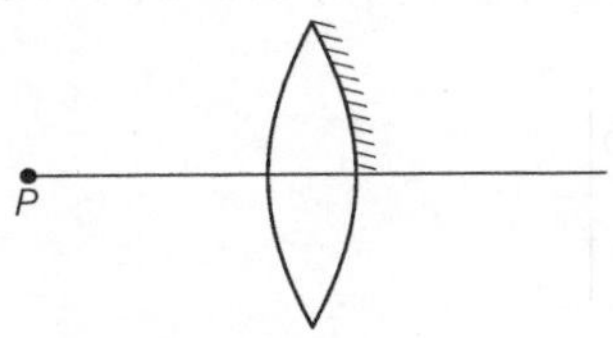

33 A pin hole camera in which the distance between hole to screen is adjustable. It is placed 10 m away from a pole casts image of the pole. Now it is shifted 1 m further away from the pole. To keep the size of image unchanged, the required percentage change in distance between hole to screen is $2n$%. Find the value of n.

34 Two thin convex lenses of focal lengths 5 cm and 4 cm sharing the same optical axis are some distance apart. This lens system is used to form image of an object. If it is found that size of the image does not depend on the distance between the first lens and the object. Find the distance d (in cm) between the lenses.

35 A lens of focal length 15 cm and aperture 12 mm is fitted in a cardboard permitting light to pass only through the lens. On the optical axis a point source of light is placed at distance 30 cm from the lens. A screen is placed at distance 15 cm from the lens. Find the radius (in mm) of light spot formed on the screen.

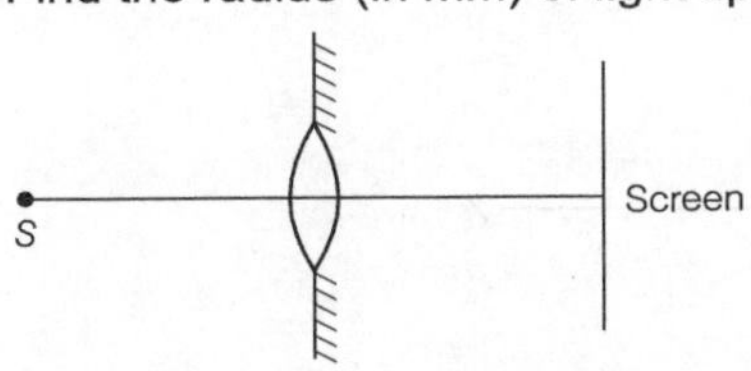

36 Two lenses L_1 and L_2 of respective focal length 10 cm and 60 cm are coaxially placed as shown in the figure. The apertures of lenses L_1 and L_2 are 1 mm and 10 mm. A point source of light is placed at distance 10 cm from lens L_1. Find the radius (in mm) of the light spot on screen due to refraction through both lenses.

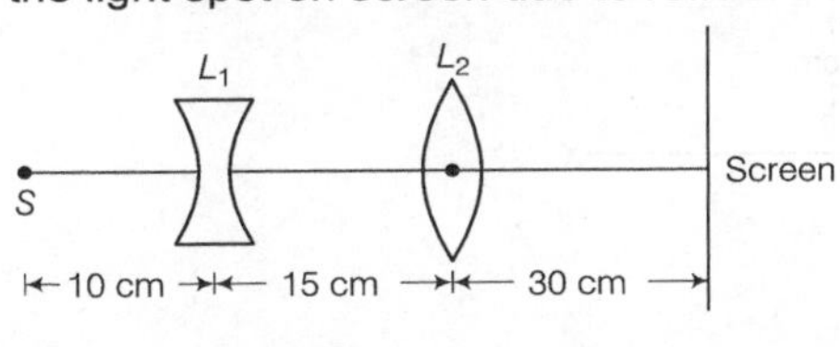

37 In the shown arrangement, a point source of light S is placed 60 cm left to a thin concave lens of concave lens of focal length 60 cm.

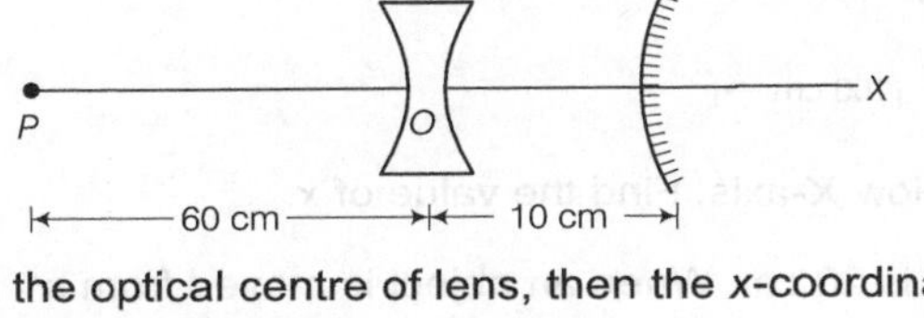

If origin is taken the optical centre of lens, then the x-coordinate of final image of point object P is $5x_0$ cm. Find the value of x_0.

Working Space

38 A rectangular slab of length $l = 40$ cm and thickness $d = 5$ cm is placed infront of a thin convex lens of focal length $f = 10$ cm. A screen is placed in the focal plane of the lens. Refractive index of material of the slab increases linearly from 1.5 to 1.6 from bottom to top face. A parallel beam of light is made to incident on the slab as shown in the figure. The beam converges at a point y_0 (in cm) above point O. Find the value of y_0.

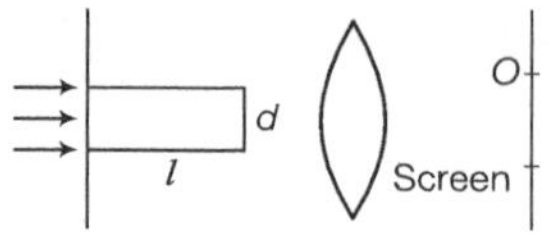

39 A thin equiconvex lens of refractive index 1.5 and focal length 20 cm in air is fitted at the month of cylindrical vessel filled with water of refractive index $\frac{4}{3}$. A parallel beam of light incident on the lens converges at the bottom of vessel. If the height of vessel is $10\,n$ cm. Find the value of n.

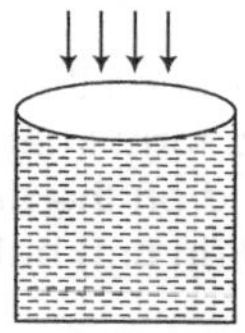

40 A rectangular slab of refractive index $\frac{3}{2}$ has a cavity inside it. The cavity has shape of a thin equiconcave lens, whose surfaces have radius of curvature $R = 20$ cm as shown in the figure. If the distance between object and final image is $5x$ cm. Find the value of x.

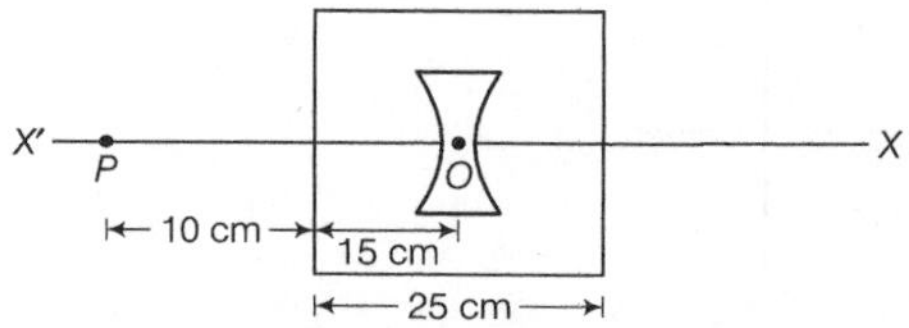

41 A thin convex lens of focal length 20 cm and aperture 5 cm is cut along the line AB. The part of the lens shown shaded in the diagram (I) is now cut and placed on the optical axis as shown in diagram (II).

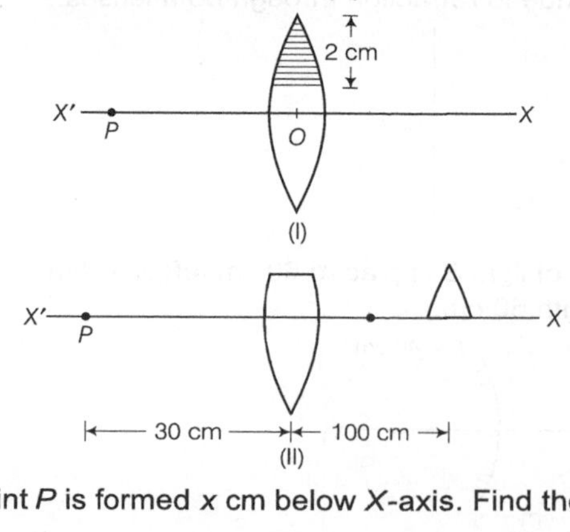

The image of point P is formed x cm below X-axis. Find the value of x.

42 The focal length of a thin biconvex lens is 20 cm. When an object is moved from a distance of 25 cm infront of it to 50 cm, the magnification of its image changes from m_{25} to m_{50}. The ratio $\frac{m_{25}}{m_{50}}$ is.

[IIT JEE 2010]

Working Space

43 A pin is placed 10 cm infront of a convex lens of focal length 20 cm and made of a matrial of refractive index 1.5. The convex surface of the lens further away from the pin is silvered and has a radius of curvature of 22 cm. Determine the position of the final image. Is the image real or virtual?

Subjective Questions

Direction *(Q. Nos. 44-51) These questions are subjective in nature, need to be solved completely on notebook.*

44 In the figure, light is incident on a thin lens as shown. The radius of curvature for both the surfaces is R. Determine the focal length of this system. **[IIT JEE 2003]**

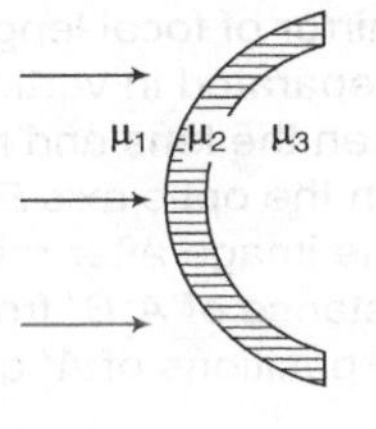

45 The radius of curvature of the convex face of a planoconvex lens is 12 cm and its $\mu = 1.5$.

i. Find the focal length of the lens. The plane face of the lens is now silvered.
ii. At what distance from the lens will parallel rays incident on the convex surface converge?
iii. Sketch the ray diagram to locate the image, when a point object is placed on the axis 20 cm from the lens.
iv. Calculate the image distance, when the object placed as in (c).

46 The convex surface a thin concave-convex lens of glass of refractive index 1.5 has a radius of curvature 20 cm. The concave surface has a radius of curvature 60 cm. The convex side is silvered and placed on a horizontal surface.

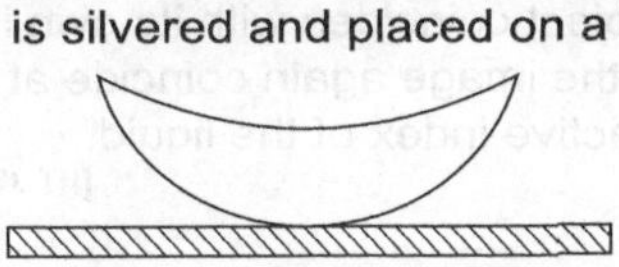

i. Where should a pin be placed on the optic axis such that, its image is formed at the same place?
ii. If the concave part is filled with water of refractive index 4/3, find the distance through which the pin should be moved, so that the image of the pin again coincides with the pin.

47 A plano-convex lens has a thickness of 4 cm. When placed on a horizontal table, with the curved surface in contact with it, the apparent depth of the bottom most point of the lens is found to be 3 cm. If the lens is inverted such that the plane face is in contact with the table, the apparent depth of the centre of the plane face is found to be 25/8 cm. Find the focal length of the lens. Assume thickness to be negligible while finding is focal length.

48 A thin plano-convex lens of focal length f is split into two halves. One of the halves is shifted along the optical axis. The separation between object and image planes is 1.8 m. The magnification of the image formed by one of the half lens is 2. Find the focal length of the lens and separation between the halves. Draw the ray diagram for image formation.

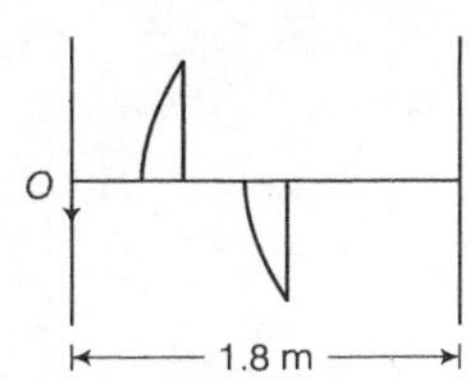

49 A thin equiconvex lens of glass of refractive index $\mu = 3/2$ and of focal length 0.3 m in air is sealed into an opening at one end of a tank filled with water ($\mu = 4/3$). On the opposite side of the lens, a mirror is placed inside the tank on the tank wall perpendicular to the lens axis as shown in the figure. The separation between the lens and the mirror is 0.8 m. A small object is placed outside the tank infront of lens. Find the position (relative to the lens) of the image of the object formed.

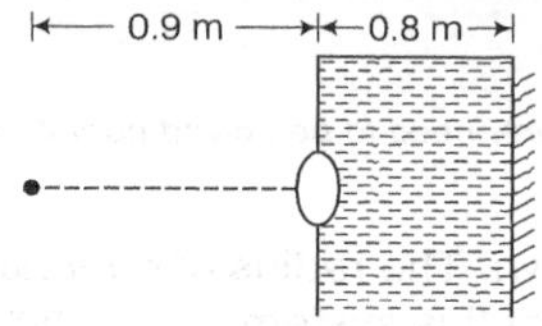

50 A convex lens of focal length 15 cm and a concave mirror of focal length 30 cm are kept with their optic axis *PQ* and *RS* parallel but separated in vertical direction by 0.6 cm as shown in the figure. The distance between the lens and mirror is 30 cm. An upright object *AB* of height 1.2 cm is placed on the optic axis *PQ* of the lens at a distance of 20 cm from the lens. If *A′B′* is the image after refraction from the lens and the reflection from the mirror, find the distance of *A′B′* from the pole of the mirror and obtain its magnification. Also, locate positions of *A′* and *B′* with respect to the optic axis *RS*. **[IIT JEE 2000]**

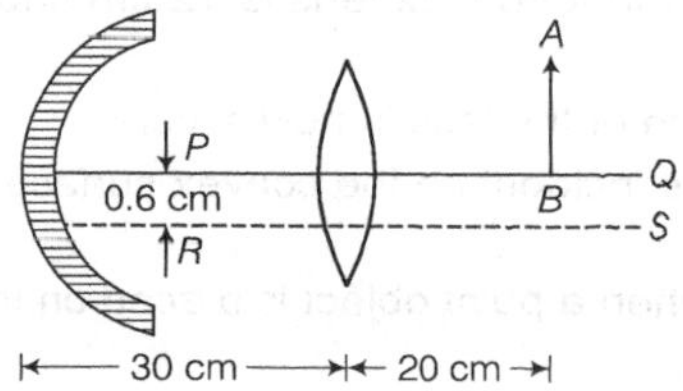

51 A thin biconvex lens of refractive index 3/2 is placed on a horizontal plane mirror as shown in the figure. The space between the lens and the mirror is then filled with water of refractive index 4/3. It is found that when a point object is placed 15 cm above the lens on its principal axis, the object coincides with its own image. On repeating with another liquid, the object and the image again coincide at a distance 25 cm from the lens. Calculate the refractive index of the liquid. **[IIT JEE 2001]**

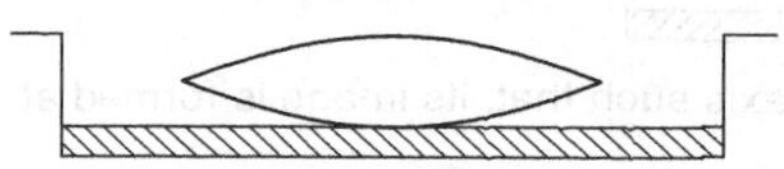

DPP-13 Prism

Conceptual Questions

Direction (*Q. Nos. 1-2*) *These questions are Conceptual in nature, need to be solved completely on notebook.*

Working Space

1 **Situation** A prism has angle of prism *A*. The prism is placed in air.

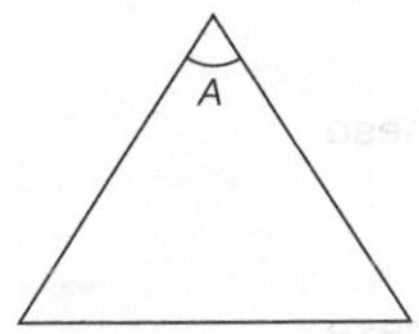

i. A ray of light incident at an angle of 60° on the face of a prism having refracting angle $A = 30°$. The ray emerges cut of the prism makes an angle 30° with the incident ray. The deviation in the ray is

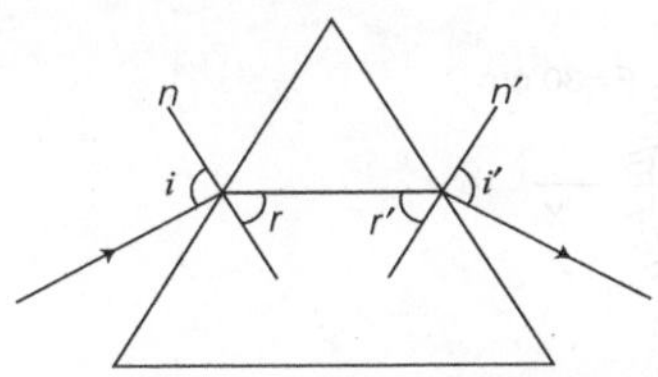

a. 30°
b. 60°
c. 90°
d. None of these

ii. In previous problem, the angle of emergence (i') is

a. 30°
b. 60°
c. zero
d. None of these

iii. In previous problem, the final emergent ray

a. emerges out perpendicular to the face through which it emerges
b. emerges out along the surface through which it emerges
c. the ray does not emerge out
d. None of the above

iv. If the minimum deviation produced by prism is equal to angle of prism. The angle of prism is 60°. The refractive index of the prism is

a. $\sqrt{2}$
b. $\sqrt{3}$
c. 2
d. 3/2

v. If angle of prism is 60°. Find the value of refractive index μ of the prism such that no ray is transmitted through the prism.

a. $\mu > (2/\sqrt{3})$
b. $\mu \leq (2/\sqrt{3})$
c. $\mu > 2$
d. None of the above

vi. One face of the prism is silvered. The angle of prism is 30°. A ray incident on unsilvered face at an angle 45° is refracted and reflected from the silver face and finally retrace its path. Draw the ray diagram.

Working Space

a. 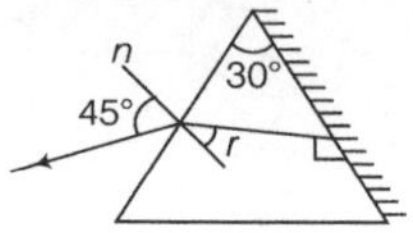

b. 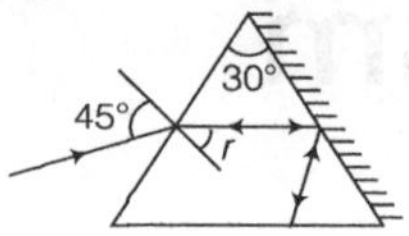

c. 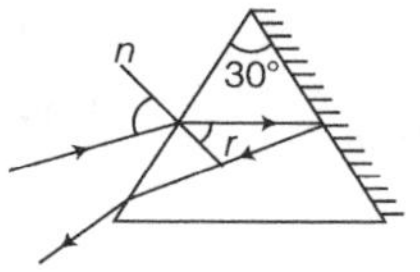

d. None of these

vii. In previous problem, the value of r is

a. 30° b. 45°

c. 60° d. None of these

viii. In previous problem, the refractive index of prism is

a. $\sqrt{3}$ b. 3/2

c. $\sqrt{2}$ d. None of these

2 **Situation** A point O is placed at the origin. The refractive index of the thin prism is $\mu = 3/2$.

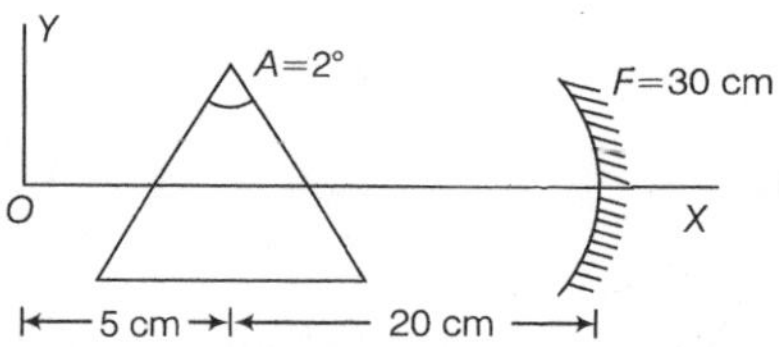

i. The coordinates of image O_1 formed due to first refraction through prism is

a. $\left(0, \frac{\pi}{36} \text{cm}\right)$ b. $\left(0, -\frac{\pi}{36} \text{cm}\right)$

c. $\left(0, \frac{\pi}{90} \text{cm}\right)$ d. $\left(175 \text{cm}, \frac{\pi}{6} \text{cm}\right)$

ii. O_1 behaves as

a. object for concave mirror

b. image for concave mirror

c. object distance for concave mirror is 25 cm in left

d. None of the above

iii. The x-coordinate of image formed by concave mirror is

a. $\pi/6$ cm b. 175 cm

c. 36 cm d. 150 cm

iv. The y-coordinate of image formed by concave mirror is

a. $\pi/36$ cm b. 6 cm

c. $\pi/6$ cm d. None of these

Only One Option Correct Type

Direction (Q. Nos. 3-20) *This section contains 18 multiple choice questions. Each question has four choices (a), (b), (c) and (d), out of which ONLY ONE is correct.*

3 A ray of light is passed through a prism of refractive angle 50°. The rotation of prism causes the ray to be deviated various amounts, the least of which is 30°. Find the refractive index of glass prism. (Take, $\sin 25° = 0.42$, $\cos 50° = 0.64$)

a. 1.41 b. 1.52 c. 1.6 d. 1.58

Working Space

4 A ray of light strikes a plane mirror at an angle of incidence 45° as shown in the figure. After reflection, the ray passes through a prism of refractive index 1.5 as shown in the figure. Through what angle must the ray be rotated, if the total deviation of the ray is to be 90°.

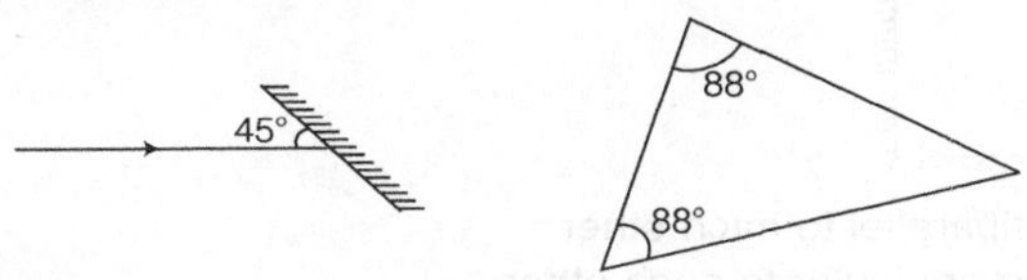

a. 1° in clockwise
b. 1° in anticlockwise
c. 2° in clockwise
d. 4° in anticlockwise

5 A prism of refracting angle 60° produces a deviation of 40°in a ray for two angle of incidence. The difference between them is 10°. Find the angles.

a. 45°, 35°
b. 35°, 25°
c. 45°, 55°
d. 60°, 70°

6 A prism of refracting angle 90° is placed in air. Find the refractive index of prism so that a ray incident on prism at any value of angle of incidence, totally reflected from face *AC*.

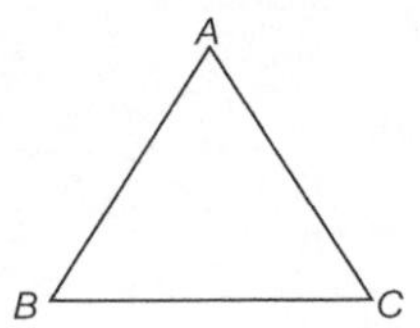

a. $\mu < \sqrt{2}$
b. $\mu > 1$
c. $\mu > \sqrt{2}$
d. $\mu > \frac{2}{\sqrt{3}}$

7 In the figure, a slab *BCDE* is in touch of prism *ABC*. Find the value of θ, so that light incident normally on face *AC* does not cross the face *BC*.

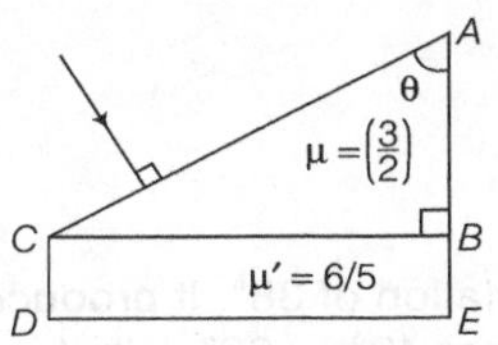

a. $\theta > 37°$
b. $\theta < 37°$
c. $\theta > 53°$
d. $\theta < 53°$

8 A prism of refractive index $\sqrt{2}$ has a angle of prism 30°. One of the refracting surface of prism is silvered. Find the angle of incidence *i* for which the ray retraces its path.

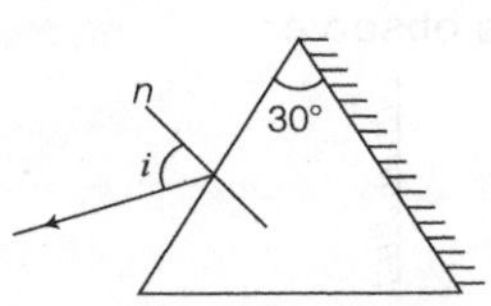

a. 30°
b. 60°
c. 45°
d. 90°

Working Space

9 In the given picture, two refractive surfaces of prism are silvered. Mark the correct option.

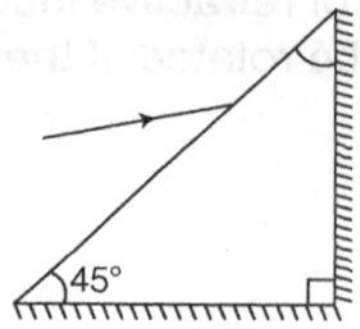

a. The incident and emergent ray are antiparallel to each other
b. The incident and emergent ray are perpendicular to each other
c. The angle between in incident and emergent ray is 120°
d. None of the above

10 The refractive index of a prism varies as $\mu = \frac{4}{3} + by$ where y is in cm, along Y-axis. Find the value of b, so that light does escape through face AB.

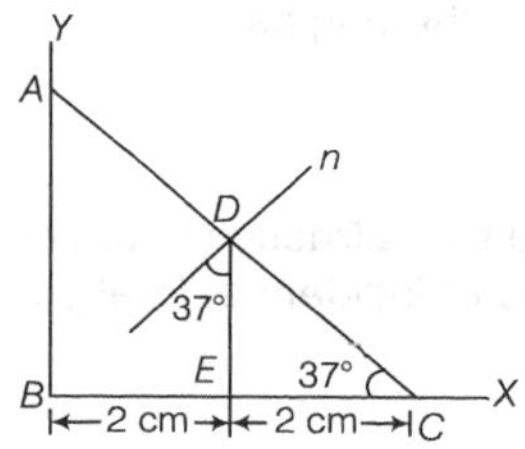

a. $b < \frac{2}{9}$
b. $b > \frac{2}{9}$
c. $b > \frac{2}{3}$
d. $b > \frac{4}{3}$

11 A beam of monochromatic light is incident at $i = 50°$ on one face of an equilateral prism, the angle of emergence is 40°, the angle of minimum deviation is

a. $< 30°$
b. $= 30°$
c. $> 30°$
d. $\leq 30°$

12 A prism of refracting angle 90° produces minimum deviation 30°. Find the refractive index of the prism.

a. $\sqrt{3}$
b. $\sqrt{2}$
c. $\sqrt{\frac{3}{2}}$
d. $\frac{3}{2}$

13 A certain prism is found to produce a minimum deviation of 38°. It produces a deviation of 44°, when the angle of incidence is either 42° or 62°. What is the angle of incidence when it is undergoing minimum deviation?

a. 30°
b. 40°
c. 49°
d. 60°

14 An observer is situated between thin prism and the plane mirror. Find the separation between images of source S seen by the observer.

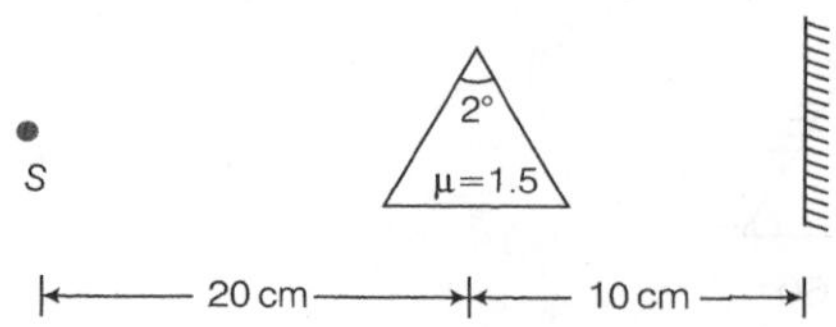

a. 60 cm
b. $\pi/9$ cm
c. $\sqrt{\left(\frac{\pi}{9}\right)^2 + (60)^2}$
d. 30 cm

Working Space

15 A thin prism of angle 2° is placed at a distance 180 cm from a screen as shown in the figure. Find the distance of a point on the screen from point *O* where the ray strikes with the screen. The refractive index of prism is 1.5.

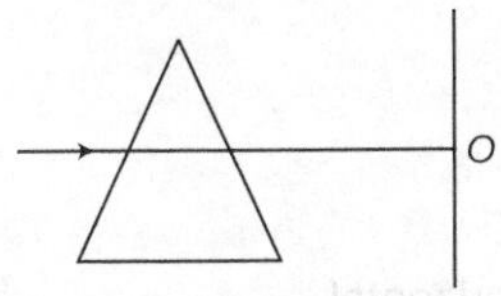

a. 1.5 cm b. 3.14 cm c. 2 cm d. 1 cm

16 Figure shows the graph of angle of deviation δ *versus* angle of incidence *i* for a light ray striking a prism. The prism angle is

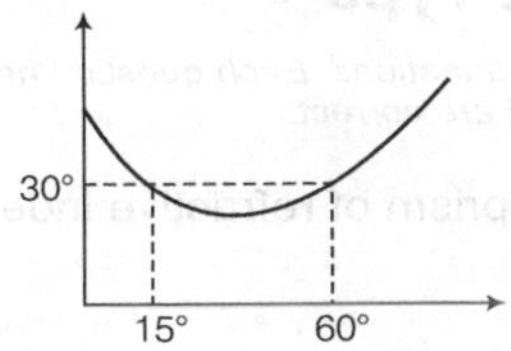

a. 30° b. 45° c. 60° d. 75°

17 A parallel beam of light is incident on the thin prism of angle of prism 3.6°. The refractive index of the prism is 1.5. The focal length of the lens is 20 cm. The co-ordinate of converging point of the beam *i* is

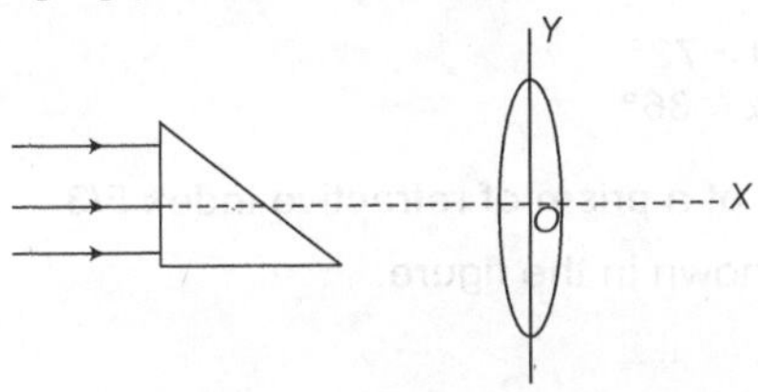

a. $\left(20\,\text{cm}, \frac{-\pi}{5}\,\text{cm}\right)$ b. $(20\,\text{cm}, 1.2\pi\,\text{cm})$

c. $\left(-20\,\text{cm}, +\frac{\pi}{5}\,\text{cm}\right)$ d. $(20\,\text{cm}, -1.2\pi\,\text{cm})$

18 An isosceles prism of angle 120° has a refractive index 1.44. Two parallel rays of monochromatic light enter the prism parallel to each other in air as shown in the figure. The rays emerging from the opposite face **[IIT JEE 1995]**

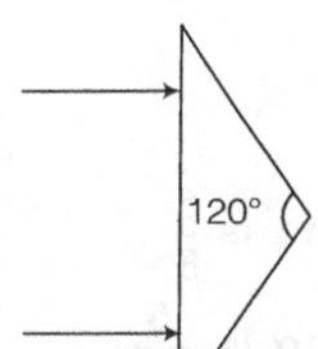

a. are parallel to each other
b. are diverging
c. make an angle $2[\sin^{-1}(0.72) - 30°]$ with each other
d. make an angle $2\sin^{-1}(0.72)$ with each other

19 A given ray of light suffers minimum deviation in an equilateral prism *P*. Additional prism *Q* and *R* of identical shape and of the same material as *P* are now added as shown in the figure. The ray will suffer **[IIT JEE 2001]**

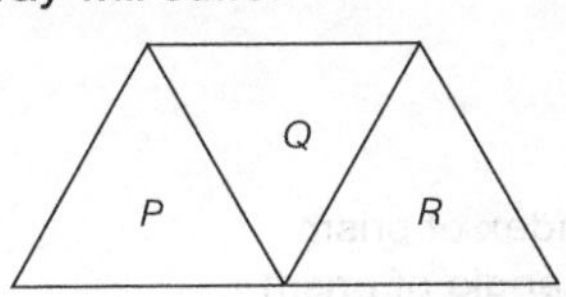

a. greater deviation b. no deviation
c. same deviation as before d. total internal reflection

20 A ray of light is incident on an equilateral glass prism placed on a horizontal table. For minimum deviation which of the following is correct? **[IIT JEE 2004]**

Working Space

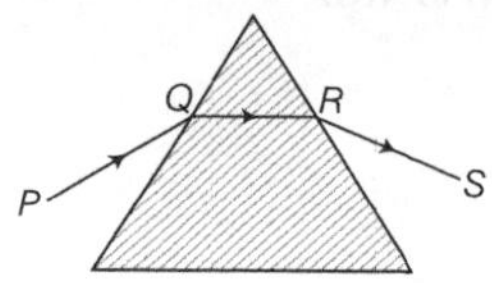

a. PQ is horizontal
b. QR is horizontal
c. RS is horizontal
d. Either PQ or RS is horizontal

One or More than One Options Correct Type

Direction *(Q. Nos. 21-29) This section contains 9 multiple choice questions. Each question has four choices (a), (b), (c) and (d), out of which ONE or MORE THAN ONE are correct.*

21 For ray diagram as shown in the case of isosceles prism of refractive index μ,

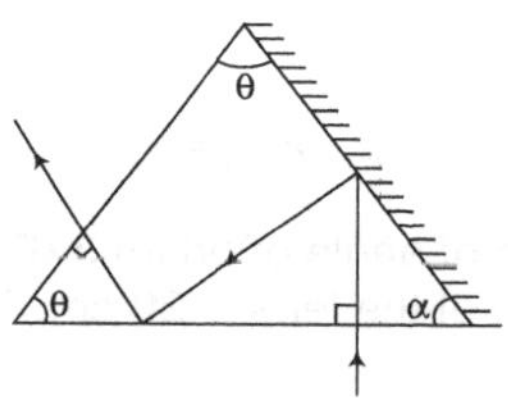

a. $\alpha = 72°$
b. $\theta = 72°$
c. $\theta = 36°$
d. $\alpha = 36°$

22 A ray of light is incident normally on face AB of a prism of refractive index 5/3 immersed in water of refractive index $\frac{4}{3}$ as shown in the figure.

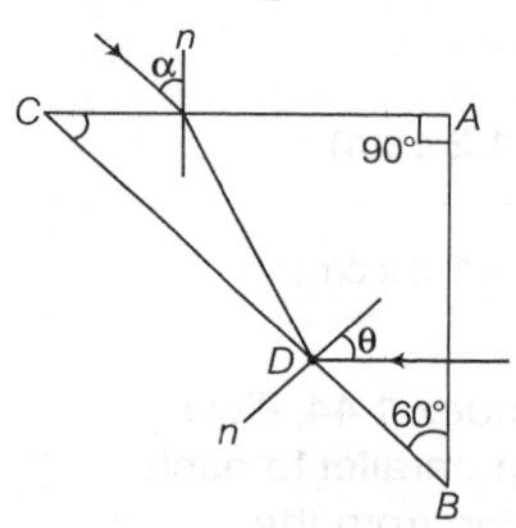

a. The value of $\sin\alpha$ is $\frac{5}{8}$

b. The value of $\sin\alpha$ is $\frac{5}{4\sqrt{3}}$

c. Total internal reflection does not take place at point D, if water is replaced by a medium of refractive index $\frac{5}{2\sqrt{3}}$

d. Total internal reflection does not take place at point D, if water is replaced by a medium whose refractive index $\frac{6}{5}$

23 For thin prism,

a. deviation does not depend on angle of incidence
b. deviation increases with increases in refractive index of prism
c. deviation is directly proportional to the refracting angle of prism
d. None of the above

Working Space

24 For refraction of light through prism placed in air,

a. for every angle of deviation, there are two angle of incidence
b. if light is passing through equilateral prism, ray is parallel to base for minimum deviation
c. there are two angle of incidence for maximum deviation
d. if refractive index of prism increases, angle of minimum deviation increases

25 Figure shows a part of prism refractive index 1.5. The incident ray suffers from minimum deviation.

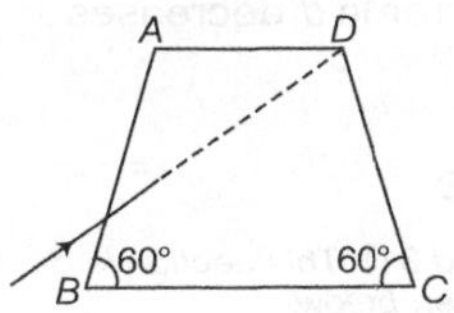

a. The angle of refraction at the surface *AB* is 30°
b. The refracting angle of prism is 60°
c. Total internal reflection takes place at surface *DC*
d. Minimum deviation is not calculated without knowing the angle of incidence

26 Two prisms are combined as shown in the figure. Light beam incident normally on the face *AB* is not enter into prism *ADC*. Then,

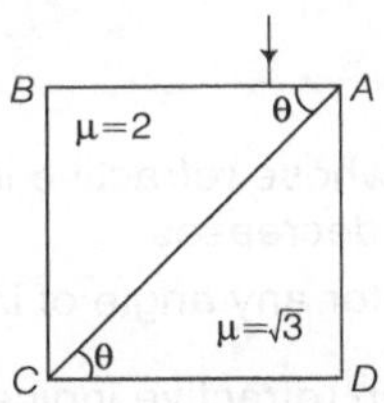

a. $\theta = 65°$
b. $\theta = 70°$
c. $\theta = 60°$
d. $\theta = 45°$

27 In the given prism, $< C = 90°$ and $\alpha \leq \theta$.

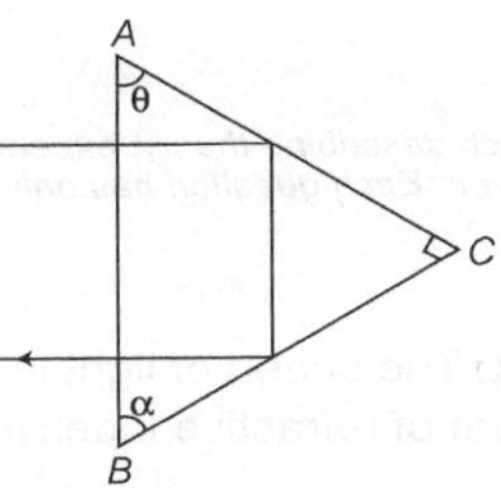

a. The minimum value of refractive index is $\sqrt{2}$
b. The minimum value of refractive index is $\sqrt{3}$
c. For refractive index of prism, $\mu = \frac{5}{3}$, it is not possible to achieve $\alpha = 30°$
d. None of the above

28 A ray *OP* of monochromatic light is incident on the face *AB* of prism *ABCD* near vertex *B* at an incident angle of 60° as shown in the figure. If the refractive index of the material of the prism is $\sqrt{3}$. Which of the following is/are correct? **[IIT JEE 2010]**

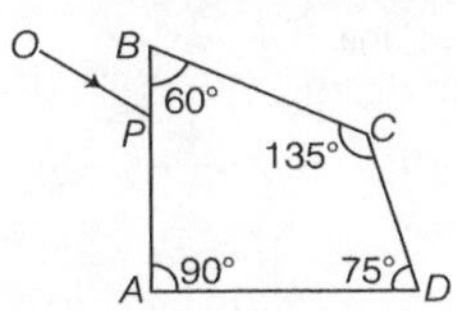

a. The ray gets totally internally reflected at face *CD*
b. The ray comes out through face *CD*
c. The angle between the incident ray and the emergent ray is 90°
d. The angle between the incident ray and the emergent ray is 120°

Working Space

29 Using the expression $2d\sin\theta = \lambda$, one calculates the values of d by measuring the corresponding angles θ in the range to 90°. The wavelength λ is exactly known and the error in θ is constant for all values of θ. As, θ increases from 0° **[JEE Advanced 2013]**

a. the absolute error in d remains constant
b. the absolute error in d increases
c. the fractional error in d remains constant
d. the fractional error in d decreases

Statement Type

Direction (*Q. Nos.* 30 *and* 31) *This section is based on Statement I and Statement II. Select the correct answer from the codes given below.*

a. Both Statement I and Statement II are correct and Statement II is the correct explanation of Statement I
b. Both Statement I and Statement II are correct but Statement II is not the correct explanation of Statement I
c. Statement I is correct but Statement II is incorrect
d. Statement II is correct but Statement I is incorrect

30 **Statement I** When a thin prism is placed in a medium, whose refractive index is lesser than that of prism deviation produced by prism decreases.

Statement II Deviation produced by thin prism is same for any angle of incidence.

31 **Statement I** For a prism, minimum deviation depends on refractive indices of prism and surrounding.

Statement II $\delta = i + i' - A$.

Comprehension Type

Direction (*Q. Nos.* 32-38) *This section contains* 3 *paragraphs, each describing theory, experiments, data, etc. Seven questions related to the paragraphs have been given. Each question has only one correct answer among the four given options (a), (b), (c) and (d).*

Passage I

The speed of light in different medium are different. The speed of light in vacuum is $c_0 = 3 \times 10^8\ \text{ms}^{-1}$. The speed of light in a medium of refractive index μ is $c = \frac{c_0}{\mu}$. The table given below presents some refractive indices of custom glass, for different wavelengths of visible light.

Colour	Wavelength in vacuum	Refractive index
Yellow	580×10^{-9} m	1.500
Yellow-orange	600×10^{-9} m	1.498
Orange	620×10^{-9} m	1.496
Orange red	640×10^{-9} m	1.494

32 Inside custom glass

a. orange and yellow light travel equally fast
b. orange light travels faster than yellow light
c. orange light travels slower than yellow light
d. None of the above

Working Space

33 For blue green light of wavelength 520 nm, the refractive index of custom glass is equal to

a. 1.494
b. 1.500
c. 1.506
d. 1.520

Passage II

As compared to yellow light waves, blue green light waves are 60 nm shorter. In the given table, reading from bottom to top, we see that each 20 nm decrease in wavelength corresponds to a 0.002 increase in refractive index μ. Since, yellow and green light differ in wavelength by 60 nm, the corresponding μ' should differ by 0.006

$$\therefore \quad \mu_{\text{green blue}} = \mu_{\text{yellow}} + 0.006 = 1.506$$

34 For visible light, which graph best expresses the refractive index of custon glass as a function of the frequency of the light.

a. 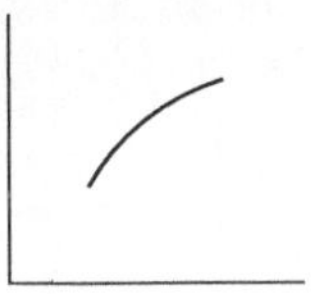

b. 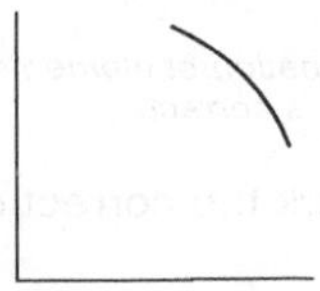

c. 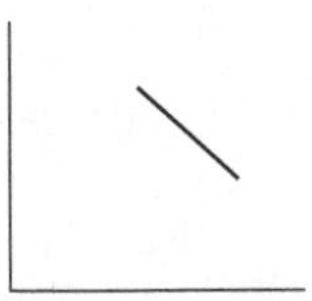

d. 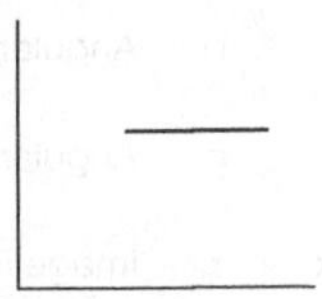

35 Which of the following phenomena happens because refractive index varies with wavelength?

a. Lens focuses light
b. A prism breaks sunlight into different colours
c. Total internal reflection ensures that light travels down a fibre-optic cable
d. All of the above

Passage III

A student wants to choose a particular wavelength λ_0 for which the ray passes through the interface *BC* of combination of two prism as shown in the figure, without deviation.

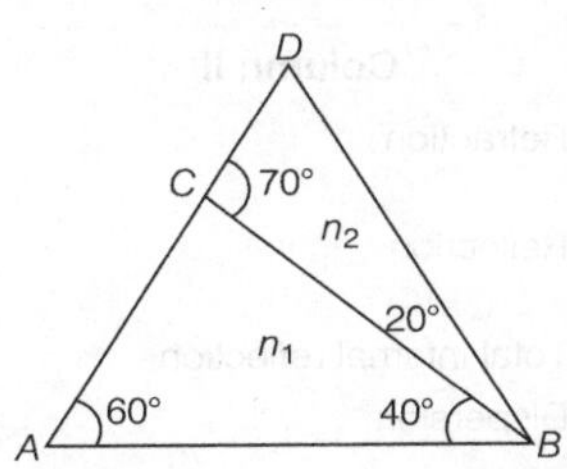

The refractive indices of prism are $n_1 = 1.2 + \dfrac{10.8 \times 10^4}{\lambda^2}$

and $n_2 = 1.45 + \dfrac{1.80 \times 10^4}{\lambda^2}$, where λ is in nm.

36 The result of student is

a. 600 nm
b. 700 nm
c. 400 nm
d. 500 nm

Working Space

37 For the wavelength λ_0,

a. the system behaves as a single prism of refracting angle of 90°
b. the system behaves as a single prism of refracting angle 60°
c. the system never be as assumed as a single prism
d. the system behaves as a single prism of refracting angle of 80°

38 For a light of wavelength λ_0, find the angle of incidence *i* on the face *AC* such that the deviation produced by the combination of prisms in minimum.

a. $\cos^{-1}\frac{3}{4}$ b. $\tan^{-1}\frac{3}{4}$
c. $\sin^{-1}\frac{3}{4}$ d. $\sec^{-1}\frac{3}{4}$

Matching List Type

***Direction** (Q. Nos. 39-42) Choices for the correct combination of elements from Column I and Column II are given as options (a), (b), (c) and (d), out of which one is correct.*

39 Match the Column I with Column II and mark the correct option from the codes given below.

	Column I		Column II
i.	If refractive index of prism material is same as that of surrounding medium	p.	Angular deviation is zero
ii.	If ray incidence along the optical axis of a thin lens from a point source	q.	Angular deviation is 180°
iii.	If ray incidence along optical axis of convex mirror from point source	r.	Image is lying on optical axis
iv.	If ray incidence along the optical axis of convex refractive surface from a point source	s.	No image is formed

Codes

	i	ii	iii	iv
a.	p,r	p,s	p,r	q,r
b.	p,s	p,r	q,r	p,r
c.	p	r	q	s
d.	s	p	q	r

40 Match the Column I with Column II and mark the correct option from the codes given below.

	Column I		Column II
i.	When light strikes a boundary between two transparent medium from denser medium	p.	Refraction
ii.	A monochromatic light incident on the surface of perfect transparent prism	q.	Reflection
iii.	Simple microscope	r.	Total internal reflection
iv.	Paraxial parallel beam of white light incident on an ideal thin lens	s.	Dispersion

Codes

	i	ii	iii	iv
a.	p,r,s	p,r	p	p,s
b.	p,s	r	p,r	s
c.	r,s	p,r	q	p
d.	p	q	r	s

Working Space

41 A ray of light falls normally on an equilateral prism of refractive index $\mu = \sqrt{3}$. Match the Column I with Column II and mark the correct option from the codes given below.

	Column I		Column II
i.	Angle i_1	p.	0°
ii.	Angle r_1	q.	90°
iii.	Angle i_2	r.	60°
iv.	Angle r_2	s.	None

Codes

	i	ii	iii	iv		i	ii	iii	iv
a.	p	p	s	r	b.	p	q	r	s
c.	s	q	p	r	d.	p	s	q	r

42 A right angled prism of refractive index μ_1 is placed in a rectangular block of refractive index μ_2, which is surrounded by a medium of refractive index μ_3, as shown in the figure, A ray of light *e* enters the rectangular block at normal incidence. Depending upon the relationships between μ_1, μ_2 and μ_3, it takes one of the four possible paths *ef*, *eg*, *eh* or *ei*. **[JEE Advanced 2013]**

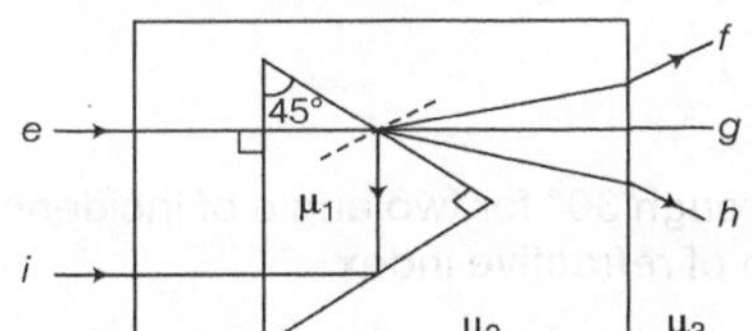

Match the paths in Column I with the conditions of refractive indices in Column II and mark the correct option from the codes given below.

	Column I		Column II
i.	$e \to f$	p.	$\mu_1 > \sqrt{2}\mu_2$
ii.	$e \to g$	q.	$\mu_1 > \mu_1$ and $\mu_2 > \mu_3$
iii.	$e \to h$	r.	$\mu_1 = \mu_2$
iv.	$e \to i$	s.	$\mu_2 < \mu_1 < \sqrt{2}\mu_2$ and $\mu_2 > \mu_3$

Codes

	i	ii	iii	iv		i	ii	iii	iv
a.	q	r	p	s	b.	p	q	s	r
c.	s	p	q	r	d.	q	r	s	p

One Integer Value Correct Type

Direction *(Q. Nos. 43-49) This section contains 7 questions. When worked out will result in an integer from 0 to 9 (both inclusive).*

43 In the given figure, $\mu_1 = 1.5$ and $\mu_2 = \dfrac{4}{3}$. A point source P is placed at 20 cm from the thin prisms combination. The focal lengths of L_1 and L_2 are 20 cm and 10 cm respectively.

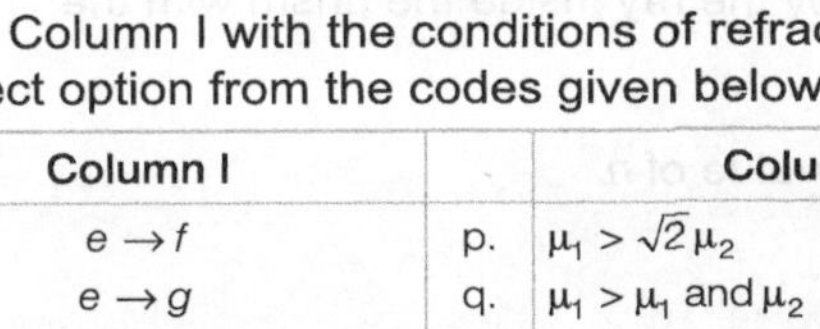

The separation between thin prisms combination and lens L_1 is 10 cm, which the separation between lenses is 70 cm. The position (in cm) of final image from the lens L_2.

Working Space

44 A beam of monochromatic light is incident on a glass regular hexagon prism. The width of the beam is such that its rays fall up the middle of adjacent sides of the prism as shown. For $\mu > \left(\frac{13}{n}\right)^{\frac{1}{2}}$, two separate parallel beams emerge out of the prism. Find the value of n.

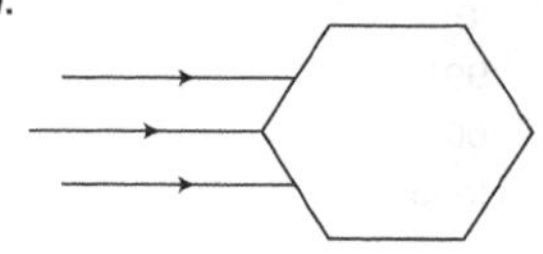

45 Due to manufacturing defect, the plane surface of a thin plano convex lens has been made tilted at a small angle $\theta = 2°$ outward from its usual place.

Lens part has focal length of $\frac{180}{\pi}$ cm and refractive index of glass used is $\mu = \frac{3}{2}$. A parallel beam of light is incident as shown in figure. Find the distance (in cm) from X-axis, where beam converges.

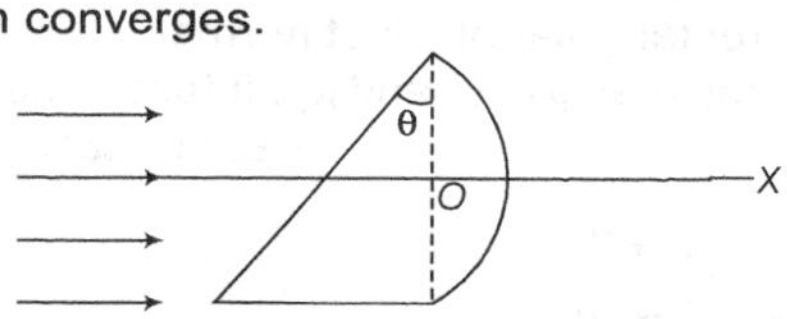

46 A prism of angle of prism 60° deviates a ray through 30° for two angle of incidence differing by 30°. Find the greatest integer value of refractive index.

47 A ray of light undergoes devlation of 30° when incident on as an equilateral prism of refractive index $\sqrt{2}$. Find the angle made by the ray inside the prism with the base of the prism.

48 The deviation in ray as shown is $\frac{\pi}{n}$. Find the value of n.

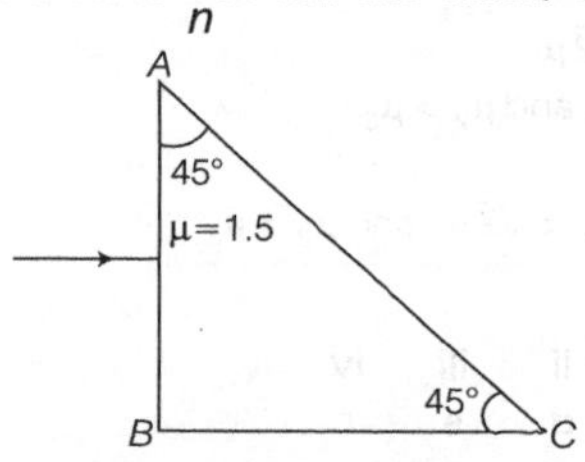

49 Figure shows an isosceles, right angled triangular glass prism has a cavity inside it. Cavity has a shape of a thin equiconvex lens whose surfaces have radius of curvature $R = 20$ cm. In the figure, S is a point source and screen is opaque sheet having small hole such that hole and source both lie on X-axis. The refractive index of prism is 1.5. If final image is formed at $\frac{160}{x}$ cm above surface BC. Find the value of x.

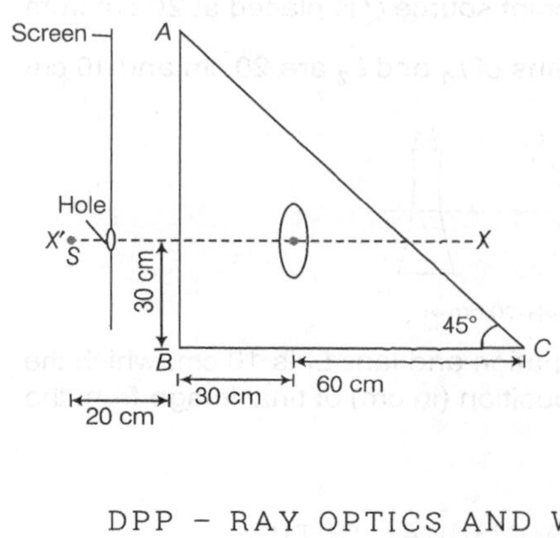

Working Space

Subjective Questions

Direction (Q. Nos. 50-53) *These questions are subjective in nature, need to be solved completely on notebook.*

50 A right angled prism is to be made by selecting a proper material and the angles A and B ($B \leq A$), as shown in the figure. It is desired that a ray of light incident on the face AB emerges parallel to the incident direction after two internal reflections.

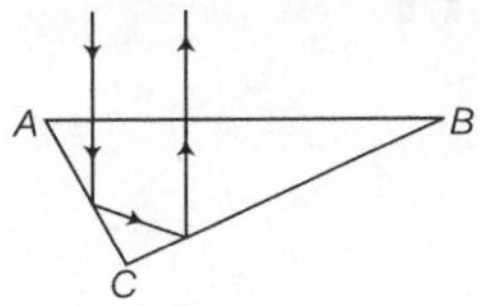

i. What should be the minimum refractive index n for this to be possible

ii. For $n = 5/3$ is it possible to achieve this with the angle B equal to 30°?

51 A right angled prism (45°-90°-45°) refractive index n has a plane of refractive index n_1 ($n_1 < n$) cemented to its diagonal face. The assembly is in air. The ray is incident on AB.

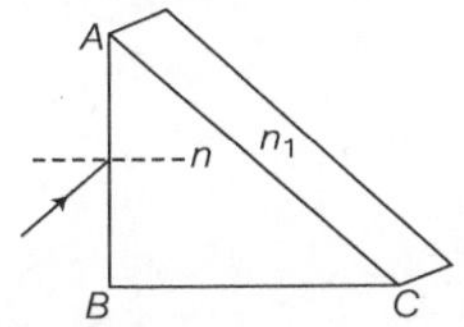

i. Calculate the angle of incidence at AB for which the ray strikes the diagonal face at the critical angle.

ii. Assuming $n = 1.352$, calculate the angle of incidence at AB for which the refracted ray passes through the diagonal face undeviated.

52 A ray of light is incident on a prism ABC of refractive index $\sqrt{3}$ as shown in the figure.

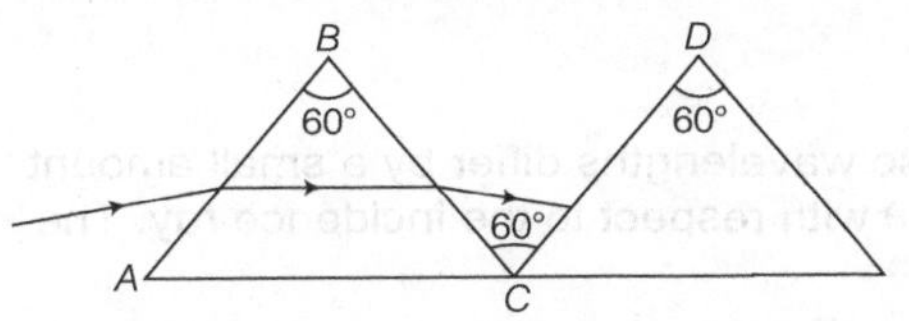

i. Find the angle of incidence for which the deviation of light ray by the prism ABC is minimum.

ii. By what angle the second identical prism must be rotated, so that the final ray suffers net minimum deviation.

DPP-14 Dispersion and Spectrum

Only One Option Correct Type

Working Space

Direction *(Q. Nos. 1-4) This section contains 4 multiple choice questions. Each question has four choices (a), (b), (c) and (d), out of which ONLY ONE is correct.*

1 A beam of light has a small wavelength spread $\delta\lambda$ about a central wavelength λ. The beam travels in vacuum until it enters a glass plate at an angle θ relative to normal to the plate as shown in the figure. The refractive index of the glass is $\mu(\lambda)$. The angular spread $\delta\theta'$ of the refracted beam is

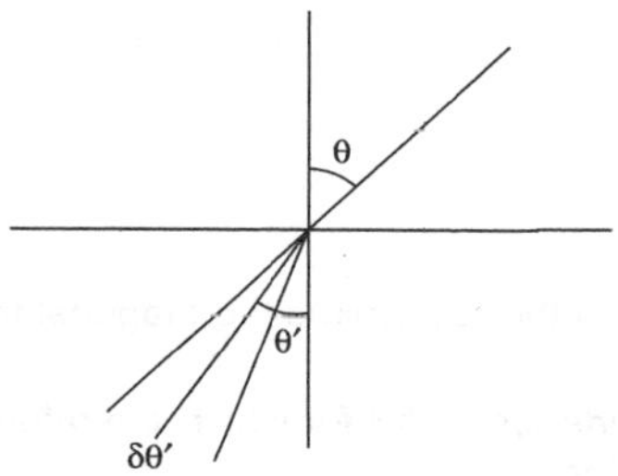

a. $\delta\theta' = \delta\theta$

b. $\delta\theta' = \dfrac{1}{\mu \cot\theta'}(\delta\lambda)\dfrac{dn}{d\lambda}$

c. $\delta\theta' = \mu \cos\theta \, \delta\theta$

d. None of the above

2 Two incident monochromatic waves whose wavelengths differ by a small amount $d\lambda$ are separated angularly at θ and $\theta + d\theta$ with respect to the incidence ray. The dispersive power is

a. $\dfrac{d\theta}{d\lambda}$

b. $\dfrac{d\theta}{\theta}$

c. $\dfrac{d\lambda}{\lambda}$

d. $\lambda\left(\dfrac{d\lambda}{d\theta}\right)$

3 It is desired to make converging achromatic combination of two lens of focal length L_1 and focal and L_2. Then

a. both lenses should be converging

b. both lenses should be diverging

c. one should be converging but other should be diverging

d. None of the above

4 The focal length of a thin lens for yellow colour is 10 cm. Its dispersive power is 0.04. Its axial chromatic aberration (longitudinal chromatic) aberration is

a. 0.04 cm

b. 0.4 cm

c. 0.2 cm

d. 0.004 cm

Working Space

One or More than One Options Correct Type

Direction *(Q. Nos. 5-7) This section contains 3 multiple choice questions. Each question has four choices (a), (b), (c) and (d), out of which ONE or MORE THAN ONE are correct.*

5 In the case of spherical mirror of large aperture,

a. chromatic aberration is present
b. spherical aberration is present
c. the image of point object is a blurred surface
d. None of the above

6 In the case of lens of large aperture.

a. Spherical aberration is present
b. Chromatic aberration is present
c. If a point object is placed at a point off the optical axis, a comet shaped image is formed by the lens on a screen placed perpendicular to optical axis
d. If a point object is placed at a point an optical axis, the shape of image formed by the lens on a screen placed perpendicular to optical axis is a circular disc

7 The refractive indices of a thin prism is 1.623 (for red) and 1.642 (for violet). The angle of prism is 4°.

a. The refractive index of prism for yellow light is 1.632
b. The dispersive power of material of prism is 0.03
c. The angular dispersion produced by prism is 0.076
d. Deviation produced by prism for mean colour is 2.528°

Matching List Type

Direction *(Q. Nos. 8 and 9) Choices for the correct combination of elements from Column I and Column II are given as options (a), (b), (c) and (d), out of which one is correct.*

8 Two thin lenses $f_1 = +10$ cm and f_2 and their respective dispersive power is ω_1 and ω_2 are co-axially placed in contact. The combination behaves as a chromatic combination. Match the Column I with Column II and mark the correct option from the codes given below.

	Column I		Column II
i.	$\omega_1 = 0.04$ and $\omega_2 = 0.02$	p.	The combination behaves as converging lens
ii.	$\omega_1 = 0.02$ and $\omega_2 = 0.04$	q.	The combination behaves as diverging lens
iii.	$\omega_1 = \omega_2 = 0.02$	r.	Not possible
iv.	$f_2 = +20$ cm	s.	The modulus of focal length of combination is 10 cm
		t.	Behaves as thin slab

Codes

	i	ii	iii	iv		i	ii	iii	iv
a.	p	q	r	s	b.	p,s	q,s	t	r
c.	q,s	p	t	r	d.	t	p	q	s

9 Match the Column I with Column II and mark the correct option from codes given below.

	Column I		Column II
i.	Spectrometer	p.	To determine focal length of lens
ii.	Displacement method	q.	Pure spectrum
iii.	Optical fibre	r.	$\frac{\omega_1}{f_1} + \frac{\omega_2}{f_2} = 0$
iv.	A chromatic combination of lenses	s.	Total internal reflection

Codes

	i	ii	iii	iv		i	ii	iii	iv
a.	p	q	r	s	b.	q	p	s	r
c.	p,q	p,q	r,s	r,s	d.	q	p	p,q,r,s	q,s

Working Space

Statement Type

Direction *(Q. Nos. 10-15) This section is based on Statement I and Statement II. Select the correct answer from the codes given below.*

a. Both Statement I and Statement II are correct and Statement II is the correct explanation of Statement I
b. Both Statement I and Statement II are correct but Statement II is not the correct explanation of Statement I
c. Statement I is correct but Statement II is incorrect
d. Statement II is correct but Statement I is incorrect

10 **Statement I** Dispersive power of a material never be negative.

Statement II The refractive index of a material for violet ray is greater than that for real ray.

11 **Statement I** Solid glass sphere, thin lens and a prism of same material placed in air have same dispersive power.

Statement II Dispersive power, $\omega = \left(\dfrac{\mu_V - \mu_r}{\mu_y - 1}\right)$.

12 **Statement I** If a glass prism is dipped in water, its dispersive power decreases.

Statement II Dispersive power depends on angle of prism.

13 **Statement I** The focal length of a spherical mirror of small aperture depends on the colour of light.

Statement II The refractive index of glass for different colours are different.

14 **Statement I** Bright emission line spectrum against a dark background is observed during a total eclipse.

Statement II Light from the chromosphere can reach the earth, while from the photosphere is completely cut off.

15 **Statement I** The line spectrum of hot starts is not so sharp.

Statement II At high temperature, the atoms in the hot starts emitting light are moving at tremendous speeds.

One Integer Value Correct Type

Direction *(Q. Nos. 16 and 17) This section contains 2 questions. When worked out will result in an integer from 0 to 9 (both inclusive).*

16 In a direction vision spectroscope there are two flint glass prism each of angle 5° and dispersive power 0.36 and two crown glass prisms of dispersive power 0.24. If net dispersion produced by the system is $7n$ minute. Find the value of n. (Take, $\mu_{crown} = 1.5$ and $\mu_{flint} = 1.68$)

17 A thin converging lens of refractive index 1.5 and dispersive power 0.01 is perfectly cemented to a plano-concave lens of refractive index 1.6 and dispersive power 0.02. For a chromatic combination of power $+5D$. If $\dfrac{1}{|r_1|} + \dfrac{1}{|r_2|} = \dfrac{1}{n}$.

Find the value of n. Here r_1 and r_2 (in cm) are the radii of curvature of the convex lens.

DPP-15 Optical Instruments

Only One Option Correct Type

Working Space

Direction (Q. Nos. 1-12) *This section contains 12 multiple choice questions. Each question has four choices (a), (b), (c) and (d), out of which ONLY ONE is correct.*

1 The focal length of lens of a camera is 10 cm and focussing range 0.6 m out to infinity. Find the range of movement necessary between lens and film,

a. 1 cm b. 2 cm
c. 3 cm d. 4 cm

2 The lens of a certain 3 cm film size camera has focal length 50 cm and diameter 10 cm. Its *f*-number is

a. 1 b. 2
c. 3 d. 5

3 In previous problem, its exposure time is $\frac{1}{400}$ s. Find its exposure time for $\frac{f}{6}$.

a. $\frac{9}{2500}$ s b. $\frac{3}{250}$ s
c. $\frac{9}{200}$ s d. $\frac{1}{100}$ s

4 A pinhole camera with 10 cm high film is to be used to take a picture of a man whose height is 2 m. The film is 20 cm from the pinhole. How far should the camera be from the man to include the full height of the tree?

a. 1 m b. 2 m
c. 3 m d. 4 m

5 The image of the Sun is focussed by achromatic thin convex lens of focal length 75 cm on a screen. The image is seen by eye at least distance of distinct vision from screen. The angular magnification is

a. -1 b. -2
c. -3 d. -4

6 The aperture of a telescope is 1.22 m. The wavelength of visible light is 500 nm. The resolving power of telescope is

a. 10^6 b. 10^7
c. 2×10^6 d. 4×10^6

7 In previous problem, if distance between the earth and moon is 4×10^8 m. The minimum separation between objects on the moon surface. So that they are just resolved is

a. 50 m b. 100 m
c. 150 m d. 200 m

Working Space

8 The near and far points of a woman are 30 cm and 180 cm. Find the power of lens she should use while reading at 25 cm.

a. 1.5 D
b. $-\frac{2}{3}D$
c. $\frac{2}{3}D$
d. 2 D

9 In previous problem, by using the lens on the eye and what maximum is clearly visible?

a. 53 cm
b. 81.82 cm
c. 40.8 cm
d. 20.3 cm

10 The image formed by an objective of a compound microscope is

a. virtual and diminished
b. real and diminished
c. real and enlarged
d. virtual and enlarged

11 An astronomical telescope has a large aperture to

a. reduce spherical aberration
b. have high resolution
c. increase span of observation
d. have low dispersion

12 Which of the following is used in optical fibres?

a. Total internal reflection
b. Scattering
c. Diffraction
d. Refraction

One or More than One Options Correct Type

Direction (Q. Nos. 13-17) *This section contains 5 multiple choice questions. Each question has four choices (a), (b), (c) and (d), out of which ONE or MORE THAN ONE are correct.*

13 An object is seen through a simple microscope of focal length of 10 cm. The least distance of distinct vision of eye is 25 cm.

a. The angular magnification for normal adjustment is 2.5
b. The angular magnification for clear vision is 3.5
c. The angular magnification for normal adjustment is 1.5
d. The angluar magnification for clear vision – 4.5

14 A small object is placed at a distance of 4 cm from a simple microscope of focal length of 5 cm.

a. A virtual image is formed at distance 20 cm from the lens
b. A real image is formed at distance 20 cm from the lens
c. The linear magnification is 5
d. The angular manification is 6.25

15 The focal lengths of objective and eyepiece of astronomical telescope is 100 cm and 2 cm respectively. The telescope is focussed to see an object 5 km from the objective. For normal adjustment,

a. the length of the tube is 102 cm
b. the length of the tube is 98 cm
c. the angular magnification is – 50
d. the angular magnification is – 0.02

Working Space

16 In previous problem, for clear vision

a. the length of tube is 102 cm
b. the length of tube is 101.85 cm
c. the angular magnification is – 50
d. the angular magnification is – 54

17 A child can adjust the focal length of his eye lens between 513 cm to 2.5 cm. His far point is infinity.

a. The distance between eye lens and retina is 2.5 cm
b. The distance between eye lens and retina is 513 cm
c. The near point is 10 cm
d. The near point is 5 cm

Comprehension Type

Direction (Q. Nos. 18 and 19) *This section contains a paragraph, which describing theory, experiments, data, etc. Two questions related to the paragraph have been given. Each question has only one correct answer among the four given options (a), (b), (c) and (d).*

Passage

In a compound microscope the objective and the eyepiece have focal length of 1 cm and 5 cm respectively and are placed at distance of $\frac{125}{6}$ cm. The compound microscope is adjusted for clear vision.

18 The position of object from objective is

a. $\frac{100}{94}$ cm
b. $\frac{95}{94}$ cm
c. 0.94 cm
d. $\frac{100}{47}$ cm

19 The total magnification is

a. – 100
b. – 50
c. – 94
d. None of these

Statement Type

Direction (Q. Nos. 20-22) *This section is based on Statement I and Statement II. Select the correct answer from the codes given below.*

a. Both Statement I and Statement II are correct and Statement II is the correct explanation of Statement I
b. Both Statement I and Statement II are correct but Statement II is not the correct explanation of Statement I
c. Statement I is correct but Statement II is incorrect
d. Statement II is correct but Statement I is incorrect

20 **Statement I** A swimmer see only hazy contours of objects, when he opens his eyes under water, while they are distinctly visible, if he is using a mask.

Statement II The refractive index of the liquid inside the eye is very close to that of water.

21 **Statement I** It is difficult to thread a needle with one eye closed.

Statement II It is difficult to estimate the distance of the object with one eye.

22 **Statement I** Red light used as a danger signal.

Statement II The scattering in red light due to fog and dust particles is least.

Matching List Type

Direction *(Q. Nos. 23-25) Choices for the correct combination of elements from Column I and Column II are given as options (a), (b), (c) and (d), out of which one is correct.*

23 A compound microsocpe consists of two converging lenses of focal length 2 cm and 10 cm. The separation between lenses is 15 cm. Match the Column I with Column II and mark the correct option from the codes given below.

	Column I		Column II
i.	Focal length of objective (in cm)	p.	$\frac{11}{4}$
ii.	For normal adjustment, object distance in (cm) from objective.	q.	3.75
iii.	For normal adjustment, numerical value of angular magnification	r.	$\frac{10}{3}$
iv.	For clear vision, numerical value of angular magnification	s.	2

Codes

	i	ii	iii	iv		i	ii	iii	iv
a.	p	q	r	s	b.	s	r	q	p
c.	s	r	p	q	d.	s	p	q	p

24 The focal lengths of objective and eyepiece of a Galilean telescope is 100 cm and 10 cm respectively. Match the Column I with Column II and mark the correct option from the codes given below.

	Column I		Column II
i.	For nomal adjustment, tube length in cm	p.	13.33
ii.	For normal adjustment, angular magnification	q.	10
iii.	It the telescope is to focus an object 4 m away from objective, tube length in cm for image formed at infinity	r.	90
iv.	Angular magnification in part (iii)	s.	123.3

Codes

	i	ii	iii	iv		i	ii	iii	iv
a.	p	q	r	s	b.	r	q	s	p
c.	p	r	s	q	d.	q	p	r	s

25 Match the Column I with Column II and mark the correct option from the codes given below.

	Column I		Column II
i.	Final image errect	p.	Simple microscope
ii.	Final image is inverted	q.	Compound microscope
iii.	Focal length of objective is larger than eye piece	r.	Astronomical telescope
iv.	Focal length of eye piece is larger than objective	s.	Galilean telescope

Codes

	i	ii	iii	iv
a.	p,s	q,r	r,s	q
b.	p,s	q,s	r,s	q
c.	p	q,s	r,s	s
d.	p,q,r	q	r	s

One Integer Value Correct Type

Working Space

Direction *(Q. Nos. 26-32) This section contains 7 questions. When worked out will result in an integer from 0 to 9 (both inclusive) .*

26 The focal lengths of eyepiece and objective of a compound microscope are 5 cm and 0.5 cm. Its angular magnification for normal adjustment is – 15. Find the separation (in cm) between eyepiece and objectve of the microscope.

27 A compound miscroscope is used to enlarge an object placed at a distance 3 cm from its objective, which consists of several convex lenses in contact and has focal length 2 cm. If a lens of focal length 10 cm is removed from the objective, find out the distance (in cm) by which the eyepiece of the microscope must be moved to refocus of the image.

28 The focal lengths of the objective and eyepiece of a compound microscope are 2 cm and 3 cm respectively. The distance between the objective and the eyepiece is $5n$ cm. For nomal adjustment, the distance of image produced by objective from objective is 12 cm. Find the value of n.

29 The diameter and focal length of condenser of a projecting camera are D and f respectively. The diameter of lens is 2 cm. The distance between the source of light of dimension 0.6 cm and lens is 40 cm. If $f = \frac{10x}{13}$ cm. Find the value of x.

30 A near sighted child, the accomodation of his eye is between $x_1 = 12$ cm and $x_2 = 60$ cm, wears spectacles through which he distinctly see remote objects. The minimum distance at which he can read a word through his spectacles is $5n$ cm. Find the value of n.

31 A far sighted man, Ram and a short sighted man Shyam, see objects through their spectacles as a man with normal eye sight. When Ram accidently wears the spectacles of Shyam, he found that he could see distinctly only infinitely remote objects. At what minimum distance y can Shyam sees small word, if he wears Ram's spectacles, if $y = \frac{25}{n}$ cm. Find the value of n.

32 The objective is taken out of a astronomical telescope adjusted for normal vision and replaced by a diaphragm with a diameter D. A screen shows a real image of diaphragm having diameter, $d = \frac{D}{25}$ at a certain distance from eyepiece. If the mode of angular magnification of the telescope was $5n$. Find the value of n.

SET - 1

Revisal Problems
for JEE Main

Only One Option Correct Type

Working Space

Direction *(Q. Nos. 1-26) This section contains 26 multiple choice questions. Each question has four choices (a), (b), (c) and (d), out of which ONLY ONE option is correct.*

1 The two head lights of a vehicle are 100 cm apart. If the diameter of pupil of eye is 0.25 cm. What is approximate distance beyond which it will not be possible to see these light separately. (The wavelength of light is 500 nm)

a. 2.1 km
b. 4.1 km
c. 8.2 km
d. None of these

2 Both mirror and point object P perform SHM with same amplitude of 5 cm and angular frequency of 4 rad s^{-1}, both are at their respective extreme position at $t = 0$. Find the maximum possible speed of image.

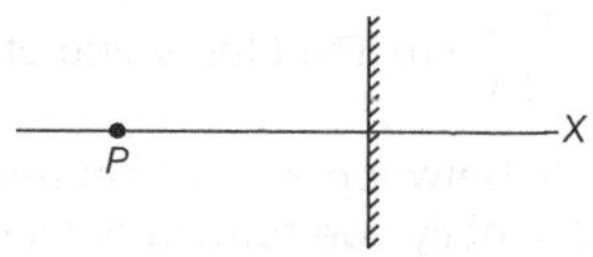

a. 20 cms^{-1}
b. 40 cms^{-1}
c. 60 cms^{-1}
d. 80 cms^{-1}

3 A converging beam of light is incident on convex mirror. The reflected beam

a. may be converging beam
b. may be a diverging beam
c. may be a parallel beam
d. All of these

4 A point object P is moving along line AC which is passing through centre of curvature of the convex mirror. The image of point object P is P'. If P is moving towards C with speed v_0. The image velocity v before striking the object with mirror.

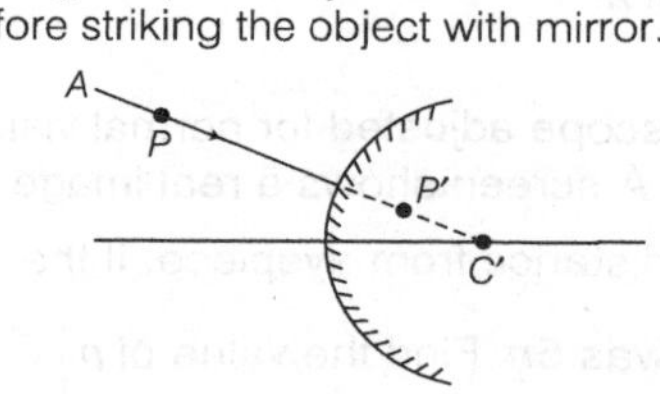

a. $> v_0$ towards C
b. $= v_0$ towards A
c. $< v_0$ towards C
d. $< v_0$ towards A

5 A point source of light S is placed at distance 10 cm from the pole of a convex mirror of focal length 10 cm. The intesity of light at the centre of screen in the absence of mirror is 49 Wm^{-2}. Find the intensity of light on screen in the presence of mirror.

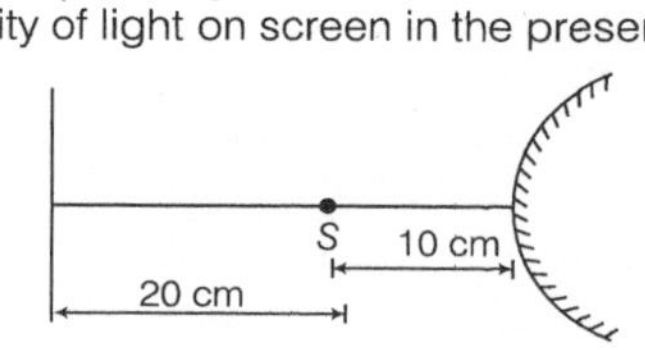

a. 49 Wm^{-2}
b. 53 Wm^{-2}
c. 80 Wm^{-2}
d. 98 Wm^{-2}

6 When an object is at distance 10 cm and 30 cm from a thin converging lens, virtual and real image of same size are formed. The power of lens is

a. $2D$
b. $4D$
c. $5D$
d. $10D$

Working Space

7 At the mouth of a cylindrical vessel, a plano convex lens of refractive index 1.5 and radius of curvature of curved surface 30 cm is filled. A point source S is at the bottom of vessel. The height of vessel is 40 cm and filled with liquid of refractive index 1.2. Find the position of image.

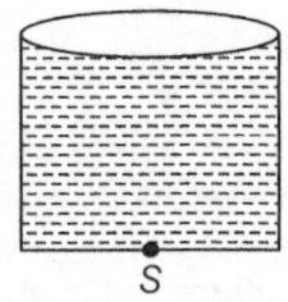

a. The image is 37.5 cm below the lens
b. The image is 56.25 below the lens
c. The image is real
d. The image is 30 cm below the lens

8 For the situation as shown in the figure, find the velocity of image formed by the mirror.

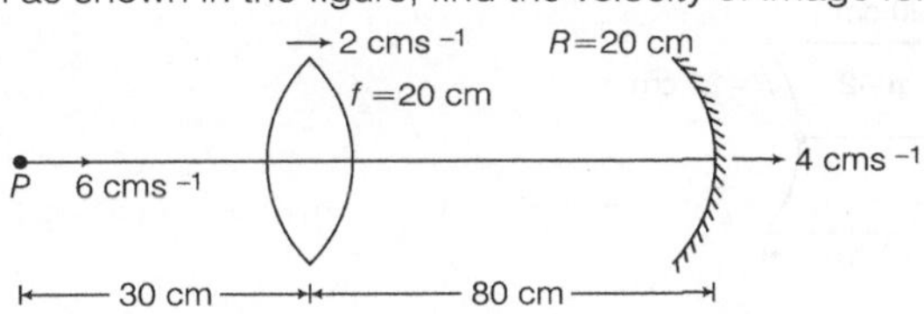

a. $10\,\text{cms}^{-1}$ in the direction of velocity of lens
b. $10\,\text{cms}^{-1}$ in the opposite direction of velocity of lens
c. zero
d. $20\,\text{cms}^{-1}$ in the direction of velocity of object

9 In the situation as shown in the figure, the focal length of the thin plane concave lens is 20 cm. A point object P is at distance of 20 cm from the lens. Find the velocity of image formed by plane mirror at the instant as shown in the figure.

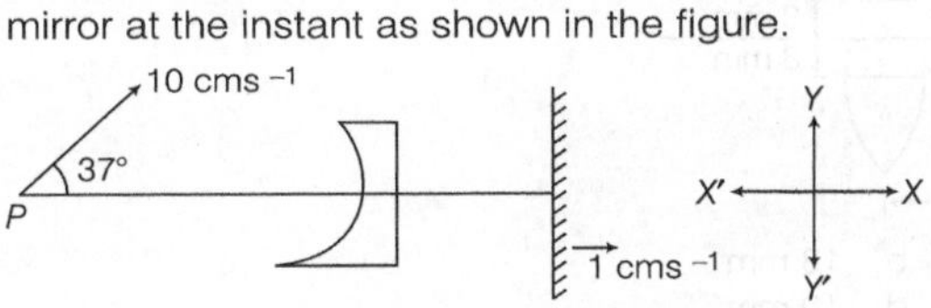

a. $(-2\hat{i} - 3\hat{j})\,\text{cms}^{-1}$
b. $(-3\hat{j})\,\text{cms}^{-1}$
c. $(-2\hat{i} + 3\hat{j})\,\text{cms}^{-1}$
d. $3\hat{j}\,\text{cms}^{-1}$

10 The converging point of incident beam is at the 20 cm from concave lens as shown in the figure are. The concave lens then moved away from convex lens with speed $1\,\text{cms}^{-1}$. The emergent ray from convex lens is

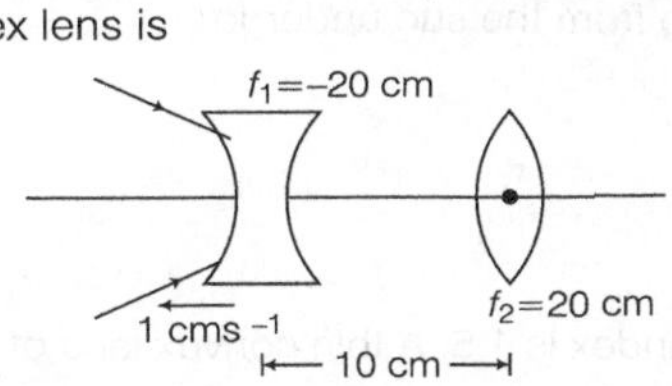

a. always converging at a single point
b. always diverging in nature
c. first converging then diverging
d. first diverging then converging

11 Four identical thin prism are combined as shown in figure. A beam of white light incident on first prism.

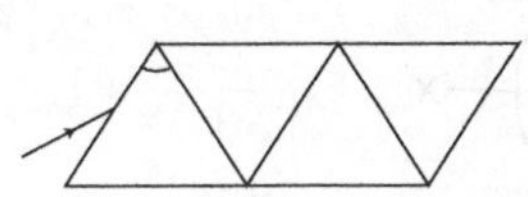

Mark the correct option.

a. The net deviation is zero
b. Angular dispersion may be non-zero
c. The net deviation may be non-zero
d. All of these

Working Space

12 Two thin prism each of angle of prism 5° and refractive index 1.5 are combined base to base as shown in the figure. Find separation between images formed due to refraction through prisms.

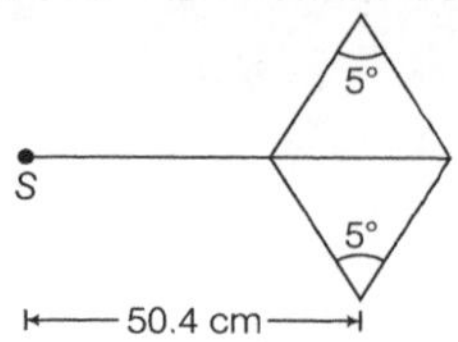

a. 2.2 cm
b. 4.4 cm
c. 6.6 cm
d. 8.2 cm

13 A thin equiconvex lens of refractive index 1.5 and radius of curvature of each surface 20 cm is fitted in a vessel whose bottom is spherical of radius 20 cm. Find the distance of image of point object *P* from bottom.

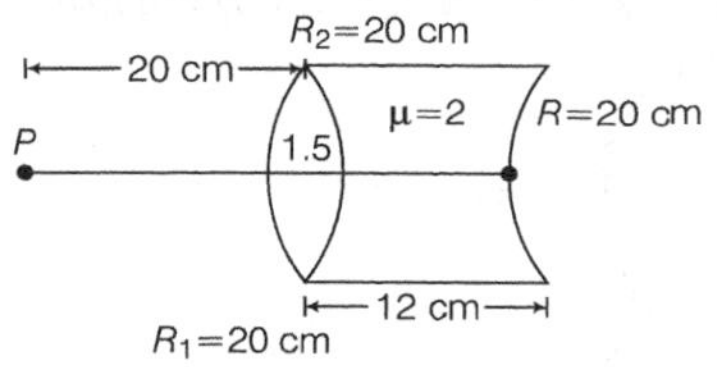

a. 12 cm
b. 30 cm
c. 24 cm
d. 40 cm

14 The two halves of a thin lens of focal length 20 cm are symmetrically separated by 6 mm. Find the separation between images formed by two halves.

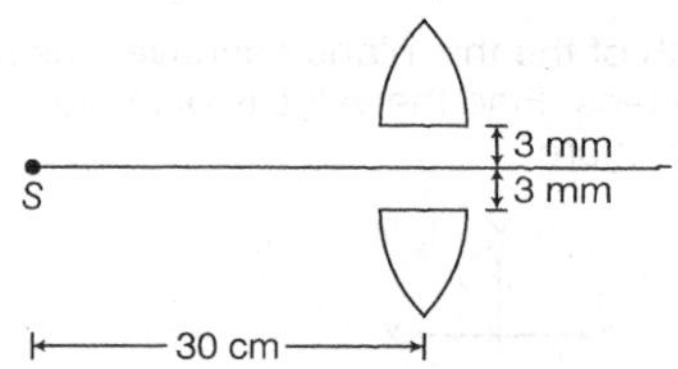

a. 12 mm
b. 18 mm
c. 21 mm
d. 10 mm

15 The critical angle for glass-water interface is

a. same on air-water inferface
b. lesser than air-glass interface $\left(\mu_w = \frac{4}{3}, \mu_g = \frac{3}{2}\right)$
c. greater than air glass interface
d. same for red and blue colour

16 Secondary rainbow is formed when light rays coming from the sun undergo

a. refraction, internal reflection and then refraction
b. only refraction
c. refraction, two internal reflections and then refraction
d. only internal reflections

17 The refracting angle of prism is $\left(\frac{14}{11}\right)^\circ$ and refractive index is 1.5. A thin convex lens of focal length 20 cm is placed at distance 30 cm from the point source *S* as shown in figure. The position of image of point source *S* is

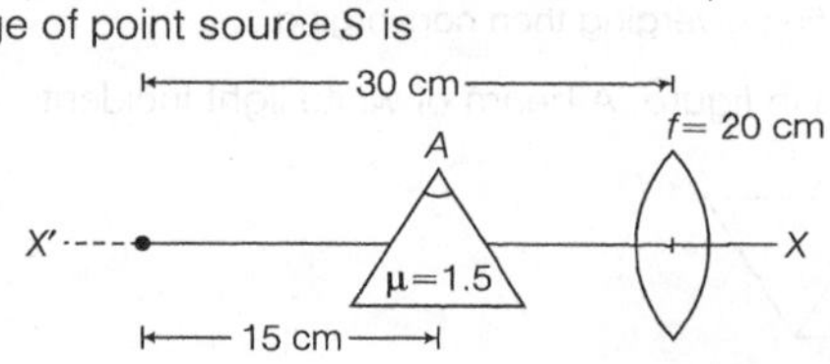

a. $\frac{1}{3}$ cm above the *X*-axis
b. $\frac{1}{3}$ cm below the *X*-axis
c. $\frac{2}{3}$ cm below the *X*-axis
d. $\frac{2}{3}$ cm above the *X*-axis

18 The radius of concave surface is 20 cm and radius of convex surface is 10 cm. The refractive index of material of lens is 1.5. If image of point object P coincides with it. Find the distance between object and lens.

Working Space

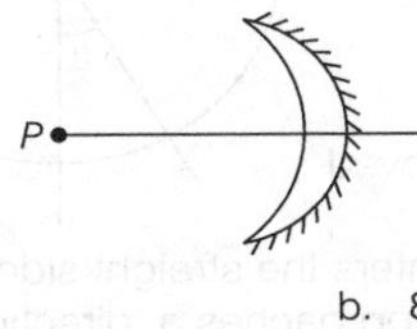

a. 4 cm
b. 8 cm
c. 12 cm
d. 16 cm

19 A point object P is approachig towards a thin and silvered equiconvex lens of focal length 20 cm and refractive index 1.5 with a speed of 2 cms^{-1} as shown in figure.

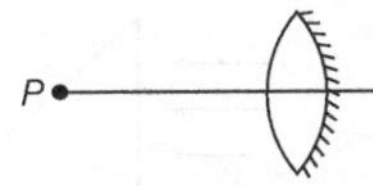

Find the speed of image when object is either away or towards the lens.

a. 1 cms^{-1} away from lens
b. 2 cms^{-1} towards the lens
c. 1 cms^{-1} towards the lens
d. 4 cms^{-1} away from lens

20 A beam of light incident on face AB of a equilateral prism of refractive index 1.5 as shown in the figure.

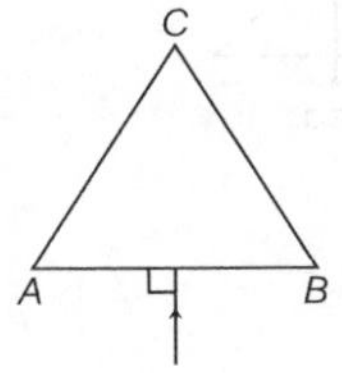

a. The beam will emerge from face AB
b. The beam will emerge from side BC
c. The beam will emerge from face AC
d. The beam will not emerge

21 A point object P is placed at 180 cm from a thin converging lens of focal length 90 cm. The lens then starts rotating about optical centre with constant angular velocity $\frac{1}{\pi}$ $rads^{-1}$ in anti-clockwise direction. The speed of image of point object P if the lens is rotated throught 1°.

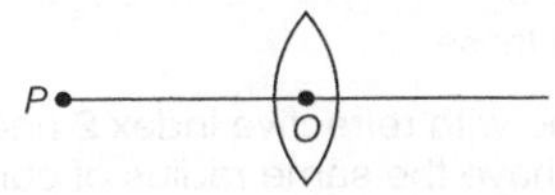

a. 1 cms^{-1}
b. 2 cms^{-1}
c. 3 cms^{-1}
d. 4 cms^{-1}

22 Find the number of images in the following arrangement

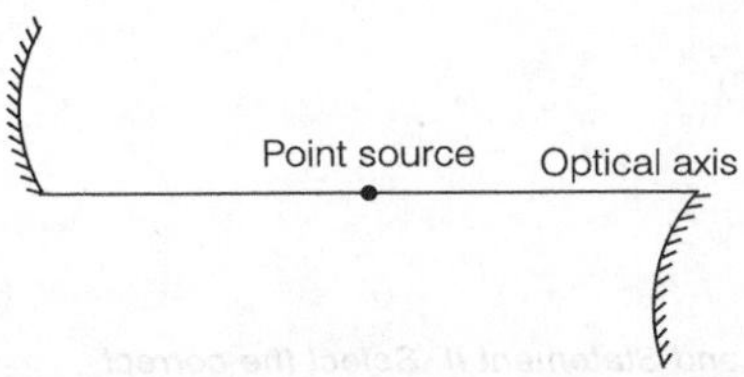

a. When object is placed at the centre of curvature of concave mirror only 2 images are formed
b. When object is placed any where in between the mirror anywhere are formed
c. If the object is placed any where only 2 images are formed
d. Both (a) and (c)

Working Space

23 A beam of white light is incident on the curved edge of a semicircular piece of glass as shown in figure.

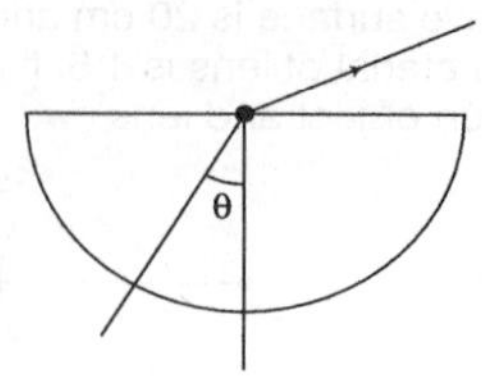

The light enters the curved surface along the normal, so it experiences no refraction.

It encounters the straight side at the centre of curvature of the curved side and refracts into the air. The incoming beam is moved clockwise. (So that the angle θ increases) such that the beam always enters along the normal to the curved side and encounters the straight side at the centre of curvature of the curved side.As the refracted beam approaches a direction parallel to the straight side, it becomes

a. red
b. blue
c. white
d. Yellow

24 A beam of light consisting of red, green and blue colours is incident on a right angle prism as shown in figure. The refractive indices of the material of the prism for the above red, green and blue wavelengths are 1.39, 1.44 and 1.47 respectively.

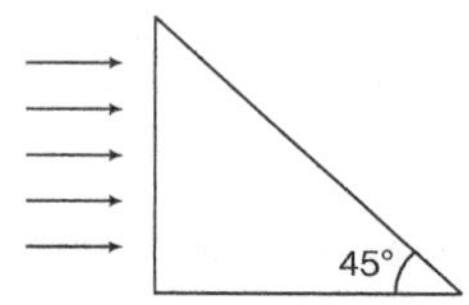

Which of the following is correct for the given condition.

a. 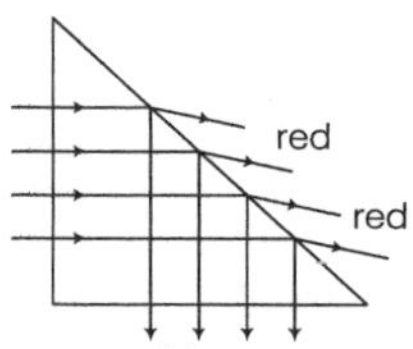

b. 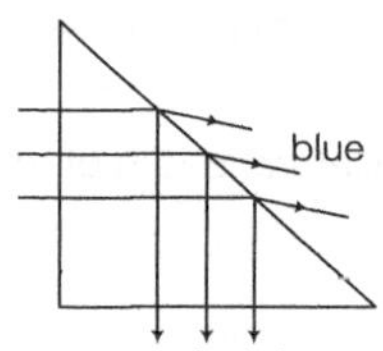

c. 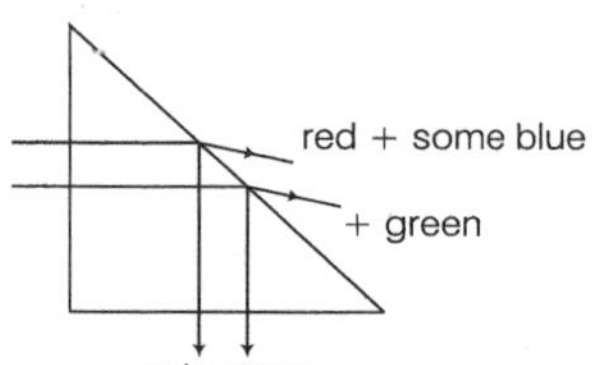

d. None of these

25 A thin converging lens of focal length $f = 1.5\,\text{m}$ is placed along Y-axis such that its optical centre coincides with the origin. A small light source S is placed at $(-2.0\,\text{m}, 0.1\,\text{m})$, where a plane mirror inclined at an angle θ, $\tan\theta = 0.3$ with the vertical, be placed such that maximum intensity is obtained at a point whose y-coordinate is 0.3 m. Also, find the x-coordinate of the final image.

a. (4 m, 0.3 m)
b. (8m, 0.6 m)
c. (1m, 0.2m)
d. None of these

26 There are two thin symmetrical lenses, one is converging with refractive index 2 and the other is diverging with refractive index 1.5. Both lenses have the same radius of curvature of 10 cm. The lenses were put close together and submerged into water. What is the focal length of this system in water. The refractive index of water is $\frac{4}{3}$.

a. 40 cm
b. $\frac{40}{3}$ cm
c. $\frac{20}{3}$ cm
d. $\frac{-40}{3}$ cm

Statement Type

***Direction** (Q. Nos. 27-35) This section is based on Statement I and Statement II. Select the correct answer from the codes given below.*

a. Both Statement I and Statement II are correct and Statement II is the correct explanation of Statement I
b. Both Statement I and Statement II are correct but Statement II is not the correct explanation of Statement I
c. Statement I is correct but Statement II is incorrect
d. Statement II is correct but Statement I is incorrect

Working Space

27 **Statement I** A ray is reflected in turn by three plane mirrors, placed mutually perpendicular to each other. The final emergent ray is anti-parallel to incident ray.

Statement II According to law of reflection, incident ray reflected ray, normal on incidence point are coplanar.

28 **Statement I** Sound like light is passing from air to water also bend towards the normal.

Statement II Air is acousfically denser than water.

29 **Statement I** Astronomers looking at distant glaxies talk about looking back in time.

Statement II Light take billion years to reach the earth from distant glaxies.

30 **Statement I** It is difficult to see through a closed glass window from well lighted room at night, but it is relatively easy when the light of room is switched OFF.

Statement II At night, when light is switched ON, the room's reflected light is much more intense than the transmitted light from outside.

31 **Statement I** When colour of light changes, the focal length of thin lens change but that of spherical mirror does not change.

Statement II Refractive index of material depends on wavelength of light while reflection of light does not depend on wavelength of light.

32 **Statement I** The distance between real image and real object is always greater or equal to four times of focal length of lens while in the case of spherical mirror, the distance between real image and real object is zero.

Statement II The image formed on screen is always real.

33 **Statement I** If a lens behaves as converging lens placed in a medium, the same lens may behave as diverging, if it is placed in other medium

Statement II The focal length of lens does not depend on surrounding medium.

34 **Statement I** The resolving power of an electron microscope is greater than that of an optical microscope.

Statement II The wavelength of electron is lesser than the wavelength of visible light.

35 **Statement I** A star will appear to twinkle, if seen from the surface of Moon

Statement II Atmosphere is not present on the moon.

Matching List Type

***Direction** (Q. No. 36) Choices for the correct combination of elements from Column I and Column II are given as options (a), (b), (c) and (d), out of which one is correct.*

36 In the case of compound microscope, due to change occur in Column I, the change in magnifying power given in Column II. Match the column I with Column II and mark the correct option from the codes given below.

	Column I		Column II
i.	I object is moved closer to objective lens (between F to $2F$)	p.	Increase
ii.	The length of tube decreased	q.	Decrease
iii.	Objective and eye piece are interchanged	r.	Will not change
iv.	Eye is not closed to eyepiece	s.	May increase or may decrease
v.	White light is replaced by red light	t.	Can not say any thing

Codes

	i	ii	iii	iv	v		i	ii	iii	iv	v
a.	p	p	r	q	q	b.	p	q	r	q	q
c.	q	p	s	t	q	d.	p	q	s	t	r

SET - 2

Revisal Problems for JEE Advanced

Only One Option Correct Type

Working Space

Direction (Q. Nos. 1-16) *This section contains 16 multiple choice questions. Each question has four choices (a), (b), (c) and (d), out of which ONLY ONE option is correct.*

1 Two plane mirrors of length 1 m each are separated by distance 1 m and a point source S is 1 m from the mirrors. Final length of reflected patch on the screen 2 m from the mirrors.

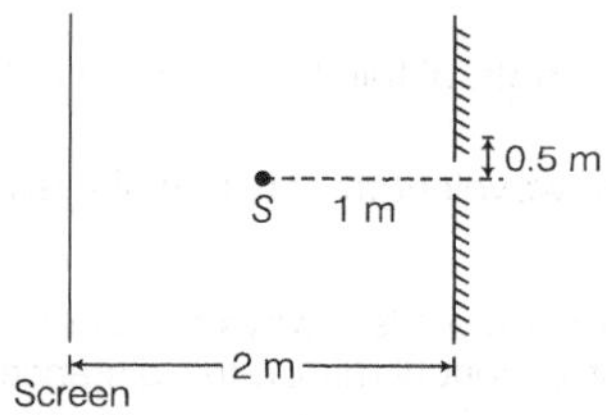

a. 1 m
b. 2 m
c. 4 m
d. 6 m

2 A circular screen of radius 20 cm and centre at point O is shown in the figure. An insect is moving on circular screen with a speed of π cms^{-1}. Its female partner is at the centre of screen. Find the frequency in which the insect coincides with the image of its female partner.

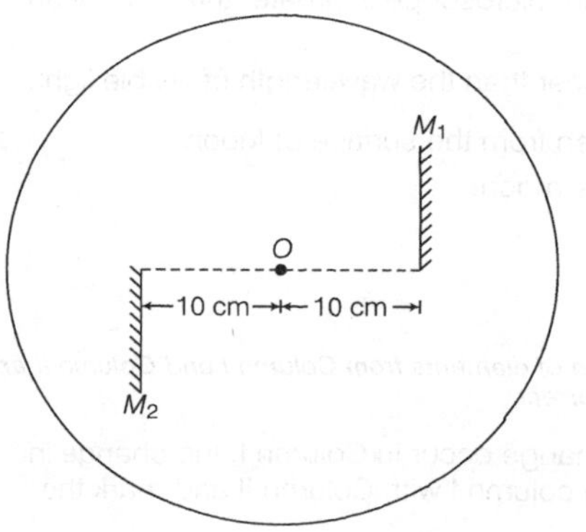

a. 1 Hz
b. 0.05 Hz
c. 0.02 Hz
d. 2 Hz

3 In the figure as shown, find the speed of second image formed by mirror M_1.

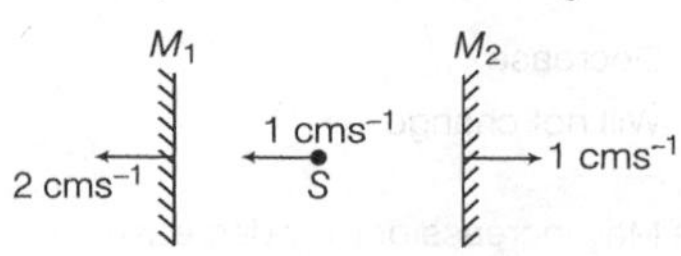

a. 3 cms^{-1}
b. 7 cms^{-1}
c. 6 cms^{-1}
d. 4 cms^{-1}

4 The refractive index of medium of a rectangular slab varies as $\mu = \sqrt{1+y}$. The ray enter from air to slab at origin in the condition of grazing incidence. The trajectory of ray inside slab is

a. straight line
b. parabola
c. ellipse
d. circle

5 The aperture of thin convex lens is 2 mm. Find the radius of circular bright spot on screen

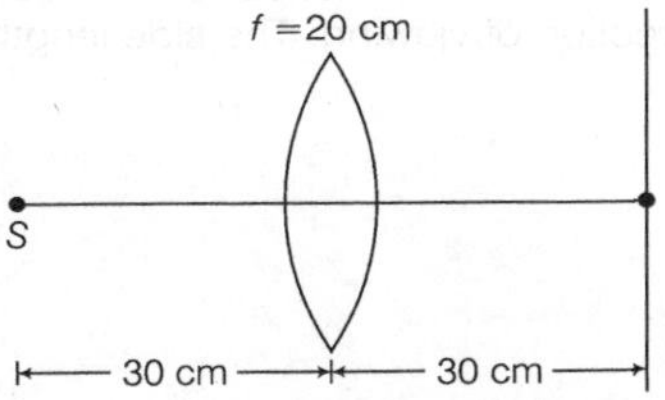

a. 1 mm b. 2 mm
c. 3 mm d. 4 mm

6 Two thin convex lenses are placed coaxially. Find the value of d, so that final emergent ray is diverging in nature.

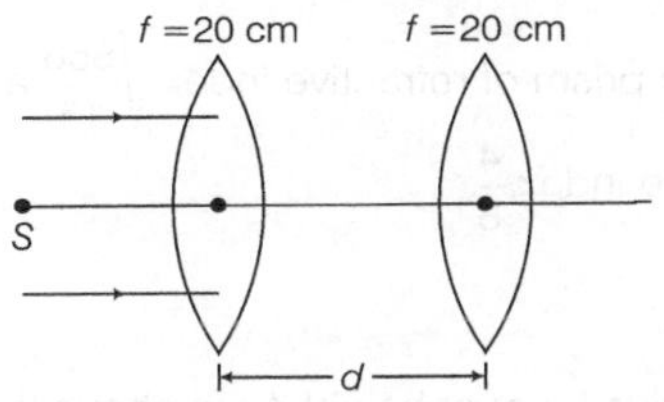

a. $d < 20$ cm b. 20 cm $< d <$ 40 cm
c. 10 cm $< d < 20$ d. None of these

7 In the figure shown, consider the first reflection at the convex mirror and second reflection at the plane mirror. The plane mirror is rotating with constant angular velocity $\omega = 2$ rads^{-1} about point O. Find the initial speed of image formed by plane mirror.

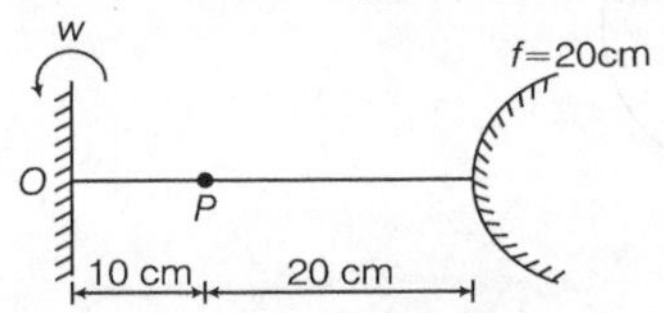

a. 40 cms^{-1} b. 80 cms^{-1}
c. 160 cms^{-1} d. 200 cms^{-1}

8 Consider a line through centre of curvature. The object P is h distance below the optical axis. The image of point object P is H distance either above or below the optical axis. Then,

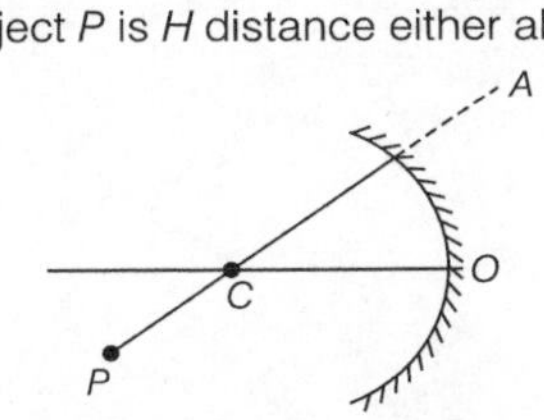

a. $h = H$ b. $h > H$
c. $h < H$ d. $h \leq H$

9 Water is injected from below so that its level rises with constant speed 2 cms^{-1}. A bird observed that fish F inside water is stationary. Find the velocity of bird. The refractive index of water is $\frac{4}{3}$

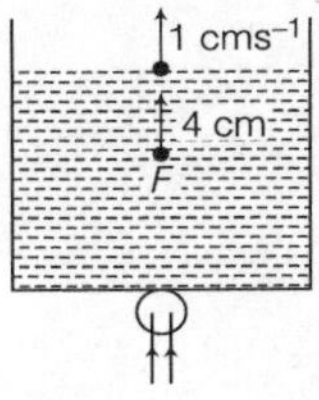

a. 3.25 cms^{-1} is downward direction b. 3.2 cms^{-1} in upward direction.
c. 4 cms^{-1} in upward direction d. 1 cms^{-1} in upward direction

Working Space

10 A glass cube has a spot at its centre. What portion of the each face of cube must be converted to make it invisible, no matter what the direction of viewing. The side length of cube is 10 cm and its refractive index is $\frac{5}{3}$.

a. Circular central part of diameter 7.5 cm
b. Circular central part of radius 5 cm
c. Circular central part of radius 3 cm
d. None of the above

11 When a plane convex lens is viewed from flat surface, its apparent depth is 2 cm. The radius of curvature of curved surface is 20 cm. When it is viewed from curved surface its apparent depth is 2.22 cm. The refractive index of lens is nearly equal to

a. 1.4 b. 1.5
c. 1.6 d. 1.7

12 Find the maximum angle of deviation produced by a prism of refractive index $\sqrt{\frac{656}{15}}$ and angle of prism 90° and is placed in water of refractive index $\frac{4}{3}$.

a. 37° b. 53°
c. 30° d. 60°

13 A ray of light incident on a transparent sphere of radius 5 cm at height 4 cm above the centre as shown in the figure. The refractive index of sphere is 1.6. Find the deviation in the ray.

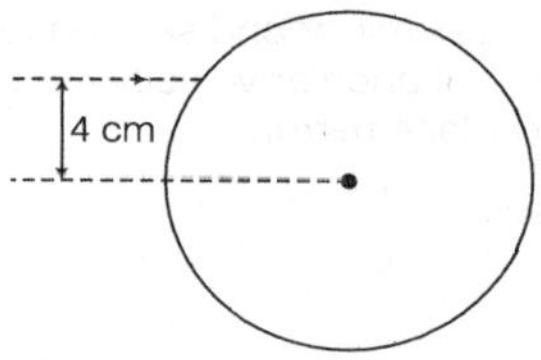

a. 30° b. 46°
c. 60° d. 90°

14 Find the deviation in ray in diagram as shown in the figure.

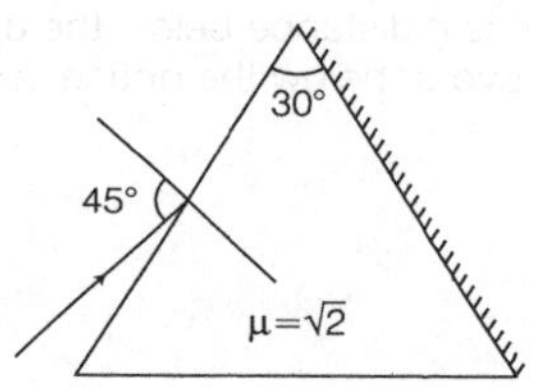

Mark the correct option.

a. The ray finally retraces its path
b. The ray will emerges from bottom faces
c. The ray will suffer total internal reflection from bottom face
d. None of the above

15 Find the position of image from the mirror formed by convex mirror as shown in the figure.

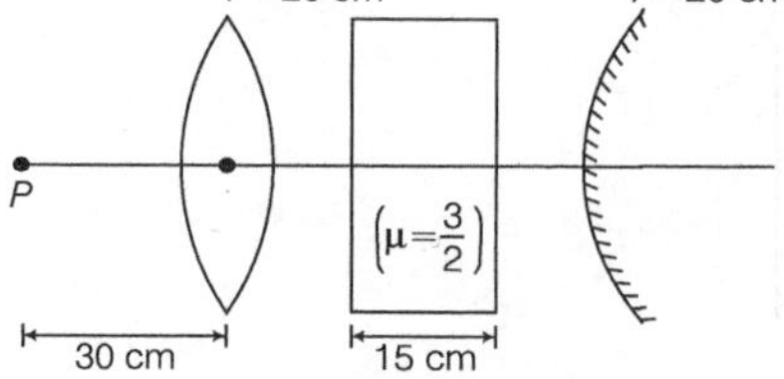

a. 5 cm b. 10 cm
c. 20 cm d. 30 cm

Working Space

16 Find the value of refractive index of prism as shown in the figure, so that total internal reflection takes place at face *AC*.

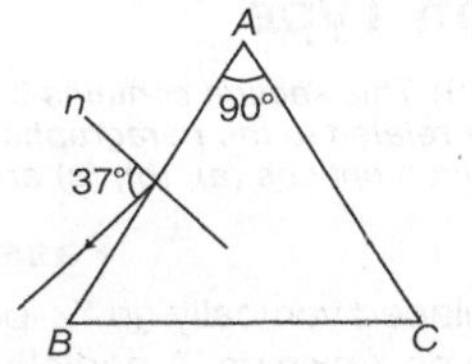

a. $\mu > \frac{\sqrt{34}}{5}$

b. $\mu \le \sqrt{\frac{34}{5}}$

c. $\mu > 1.5$

d. $\mu \le 1.5$

One or More than One Options Correct Type

Direction (Q. Nos. 17-21) ***This section contains 5 multiple choice questions. Each question has four choices (a), (b), (c) and (d), out of which ONE or MORE THAN ONE are correct.***

17 The distance between a light source and screen is 1.2 m. When a convex lens is moved along the optical axis, sharp image of source can be obtained for two lens positions. The area ratio on images on the screen is 1 : 9.

a. The focal length of lens is 22.5 cm

b. The smaller image is more bright than larger image

c. The smaller image is a less brighter than larger image

d. None of the above

18 The refractive index of the medium with in a certain region $x > 0$, $y > 0$, changes with *y*. A ray of light ravelling in *x*-direction in air strike the medium at grazing incidence and moves along a circular are of radius *R*.

a. The refractive index of medium is $\mu = \frac{R}{R-y}$

b. If the maximum value of refractive index is 2

c. The refractive index of medium is $\frac{R}{R+y}$

d. None of the above

19 The rainbow is formed when light rays coming from the Sun undergo

a. a refraction, internal reflection and then refraction

b. refraction, two internal reflection and then refraction

c. only two refraction

d. only two internal reflection

20 The cross-section of a prism is shown in the figure. The minimum value of *y* for which the ray gets refracted from surface *OB* is

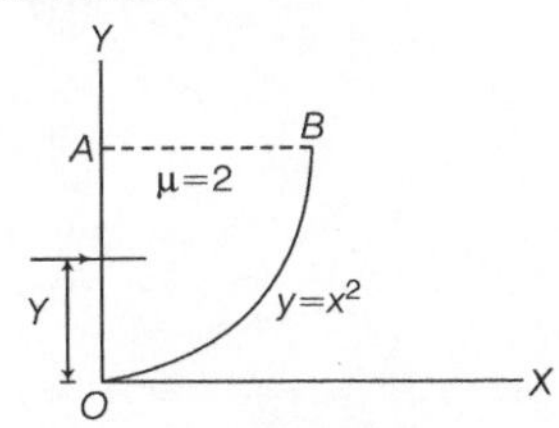

a. 0.75 m

b. 1 m

c. 2 m

d. non-zero

21 A thin convex lens of focal length 30 cm is placed on origin whose optical axis coinciding with *X*-axis. Take, *x* and *y* in cm

a. A ray of light whose equation is $y = 1$ is incident on the lens in position *x*-direction, the equation of refracted ray is $\frac{x}{30} + y = 1$

b. A light ray $y + \frac{1}{39}x + 1 = 0$ is incidont on the lens in positive *x*-direction, the equation of refracted ray is $\frac{x}{130} - y - 1 = 0$

c. For a ray $y = 1$, the refracted ray intersect the *X*-axis at (30 cm, 0)

d. None of the above

Comprehension Type

Working Space

Direction (*Q. Nos. 22-29*) *This section contains 3 paragraphs, each describing theory, experiments, data, etc. Eight questions related to the paragraphs have been given. Each question has only one correct answer among the four given options (a), (b), (c) and (d).*

Passage I

A plane mirror is placed vertically (in *YZ*-plane) as shown in figure. Vertically upward direction is talcum as *Y*-axis us. A particle of mass *m* is projected in vertical plane *XY* with speed $20\sqrt{2}$ with *X*-axis. At the same time, another particle of same mass projected upward with speed 20 ms^{-1} from print point *A* (40 m 0,0).

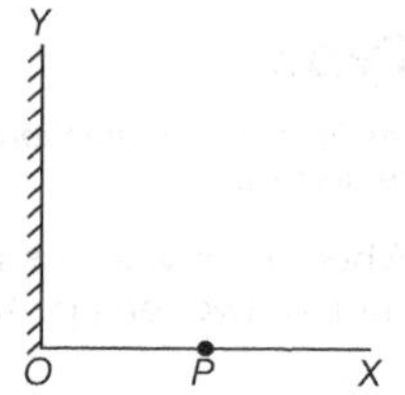

22 Find the magnitude of relative velocity between images of particles at, $t = 1$ s.

a. 10 ms^{-1} b. 20 ms^{-1}
c. 30 ms^{-1} d. zero

23 If collision between particles is completely inelastic. Find the velocity of image of combined mass just after collision.

a. 5 ms^{-1} towards negative *X*-axis
b. 10 ms^{-1} towards positive *X*-axis
c. 10 ms^{-1} towards positive *Y*-axis
d. 10 ms^{-1} towards positive *X*-axis

Passage II

An equilateral prism *AOC* is placed in air with its side by lying horizontally along *X*-axis as shown in the figure. A ray given by $x + \sqrt{3y} - 10 = 0$ in incident at a point *P* of face *AO* of prism.

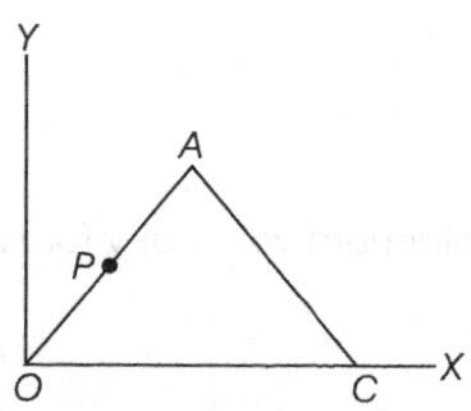

24 Find the value of μ for which the ray grazes on face *AC*.

a. $\sqrt{3}$ b. 1.5
c. $\frac{2}{\sqrt{3}}$ d. $\sqrt{2}$

25 For $\mu = 1.5$

a. the ray refracted from face *AC*
b. the ray totally reflected from face *AC*
c. the ray is at grazing emergence
d. None of the above

26 Find the equation of ray coming out of prism of face *OC* is silvered. (Take, $\mu = 1.5$)

a. $x + \sqrt{3y} - 10 = 0$
b. $x - \sqrt{3y} + 10 = 0$
c. $\sqrt{3x} + y - 10 = 0$
d. $x + \frac{1}{\sqrt{3}}y - 10 = 0$

Working Space

Passage III

A long rectangular slab of trans parent medium of thickness d has refractive index $\mu = \sqrt{1 + e^{x/d}}$.

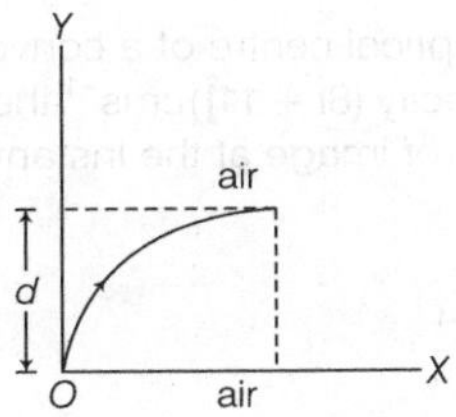

27 The equation of ray is

a. $y = e^{x/d}$ b. $y = 2d\,(1 - e^{-\frac{x}{2d}})$

c. $e^{-\frac{x}{2d}}$ d. None of these

28 Find x-coordinate of the point A, where the ray intersect the upper surface of the slab air boundary.

a. $2d/n^2$ b. d/n^2

c. d/n^3 d. d

29 Find the refractive index of the medium at point A.

a. $\sqrt{2}$ b. $\sqrt{3}$

c. 1.5 d. $\sqrt{5}$

Matching List Type

Direction *(Q. No. 30) Choices for the correct combination of elements from Column I and Column II are given as options (a), (b), (c) and (d), out of which one is correct.*

30 An object P is placed 30 cm.

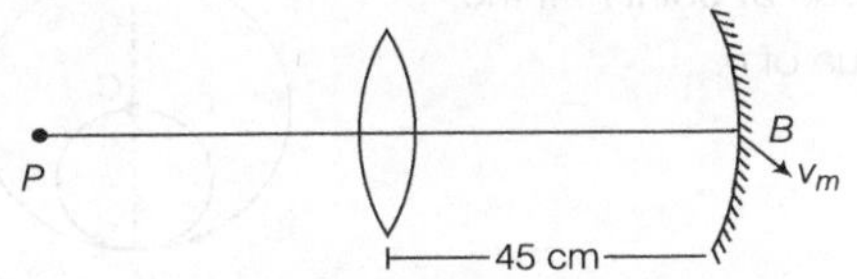

From a convex lens of focal length 10 cm. A concave mirror of local length 30 cm starts moving towards right from point B and has velocity $v_m = \frac{(45 - x)^2}{300}$ cms^{-1} at any point on the axis. Here, x is the separation between the lens and the mirror. Match the Column I with Column II and mark the correct option from the codes given below.

	Column I		Column II
i.	The speed in cms^{-1} of image formed by mirror with respect to mirror is	p.	3
ii.	The speed in cms^{-1} of image formed by mirror for $x = 15$ cm	q.	6
iii.	The mode of magnification by one refraction and reflection through mirror for $x = 30$ cm	r.	4
iv.	The net magnification for one refraction and one reflection for $x = 40$ cm	s.	1

Codes

	i	ii	iii	iv		i	ii	iii	iv
a.	p	q	s	p	b.	p	q	s	r
c.	p	q	r	s	d.	p	q	p	s

One Integer Value Correct Type

Working Space

Direction *(Q. Nos. 31-36) This section contains 6 questions. When worked out will result in an integer from 0 to 9 (both inclusive).*

31 A point object P is located at a distance 15 cm from the optical centre of a convex lens of focal length 10 cm on its opitcal axis is moving with a velocity $(8\hat{i} + 11\hat{j})\text{cms}^{-1}$ the lens is moving with a velocity of $(4\hat{i} + 2\hat{j})\text{cms}^{-1}$. If **v** is the velocity of image at the instant as shown in the figure. The value of $|\mathbf{v}| = n\sqrt{41}\,\text{cms}^{-1}$.
Find the value of n.

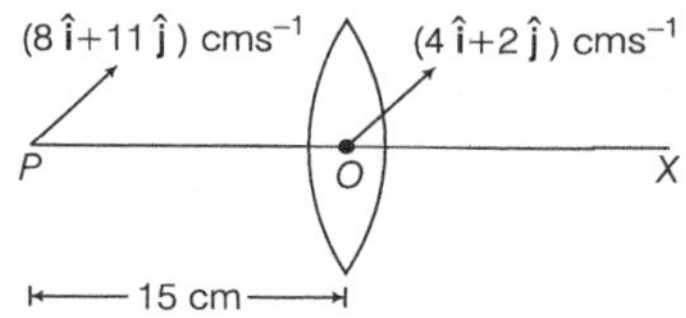

32 A point source of light is moving with constant velocity $\sqrt{2}\ \text{cms}^{-1}$ inside a thin spherical shell of radius $R = 10$ cm which is filled with a liquid of refractive index $\mu = \sqrt{2}$. If at $t = 0$, the source is at the centre of the sphere, then at what minimum time (in second) a thin dark ring will be visible for an observer outside the sphere.

33 A thin concave lens having focal length 20 cm is cut into two equal parts 1 cm above the principal axis. The lower portion is placed with optical axis at origin and upper portion is placed at (10 cm, 0). A point object P is placed at (– 20 cm, 0). If y-coordinate of image is $\frac{-y}{2}$, find the value of y^2.

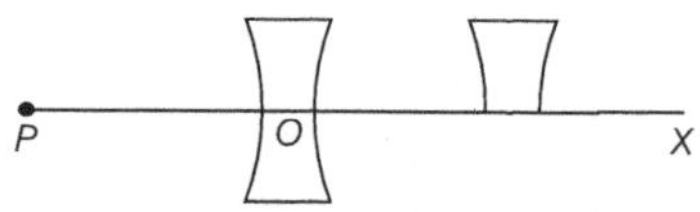

34 A transparent sphere of radius 10 cm has a non-concentric cavity of radius 5 cm. Find the value of refractive index of the sphere, if a parallel paraxial beam of light incident on sphere focussed at point P. If the refractive index of sphere is $\mu = \frac{13}{n}$. Find the value of n.
(Take, $\sqrt{5} = 2.2$)

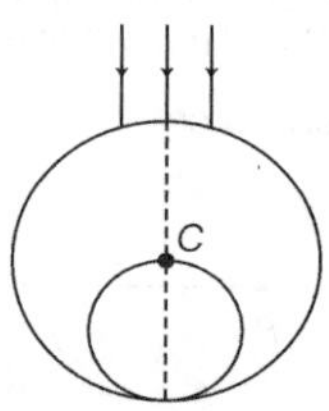

35 Bottom of a glass beaker is made of a thin planoconvex lens whose plane surface is silvered and radius of curved surface is 20 cm as shown in the figure. It is found that the image of a floating object P on the surface of water coincides with it. If the depth of water is $40x$ cm. Find the value of x.

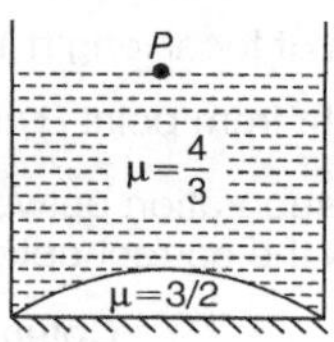

36 A ray is incident normally on a right angle prism whose refractive index is $\sqrt{3}$ and prism angle $\alpha = 30°$ after crossing prism ray passes through glass hemisphere. It strikes the glass hemisphere at $\frac{R}{2}$ distance from principal axis. If angle of deviation is $\frac{\pi}{n}$. Find the value of n.

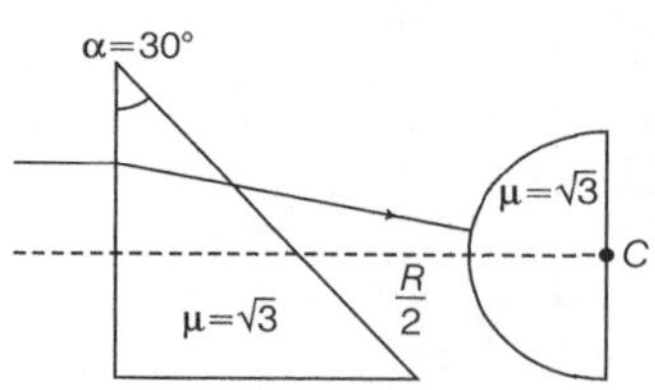

JEE Main & AIEEE Archive

(Compilation of Last 13 Years' Questions)

1 A thin convex lens made from crown glass $\left(\mu = \frac{3}{2}\right)$ has focal length f. When it is measured in two different liquids having refractive indices $\frac{4}{3}$ and $\frac{5}{3}$. It has the focal lengths f_1 and f_2, respectively. The correct relation between the focal lengths is **(JEE Main 2014)**

a. $f_1 = f_2 < f$
b. $f_1 > f$ and f_2 becomes negative
c. $f_2 > f$ and f_1 becomes negative
d. f_1 and f_2 both become negative

2 A green light is incident from the water to the air-water interface at the critical angle (C). Select the correct statement. **(JEE Main 2014)**

a. The entire spectrum of visible light will come out of the water at an angle of 90° to the normal
b. The spectrum of visible light, whose frequency is less than that of green light will come out of the air medium
c. The spectrum of visible light, whose frequency is more than that of green light will come out to the air medium
d. The entire spectrum of visible light will come out of the water at various angles to the normal

3 Diameter of a plano-convex lens is 6 cm and thickness at the centre is 3 mm. If speed of light in material of lens is $2 \times 10^8\ \text{ms}^{-1}$, the focal length of the lens is **(JEE Main 2013)**

a. 15 cm
b. 20 cm
c. 30 cm
d. 10 cm

4 The graph between angle of deviation (δ) and angle of incidence (i) for a triangular prism is represented by **(JEE Main 2013)**

a. 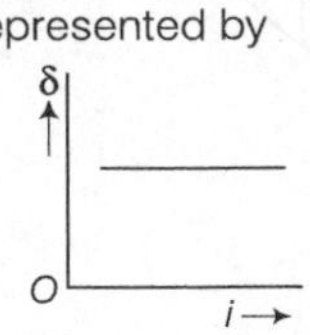

b. 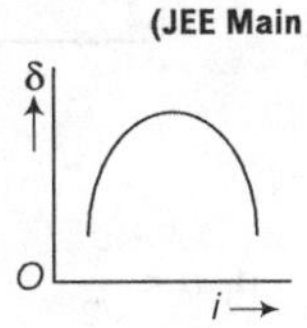

c. 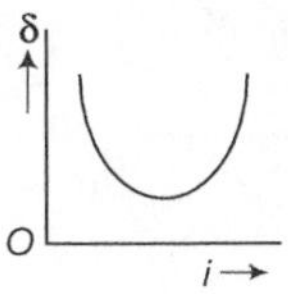

d. 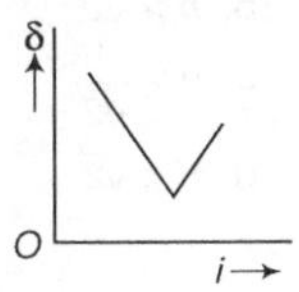

5 An object 2.4 m infront of a lens forms a sharp image on a film 12 cm behind the lens. A glass plate 1 cm thick, of refractive index 1.50 is interposed between lens and film with its plane faces parallel to film. At what distance (from lens) should object shifted to be in sharp focus on film? **(AIEEE 2012)**

a. 7.2 m
b. 2.4 m
c. 3.2 m
d. 5.6 m

6 When monochromatic red light is used instead of blue light in a convex lens, its focal length will **(AIEEE 2011)**

a. not depend on colour of light
b. increase
c. decrease
d. remain same

7 Let the zx-plane be the boundary between two transparent media. Medium 1 in $z \geq 0$ has a refractive index of $\sqrt{2}$ and medium 2 with $z < 0$ has a refractive index of $\sqrt{3}$. A ray of light in medium 1 given by the vector $\mathbf{A} = 6\sqrt{3}\,\hat{\mathbf{i}} + 8\sqrt{3}\,\hat{\mathbf{j}} - 10\,\hat{\mathbf{k}}$ is incident on the plane of separation. The angle of refraction in medium 2 is **(AIEEE 2011)**

a. 45°
b. 60°
c. 75°
d. 30°

8 A car is fitted with a convex side-view mirror of focal length 20 cm. A second car 2.8 m behind the first car is overtaking the first car at a relative speed of $15\ \text{ms}^{-1}$. The speed of the image of the second car as seen in the mirror of the first one is **(AIEEE 2011)**

a. $\frac{1}{15}\ \text{ms}^{-1}$
b. $10\ \text{ms}^{-1}$
c. $15\ \text{ms}^{-1}$
d. $\frac{1}{10}\ \text{ms}^{-1}$

9 As, the beam enters the medium, it will **(AIEEE 2010)**

a. diverge
b. converge
c. diverge near the axis and converge near the periphery
d. travel as a cylindrical beam

10 The initial shape of the wavefront of the beam is **(AIEEE 2010)**

a. convex
b. concave
c. convex near the axis and concave near the periphery
d. planar

11 The speed of light in the medium is **(AIEEE 2010)**

a. minimum on the axis of the beam
b. the same everywhere in the beam
c. directly proportional to the intensity I
d. maximum on the axis of the beam

12 In an optics experiment, with the position of the object fixed, a student varies the position of a convex lens and for each position, the screen is adjusted to get a clear image of the object. A graph between the object distance u and the image distance v, from the lens is plotted using the same scale for the two axes. A straight line passing through the origin and making an angle of 45° with the x-axis meets the experimental curve at P. The coordinates of P will be **(AIEEE 2009)**

a. $(2f, 2f)$ b. $\left(\frac{f}{2}, \frac{f}{2}\right)$

c. (f, f) d. $(4f, 4f)$

13 A transparent solid cylinder rod has a refractive index of $\frac{2}{\sqrt{3}}$. It is surrounded by air. A light ray is incident at the mid-point of one end of the rod as shown in the figure.

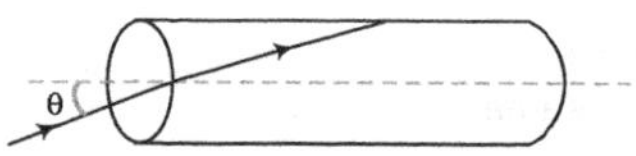

The incident angle θ for which the light ray grazes along the wall of the rod is **(AIEEE 2009)**

a. $\sin^{-1}\left(\frac{1}{2}\right)$

b. $\sin^{-1}\left(\frac{\sqrt{3}}{2}\right)$

c. $\sin^{-1}\left(\frac{2}{\sqrt{3}}\right)$

d. $\sin^{-1}\left(\frac{1}{\sqrt{3}}\right)$

14 A student measures the focal length of a convex lens by putting an object pin at a distance u from the lens and measuring the distance v of the image pin. The graph between u and v plotted by the student should look like **(AIEEE 2008)**

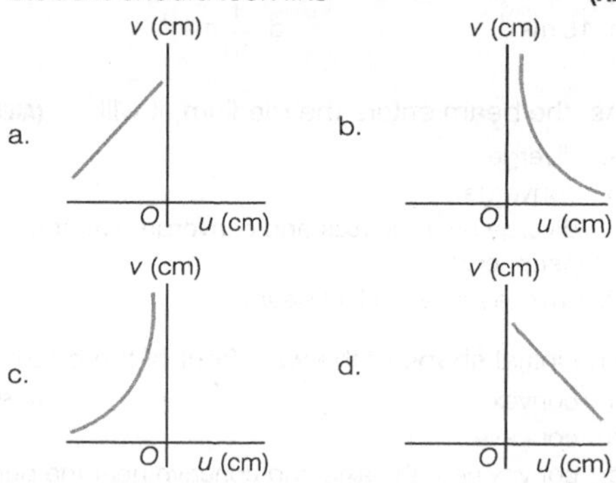

15 Two lenses of power $-15\ D$ and $+5\ D$ are in contact with each other. The focal length of the combination is **(AIEEE 2007)**

a. -20 cm b. -10 cm

c. $+20$ cm d. $+10$ cm

16 The refractive index of glass is 1.520 for red light and 1.525 for blue light. Let D_1 and D_2 be angles of minimum deviations for red and blue light respectively in a prism of this glass. Then, **(AIEEE 2006)**

a. $D_1 < D_2$

b. $D_1 = D_2$

c. $D_1 > D_2$

d. Can be less than or greater than D_2 depending upon the angle of prism

17 A fish looking up through the water sees the outside world, contained in a circular horizon. If the refractive index of water is $4/3$ and the fish is 12 cm below the water surface, the radius of this circle (in cm) is **(AIEEE 2005)**

a. $36\sqrt{7}$

b. $\frac{36}{\sqrt{7}}$

c. $36\sqrt{5}$

d. $4\sqrt{5}$

18 Two point white dots are 1 mm apart on a black paper. They are viewed by eye of pupil diameter 3 mm. Approximately, what is the maximum distance at which these dots can be resolved by the eye? (Take, wavelength of light = 500 nm) **(AIEEE 2005)**

a. 5 m b. 1 m

c. 6 m d. 3 m

19 A thin glass (refractive index 1.5) lens has optical power of -5 D in air. Its optical power in a liquid medium with refractive index 1.6 will be **(AIEEE 2005)**

a. 1 D b. -1 D

c. 25 D d. -25 D

20 A light ray is incident perpendicular to one face of a 90° prism and is totally internally reflected at the glass-air interface. If the angle of reflection is 45°, we conclude that for the refractive index n as **(AIEEE 2004)**

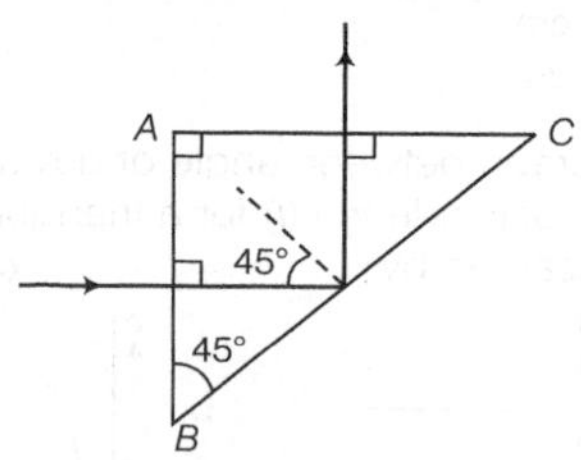

a. $n < \frac{1}{\sqrt{2}}$

b. $n > \sqrt{2}$

c. $n > \frac{1}{\sqrt{2}}$

d. $n < \sqrt{2}$

JEE Archive

21 A plano-convex lens of refractive index 1.5 and radius of curvature 30 cm is silvered at the curved surface. Now, this lens has been used to form the image of an object. At what distance from this lens, an object be placed in order to have a real image of the size of the object? **(AIEEE 2004)**

a. 20 cm b. 30 cm
c. 60 cm d. 80 cm

22 The angle of incidence at which reflected light is totally polarised for reflection from air to glass (refractive index n), is

a. $\sin^{-1}(n)$ b. $\sin^{-1}\left(\frac{1}{n}\right)$
c. $\tan^{-1}\left(\frac{1}{n}\right)$ d. $\tan^{-1}(n)$

23 The image formed by an objective of a compound microscope is **(AIEEE 2003)**

a. virtual and diminished
b. real and diminished
c. real and enlarged
d. virtual and enlarged

24 The earth radiates in the infra-red region of the spectrum. The spectrum is correctly given by **(AIEEE 2003)**

a. Rayleigh Jeans law
b. Planck's law of radiation
c. Stefan's law of radiation
d. Wien's law

25 To get three images of a single object, one should have two plane mirrors at an angle of **(AIEEE 2003)**

a. 60° b. 90°
c. 120° d. 30°

26 If two mirrors are kept at 60° to each other, then the number of images formed by them is **(AIEEE 2002)**

a. 5 b. 6
c. 7 d. 8

27 Wavelength of light used in an optical instrument are $\lambda_1 = 4000$ Å and $\lambda_2 = 5000$ Å, then ratio of their respective resolving powers (corresponding to λ_1 and λ_2) is **(AIEEE 2002)**

a. 16 : 25 b. 9 : 1
c. 4 : 5 d. 5 : 4

28 An astronomical telescope has a large aperture to **(AIEEE 2002)**

a. reduce spherical aberration
b. have high resolution
c. increase span of observation
d. have low dispersion

29 Which of the following is used in optical fibres? **(AIEEE 2002)**

a. Total internal reflection
b. Scattering
c. Diffraction
d. Refraction

JEE Advanced & IIT JEE Archive

(Compilation of Last 10 Years' Questions)

1 A point source S is placed at the bottom of a transparent block of height 10 mm and refractive index 2.72. It is immersed in a lower refractive index liquid as shown in the figure. It is found that the light emerging from the block to the liquid forms a circular bright spot of diameter 11.54 mm on the top of the block. The refractive index of the liquid is

(2014 Adv., Only One Option Correct Type)

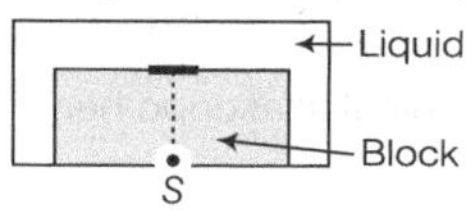

a. 1.21 b. 1.30
c. 1.36 d. 1.42

2 Four combinations of two thin lenses are given in Column I. The radius of curvature of all curved surfaces is r and the refractive index of all the lenses is 1.5. Match lens combinations in Column I with their focal length in Column II and select the correct answer using the code given below the columns. **(2014 Adv. Matching Type)**

	Column I		Column II
i.		p.	$2r$
ii.		q.	$r/2$
iii.		r.	$-r$
iv.		s.	r

Codes

	i	ii	iii	iv		i	ii	iii	iv
a.	q	s	r	p	b.	r	p	s	q
c.	s	q	p	r	d.	p	r	q	s

3 A transparent thin film of uniform thickness and refractive index $n_1 = 1.4$ is coated on the convex spherical surface of radius R at one end of a long solid glass cylinder of refractive index $n_2 = 1.5$, as shown in the figure.

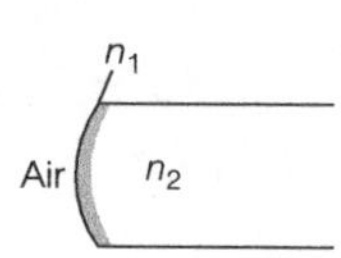

Rays of light parallel to the axis of the cylinder traversing through the film from air to glass get focused at distance f_1 from the film, while rays of light traversing from glass to air get focused at distance f_2 from the film. Then,

(2014 Adv., One or More than One Options Correct Type)

a. $|f_1| = 3R$ b. $|f_1| = 2.8R$
c. $|f_2| = 2R$ d. $|f_2| = 1.4R$

4 The image of an object, formed by a plano-convex lens at a distance of 8 m behind the lens, is real and is one-third the size of the object. The wavelength of light inside the lens is 2/3 times the wavelength in free space. The radius of the curved surface of the lens is

(2013 Adv, Only One Option Correct Type)

a. 1 m b. 2 m c. 3 m d. 6 m

5 A ray of light travelling in the direction $\frac{1}{2}(\hat{\mathbf{i}} + \sqrt{3}\hat{\mathbf{j}})$ is incident on a plane mirror. After reflection, it travels along the direction $\frac{1}{2}(\hat{\mathbf{i}} - \sqrt{3}\hat{\mathbf{j}})$. The angle of incidence Is **(2013 Adv., Only One Option Correct Type)**

a. 120° b. 45° c. 60° d. 75°

6 A right angled prism of refractive index μ_1 is placed in a rectangular block of refractive index μ_2, which is surrounded by a medium of refractive index μ_3, as shown in the figure. A ray of light e enters the rectangular block at normal incidence. Depending upon the relationships between μ_1, μ_2 and μ_3, it takes one of the four possible paths ef, eg, eh or ei.

(2013 Adv., Matching Type)

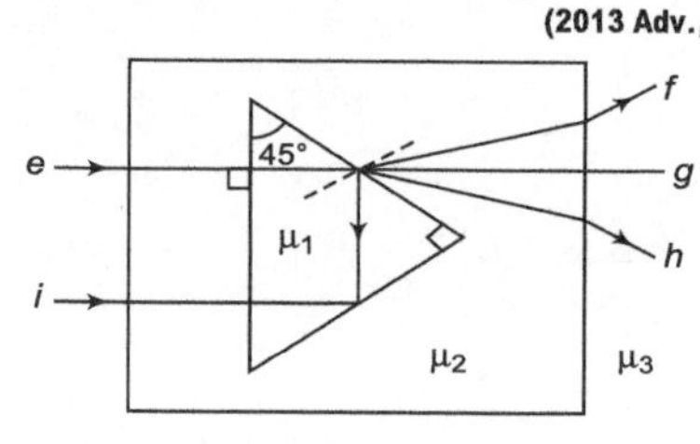

Match the paths in Column I with conditions of refractive indices in Column II and select the correct answer using the codes given below the columns.

	Column I		Column II
i.	$e \to f$	p.	$\mu_1 > \sqrt{2}\mu_2$
ii.	$e \to g$	q.	$\mu_1 > \mu_1$ and $\mu_2 > \mu_3$
iii.	$e \to h$	r.	$\mu_1 = \mu_2$
iv.	$e \to i$	s.	$\mu_2 < \mu_1 < \sqrt{2}\mu_2$ and $\mu_2 > \mu_3$

Codes

	i	ii	iii	iv		i	ii	iii	iv
a.	q	r	p	s	b.	p	q	s	r
c.	s	p	q	r	d.	q	r	s	p

JEE Archive

7 A biconvex lens is formed with two thin plano-convex lenses as shown in the figure. Refractive index n of the first lens is 1.5 and that of the second lens is 1.2. Both the curved surfaces are of the same radius of curvature $R = 14$ cm. For this biconvex lens, for an object distance of 40 cm, the image distance will be **(2012, Only One Option Correct Type)**

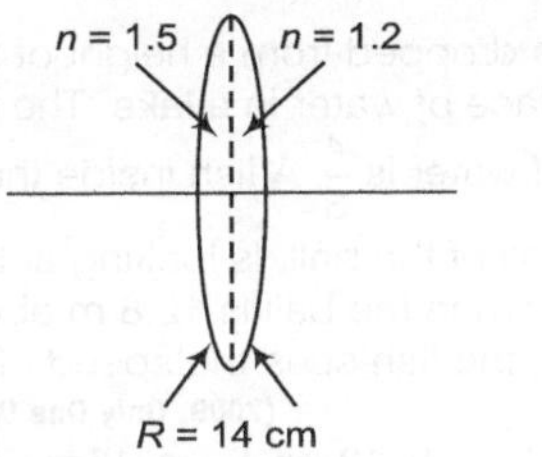

a. – 280.0 cm b. 40.0 cm
c. 21.5 cm d. 13.3 cm

Passage for Q. Nos. (8 and 9)

Most materials have the refractive index, $n > 1$. So, when a light ray from air enters a naturally occurring material, then by Snell's law, $\frac{\sin \theta_1}{\sin \theta_2} = \frac{n_2}{n_1}$, it is understood that the refracted ray bends towards the normal. But it never emerges on the same side of the normal as the incident ray. According to electromagnetism, the refractive index of the medium is given by the relation, $n = \left(\frac{c}{v}\right) = \pm \sqrt{\varepsilon_r \mu_r}$, where c is the speed of electromagnetic waves in vacuum, v its speed in the medium, ε_r and μ_r are the relative permittivity and permeability of the medium respectively. In normal materials, both ε_r and μ_r are positive, implying positive n for the medium. When both ε_r and μ_r are negative, one must choose the negative root of n. Such negative refractive index materials can now be artificially prepared and are called meta-materials. They exhibit significantly different optical behaviour, without violating any physical laws. Since n is negative, it results in a change in the direction of propagation of the refracted light. However, similar to normal materials, the frequency of light remains unchanged upon refraction even in meta-materials.
(2012, Comprehension Type)

8 Choose the correct statement.
a. The speed of light in the meta-material is $v = c|n|$
b. The speed of light in the meta-material is $v = \frac{c}{|n|}$
c. The speed of light in the meta-material is $v = c$
d. The wavelength of the light in the meta-material (λ_m) is given by $\lambda_m = \lambda_{air}|n|$, where λ_{air} is the wavelength of the light in air.

9 For light incident from air on a meta-material, the appropriate ray diagram is

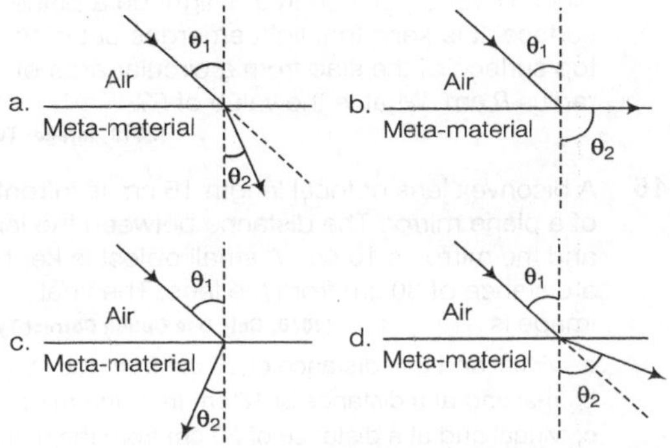

10 A light ray travelling in glass medium is incident on glass-air interface at an angle of incidence θ. The reflected (R) and transmitted (T) intensities, both as function of θ, are plotted. The correct sketch is
(2011, Only One Option Correct Type)

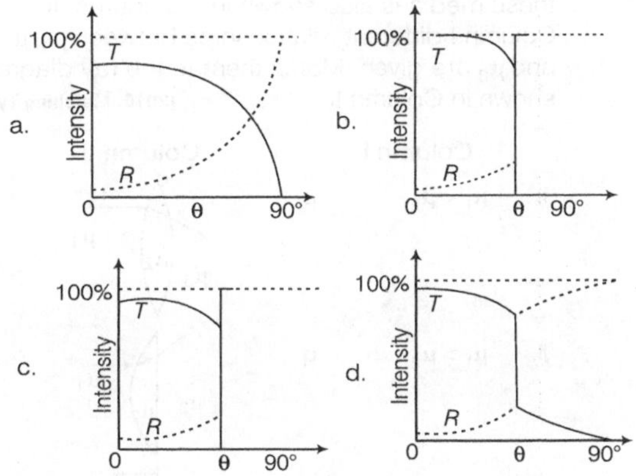

11 Water (with refractive index $= 4/3$) in a tank is 18 cm deep. Oil of refractive index $\frac{7}{4}$ lies on water making a convex surface of radius of curvature $R = 6$ cm as shown. Consider oil to act as a thin lens. An object S is placed 24 cm above water surface. The location of its image is at x cm above the bottom of the tank. Then, x is
(2011, Integer Type)

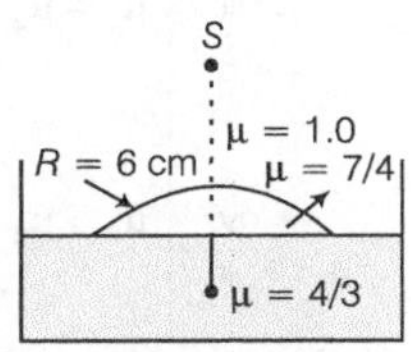

12 The focal length of a thin biconvex lens is 20 cm. When an object is moved from a distance of 25 cm in front of it to 50 cm, the magnification of its image changes from m_{25} to m_{50}. The ratio $\frac{m_{25}}{m_{50}}$ is
(2010, Integer Type)

13 Image of an object approaching a convex mirror of radius of curvature 20 m along its optical axis is observed to move from $\frac{25}{3}$ m to $\frac{50}{7}$ m in 30 s. What is the speed of the object in km h^{-1}? **(2010, Integer Type)**

14 A large glass slab $\left(\mu = \frac{5}{3}\right)$ of thickness 8 cm is placed over a point source of light on a plane surface. It is seen that light emerges out of the top surface of the slab from a circular area of radius R cm. What is the value of R? **(2010, Integer Type)**

15 A biconvex lens of focal length 15 cm is infront of a plane mirror. The distance between the lens and the mirror is 10 cm. A small object is kept at a distance of 30 cm from the lens. The final image is **(2010, Only One Option Correct Type)**

a. virtual and at a distance of 16 cm from the mirror
b. real and at a distance of 16 cm from the mirror
c. virtual and at a distance of 20 cm from the mirror
d. real and at a distance of 20 cm from the mirror

16 Two transparent media of refractive indices μ_1 and μ_3 have a solid lens shaped transparent material of refractive index μ_2 between them as shown in figures in Column II. A ray traversing these media is also shown in the figures. In Column I different relationships between μ_1, μ_2 and μ_3 are given. Match them to the ray diagram shown in Column II. **(2010, Matching Type)**

	Column I		Column II
i.	$\mu_1 < \mu_2$	p.	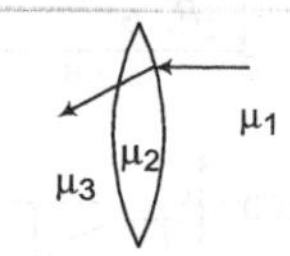
ii.	$\mu_1 > \mu_2$	q.	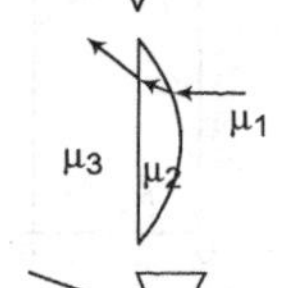
iii.	$\mu_2 = \mu_3$	r.	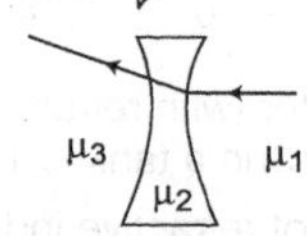
iv.	$\mu_2 > \mu_3$	s.	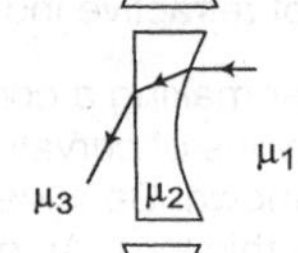
		t.	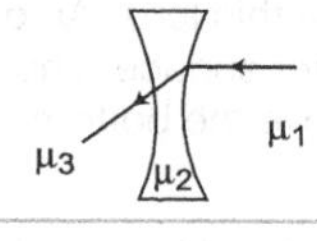

Codes

	i	ii	iii	iv
a.	p,r	q,s,t	p,r,t	q,s
b.	q	p,q	p,s	p,q
c.	r	s,t	r,q	r,s
d.	s,t	r	q,t	t,p

17 A ray OP of monochromatic light is incident on the face AB of prism $ABCD$ near vertex B at an incident angle of 60° (see figure).

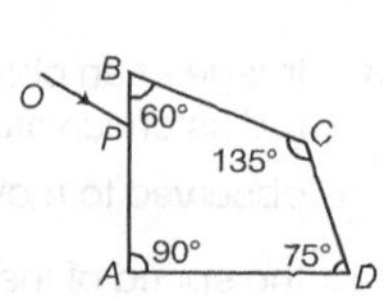

If the refractive index of the material of the prism is $\sqrt{3}$, which of the following is (are) correct? **(2010, One or More than One Options Correct Type)**

a. The ray gets totally internally reflected at face CD
b. The ray comes out through face AD
c. The angle between the incident ray and the emergent ray is 90°
d. The angle between the incident ray and the emergent ray is 120°

18 A ball is dropped from a height of 20 m above the surface of water in a lake. The refractive index of water is $\frac{4}{3}$. A fish inside the lake, in the line of fall of the ball, is looking at the ball. At an instant, when the ball is 12.8 m above the water surface, the fish sees the speed of ball as **(2009, Only One Option Correct Type)**

a. 9 ms^{-1} b. 12 ms^{-1} c. 16 ms^{-1} d. 21.33 ms^{-1}

19 A student performed the experiment of determination of focal length of a concave mirror by u-v method using an optical bench of length 1.5 m. The focal length of the mirror used is 24 cm. The maximum error in the location of the image can be 0.2 cm. The 5 sets of (u, v) values recorded by the student (in cm) are : (42, 56), (48, 48), (60, 40), (66, 33), (78, 39). The data set(s) that cannot come from experiment and is (are) incorrectly recorded, is (are) **(2009, One or More than One Options Correct Type)**

a. (42, 56) b. (48, 48) c. (66, 33) d. (78, 39)

20 An optical component and an object S placed along its optic axis are given in Column I. The distance between the object and the component can be varied. The properties of images are given in Column II. Match all the properties of images from Column II with the appropriate components given in Column I. **(2008, Matching Type)**

	Column I		Column II
i.	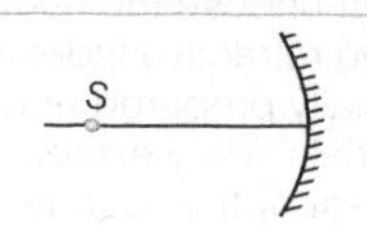	p.	Real image
ii.	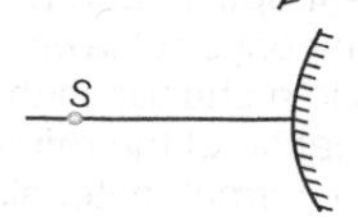	q.	Virtual image
iii.	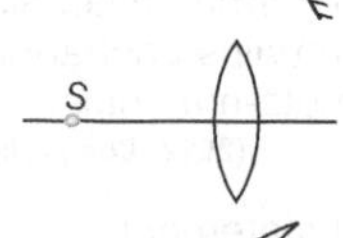	r.	Magnified image
iv.	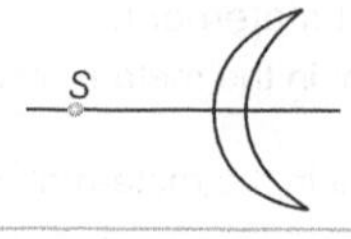	s.	Image at infinity

Codes

	i	ii	iii	iv
a.	p,q,r	q	p,q,r	p,q,r
b.	p,q	q	r	r,p
c.	r	q	p	s
d.	p,q,r,s	q	p,q,r,s	p,q,r,s

21 Two beams of red and violet colours are made to pass separately through a prism (angle of the prism is 60°). In the position of minimum deviation, the angle of refraction will be **(2008, Only One Option Correct Type)**

a. 30° for both the colours
b. greater for the violet colour
c. greater for the red colour
d. equal but not 30° for both the colours

22 A light beam is travelling from Region I to Region IV (Refer figure). The refractive index in Regions I, II, III and IV are $n_0, \frac{n_0}{2}, \frac{n_0}{6}$ and $\frac{n_0}{8}$, respectively. The angle of incidence θ for which the beam just misses entering Region IV is **(2008, Only One Option Correct Type)**

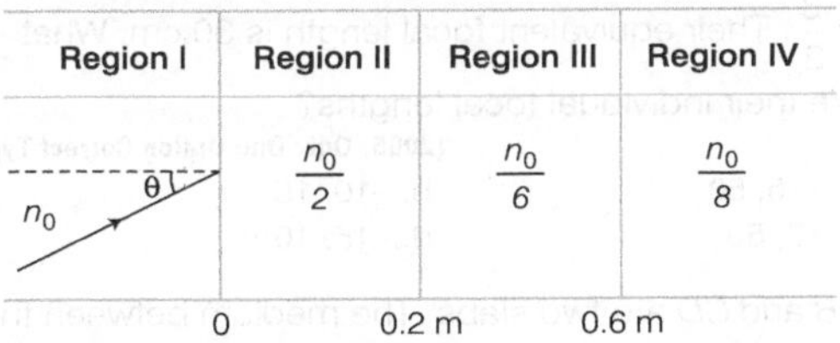

a. $\sin^{-1}\left(\frac{3}{4}\right)$
b. $\sin^{-1}\left(\frac{1}{8}\right)$
c. $\sin^{-1}\left(\frac{1}{4}\right)$
d. $\sin^{-1}\left(\frac{1}{3}\right)$

23 A ray of light travelling in water is incident on its surface open to air. The angle of incidence is θ, which is less than the critical angle. Then, there will be **(2007, Only One Option Correct Type)**

a. only a reflected ray and no refracted ray
b. only a refracted ray and no reflected ray
c. a reflected ray and a refracted ray and the angle between them would be less than $180° - 2\theta$
d. a reflected ray and a refracted ray and the angle between them would be greater than $180° - 2\theta$

24 In an experiment to determine the focal length (f) of a concave mirror by the u-v method, a student places the object pin A on the principal axis at a distance x from the pole P. The student looks at the pin and its inverted image from a distance keeping his/her eye in line with PA. When the student shifts his/her eye towards left, the image appears to the right of the object pin. Then **(2007, Only One Option Correct Type)**

a. $x < f$
b. $f < x < 2f$
c. $x = 2f$
d. $x > 2f$

Passage for Q. Nos. (25-27)

The figure shows a surface XY separating two transparent media, medium-1 and medium-2. The lines ab and cd represent wavefronts of a light wave travelling in medium-1 and incident on XY. The lines ef and gh represent wavefronts of the light wave in medium-2 after refraction. **(2007, Comprehension Type)**

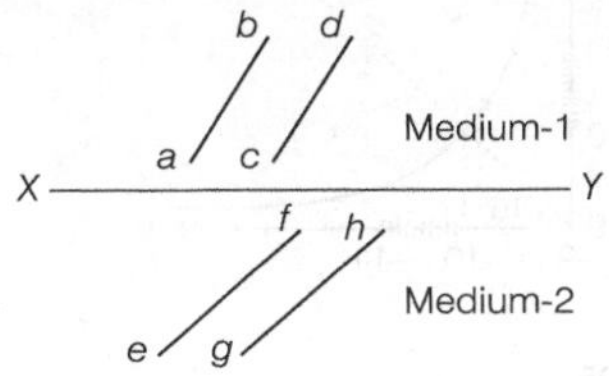

25 Light travels as a

a. parallel beam in each medium
b. convergent beam in each medium
c. divergent beam in each medium
d. divergent beam in one medium and convergent beam in the other medium

26 The phases of the light wave at c, d, e and f are ϕ_c, ϕ_d, ϕ_e and ϕ_f respectively. It is given that $\phi_c \neq \phi_f$

a. ϕ_c cannot be equal to ϕ_d
b. ϕ_d can be equal to ϕ_e
c. $(\phi_d - \phi_f)$ is equal to $(\phi_c - \phi_e)$
d. $(\phi_d - \phi_c)$ is not equal to $(\phi_f - \phi_e)$

27 Speed of light is

a. the same in medium-1 and medium-2
b. larger in medium-1 than in medium-2
c. larger in medium-2 than in medium-1
d. different at b and d

28 **Statement I** The formula connecting u, v and f for a spherical mirror is valid only for mirrors whose sizes are very small compared to their radii of curvature. **(2007, Statement Type)**

Statement II Laws of reflection are strictly valid for plane surfaces, but not for large spherical surfaces.

a. Both Statement I and Statement II are correct and Statement II is the correct explanation of Statement I
b. Both Statement I and Statement II are correct but Statement II is not the correct explanation of Statement I
c. Statement I is correct but Statement II is incorrect
d. Statement II is correct but Statement I is incorrect

29 A point object is placed at a distance of 20 cm from a thin plano-convex lens of focal length 15 cm. The plane surface of the lens is now silvered. The image created by the system is at **(2006, Only One Option Correct Type)**

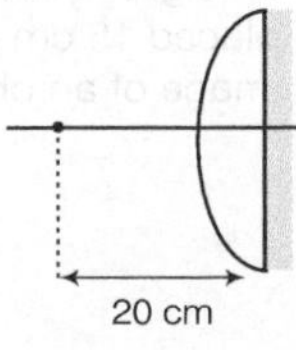

a. 60 cm to the left of the system
b. 60 cm to the right of the system
c. 12 cm to the left of the system
d. 12 cm to the right of the system

30 The graph between object distance u and image distance v for a lens is given below. The focal length of the lens is **(2006, Only One Option Correct Type)**

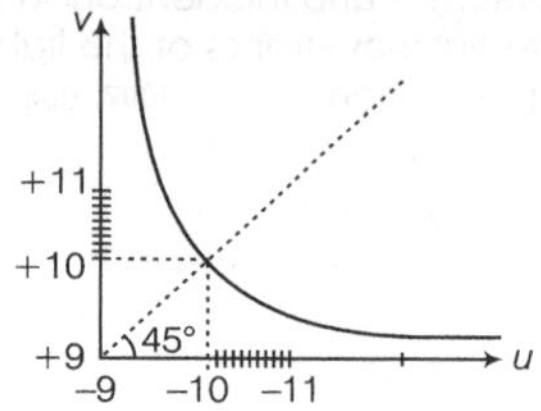

a. 5 ± 0.1
b. 5 ± 0.05
c. 0.5 ± 0.1
d. 0.5 ± 0.05

31 A biconvex lens of focal length f forms a circular image of radius r of sun in focal plane. Then which option is correct? **(2006, Only One Option Correct Type)**

a. $\pi r^2 \propto f$
b. $\pi r^2 \propto f^2$
c. If lower half part is convered by black sheet, then area of the image is equal to $\pi r^2/2$
d. If f is doubled, intensity will increase

32 Some laws/processes are given in Column I. Match these with the physical phenomena given in Column II. **(2006, Matching Type)**

	Column I		Column II
i.	Intensity of light received by lens	p.	radius of aperture (R)
ii.	Angular magnification	q.	dispersion of lens
iii.	Length of telescope	r.	focal length f_o, f_e
iv.	Sharpness of image	s.	spherical aberration

Codes

	i	ii	iii	iv
a.	p	r	r	p,q,r
b.	q	p	s	r,p
c.	s	s	q,s	r,s
d.	r	r	p	q

33 A container is filled with water ($\mu = 1.33$) upto a height of 33.25 cm. A concave mirror is placed 15 cm above the water level and the image of an object placed at the bottom is formed 25 cm below the water level. The focal length of the mirror is **(2005, Only One Option Correct Type)**

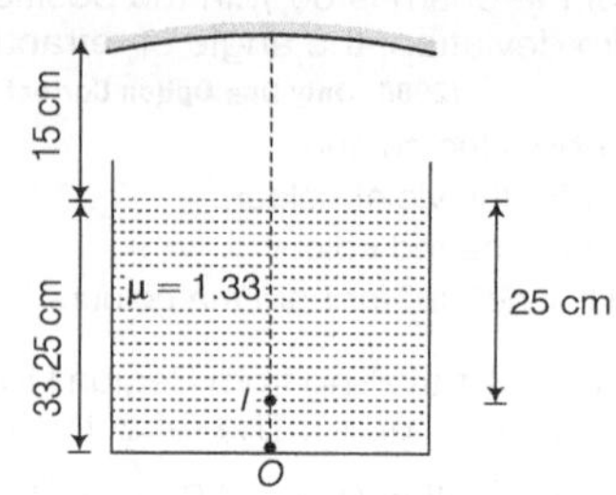

a. 10 cm
b. 15 cm
c. 20 cm
d. 25 cm

34 A convex lens is in contact with concave lens. The magnitude of the ratio of their focal length is $\frac{3}{2}$. Their equivalent focal length is 30 cm. What are their individual focal lengths? **(2005, Only One Option Correct Type)**

a. −75, 50
b. −10, 15
c. 75, 50
d. −15, 10

35 AB and CD are two slabs. The medium between the slabs has refractive index 2. Find the minimum angle of incidence of Q, so that the ray is totally reflected by both the slabs. **(2005, Subjective Type)**

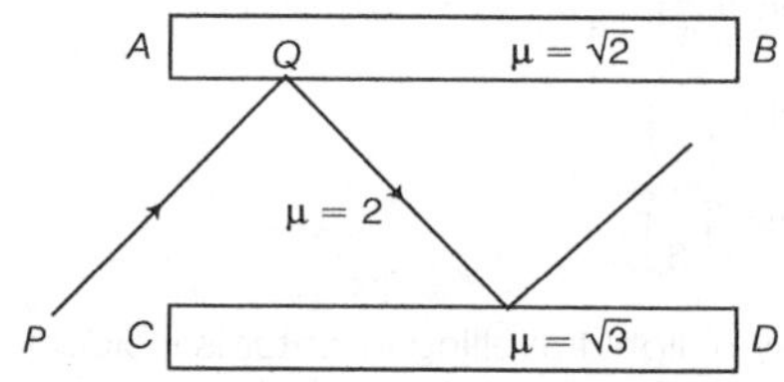

36 A ray of light is incident on a prism ABC of refractive index $\sqrt{3}$ as shown in the figure. **(2005, Subjective Type)**

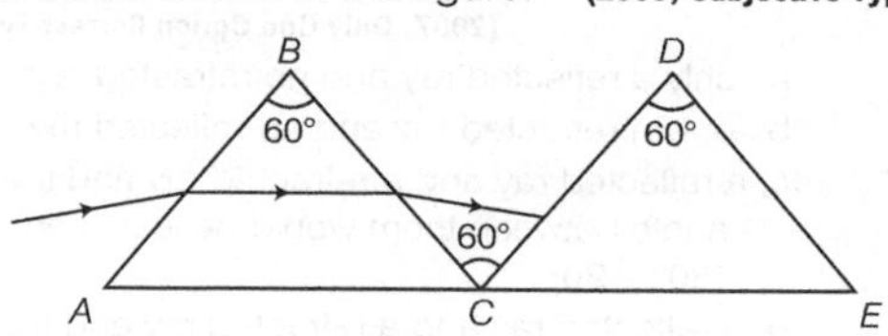

i. Find the angle of incidence for which the deviation of light ray by the prism ABC is minimum.
ii. By what angle the second identical prism must be rotated, so that the final ray suffers net minimum deviation.

2. Wave Optics

DPP

DAILY PRACTICE PROBLEMS

DPP-1 Electomagnetic Waves

Only One Option Correct Type

Working Space

Direction *(Q. Nos. 1-16) This section contains 16 multiple choice questions. Each question has four choices (a), (b), (c) and (d), out of which ONLY ONE is correct.*

1 A linearly polarised electromagnetic wave given as, $\mathbf{E} = E_0 \sin(kz - \omega t)\,\hat{\mathbf{i}}$ is incident from vacuum on the surface of glass normally. The reflected wave is given as

a. $\mathbf{E}_r = -E_0 \sin(kz + \omega t)\,\hat{\mathbf{i}}$

b. $\mathbf{E}_r = -E_0 \cos(kz - \omega t)\,\hat{\mathbf{i}}$

c. $\mathbf{E}_r = -E_0 \cos(kz + \omega t)\,\hat{\mathbf{i}}$

d. $\mathbf{E}_r = -E_0 \sin(kz - \omega t)\,\hat{\mathbf{i}}$

2 When light energy of 60 W incident normally on a plane mirror. The total momentum delivered is

a. 20×10^{-7} kg ms^{-1} b. 2×10^{-7} kg ms^{-1}

c. 60 kg ms^{-1} d. 60×10^{-8} kg ms^{-1}

3 The electric field intensity produced by the radiation coming from the 100 W bulb at a distance 1 m is E. The electric field at distance 2 m is

a. E b. $2E$

c. $\frac{E}{2}$ d. $\frac{E}{4}$

4 If $\hat{\mathbf{E}}$ and $\hat{\mathbf{B}}$ represent unit vectors along electric field vector and magnetic field vector. The unit vector along direction of propagation of electromagnetic wave is $\hat{\mathbf{c}}$. Then,

a. $\hat{\mathbf{c}} = (\hat{\mathbf{E}} \cdot \hat{\mathbf{B}})\hat{\mathbf{B}}$ b. $\hat{\mathbf{c}} = (\hat{\mathbf{E}} \times \hat{\mathbf{B}})$

c. $\hat{\mathbf{c}} = (\hat{\mathbf{B}} \times \hat{\mathbf{E}})$ d. None of these

5 Intensity of solar radiation on the surface of the earth is 1400 Wm^{-2}. The amplitude of electric field and magnetic field in solar radiation is

a. 1.03×10^3 N/C, 3.4 μT b. 1.02×10^3 N/C, 3 μT

c. 5×10^3 N/C, 3.8 μT d. None of these

6 If space between plates of parallel capacitor is filled with poorly conducting material. If an AC source is connected across the capacitor, current through material is

a. only conduction current
b. only displacement current
c. displacement current as well as conduction current
d. None of the above

7 During the propagation of electromagnetic waves in a medium **[JEE Main 2014]**

a. electric energy density is double of the magnetic energy density
b. electric energy density is half of the magnetic energy density
c. electric energy density is equal to the magnetic energy density
d. Both electric and magnetic energy densities are zero

Working Space

8 The magnetic field in a travelling electromagnetic wave has a peak value of 20 nT. The peak value of electric field strength is **[JEE Main 2013]**

a. $3\ Vm^{-1}$
b. $6\ Vm^{-1}$
c. $9\ Vm^{-1}$
d. $12\ Vm^{-1}$

9 An electromagnetic wave in vacuum has the electric and magnetic fields **E** and **B**, which are always perpendicular to each other. The direction of polarisation is given by **X** and that of wave propagation by **k**. Then **[AIEEE 2012]**

a. $\mathbf{X} \parallel \mathbf{B}$ and $\mathbf{k} \parallel \mathbf{B} \times \mathbf{E}$
b. $\mathbf{X} \parallel \mathbf{E}$ and $\mathbf{k} \parallel \mathbf{E} \times \mathbf{B}$
c. $\mathbf{X} \parallel \mathbf{B}$ and $\mathbf{k} \parallel \mathbf{E} \times \mathbf{B}$
d. $\mathbf{X} \parallel \mathbf{E}$ and $\mathbf{k} \parallel \mathbf{B} \times \mathbf{E}$

10 The rms value of the electric field of the light coming from the Sun is $720\ NC^{-1}$. The average total energy density of the electromagnetic wave is **[AIEEE 2006]**

a. $4.58 \times 10^{-6}\ Jm^{-3}$
b. $6.37 \times 10^{-9}\ Jm^{-3}$
c. $81.35 \times 10^{-12}\ Jm^{-3}$
d. $3.3 \times 10^{-3}\ Jm^{-3}$

11 An electromagnetic wave of frequency $v = 3.0$ MHz passes from vacuum into a dielectric medium with permittivity $\varepsilon = 4.0$. Then **[AIEEE 2004]**

a. wavelength is doubled and the frequency remains unchanged
b. wavelength is doubled and frequency becomes half
c. wavelength is halved and frequency remains unchanged
d. wavelength and frequency both remains unchanged

12 A radiation of energy E falls normally on a perfectly reflecting surface. The momentum transferred to the surface is **[AIEEE 2004]**

a. $\frac{E}{c}$
b. $\frac{2E}{c}$
c. Ec
d. $\frac{E}{c^2}$

13 Which of the following radiations has the least wavelength? **[AIEEE 2003]**

a. γ-rays
b. β-rays
c. α-rays
d. X-rays

14 Which of the following are not electromagnetic waves? **[AIEEE 2003]**

a. Cosmic rays
b. γ-rays
c. β-rays
d. X-rays

15 Electromagnetic waves are transverse in nature is evident by

a. polarisation
b. interference
c. reflection
d. diffraction

16 The relative permittivity and relative permeability of a medium are ε_r and μ_r respectively. Find the absolute refractive index of the medium.

a. $\frac{1}{\sqrt{\mu_r \varepsilon_r}}$
b. $\sqrt{\mu_r \varepsilon_r}$
c. $\sqrt{\frac{\varepsilon_r}{\mu_r}}$
d. $\mu_r \varepsilon_r$

One or More than One Options Correct Type

Direction *(Q. Nos. 17-21) This section contains 5 multiple choice questions. Each question has four choices (a), (b), (c) and (d), out of which ONE or MORE THAN ONE are correct.*

17 To dissociate a molecule of carbon monoxide into carbon and oxygen atoms, 11 eV energy is required.

a. The maximum wavelength required is 112.91 nm
b. The electromagnetic radiation required in visible region
c. The electromagnetic radiation required in ultraviolet-region
d. The electromagnetic radiation required in microwave region

Working Space

18 If electromagnetic wave is given as $\mathbf{E} = E_0 \sin(\omega t - kz)\,\hat{\mathbf{i}}$. Then, electromagnetic wave in the form of magnetic field is

a. $\mathbf{B} = \frac{E_0}{c}\sin(\omega t - kz)\,\hat{\mathbf{i}}$
b. $\mathbf{B} = \frac{E_0}{c}\sin(\omega t - kz)\,\hat{\mathbf{j}}$
c. $\mathbf{B} = \frac{E_0}{c}\cos(\omega t - kz)\,\hat{\mathbf{j}}$
d. $\mathbf{B} = -\frac{E_0}{c}\sin(\omega t - kz)\,\hat{\mathbf{j}}$

19 If an electromagnetic wave is travelling along X-axis. Then,

a. $\mathbf{E}\cdot\hat{\mathbf{i}} = 0$
b. $\mathbf{E}\cdot\hat{\mathbf{i}} \neq 0$
c. $\mathbf{B}\cdot\hat{\mathbf{i}} = 0$
d. $\mathbf{B}\cdot\hat{\mathbf{i}} \neq 0$

20 An electromagnetic wave is generated due to oscillation of a charged particle with a frequency 10^9Hz.

a. The wavelength of electromagnetic wave in vacuum is 30 cm
b. This wave is radio wave
c. This wave is microwave
d. This wave is light wave

21 Mark the correct option(s).

a. A charged particle moving with constant speed may produce electromagnetic wave
b. A charged particle moving with constant velocity may be produce electromagnetic wave
c. A charged particle moving an circular path produces electromagnetic wave
d. A charged particle moving under gravity produces electromagnetic wave

Statement Type

***Direction** (Q. Nos. 22-25) This section is based on Statement I and Statement II. Select the correct answer from the codes given below.*

a. Both Statement I and Statement II are correct and Statement II is the correct explanation of Statement I
b. Both Statement I and Statement II are correct but Statement II is not the correct explanation of Statement I
c. Statement I is correct but Statement II is incorrect
d. Statement II is correct but Statement I is incorrect

22 **Statement I** The antenna of portable radio should be parallel to electric and magnetic part of the electromagnetic wave.

Statement II Electromagnetic wave is plane polarised.

23 **Statement I** Microwave oven heats up a food item containing water most sufficiently.

Statement II The frequency of the microwave matches the resonant frequency of water molecules.

24 **Statement I** A variable frequency alternating current source is connected to a parallel plate capacitor. When the frequency of AC source increases, displacement current decreases.

Statement II Reactance of capacitor is $X_C = \frac{1}{\omega C}$.

25 **Statement I** If a laser beam incident on a tiny ball, it may suspended freely in vacuum.

Statement II Tails of comets are due to solar radiation.

Working Space

Matching List Type

Direction *(Q. Nos. 26-30) Choices for the correct combination of elements from Column I and Column II are given as options (a), (b), (c) and (d), out of which one is correct.*

26 Match the Column I with Column II and mark the correct option from the codes given below.

	Column I		Column II
i.	Electromagnetic wave used in communication satellite	p.	Ultraviolet
ii.	Electromagnetic wave used to kill germs in water purifies	q.	Infrared
iii.	Electromagnetic wave used to detect leakage of oil in underground pipelines	r.	Microwaves
iv.	Electromagnetic wave used to improve visibilty in run ways during fog and mist condition	s.	X-ray

Codes

	i	ii	iii	iv		i	ii	iii	iv
a.	p	q	r	s	b.	r	p	s	q
c.	r	s	p	q	d.	s	p	q	r

27 Match the Column I with Column II and mark the correct option from the codes given below.

	Column I		Column II
i.	Production of radio wave	p.	Vibration of atoms and molecules
ii.	Production of microwave	q.	Klystron valve or magnetron valve
iii.	Production of infrared	r.	Rapid acceleration and deceleration of electrons in aerials
iv.	Light	s.	Electrons in atom jump from higher orbit to lower orbit

Codes

	i	ii	iii	iv		i	ii	iii	iv
a.	r	q	p	s	b.	p	q	r	s
c.	q	r	p	s	d.	p	q	s	r

28 Match the Column I with Column II and mark the correct option from the codes given below.

	Column I		Column II
i.	Detection of radio wave	p.	Human eye
ii.	Detection of microwave	q.	Point contact diodes
iii.	Detection of infrared	r.	Bolometer
iv.	Detection of light	s.	Receivers aerials

Codes

	i	ii	iii	iv		i	ii	iii	iv
a.	s	q	r	p	b.	p	q	r	s
c.	q	p	s	r	d.	p	q	s	r

29 Match the Column I with Column II and mark the correct option from the codes given below.

	Column I		Column II
i.	Radio wave	p.	$\lambda < 10^{-3}$nm
ii.	Microwave	q.	$1\text{nm} > \lambda > 10^{-3}$nm
iii.	X-ray	r.	> 0.1m
iv.	Gamma ray	s.	$0.1\text{ m} > \lambda > 1\text{mm}$

Codes

	i	ii	iii	iv		i	ii	iii	iv
a.	p	q	r	s	b.	q	p	s	r
c.	r	s	q	p	d.	p, r	q, s	q, r	p, s

Working Space

30 Match the Column I (electromagnetic wave type) with Column II (Its association/application) and mark the correct option from the codes given below.

	Column I		Column II
i	Infrared waves	p.	To treat muscular strain
ii	Radio waves	q.	For broadcasting
iii	X-rays	r.	To detect fracture of bones
iv	Ultraviolet	s.	Absorbed by the ozone layer of the atmosphere

Codes

	i	ii	iii	iv		i	ii	iii	iv
a.	s	r	q	p	b.	p	q	s	r
c.	r	q	p	s	d.	p	q	r	s

One Integer Value Correct Type

Direction (Q. Nos. 31-35) *This section contains 5 questions. When worked out will result in an integer from 0 to 9 (both inclusive).*

31 A parallel plate capacitor with plate area A is charged as such it has charge $q = I_0 t$ at instant t. Consider a plane surface of area $\frac{A}{2}$ parallel to plates and drawn symmetrically between plates. If displacement current through the area is $\frac{I_0}{n}$.

Find the value of n.

32 A 6×10^{-12}F parallel plate capacitor has circular plates of area 10^{-2}m^2. It is charged with a charge of 42×10^{-11}C charge. Now, it plates are connected through a resistor of resistance 2Ω. The rate of change of electric field just after connecting with resistor is 395.48×10^{2n} $\text{Vm}^{-1}\text{s}^{-1}$. Find the value of n.

33 A circular cross-section conductor of radius 1.5 mm carries a current $i = 5.5 \sin (4 \times 10^{10} t)\ \mu$A. The conductivity of conductor is 35×10^6 s/m and relative permittivity is 1. If the amplitude of displacement current density is $\frac{7.86}{10^n}\ \mu\ \text{Am}^{-2}$. Find the value of n.

34 In a material for which conductivity $\sigma = 5$ s/m and relative permittivity, $\varepsilon_r = 1$, electric field is $E = E_0 \sin \omega t$. For $\omega = 565 \times 10^n$ rads^{-1}, displacement current density and conduction current density are same. Find the value of n.

35 A coaxial cylindrical capacitor with inner radius 5 mm, outer radius 6 mm and length 50 cm has dielectric for which $\varepsilon_r = 6.7$. An AC voltage $V = 250 \sin (377\,t)$ V is applied across capacitor. If displacement current amplitude is $\frac{9.63}{10^n}$ A. Find the value of n.

DPP-2 Interference of Light-I

Conceptual Questions

Working Space

Direction (*Q. Nos.* 1 *and* 2) *These questions are conceptual in nature, need to be solved completely on notebook.*

1 Situation A beam of light consisting of two wavelengths is used to obtain interference fringes in Young's double slits experiment, the separation between slits is 2 mm and the distance between the plane of the slits and screen is 120 cm.

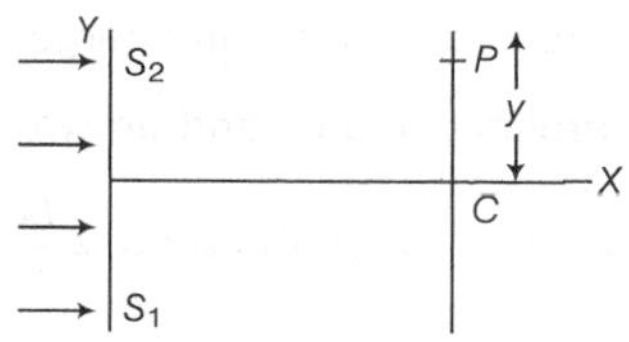

i. The shape of central bright fringe on the screen is
   a. straight line
   b. hyperbolic
   c. circular
   d. None of these

ii. If incident beam is monochromatic, the shape of fringes on the screen is
   a. hyperbolic but central fringe is straight line along Z-axis
   b. circular
   c. straight line
   d. hyperbolic but central bright fringe is straight line along Y-axis

iii. If S_1 and S_2 are broad sources of monochromatic light, the broadness of slits S_1 and S_2 are along Z-axis the shape of fringes on the screen is
   a. hyperbolic
   b. hyperbolic but central bright fringes is straight line
   c. straight line parallel to Y-axis
   d. straight line parallel to Z-axis

iv. The path difference at point P is
   a. $\Delta x = \frac{yd}{D}$
   b. $\Delta x = \frac{yD}{d}$
   c. $\Delta x = \frac{yd}{D} + \bar{2}$
   d. None of these

v. Light consists of wavelength $\lambda_1 = 650$ nm and 520 nm. If P is point of third maxima for wavelength $\lambda_1 = 650$ nm. Find the value of y.
   a. 2×10^{-3}m
   b. 1.17×10^{-3}m
   c. 1.56×10^{-3}m
   d. None of these

vi. In preview problem, the order of fringes coincide are
   a. 4th and 5th
   b. 8th and 10th
   c. 4th and 10
   d. None of these

vii. In previous problem, the least distance from the central maximum, when the bright fringes due to both the wavelength coincide is
   a. $y = 1.17 \times 10^{-3}$ m
   b. $y = 1.56 \times 10^{-3}$ m
   c. $y = 6.5 \times 10^{-3}$ m
   d. $y = 5.2 \times 10^{-3}$ m

Working Space

2 **Situation** A source of light of wavelength $\lambda = 500$ mm is placed in situation as shown in the figure. Considering interference of direct and reflected rays.

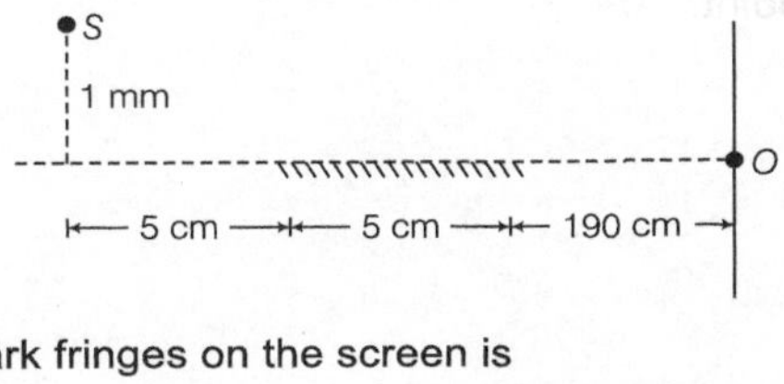

i. The shape of dark fringes on the screen is

a. hyperbolic
b. straight line
c. circular
d. None of these

ii. The position of image of S by the mirror is

a. 2 mm below S
b. 1 mm below
c. 4 mm below S
d. None of these

iii. The fringe width on the screen is

a. 0.5 mm
b. 2 mm
c. 40 mm
d. None of these

iv. The length of fringe pattern on the screen is

a. 5 cm
b. 2 cm
c. 4 cm
d. None of these

v. The number of bright fringe on the screen is

a. 20
b. 80
c. 40
d. 39

vi. The path difference at point O is

a. zero
b. $\frac{\lambda}{2}$
c. λ
d. None of these

vii. The point O is a point of

a. bright fringe
b. dark fringe
c. either bright or dark fringe
d. None of these

Only One Option Correct Type

Direction (Q. Nos. 3-42) *This section contains 40 multiple choice questions. Each question has four choices (a), (b), (c) and (d), out of which ONLY ONE is correct.*

3 If the first order minima in Young's double slit experiment occurs direct by infront of one of the slits. If distance between slits is d and the distance between slits and screen is D, then the wavelength of used light is

a. $\frac{d^2}{2D}$
b. $\frac{d^2}{D}$
c. $\frac{d^2}{3D}$
d. $\frac{d^2}{4D}$

4 Consider an Young's double slit experiment that has different slits width as a result, amplitudes of waves from two slits are I_0 and $4I_0$, then the intensity at a point on screen, where phase difference between waves is $\frac{2\pi}{3}$ is

a. I_0
b. $2I_0$
c. $3I_0$
d. $4I_0$

Working Space

5 Three light waves combine at a certain point, where their electric field components are $E_1 = E_0 \cos \omega t$, $E_2 = E_0 \sin(\omega t + 53°)$, $E_3 = E_0 \sin(\omega t - 37°)$. Find their resultant component $E(t)$ at that point.

a. $\frac{\sqrt{85}}{5} \sin\left(\omega t + \tan^{-1}\frac{6}{7}\right)$

b. $E_0 \sin(\omega t + 60°)$

c. $E = \frac{\sqrt{85}}{5} \sin\left(\omega t + \tan^{-1}\frac{7}{6}\right)$

d. None of the above

6 Find a unit vector along ray for a wavefront, whose equation is $x^2 + y^2 - z = 6$ at the point (2, 3, 7).

a. $\frac{4\hat{i} + 6\hat{j} - \hat{k}}{\sqrt{53}}$

b. $\frac{2\hat{i} + 2\hat{j} - \hat{k}}{3}$

c. $\frac{4\hat{i} + 6\hat{j} - \hat{k}}{\sqrt{53}}$

d. None of these

7 A wavefront AB passing through a system C emerges as DE as shown in the figure. The system C could be

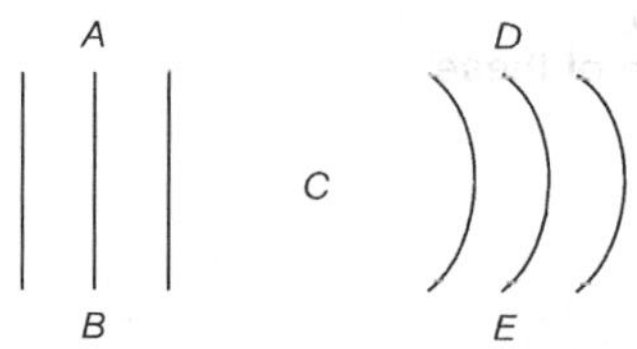

a. converging lens

b. diverging lens

c. concave mirror

d. convex mirror

8 The wavefront of light beam is given by the equation $2x + y + 2z = C$, where C is arbitrary constant. The angle made by the direction of light with X-axis is

a. $\cos^{-1}\left(\frac{2}{3}\right)$

b. $\sin^{-1}\left(\frac{2}{3}\right)$

c. $\cos^{-1}\left(\frac{1}{3}\right)$

d. $\cos^{-1}\left(\frac{2}{5}\right)$

9 If two independent sodium vapour lamps are used as sources in Young's double slit experiment. Then,

a. interference pattern is disappeared

b. only dark fringes are formed

c. interference pattern becomes more sharp

d. None of the above

10 Following transverse waves,

$$y_1 = 2\sin(100\pi t - 5.3x),$$

$$y_2 = 2\sqrt{2}\sin\left(100\pi t - 5.3x + \frac{\pi}{4}\right)$$

and

$$y_3 = \cos(100\pi t - 5.3x)$$

superpose in a homogenous medium.

Find the resultant amplitude at $x = 0$.

a. 1 unit

b. 4 unit

c. 5 unit

d. 2 unit

Working Space

11 A monochromatic light of wavelength is incident at an angle α with the normal to a plane containing two slits of separation d. Find the minimum value of α, so that a minima is formed at centre of screen.

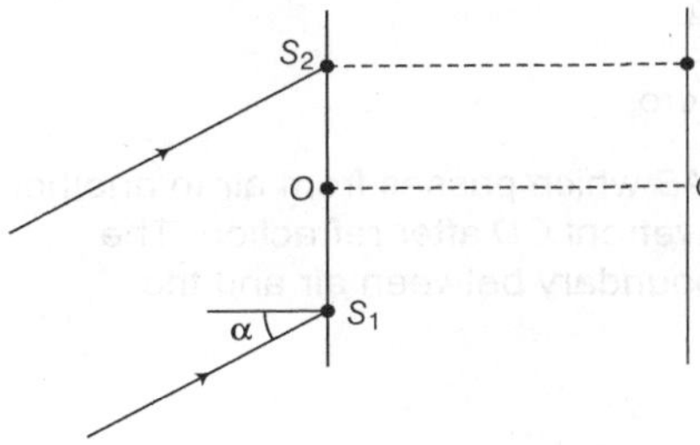

a. $\cos^{-1}\left(\frac{\lambda}{d}\right)$ b. $\sin^{-1}\left(\frac{\lambda}{2d}\right)$

c. $\sin^{-1}\left(\frac{\lambda}{d}\right)$ d. $\sin^{-1}\left(\frac{\lambda}{4d}\right)$

12 In Young's double slit experiment, the slit S_1 and S_2 are illuminated with coherent microwave sources, each of frequency 10^6 Hz. The initial phase difference between them is zero. The separation between slits is d = 150m. The intensity on screen due to each source is 1 Wm^{-2}. Find intensity on screen for $\theta = 90°$.

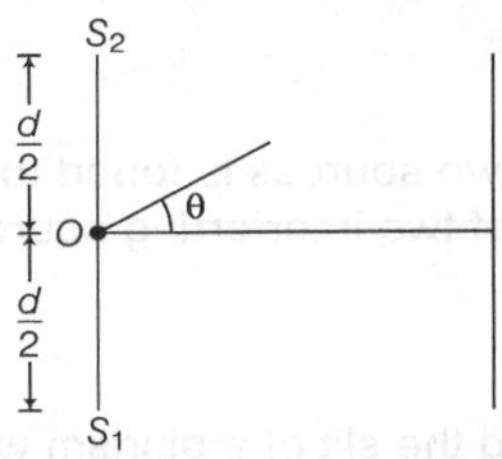

a. 1Wm^{-2} b. 2 Wm^{-2}

c. 3 Wm^{-2} d. 4 Wm^{-2}

13 Slits in Young's double slit experiment are at 1 mm separation. They are illuminated by white light (400 nm to 700 nm) on screen 1 m away interference pattern is formed. If a pinhole is made on this screen at a distance 2 mm from the central white fringe, some wavelengths will be absent in the transmitted light. The absent wavelengths are

a. 571.4 nm and 444.4 nm b. 500 nm and 450 nm

c. 402 nm, 507 nm and 609 nm d. 550 nm, 600 nm and 680 nm

14 In Young's double slits experiment, when a transparent sheet of refractive index 1.4 is introduced in the path of one of the interferring waves of wavelength 500 nm, the intensity at the central position of the screen remains unchanged. The minimum thickness of the sheet is

a. 2 nm b. 200 nm

c. 1250 nm d. 2500 nm

15 The slits in a double slit experiment are illuminated by light of wavelength 600 nm. A thin transparent sheet of thickness t is placed infront of one of this slits. The number of fringe shifting, $N = 20\mu - 20$, where μ is refractive index of the transparent sheet of glass. For $2 \leq \mu \leq 1$, the value of t is

a. 24×10^{-6}m b. 4.8 mm

c. 4.5 mm d. 24 mm

Working Space

16 If path difference at a point is λ in YDSE, the intensity is I_0. Find the intensity at a point, where path difference is 5λ.

a. I_0 b. $\frac{I_0}{2}$

c. $5I_0$ d. zero

17 The following figure represents a wavefront *AB* which passes from air to another transparent medium and produces a new wavefront *CD* after refraction. The refractive index of the medium is (*PQ* is the boundary between air and the medium).

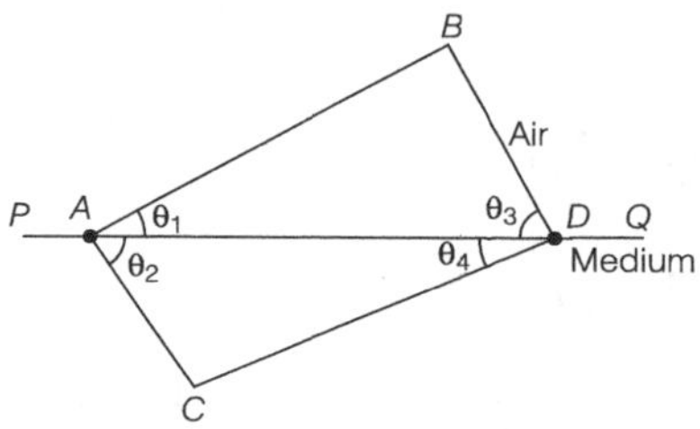

a. $\frac{\cos\theta_1}{\cos\theta_4}$ b. $\frac{\cos\theta_4}{\cos\theta_1}$

c. $\frac{\sin\theta_1}{\sin\theta_4}$ d. $\frac{\sin\theta_2}{\sin\theta_3}$

18 In YDSE, the amplitude of intensity variation of the two sources is found to be 5% of the average intensity. The ratio of the intensities of two interferring sources is

a. 2564 b. 1089

c. 1681 d. 869

19 When interference pattern is obtained by illuminating the slit of a biprism with white light,

a. the central band will be red

b. the interference band next to the central band will be red

c. the interference band next to the central band will be violet

d. a completely dark band cannot be observed

20 The figure shows a plane wavefront at a time *t* and at a time t_1. In the time interval $(t_1 - t)$ the wavefront must have passed through

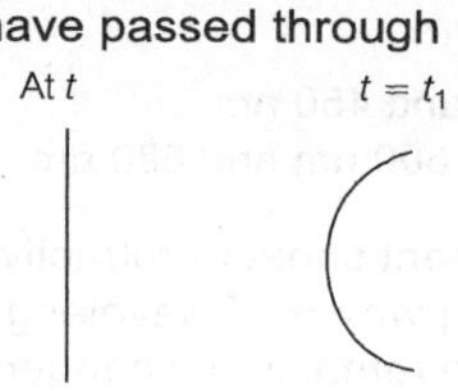

a. a prism

b. a prism and a convex lens

c. a convex lens

d. a plane mirror and a concave lens

21 In a two slit experiment with monochromatic light, fringes are obtained on a screen placed at some distance from the slits. If screen is moved by 5×10^{-2} m towards the slits, then change in fringe width is 3×10^{-5} m. If the distance between slits is 10-3 m, then wavelength of the light used will be

a. 4000 Å b. 6000 Å

c. 5890 Å d. 8000 Å

Working Space

22 Figure shows two coherent sources S_1 and S_2 that emits radio waves of wavelength λ in all direction. The sources are exactly in phase and are separated by a distance equal to 1.5 λ. The vertical broken line is the perpendicular bisector of the distance between the sources. If we start at the indicated start point and travel along path as shown in the figure.

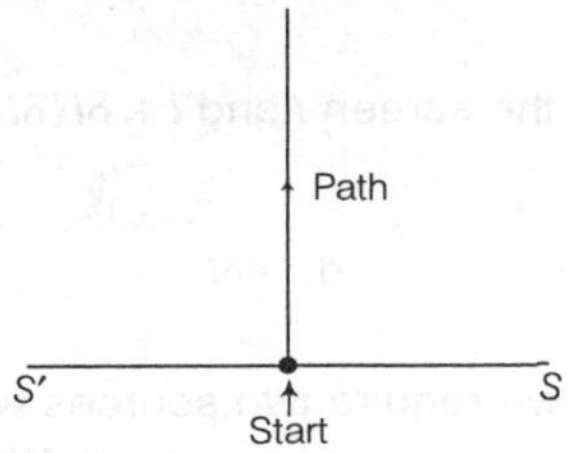

a. At every point of path, maxima is found
b. Maxima and minima are found alternatively
c. At every point, only minima is found
d. A number of points are found, where intensity is zero

23 The propagation of wavefront from air to glass as shown in the figure. As, the wave moves into the glass, a Huygens' wavelet at point *e* will expand to pass through point *C* at a distance of λ_1 (wavelength of light in air) from point *e*. In the same time interval, a Huygens' wavelet a point *h* will expand to pass through point *g* in glass. The distance *hg* is

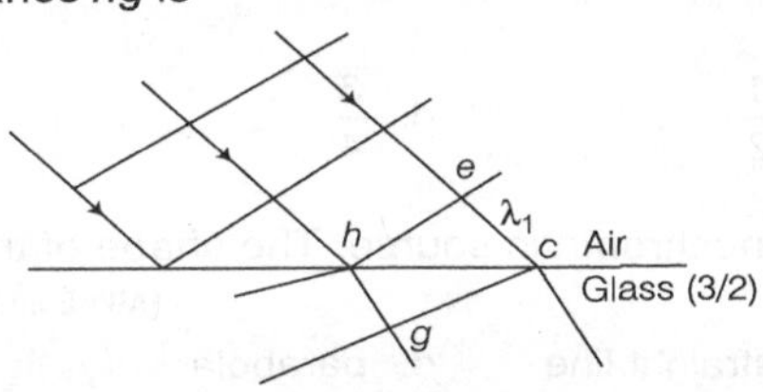

a. λ_1
b. $\frac{2\lambda_1}{3}$
c. $\frac{3\lambda_1}{2}$
d. Cannot be determined

24 In a Young's double slit experiment, the slits are illuminated by monochromatic light. The entire set up is immersed in pure water. Which one of the following act can not be restore the original fringe width?
a. Bringing the slits close together
b. Moving the screen away from the slit plane
c. Replacing the incident light of a longer wavelength
d. Introducing a thin transparent slab infront of one of the slits

25 Two coherent sources *A* and *B* emitting light of wavelength λ are placed at position $(-D, 0)$ and $(-D, 3\lambda)$ respectively. Here, $D >> \lambda$. The number of minima on *Y*-axis and maxima on *X*-axis are respectively.

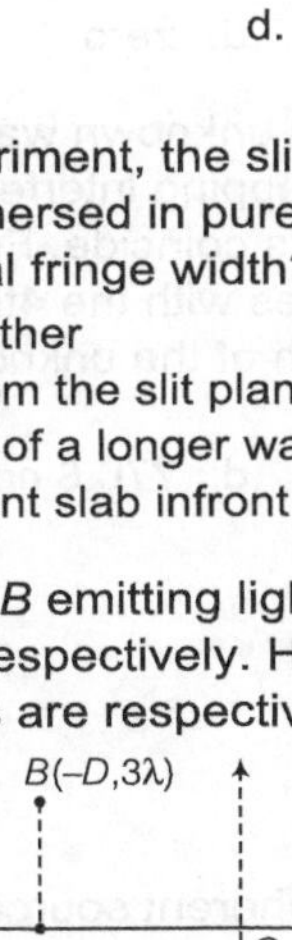

a. 6, 5
b. 5, 6
c. 4, 3
d. 3, 4

Working Space

26 In a Young's double slit experiment, if the slits are of unequal width,

a. fringes will not be formed
b. the positions of minimum intensity will not be completely dark
c. bright fringe will not be formed at the centre of the screen
d. distance between two consecutive bright fringes will not be equal to the distance between two consecutive dark fringes

27 In a YDSE set up the intensity due to 2 beams on the screen I and $I + \delta I (\delta I << I)$. The intensity at minima is

a. zero b. $\frac{4(\delta I)^2}{I}$ c. $\frac{(\delta I)^2}{4I}$ d. $4\delta I$

28 To demonstrate the phenomenon of interference, we require two sources which emit radiations of **[AIEEE 2003]**

a. nearly the same frequency
b. the same frequency
c. different wavelength
d. the same frequency and having a definite phase relationship

29 In a Young's double slit experiment, the intensity at a point where the path difference is $\frac{\lambda}{6}$ (λ being the wavelength of the light used) is I. If I_0 denotes the maximum intensity, then $\frac{I}{I_0}$ is equal to

a. $\frac{1}{\sqrt{2}}$ b. $\frac{\sqrt{3}}{2}$ c. $\frac{1}{2}$ d. $\frac{3}{4}$

30 A Young's double slit experiment uses a monochromatic source. The shape of the interference fringes formed on a screen is **[AIEEE 2005]**

a. hyperbola b. circle c. straight line d. parabola

31 The maximum number of possible interference maxima for slit separation equal to twice the wavelength in Young's double slit experiment, is **[AIEEE 2004]**

a. infinite b. five c. three d. zero

32 A mixture of light consisting of wavelength 590 nm and an unknown wavelength, illuminates Young's double slit and gives rise to two overlapping interference patterns on the screen. The central maximum of both lights coincide. Further, it is observed that the third bright fringe of known light coincides with the 4th bright fringe of the unknown light. From this data, the wavelength of the unknown light is **[AIEEE 2009]**

a. 393.4 nm b. 885.0 nm c. 442.5 nm d. 776.8 nm

33 The initial shape of the wavefront of the beam is **[AIEEE 2010]**

a. convex
b. concave
c. convex near the axis and concave near the periphery
d. planar

34 In a Young's double slit experiment, the two slits act as coherent sources of waves of equal amplitude A and wavelength λ. In another experiment with the same arrangement, the two slits are made to act as incoherent sources of waves of same amplitude and wavelength. If the intensity at the middle point of the screen in the first case is I_1 and in the second case is I_2, then the ratio $\frac{I_1}{I_2}$ is **[AIEEE 2011]**

a. 4 b. 2 c. 1 d. 0.5

Working Space

35 At two points P and Q on screen in Young's double slit experiment, waves from slits S_1 and S_2 have a path difference of 0 and $\frac{\lambda}{4}$, respectively. The ratio of intensities at P and Q will be **[AIEEE 2011]**

a. 3 : 2
b. 2 : 1
c. $\sqrt{2}:1$
d. 4 : 1

36 In Young's double slit experiment, one of the slits is wider than other, so that amplitude of the light from one slit is double of that from other slit. If I_m is the maximum intensity, the resultant intensity I when they interfere at phase difference ϕ, is given by **[AIEEE 2012]**

a. $\frac{I_m}{9}(4+5\cos\phi)$
b. $\frac{I_m}{3}\left(1+2\cos^2\frac{\phi}{2}\right)$
c. $\frac{I_m}{5}\left(1+4\cos^2\frac{\phi}{2}\right)$
d. $\frac{I_m}{9}\left(1+8\cos^2\frac{\phi}{2}\right)$

37 In Young's double slit experiment intensity at a point is $\frac{1}{4}$ of the maximum intensity. Angular position of this point is **[IIT JEE 2005]**

a. $\sin^{-1}\left(\frac{\lambda}{d}\right)$
b. $\sin^{-1}\left(\frac{\lambda}{2d}\right)$
c. $\sin^{-1}\left(\frac{\lambda}{3d}\right)$
d. $\sin^{-1}\left(\frac{\lambda}{4d}\right)$

38 In Young's double slit experiment is carried out by using green, red and blue light, one colour at a time. The fringe widths recorded are β_G, β_R and β_B respectively. Then, **[IIT JEE 2012]**

a. $\beta_G > \beta_B > \beta_R$
b. $\beta_B > \beta_G > \beta_R$
c. $\beta_R > \beta_B > \beta_G$
d. $\beta_R > \beta_G > \beta_B$

39 In the ideal double slit experiment, when a glass-plate (refractive index 1.5) of thickness t is introduced in the path of one of the interferring beams (wavelength λ), the intensity at the position where the central maximum occurred previously remains unchanged. The minimum thickness of the glass-plate is **[IIT JEE 2002]**

a. 2λ
b. $2\lambda/3$
c. $\lambda/3$
d. λ

40 In a Young's double slit experiment, 12 fringes are observed to be formed in a certain segment of the screen, when light of wavelength 600 nm is used. If the wavelength of light is changed to 400 nm, number of fringes observed in the same segment of the screen is given by **[IIT JEE 2001]**

a. 12
b. 18
c. 24
d. 30

41 Two beams of light having intensities I and $4I$ interfere to produce a fringe pattern on a screen. The phase difference between the beams is $\lambda/2$ at point A and π at point B. Then, the difference between resultant intensities A and B is **[IIT JEE 2001]**

a. $2I$
b. $4I$
c. $5I$
d. $7I$

42 In a double slit experiment instead of taking slits of equal widths, one slit is made twice as wide as the other, then in the interference pattern **[IIT JEE 2000]**

a. the intensities of both the maxima and the minima increases
b. the intensity of the maxima increases and the minima has zero intensity
c. the intensity of maxima decreases and that of minima increases
d. the intensity of maxima decreases and the minima has zero intensity

DPP-3 Interference of Light-II

One or More than One Options Correct Type

Working Space

Direction *(Q. Nos. 1-12) This section contains 12 multiple choice questions. Each question has four choices (a), (b), (c) and (d), out of which ONE or MORE THAN ONE are correct.*

1 In light wave,

a. electric field and magnetic field are in same phase
b. electric field, magnetic field and propagation of wave, all are mutually perpendicular to each other
c. light wave is electromagnetic wave
d. the speed of light in vacuum is independent of reference frame

2 An sustained interference pattern is formed on the screen, when light from two different monochromatic sources are allowed to interfere. Then, it is true that

a. phase difference between sources should be independent of time
b. frequencies of light from two sources may be different
c. sources may be incoherent
d. sources may be in different medium

3 In Young's double slit experiment, the interference pattern is found to have an intensity ratio between the bright and dark fringes is 9. This implies that

a. the intensity at screen due to two slits may be 5 unit and 4 unit respectively
b. the intensities at screen due to two slits may be 4 unit and 1 unit respectively
c. the amplitude ratio is 3
d. the amplitude ratio is 2

4 If Young's double slit experiment is performed with white light, then

a. central bright fringe is white
b. the bright fringe next to central bright fringe is violet
c. the first order fringes are coloured
d. None of the above

5 n light beam each of intensity I_0 are superposed. Then, mark the correct option.

a. If sources are coherent, minimum intensity is zero
b. If sources are coherent, maximum intensity is nI_0
c. If sources are coherent, maximum intensity is n^2I_0
d. If sources are incoherent, resultant intensity is nI_0

6 When a thin transparent sheet of refractive index, $\mu = \frac{3}{2}$ is placed near one of the slits in Young's double slit experiment, intensity at the centre reduces to 50% of the maximum intensity. The possible thickness of the transparent sheet is (Assume that, wavelength of light used is 600 nm)

a. 300 nm
b. 600 nm
c. 900 nm
d. 1500 nm

7 A transparent thin sheet of refractive index $\mu\,(\mu > 1)$ and thickness t is inserted infront of upper slit of YDSE apparatus. The wavelength of light used is λ. Assume that, there is no absorption of light by the slab.
Mark the correct option(s).

a. Maxima may be formed at the centre of screen
b. Central maxima is not formed at the centre of screen
c. Minima may be formed at the centre of screen
d. If slits are identical and the thickness of the transparent sheet is $\frac{5\lambda}{2(\mu - 1)}$, intensity at the centre of screen is zero

8 In Young's double slit experiment, the separation between slits is 2 mm, the distance between the slit and screen is 1m. In the interference pattern, there is a maxima exactly infront of one of the slits. Then, the possible value of wavelength(s) used in experiment are

a. 400 nm
b. 500 nm
c. 666.67 nm
d. 200 nm

9 A thin transparent sheet of thickness 0.02 nm having refractive index 1.45 is pasted across one of the slit in a Young's double slit experiment. The paper transmits 25% of light falling on it. The wavelength of light used is 600 nm. Then, mark the correct options.

a. The ratio of maximum and minimum amplitude in fringe pattern is 3
b. The position of central maxima from centre of screen is 9 mm
c. The total number of fringe crossing through the centre of screen, if an identical transparent sheet is pasted on the other slit is 15
d. The ratio of maximum and minimum intensity is 9

10 In an Young's double slit experiment, a monochromatic light of wavelength 500 nm is used as shown in the figure.

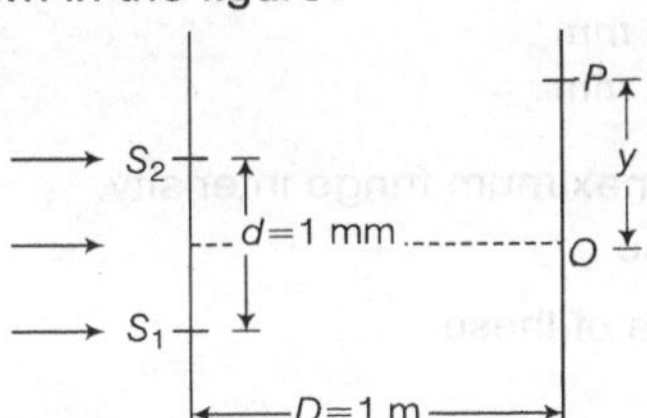

Mark the correct option(s).

a. If intensity due to each slits on screen is $2\,\text{Wm}^{-2}$, the intensity at point P for $y = 0.5\,\text{mm}$ is $8\,\text{Wm}^{-2}$
b. Angular fringe width is 5×10^{-4} rad
c. The position of first minima is 0.25 mm
d. The fringe width decreases, when the experiment is performed in water

11 In the given figure, the two light waves that are represented by the rays have wavelength 550.0 mm before entering media 1 and 2. They also have equal amplitudes and are in phase. Medium 1 is now just air, and medium 2 is transparent plastic layer of index of refraction 1.600 and thickness 2.600 μm.

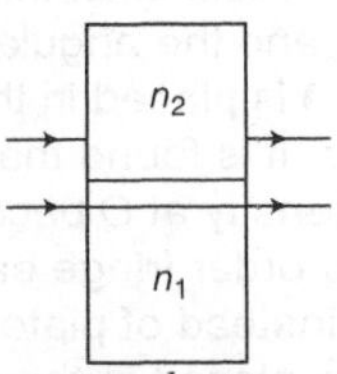

Then, mark the correct options.

a. The phase difference of emerging waves in wavelengths is 2.84
b. The effective phase difference of emerging waves in wavelengths is 17.8
c. The effective phase difference of emerging waves in wavelengths is 0.84
d. The effective phase difference of emerging waves in wavelengths is 2.84

Working Space

12 A transparent slab of thickness t and refractive index μ is inserted infront of upper slit of Young's double slit experiment apparatus. The wavelength of light used is λ. Assume that, there is no absorption of light by the slab.
Mark the correct option(s).

a. The intensity of dark fringes will be zero, if slits are identical
b. The change in optical path due to insertion of plate is μt
c. The change in optical path due to insertion of plate $(\mu - 1)t$
d. For making intensity zero at centre of screen, the thickness can be $5\lambda/2(\mu - 1)$

Comprehension Type

***Direction** (Q. Nos. 13-24) This section contains 4 paragraphs, each describing theory, experiments, data, etc. Twelve questions related to the paragraphs have been given. Each question has only one correct answer among the four given options (a), (b), (c) and (d).*

Passage I

The Young's double slit experiment is done in a medium of refractive index 4/3. A light of 600 nm wavelength is falling on the slits have 0.45 mm separation. The lower slit S_2 is covered by a thin glass sheet of thickness 10.4 mm and refractive index 1.5. The interference pattern is observed on a screen placed at 1.5 m. The interference pattern is observed on a screen placed 1.5 m from the slits as shown in the figure.

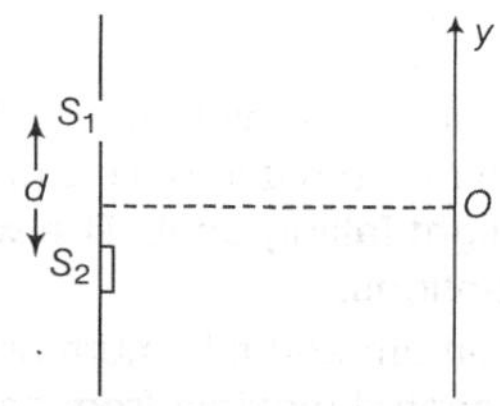

(All wavelengths in this problem are for the given medium of refractive index 4/3. Ignore dispersion.)

13 Find the location of central maxima (bright fringe with zero path difference) on the Y-axis.

a. 5.22 mm
b. 4.33 mm
c. 3.43 mm
d. 3.33 mm

14 Find the light intensity at point O relative to the maximum fringe intensity.

a. half of I_{max}
b. Same
c. $\frac{3}{4}$ of I_{max}
d. None of these

15 Now, if 1600 nm light is replaced by white light of range 400 to 700 nm, find the wavelengths of the light that from maxima exactly at point O.

a. 650 nm, 433.3 nm
b. 665 nm, 500 nm
c. 675 nm, 550 nm
d. 685 nm, 575 nm

Passage II

In Young's double slit experiment, light of wavelength 6000 Å is incident on the double slits. Intensity of central maximum, formed at O, is I_0, and the angular fringe width is 0.1° If a thin transparent plate A is placed in the path of one of the interferring beams, it is found that smallest thickness of the plate for which intensity at O becomes half of maximum is 250 nm. Also, the zero order fringe earlier at O now forms 0.5 mm from O. However instead of plate A, if a plate B made of another material is placed in the path of one of the interfering beams, smallest thickness of B for which intensity at O becomes zero is 300 nm. (Neglect absorption of light by the plate.)

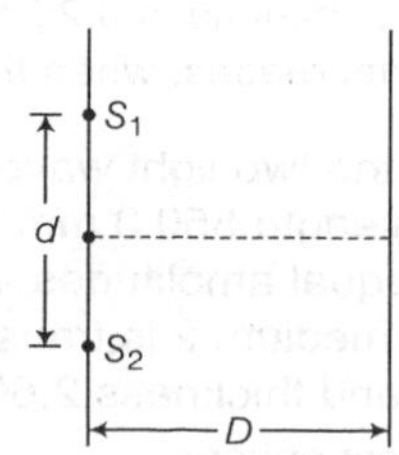

16 Refractive index of plate A is

a. 14
b. 1.5
c. 1.6
d. 1.82

17 Refractive index of plate B is

a. 2
b. 1.8
c. 1.6
d. 1.48

18 Distance of screen from the double slits is

a. 1.1 m
b. 1.4 m
c. 0.8 m
d. 1.5 m

Passage III

Young's interference experiment has a set-up as shown in the figure. Consider that a source of monochromatic light of wavelength 5000 Å is placed at S. S_1 and S_2 are equidistant from S and have equal widths. A wavefront proceeds from S, and S_1 and S_2, being points on this wavefront, emit light waves, which superimpose to result in an interference pattern on the screen kept at a position (1). O is a point on the screen equidistant from S_1 and S_2. P is a point on the screen 1 mm from O at which path difference between waves from S_1 and S_2 is 11250 Å. Position of any point on the screen can be expressed by θ. The path difference is zero at O and a central bright fringe of intensity I_0 is obtained at O. If the screen is now shifted to a new position (2) [not shown in figure]. So that D changes, the fringe width is found to be 50% more than its earlier value and the angular fringe width is 0.1°. (Assume, D to be large and θ is very small so that, $\sin\theta \approx \tan\theta \approx \theta$).

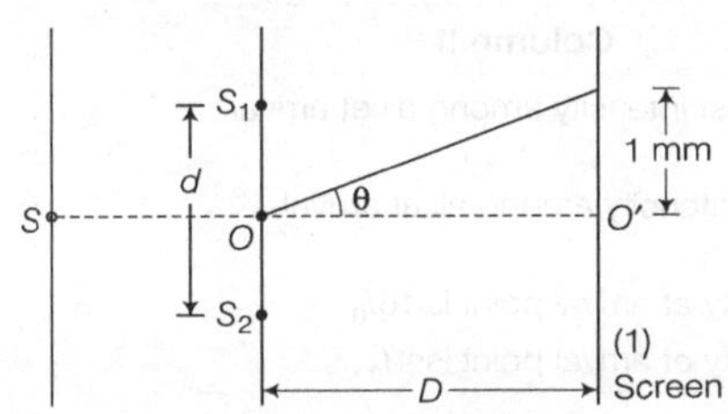

19 Intensity at P when the screen is at position (1) is

a. $2I_0$
b. I_0
c. $I_0/2$
d. $I_0/4$

20 Fringe width (in mm) when screen is at position (1) will be

a. 0.92
b. 0.66
c. 0.44
d. 0.25

21 Distance d between S_1 and S_2(in mm) is

a. 0.29
b. 0.49
c. 0.99
d. 0.66

Passage IV

In Young's double slit experiment, a parallel beam of light of wavelength 500 nm is incident on the slits at an angle of 30°. x and y are two transparent sheet each of refractive index 1.5. The thickness of x is 10 μm and that y is 5 μm. The light coming from x and y have intensities 1Wm^{-2} and 9 Wm^{-2} on the screen.

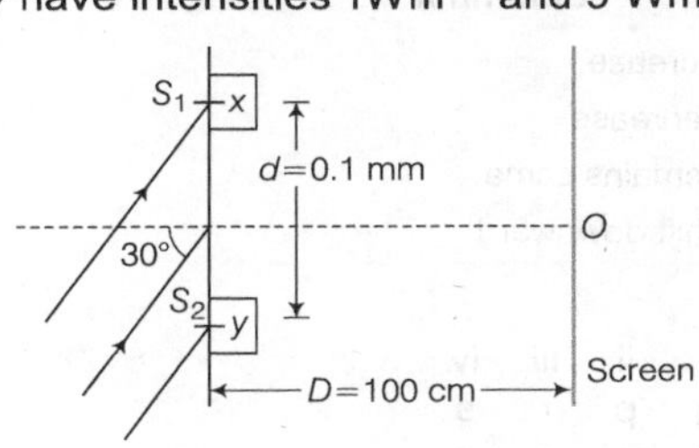

Working Space

22 The intensity at point O is

a. $1\ Wm^{-2}$
b. $4\ Wm^{-2}$
c. $16\ Wm^{-2}$
d. $36\ Wm^{-2}$

23 The position of central maximum from centre of screen is

a. 52.5 cm below point O
b. 2 mm above point O
c. 52.5 cm above point O
d. at point O

24 Fringe width on the screen is

a. 5×10^{-3} cm
b. 1 mm
c. 3×10^{-3} cm
d. None of these

Matching List Type

***Direction** (Q. Nos. 25-30) Choices for the correct combination of elements from Column I and Column II are given as options (a), (b), (c) and (d), out of which one is correct.*

25 Each of four pairs of light waves given in Column I arrives at a certain point on a screen. The waves have same wavelength. At the arrival point, their intensities and phase differences are given in Column I. Match the Column I with Column II and mark the correct options from the codes given below.

	Column I		Column II
i.	$4I_0, 6I_0$ and π radian	p.	Greatest intensity among all at arrival point
ii.	$9I_0, 25I_0$ and π radian	q.	Least intensity among all at arrival point
iii.	$81I_0, 49I_0$ and 3π radian	r.	Intensity at arrival point is $16I_0$
iv.	$4I_0, 4I_0$ and 0 radian	s.	Intensity at arrival point is $4I_0$

Codes

	i	ii	iii	iv		i	ii	iii	iv
a.	q	s	p	r	b.	p	q	r	s
c.	s	q	p	r	d.	q	p	s	r

26 In the Young's double slit experiment as shown in the figure, Δx is the path difference between waves at a point on screen. Now, a thin transparent glass sheet is introduced infront of S_1 which absorbed 75% of light incident of it.

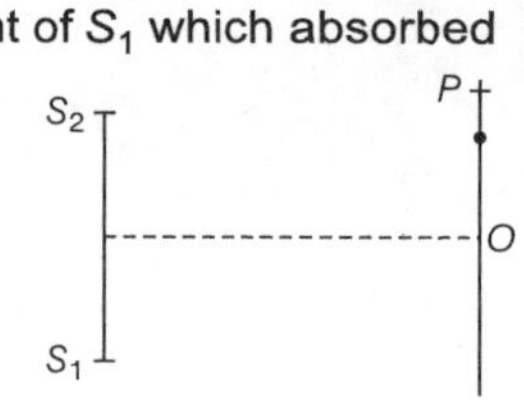

Match the Column I with Column II and mark the correct options from the codes given below.

	Column I		Column II
i.	Intensity of maxima will	p.	Increase
ii.	Intensity of minima will	q.	Decrease
iii.	Fringe width will	r.	Remains same
iv.	Fringe pattern will	s.	Shift downward

Codes

	i	ii	iii	iv		i	ii	iii	iv
a.	p	q	r	s	b.	q	p	r	s
c.	r	p	q	s	d.	q	p	s	r

Working Space

27 Match the Column I with Column II and mark the correct option from the codes given below.

	Column I		Column II
i.	Planar wavefront incident on a thin convex lens placed in air	p.	After reflection or refraction planar wave
ii.	Planar wavefront incident on a prism placed in air incidence side	q.	Spherical wavefront having centre towards incident side
iii.	Planar wavefront incident on a concave mirror of small aperture	r.	Spherical wavefront having centre on other side
iv.	Planar wavefront incident on convex mirror	s.	Planar wavefront parallel to incident wavefront

Codes

	i	ii	iii	iv
a.	p	r	s	q
b.	q	p	r	s
c.	r	p	q	r
d.	p	s	q	r

28 In the YDSE apparatus as shown in figure, Dx is the path difference between S_2P and S_1P. Now, a glass slab is introduced infront of S_2, then match the Column I with Column II and mark the correct option from the codes given below.

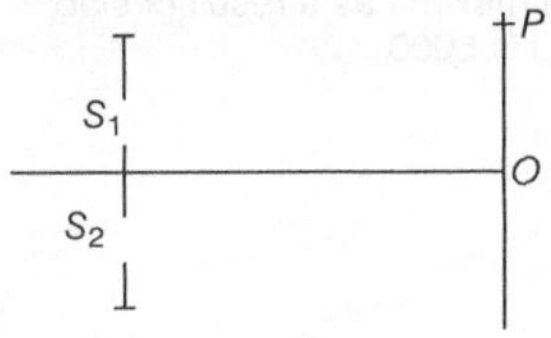

	Column I		Column II
i.	Dx at P will	p.	Increase
ii.	Fringe width will	q.	Decrease
iii.	Fringe pattern will	r.	Remains same
iv.	Number of fringes between O and P will	s.	Shift upward
		t.	Shift downward

Codes

	i	ii	iii	iv
a.	p	q	r	s
b.	p	r	t	r
c.	t	q	s	t
d.	p	r	s	t

29 Match the Column I with Column II and mark the correct option from the codes given below.

	Column I		Column II
i.	For a sustained interference pattern of light	p.	Amplitude of waves from the two sources must be same
ii.	Newton's corpuscular theory	q.	Could not explain the phenomenon of interference of light
iii.	Huygens' wave theory	r.	Coherent source is required
iv.	For maximum contrast in the sustained interference pattern due to YDSE	s.	Could not explain photoelectric effect

Codes

	i	ii	iii	iv
a.	r	q	s	p,r
b.	p,q	q,s	r	p
c.	p	q	r	p,s
d.	p,r	p,s	q,r	p,r

Working Space

30 In a YDSE set-up, light of wavelength of light 4000 Å is used. Distance of screen from the slit is 2 m and distance between the slits is 1 mm. There are four slabs, slab 1 (thickness 2 mm, refractive index 2), slab 2 (thickness 1 mm, refractive index 3), slab 3 (thickness 4 mm, refractive index 3/2), slab 4 (thickness 3 mm, refractive index 5/3).

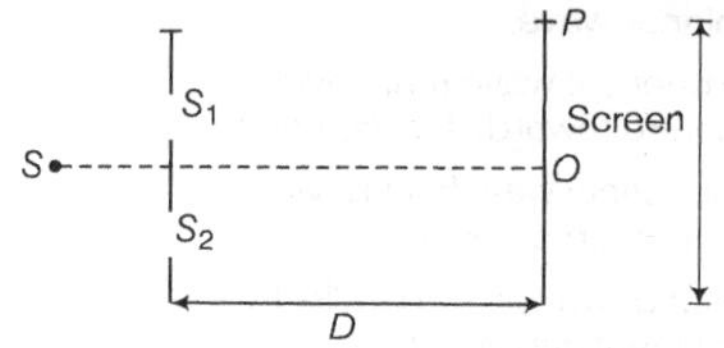

Match the Column I with Column II and mark the correct option from the codes given below.

	Column I		Column II
i.	If slab 1 is placed infront of slit S_1	p.	Central maxima is at C
ii.	If slab 2 is placed infront of slit S_2 along with condition (i)	q.	Central maxima is above C
iii.	If slab 3 is placed behind slit S_1 along with condition (ii)	r.	Fringe width = 0.8 mm
iv.	If slab 4 is placed behind slit S_2 along with condition (iii)	s.	Number of fringes crossing the central maxima as a result of slab placing is 5000

Codes

	i	ii	iii	iv
a.	q,r,s	p,r	q,r,s	p,r
b.	p,q,r	p,r	q	p
c.	q	r	s	p
d.	s	p	r	s

DPP-4 Interference of Light-III

Statement Type

Working Space

Direction (*Q. Nos.* 1-5) *This section is based on Statement I and Statement II. Select the correct answer from the codes given below.*

a. Both Statement I and Statement II are correct and Statement II is the correct explanation of Statement I
b. Both Statement I and Statement II are correct but Statement II is not the correct explanation of Statement I
c. Statement I is correct but Statement II is incorrect
d. Statement II is correct but Statement I is incorrect1

1 **Statement I** The colour of a monochromatic light wave is different in different medium.

Statement II The wavelength of monochromatic light waves in different medium are different.

2 **Statement I** If light wave superposed in a certain section of space and extinguish each other. The light energy is converted into other form of energy.

Statement II Light wave is electromagnetic wave.

3 **Statement I** In Young's double slit experiment, the interference pattern disappears when one of the slits is closed.

Statement II Interference occurs due to the superposition of light waves from two coherent sources.

4 **Statement I** In Young's double slit experiment having two identical slit. If width of one of the slits is increased, then intensities of both bright and dark fringes will increase.

Statement II On increasing the width of one slit in YDSE, the will be no point on the screen which is physically dark (zero intensity).

5 **Statement I** If the optical path between points A and B in first medium is same as the optical path between two points D and E in second medium. Then, time taken by light to travel from point A to B is same as that of from point D to E.

Statement II The speed of light in denser medium is same as that of in rarer medium.

One Integer Value Correct Type

Direction (*Q. Nos.* 6-14) *This section contains 9 questions. When worked out will result in an integer from 0 to 9 (both inclusive).*

6 In Young's double slit experiment, one of the slits is covered by a thin film of thickness $t = 0.04$ mm and refractive index 1.5 for wavelength 400 nm and 1.4 for wavelength 600 nm. A beam of light consisting two wavelengths $\lambda_1 = 400$ nm and $\lambda_2 = 600$ nm falls normally on the plane of the slits. Find the distance between two central maxima (in mm).

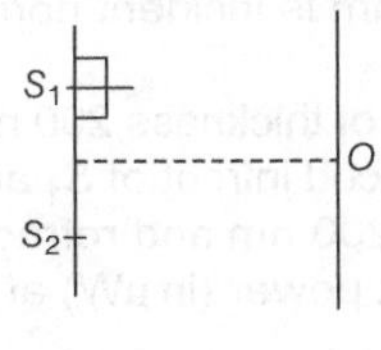

Working Space

7 In a Young's double slit experiment, the two slits are illuminated by a monochromatic light source S of wavelength 400 nm.

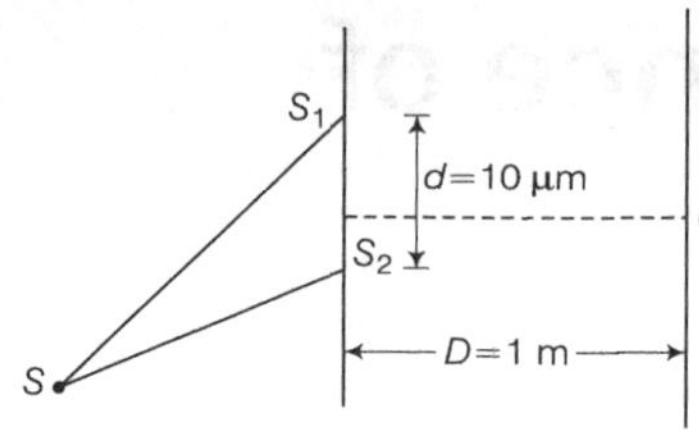

If $SS_1 = 10\ \mu m$ and SS_2 is $50\ \mu m$, find the number of maxima formed on screen.

8 Two coherent light sources S_1 and S_2 each of wavelength λ are separated by a distance $d = 2\lambda$. Find the number of minima formed on the screen as shown in the figure.

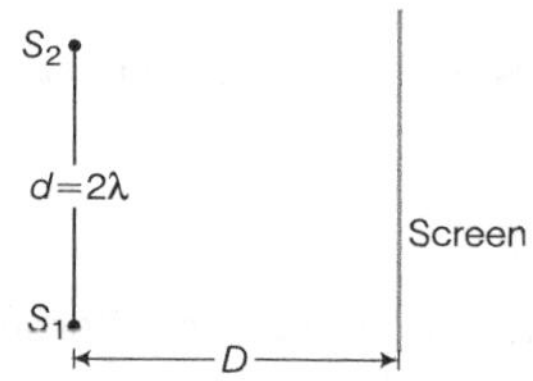

9 Consider the situation as shown in the figure, two slits S_1 and S_2 are placed symmetrically about the line OP which is perpendicular to screen and bisector to line joining the slits. The space between screen and slits is filled with a liquid of refractive index $\mu_3 = 1.8$. A thin transparent sheet of thickness $6\ \mu m$ and refractive index $\mu_1 = 1.2$ is placed infront of slit S_1 and another thin transparent sheet of thickness $12\ \mu m$ and refractive index 1.4 is placed infront of slit S_2. If first maxima is formed at centre C of the screen, the wavelength in air is $100x$ nm. Find the value of x.

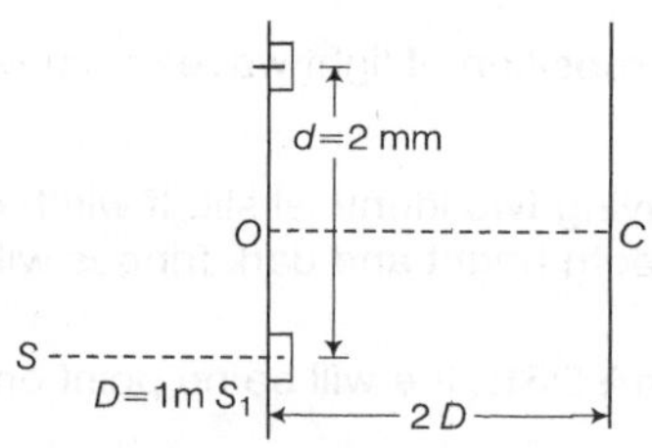

10 In Young's double slit experiment, how many maxima can be obtained on screen if $d = 3\lambda$.

11 In Young's double slit experiment, two parallel slits are illuminated by two wavelengths 400 nm and 700 nm. The fourth dark fringe resulting from 400 nm wavelength of light coincides on screen with nth order bright fringe of 700 nm wavelength. Find the value of n.

12 Two circular aperture S_1 and S_2 of radii 1 mm and 2 mm. A parallel beam of light $\frac{7}{22}\ Wm^{-2}$ and wavelength 600 nm is incident normally as shown in the figure.

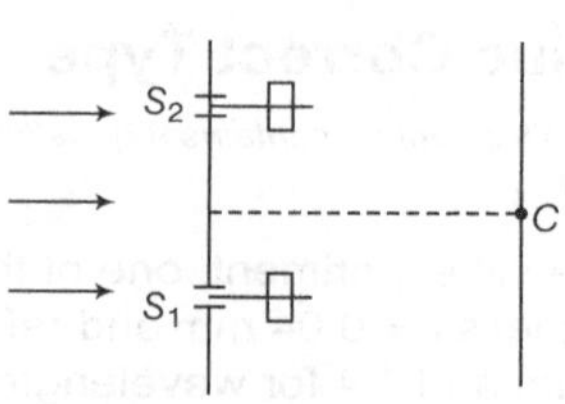

A transparent film of thickness 200 nm having refractive index $\mu_1 = 2$ is placed infront of S_1 and other transparent film of thickness 200 nm and refractive index 1.5 is placed infront of other slit. Find the resultant power (in μW) at the centre of screen.

Working Space

13 Two identical monochromatic light sources S_1 and S_2 intensity 10^{-15} Wm^{-2} produce wavelength of light 400 nm. A glass of thickness 3×10^{-6} m is placed infront of S_1 whose refractive index $\mu = 1 + \sqrt{x}$, where x is in micrometre. Find intensity (in femto Wm^{-2}) at the centre of screen.

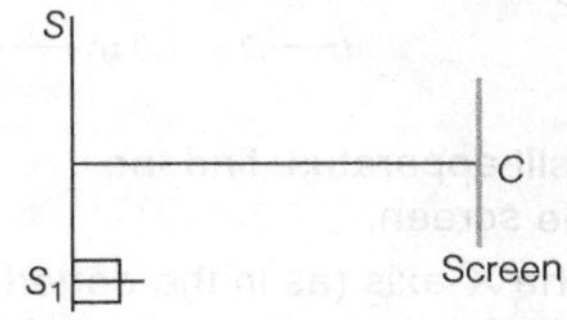

14 In Young's double slit experiment, the source consists of two wavelengths $\lambda_1 = 400$ nm and $\lambda_2 = 400.2$ nm. Find the distance (in cm) from the centre of screen where the fringes disappear, if $d = 10$ mm and $D = 100$ cm.

Subjective Questions

Direction *(Q. Nos. 15-23) These questions are subjective in nature, need to be solved completely on notebook.*

15 In a Young's double slit experiment, two wavelengths of 500 nm and 700 nm were used. What is the minimum distance from the central maximum where their maximas coincide again? Take, $D/d = 10^3$. Symbols have their usual.

16 A vessel *ABCD* of 10 cm width has two small slits S_1 and S_2 sealed with identical glass plates of equal thickness. The distance between the slits is 0.8 mm. *POQ* is the line perpendicular to the plane *AB* and passing through *O*, the middle point of S_1 and S_2. A monochromatic light source is kept at *S*, 40 cm below *P* and 2 m from the vessel, to illuminate the slits as shown in the figure. Calculate the position of the central bright fringe on the other wall *CD* with respect to the line *OQ*. Now, a liquid is poured into the vessel and filled upto *OQ*. The central bright fringe is found to be at *Q*. Calculate the refractive index of the liquid.

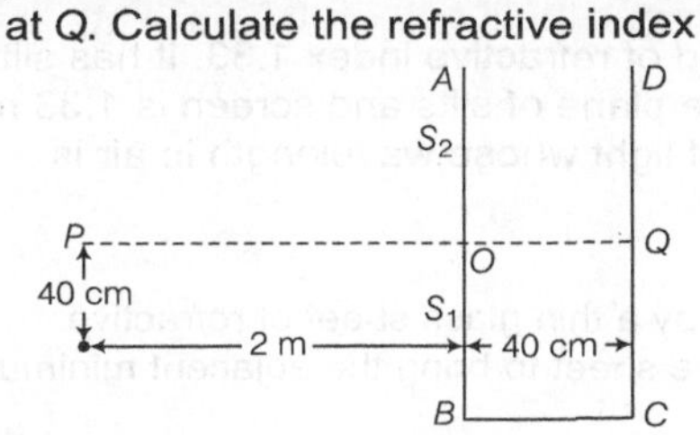

17 In Young's double slit experiment is done in a medium of refractive index 4/3. A light of 600 nm wavelength is falling on the slits having 0.45 mm separation. The lower slit S_2 is covered by a thin glass sheet of thickness 10.4 μm and refractive index 1.5. The interference pattern is observed on a screen placed 1.5 m from the slits as shown in the figure.

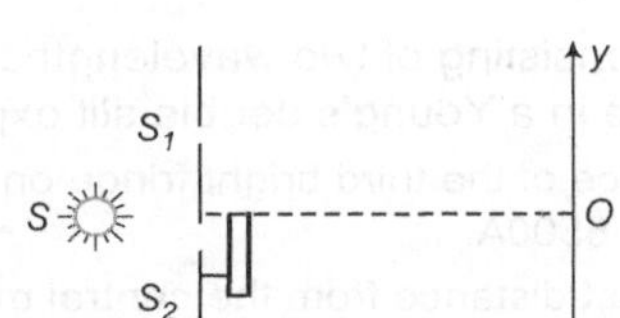

i. Find the location of central maximum (bright fringe with zero path difference) on the *Y*-axis.

ii. Find the light intensity of point *O* relative to the maximum fringe intensity.

iii. Now, if 600 nm light is replaced by white light of range 400 to 700 nm, find the wavelengths of the light that form maxima exactly at point *O*.

(All wavelengths in the problem are for the given medium of refractive index 4.3. Ignore dispersion).

Working Space

18 A coherent parallel beam of microwaves of wavelength $\lambda = 0.5$ mm falls on a Young's double slit apparatus. The separation between the slits is 1.0 mm. The intensity of microwaves is measured on a screen placed parallel to the plane of the slits at a distance of 1.0 m from it as shown in the figure.

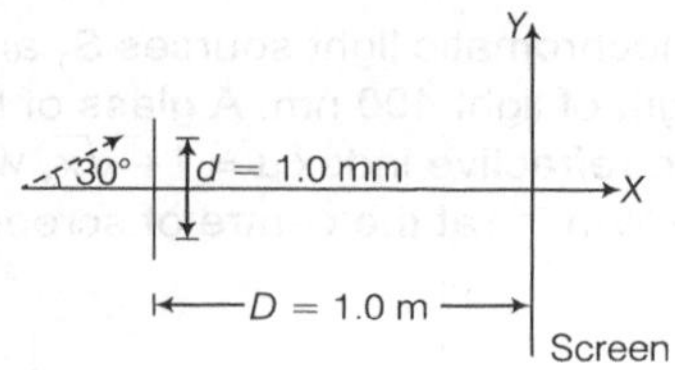

i. If the incident beam falls normally on the double slit apparatus, find the y-coordinates of all the interference minima on the screen.

ii. If the incident beam makes an angle of 30° with the X-axis (as in the dotted arrow shown in figure), find the y-coordinates of the first minima on either side of the central maximum.

19 In a Young's experiment, the upper slit is covered by a thin glass plate of refractive index 1.4, while the lower slit is covered by another glass plate, having the same thickness as the first one but having refractive index 1.7. Interference pattern is observed using light of wavelength 5400 Å. It is found that the point P on the screen, where the central maximum ($n = 0$) fall before the glass plates were inserted, now has 3/4 the original intensity. It is further observed that what used to be the fifth maximum earlier lies below the point P while the sixth minima lies above P. Calculate the thickness of glass plate. (Absorption of light by glass plate may be neglected)

20 In Young's experiment, the source is red light of wavelength 7×10^{-7} m. When a thin glass plate of refractive index 1.5 at this wavelength is put in the path of one of the interferring beams, the central bright fringe shifts by 10^{-3} m to the position previously occupied by the 5th bright fringe.

i. Find the thickness of the plate. When the source is now changed to green light of wavelength 5×10^{-7} m, the central fringe shifts to a position initially occupied by the 5th bright fringe due to red light.

ii. Find the refractive index of glass for green light, estimate the change in fringe width due to the change in wavelength.

21 A double slit apparatus is immersed in a liquid of refractive index 1.33. It has slit separation of 1 mm and distance between the plane of slits and screen is 1.33 m. The slits are illuminated by a parallel beam of light whose wavelength in air is 6300 Å.

i. Calculate the fringe width.

ii. One of the slits of the apparatus is covered by a thin glass sheet of refractive index 1.53. Find the smallest thickness of the sheet to bring the adjacent minimum as the axis.

22 A beam of light consisting of two wavelengths, 6500Å and 5200Å is used to obtain interference fringe in a Young's double slit experiment.

i. Find the distance of the third bright fringe on the screen from the central maximum for wavelength 6500Å.

ii. What is the least distance from the central maximum where the bright fringes due to both the wavelengths coincide?

The distance between the slits is 2 mm and the distance between the plane of the slits and the screen is 120 cm.

23 In Young's double slit experiment using monochromatic light, the fringe pattern shifts by a certain distance on the screen when a mica sheet of refractive index 1.6 and thickness 1.964 microns is introduced in the path of one of the interfering waves. The mica sheet is then removed and the distance between the slits and the screen is doubled. It is found that the distance between successive maxima (or minima) now is the same as the observed fringe shift upon the introudction of the mica sheet. Calculate the wavelength of the monochromatic light used in the experiment.

DPP-5 Interference of Light-IV

Conceptual Type

Working Space

Direction *(Q. Nos. 1-3) These questions are conceptual in nature, need to be solved completely on notebook.*

1 **Situation** White light may be considered to have wavelength from 400 mm to 750 mm. If an oil film of thickness $t = 10^{-4}$ cm. The light incidence normally the refractive index of oil is $\mu = 1.4$.

i. If interference is consider in reflected light, the path difference is
   a. $\Delta x = 2\mu t$
   b. $\Delta x = 2\mu t - \lambda/2$
   c. $\Delta x = \lambda/2$
   d. None of the above

ii. The wavelengths of visible light for weak reflection are
   a. 400 nm, 466.7 nm, 560 nm, 700 nm
   b. 430 nm, 509.1 nm, 622.2 nm
   c. 400 nm, 509.1 nm , 622.2 nm
   d. None of the above

iii. The wavelength of visible light for strong reflection are
   a. 400 nm, 466.7 nm, 560 nm, 700 nm
   b. 430.8 nm, 509.1 nm, 622.2 nm
   c. 400 nm, 509.1 nm, 622.2 nm
   d. None of the above

iv. If interference is considered is transmitted light, the strongly transmitted wavelength are
   a. 400 nm, 466.7 nm, 560 nm, 700 nm
   b. 430.8 nm, 509.1 nm, 622.2 nm
   c. 400 nm, 509.1 nm, 622.2 nm
   d. None of the above

v. For destructive interference, the wavelength are
   a. 400 nm, 466.7 nm, 560 nm
   b. 430.8 nm, 509.1 nm 622.2 nm
   c. 400 nm, 509.1 nm, 622.2 nm
   d. None of the above

vi. At what maximum width δ_{max} of the slit, the interference fringes on the screen are still observed sufficiently sharp.
   a. $\frac{\lambda}{2\alpha}\left(1+\frac{r}{b}\right)$ b. $\frac{\lambda}{4\alpha}\left(1+\frac{r}{b}\right)$
   c. 43×10^{-6} m d. None of these

Working Space

2 **Situation** Illustrates the interference experiment with Fresnel mirros. The angle between the mirror is $\alpha = 12'$, the distances from the mirros' intersection line to the narrow slit S and the screen Sc are equal to $r = 10.0$ cm and $b = 130$ cm respectively. The wavelength of light is $\lambda = 0.55\ \mu m$.

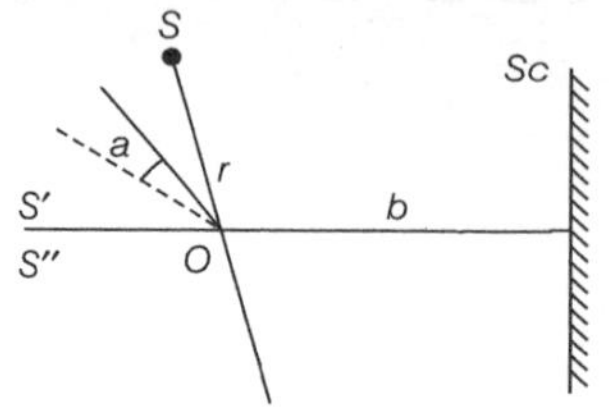

i. Find the separation betweem images formed by Fresnel mirros.
   a. $d = r\alpha$ b. $d = 2r\alpha$
   c. $d = r\alpha/2$ d. $d = 4r\alpha$

ii. Find the distance between images and screen.
   a. $D = b + r$ b. $D = b + r\alpha$
   c. $D = b + 2r$ d. None of these

iii. Find the fringe width.
   a. $\beta = \frac{D\lambda}{d} = \left\{\frac{(b+r)}{2r\alpha}\right\}\lambda$ b. $\beta = 1.1$ mm
   c. $\beta = 2.2$ mm d. None of these

iv. Find the length of fringe pattern.
   a. $l = 2b\alpha$ b. $l = b\alpha$
   c. $l = 3b\alpha$ d. None of these

v. Find the number of possible maxima.
   a. 8 b. 4.15
   c. 8.30 d. 9

vi. Find the shift of the interference pattern on the screen, when the slit is displaced by $\delta l = 1.0$ mm along the arc of radius r with centre at the point O.
   a. b/r b. $(b/r)_{\delta l}$
   c. $(2b/r)_{\delta l}$ d. 13 mm

3 **Situation** Two plane mirror a sources S of light emitting monochromatic rays of wavelength λ and a screen are arrange as shown in the figure.

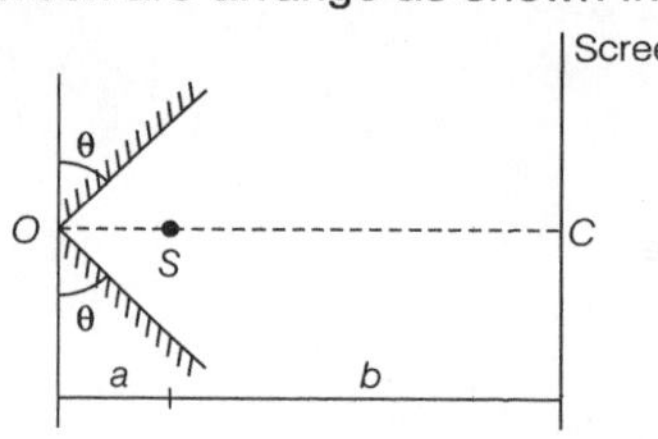

i. The perpendicular distance of each mirror from source S is
   a. $a\cos\theta = a$ b. $a\sin\theta = a\theta$
   c. $a\tan\theta = a\theta$ d. None of these

ii. If image of S formed by mirror are S_1 and S_2, then SS_1 and SS_2 are
   a. $SS_1 = SS_2 = 2a\cos\theta \approx 2a$
   b. $SS_1 = SS_2 = 2a\sin\theta \approx 2a\theta$
   c. $SS_1 = SS_2 = 2a\tan\theta \approx 2a\theta$
   d. None of the above

iii. The separation between images s_1 and s_2 is
   a. $d = S_1S_2 = 2a\theta$ b. $d = 2a$
   c. $d = 4a\theta$ d. None of these

iv. The fringe width on the screen is

a. $D = a + b$

b. $\frac{2(a+b)\lambda}{4a\theta}$

c. $D = a + 2b$

d. None of these

v. The point C of screen is a point of

a. central bright fringe
b. first dark fringe
c. neither a point of maxima nor a point of minima
d. None of the above

vi. The shape of fringe on the screen is

a. hyperbolic but central fringe is straight line
b. straight line
c. circular
d. None of the above

Only One Option Correct Type

***Direction** (Q. Nos. 4-12) These section contains 9 multiple choice questions. Each question has four choices (a), (b), (c) and (d), out of which ONLY ONE is correct.*

4 Due to interference between direct and reflected light from mirror, maxima is formed at point P. By what minimum distance mirror is shifted downward to find minima at point P. (Assume that, wavelength of light is 600 mm)

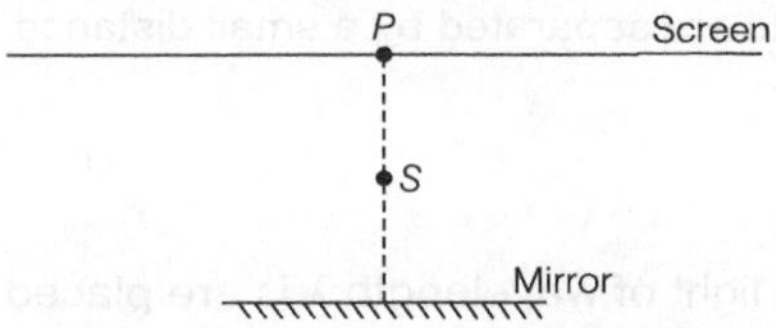

a. 100 nm

b. 200 nm

c. 300 nm

d. 400 nm

5 Upper half of the thin lens has refractive index μ and lower half has refractive index $\mu - \delta\mu (\delta\mu << \mu)$ the shape the fringe on the screen is

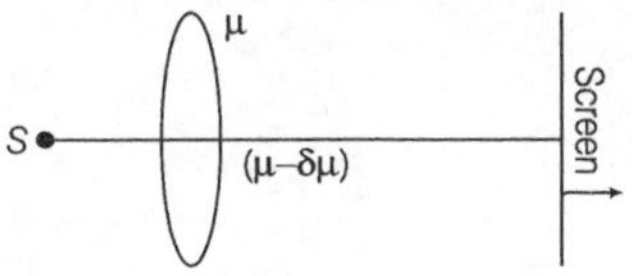

a. hyperbolic
b. straight line
c. circular
d. No sufficient information

6 A flat piece of glass is held stationary and horizontal above the flat top and of a 10 cm long vertical metal rod that has a rigidly fixed lower end. The thin film of air between the rod and the glass is observed to be bright by reflected light when it is illuminated by light of wavelength 500 nm. As the temperature is slowly increased by 25.0° C, the film changes from bright to dark and back to bright 200 times. What is the coefficient of linear expansion of the metal?

a. $25 \times 10^{-4} / °C$

b. $10^{-5} / °C$

c. $5 \times 10^{-5} / °C$

d. None of these

Working Space

7 M_1 and M_2 are plane mirrors and kept parallel to each other. At point O there will be a maxima for wavelength. Light from monochromatic source S of wavelength λ is not reaching directly on the screen. The λ is: $[D >> d, d >> \lambda]$

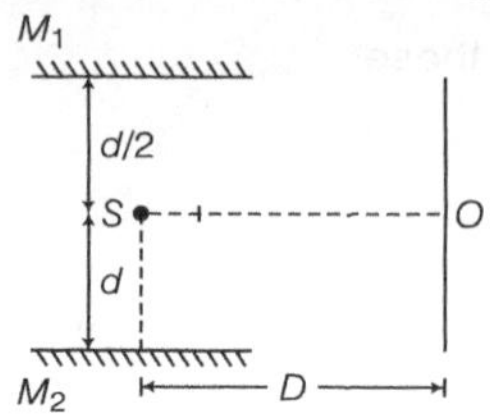

a. $\frac{3d^2}{D}$

b. $\frac{3d^2}{2D}$

c. $\frac{d^2}{D}$

d. $\frac{2d^2}{D}$

8 A thin convex lens of focal length 10 cm is cut along its diameter and the two halves are joined together such that the angle between their planes is 170°. They are placed symmetrically with respect to axis OO' as shown. To the left of the lens system a monochromatic source of light S of wavelength λ is placed at a distance of 15 cm from the lens system. A screen AB is placed at 30 m from the lens system, towards the right.

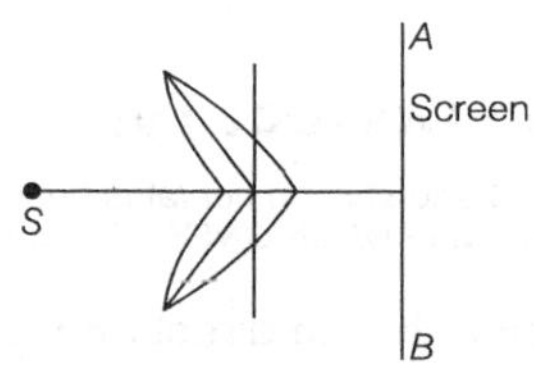

a. Image of point source S formed by both parts are separated by a small distance
b. The shape of fringe on screen in circular
c. No fringe pattern is formed on the screen
d. None of the above

9 Two coherent point sources A and B emitting light of wavelength λ is are placed at position $(-D, 0)$ and $(-D, 3\lambda)$ respectively. Here, $D >> l$. The shape of fringes on the screen placed along Y-axis (in YZ-plane) is

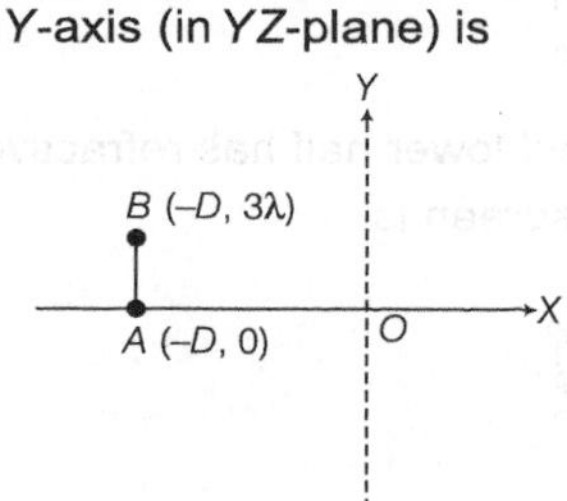

a. straight line
b. circular
c. hyperbolic
d. parabolic

10 Two coherent point sources S_1 and S_2 are separated by a small distance d as shown in the figure. The fringes obtained on the screen will be **[JEE Main 2013]**

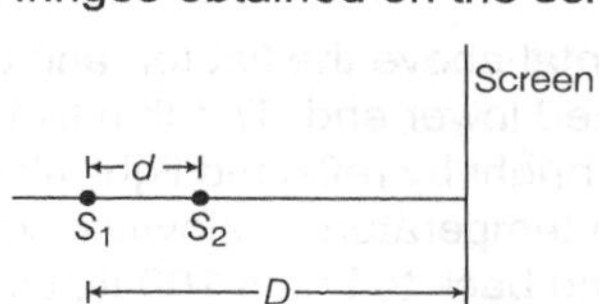

a. points
b. straight line
c. semi-circle
d. concentric circles

Working Space

11 In the adjacent diagram, *CP* represents a wavefront and *AO* and *BP*, the corresponding two rays. Find the condition of θ for constructive interference at *P* between the ray *BP* and reflected by ray *OP*.

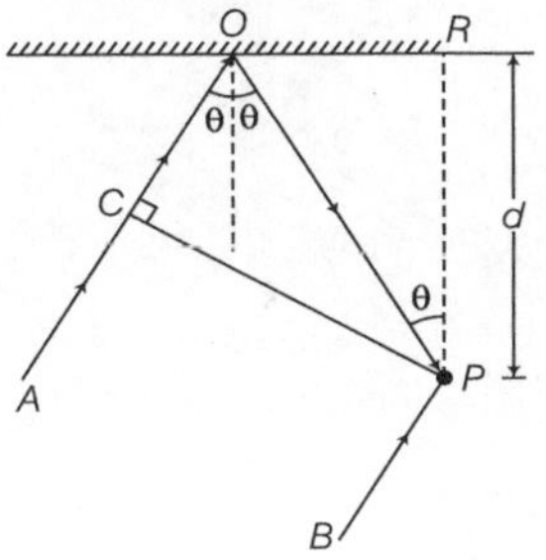

a. $\cos\theta = \frac{3\lambda}{2d}$

b. $\cos\theta = \frac{\lambda}{4d}$

c. $\sec\theta - \cos\theta = \frac{\lambda}{d}$

d. $\sec\theta - \cos\theta = \frac{4\lambda}{d}$

12 A thin slice is cut out of a glass cylinder along a plane parallel to its axis. The slice is placed on a flat plate as shown in the figure. The observed interference fringes from this combination shall be

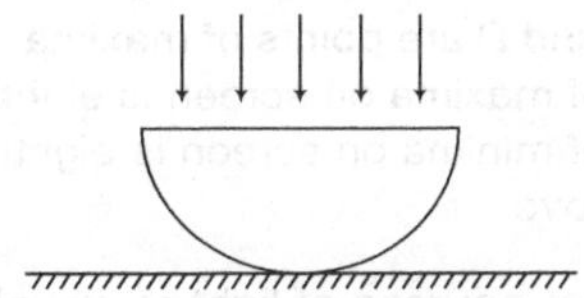

a. straight
b. circular
c. equally spaced
d. having fringe spacing which increases as we go outwards

One or More than One Options Correct Type

Direction *(Q. Nos. 13-18) This section contains 6 multiple choice questions. Each question has four choices (a), (b), (c) and (d), out of which ONE or MORE THAN ONE are correct.*

13 Mark the correct options (s).

a. A very thin soap film (refractive index $\frac{4}{3}$), whose thickness is much less than wavelength of visible light appears dark

b. A thin soap film of refractive index $\frac{4}{3}$ and thickness $\frac{3\lambda}{8}$ appears dark

c. A thin soap film of refractive index $\frac{4}{3}$ placed on glass of refractive index $\frac{3}{2}$ appears bright for thickness 3λ

d. None of the above

14 In the given arrangement,

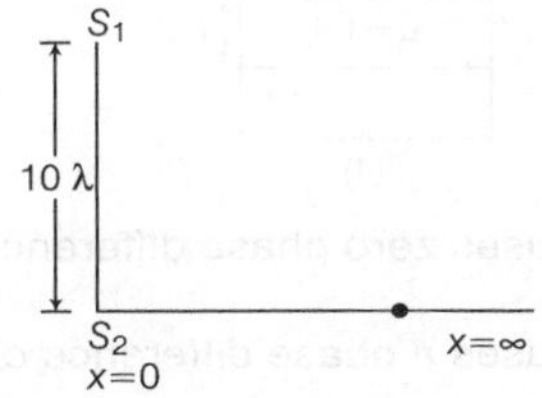

a. Number of minima is 10 for $(0 \le x < \infty)$
b. Number of maxima is 11 for $(0 \le x \le \infty)$
c. The central maxima is not obtained
d. The tenth of maxima is found at $x = 0$

Working Space

15 Each source produces intensity I_0 at the point P. The phase difference between successive sources is 60°. The resultant intensity at point P is

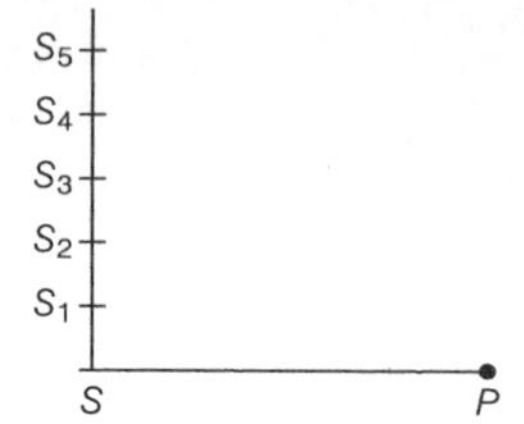

a. $6I_0$ if sources are coherent
b. zero if sources are coherent
c. $6I_0$ if sources are incoherent
d. $36I_0$ if sources are coherent

16 Two point coherent sources separated by $d = 2\lambda$ emits light of wavelength λ in phase. A circular wire of radius R ($R >> \lambda$) has its centre at point O.

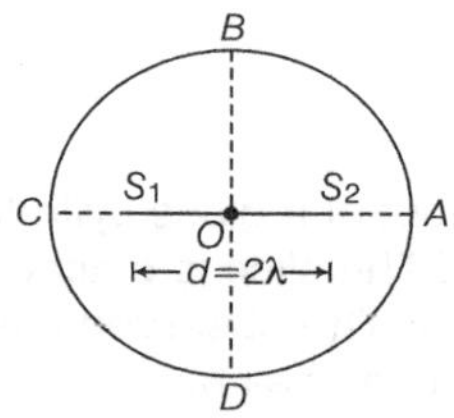

a. Points A, B, C and D are points of maxima
b. Total number of maxima on screen is eight
c. Total number of minima on screen is eight
d. None of the above

17 Two point coherent sources of light of wavelength λ, each placed as shown in the figure. The initial phase difference between the sources is zero ($D >> d$).

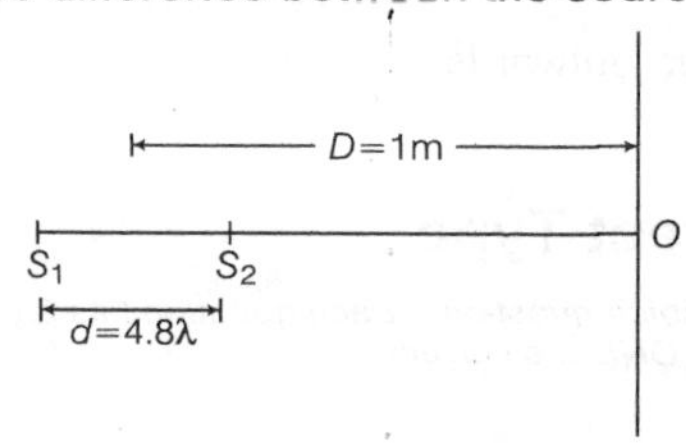

a. The shape of fringes on screen is circular
b. First minima on screen is at distance from point O is 10 cm
c. Total minima on screen is 10
d. Point O is a point of neither minima nor maxima

18 Consider a film of thickness L_0 as shown in four different cases below. Notice the observation of film with perpendicularly falling light. Mark the correct statement(s). (Take, $L = \frac{\lambda}{4}$)

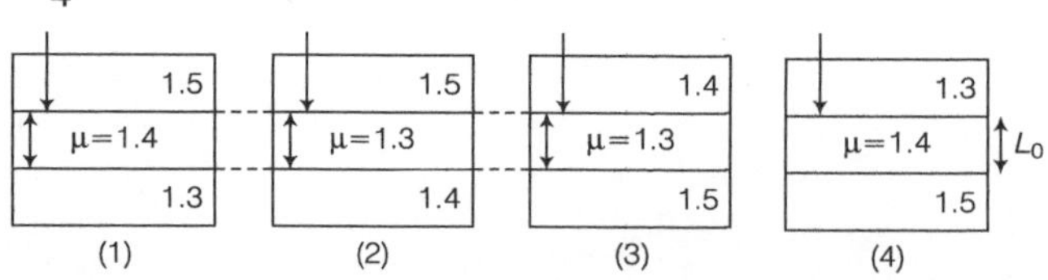

a. For (1) and (4), the reflection at film interfaces causes zero phase difference for two reflected rays
b. For (2) and (3), the reflection at film interfaces causes a phase difference of π reflected rays
c. For (1), the film will appears dark, if it is observed through reflected rays from film interfaces
d. For (4), the film will apears dark, if it is observed through reflected rays from film interfaces.

Comprehension Type

Working Space

Direction *(Q. Nos. 19-24) This section contains 2 paragraphs, each describing theory, experiments, data, etc. Six questions related to the paragraphs have been given. Each question has only one correct answer among the four given options (a), (b), (c) and (d).*

Passage I

A student wants to choose a particular wavelength λ_0 for which the ray passes through the interface BC of combination of two prism as shown in the figure, without deviation. The refractive indices of prism are

$$\mu_1 = 1.2 + \frac{10.8 \times 10^4}{\lambda^2} \text{ and } \mu_2 = 1.45 + \frac{1.80 \times 10^2}{\lambda^2}$$

where λ is in nm.

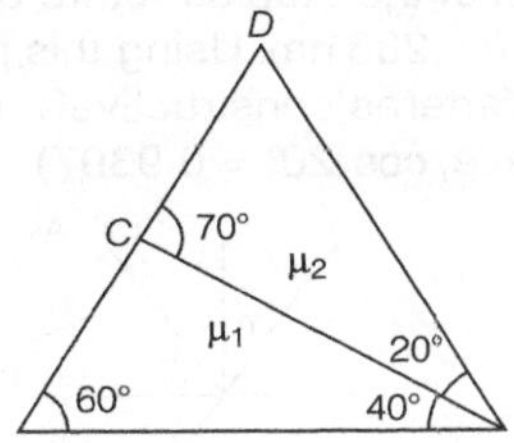

19 The result of student is

a. 600 nm b. 700 nm c. 400 nm d. 500 nm

20 For the wavelength λ_0,

a. The system behaves as a single prism of reracting angle of 90°
b. The system behaves as a single prism of refracting angle 60°
c. The system never be assumed as a single prism
d. The system behaves as a single prism of angle of prism 80°

21 For light wavelength λ_0, find the angle of incidence i on the face AC such that the deviation produced by the combination of prisms is minimum.

a. $\cos^{-1}\frac{3}{4}$ b. $\tan^{-1}\frac{3}{4}$ c. $\sin^{-1}\frac{3}{4}$ d. $\sec^{-1}\frac{3}{4}$

Passage II

Interference from light reflected by step structures or partially overlapping scales produces the iridescent colors seen in many butterflies, moths, birds, and fish. A stunning example is the shimmering blue of the Morpho butterfly. Figure (a), shows the Morpho wing as viewed under an electron miscroscope. The tree-like structures that project up from the top surface of the wing are made of transparent material. Light is thus reflected from a series of steps. Let us concentrate on two rays reflected from the tops of successive steps of thickness t_1 with spacing t_2 betwen the steps (Fig. b). Both are inverted on reflection, so there is no relative phase difference due to reflection. At normal incidence, the path difference is $2(t_1 + t_2)$. However, the ray passes through a thickness t_1 of the step, where the index of refraction is $n = 1.5$. We cannot find the wavelength for constructive interference simply by setting the path difference equal to a whole number of wavelengths.

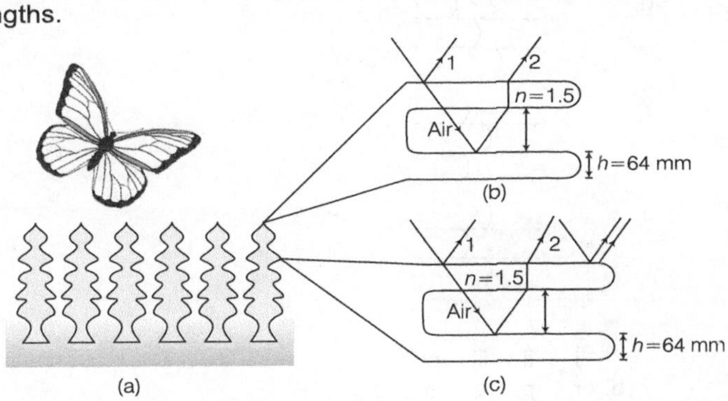

22 For normal incidence, the path difference is

a. $2(t_2 + nt_1)$ b. $2(t_2 + nt_1 - 1/2)$ c. $2(t_2 + t_1 - 1/2)$ d. $2(t_1 + t_2)$

Working Space

23 The dominant wavelength in the light we see when looking at butterfly wing at normal incidence is

a. 223 nm b. 446 nm
c. 400 nm d. 700 nm

24 A simplified model of the step structure of the wing of the Morpho butterfly is that the step height is $h = 233$ nm. Using this model, find the wavelength of reflected visible light the inferferes constructively, if the wing is viewed at an angle $\theta = 20^\circ$ to the normal. (Take, $\cos 20^\circ = 0.9397$)

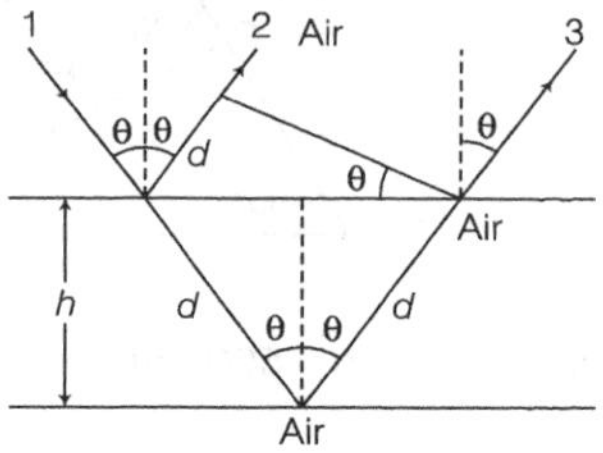

a. 419 nm b. 446 nm
c. 209.5 nm d. 223 nm

Matching List Type

Direction *(Q. Nos. 25 and 26) Choices for the correct combination of elements from Column I and Column II are given as options (a), (b), (c) and (d), out of which one is correct.*

25 For the situation as shown in the figure Column I, relates the values of μ_1, μ_2 and μ_3 and Column II shows the possible shapes of wavefronts. Match the Column I with Column II and mark the correct option from the codes given below.

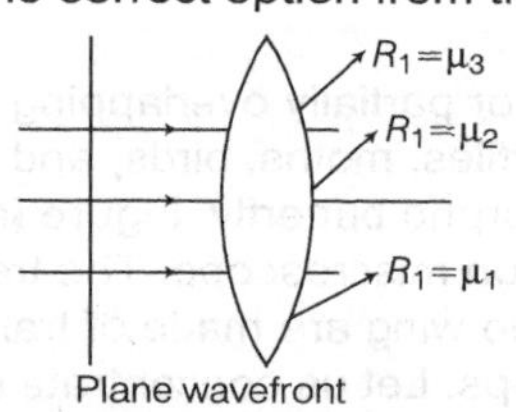

	Column I		Column II
i.	$\mu_1 = 1.4, \mu_2 = 1.5,$ $\mu_3 = 1.6$	p.	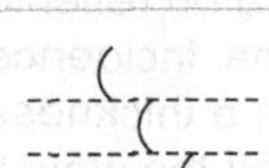
ii.	$\mu_1 = 1.6, \mu_2 = 1.5,$ $\mu_3 = 1.4$	q.	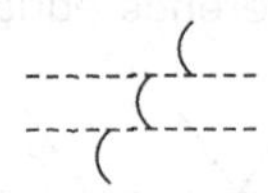
iii.	$\mu_1 = 1.7, \mu_2 = 1.5,$ $\mu_3 = 1.7$	r.	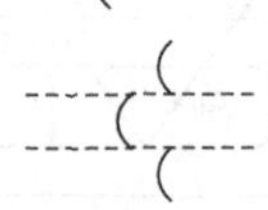
iv.	$\mu_1 = 1.3, \mu_2 = 1.5,$ $\mu_3 = 1.3$	s.	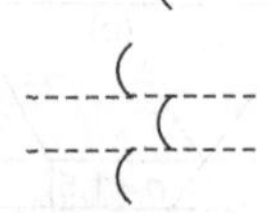

Codes

	i	ii	iii	iv		i	ii	iii	iv
a.	p	s	q	r	b.	r	p	q	s
c.	q	p	r	s	d.	q	p	s	r

26 For the situation as shown in the figure, assume thickness of films to be very small compared to wavelength of incident light. Match the Column I with Column II and mark the correct option from the codes given below.

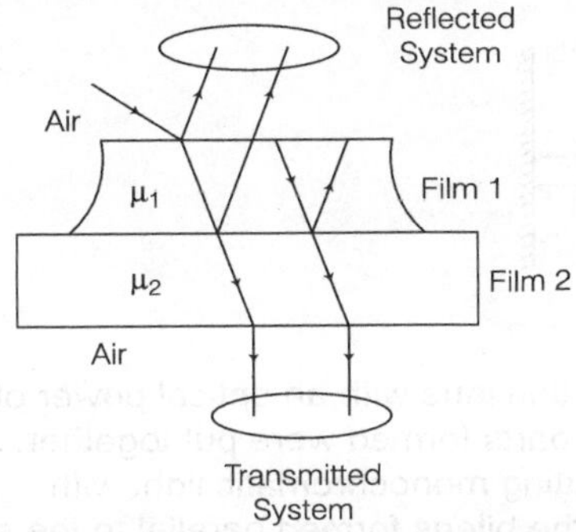

	Column I		Column II
i.	$\mu_1 = 1.5$ and $\mu_2 = 1.5$	p.	Film 1 appears shiny from the reflected system
ii.	$\mu_1 = 2$ and $\mu_2 = 1.5$	q.	Film 1 appears dark from the reflected system
iii.	$\mu_1 = \frac{4}{3}$ and $\mu_2 = 1.5$	r.	Film 1 appears shiny from the transmitted system
iv.	$\mu_1 \neq \mu_2$	s.	Film 1 appears dark from the transmitted system

Codes

	i	ii	iii	iv
a.	q,r	q,s	p,s	p,r,q,s
b.	q	p	r	s
c.	p,q	s,q	s,p	r,p,q,s
d.	r,q	s,p	p	q,p,r,s

One Integer Value Correct Type

Direction *(Q. No. 27-32) This section contains 6 questions. When worked out will result in an integer from 0 to 9 (both inclusive).*

27 A parallel beam of light of wavelength 600 nm is incident on a film of refractive index $\sqrt{3}$. The angle of incidence is 60°. The minimum thickness of the film for maximum reflection is 100x nm. Find the value of x.

28 The figure shows the modified Fresnel biprism experiment with monochromatic source of wavelength 500 nm. The refractive angle of the glass prism is 2°. Find the number of bright fringes formed.

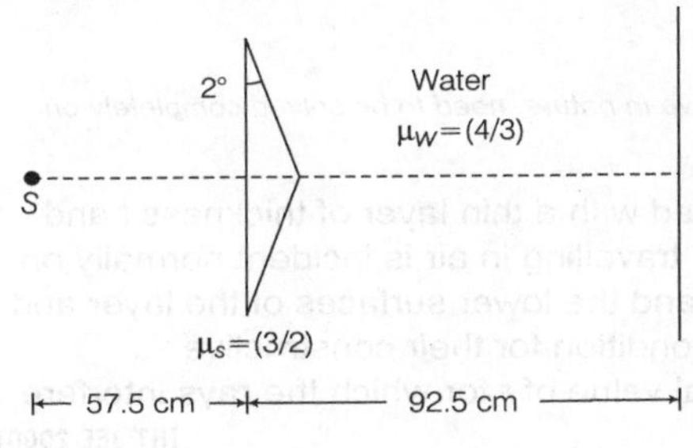

Working Space

29 Consider interference of direct and reflected light in the situation as shown in the figure. Find the number of minima that will be formed on the screen.

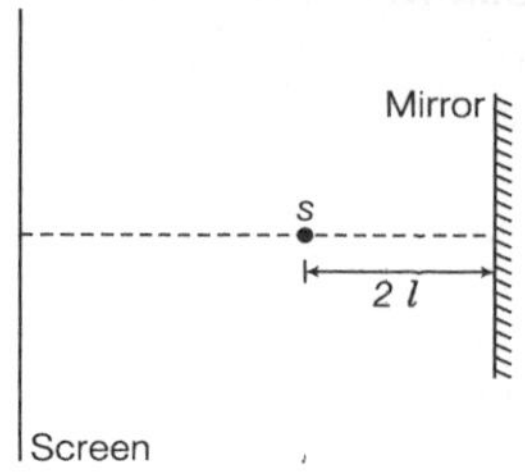

30 A strip of wdith h = 1.00 mm was cut of from a thin lens with an optical power of P = + 2.00 D along its diameter. Next the lens parts formed were put together. A source S in the form of a luminescent slits emitting monochromatic light with λ_0 = 500 nm was placed in the focal plane of the bilens formed parallel to the slit as shown in the figure. A screen was placed after the bilens at a distance of b = 1.00 m from it. Determine the maximum number N of fringes that can be obeserved in this case.

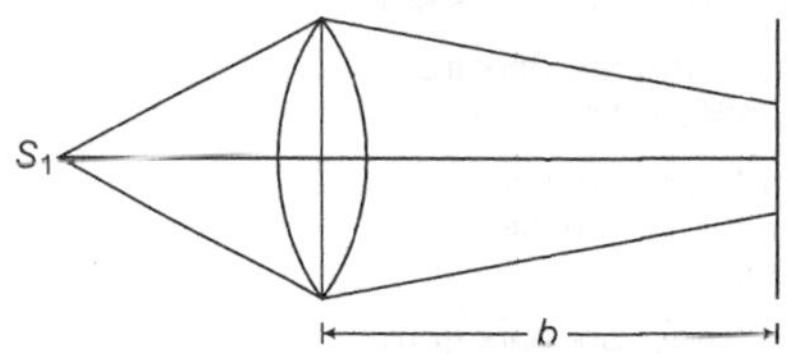

31 The arrangement for a mirror experiment as shown in the figure. S is a point source of wavelength 600 nm D and C represent the two ends of a mirror placed horizontally, and LOM represents the screen. If number of maxima on screen is 11 N. Find the value of N.

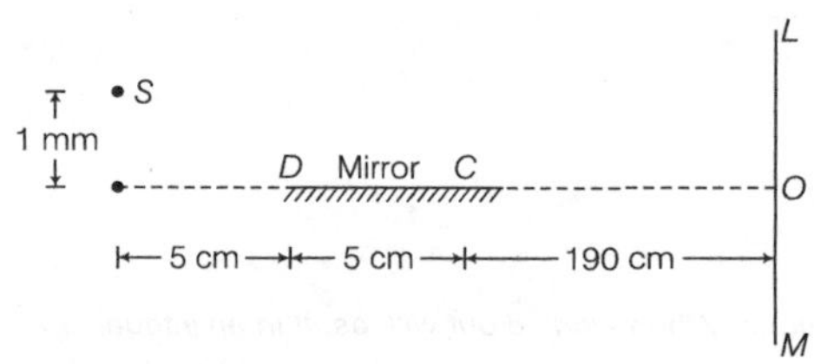

32 A broad source of light of wavelength of 6000 Å illuminates normally two glass plates 12 cm long that meet at one end and are separated by a wire 48×10^{-3} mm in diameter at the end. The number of bright fringe formed over the 12 cm distance is 5 n. Find the value of n.

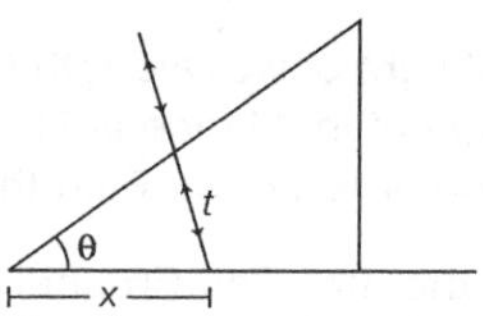

Subjective Questions

Direction (Q. Nos. 33-36) *These questions are subjective in nature, need to be solved completely on notebook.*

33 A glass plate of refractive index 1.5 is coated with a thin layer of thickness t and refractive index 1.8. Light of wavelength λ' travelling in air is incident normally on the layer. It is partly reflected at the upper and the lower surfaces of the layer and the two reflected rays interfere. Write the condition for their constructive interference. If λ = 648 mm, obtain the local value of r for which the rays interfere constructively. **[IIT JEE 2000]**

Working Space

34 A point source S emitting light of wavelength 600 nm is placed at a very small height h above a flat reflecting surface AB as shown in the figure. The intensity of the reflected light is 36% of the incident intensity. Interference fringes are observed on a screen placed parallel to the reflecting surface at a very large distance D from it.

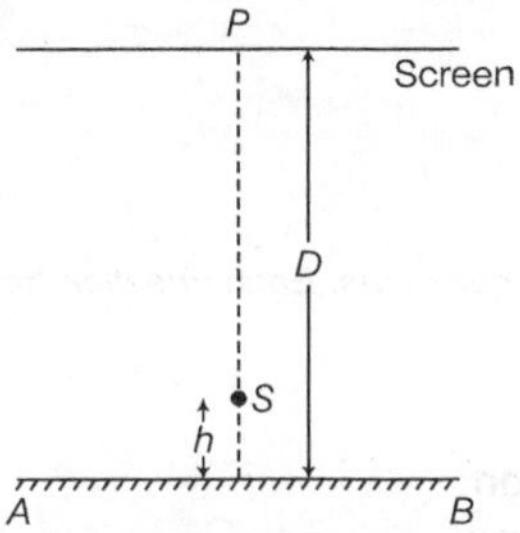

i. What is the shape of the interference fringes on the screen ?

ii. Calculate the ratio of the minimum to the maximum intensities in the inferference fringes formed near the point P as shown in the figure.

iii. If the intensity at point P corresponds to a maximum, calculate the minimum distance through which the reflecting surface AB should be shifted so that the intensity at P again becomes maximum.

35 A narrow monochromatic beam of light of intensity I is incident on a glass plate as shown in figure. Another identical glass plate as kept close to the first one and parallel to it. Each glass plate reflects 25% of the light incident on its and transmits the remaining. Find the ratio of the minimum and maximum intensities in the interference pattern formed by the two beams, obtained after only reflection at each other.

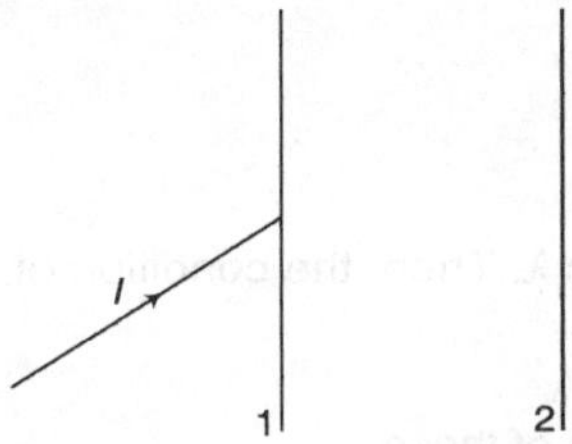

36 In a modified Young's double slit experiment, monochromatic uniform and parallel beam of light of wavelength 6000 Å and intensity $(10/\pi)$ Wm^{-2} is inclined normally on two aperture A and B of radii 0.001 m and 0.002 m respectively. A perfectly transparent film of thickness 2000 Å and refractive index 1.5 for the wavelength of 6000 Å is placed in front of aperture A (see figure). Calculate the power (in W) received at the focal spot F of the lens. The lens is symmetrically placed with respect to the apertures. Assume that, 10% of the power received by each aperture goes in the original direction and is brought to the focal spot.

DPP-6 Diffraction of Light

Only One Option Correct Type

Working Space

Direction *(Q. Nos. 1-11) This section contains 11 multiple choice questions. Each question has four choices (a), (b), (c) and (d), out of which ONLY ONE is correct.*

1 Bending of light at the edge of obstacles is

a. reflection b. refraction
c. polarisation d. diffraction

2 A parallel beam of fast moving electrons is incident normally on a narrow slit. If the speed of the electrons is decreased, the angular width of central maxima

a. increases b. decreases
c. remains same d. None of these

3 In Fresnel diffraction, if distance between the disc and the screen is decresed, the intensity of central bright spot

a. remains constant b. increases
c. decreases d. None of these

4 If blue light is used instead of red light to observe diffraction pattern from single slit. Then, diffraction pattern

a. will expand
b. will contract
c. will be more clear
d. will disappear

5 If width of slit is b and wavelength of used light is λ. Then, the condition of diffraction is

a. $b \approx \lambda$ b. $b >> \lambda$
c. $b << \lambda$ d. None of these

6 A screen is placed at a distance of 1 m from a circular hole illuminated by a parallel monochromatic light of wavelength 490 nm. Find the radius of fourth half period zone.

a. 0.12 cm b. 0.14 cm
c. 0.16 cm d. 0.18 cm

7 If the radius of first zone is 0.3 mm for wavelength 500 nm. Then,

a. first focal length is 18 cm
b. first focal length is 6 cm
c. first focal length is 3.6 cm
d. first focal length is 10 cm

8 A narrow slit of width 0.02 mm is illuminated by a parallel beam of light. The diffraction pattern is observed through a telescope. It is found that to reach the first minima, the telescope has to be rotated through $1°30'$ from the direction of the direct ray. Calculate the wavelength used.

a. 400 nm b. 523.3 nm
c. 600 nm d. 700 nm

Working Space

9 Yellow light is used in a single slit diffraction experiment with slit width of 0.6 mm. If yellow light is replaced by X-rays, then the observed pattern will reveal

a. that the central maximum is narrower
b. more number of fringes
c. less number of fringes
d. No diffraction pattern

10 A parallel monochromatic beam of light is incident normally on a narrow slit. A diffraction pattern is formed on a screen placed perpendicular to the direction of the incident beam. At the first minima of the diffraction pattern, the phase difference between the rays coming from the two edges of the slit is

a. zero
b. $\pi/2$
c. π
d. 2π

11 A narrow slit of width 1 mm is illuminated by monochromatic light of wavelength 600 nm. The distance between the first minima on either side of a screen at a distance of 2 m is

a. 1.2 cm
b. 1.2 mm
c. 2.4 cm
d. 2.4 mm

One or More than One Option Correct Type

***Direction** (Q. Nos. 12 and 13) This section contains 2 multiple choice questions. Each question has four choices (a), (b), (c) and (d), out of which ONE or MORE THAN ONE are correct.*

12 Mark the correct option(s).

a. In an interference pattern, the minima are usually almost perfectly dark while in a diffraction pattern they are not so
b. In an interference pattern, all the maxima are of same intensity but not so in the diffraction pattern
c. In interference fringes are usually equispaced but diffraction fringes are never equally spaced
d. None of the above

13 In the case of double slit diffraction, if slit width increases,

a. central peak becomes sharper
b. fringe spacing does not change
c. less interference maxima fall within the central diffraction maximum
d. None of the above

Comprehension Type

***Direction** (Q. Nos. 14 and 15) This section contains a paragraph, describing theory, experiments, data, etc. Two questions related to the paragraph have been given. Each question has only one correct answer among the four given options (a), (b), (c) and (d).*

Passage

A parallel beam of monochromatic light of wavelength 500 nm is normally incident on a single slit. The central maximum spreads out at 30° on both sides of the direction of incident light.

14 Calculate the slit width.

a. 10^{-3} mm
b. 10^{-6} mm
c. 2×10^{-3} mm
d. 5×10^{-3} mm

Working Space

15 For what width of slit the central maximum would spread out to 90° from the direction of incident light?

a. 10^{-3} mm   b. 250 nm
c. 500 nm   d. 1000 nm

Matching List Type

Direction (Q. No. 16) *Choices for the correct combination of elements from Column I and Column II are given as options (a), (b), (c) and (d), out of which one is correct.*

16 Consider the diffraction through single slit of width b and wavelength of light λ. Match the Column I with Column II and mark the correct option from the codes given below.

	Column I		Column II
i.	$\lambda << b$	p.	Diffraction pattern is found on screen
ii.	$\lambda < b$	q.	No diffraction pattern is found on screen
iii.	$\lambda = b$	r.	A sharp bright image of slit is formed on screen
iv.	$\lambda > b$	s.	Central maxima will extend over whole screen

Codes

	i	ii	iii	iv
a.	q,r	p	q,s	q
b.	r	p	q	s
c.	p	q	r	s
d.	q,r	p,q	s	q,s

Statement Type

Direction (Q. Nos. 17 and 18) *This section is based on Statement I and Statement II. Select the correct answer from the codes given below.*

a. Both Statement I and Statement II are correct and Statement II is the correct explanation of Statement I
b. Both Statement I and Statement II are correct but Statement II is not the correct explanation of Statement I
c. Statement I is correct but Statement II is incorrect
d. Statement II is correct but Statement I is incorrect

17 **Statement I** Two objects of equal intensity will be just resolved by an optical instrument, if the central diffraction maximum of one lies at the first minimum of the other.

Statement II Diffraction can be explained on the basis of Huygens' theory.

18 **Statement I** In single slit diffraction, if slit is made narrower, central maximum becomes wider.

Statement II Presence of light in shadow is due to diffraction.

One Integer Value Correct Type

Working Space

Direction *(Q. Nos. 19-22) This section contains 4 questions. When worked out will result in an integer from 0 to 9 (both inclusive).*

19 In Fraunhoffer diffraction due to a narrow slit, a screen is placed 2 m away from the lens to obtain the pattern. If the slit width is 0.2 mm and the first minima is 5 mm on either side of the central maximum, find the wave of light. If your answer is $n \times 10^{-7}$ m, find the value of n.

20 A linear aperture whose width 0.2 mm is placed immediately infront of a lens of focal length 60 cm. This aperture is illuminated normally by a parallel beam whose wavelength is 500 nm. The distance between the centre and first dark band of diffraction pattern on a screen placed 60 cm from lens is $3x \times 10^{-2}$ cm. Find the value of x.

21 The diameter of the first ringe of a zone plate is $\sqrt{2}$ mm. If a plane wave of wavelength 500 nm falls on the plate, where should the screen be placed so that light is focussed to a brightest spot. (Give answer in metre).

22 An object is placed at 20 cm from a zone plate and the brightest image is situated at 20 cm from the zone plate with light of wavelength 400 nm. If the number of Fresnel's zones in a radius of 10 mm that plate is 500 x. Find the value of x.

Subjective Questions

Direction *(Q. No. 23) These questions are subjective in nature, need to be solved completely on notebook.*

23 Angular width of central maximum in the Fraunhoffer diffraction pattern of a slit is measured. The slit is illuminated by light of wavelength 6000 Å. When the slit is illuminated by light of another wavelength, the angular width decreases by 30%. Calculate the wavelength of the light. The same decrease in the angular width of central maximum is obtained when the original apparatus is immersed in a liquid. Find the refractive index of the liquid.

DPP-7 Polarisation of Light

Only One Option Correct Type

Working Space

Direction (Q. Nos. 1-10) This section contains 10 multiple choice questions. Each question has four choices (a), (b), (c) and (d), out of which ONLY ONE is correct.

1 Which of the following phenomena can be demonstrated by light but not with sound wave in an air column?

a. Reflection b. Diffraction
c. Refraction d. Polarisation

2 A light wave has amplitude A and angle between analyser and polariser is 45°. Light is reflected by analyser has amplitude

a. $\frac{A}{2}$ b. $\frac{A}{\sqrt{2}}$
c. $2A$ d. $\sqrt{3}A$

3 An electromagnetic wave travels in vacuum along x-direction, is $\mathbf{E} = (E_1\hat{\mathbf{j}} + E_2\hat{\mathbf{k}}) \sin (kx - \omega t)$. The corresponding magnetic field is

a. $\mathbf{B} = \frac{1}{c}(E_1\hat{\mathbf{j}} + E_2\hat{\mathbf{k}}) \sin (kx + \omega t)$

b. $\mathbf{B} = \frac{1}{c}(-E_2\hat{\mathbf{j}} + E_1\hat{\mathbf{k}}) \sin (kx - \omega t)$

c. $\mathbf{B} = \frac{1}{c}(E_1\hat{\mathbf{j}} - E_2\hat{\mathbf{k}}) \sin (kx - \omega t)$

d. $\mathbf{B} = \frac{1}{c}(-E_1\hat{\mathbf{j}} + E_2\hat{\mathbf{k}}) \cos (kx - \omega t)$

4 In previous problem,

a. electromagnetic wave is non-polarised
b. electromagnetic wave is plane polarised
c. electromagnetic wave is circularly polarised
d. electromagnetic wave is elliptically polarised

5 Unpolarised light is equivalent to

a. the superposition of two mutually perpendicular identical unpolarised light
b. linearly polarised light
c. circularly polarised light
d. None of the above

6 If an unpolarised light is converted into plane polarised light, its intensity becomes

a. half b. one-fourth
c. one-third d. one-eighth

7 Two beams, A and B of plane polarised light with mutually perpendicular planes of polarisation are seen through a polaroid. From the position, when the beam A has maximum intensity (and beam B has zero intensity), a rotation of polaroid through 30° makes the two beams appear equally bright. If the initial intensities of the two beams are I_A and I_B respectively, then I_A/I_B equals **[JEE Main 2014]**

a. 3 b. 3/2 c. 1 d. 1/3

8 A beam of unpolarised light of intensity passed through a polaroid *A* and *B* through another polaroid *B* which oriented so that its principal plane makes angle of 45° relative to that of *A*. The intensity of the emergent light is **[JEE Main 2014]**

a. I_0

b. $\frac{I_0}{2}$

c. $\frac{I_0}{4}$

d. $\frac{I_0}{8}$

9 A polariser is used to

a. produce a wave
b. produce electromagnetic wave
c. produce transverse wave
d. produce polarised light

10 Which of the following cannot be polarised?

a. Light wave

b. Transverse wave

c. Radio wave

d. Ultrasonic wave

One or More than One Options Correct Type

Direction *(Q. No. 11) This section contains a multiple choice question. Each question has four choices (a), (b), (c) and (d), out of which ONE or MORE THAN ONE are correct.*

11 When sunlight is incident on water surface at an angle of incidence 53°, the reflected ray is found to be completely polarised. Then,

a. the angle of refraction is 37°

b. the refractive index of water is $\frac{3}{2}$

c. the angle of refraction is 53°

d. the refractive index of water is $\frac{4}{3}$

Comprehension Type

Direction *(Q. Nos. 12-17) This section contains 3 paragraphs, each describing theory, experiments, data, etc. Six questions related to the paragraphs have been given. Each question has only one correct answer among the four given options (a), (b), (c) and (d).*

Passage I

A beam of plane polarised light is incident normally on a polariser of cross-sectional area 3×10^{-4} m^2, which rotates about the axis of the ray with an angular velocity 10π rads^{-1}. The power of incident ray is 10^{-3}W.

12 Find the intensity of emergent beam.

a. $\frac{1}{3}$ Wm^{-2}

b. $\frac{5}{3}$ Wm^{-2}

c. $\frac{4}{3}$ Wm^{-2}

d. $\frac{10}{3}$ Wm^{-2}

13 Find the energy of light passing through the polariser per revolution.

a. 10^{-4} J/rev

b. 10 J/rev

c. $\frac{5}{3}$ J/rev

d. None of these

Working Space

Passage II

Unpolarised light of intensity 32 Wm^{-2} passes through three polarisers such that the transmission axis of the last polariser is crossed with the first. The intensity of the emerging light is 3 Wm^{-2}.

14 Find the angle between the transmission of first two polarisers.

a. 30°
b. 60°
c. 90°
d. 45°

15 At what angle transmitted intensity is maximum?

a. 30°
b. 45°
c. 60°
d. 90°

Passage III

A mixture of plane polarised light and unpolarised light falls normally on a polarising sheet about the direction of the incident beam, the transmitted intensity varies by a factor 4.

16 Find the ratio of the intensities of I_P and I_0 respectively, of the polarised and unpolarised components in incident beam.

a. 1 : 2
b. 1 : 4
c. 3 : 2
d. 3 : 4

17 If the axis of the polarising sheet is fixed at an angle 45° with the direction, when the transmitted intensity is maximum. Then, find the total intensity of the transmitted beam in terms of I_0.

a. $\frac{3}{2}I_0$
b. $\frac{3}{4}I_0$
c. $\frac{6}{5}I_0$
d. $\frac{5}{4}I_0$

Matching List Type

Direction *(Q. No. 18) Choices for the correct combination of elements from Column I and Column II are given as options (a), (b), (c) and (d), out of which one is correct.*

18 Suppose an unpolarised light wave travels along the X-axis. The electric field at any instant is in the YZ-plane.

$$\mathbf{E} = E_1 \sin(\omega t - kx + \phi)\,\hat{\mathbf{j}} + E_2 \sin(\omega t - kx)\,\hat{\mathbf{k}}$$

Some informations about ϕ in Column I and corresponding results given in Column II. Match the Column I with Column II and mark the correct option from the codes given below.

	Column I		Column II
i.	$\phi = 0$	p.	Circularly polarised
ii.	$\phi = \pi$	q.	Elliptically polarised
iii.	$\phi = \frac{\pi}{2}$ and $E_1 = E_2$	r.	Linearly polarised
iv.	$\phi = \frac{\pi}{2}$ and $E_1 \neq E_2$	s.	Plane polarised

Codes

	i	ii	iii	iv
a.	r,s	r s	p	q
b.	r,s	p	q	r
c.	p,s	q,s	r	s
d.	r,s	r,s	p,s	q,s

Working Space

Statement Type

Direction (*Q. No. 19-21*) *This section is based on Statement I and Statement II. Select the correct answer from the codes given below.*

a. Both Statement I and Statement II are correct and Statement II is the correct explanation of Statement I

b. Both Statement I and Statement II are correct but Statement II is not the correct explanation of Statement I

c. Statement I is correct but Statement II is incorrect

d. Statement II is correct but Statement I is incorrect

19 **Statment I** Sungoggles made of polaroids are preferred over coloured glasses.

Statement II Polarisation only takes place in transverse waves.

20 **Statement I** Reflection can result in plane polarised light, if the light is incident on the interface from the side with higher refractive index.

Statement II Polarisation by reflection occurs when the angle of incidence is the Brewster's angle.

21 **Statement I** Polarised light can be used to study the helical surface of a nucleic acids.

Statement II Optical axis is a direction and not any particular line in the crystal.

Integer Type

Direction (*Q. Nos. 22-24*) *This section contains 3 questions. When worked out will result in an integer from 0 to 9 (both inclusive).*

22 At what angle, a beam of light be incident on the surface of glass, which has refractive index $\sqrt{3}$, so that the reflected light is plane polarised. If your answer is $\frac{\pi}{n}$. Find the value of n.

23 If a phase difference $\phi = \frac{6\pi}{10^n}$ is introduced by a quartz plate of thickness 20×10^{-6} n between ordinary and extraordinary light. The refractive index of the plate for ordinary light and extraordinary light are 1.5443 and 1.5533 respectively and wavelength is 600 nm. Find the value of n.

24 The thickness of a quarter wave plate for light of wavelength 400 nm for which refractive index for ordinary and extraordinary light areas 1.5632 and 1.5541 is $\frac{0.1099}{10^n}$ metre. Find the value of n.

SET - 1

Revisal Problems for JEE Main

Only One Option Correct Type

Working Space

Direction *(Q. Nos. 1-20) This section contains 20 multiple choice questions. Each question has four choices (a), (b), (c) and (d), out of which ONLY ONE is correct.*

1 In Young's double slit experiment, the separation between slits is 0.2 mm and they are 1 m from the screen. Find the distance between 8th maxima and 12th minima, if wavelength of light used is 500 nm.

a. 2 mm
b. 2.5 mm
c. 8.75 mm
d. 12.75 mm

2 Intensity on screen due to each slit is I_0 in Young's double slit experiment.

The intensity at a point on screen where path difference between waves $\frac{\lambda}{4}$ is

a. I_0
b. $2I_0$
c. $3I_0$
d. $4I_0$

3 Light waves from two coherent sources of amplitude ratio 9 : 1 produce interference. Find the ratio of maximum and minimum intensities.

a. 5 : 4
b. 9 : 1
c. 25 : 16
d. 5 : 8

4 In Young's double slit experiment, intensity at the centre of screen is I_0. Find the minimum distance from centre of screen, when intensity is $\frac{I_0}{4}$.

a. $\frac{\lambda D}{3d}$
b. $\frac{\lambda D}{d}$
c. $\frac{\lambda D}{2d}$
d. $\frac{\lambda D}{8d}$

5 In two slits interference pattern, the intensity at a point on screen can be represented as

a. $I = A + Bx$
b. $I = A + B\cos^2\frac{kx}{2}$
c. $I = A + B\cos^3\frac{kx}{2}$
d. $I = A + B\cos^4\frac{kx}{2}$

where A, B and k are constant and x is the linear distance of this point from centre of screen.

6 If interferring waves are polarised, then they

a. must be in same state of polarisation
b. must not be in same state of polarisation
c. may not be in same state of polarisation
d. None of the above

7 If angle of Fresnel's biprism increases

a. fringes may be disappear
b. fringes will be more visible
c. fringe width will increase
d. None of the above

8 In an interference pattern of Young's double slit experiment, we observe the 12th order maxima for wavelength 600 nm at a point on the screen. What order will be visible at the same point, if the source is replaced by light of wavelength 480 nm.

a. 4th
b. 10th
c. 15th
d. 20th

Working Space

9 A thin convex lens of refractive index 1.5 is coated with a thin film of refractive index 1.2 in order to reduce the reflection from its surface at $\lambda = 480$ nm. Find the minimum thickness of the film which will minimise the intensity of the reflected light.

a. 10^{-7} m b. 10^{-6} m
c. 2×10^{-7} m d. 3×10^{-8} m

10 Calculate the minimum thickness of the film ($\mu = 1.4$) in which interference of violet component ($\lambda = 400$ nm) of incident light can take place by reflection.

a. 714.3 Å b. 4000 Å
c. 200 Å d. 314 Å

11 Newton's rings are observed normally in reflected light of wavelength 5.9×10^{-7} m. The radius of 10th dark ring is 0.25 cm. Find the thickness of the film.

a. 10^{-6} m b. 2×10^{-6} m
c. 3×10^{-6} m d. 4×10^{-6} m

12 A beam of light travelling in water strikes a glass plate which is also immersed in water. When the angle of incidence is 60°, the reflected beam is found to be plane polarised. Find the refractive index of glass, if refractive index of water is $\sqrt{2}$.

a. $\sqrt{3}$ b. $\sqrt{\frac{3}{2}}$
c. $\sqrt{6}$ d. None of these

13 A thin mica sheet ($\mu = 1.6$) of 7 μm thickness introduced in the path of one of the interfering beams in a Fresnel's biprism arrangement shifts the central fringe to a position normally occupied by 7th bright fringe from the centre of screen. Find the wavelength of light used.

a. 400 nm b. 500 nm
c. 600 nm d. 700 nm

14 Interference fringes are produced by Fresnel's biprism in the focal plane of eyepiece 2 m away from the slit. The two images of the slit that are formed for each of two positions of a convex lens placed between the biprism and eyepiece are found to be separated by 4.5 mm and 2.9 mm, respectively. If the width of the interference fringes be 0.326 mm. Find the wavelength of light used.

a. 4000 Å b. 5870 Å
c. 6000 Å d. 7680 Å

15 A Fresnel's biprism arrangement is set with sodium light of wavelength 5893 Å and in the field of view of the eyepiece 62 fringes are observed. How many fringes shall we get in the same field of view, if we replace the source by light of wavelength 5461 Å?

a. 50 b. 55
c. 58 d. 67

16 A thin slice is cut out of a glass cylinder along a plane parallel to its axis as shown in figure. The slice is placed on a flat glass plate with the curved surface downward. A parallel beam of monochromatic light is incident normally from the top. Then,

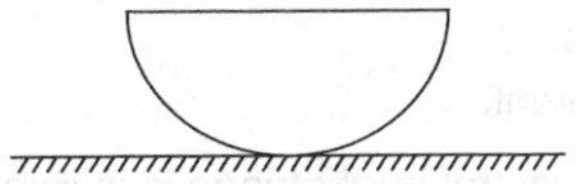

a. the fringes are circular
b. the line of contact of the cylindrical glass piece and the glass plate appears dark
c. the fringe spacing increases as we go outward
d. None of the above

17 To detect central bright fringe in Fresnel's biprism experiment.

a. White light is used
b. Monochromatic light is used
c. Separation between source and biprism should be large
d. Angle of prism should be very small

Working Space

18 The amplitude of oscillation of electric field is 18 N/C. The amplitude of oscillation of magnetic field is

a. 6×10^{-8} T
b. 3×10^{-8} T
c. 18×10^{-8} T
d. None of these

19 The electric field in a electromagnetic wave is $E = 50 \sin\left(\omega t - \frac{\omega}{c} x\right)$ in SI. Find mean energy density.

a. 24.15×10^{-9} Jm^{-3}
b. 22.13×10^{-9} Jm^{-3}
c. 29.16×10^{-10} Jm^{-3}
d. 28.15×10^{-8} Jm^{-3}

20 A beam of circularly polarised light is completely absorbed by an object on which it falls. If E_0 represents absorbed energy and f represents frequency. Then, angular momentum transferred to the object is

a. $\frac{E_0}{4\pi f}$
b. $\frac{E_0}{2\pi f}$
c. $\frac{E_0}{\pi f}$
d. $\frac{E_0}{f}$

Statement Type

Direction (Q. Nos. 21-27) *This section is based on Statement I and Statement II. Select the correct answer from the codes given below.*

a. Both Statement I and Statement II are correct and Statement II is the correct explanation of Statement I
b. Both Statement I and Statement II are correct but Statement II is not the correct explanation of Statement I
c. Statement I is correct but Statement II is incorrect
d. Statement II is correct but Statement I is incorrect

21 **Statement I** In a Young's double slit experiment, source is white but one hole is covered by a green filter and other by orange filter. Then, there is no interference fringes on the screen.

Statement II For fringe pattern on the screen, wavelength from both holes should be same.

22 **Statement I** Huygens' principle is valid for sound wave.

Statement II Sound is longitudinal wave in gaseous medium.

23 **Statement I** The diffraction of sound waves is more evident in daily life than that of light wave.

Statement II Diffraction effects are seen, if there are width of slits $b \sim \lambda$.

24 **Statement I** In the case of Newton's ring formation, the diameters of bright rings are proportional to the square roots of the odd natural numbers.

Statement II The contact point of planoconvex lens and plate during formation of Newton's ring, is a point of maxima.

25 **Statement I** Energy remains conserved in interference.

Statement II For interference sources should be coherent.

26 **Statement I** In the case of Lloyd's mirror experiment, central bright fringe is invisible.

Statement II Due to reflection of light from mirror, change in path difference is $\frac{\lambda}{2}$.

27 **Statement I** A narrow source is necessary for good contrast in the case of biprism experiment.

Statement II If extended source is used in biprism experiment, poor contrast is obtained.

SET - 2

Revisal Problems for JEE Advanced

Only One Option Correct Type

Working Space

***Direction** (Q. Nos. 1-8) This section contains 8 multiple choice questions. Each question has four choices (a), (b), (c) and (d), out of which ONLY ONE option is correct.*

1 A parallel beam of monochromatic light of intensity I_0 is incident on a glass plate, 25% of light is reflected by upper surface and 50% of light is reflected from lower surface. The ratio of minimum and maximum intensity in interference region of reflected light is

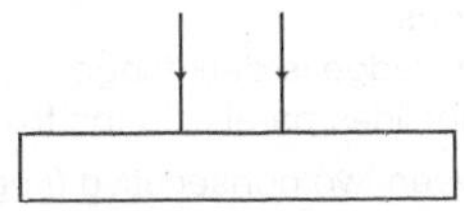

a. $\dfrac{\left(\frac{1}{2}-\sqrt{\frac{3}{8}}\right)^2}{\left(\frac{1}{2}+\sqrt{\frac{3}{8}}\right)^2}$ b. $\dfrac{(1-\sqrt{2})^2}{(1+\sqrt{2})^2}$ c. $\dfrac{3}{8}$ d. $\dfrac{4}{9}$

2 In Young's double slit experiment, one of the slits is covered with a transparent sheet of thickness 3.6×10^{-3} cm due to which position of central fringe shifts to a position originally occupied by 30th fringe. Find the refractive index of sheet.

a. 1.5 for $\lambda = 600$ nm b. 1.2 for $\lambda = 600$ nm
c. 1.5 for $\lambda = 400$ nm d. 1.3 for $\lambda = 500$ nm

3 A slit (wavelength $\lambda = 600$ nm) of adjustable width is used to illuminate two points 0.1 mm apart. Initially, the slit is closed and then gradually it is opened. The width at which the fringes on a screen 2 m apart will just disappear is

a. 1 mm b. 2 mm
c. 10 mm d. 12 mm

4 A slit of width 1.2 mm is used as a source to illuminate two point sources 0.1 mm apart on an opaque screen. The maximum permissible distance between the slit source and the screen for coherence is ($\lambda = 600$ nm)

a. 0.1 m b. 0.2 m
c. 0.25 m d. 0.3 m

5 The headlights of a truck are 1.5 m apart. How far away could the truck and the headlights just be resolved by a person's eye. (Take, wavelength of light is 550 nm, diameter of people is 7 mm and refractive index of viterous humor in eye is 1.36).

a. 10 km b. 21.28 km
c. 20 km d. None of these

6 A mixture of red (630 nm) and blue (490 nm) light pass through a 5500 slits/cm grating and falls on a screen located 1 m away. The separation in colours in first order is

a. 2 cm b. 3 cm
c. 5 cm d. 8 cm

7 A student wants to find the radius of a human hair by placing it between two flat glass plates, illuminating the plates with light of wavelength 500 nm and counting the number of bright fringes produced along plates. He finds 125 bright fringes between the edge of the plates and the hair. What is the diameter of the hair ?

a. 15.625×10^{-6} m b. 31.25×10^{-6} m
c. 6×10^{-6} m d. None of these

8 Find the maximum number of dark fringes that could be appear on a screen in a Young's double slit arrangement for $d = 0.1$ mm and $\lambda = 400$ nm.

a. 400 b. 500
c. 501 d. 449

One or More than One Options Correct Type

Working Space

Direction (Q. Nos. 9-16) ***This section contains 8 multiple choice questions. Each question has four choices (a), (b), (c) and (d), out of which ONE or MORE THAN ONE are correct.***

9 A thin wedge of angle θ (in radian) and refractive index μ is seen normally by reflected sodium light of wavelength λ.

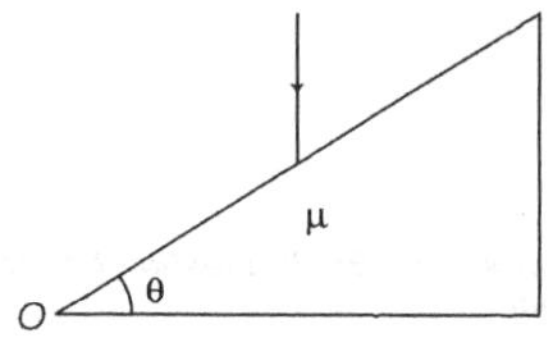

Mark the correct options.

a. The point O of the wedge is dark fringe

b. Fringes are straight lines parallel to the thin edge of the wedge

c. The spacing between two consecutive fringes is $\frac{\lambda}{2\mu\theta}$

d. None of the above

10 Monochromatic light waves of wavelength λ from two coherent sources fall on a large plane screen. One of the waves is emanated from a point source S located at distance L from the screen and other one is a broad plane wave as shown in the figure.

Mark the correct options.

a. Central point of screen is bright spot

b. The shape of fringe on screen is circular

c. The distance between nth and $(n-1)$th bright fringe is $\sqrt{2\lambda l}(\sqrt{n}-\sqrt{n-1})$

d. None of the above

11 Two identical slits each of width a are placed d distance apart from centre to centre in Young's double slit experiment. The distance between slits and screen is D. (Take, wavelength of light λ)

a. The difference of phase differences between rays from the bottom and top of the slits is $\frac{2\pi ad}{\lambda D}$

b. The difference of phase differences between rays from the bottom and top of the slits is $\frac{\pi ad}{2\lambda D}$

c. The maximum value of a for which interference fringes continue to be sharp is $\frac{D\lambda}{2d}$

d. The maximum value of a for which interference fringes continue to be sharp is $\frac{D\lambda}{d}$

12 A system as shown in figure, consists of two coherent sources S_1 and S_2 located in a certain plane. Sources are separated by a distance d apart. The wavelength of light used is λ. The oscillation of S_2 lag in phase behind the oscillations of S_1 by $\phi_0 (\phi_0 < \pi)$

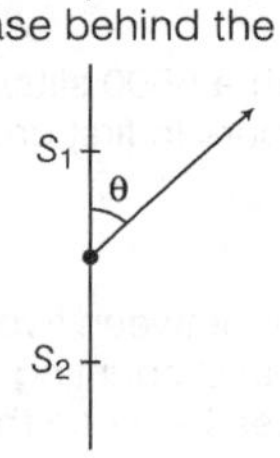

Mark the correct options.

a. For maximum intensity in the direction $\theta = \pi$, $\frac{2\pi}{\lambda}d = \phi_0 + 2m\pi$, $m = 0, 1, 2, \ldots$

b. For minimum intensity in the direction $\theta = 0$, $\frac{2\pi}{\lambda}d = (2m+1)\pi - \phi_0$

c. For maximum intensity in the direction $\theta = \pi$, $\frac{2\pi}{\lambda}d = \phi_0 + (2m+1)\pi$

d. The condition under which the radiation intensity is maximum in direction $\theta = \pi$ and minimum in direction $\theta = 0°$ is $\phi_0 = \frac{\pi}{2}$ and $\frac{d}{\lambda} = m + \frac{1}{4}$, where, $m = 0, 1, 2, 3, \ldots$

Working Space

13 In the Young's double slit experiment, the interference pattern is found to have an intensity ratio between the bright and dark fringes as 9. This implies that

a. the intensities at the screen due to the two slits are 5 units and 4 units respectively
b. the intensities at the screen due to the two slits are 4 units and 1 unit respectively
c. the amplitude ratio is 3
d. the amplitude ratio is 2

14 A light source, which emits two wavelengths $\lambda_1 = 400$ nm and $\lambda_2 = 600$ nm, is used in a Young's double-slit experiment. If recorded fringe widths for λ_1 and λ_2 are β_1 and β_2 and the number of fringes for them within a distance y on one side of the central maximum are m_1 and m_2, respectively, then

a. $\beta_2 > \beta_1$
b. $m_1 > m_2$
c. from the central maximum, 3rd maximum of λ_2 overlaps with 5th minimum of λ_1
d. the angular separation of fringes of λ_1 is greater than λ_2

15 In an interference arrangement similar to Young's double-slit experiment, the slits S_1 and S_2 are illuminated with coherent microwave sources, each of frequency 10^6 Hz. The sources are synchronised to have zero phase difference. The slits are separated by distance

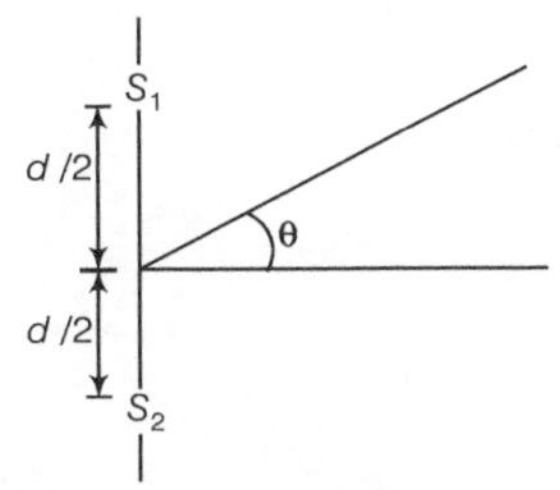

$d = 150.0$ m. The intensity $I(\theta)$ is measured as a function of θ, where θ is defined as shown. If I_0 is the maximum intensity, then $I(\theta)$ for $0 \le \theta \le 90°$ is given by

a. $I(\theta) = I_0/2$ for $\theta = 30°$
b. $I(\theta) = I_0/4$ for $\theta = 90°$
c. $I(\theta) = I_0$ for $\theta = 0°$
d. $I(\theta)$ is constant for all values of θ

16 In a Young's double slit experiment, the separation between the two slits is d and the wavelength of the light is λ. The intensity of light falling on slit 1 is four times the intensity of light falling on slit 2. Choose the correct options.

a. If $d = \lambda$, the screen will contain only one maximum
b. If $\lambda < d < 2\lambda$, at least one more maximum (besides the central maximum) will be observed on the screen.
c. If the intensity of light falling on slit 1 is reduced so that it becomes equal to that of slit 1, the intensities of the observed dark and bright fringes will increase.
d. If the intensity of light falling on slit 2 is increased so that it becomes equal to that of slit 1, the intensities of the observed dark and bright fringes will increase.

Comprehension Type

Direction (Q. Nos. 17-25) *This section contains 3 paragraphs, each describing theory, experiments, data, etc. Nine questions related to the paragraphs have been given. Each question has only one correct answer among the four given options (a), (b), (c) and (d).*

Passage I

Fermat's principle is applicable in discussion of laws of optics on the basis of geometrical tracing Fermat's principle is stated as, a ray of light travelling from one point to another by any number of reflections and refractions follows a path for which, compared with all other neighbouring paths, the time taken is either a minimum or a maximum or stationary.

17 When light reflects from a plane mirror, it follows a path of

a. least time
b. maximum time
c. stationary time
d. either of maximum time or of minimum time

Working Space

18 A ray of light travels from point A to point B. Refraction takes place at plane surface (as shown in figure).

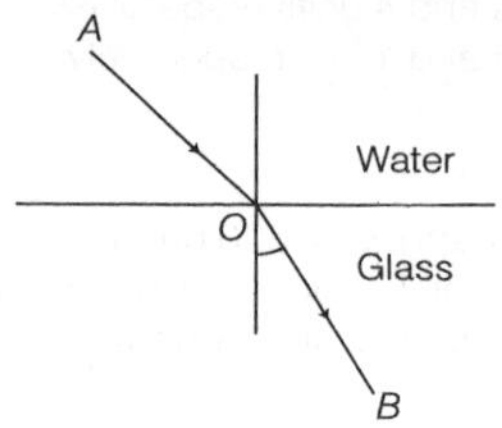

The optical path from A to B is a path of

a. least time
b. maximum time
c. either maximum time or minimum time
d. stationary time

19 Light from a point A passes through a point B after reflection from mirror (as shown in the figure) at point O. The path AOB a path of

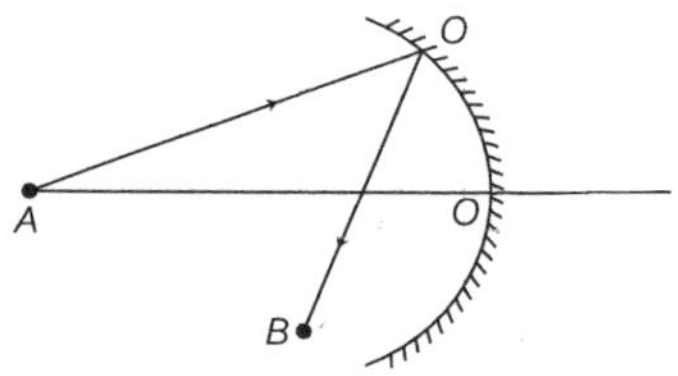

a. least time
b. maximum time
c. stationary time
d. either maximum or minimum time

Passage II

Two point coherent sources with the same phases of oscillation are shown in the figure.

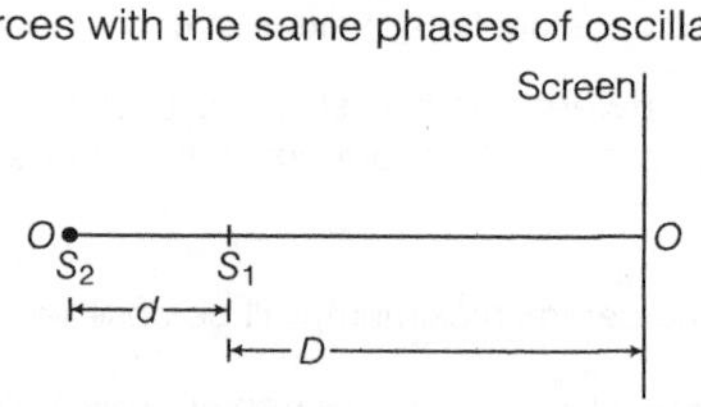

Here, $D >> \lambda$ and $d = n\lambda >> \lambda_3$, where n is an integer.

20 The shape of fringe on screen is

a. straight line
b. circle
c. hyperbola
d. parabola

21 What is the distance on the screen from the perpendicular to the nearest bright band?

a. $\sqrt{\dfrac{2D(D+n\lambda)}{n}}$
b. $\sqrt{\dfrac{D(D+n\lambda)}{n}}$
c. $\sqrt{2nD(D+n\lambda)}$
d. None of these

22 Find the radius of ith bright ring if $D = d = n\lambda$, $n >> l$ and $i = n, n-1, n-2$, etc.

a. $\dfrac{\lambda}{3i}\sqrt{(9n^2-i^2)(n^2-i^2)}$
b. $\dfrac{\lambda}{2i}\sqrt{(9n^2-i^2)(n^2-i^2)}$
c. $\dfrac{\lambda}{4i}\sqrt{(9n^2-2i^2)(2n^2-3i^2)}$
d. None of the above

Working Space

Passage III

The figure shows a surface XY separating two transparent media, medium 1 and medium 2. The lines ab and cd represent wavefronts of a light wave travelling in medium 1 and incident on XY. The lines ef and gh represent wavefronts of the light wave in medium-2 after refraction. **[IIT JEE 2007]**

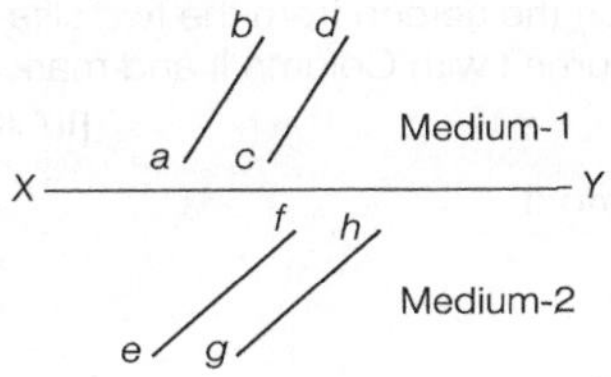

23 Light travels as a

a. parallel beam in each medium
b. convergent beam in each medium
c. divergent beam in each medium
d. divergent beam in one medium and convergent beam in the other medium

24 The phases of the light wave at c, d, e and f are ϕ_c, ϕ_d, ϕ_e and ϕ_f respectively. It is given that $\phi_c \neq \phi_f$

a. ϕ_c cannot be equal to ϕ_d
b. ϕ_d can be equal to ϕ_e
c. $(\phi_d - \phi_f)$ is equal to $(\phi_c - \phi_e)$
d. $(\phi_d - \phi_c)$ is not equal to $(\phi_f - \phi_e)$

25 Speed of light is

a. the same in medium-1 and medium-2
b. larger in medium-1 than in medium-2
c. larger in medium-2 than in medium-1
d. different at b and d

Matching List Type

***Direction** (Q. Nos. 26 and 27) Choices for the correct combination of elements from Column I and Column II are given as options (a), (b), (c) and (d), out of which one is correct.*

26 A convex lens of focal length 0.25 m was cut along the diameter into two identical halves and was again set leaving a thin gap of width $a = 1$ mm. A narrow slit emitting monochromatic light of wavelength 600 nm was placed at the focus of the lens. A screen was placed behind the lens at a distance $b = 0.5$ m. Match the Column I with Column II and mark the correct option from the codes given below.

	Column I		Column II
i.	Position of images (in cm) formed by two halves from the source.	p.	13
ii.	The fringe width in mm	q.	infinity
iii.	Length of fringe pattern on screen in mm	r.	0.15
iv.	Number of bright fringe on screen	s.	2

Codes

	i.	ii.	iii.	iv.
a.	q	r	s	p
b.	p	q	r	s
c.	p,q	r,s	p,q	s
d.	s	p	q	p

27 Column I shows four situations of standard Young's double slit arrangement with the screen placed far away from the slits S_1 and S_2. In each of these cases $S_1P_0 = S_2P_0$, $S_1P_1 - S_2P_1 = \frac{\lambda}{4}$ and $S_1P_2 - S_2P_2 = \lambda/2$, where λ is the wavelength of the light used. In the cases *B*, *C* and *D*, a transparent sheet of refractive index μ and thickness t is pasted on slit S_2. The thickness of the sheets are different in different cases. The phase difference between the light waves reaching a point *P* on the screen from the two slits is denoted by $\delta(P)$ and the intensity by $I(P)$. Match the Column I with Column II and mark the correct option from the codes given below. **[IIT JEE 2009]**

	Column I			Column II
i.		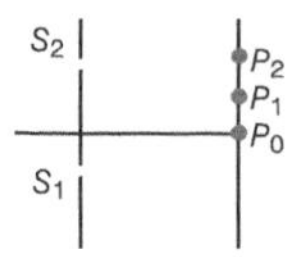	p.	$\delta(P_0) = 0$
ii.	$(\mu - 1)t = \frac{\lambda}{4}$	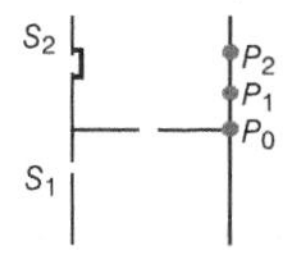	q.	$\delta(P_1) = 0$
iii.	$(\mu - 1)t = \frac{\lambda}{2}$	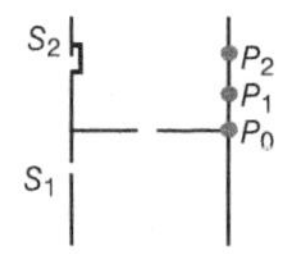	r.	$I(P_1) = 0$
iv.	$(\mu - 1)t = \frac{3\lambda}{4}$	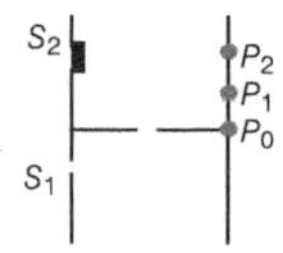	s.	$I(P_0) > I(P_1)$
			t.	$I(P_2) > I(P_1)$

Codes

	i.	ii.	iii.	iv.
a.	q	r	s	p
b.	p,s	q	t	r,s,t
c.	p,q	r,s	p,q	s
d.	s	p	q	p

One or More than One Options Correct Type

Direction *(Q. Nos. 28-32) This section contains 5 multiple choice questions. Each question has four choices (a), (b), (c) and (d), out of which ONE or MORE THAN ONE are correct.*

28 A thin convex lens of focal length 0.2 m is cut along its diameter and two halves are displaced by 1 cm along the optical axis. A monochromatic light source of wavelength 4760 Å is placed at a distance 0.4 m from the first half as shown in figure.

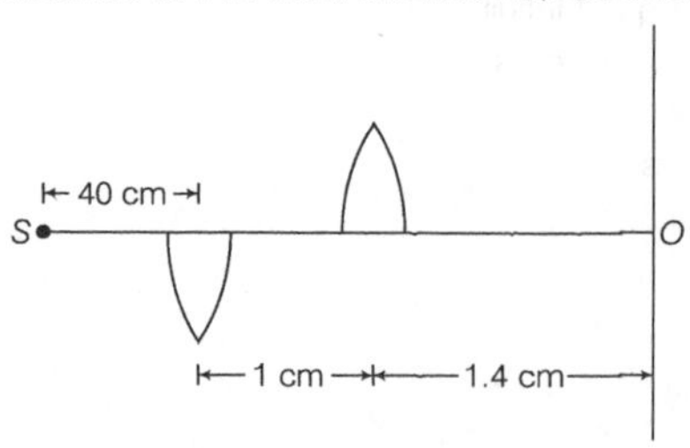

If the position of first maxima from point *O* is $\frac{2236}{n} \times 10^{-4}$ m. Find the value of *n*

Working Space

Working Space

29 Two identical laser beams *A* and *B* of same intensity and wavelength λ fall on a plane screen. The directions of the beam propagations make angles $\theta_1 = 53°$ and $\theta_2 = 37°$ with the normal to the screen and lie in the same plane as shown in the figure. The distance between adjacent fringes is $\beta = n\lambda$. Find the value of *n*.

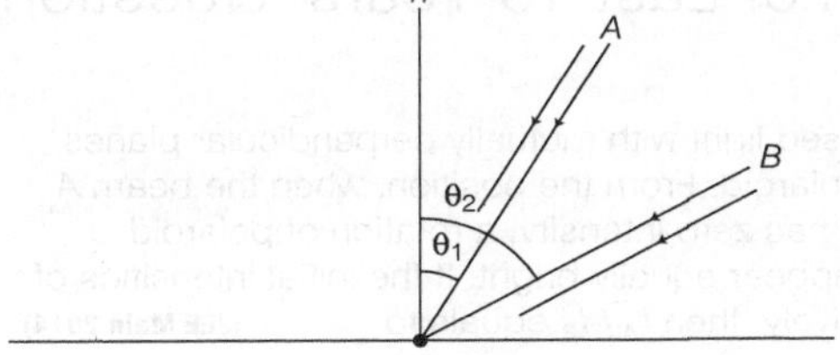

30 The distance from the source to the Fresnel's biprism is 1 m and from the biprism to the screen is 4 m. The angle of biprism is 0.0002 radian and its refractive index is 1.5. The wavelength of light is 600 nm. Find the number of bright fringes formed on the screen.

31 A parallel beam of light of wavelength 450 nm is incident normally on the surface of a wedge made up of glass of refractive index $\mu = 1.5$. The angle between wedge faces is $\left(\frac{1}{20}\right)^\circ$. If the distance between two successive axis is $43n \times 10^{-6}$ m.

Find the value of *n*.

32 Two slits S_1 and S_2 are on a plane inclined at an angle of 45° with horizontal as shown in the figure. The wavelength of light used is 500 nm. The fringe width on screen is $n \times 10^{-4}$m. Find the value of *n*.

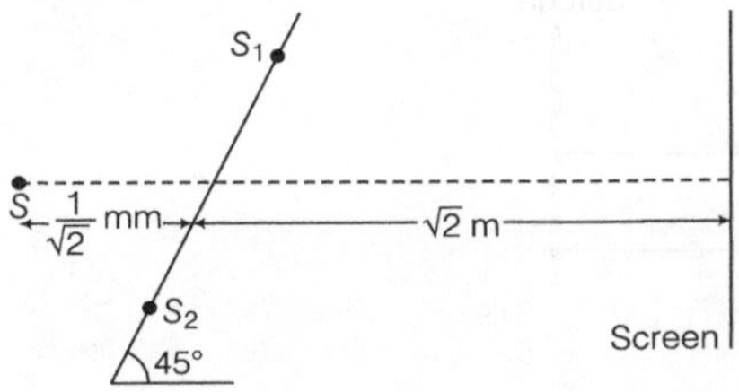

JEE Main & AIEEE Archive

(Compilation of Last 13 Years' Questions)

1 Two beams A and B, of plane polarised light with mutually perpendicular planes of polarisation are seen through a polaroid. From the position, when the beam A has maximum intensity and beam B has zero intensity, a rotation of polaroid through 30° makes the two beams appear equally bright. If the initial intensities of the two beams are I_A and I_B respectively, then I_A / I_B equals to **(JEE Main 2014)**

a. 3 b. 3/2
c. 1 d. 1/3

2 A beam of unpolarised light of intensity I_0 is passed through a polaroid A and then through another polaroid B which is oriented so that its principal plane makes an angle of 45° relative to that of A. The intensity of the emergent light is **(JEE Main 2013)**

a. I_0 b. $\frac{I_0}{2}$
c. $\frac{I_0}{4}$ d. $\frac{I_0}{8}$

3 Two coherent point sources S_1 and S_2 are separated by a small distance d as shown in the figure. The fringes obtained on the screen will be **(JEE Main 2013)**

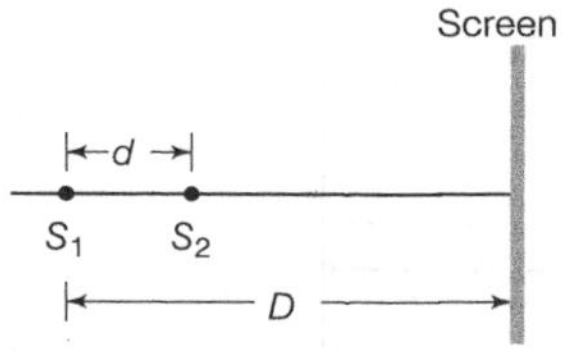

a. points
b. straight lines
c. semi-circle
d. concentric circles

4 In Young's double slit experiment, one of the slits is wider than other, so that amplitude of the light from one slit is double of that from other slit. If I_m is the maximum intensity, the resultant intensity I, when they interfere at phase difference ϕ, is given by **(AIEEE 2012)**

a. $\frac{I_m}{9}(4 + 5\cos\phi)$
b. $\frac{I_m}{3}\left(1 + 2\cos^2\frac{\phi}{2}\right)$
c. $\frac{I_m}{5}\left(1 + 4\cos^2\frac{\phi}{2}\right)$
d. $\frac{I_m}{9}\left(1 + 8\cos^2\frac{\phi}{2}\right)$

5 At two points P and Q on screen in Young's double slit experiment, waves from slits S_1 and S_2 have a path difference of 0 and $\frac{\lambda}{4}$, respectively. The ratio of intensities at P and Q will be **(AIEEE 2011)**

a. 3 : 2 b. 2 : 1
c. $\sqrt{2}$: 1 d. 4 : 1

6 A beaker contains water upto a height h_1 and kerosene of height h_2 above water so that the total height of (water + kerosene) is $(h_1 + h_2)$. Refractive index of water is μ_1 and that of kerosene is μ_2. The apparent shift in the position of the bottom of the beaker, when viewed from above is **(AIEEE 2011)**

a. $\left(1-\frac{1}{\mu_1}\right)h_2+\left(1-\frac{1}{\mu_2}\right)h_1$

b. $\left(1+\frac{1}{\mu_1}\right)h_1+\left(1+\frac{1}{\mu_2}\right)h_2$

c. $\left(1-\frac{1}{\mu_1}\right)h_1+\left(1-\frac{1}{\mu_2}\right)h_2$

d. $\left(1+\frac{1}{\mu_1}\right)h_2-\left(1+\frac{1}{\mu_2}\right)h_1$

7 **Statement I** On viewing the clear, blue portion of the sky through a Calcite crystal, the intensity of transmitted light varies as the crystal is rotated.

Statement II The light coming from the sky is polarised due to scattering of sun light by particles in the atmosphere. The scattering is largest for blue light. **(AIEEE 2011)**

a. Both Statement I and Statement II are correct and Statement II is the correct explanation of Statement I
b. Both Statement I and Statement II are correct but Statement II is not the correct explanation of Statement I
c. Statement I is correct but Statement II is incorrect
d. Statement II is correct but Statement I is incorrect

8 In a Young's double slit experiment, the two slits act as coherent sources of waves of equal amplitude A and wavelength λ. In another experiment with the same arrangement, the two slits are made to act as incoherent sources of waves of same amplitude and wavelength. If the intensity at the middle point of the screen in the first case is I_1 and in the second case is I_2, then the ratio $\frac{I_1}{I_2}$ is **(AIEEE 2011)**

a. 4 b. 2
c. 1 d. 0.5

9 This question has a paragraph followed by two Statements, Statement I and Statement II. Of the given four alternatives after the Statements, choose the one that describes the statements. A thin air film is formed by putting the convex surface of a plane-convex lens over a plane glass plate. With monochromatic light, this film gives an interference pattern due to light reflected from the top (convex) surface and the bottom (glass plate) surface of the film. **(AIEEE 2011)**

Statement I When light reflects from the air-glass plate interface, the reflected wave suffers a phase change of π.

Statement II The centre of the interference pattern is dark.

a. Both Statement I and Statement II are correct and Statement II is the correct explanation of Statement I
b. Both Statement I and Statement II are correct but Statement II is not the correct explanation of Statement I
c. Statement I is correct but Statement II is incorrect
d. Statement II is correct but Statement I is incorrect

10 A mixture of light, consisting of wavelength 590 nm and an unknown wavelength, illuminates Young's double slit and gives rise to two overlapping interference patterns on the screen. The central maximum of both lights coincide. Further, it is observed that the third bright fringe of known light coincides with the 4th bright fringe of the unknown light. From this data, the wavelength of the unknown light is **(AIEEE 2009)**

a. 393.4 nm b. 885.0 nm
c. 442.5 nm d. 776.8 nm

11 In a Young's double slit experiment, the intensity at a point, where the path difference is $\frac{\lambda}{6}$ (λ is being the wavelength of the light used) is I. If I_0 denotes the maximum intensity, then I/I_0 is equal to **(AIEEE 2007)**

a. $\frac{1}{\sqrt{2}}$ b. $\frac{\sqrt{3}}{2}$

c. $\frac{1}{2}$ d. $\frac{3}{4}$

12 A Young's double slit experiment uses a monochromatic source. The shape of the interference fringes formed on a screen is

a. hyperbola b. circle **(AIEEE 2005)**

c. straight line d. parabola

13 When an unpolarised light of intensity I_0 is incident on a polarising sheet, the intensity of the light, which does not get transmitted is

(AIEEE 2005)

a. $\frac{1}{2}I_0$ b. $\frac{1}{4}I_0$

c. zero d. I_0

14 The maximum number of possible interference maxima for slit-separation equal to twice the wavelength in Young's double-slit experiment, is

(AIEEE 2004)

a. infinite b. five

c. three d. zero

15 To demonstrate the phenomenon of interference, we require two sources for which emit radiations of **(AIEEE 2003)**

a. nearly the same frequency
b. the same frequency
c. different wavelength
d. the same frequency and having a definite phase relationship

JEE Advanced & IIT JEE Archive

(Compilation of Last 10 Years' Questions)

1 A light source, which emits two wavelengths $\lambda_1 = 400$ nm and $\lambda_2 = 600$ nm, is used in a Young's double-slit experiment. If recorded fringe widths for λ_1 and λ_2 are β_1 and β_2 and the number of fringes for them within a distance y on one side of the central maximum are m_1 and m_2, respectively, then

(2014 Adv., One or More than One Options Correct Type)

a. $\beta_2 > \beta_1$

b. $m_1 > m_2$

c. from the central maximum, 3rd maximum of λ_2 overlaps with 5th minimum of λ_1

d. the angular separation of fringes of λ_1 is greater than λ_2

2 Using the expression $2d \sin\theta = \lambda$, one calculates the values of d by measuring the corresponding angles θ in the range 0 to 90°. The wavelength λ is exactly known and the error in θ is constant for all values of θ. As θ increases from 0°

(2013 Adv., One or More than One Options Correct Type)

a. the absolute error in d remains constant

b. the absolute error in d increases

c. the fractional error in d remains constant

d. the fractional error in d decreases

3 In the Young's double slit experiment using a monochromatic light of wavelength λ the path difference (in terms of an integer n) corresponding to any point having half the peak intensity is **(2013 Adv., Only One Option Correct Type)**

a. $(2n+1)\frac{\lambda}{2}$

b. $(2n+1)\frac{\lambda}{4}$

c. $(2n+1)\frac{\lambda}{8}$

d. $(2n+1)\frac{\lambda}{16}$

4 In the determination of Young's modulus $\left(Y = \frac{4MLg}{\pi l d^2}\right)$ by using Searle's method, a wire of length $L = 2$ m and diameter $d = 0.5$ mm is used. For a load $M = 2.5$ kg, an extension $l = 0.25$ mm in the length of the wire is observed. Quantities d and l are measured using a screw gauge and a micrometer, respectively. They have the same pitch of 0.5 mm. The number of divisions on their circular scale is 100. The contributions to the maximum probable error of the Y measurement is

(2012, Only One Option Correct Type)

a. due to the errors in the measurements of d and l are the same

b. due to the error in the measurement of d is twice that due to the error in the measurement of l

c. due to the error in the measurement of l is twice that due to the error in the measurement of d

d. due to the error in the measurement of d is four times that due to the error in the measurement of l

5 Column I shows four situations of standard Young's double slit arrangement with the screen placed far away from the slits S_1 and S_2. In each of these cases $S_1P_0 = S_2P_0$, $S_1P_1 - S_2P_1 = \frac{\lambda}{4}$ and $S_1P_2 - S_2P_2 = \lambda/2$, where λ is the wavelength of the light used. In the cases B, C and D, a transparent sheet of refractive index μ and thickness t is pasted on slit S_2. The thickness of the sheets are different in different cases. The phase difference between the light waves reaching a point P on the screen from the two slits is denoted by $\delta(P)$ and the intensity by $I(P)$.

JEE Archive

Match each situation given in Column I with the statement(s) in Column II valid for that situation. **(2009, Matching Type)**

	Column I		Column II
i.	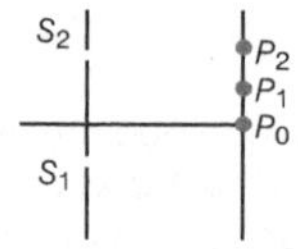	p.	$\delta(P_0) = 0$
ii.	$(\mu - 1)t = \frac{\lambda}{4}$ 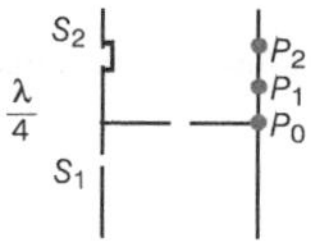	q.	$\delta(P_1) = 0$
iii.	$(\mu - 1)t = \frac{\lambda}{4}$ 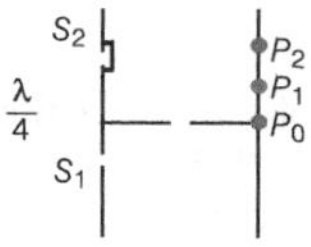	r.	$I(P_1) = 0$
iv.	$(\mu - 1)t = \frac{3\lambda}{4}$ 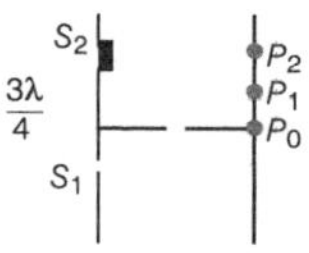	s.	$I(P_0) > I(P_1)$
		t.	$I(P_2) > I(P_1)$

Codes

	i	ii	iii	iv
a.	p,s	q	t	r,s,t
b.	q	p	q	s,t
c.	r	q	s	r
d.	s	r	p	t

6 Young's double slit experiment is carried out by using green, red and blue light, one colour at a time. The fringe widths recorded are β_G, β_R and β_B respectively. Then, **(2012, Only One Option Correct Type)**

a. $\beta_G > \beta_B > \beta_R$ b. $\beta_B > \beta_G > \beta_R$
c. $\beta_R > \beta_B > \beta_G$ d. $\beta_R > \beta_G > \beta_B$

7 In Young's double slit experiment intensity at a point is (1/4) of the maximum intensity. Angular position of this point is **(2005, Only One Option Correct Type)**

a. $\sin^{-1}\left(\frac{\lambda}{d}\right)$ b. $\sin^{-1}\left(\frac{\lambda}{2d}\right)$
c. $\sin^{-1}\left(\frac{\lambda}{3d}\right)$ d. $\sin^{-1}\left(\frac{\lambda}{4d}\right)$

8 In a YDSE bi-chromatic light of wavelengths 400 nm and 560 nm are used. The distance between the slits is 0.1 mm and the distance between the plane of the slits and the screen is 1 m. The minimum distance between two successive regions of complete darkness is **(2004, Only One Option Correct Type)**

a. 4 mm b. 5.6 mm
c. 14 mm d. 28 mm

Answers *with* Explanations

Answers

1. Geometrical Optics

DPP-1 Plane Mirror - I

1. (i) *a,b* (ii) *a,b,c,d* (iii) *a,b,c* (iv) *b,c,d* (v) *a,b,c* (vi) *b* (vii) *a,b,c* (viii) *a* (ix) *a,b,c,d* (x) *a* (xi) *a* (xii) *b* (xiii) *c* (xiv) *a* (xv) *c* (xvi) *a,b,c* (xvii) *a,c,d* (xviii) *b* (xix) *a* (xx) *b* (xxi) *b* (xxii) *a*
2. (i) *c* (ii) *a* (iii) *c* (iv) *b* (v) *b* (vi) *b,c,d* (vii) *a,b*
3. (i) *a* (ii) *d* (iii) *a* (iv) *c*
4. (i) *a* (ii) *b* (iii) *b* (iv) *b* (v) *b* (vi) *b* (vii) *c* (viii) *b*
5. (i) *b* (ii) *c*
6. (i) *a* (ii) *a,b* (iii) *c*
7. (i) *a* (ii) *b* (iii) *b* (iv) *c* (v) *b* (vi) *b* (vii) *a,c* (viii) *c* (ix) *c* (x) *b* (xi) *c* (xii) *a,c* (xiii) *a,c* (xiv) *a*
8. (i) *a,b,c,d* (ii) *c* (iii) *b* (iv) *c* (v) *b* (vi) *a* (vii) *b* (viii) *c* (ix) *d* (x) *b* (xi) *b* (xii) *b*
9. (i) *b* (ii) *b* (iii) *b* (iv) *c* (v) *c* (vi) *a* **10.** (i) *c* (ii) *a* (iii) *d* (iv) *d* (v) *b* **11.** (i) *a* (ii) *a,b* (iii) *c*
12. (i) *a* (ii) *b* **13.** (i) *b* (ii) *c* (iii) *b* **14.** *b* **15.** *a* **16.** *b* **17.** *c* **18.** *a* **19.** *d*
20. *c* **21.** *d* **22.** *b* **23.** *b* **24.** *b* **25.** *a* **26.** *d* **27.** *a* **28.** *d* **29.** *a*
30. *b* **31.** *b* **32.** *d* **33.** *d* **34.** *a* **35.** *d* **36.** *b* **37.** *a* **38.** *a,c* **39.** *a,d*
40. *a,c* **41.** *a,b,c*

DPP-2 Plane Mirror - II

1. *a* **2.** *b* **3.** *c* **4.** *d* **5.** *b* **6.** *b* **7.** *a* **8.** *b* **9.** *b* **10.** *b*
11. *c* **12.** 4 **13.** 3 **14.** 2 **15.** 6 **16.** 6 **17.** 2 **18.** 5 **19.** 4 **20.** 2
21. 1 **22.** 2 **23.** 5 **24.** 5 **25.** 9

DPP-3 Spherical Mirror - I

1. (i) *c* (ii) *b* (iii) *b* (iv) *c* (v) *b* **2.** (i) *b* (ii) *b* (iii) *b*
3. (i) *a* (ii) *d* (iii) *a,b,c* (iv) *b* (v) *b* (vi) *a* (vii) *a,b,c,d* (viii) *b* (ix) *a* (x) *b* (xi) *b*
4. (i) *a* (ii) *b* (iii) *a* (iv) *b* **5.** (i) *a* (ii) *a* (iii) *a,b,c* (iv) *a,b,c* (v) *a* (vi) *a* **6.** (i) *b* (ii) *b* (iii) *b*
7. *b* **8.** *c* **9.** *b* **10.** *c* **11.** *a* **12.** *d* **13.** *b* **14.** *b* **15.** *c* **16.** *b*
17. *b* **18.** *c* **19.** *b* **20.** *c* **21.** *a* **22.** *a* **23.** *a* **24.** *c* **25.** *d* **26.** *b*
27. *a,b,c,d* **28.** *a,d* **29.** *a,c* **30.** *a,b* **31.** *b,c* **32.** *a,b*

DPP-4 Spherical Mirror - II

1. *c* **2.** *d* **3.** *b* **4.** *a* **5.** *b* **6.** *c* **7.** *b* **8.** *b* **9.** *a* **10.** *a*
11. 3 **12.** 4 **13.** 3 **14.** 2 **15.** 7 **16.** 4 **17.** 1 **18.** 1 **19.** 8 **20.** 2
21. 6 **22.** 5 **23.** 9

DPP-5 Refraction of Light - I

1. (i) *a* (ii) *a,b,c* (iii) *a* (iv) *c* (v) *a* (vi) *b* (vii) *b* (viii) *b*
2. (i) *a* (ii) *a,c* (iii) *a* (iv) *a* (v) *b* (vi) *a,b,c,d*
3. (i) *a,b* (ii) *a* (iii) *c* (iv) *c* (v) *a* (vi) *a,b* (vii) *b* (viii) *b*
4. (i) *b* (ii) *b,c* (iii) *a* (iv) *a* (v) *a* (vi) *b* (vii) *a* (viii) *c* (ix) *a,b* (x) *a* (xi) *b*
5. (i) *b* (ii) *a* (iii) *a* (iv) *a* (v) *a* (vi) *b*
6. (i) *a* (ii) *b* (iii) *a* (iv) *a* (v) *b* (vi) *b* (vii) *b* (viii) *b* **7.** (i) *b* (ii) *c* (iii) *d* (iv) *a* **8.** *d* **9.** *b*
10. *b* **11.** *c* **12.** *a* **13.** *a* **14.** *a* **15.** *c* **16.** *b* **17.** *c* **18.** *a* **19.** *c*
20. *c* **21.** *a* **22.** *c* **23.** *c* **24.** *b* **25.** *b*

DPP-6 Refraction of light - II

1. *a,b* **2.** *a,b* **3.** *a,b,c* **4.** *a,b* **5.** *a,b* **6.** *b* **7.** *c* **8.** *b* **9.** *c* **10.** *a*
11. *b* **12.** *b* **13.** *b* **14.** *c* **15.** *c* **16.** *c* **17.** 4 **18.** 2 **19.** 5 **20.** 3
21. 2 **22.** 4 **23.** 2

DPP-7 Image Formation due to Refraction through Spherical Surface-I

1. (i) *c* (ii) *b* (iii) *b** **2.** (i) *c* (ii) *a* (iii) *b* **3.** (i) *b* (ii) *a,c* (iii) *b* (iv) *a* (v) *b* (vi) *b* (vii) *b*
4. *a* **5.** *b* **6.** *c* **7.** *a* **8.** *c* **9.** *b* **10.** *c* **11.** *c* **12.** *d* **13.** *a*
14. *d* **15.** *b* **16.** *b* **17.** *d* **18.** *a* **19.** *c* **20.** *c* **21.** *a,c* **22.** *a,b,c* **23.** *b,c*

DPP-8 Image Formation due to Refraction through Spherical Surface-II

1. *a,d* **2.** *a,c,d* **3.** *a,b,c* **4.** *a,c,d* **5.** *a,c* **6.** *b* **7.** *b* **8.** *a* **9.** *a* **10.** *b*
11. 6 **12.** 3 **13.** 2 **14.** 5 **15.** 5 **16.** 2 **17.** 4 **18.** 2 **19.** 2
20. $\left(\frac{4}{3}\right)$ **21.** $\left(\frac{\mu_3 R}{\mu_3 - \mu_1}\right)$ **22.** 6.06 m

DPP-9 Image Formation due to Refraction through Plane Surface-III

1. (i) *a,c* (ii) *a,c* (iii) *d* (iv) *b* (v) *c* (vi) *b* **2.** (i) *b* (ii) *c* (iii) *b* (iv) *b* (v) *b* (vi) *b*
3. *c* **4.** *d* **5.** *b* **6.** *c* **7.** *a* **8.** *b* **9.** *b* **10.** *a* **11.** *b* **12.** *c*
13. *d* **14.** *b* **15.** *a* **16.** *b* **17.** *c* **18.** *a,b,c,d* **19.** *a,c* **20.** *a,c* **21.** *a,c*
22. *b* **23.** *a* **24.** *a* **25.** *a* **26.** *b* **27.** *b* **28.** *b* **29.** 5 **30.** 9 **31.** 5
32. 9

DPP-10 Lens-I

1. (i) *a* (ii) *c* (iii) *c* (iv) *b* **2.** (i) *a*(ii) *b* (iii) *c* (iv) *b* (v) *a* (vi) *a*
3. (i) *a* (ii) *a* (iii) *a* (iv) *a* (v) *b* (vi) *a* (vii) *b* (viii) *a* (ix) *a* (x) *b* **4.** (i) *a* (ii) *a* (iii) *a* (iv) *a* (v) *b*
5. *b* **6.** *b* **7.** *b* **8.** *d* **9.** *c* **10.** *a* **11.** *c* **12.** *b* **13.** *c* **14.** *d*
15. *a* **16.** *a* **17.** *a* **18.** *a* **19.** *a* **20.** *b* **21.** *a* **22.** *c* **23.** *d* **24.** *b*
25. *a* **26.** *c* **27.** *a* **28.** *a* **29.** *b* **30.** *d* **31.** *c* **32.** *b* **33.** *b* **34.** *c*
35. *b* **36.** *b* **37.** *d* **38.** *c* **39.** *b* **40.** *b* **41.** *b* **42.** *d* **43.** *a* **44.** *c*

DPP-11 Lens-II

1. *a,b,c* **2.** *a,b* **3.** *a,b* **4.** *a,b,c* **5.** *a,d* **6.** *a,b* **7.** *a,b* **8.** *a,b,c* **9.** *a,b,c*

DPP-12 Lens-III

1. *b* **2.** *a* **3.** *b* **4.** *b* **5.** *a* **6.** *b* **7.** *d* **8.** *b* **9.** *a* **10.** *b*
11. *b* **12.** *a* **13.** *a* **14.** *a* **15.** *d* **16.** *b* **17.** *a* **18.** *c* **19.** *a* **20.** *d*
21. *d* **22.** *a* **23.** *a* **24.** *c* **25.** 4 **26.** 5 **27.** 2 **28.** 9 **29.** 5 **30.** 5
31. 9 **32.** 4 **33.** 5 **34.** 9 **35.** 2 **36.** 8 **37.** 4 **38.** 8 **39.** 4 **40.** 5
41. 1 **42.** 6 **43.** 11 **44.** $\frac{\mu_3 R}{\mu_3 - \mu_1}$ **45.** (i) + 24 cm (ii) –12 cm (iv) –30 cm
46. (i) 15 cm (ii) 1.16 cm **47.** $f = 75$ cm **48.** $f = 0.4$ m, $m_1 = -2$, $m_2 = -\frac{1}{2}$ **49.** 0.9
50. $A'B' = -1.8$ cm **51.** $\mu = 1.6$

DPP-13 Lens-III

1. (i) *a* (ii) *c* (iii) *a* (iv) *b* (v) *c* (vi) *a* (vii) *a* (viii) *c* **2.** (i) *a* (ii) *a,c* (iii) *b* (iv) *c*
3. *b* **4.** *b* **5.** *c* **6.** *c* **7.** *b* **8.** *c* **9.** *a* **10.** *b* **11.** *a* **12.** *c*
13. *c* **14.** *a* **15.** *b* **16.** *b* **17.** *a* **18.** *c* **19.** *c* **20.** *b* **21.** *a,c,d* **22.** *a,c*
23. *a,b,c* **24.** *b,c,d* **25.** *a,b* **26.** *a,b* **27.** *a,c* **28.** *c* **29.** *a,b,c* **30.** *b* **31.** *b* **32.** *b*
33. *c* **34.** *a* **35.** *b* **36.** *a* **37.** *b* **38.** *c* **39.** *b* **40.** *a* **41.** *a* **42.** *d*
43. 5 **44.** 4 **45.** 1 **46.** 3 **47.** 0 **48.** 2 **49.** 3 **50.** (i) $\sqrt{2}$ (ii) $0.5 < 0.6 < 0.86$
51. (i) $i_1 = \sin^{-1}\left\{\frac{1}{2}\sqrt{(n^2 - n_1^2)} - n_1\right\}$ (ii) $i = 73°$ **52.** 2 cm, $\mu = 1.0016$
53. (i) $i_1 = 60°$ (ii) The second prism should be rotated by 60° (Anti-clock wise)

DPP-14 Dispersion and Spectrum

1. *b* **2.** *b* **3.** *c* **4.** *b* **5.** *b,c* **6.** *a,b,c,d* **7.** *a,b,c,d* **8.** *c* **9.** *b* **10.** *a*
11. *a* **12.** *c* **13.** *d* **14.** *a* **15.** *a* **16.** 7 **17.** 5

DPP-15 Optical Instruments

1. *b* 2. *d* 3. *a* 4. *d* 5. *c* 6. *c* 7. *d* 8. *c* 9. *b* 10. *c*
11. *b* 12. *a* 13. *a,b* 14. *a,c,d* 15. *a,c* 16. *b,d* 17. *a,d* 18. *a* 19. *c* 20. *a*
21. *a* 22. *a* 23. *b* 24. *b* 25. *a* 26. 7 27. 9 28. 3 29. 9 30. 3
31. 2 32. 5

Revisal Problems (JEE Main)

1. *b* 2. *c* 3. *d* 4. *d* 5. *b* 6. *c* 7. *a* 8. *b* 9. *b* 10. *a*
11. *a* 12. *b* 13. *a* 14. *b* 15. *c* 16. *c* 17. *b* 18. *b* 19. *a* 20. *c*
21. *b* 22. *d* 23. *b* 24. *a* 25. *a* 26. *b* 27. *b* 28. *d* 29. *a* 30. *a*
31. *a* 32. *b* 33. *b* 34. *a* 35. *d* 36. *a*

Revisal Problems (JEE Advanced)

1. *d* 2. *b* 3. *a* 4. *b* 5. *a* 6. *b* 7. *c* 8. *b* 9. *b* 10. *d*
11. *b* 12. *b* 13. *b* 14. *a* 15. *b* 16. *a* 17. *a,b* 18. *a,b* 19. *a,b* 20. *a,d*
21. *a,b,c* 22. *b* 23. *d* 24. *c* 25. *b* 26. *a* 27. *b* 28. *a* 29. *d* 30. *a*
31. 4 32. 5 33. 1 34. 5 35. 4 36. 3

JEE Main & AIEEE Archive

1. *b* 2. *b* 3. *c* 4. *c* 5. *a* 6. *b* 7. *a* 8. *a* 9. *b* 10. *d*
11. *c* 12. *a* 13. *d* 14. *c* 15. *b* 16. *a* 17. *b* 18. *a* 19. *a* 20. *b*
21. *a* 22. *d* 23. *c* 24. *a* 25. *b* 26. *a* 27. *d* 28. *b* 29. *a*

JEE Advanced & IIT JEE Archive

1. *c* 2. *a* 3. *a,c* 4. *c* 5. *a* 6. *d* 7. *b* 8. *b* 9. *c* 10. *c*
11. 2 12. 6 13. 3 14. 6 15. *b* 16. *a* 17. *a,b,c* 18. *c* 19. *c,d* 20. *a*
21. *a* 22. *b* 23. *c* 24. *b* 25. *a* 26. *c* 27. *b* 28. *c* 29. *c* 30. *b*
31. *b* 32. *a* 33. *c* 34. *d* 35. 60°
36. *(i)* $i_1 = 60°$ *(ii)* The second prism should be rotated by 60° (anti-clockwise).

2. Wave Optics

DPP-1 Electromagnetic Waves

1. *a* 2. *b* 3. *c* 4. *b* 5. *a* 6. *c* 7. *c* 8. *b* 9. *b* 10. *a*
11. *c* 12. *b* 13. *a* 14. *c* 15. *a* 16. *b* 17. *a,c* 18. *b,c* 19. *a,c* 20. *a,b*
21. *a,c,d* 22. *a* 23. *d* 24. *d* 25. *b* 26. *b* 27. *a* 28. *a* 29. *c* 30. *d*
31. 2 32. 6 33. 3 34. 9 35. 5

DPP-2 Interference of light-I

1. *(i) a (ii) d, (iii) d (iv) a (v) b (vi) c,b (vii) b*
2. *(i) a (ii) a, (iii) a (iv) b (v) c (vi) b (vii) b*
3. *b* 4. *c* 5. *a* 6. *a* 7. *b* 8. *a* 9. *a* 10. *c*
11. *b* 12. *b* 13. *a* 14. *c* 15. *a* 16. *a* 17. *c* 18. *c* 19. *c* 20. *b*
21. *b* 22. *a* 23. *b* 24. *d* 25. *a* 26. *b* 27. *c* 28. *d* 29. *b* 30. a
31. *b* 32. *c* 33. *d* 34. *b* 35. *b* 36. *d* 37. *c* 38. *d* 39. *a* 40. *b*
41. *b* 42. *d*

DPP-3 Interference of Light-II

1. *a,b,c,d* 2. *a,d* 3. *b,d* 4. *a,b,c* 5. *a,c,d* 6. *a,c,d* 7. *a,b,c,d* 8. *a,b,c* 9. *a,b,c,d*
10. *a,b,c,d* 11. *a,c* 12. *a,c,d* 13. *b* 14. *b* 15. *a* 16. *c* 17. *a* 18. *a*
19. *c* 20. *c* 21. *a* 22. *c* 23. *a* 24. *a* 25. *a* 26. *b* 27. *c* 28. *b*
29. *a* 30. *a*

DPP-4 Interference of Light - III

1. *d* 2. *d* 3. *a* 4. *b* 5. *c* 6. 1 7. 5 8. 4 9. 8 10. 5
11. 2 12. 7 13. 4 14. 4 15. 3.5 mm 16. 2 m

17. *(i)* $y = 4.33\,\mu m$ *(ii)* $I = \frac{3}{4} I_{max}$

18. *(i)* $y_1 = \pm 0.26\,m$, $y_2 = \pm 1.13\,m$ *(ii)* $y_1 = 0.26\,m$, $y_2 = 1.33\,m$

19. $t = 9.3\,\mu m$

20. *(i)* 7×10^{-6} m *(ii)* $\mu' = 1.6$, $\Delta\omega = -5.71 \times 10^{-5}$ m

21. *(i)* $\omega = 0.63$ mm *(ii)* $t = 1.5790\,\mu m$

22. *(i)* 1.17 mm *(ii)* 1.56 mm

23. 5892 Å

DPP-5 Interference of Light - IV

1. *(i) b (ii) a (iii) b (iv) a (v) b (vi) b,c*
2. *(i) b (ii) a (iii) a,b (iv) a (v) d (vi) b,d*
3. *(i) a (ii) a (iii) c (iv) c (v) a (vi) a*

4. *c* 5. *c* 6. *b* 7. *b* 8. *c* 9. *b* 10. *d* 11. *b* 12. *a*
13. *a,b,c* 14. *a,c,d* 15. *b,c* 16. *a,b,c* 17. *a,b,c,d* 18. *a,b* 19. *a* 20. *b* 21. *c*
22. *a* 23. *a* 24. *c* 25. *d* 26. *a* 27. 2 28. 71 29. 7 30. 9 31. 3
32. 8 33. $t_{max} = 90$ nm 34. *(ii)* $\frac{1}{16}$ *(iii)* 300 mm 35. $\frac{1}{49}$ 36. 7×10^{-6} W

DPP-6 Diffraction of Light

1. *d* 2. *a* 3. *c* 4. *b* 5. *a* 6. *b* 7. *a* 8. *b* 9. *d* 10. *d*
11. *d* 12. *a,b,c* 13. *a,b,c* 14. *a* 15. *c* 16. *a* 17. *b* 18. *b* 19. 5 20. 5
21. 1 22. 5 23. *(i)* $\lambda' = 42000$ Å *(ii)* $\mu \approx 1.43$

DPP-7 Polarisation of Light

1. *d* 2. *b* 3. *b* 4. *b* 5. *a* 6. *a* 7. *d* 8. *c* 9. *d* 10. *d*
11. *a,d* 12. *b* 13. *a* 14. *a* 15. *b* 16. *c* 17. *d* 18. *a* 19. *b* 20. *a*
21. *b* 22. 3 23. 2 24. 7

Revisal Problems (JEE Main)

1. *c* 2. *b* 3. *c* 4. *a* 5. *b* 6. *b* 7. *a* 8. *c* 9. *a* 10. *a*
11. *c* 12. *c* 13. *c* 14. *b* 15. *d* 16. *b* 17. *a* 18. *a* 19. *b* 20. *b*
21. *a* 22. *b* 23. *a* 24. *c* 25. *b* 26. *b* 27. *b*

Revisal Problems (JEE Advanced)

1. *a* 2. *a* 3. *d* 4. *b* 5. *b* 6. *d* 7. *a* 8. *b* 9. *a,b,c*
10. *a,b,c* 11. *a,c* 12. *a,b,d* 13. *b,d* 14. *a,b,c* 15. *a,b,c* 16. *a,b* 17. *a* 18. *a* 19. *b*
20. *b* 21. *a* 22. *b* 23. *a* 24. *c* 25. *b* 26. *a* 27. *b* 28. 5 29. 5
30. 5 31. 4 32. 7

JEE Main & AIEEE Archive

1. *d* 2. *c* 3. *d* 4. *d* 5. *b* 6. *c* 7. *c* 8. *b* 9. *b* 10. *c*
11. *d* 12. *d* 13. *a* 14. *b* 15. *d*

JEE Advanced & IIT JEE Archive

1. *a,b,c* 2. *d* 3. *b* 4. *b* 5. *a* 6. *d* 7. *c* 8. *d*

1. Geometrical Optics

DPP-1 Plane Mirror-I

1. (i) (a,b)

(ii) (a,b,c,d)

(iii) (a,b,c) According to Newton's corpuscular model of light, light consists of minute particles known as corpuscles.
Corpuscles are elastic in nature. They move in homogeneous medium without deviation.

(iv) (b,c,d)	(v) (a,b,c)	(vi) (b)
(vii) (a,b,c)	(viii) (a)	(ix) (a,b,c,d)
(x) (a)	(xi) (a)	(xii) (b)
(xiii) (c)	(xiv) (a)	(xv) (c)
(xvi) (a,b,c)	(xvii) (a,c,d)	(xviii) (b)
(xix) (a)	(xx) (b)	(xxi) (b)
(xxii) (a)		

2. (i) (c) $\hat{\mathbf{e}}_1 = |\hat{\mathbf{e}}_1|\sin\theta\hat{\mathbf{i}} - |\hat{\mathbf{e}}_1|\cos\theta\hat{\mathbf{j}} = \sin\theta\hat{\mathbf{i}} - \cos\theta\hat{\mathbf{j}}$

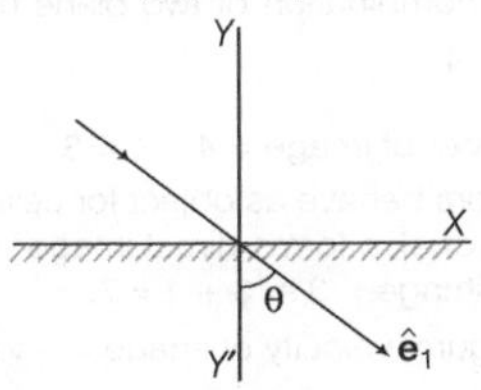

(ii) (a) Since, the mirror is in XY-plane. So, normal on incidence point is along Y-axis. Thus, unit vector along normal on incidence point is $\hat{\mathbf{j}}$.

(iii) (c) $\hat{\mathbf{e}}_1 \cdot \hat{\mathbf{n}} = |\hat{\mathbf{e}}_1|\,|\hat{\mathbf{n}}|\cos(\pi - \theta) = -\cos\theta$

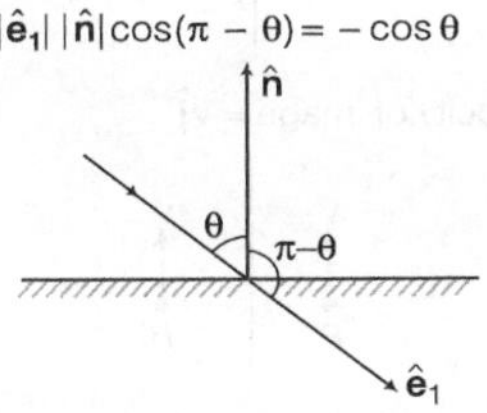

(iv) (b) $\because -\cos\theta = (\hat{\mathbf{e}}_1 \cdot \hat{\mathbf{n}})$

$\therefore \quad \cos\theta = |\hat{\mathbf{e}}_1 \cdot \hat{\mathbf{n}}|$

(v) (b) $\because$ Angle of incidence = angle of reflection

$\therefore \quad \cos\alpha = \cos\theta = |\hat{\mathbf{e}}_1 \cdot \hat{\mathbf{n}}|$

(vi) (b,c,d) (b) The unit vector along reflected ray is

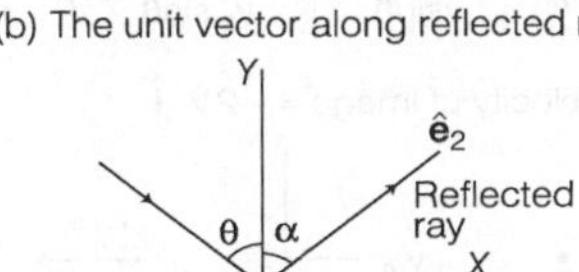

$$\hat{\mathbf{e}}_2 = e_2\sin\alpha\hat{\mathbf{i}} + e_2\cos\alpha\hat{\mathbf{j}}$$
$$= \sin\alpha\hat{\mathbf{i}} + \cos\alpha\hat{\mathbf{j}}$$

(c) $\because \alpha = \theta$

$\therefore \quad \hat{\mathbf{e}}_2 = \sin\theta\hat{\mathbf{i}} + \cos\theta\hat{\mathbf{j}}$

(d) $\because \hat{\mathbf{e}}_1 - 2(e_1 \cdot \hat{\mathbf{n}})\hat{\mathbf{n}} = \sin\theta\hat{\mathbf{i}} - \cos\theta\hat{\mathbf{j}} + 2\cos\theta\hat{\mathbf{j}}$

$$= \sin\theta\hat{\mathbf{i}} + \cos\theta\hat{\mathbf{j}} = \hat{\mathbf{e}}_2.$$

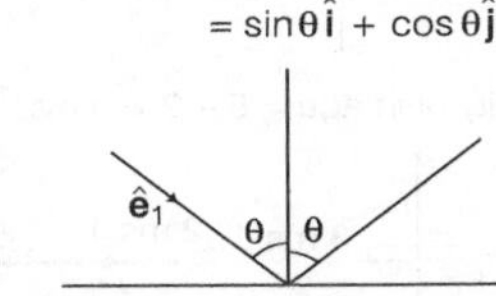

(vii) (a,b) The component of unit vector along incident ray along Y-axis is $-\cos\theta$, but that of reflected ray is $\cos\theta$.

3. (i) (a) The unit vector along incident ray is

$$\hat{\mathbf{e}}_1 = \frac{3\hat{\mathbf{i}} - 4\hat{\mathbf{k}}}{\sqrt{3^2 + (-4)^2}} = \frac{3}{5}\hat{\mathbf{i}} - \frac{4}{5}\hat{\mathbf{k}}.$$

(ii) (d) The plane mirror is XY-plane. So, normal is along Z-axis. So, unit vector along normal is $\hat{\mathbf{k}}$.

(iii) (a) If θ is the angle of incidence, then

$$\cos\theta = |\hat{\mathbf{e}}_1 \cdot \hat{\mathbf{n}}| = \left|\left(\frac{3}{5}\hat{\mathbf{i}} - \frac{4}{5}\hat{\mathbf{k}}\right)\cdot\hat{\mathbf{k}}\right| = \frac{4}{5}$$

$\therefore \quad \theta = 37°$

(iv) (c) $\hat{\mathbf{e}}_2 = e_2\sin\theta\hat{\mathbf{i}} + e_2\cos\theta\hat{\mathbf{k}}$

$$= \sin 37°\hat{\mathbf{i}} + \cos 37°\hat{\mathbf{k}} = \frac{3}{5}\hat{\mathbf{i}} + \frac{4}{5}\hat{\mathbf{k}}.$$

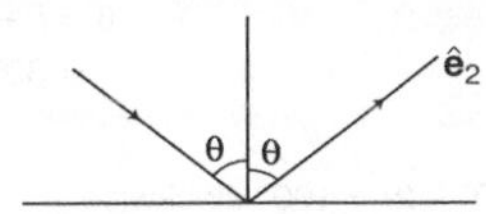

4. (i) (a) $\hat{\mathbf{e}}_1 = \dfrac{\hat{\mathbf{i}} + \hat{\mathbf{j}} + \hat{\mathbf{k}}}{\sqrt{1^2 + 1^2 + 1^2}} = \dfrac{\hat{\mathbf{i}} + \hat{\mathbf{j}} + \hat{\mathbf{k}}}{\sqrt{3}}.$

(ii) (b) First plane mirror is in YZ-plane.
So, the unit vector along normal on incidence point is $\hat{\mathbf{i}}$.

(iii) (b) Due to reflection, the normal component becomes opposite.
$\therefore$ The unit vector along reflected ray from first mirror is

$$\hat{\mathbf{e}}_2 = \frac{-\hat{\mathbf{i}} + \hat{\mathbf{j}} + \hat{\mathbf{k}}}{\sqrt{3}}.$$

(iv) (b) The reflected ray from first mirror is incident ray for second mirror.
$\therefore$ The unit vector along incident ray on second mirror is

$$\frac{-\hat{\mathbf{i}} + \hat{\mathbf{j}} + \hat{\mathbf{k}}}{\sqrt{3}}.$$

(v) (b) Since, normal on second mirror is along Y-axis.
$\therefore$ The unit vector along reflected ray from second mirror is

$$\frac{-\hat{\mathbf{i}} - \hat{\mathbf{j}} + \hat{\mathbf{k}}}{\sqrt{3}}.$$

(vi) (b) The reflected ray from second plane mirror is incident ray for third mirror. The unit vector along normal on third mirror is $\hat{\mathbf{k}}$.
$\therefore$ The unit vector along reflected ray from third mirror is

$$\frac{-\hat{\mathbf{i}} - \hat{\mathbf{j}} - \hat{\mathbf{k}}}{\sqrt{3}}.$$

(vii) (c) Incidence ray and final emergent ray are opposite to each other. Hence, angle between them is 180°.

(viii) (b) $\because \delta + 60° + 60° = 180°$

$\therefore \quad \delta = 60°$

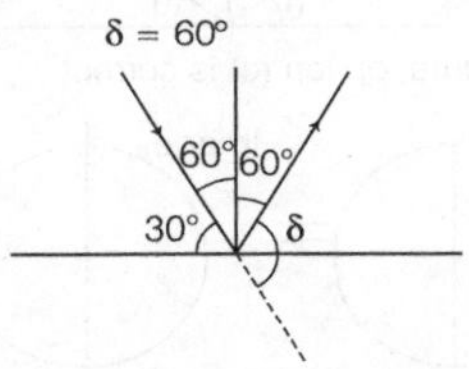

5. (i) (b) Incidence ray and final emergent ray are anti-parallel to each other. So, the deviation is 180°.

(ii) (c) $\because \delta = 2\pi - 2\theta$,
where, θ is angle between mirrors or $\pi = 2\pi - 2\theta$

$\therefore \quad \theta = \pi/2 = 90°$.

6. (i) (a)

(ii) (a,b) From figure,

$$2i = 90° - r$$

and $$2r = 90° - i$$

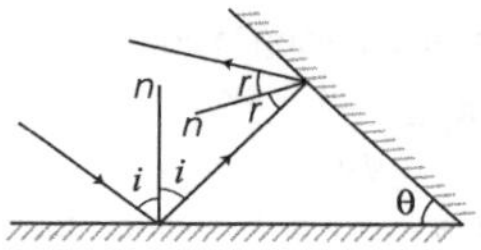

(iii) (c) $\because$ $2i = 90° - r$ or $i = 45° - r/2$

$\because$ $2r = 90° - i = 90° - 45° + \dfrac{r}{2}$

or $\dfrac{3r}{2} = 45°$ or $r = 30°$

$\therefore$ $2i = 90° - r = 90° - 30° = 60°$

$\therefore$ $i = 30°$.

From figure, $90° - i + 90° - r + \theta = 180°$

$\therefore$ $\theta = i + r = 60°$

$\therefore$ $i = 30°$

7. (i) (a)

(ii) (b) $\delta_1 = 180° - 2i = 100°$ clockwise.

(iii) (b) From figure, angle of incidence is 10°

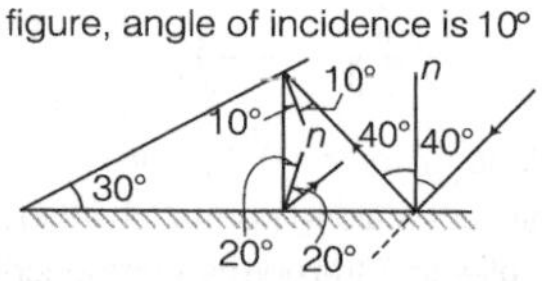

(iv) (c) $\delta_2 = 180° - 2i = 180° - 20° = 160°$ anti-clockwise.

(v) (b) From figure, angle of incidence is 20°.

(vi) (b) $\delta_3 = 180° - 2i = 180° - 40° = 140°$ anti-clockwise.

(vii) (a,c) $\therefore \delta = \delta_2 + \delta_3 - \delta_1 = 160° + 140° - 100°$

$= 200°$ anti-clockwise

$= 360° - 200° = 160°$ clockwise

(viii) (c) 11 : 59 : 60

8 : 40 : 10

3 : 19 : 50

(ix) (c) From figure,

The coordinates of image is $(-x, y, z)$.

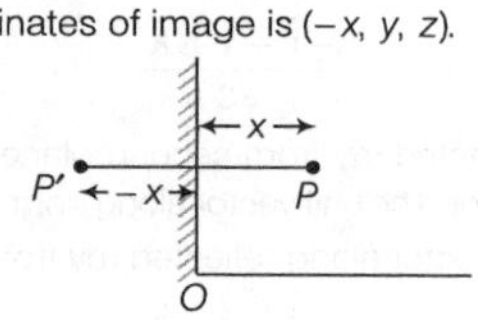

(x) (b)

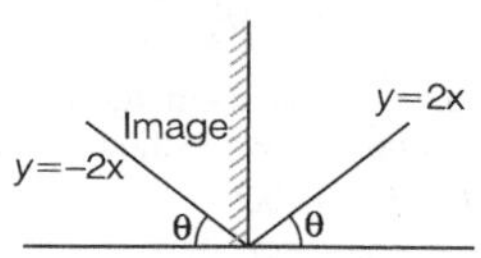

(xi) (c) From figure, option (c) is correct.

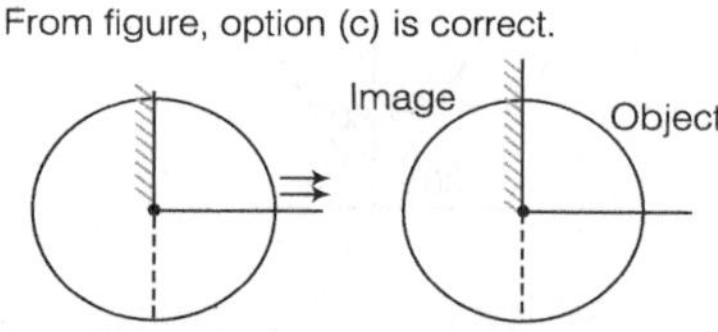

(xii) (a,c) From figure, image of the particle P executes SHM with amplitude 2 mm and time period 4s.

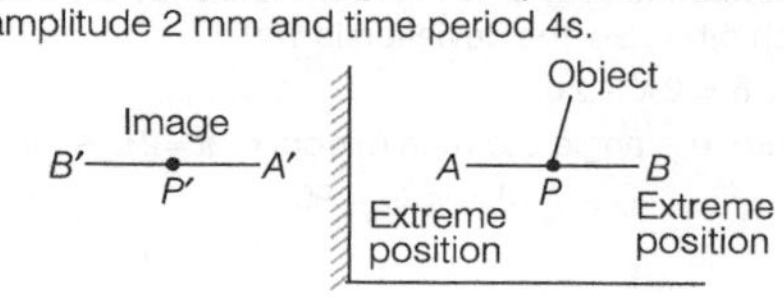

(xiii) (a,c)

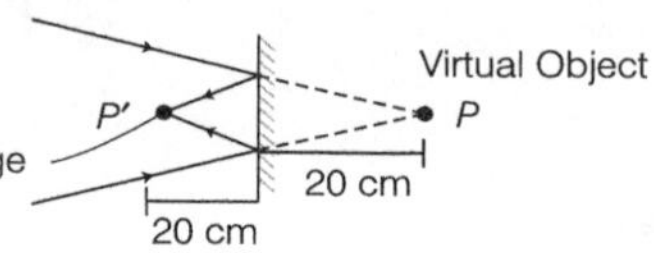

(xiv) (a) From figure,

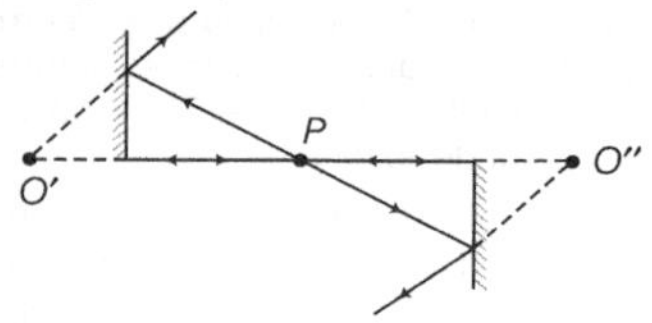

8. (i) (a,b,c,d)

(ii) (c) Here, $n = \dfrac{360°}{\theta} = \dfrac{360°}{60°} = 6$

$\therefore$ Number of images $= n - 1 = 6 - 1 = 5$

(iii) (b)

(iv) (c) Due to combination of two plane mirrors fixed in walls is $n = \dfrac{360°}{90°} = 4$

$\therefore$ The number of image $= 4 - 1 = 3$

These mirrors behave as object for ceiling mirror O.

Ceiling mirror also forms direct image of object. Thus, total number of image $= 3 + 3 + 1 = 7$.

(v) (b) From figure, velocity of image $= -v\hat{i}$.

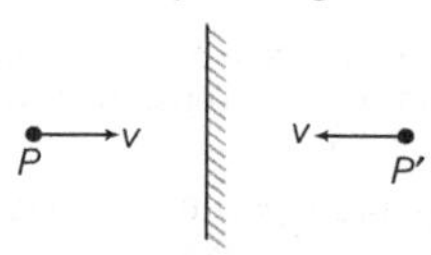

(vi) (a) The velocity of image $= v\hat{j}$

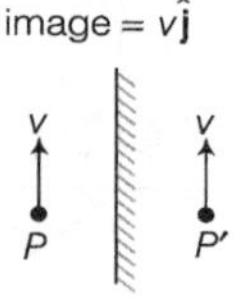

(vii) (b) The velocity of image $= -v\sin\theta\,\hat{i} + v\cos\theta\,\hat{j}$

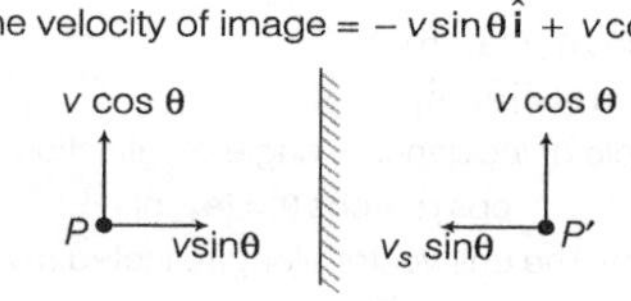

(viii) (c) The velocity of image $= -2v_m\hat{i}$

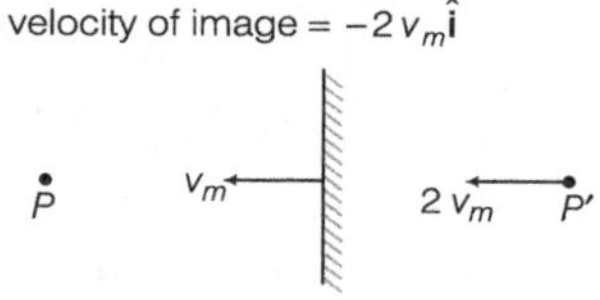

(ix) (d)

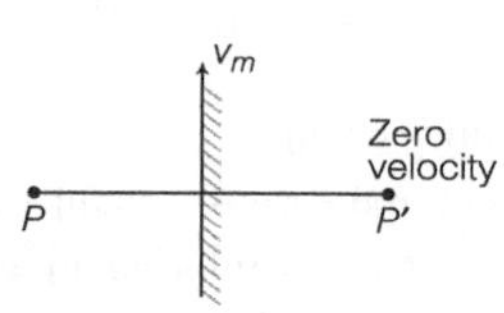

(x) (b) The velocity of image $= 6 - 2 = 4\ ms^{-1} = 4\hat{i}$

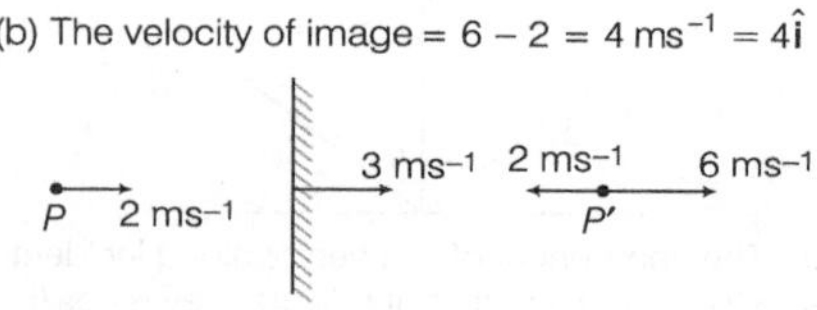

(xi) (b)

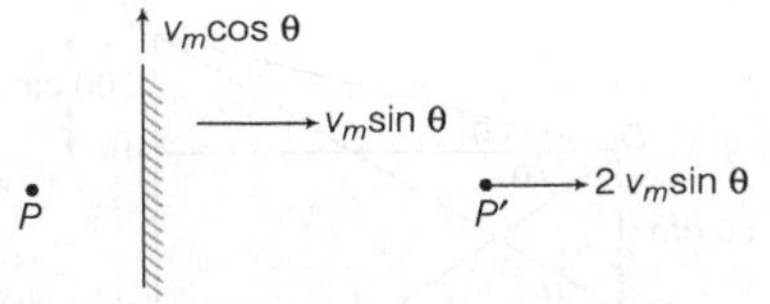

(xii) (b) $\mathbf{v}_{image} = (2v_m\sin\theta - v\cos\alpha)\hat{\mathbf{i}} + v\sin\alpha\hat{\mathbf{j}}$

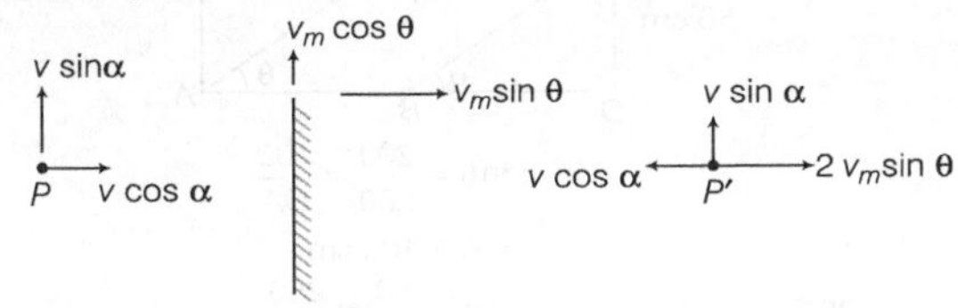

9. (i) (b) $PP_1 = 10 + 10 = 20$ cm

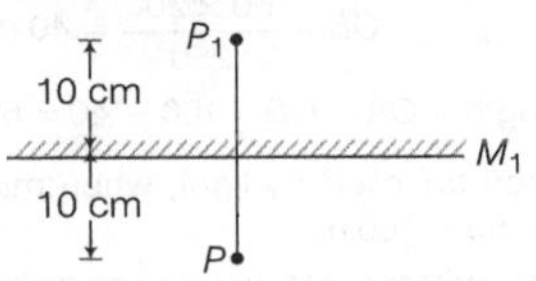

(ii) (b) $PP_1' = 10 + 10 = 20$ cm

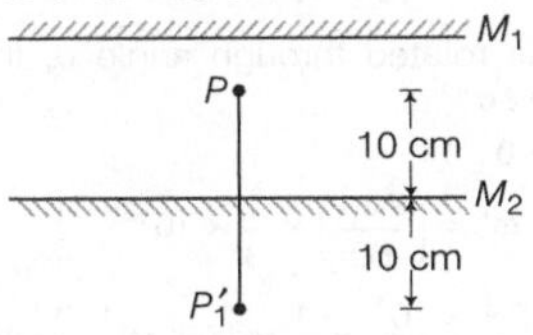

(iii) (b) For second image by mirror M_1, the first image formed by mirror M_2 behaves as object.

∴ Distance of second image from $P = 10 + 20 + 10 = 40$ cm.

(iv) (c) The velocity of first image $= 3\hat{\mathbf{i}} - 4\hat{\mathbf{j}}$d

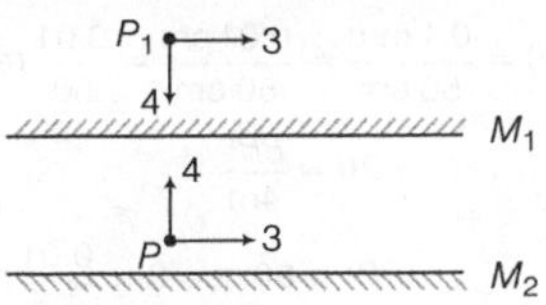

(v) (c) $\mathbf{v}_{image} = 3\hat{\mathbf{i}} - 4\hat{\mathbf{j}}$

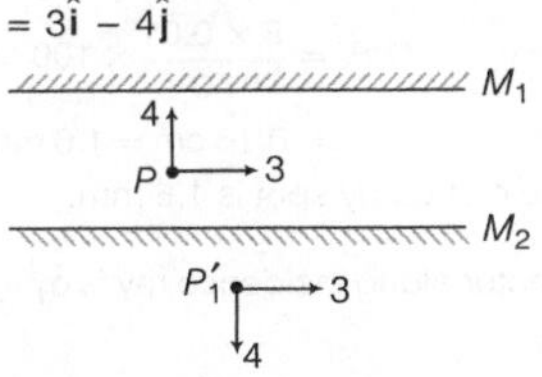

(vi) (a) For second image by M_1, P_1' behaves as object

$$\mathbf{v}_{image} = 3\hat{\mathbf{i}} + 4\hat{\mathbf{j}}$$

10. (i) (c) $n = \dfrac{360°}{90°} = 4$

∴ Numbers of image $= 4 - 1 = 3$

(ii) (a)

(iii) (d) $\mathbf{v}_{image} = 3\hat{\mathbf{i}} - 4\hat{\mathbf{j}}$

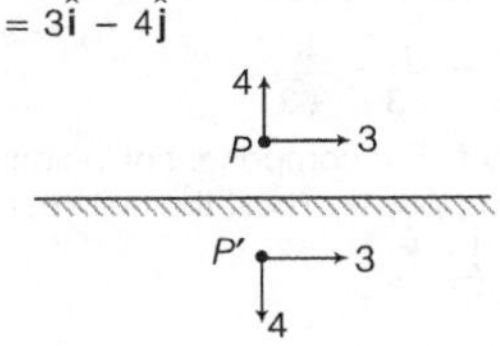

(iv) (d) $\mathbf{v}_{image} = -3\hat{\mathbf{i}} + 4\hat{\mathbf{j}}$

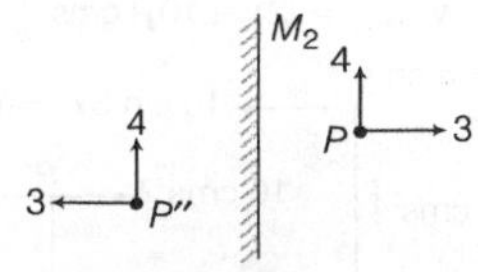

(v) (b) If P''' is image due to M_2.

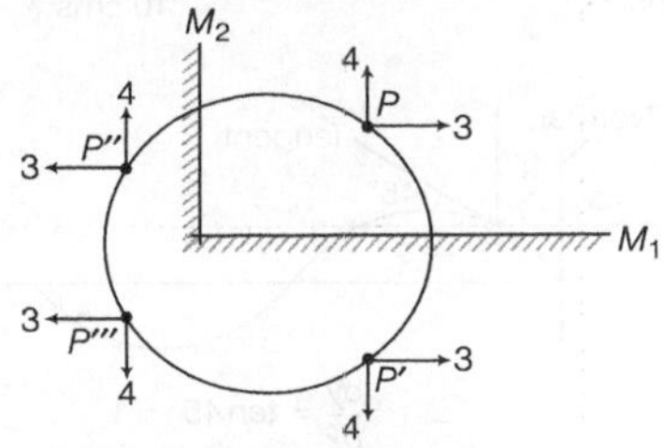

Then, P' behaves as object for P'''.

∴ $\mathbf{v}_{image} = -3\hat{\mathbf{i}} - 4\hat{\mathbf{j}}$

11. (i) (a)

(ii) (a,b)

(iii) (c) From similarity of triangle,

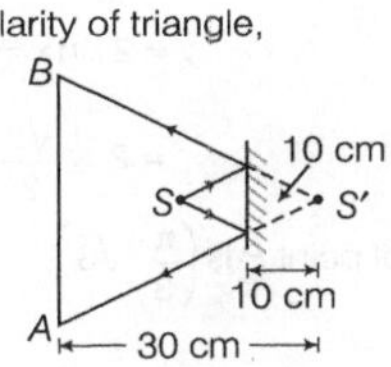

$$\frac{AB}{30} = \frac{10}{10}$$

∴ $AB = 30$ cm

12. (i) (a)

(ii) (b) From similarity of triangle,

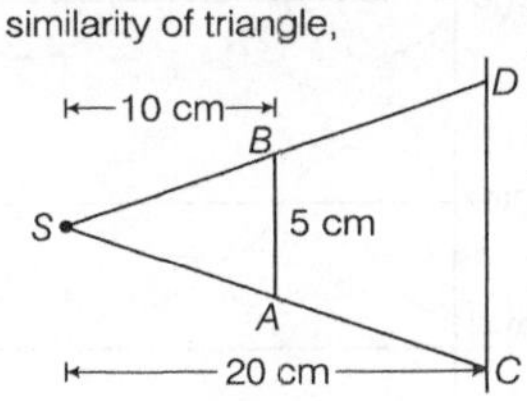

$$\frac{CD}{20} = \frac{AB}{10}$$

∴ $CD = \dfrac{5}{10} \times 20 = 10$ cm

13. (i) (b) $\dfrac{24}{2} \times \dfrac{16}{2} = 12\text{ cm} \times 8\text{ cm}$

(ii) (c) Length $= \dfrac{24}{2} = 12$ cm

Breadth $= \dfrac{b+a}{2} = \dfrac{16+24}{2} = 20$ cm

(iii) (b) Length $= \dfrac{24}{2} = 12$ cm

Breadth $= \dfrac{b-a}{2} = \dfrac{16-4}{2} = 6$ cm

14. (b) The velocity of image of P_1 is $v_{image} = 4 - 4 = 0$

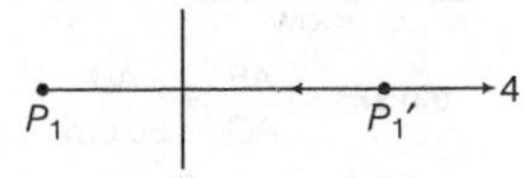

∴ Velocity of image with respect to P_2 is

$$\mathbf{v}_{ref} = 0 - (-8\hat{\mathbf{i}}) = (8\text{ cms}^{-1})\hat{\mathbf{i}}$$

15. (a) From figure,

$$\mathbf{v}_{image} = (2\hat{\mathbf{i}} - 10\hat{\mathbf{j}})\ \text{cms}^{-1}$$

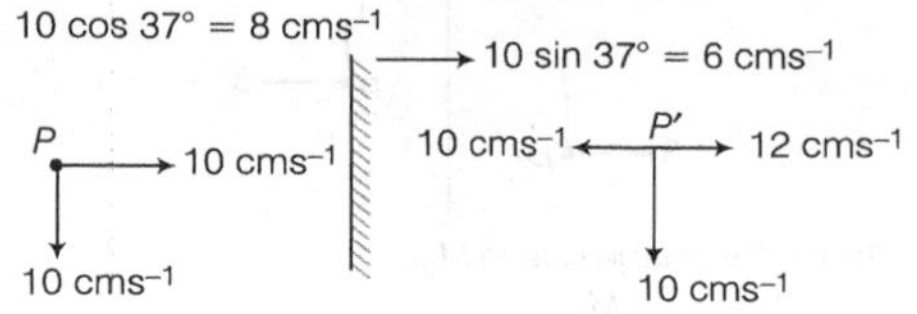

16. (b)

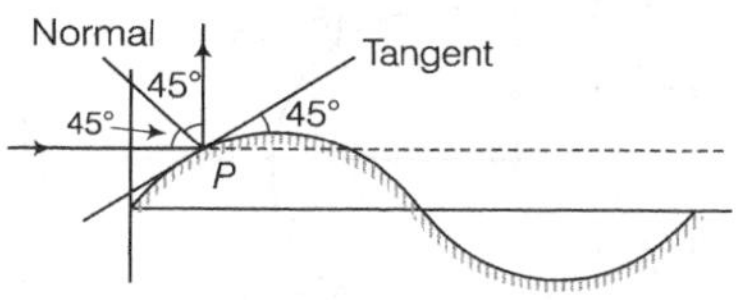

Hence, $\frac{dy}{dx} = \tan 45° = 1$

or $2\cos x = 1$

or $\cos x = \frac{1}{2}$

$\Rightarrow$ $x = \frac{\pi}{3}$

$\therefore$ $y = 2\sin x = 2\sin\frac{\pi}{3}$

$= 2 \times \frac{\sqrt{3}}{2} = \sqrt{3}$

$\therefore$ The coordinate of point P is $\left(\frac{\pi}{3}, \sqrt{3}\right)$.

17. (c)

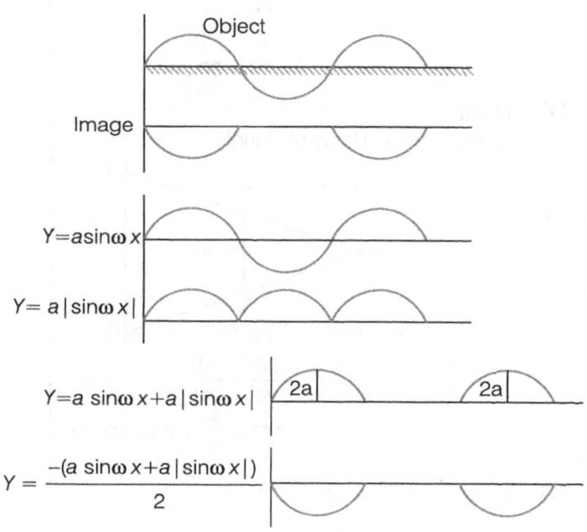

18. (a) From figure,

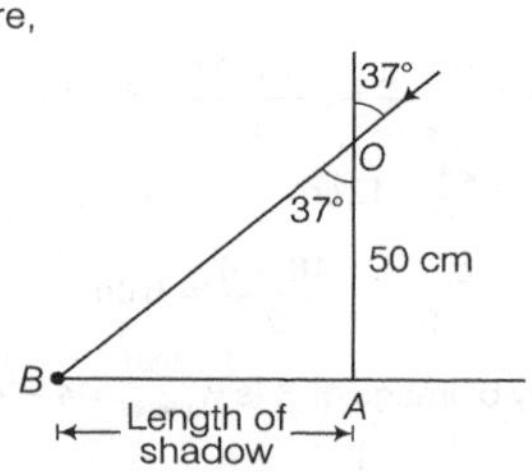

$$\tan 37° = \frac{AB}{AO} = \frac{AB}{50\ \text{cm}}$$

$\therefore$ $AB = 50 \times \frac{3}{4} = 37.5$ cm

19. (d) From figure,

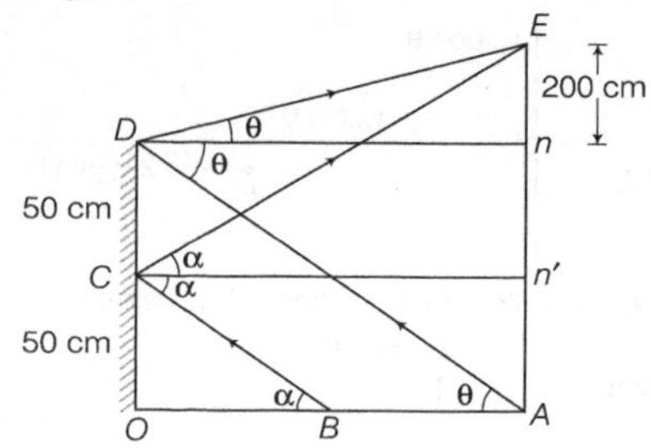

$$\tan\theta = \frac{200}{200} = \frac{100}{OA}$$

$\therefore$ $OA = 100$ cm

and $\tan\alpha = \frac{50}{OB} = \frac{250}{200}$

$\therefore$ $OB = \frac{50 \times 200}{250} = 40$ cm

$\therefore$ Required length $= OA - OB = 100 - 40 = 60$ cm.

20. (c) The distance travelled by light, when mirror rotated through angle α is $50 + 50 = 100$ m.

$\therefore$ Time taken by mirror to rotate through angle is

$$t = \frac{100}{3 \times 10^8} = \frac{1}{3} \times 10^{-6}\ \text{s}$$

If the mirror is rotated through angle α, the image is rotated through angle 2α.

$\therefore$ $2\alpha = \theta$

Here, $\alpha = \omega t = \left(\frac{2\pi n}{60}\right) \times \frac{1}{3} \times 10^{-6}$

$\therefore$ $\alpha = \frac{2\pi \times 10^4}{60} \times \frac{1}{3} \times 10^{-6} \times \frac{180}{\pi} \times 60\ \text{min} = 1.2\ \text{min}$

$\therefore$ $\theta = 2\alpha = 2.4$ min.

21. (d) If the mirror is rotated through an angle θ, spot rotated through angle 2θ.

Here, $\theta = \frac{0.1\ \text{mm}}{50\ \text{cm}} = \frac{0.01\ \text{cm}}{50\ \text{cm}} = \frac{0.01}{50}$ rad

From figure, $2\theta = \frac{PP'}{4\text{m}}$

$\therefore$ $PP' = 8\theta\ \text{m} = 8 \times \frac{0.01}{50}$ m

$\therefore$ $PP' = \frac{8 \times 0.01}{50} \times 100$ cm

$= 0.16$ cm $= 1.6$ mm

Thus, distance moved by spot is 1.6 mm.

22. (b) The unit vector along incidence ray is $\hat{\mathbf{e}}_1 = \frac{\hat{\mathbf{i}} + \hat{\mathbf{j}} - \hat{\mathbf{k}}}{\sqrt{3}}$

The unit vector along normal an incidence point is $\hat{\mathbf{n}} = \frac{\hat{\mathbf{i}} + \hat{\mathbf{j}}}{\sqrt{2}}$

$\because$ $\hat{\mathbf{e}}_2 = \hat{\mathbf{e}}_1 - 2(\hat{\mathbf{e}}_1 \cdot \hat{\mathbf{n}})\hat{\mathbf{n}}$

$$= \frac{\hat{\mathbf{i}} + \hat{\mathbf{j}} - \hat{\mathbf{k}}}{\sqrt{3}} - 2\left(\frac{1}{\sqrt{6}} + \frac{1}{\sqrt{6}}\right)\frac{(\hat{\mathbf{i}} + \hat{\mathbf{j}})}{\sqrt{2}}$$

$$= \frac{\hat{\mathbf{i}} + \hat{\mathbf{j}} - \hat{\mathbf{k}}}{\sqrt{3}} - \frac{2}{\sqrt{3}}\hat{\mathbf{i}} - \frac{2}{\sqrt{3}}\hat{\mathbf{j}}$$

$$= \frac{-\hat{\mathbf{i}}}{\sqrt{3}} - \frac{\hat{\mathbf{j}}}{\sqrt{3}} - \frac{\hat{\mathbf{k}}}{\sqrt{3}}$$

Alternate method The component of incident ray normal to the mirror becomes opposite. Thus, the unit vector along reflected ray is $\hat{\mathbf{e}}_2 = \frac{-\hat{\mathbf{i}} - \hat{\mathbf{j}} - \hat{\mathbf{k}}}{\sqrt{3}}$.

23. (b) $\therefore \tan 45° = \dfrac{x}{h}$

$\therefore \qquad x = h$

Also, $\qquad \tan\alpha = \dfrac{L-x}{H} = \dfrac{2h-h}{H} = \dfrac{h}{H}$

or $\qquad \tan 45° = \dfrac{h}{H}$

$\therefore \qquad h = H$

$\therefore \qquad H = h = 2\,\text{m}.$

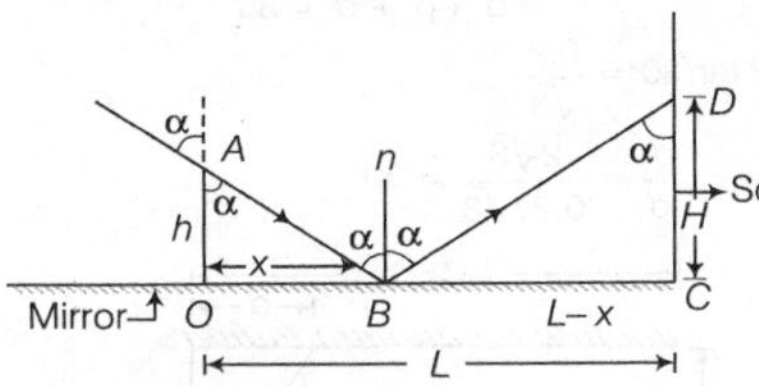

24. (b)

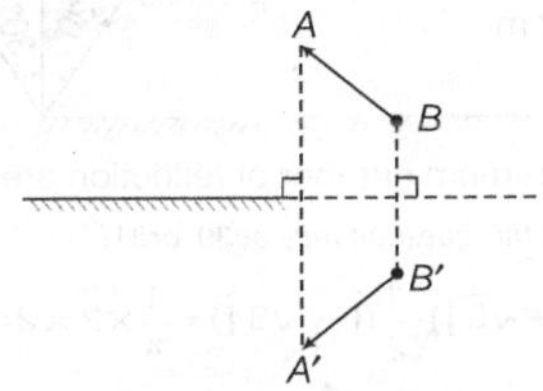

25. (a) $\because \quad v = r\omega$

$\therefore \qquad \omega = \dfrac{v}{r} = \dfrac{80\pi}{10} = 8\pi$

The angular speed of mirror, at $t = 1\text{s}$ is $\omega' = \dfrac{\omega}{2} = 4\pi$

$\therefore \qquad \omega' = \alpha t$

$\therefore \qquad \alpha = \dfrac{\omega'}{t} = \dfrac{4\pi}{1} = 4\pi$

$\therefore \qquad \alpha = 4\pi \text{ rads}^{-1}.$

26. (d) From figure, $O'F = r =$ radius of mirror

In ΔOSD and DFB, $\quad \tan\theta = \dfrac{r}{L} = \dfrac{FB}{2L}$

$\therefore \qquad FB = 2r$

$\therefore$ The radius of reflected patch $= R = O'B = O'F + FB = 3r$

$\therefore$ The area of reflected path patch $= A = \pi R^2 = \pi(3r^2) = 9\pi r^2$

$= 9 \times 1 = 9\,\text{m}^2.$

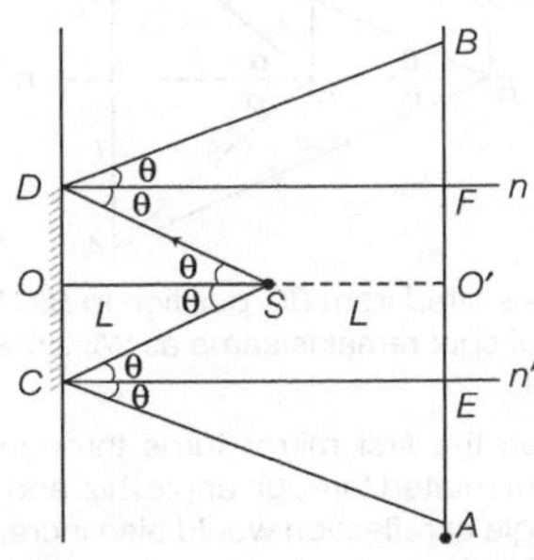

27. (a) From figure, $\alpha + 30° = 90° - \alpha$ or $2\alpha = 60°$

$\therefore \qquad \alpha = 30°$

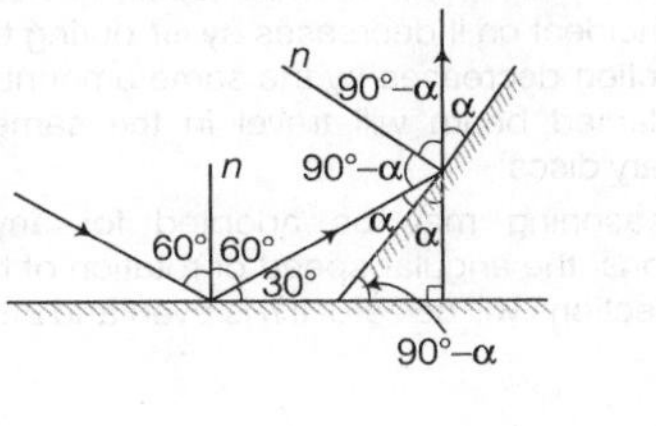

28. (d) From figure, the angle between reflected rays,

$= 30° + 30° = 60°$

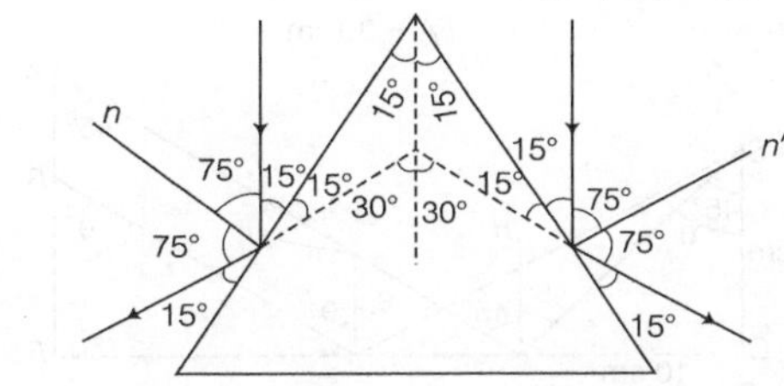

29. (a) Thus, the relative velocity is $v_{rel} = 2\sin 30° - (-2\sin 30°)$

$= 4\sin 30° = 2\text{ cms}^{-1}$

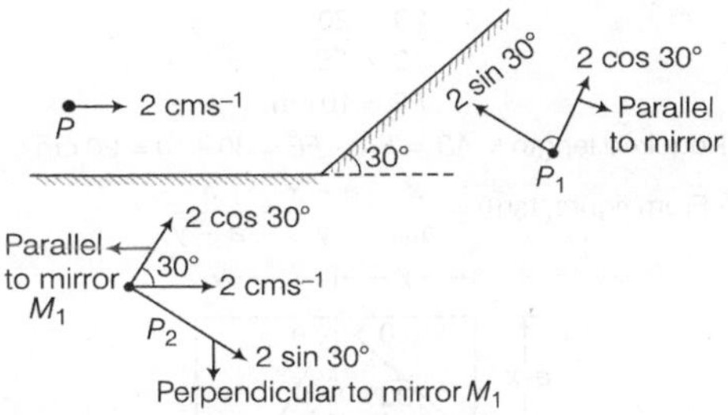

in perpendicular direction of mirror M_2.

30. (b) P_1 behaves as object for mirror M_2.

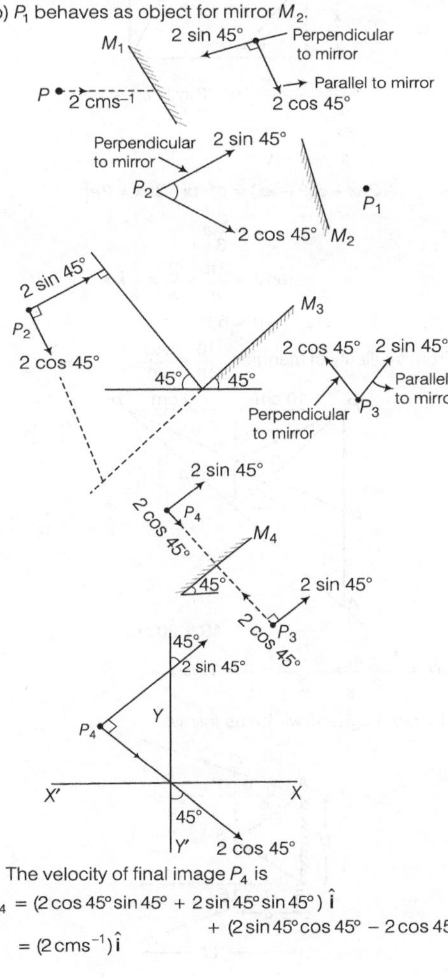

$\therefore$ The velocity of final image P_4 is

$v_4 = (2\cos 45° \sin 45° + 2\sin 45° \sin 45°)\,\hat{\mathbf{i}} + (2\sin 45° \cos 45° - 2\cos 45° \cos 45°)\,\hat{\mathbf{j}}$

$= (2\text{cms}^{-1})\,\hat{\mathbf{i}}$

31. (b) From triangle DAF, $\tan 45^\circ = \frac{FA}{DF} = \frac{FA}{30}$

$\therefore \qquad FA = 30 \text{ cm}$

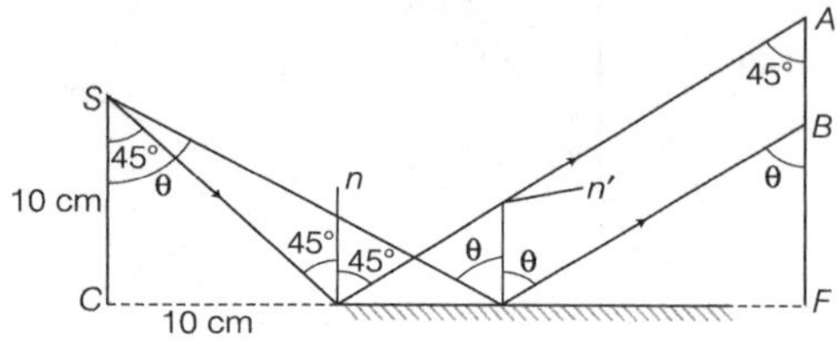

From triangle SCE and BEF,

$$\tan\theta = \frac{EC}{SC} = \frac{EF}{FB}$$

or $\qquad \frac{20}{10} = \frac{20}{FB}$

$\therefore \qquad FB = 10 \text{ cm}$

$\therefore$ Required length $= AB = FA - FB = 30 - 10 = 20$ cm.

32. (d) From figure, $\tan\theta = \frac{x}{a_{12}} = \frac{a-x}{y} = \frac{a}{a-y}$

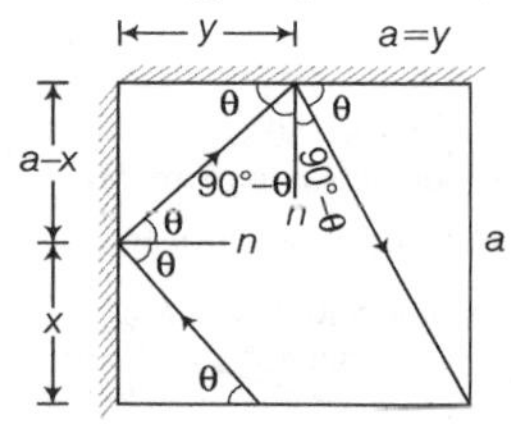

$\therefore \qquad \frac{2x}{a} = \frac{a-x}{y}$ or $2xy = a^2 - x$ ...(i)

and $\qquad \frac{2x}{a} = \frac{a}{a-y}$ or $2ax - 2xy = a^2$

or $\qquad 2ax - (a^2 - ax) = a^2$ or $3ax = 2a^2$

or $\qquad x = \frac{2}{3}a$

$\therefore \qquad \tan\theta = \frac{2\pi}{a} = \frac{2}{a} \times \frac{2}{3}a = \frac{4}{3}$

$\therefore \qquad \theta = 53^\circ$

33. (d) From similarity of triangles, $\frac{10}{10} = \frac{AB}{20}$

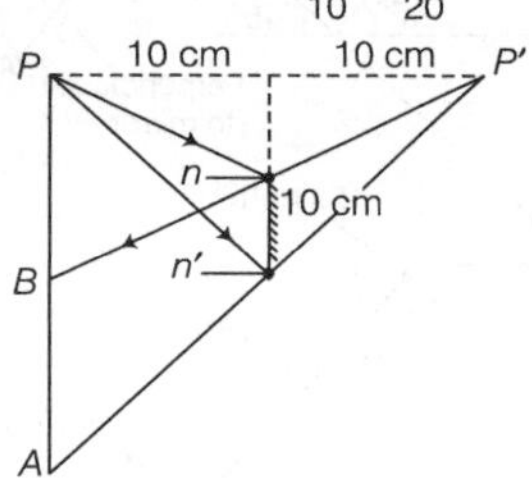

$\therefore \qquad AB = 20$ cm

34. (a) $\because d = \frac{b-a}{2} = \frac{6-4}{2} = \frac{2}{2} = 1$ inch.

35. (d) The ray diagrams will be as follows

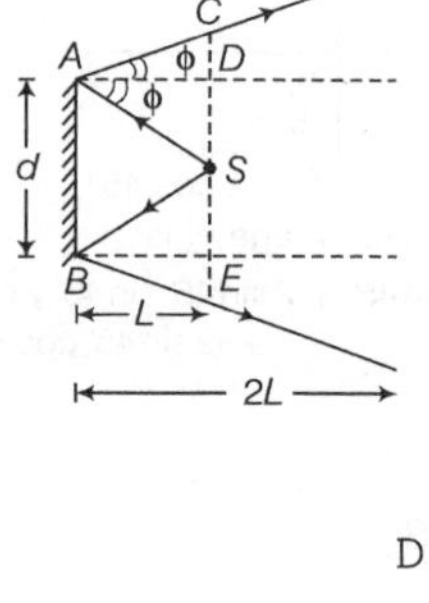

$$HI = AB - d$$

$$DS = CD = \frac{d}{2}$$

Since, $\qquad AH = 2AD$

$\therefore \qquad GH = 2CD = 2\frac{d}{2} = d$

Similarly, $\qquad IJ = d$

$\therefore \qquad GJ = GH + HI + IJ$

$$= d + d + d = 3d$$

36. (b) $d = 0.2 \tan 30^\circ = \frac{0.2}{\sqrt{3}}$

$$\frac{l}{d} = \frac{2\sqrt{3}}{0.2/\sqrt{3}} = 30$$

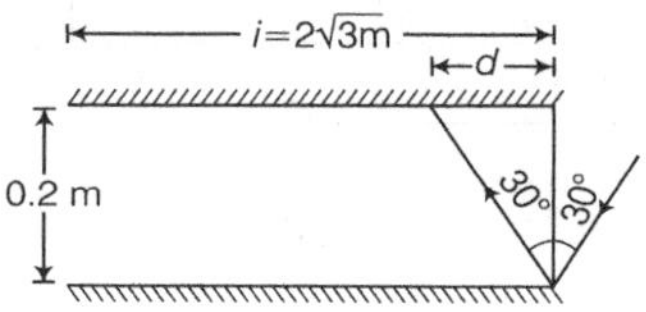

Therefore, maximum number of reflection are 30.

Note *Answer of this question may be 30 or 31 l.*

37. (a) Here, $\frac{1}{2}(\hat{i} + \sqrt{3}\hat{j}) \cdot \frac{1}{2}(\hat{i} - \sqrt{3}\hat{j}) = \frac{1}{4} \times 2 \times 2\cos\theta$

$$\cos\theta = \frac{-1}{2}$$

$\therefore \qquad \theta = 120^\circ$

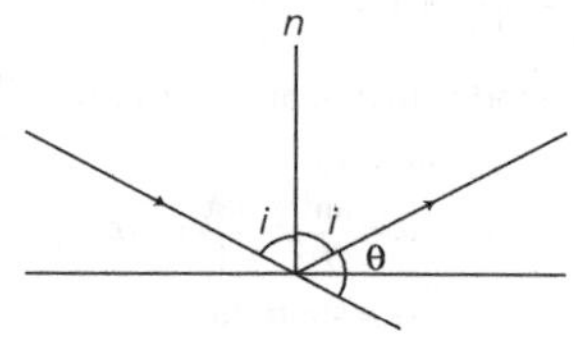

From figure,

$$\theta + 2i = 180^\circ$$

$\therefore \qquad i = 30^\circ$

38. (a,c) The length of spot on the wall is $SA = 2L$.

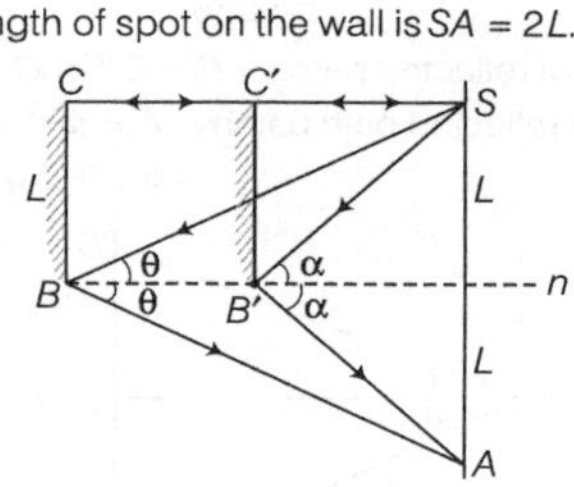

when mirror is shifted from BC position to $B'C'$ position, the size and position of spot remains same as SA. Hence, options (a) and (c) are correct.

39. (a,d) (a) When the first mirror turns through an angle ωt, the reflected beam rotated through angle $2\omega t$ and if the mirror did not rotate, the angle of reflection would also increased by $2\omega t$. Thus, after two reflections the beam would turn through $2\omega t$ as compared with the case of stationary mirrors.

Since, the second mirror does rotate, however the angle of the beam incident on it decreases by ωt during the time t. The angle of reflection decreases by the same amount and for this reason the reflected beam will travel in the same direction as with stationary discs.

This reasoning may be adopted for any two consecutive reflections, the angular speed of rotation of the beam subjected to n reflections will be zero, if n is even and 2ω if n is odd.

40. (a,c) The speed of image S_1 formed by mirror M_1 is $v_1 = 2v_o$. The speed of image S_1' formed by mirror M_2 is $v_1' = 2v_o$. For image S_2 formed by mirror M_1, image S_1' behaves as object.

So, the speed of image S_2 is $v_2 = 2v_o - 2v_o = 0$.

Similarly, the speed of image $S_2' = 2v_o - 2v_o = 0$

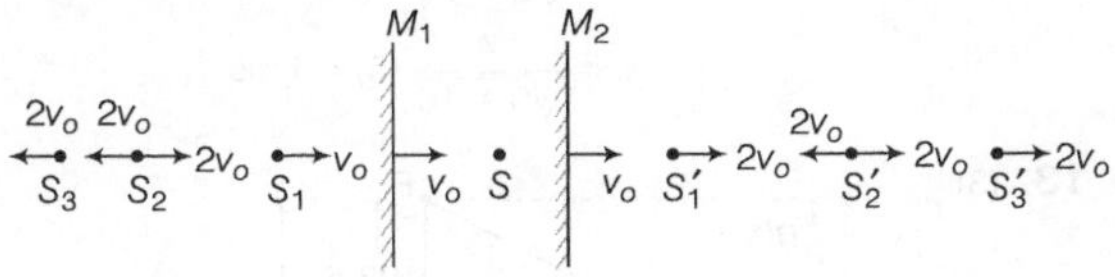

For image S_3 formed by mirror M_1,

S_2' behaves as object. So, the speed of image S_3 is

$$v_3 = 2v_o - 0 = 2v_o$$

Thus, the speed of nth image is zero, if n is even and the speed of nth image is $2v_o$, if n is odd.

41. (a,b,c) (a) When mirror is rotating, image is also rotating. Hence, locus is circle.

(b) If mirror is rotated through angle θ, the image is rotated through angle 2θ.

$\therefore$ The angular velocity of image $= \omega = 2\frac{d\theta}{dt} = 2 \times 12t$
$= 24t = 24 \text{ rads}^{-1}$

(c) $\because v = r\omega = 5 \times 24 = 120 \text{ ms}^{-1}$.

DPP-2 **Plane Mirror-II**

1. (a) A specular reflection takes place from polished half. Therefore, Dhanad sees a high percentage of light that hits the polished part of the coin. By contrast, light hitting the unpolished half undergoes more diffused reflection bouncing off the coin in all different directions. So, a smaller percentage of that light reaches Dhanad's eye. Thus, polished half looks to him more bright than unpolished half.

2. (b) Since, unpolished half reflects light diffusely, some of light reaches Rishu's eye. Thus, he sees unpolished part more bright than polished half.

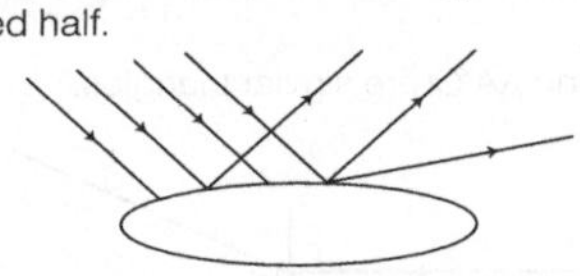

3. (c) If the size of nooks on the wall is negligible with respect to wavelength of wave, specular reflection takes place. If the size of nook are comparable to wavelength, then diffused reflection takes place.

4. (d) The beam reflected from first mirror forms an angle 2θ with the incident beam (Here, θ is the angle of incidence).

During the time t, the mirror will turn through an angle ωt and new angle of incidence will become equal to $\theta + \omega t$, as will the angle of reflection. Therefore, the angle between the incident and reflected beam will increase by $2\omega t$.

Thus, the reflected beam will turn through an angle $2\omega t$. So, the angle of incidence on second mirror is $\phi + 2\omega t$, where ϕ is angle of incidence with stationary disc. But the mirror also revolves through the angle ωt during the time t and therefore angle of incidence becomes $\phi + 3\omega t$. The angle of reflection will be the same.

Thus, after two reflections the beam will through the same angle $3\omega t$ from its direction with stationary mirrors. In similar way, after three reflections, the beam will turn through $5\omega t$ and after n reflections through mirrors, the beam will turn through angle $(2n-1)2\omega t$. In this way, angular velocity of beam is

$$\omega' = (2n-1)2\omega.$$

For five times reflections,

$$\omega' = (2 \times 5 - 1)\,2\omega = 18\omega = 18 \times 1 = 18 \text{ rads}^{-1}.$$

5. (b) The images by both mirrors moves away from the source with a speed $2v_o$. For second image S_2, the first image S_1' of other mirror behaves as object.

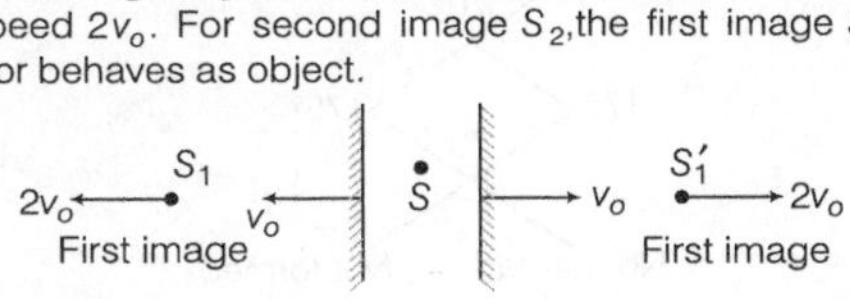

Thus, the speed of image S_2 is $4v_o$. Similarly, the speed of nth image is $2nv_o$.

6. (b)

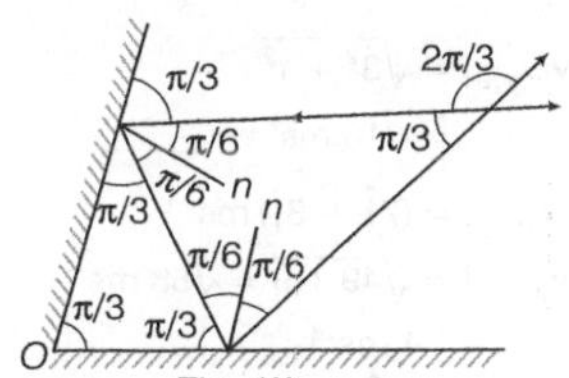

Fig. (A)

From Fig. (A), the deviation in ray is $\frac{2\pi}{3}$.

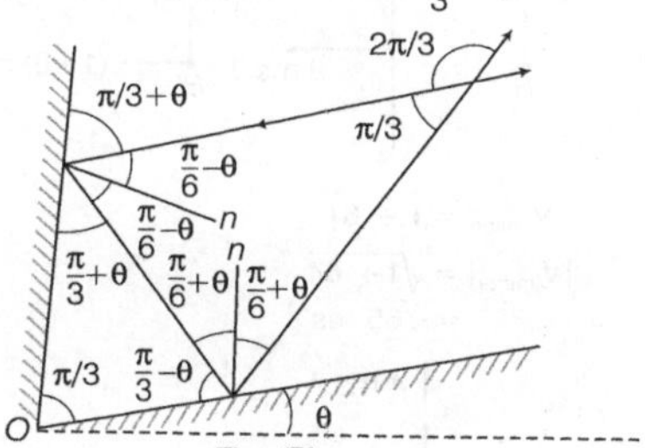

Fig. (B)

From Fig. (B), the deviation in ray is $\frac{2\pi}{3}$.

Hence, Statement I is correct.

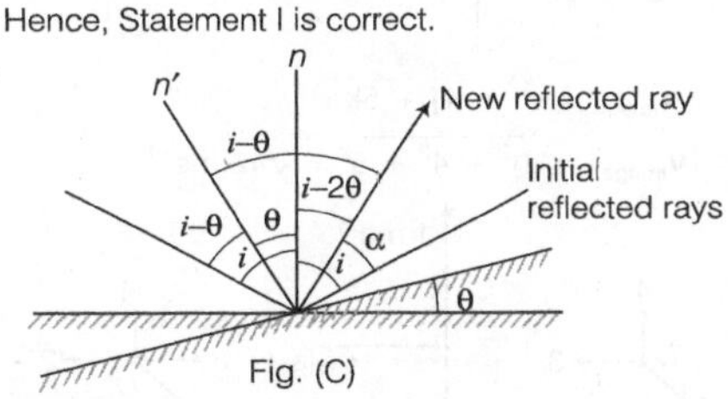

Fig. (C)

From Fig. (C),

$$i - 2\theta + \alpha = i$$

$$\therefore \qquad \alpha = 2\theta$$

Thus, when mirror is rotated through angle θ, reflected ray rotated through angle 2θ. Hence, Statement II is correct.

7. (a) Statement II is exact reason of Statement I. Hence, option (a) is correct.

8. (b) When mirror is rotated about an axis perpendicular to plane mirror, normal on incident point does not rotate. So, angle of reflection does not change. Hence, Statement I is correct. Statement II is also correct.

9. (b) The force of intraction between photon and mirror is normal to mirror. So, momentum of photon along surface of mirror remains conserved.

10. (b) The shadow will be equally distinct everywhere only if point sources of light is used. The separate sections of extended source throw shadows that are superimposed on one another. The boundary of the shadow will be the sharper, the smaller is the distance from the object to the surface on which the shadow is formed. Since, the distances between the boundaries of the shadows produced by the various sections of the source will be minimum. It is for this reason that a man's legs give sharper shadow than his head.

11. (c)

(i) $\because$ $\quad \mathbf{v}_A = (3\hat{i} + \hat{j})\,\text{ms}^{-1}$

$\therefore$ $\quad \mathbf{v}_{\text{image}} = (-3\hat{i} + \hat{j})\,\text{ms}^{-1}$

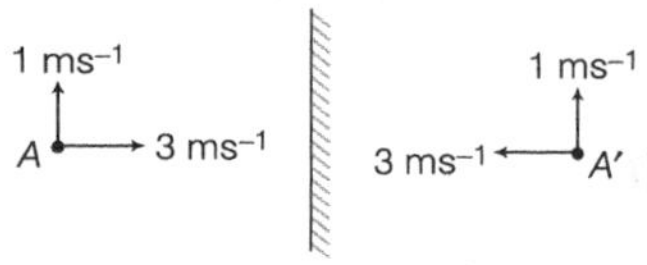

$\therefore$ $\quad |\mathbf{v}_{\text{image}}| = \sqrt{3^2 + 1^2}$

$= \sqrt{10}\,\text{ms}^{-1}$

(ii) $\therefore$ $\quad \mathbf{v}_{\text{image}} = (7\hat{i} + 3\hat{j})\,\text{ms}^{-1}$

$\therefore$ $\quad |\mathbf{v}_{\text{image}}| = \sqrt{49 + 9} = \sqrt{58}\,\text{ms}^{-1}$

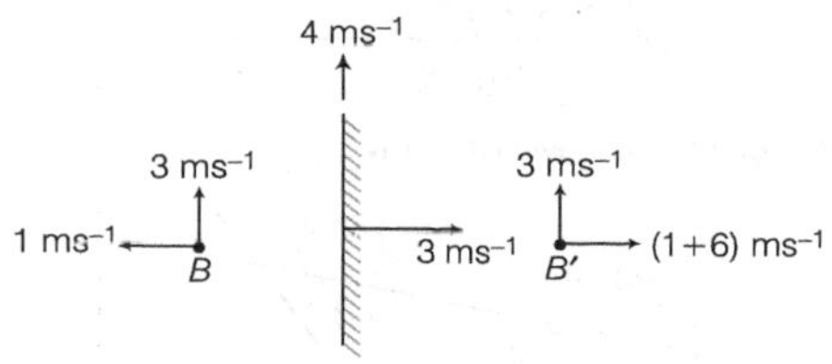

(iii) $\therefore$ $\quad \mathbf{v}_{\text{image}} = \hat{i} + 8\hat{j}$

$\therefore$ $\quad |\mathbf{v}_{\text{image}}| = \sqrt{1 + 64}$

$= \sqrt{65}\,\text{ms}^{-1}$

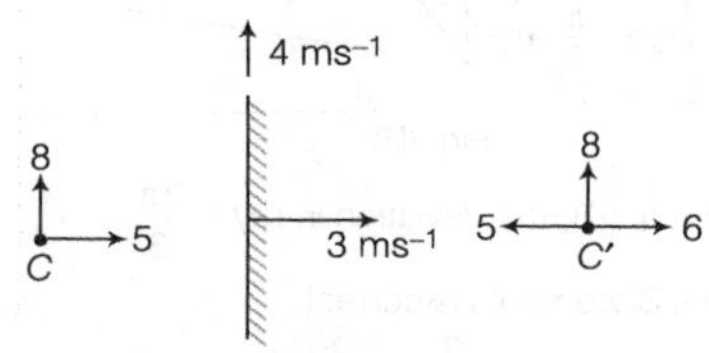

(iv) $\therefore$ $\quad \mathbf{v}_{\text{image}} = -\hat{i} + 4\hat{j} + 5\hat{k}$

$\therefore$ $\quad |\mathbf{v}_{\text{image}}| = \sqrt{1^2 + 4^2 + 5^2} = \sqrt{42}\,\text{ms}^{-1}$

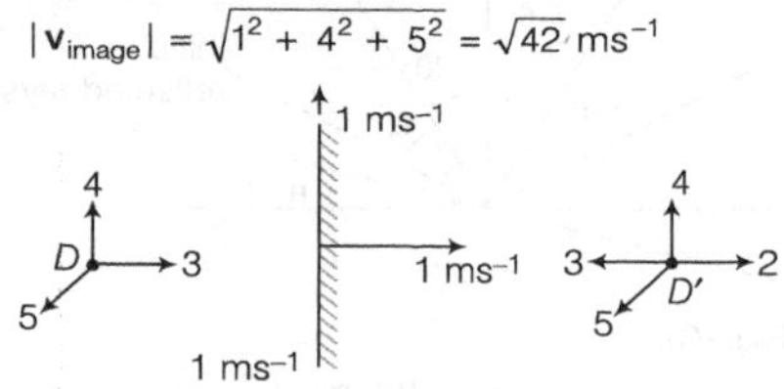

Hence, (i) $\to$ (q), (ii) $\to$ (r), (iii) $\to$ (s), (iv) $\to$ (p).

12. (4) Let the length of shadow is l.

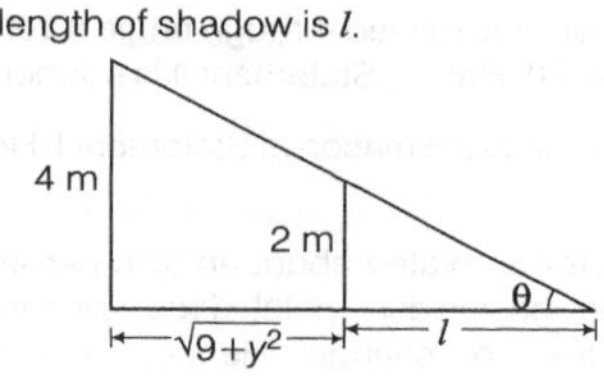

$\therefore$ $\quad \tan\theta = \dfrac{4}{\sqrt{9 + y^2} + l} = \dfrac{2}{l}$

or $\quad 2l = \sqrt{9 + y^2} + l$

or $\quad l = \sqrt{9 + y^2}$

or $\quad \dfrac{dL}{dt} = \dfrac{1}{2\sqrt{9 + y^2}}\, 2y\,\dfrac{dy}{dt}$

$= \dfrac{y}{\sqrt{9 + y^2}} \times 5$

$= \dfrac{4}{\sqrt{9 + 16}}(5) = 4\,\text{ms}^{-1}.$

13. (3)

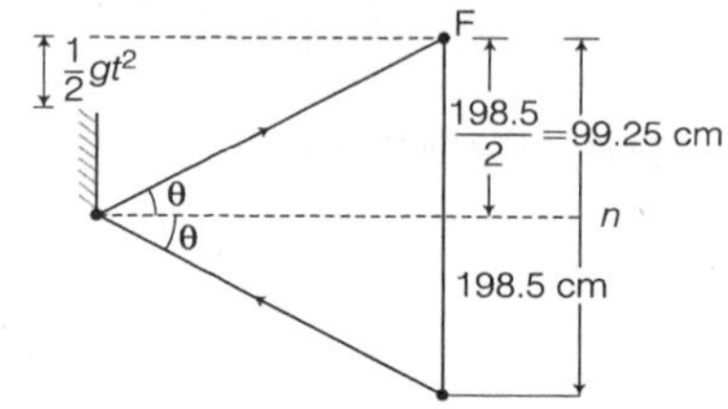

From figure,

$$\frac{1}{2}gt^2 + 50\text{ cm} = 99.25\text{ cm}$$

or $\quad \dfrac{1}{2} \times 10 \times t_m^2 + 50\text{ cm} = 99.25\text{ cm}$

or $\quad \dfrac{1}{2} \times 10 \times 100t^2\text{cm} + 50\text{ cm} = 99.25\text{ cm}$

$\therefore$ $\quad t^2 = 1015 \times 10^{-3}\,\text{s}^2$

$\therefore$ $\quad n = 3.$

14. (2) $\Delta\, A'E_1A$ and $\Delta A'CI$ are similar triangles.

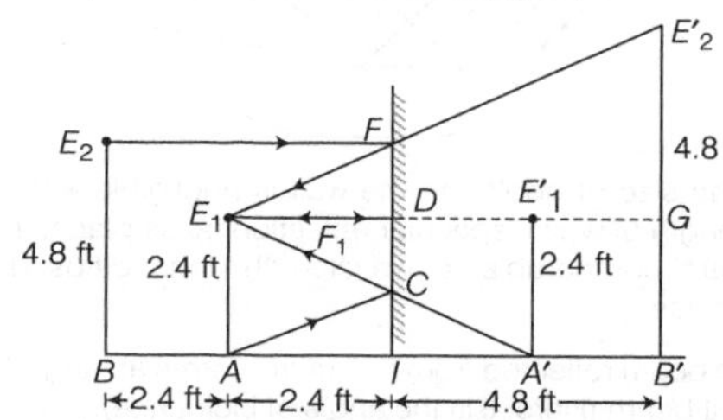

$\therefore$ $\quad \dfrac{IC}{A'I} = \dfrac{AE_1}{AA'}$ or $\dfrac{IC}{2.4} = \dfrac{2.4}{4.8}$

$\therefore$ $\quad IC = 1.2$ ft

$\Delta GE_2'E_1$ and ΔDFE_1 are similar triangles.

$\therefore$ $\quad \dfrac{DF}{DE_1} = \dfrac{E_2'G}{GE_1} = \dfrac{B'E_2' - B'G}{2.4 + 4.8}$

or $\quad \dfrac{DF}{2.4} = \dfrac{4.8 - 2.4}{7.2}$

$\therefore$ $\quad DF = \dfrac{2.4 \times 2.4}{7.2} = 0.8$ ft

Required length of mirror $= CF = CD + DF$

$= ID - IC + DF$

$= 2.4 - 1.2 + 0.8 = 2$ ft

15. (6) So, number of images = 6

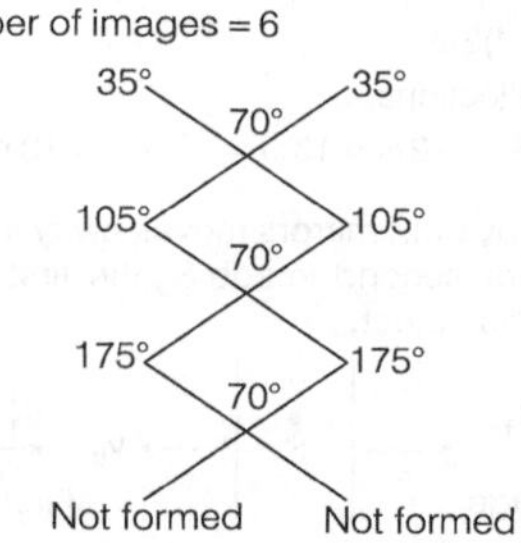

16. (6) The number of images formed by combination of two plane mirrors.

$$= \frac{360^\circ}{60^\circ} - 1 = 5$$

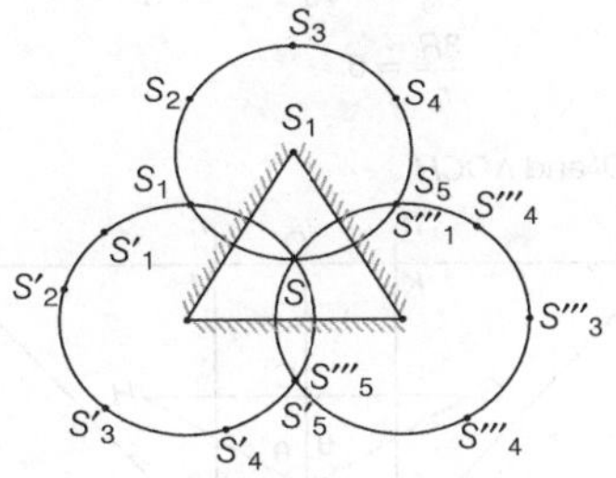

Total number of images $= 5 \times 3 - 3 = 12 = 2n$

$\therefore \quad n = 6.$

17. (2) From figure, $\sin 37^\circ = \dfrac{h}{R}$

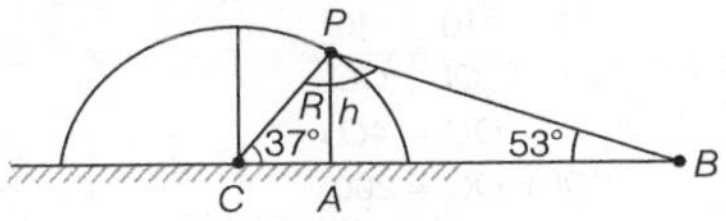

$\therefore \quad h = R \sin 37^\circ$

$$= 10 \times \frac{3}{5} = 6 \text{ cm}$$

Also, $\cos 37^\circ = \dfrac{CA}{R}$

$\therefore \quad CA = R\cos 37^\circ$

$$= 10 \times \frac{4}{5} = 8 \text{ cm}$$

and $\tan 53^\circ = \dfrac{h}{AB}$

$\therefore \quad AB = \dfrac{h}{\tan 53^\circ} = \dfrac{6}{4/3}$

$$= \frac{18}{4} = 4.5 \text{ cm}$$

Thus, minimum radius of mirror $= CB$

$= CA + CB$

$= 8 + 4.5$

$= 12.5$ cm.

18. (5) From figure, visible portion is $OA = 10$ cm $= 2h$

$\therefore \quad h = 5$

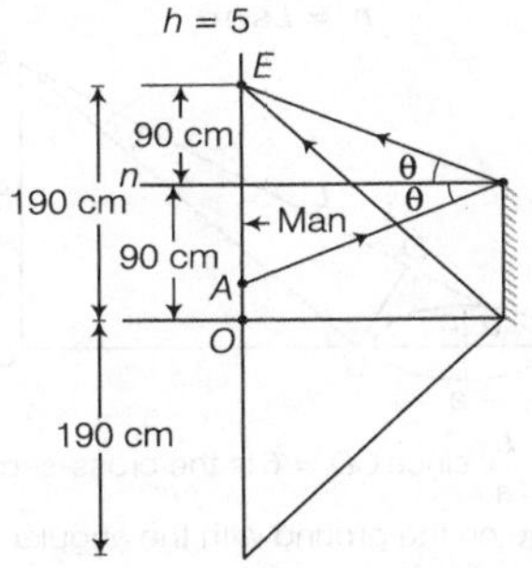

19. (4)

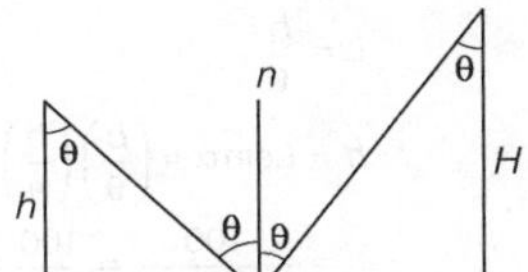

Fig. (A)

From Fig (A), $\tan\theta = \dfrac{x}{h} = \dfrac{y}{H}$

$\therefore \quad \dfrac{x}{y} = \dfrac{h}{H} = \dfrac{1.8}{5.4} = \dfrac{1}{3}$

$\therefore \quad y = 3x$

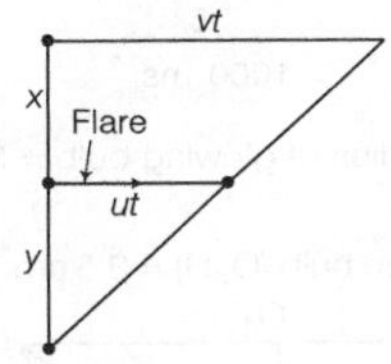

Fig. (B)

From Fig. (B)

$$\frac{vt}{ut} = \frac{x+y}{y} \quad \text{or} \quad \frac{v}{u} = \frac{x+y}{y}$$

$$\therefore \quad v = \left(1 + \frac{x}{y}\right)u = \left(1 + \frac{1}{3}\right)2.4$$

$$= \frac{4}{3} \times 3 = 4 \text{ ms}^{-1}$$

20. (2)

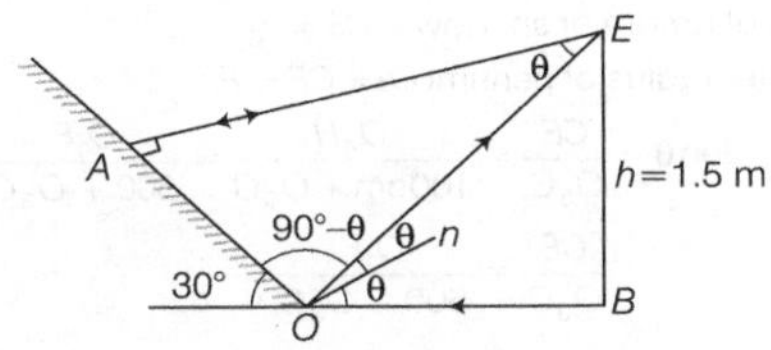

Here, $2\theta + 90^\circ - \theta + 30^\circ = 180^\circ$

or $\theta + 30^\circ = 90^\circ$

or $\theta = 60^\circ$

In ΔEOA, $\sin\theta = \dfrac{OA}{OE}$

$\therefore \quad OA = OE\sin\theta$

and $\sin 2\theta = \dfrac{1.5}{OE}$ (in ΔOBE)

$\therefore \quad OA = \dfrac{2}{\sin 2\theta}\sin\theta$

$$= \frac{2\sin 60^\circ}{\sin 120^\circ} = 2\text{m}$$

$\therefore$ Minimum length of mirror = 2m.

21. (1) Here, $P_A = \dfrac{P_s A}{4\pi r^2}$

where, r is distance of mirror A from source S and A is area of mirror.

Here, $P_B = \dfrac{P_s A\cos 60^\circ}{4\pi r'^2} = \dfrac{P_s A\cos 60^\circ}{\left(\dfrac{r}{\cos 60^\circ}\right)^2}$

$$= \frac{P_s A}{4\pi r^2}\cos^3 60^\circ = P_A \cos^3 60^\circ = 1\,\text{W}.$$

22. (2) $\sin 37^\circ = \dfrac{10}{r}$

$\therefore \quad r = \dfrac{10}{\frac{3}{5}} = \dfrac{50}{3}$ m

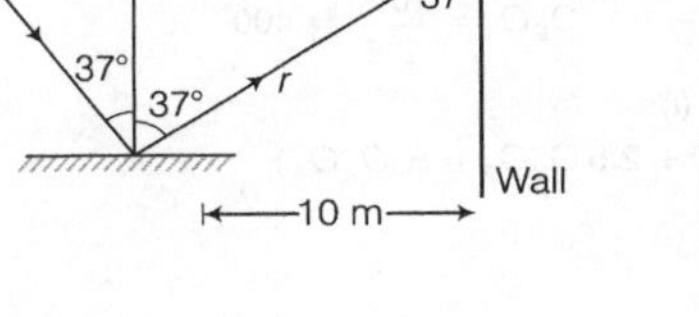

$\therefore$ The angular speed of spot

$$= 2\omega = 36 = \frac{v \sin 37°}{r}$$

$$\therefore \quad v = \frac{36 \times 50/3}{\frac{3}{5}} = \frac{36 \times 50 \times 5}{9}$$

$$= 1000 \text{ ms}^{-1}.$$

23. (5) Here, O_1F = radius of glowing bulb = 5 cm and IH is diameter of opaque bulb.

$\therefore$ Radius of opaque bulb (O_4H) = 2.5 cm

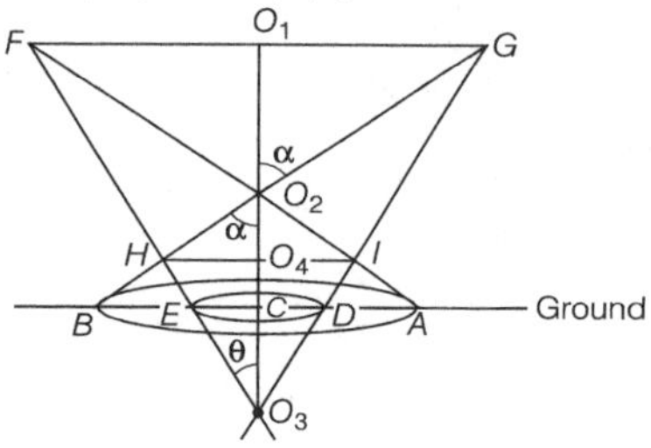

Radius of umbra or shadow = $CE = r_0$

and outer radius of penumbra = $CB = R$

Here, $\tan\theta = \frac{CE}{O_3C} = \frac{O_4H}{100\text{cm} + O_3C} = \frac{O_1F}{500 + O_3C}$

$$\Rightarrow \frac{CE}{O_3C} = \frac{O_1F}{500 + O_3C}$$

$$\Rightarrow \frac{r_0}{O_3C} = \frac{5}{500 + O_3C}$$

$$\Rightarrow 500r_0 + (O_3C)\, r_0 = 5(O_3C) \quad \ldots\text{(i)}$$

Also, $\frac{CE}{O_3C} = \frac{O_4H}{100 + O_3C}$

$$\Rightarrow \frac{r_0}{O_3C} = \frac{2.5}{100 + O_3C}$$

$$\Rightarrow 2.5\, O_3C = 100r_0 + r_0(O_3C)$$

$$\Rightarrow 2.5\, O_3C = 100r_0 + 5(O_3C) - 500\, r_0$$

$$\Rightarrow 2.5\, O_3C = 400r_0$$

$$\Rightarrow O_3C = \frac{400\, r_0}{2.5} = 160r_0$$

From Eq. (i),

$$500r_0 + O_3Cr_0 = 5(O_3C)$$

$$\Rightarrow 500r_0 + 160\, r_0^2 = 5 \times 160\, r_0$$

$$\Rightarrow 160\, r_0^2 = 800\, r_0 - 500r_0 = 300\, r_0$$

$$\Rightarrow r_0 = \frac{300}{160} = \frac{15}{8} = 1.875 \text{ cm}$$

Also, $\tan\alpha = \frac{O_4H}{O_2O_4} = \frac{CB}{100 + O_2O_4} = \frac{O_1G}{400 - O_2O_4}$

$$\frac{2.5}{O_2O_4} = \frac{R}{100 + O_2O_4}$$

$$\Rightarrow (2.5)(100 + O_2O_4) = R(O_2O_4)$$

$$\Rightarrow 250 + 2.5O_2O_4 = R(O_2O_4) \quad \ldots\text{(i)}$$

and $\frac{2.5}{O_2O_4} = \frac{5}{400 - O_2O_4}$

$$\Rightarrow 5O_2O_4 = 1000 - 2.5O_2O_4$$

$$\Rightarrow 2.5O_2O_4 = 1000$$

$$\Rightarrow O_2O_4 = \frac{1000}{2.5} = 400$$

From Eq. (i),

$$250 + 2.5\, O_2O_4 = R(O_2O_4)$$

$$\Rightarrow 250 + 2.5 \times 400 = 400R$$

$$\Rightarrow R = 3.125$$

$$\Rightarrow \frac{R}{r_0} = \frac{3.125}{15} \times 8 = \frac{5}{3}$$

$$\therefore \frac{3R}{r_0} = 5.$$

24. (5) In ΔEOI and ΔOCH,

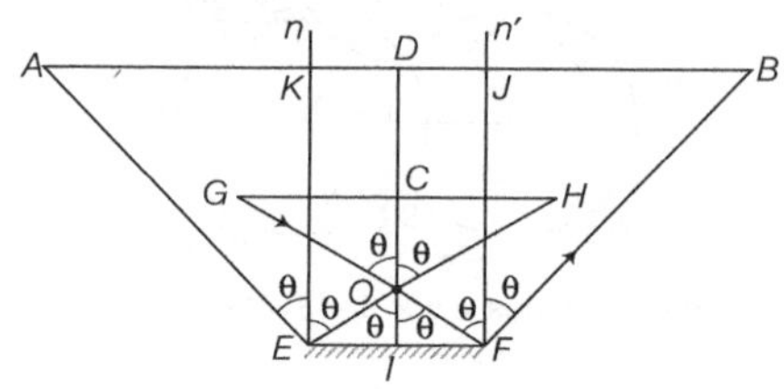

$$\tan\theta = \frac{IE}{OI} = \frac{CH}{OC}$$

$$\Rightarrow \frac{10}{OI} = \frac{40}{OC}$$

$$\Rightarrow OC = 4OI$$

$$\because OI + OC = 200$$

or $5OI = 200$

$\therefore$ $OI = 40$ cm

and $OC = 4OI$

$$= 4 \times 40 = 160 \text{ cm}.$$

In ΔAKE, $\tan\theta = \frac{AK}{EI} = \frac{AK}{300 \text{ cm}}$ or $\frac{IE}{OI} = \frac{AK}{300}$

or $\frac{10}{40} = \frac{AK}{300}$

$\therefore$ $AK = \frac{1}{4} \times 300 = 75$ cm

But $AK = JB = 75$ cm

$\therefore$ The length off reflected patch

$$= AK + KJ + JB$$

$$= 75 + 20 + 75$$

$$= 170 \text{ cm} = 34x$$

$$\therefore x = \frac{170}{34} = \frac{10}{2} = 5.$$

25. (9) From figure, $\sin\alpha = \frac{h}{L}$

$\therefore$ $h = L\sin\alpha$

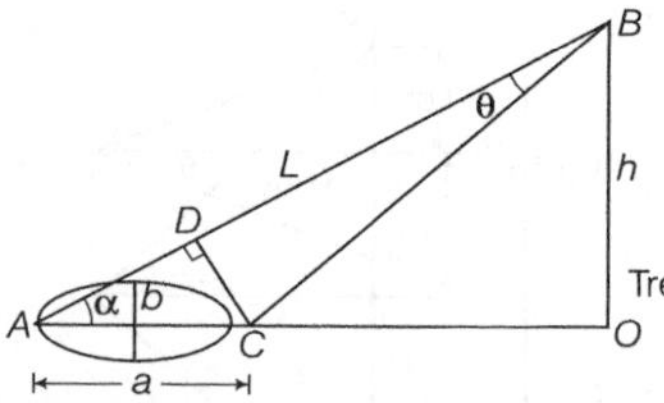

while, $\sin\alpha = \frac{b}{a}$, since $CD = 6$ is the cross-sectional diameter of the light come on the ground with the angular dimension of the sun's disc θ_L we obtain.

$$L = \frac{b}{\theta}$$

$$\therefore h = L\sin\alpha = \left(\frac{b}{\theta}\right)\left(\frac{b}{a}\right) = \frac{b^2}{a\theta}$$

$$= \frac{100}{12 \times \frac{1}{108}} = \frac{100}{12} \times 108$$

$$= 900 \text{ cm} = 9 \text{ m}.$$

DPP-3 **Spherical Mirror- I**

1. (i) (c) $u = -30$ cm, $f = -20$ cm

According to the mirror formula,

$$\frac{1}{v} + \frac{1}{u} = \frac{1}{f}$$

$$\Rightarrow \quad \frac{1}{v} - \frac{1}{30} = -\frac{1}{20}$$

$$\Rightarrow \quad \frac{1}{v} = \frac{1}{30} - \frac{1}{20} = \frac{2-3}{60}$$

$$\therefore \quad v = -60 \text{ cm}$$

(ii) (b) $m = \frac{-v}{u} = -\frac{(-60)}{(-30)} = -2$

(iii) (b) $m = \frac{l'}{l}$

$\therefore \quad l' = ml = -2 \times 2 = -4$ cm

(iv) (c)

(v) (b)

2. (i) (b) The converging point is behind the mirror. So, it behaves as virtual object.

(ii) (b) The new converging point is the image point.

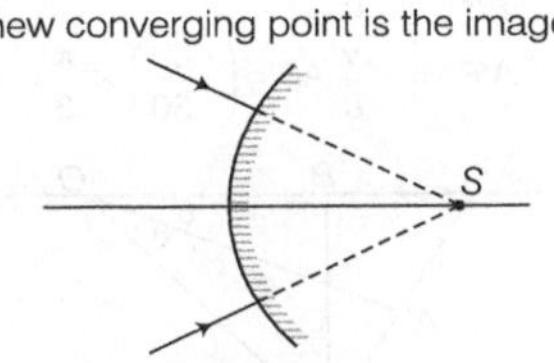

$\because \quad u = +20$ cm, $f = +5$ cm

$$\frac{1}{v} + \frac{1}{u} = \frac{1}{f}$$

or $$\frac{1}{v} + \frac{1}{20} = \frac{1}{5}$$

or $$\frac{1}{v} = \frac{1}{5} - \frac{1}{20} = \frac{3}{20}$$

$$\therefore \quad v = \frac{20}{3} \text{ cm.}$$

(iii) (b)

3. (i) (a) Since, image is formed on screen. So, it must be real.

(ii) (d) Here, $u = -x$, $m = \frac{-v}{u}$ or $-3 = \frac{-v}{-x}$

$\therefore \quad v = -3x$

(iii) (a,b,c) (iv) (b) (v) (b)

(vi) (a) (vii) (a,b,c,d)

(viii) (b) $\because \quad \frac{1}{3x} + \frac{1}{x} = \frac{2}{R}$ ...(i)

and $\frac{1}{2(x-6)} + \frac{1}{x-6} = \frac{2}{R}$ ...(ii)

From Eqs. (i) and (ii), we get $R = 72$ cm

(ix) (a) $\because \quad \frac{1}{3x} + \frac{1}{x} = \frac{2}{R} = \frac{2}{72} = \frac{1}{36}$

or $$\frac{1+3}{3x} = \frac{1}{36}$$

$$\therefore \quad x = \frac{36 \times 4}{3} = 48 \text{ cm}$$

(x) (b) $\because \quad v = -3x = -48 \times 3$
$= -144$ cm

and $v' = -2(x+6) = -2(54)$
$= -108$ cm

$\therefore$ Separation between images $= |v' - v| = 36$ cm.

(xi) (b) The shifting in screen = Shifting in image = 36 cm.

4. (i) (a) For concave mirror, $u = -30$ cm, $f = -20$ cm

$$\because \quad \frac{1}{v} + \frac{1}{u} = \frac{1}{f}$$

or $$\frac{1}{v} - \frac{1}{30} = -\frac{1}{20}$$

or $$\frac{1}{v} = \frac{1}{30} - \frac{1}{20} = \frac{2-3}{60} = -\frac{1}{60}$$

$$\therefore \quad v = -60 \text{ cm.}$$

(ii) (b) $\because \quad \frac{1}{v} + \frac{1}{u} = \frac{1}{f}$

or $$-\frac{1}{v^2}\frac{dv}{dt} - \frac{1}{u^2}\frac{du}{dt} = 0$$

$$\therefore \quad \frac{dv}{dt} = -\frac{v^2}{u^2}\frac{du}{dt}$$

$$\therefore \quad v_{image} = -\left(\frac{-60}{-30}\right)^2 \times 2 \text{ cms}^{-1}$$

$$= -8 \text{ cms}^{-1}$$

(iii) (a) The image formed by concave mirror behaves as object for plane mirror.

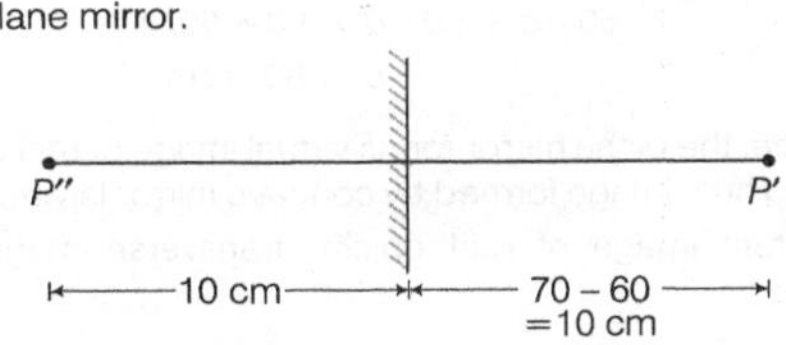

(iv) (b)

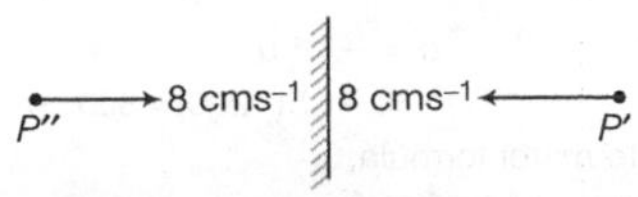

5. (i) (a)

(ii) (a) For paraxial rays, $f = R/2$, $R = 2f$

(iii) (a,b,c) From figure,

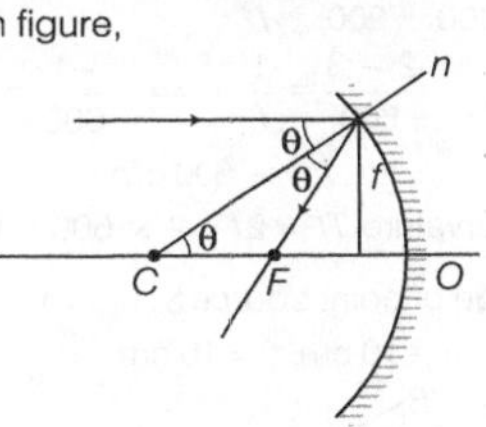

$$\sin\theta = \frac{f}{R} = \frac{f}{2f} = \frac{1}{2}$$

$$\therefore \quad \theta = 30°.$$

(iv) (a,b,c) From figure of solution of Eq. (iii),

$$\frac{R}{\sin(\pi - 2\theta)} = \frac{CF}{\sin\theta}$$

$$\therefore \quad CF = R/2 \sec\theta$$

$$CF = \frac{R}{2}\sec 30°$$

$$= R/\sqrt{3} = 2f/\sqrt{3}.$$

(v) (a) $\because \quad \sin\theta = \frac{\sqrt{3}f}{R} = \frac{\sqrt{3}f}{2f} = \frac{\sqrt{3}}{2}$

$$\therefore \quad \theta = 60°$$

$$\therefore \quad CF' = \frac{R}{2}\sec\theta = \frac{R}{2}\sec 60° = R$$

(vi) (a) $FF' = R - \frac{2f}{\sqrt{3}} = 2f - \frac{2f}{\sqrt{3}}$

$$= 2f\left(1 - \frac{1}{\sqrt{3}}\right) = f\left(2 - \frac{2}{\sqrt{3}}\right).$$

6. (i) (b) Two images, one due to plane mirror and other due to convex mirror.

(ii) (b) For convex mirror, $u = -50$ cm,

$$f = +25 \text{ cm}$$

$$\frac{1}{v} + \frac{1}{u} = \frac{1}{f}$$

or $$\frac{1}{v} - \frac{1}{50} = \frac{1}{25}$$

or $$\frac{1}{v} = \frac{1}{25} + \frac{1}{50} = \frac{2+1}{50}$$

$\therefore$ $$v = 50/3 \text{ cm}$$

(iii) (b) Images formed by both mirrors should be at as the same place.

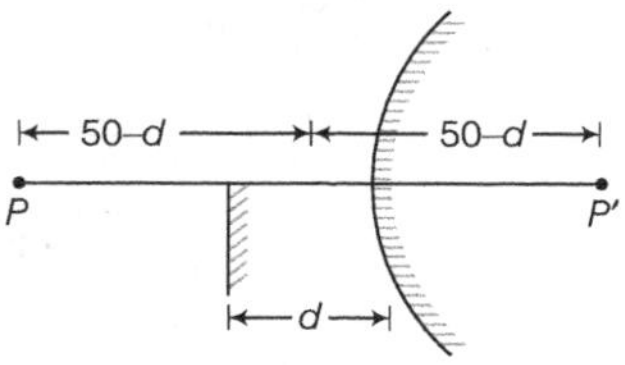

From figure,

$$50 - d + 50 - d = 50 + 50/3$$

$\therefore$ $$d = 50/3 \text{ cm.}$$

7. (b) Since, the plane mirror forms virtual image of real object same in size. Thus, image formed by concave mirror is virtual.

For virtual image of real object, transverse magnification is positive.

$\therefore$ $$m = +1.5 = -\frac{v}{u}$$

$\therefore$ $$u = -1.5u$$
$$= -1.5(-200) = 300 \text{ cm}$$

According to mirror formula,

$$\frac{1}{v} + \frac{1}{u} = \frac{1}{f}$$

$\Rightarrow$ $$\frac{1}{300} - \frac{1}{200} = \frac{1}{f}$$

$\Rightarrow$ $$\frac{2-3}{600} = \frac{1}{f} \Rightarrow \frac{-1}{600} = \frac{1}{f}$$

$\therefore$ $$f = -600 \text{ cm}$$

$\therefore$ Radius of curvature, $R = 2f = 2 \times 600 = 1200$ cm $= 12$ m.

8. (c) If S' is image of point source S.

$\because$ $$u = -10 \text{ cm}, f = 10 \text{ cm}$$

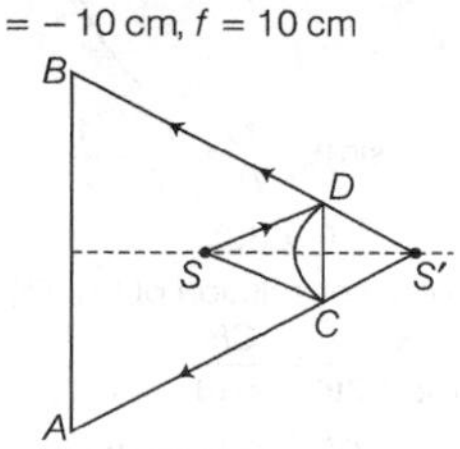

$$\frac{1}{v} + \frac{1}{u} = \frac{1}{f}$$

or $$\frac{1}{v} - \frac{1}{10} = \frac{1}{10}$$

or $$\frac{1}{v} = \frac{1}{10} + \frac{1}{10} = \frac{1}{5}$$

$\therefore$ $$v = 5 \text{ cm}$$

$S'CD$ and $S'AB$ are similar triangles.

$\therefore$ $$\frac{AB}{25 \text{ cm}} = \frac{CD}{5 \text{ cm}} = \frac{1 \text{ mm}}{5 \text{ cm}}$$

$\therefore$ $$AB = (1 \text{ mm}) \frac{25 \text{ cm}}{5 \text{ cm}} = 5 \text{ mm}$$

9. (b) Here, $OA = OP\cos\theta = OP = 30$ cm

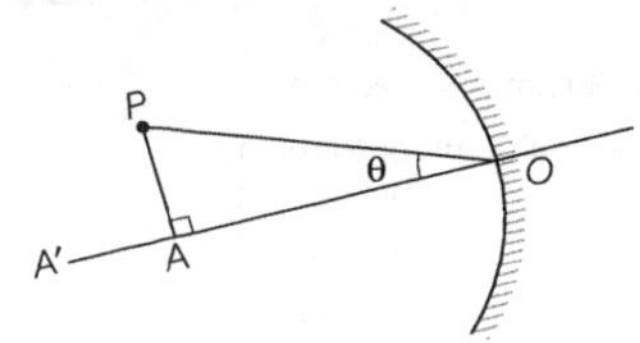

and $$AP = OP\sin\theta = OP \times \theta$$
$$= 30 \times 2 \times \frac{\pi}{180} = \frac{\pi}{3} \text{ cm}$$

Here, $u = OA = -30$ cm, $f = -20$ cm

$\because$ $$\frac{1}{v} + \frac{1}{u} = \frac{1}{f}$$

or $$\frac{1}{v} - \frac{1}{30} = -\frac{1}{20}$$

or $$\frac{1}{v} = \frac{1}{30} - \frac{1}{20} = \frac{2-3}{60}$$

$\therefore$ $$v = -60 \text{ cm}$$

$\therefore$ $$OA' = 60 \text{ cm}$$

$\because$ $$\frac{A'P'}{AP} = -\frac{v}{u}$$

$\therefore$ $$A'P' = -\frac{v}{u} AP = \left(\frac{-60}{-30}\right) \times \frac{\pi}{3} = -\frac{2\pi}{3} \text{ cm}$$

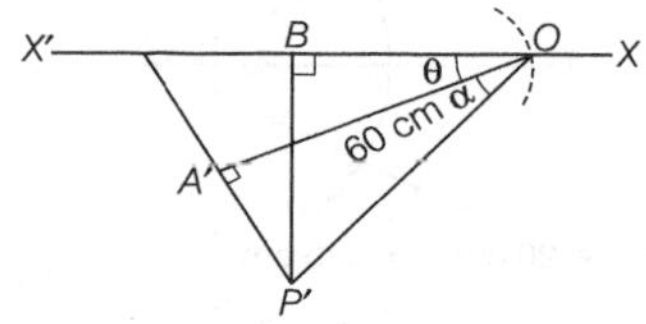

From figure, $$\tan\alpha = \frac{A'P'}{OA'} = \frac{2\pi}{3 \times 60} = \frac{\pi}{90}$$

But $$\tan\alpha = \alpha = \sin\alpha = \frac{A'P'}{OP'}$$

$\therefore$ $$OP' = \frac{2\pi}{3 \times \pi/90} = 60 \text{ cm}$$

$\therefore$ $$\sin(\alpha + \theta) = \frac{BP'}{OP'}$$

$\therefore$ $$BP' = OP'(\alpha + \theta) = 60\left(\frac{\pi}{90} + \frac{\pi}{90}\right)$$
$$= 60 \times \frac{2\pi}{90} = \frac{4\pi}{3} \text{ cm}$$

and $$OB = OP'\cos(\alpha + \theta) = OP' = 60 \text{ cm}$$

$\therefore$ Coordinates of image $= \left(-60 \text{ cm}, -\frac{4\pi}{3} \text{ cm}\right)$

10. (c) Here, $u = -30$ cm, $f = -20$ cm

$\because$ $$\frac{1}{v} + \frac{1}{u} = \frac{1}{f}$$

$\Rightarrow$ $$\frac{1}{v} - \frac{1}{30} = -\frac{1}{20}$$

$\Rightarrow$ $$\frac{1}{v} = \frac{1}{30} - \frac{1}{20} = \frac{2-3}{60}$$

$\therefore$ $$v = -60 \text{ cm}$$

$\therefore$ Transverse magnification is $m = \frac{-v}{u} = -2$.

and longitudinal magnification is $m_l = \frac{-v^2}{u^2} = -u$.

The magnification in sides parallel to optical axis is m_l and the magnification in sides perpendicular to optical axis is m.

$\therefore$ $$\frac{l'}{l} = \frac{-v^2}{u^2}$$

$\therefore \quad l' = -4l = -4$ mm

and $\quad \dfrac{b'}{b} = m = -2$

$b' = -2 \times 1 = -2$ mm

$\therefore$ Volume of image $= l' \times b' \times b'$

$= 4 \times (2)(2) = 16\ \text{mm}^3$

Thus, the magnification in the volume is 16.

11. (a) If mirror is at left extreme position, then

$u = -15$ cm, $f = +10$ cm

$\therefore \quad \dfrac{1}{v} + \dfrac{1}{u} = \dfrac{1}{f}$

or $\quad \dfrac{1}{v} - \dfrac{1}{15} = \dfrac{1}{10}$

$\Rightarrow \quad \dfrac{1}{v} = \dfrac{1}{15} + \dfrac{1}{10} = \dfrac{2+3}{30} = \dfrac{5}{30}$

$\therefore \quad v = 6$ cm

If mirror is at right extreme position.

Here, $\quad u' = -25$ cm

$\Rightarrow \quad f' = +10$ cm

$\therefore \quad \dfrac{1}{v'} + \dfrac{1}{u'} = \dfrac{1}{f'}$

or $\quad \dfrac{1}{v'} - \dfrac{1}{25} = \dfrac{1}{10}$

or $\quad \dfrac{1}{v'} = \dfrac{2+5}{50} = \dfrac{7}{50}$

$\therefore \quad v' = \dfrac{50}{7}$ cm

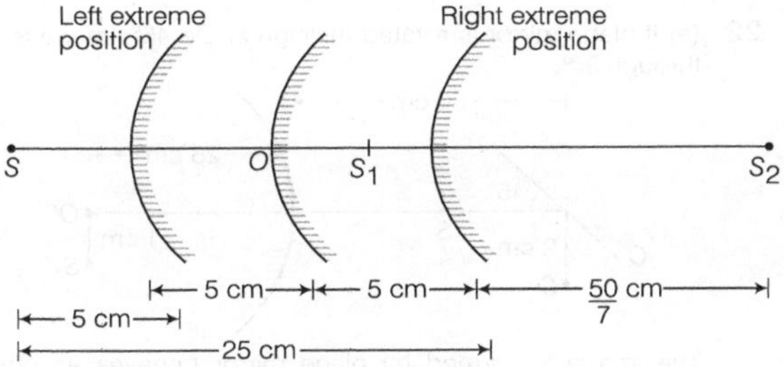

$\therefore$ The range of vibration of image,

$= S_1S_2 = \dfrac{50}{7} + 10 - 6$

$= \dfrac{50 + 70 - 42}{7} = 11.14$ cm

12. (d) For point A,

$u = -10$ cm, $f = +10$ cm

$\because \quad \dfrac{1}{v} + \dfrac{1}{u} = \dfrac{1}{f}$

or $\quad \dfrac{1}{v} - \dfrac{1}{10} = \dfrac{1}{10}$

or $\quad \dfrac{1}{v} = \dfrac{2}{10}$

$\therefore \quad v = +5$ cm

For point B,

$u' = -15$ cm, $f' = +10$ cm

$\therefore \quad \dfrac{1}{v'} + \dfrac{1}{u'} = \dfrac{1}{f'}$

$\Rightarrow \quad \dfrac{1}{v'} - \dfrac{1}{15} = \dfrac{1}{10}$

$\Rightarrow \quad \dfrac{1}{v'} = \dfrac{1}{10} + \dfrac{1}{15} = \dfrac{3+2}{30}$

$\therefore \quad v' = 6$ cm

The length of image of $AB = |v' - v|$

$= 6 - 5 = 1$ cm

13. (b) Here, $u = -10$ cm, $f = +10$ cm

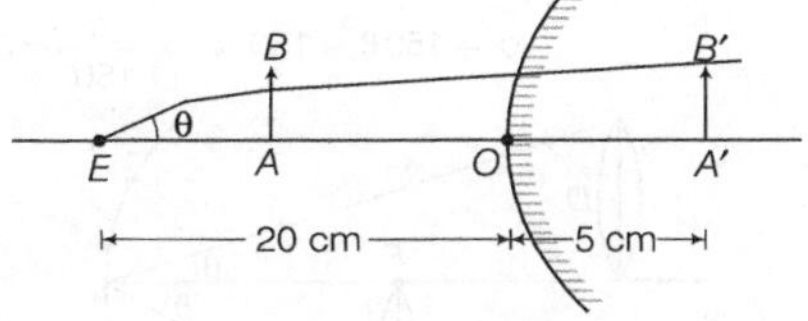

$\because \quad \dfrac{1}{v} + \dfrac{1}{u} = \dfrac{1}{f}$

$\therefore \quad v = +5$ cm

Here, $\quad \tan\theta = \dfrac{A'B'}{EA'} = \dfrac{1\text{ cm}}{25\text{ cm}}$

$\therefore \quad \theta = \dfrac{1}{25} = 0.04$ rad

$= 0.04 \times \dfrac{180^\circ}{\pi} = 2.3^\circ$

14. (b) Here, $\quad u = -20$cm

$f = +10$ cm

$\because \quad \dfrac{1}{v} + \dfrac{1}{u} = \dfrac{1}{f}$

$\Rightarrow \quad \dfrac{1}{v} - \dfrac{1}{20} = \dfrac{1}{10}$

or $\quad \dfrac{1}{v} = \dfrac{1}{20} + \dfrac{1}{10} = \dfrac{1+2}{20}$

$\therefore \quad v = \dfrac{20}{3}$ cm

For upright,

$\dfrac{L'}{L} = -\dfrac{v}{u} = -\dfrac{20}{3(-2^\circ)}$

$\therefore \quad L' = \dfrac{L}{3}$

For along the optical axis,

$\dfrac{L''}{L} = \dfrac{-v^2}{u^2}$

or $\quad L'' = -\dfrac{1}{9}L$

$\therefore \quad \text{Ratio} = \dfrac{\frac{L}{3}}{\frac{L}{9}} = 3:1$

15. (c) Here, $\quad u = -10$ cm

$f = +10$ cm

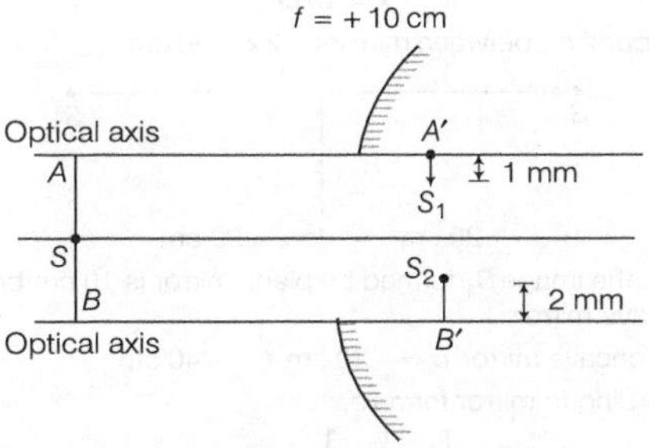

$\because \quad \dfrac{1}{v} + \dfrac{1}{u} = \dfrac{1}{f}$

or $\quad \dfrac{1}{v} - \dfrac{1}{10} = \dfrac{1}{10}$

$\therefore \quad v = 5$ cm

$\therefore \quad m = -\dfrac{v}{u} = \dfrac{1}{2}$

$\therefore \quad S_1S_2 = 6 - 3 = 3$ mm

16. (b) From figure, $\tan\theta = \theta = \dfrac{d}{f} = \dfrac{d}{150}$

$\therefore$ $$d = 150\theta = 150 \times 1 \times \frac{\pi}{180^\circ} = 2.62 \text{ cm}$$

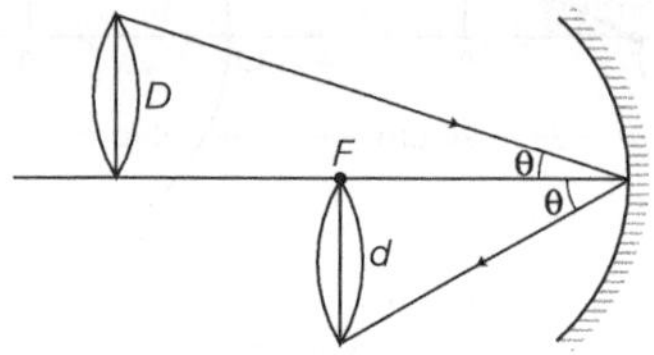

17. (b) The rays from distant star are parallel to each other. Hence, they converge at the focus.

18. (c) $u = -20$ cm, $f = +20$ cm

$\because$ $$\frac{1}{v} + \frac{1}{u} = \frac{1}{f} \text{ or } \frac{1}{v} - \frac{1}{20} = \frac{1}{20}$$

$\therefore$ $$v = +10 \text{ cm}$$

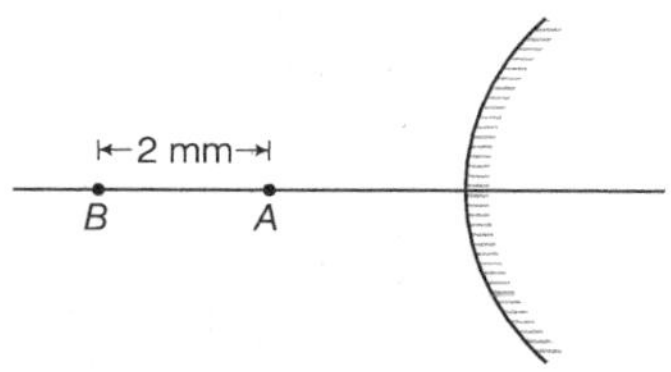

$\therefore$ Longitudinal magnification $= \dfrac{A'B'}{AB} = \dfrac{-v^2}{u^2}$

$\therefore$ $$A'B' = -\frac{100}{400}(2\text{ mm}) = -0.5 \text{ mm}$$

Thus, the length of image is 0.5 mm.

19. (b) The image formed by plane mirror behaves as object for concave mirror. If the image formed by plane mirror is at the focus of concave mirror, then after reflection from concave mirror, beam becomes parallel to optical axis.

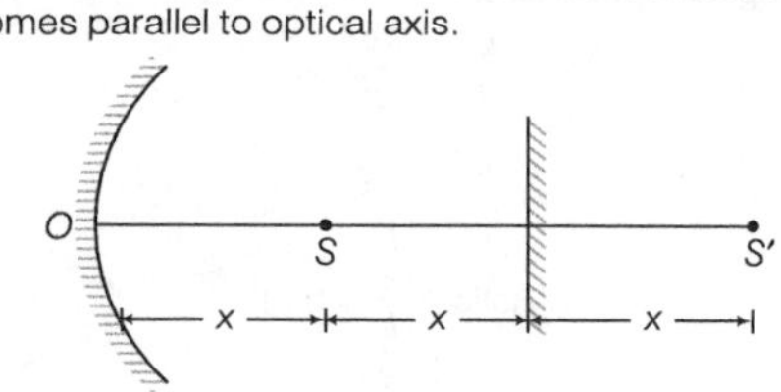

From figure, $3x = f = 30$ cm

$\therefore$ $x = 10$ cm

$\therefore$ Separation between mirrors $= 2x = 20$ cm.

20. (c)

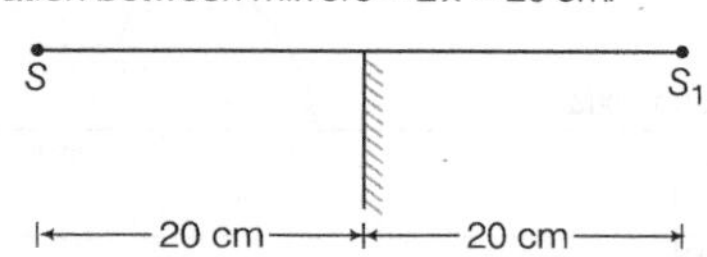

Thus, the image S_1 formed by plane mirror is 10 cm behind the concave mirror.

For concave mirror, $u = -30$ cm, $f = -40$ cm

According to mirror formula,

$$\frac{1}{v} + \frac{1}{u} = \frac{1}{f}$$

or $$\frac{1}{v} + \frac{1}{-30} = -\frac{1}{40}$$

or $$\frac{1}{v} = \frac{1}{30} - \frac{1}{40} = \frac{4-3}{120} = \frac{1}{120}$$

$\therefore$ $$v = 120 \text{ cm}$$

Thus, the separation between images is

$$120 - 10 = 110 \text{ cm}$$

21. (a) For mirror M_1, $\mu = -30$ cm, $f = -20$ cm

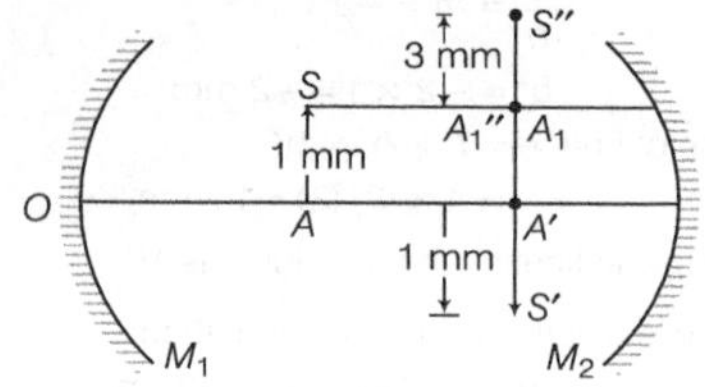

$\because$ $$\frac{1}{v} + \frac{1}{u} = \frac{1}{f} \text{ or } \frac{1}{v} - \frac{1}{30} = -\frac{1}{20}$$

$\therefore$ $$v = -60 \text{ cm}$$

$$m = \frac{A'S'}{AS} = \frac{-v}{u}$$

or $$A'S' = -\frac{(-60)}{(-30)}(1\text{ mm}) = -2\text{ mm}$$

S' behaves as object for mirror M_2.

$\therefore$ $$u' = -10 \text{ cm}, f' = -5 \text{ cm}$$

$\because$ $$\frac{1}{v'} + \frac{1}{u'} = \frac{1}{f'} \text{ or } \frac{1}{v'} - \frac{1}{10} = \frac{-1}{5}$$

or $$\frac{1}{v'} = \frac{1}{10} - \frac{1}{5} = \frac{1-2}{10}$$

$\therefore$ $$v' = -10\text{cm}$$

$\therefore$ $$m' = \frac{A_1''S''}{A_1S'} = -\frac{v'}{u'} = \frac{-(-10)}{-10} = -1$$

$\therefore$ $$A_1''S'' = -1(-3) = 3 \text{ cm}$$

Thus, coordinates of final image S'' is (60 cm, 4 mm).

22. (a) If plane mirror is rotated through angle 45°, image is rotated through 90°.

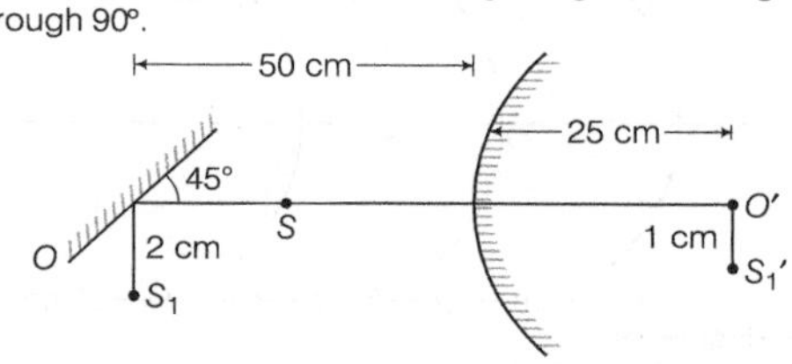

The image S_1 formed by plane mirror behaves as object for convex mirror.

For convex mirror, $u = -50$ cm, $f = +50$ cm

$\because$ $$\frac{1}{v} + \frac{1}{u} = \frac{1}{f} \text{ or } \frac{1}{v} - \frac{1}{50} = \frac{1}{50}$$

or $$\frac{1}{v} = \frac{1}{50} + \frac{1}{50} = \frac{2}{50} = \frac{1}{25}$$

$\therefore$ $$v = +25 \text{ cm}$$

$\because$ $$m = \frac{O'S_1'}{OS_1} = -\frac{v}{u} = \frac{-25}{-50} = \frac{1}{2}$$

$\therefore$ $$O'S_1' = \frac{1}{2}(-2) = -1 \text{ cm}$$

Thus, coordinates of image are (75 cm, −1 cm).

23. (a) $u = -10$ cm, $f = +10$ cm

$\because$ $$\frac{1}{v} + \frac{1}{u} = \frac{1}{f} \text{ or } \frac{1}{v} - \frac{1}{10} = \frac{1}{10}$$

or $$\frac{1}{v} = \frac{2}{10}$$

$\because$ $$v = +5 \text{ cm}$$

$\because$ $$\frac{1}{v} + \frac{1}{u} = \frac{1}{f} \text{ or } -\frac{1}{v^2}\frac{dv}{dt} - \frac{1}{u^2}\frac{du}{dt} = 0$$

$\therefore$ $$\frac{dv}{dt} = -\frac{v^2}{u^2}\frac{du}{dt}$$

$\therefore$ $$v_{\text{image}} = -\left(\frac{5}{-10}\right)^2 \times 4 \text{ cms}^{-1} = \frac{-4}{4} = -1 \text{ cms}^{-1}$$

Thus, image is moving leftward with a speed of 1 cms^{-1}.

24. (c) Here, $u = -x$ and $v = +y$

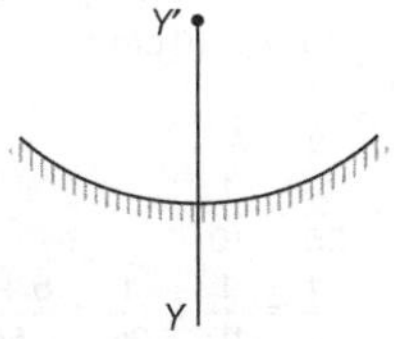

$\because \quad \frac{1}{v} + \frac{1}{u} = \frac{1}{F}$ or $\frac{1}{y} - \frac{1}{x} = -\frac{1}{f}$

$$-\frac{1}{y^2}\frac{dy}{dt} + \frac{1}{x^2}\frac{dx}{dt} = 0$$

or $$\frac{dy}{dt} = \frac{y^2}{x^2}\frac{dx}{dt} = \left(\frac{f}{f-x}\right)^2 \frac{dx}{dt}$$

$\therefore$ $$v_{image} = \left(\frac{f}{f-x}\right)^2 \sqrt{2g\left(\frac{f}{2} - x\right)}$$

For maximum value of v_{image},

$$\frac{d\,v_{image}}{dx} = 0$$

$\Rightarrow$ $$x = \frac{f}{3}$$

$\therefore$ $$(v_{image})_{max} = \frac{3}{4}\sqrt{3fg} = \frac{3}{4}\sqrt{3 \times 0.3 \times 10}$$

$$= \frac{9}{4}\,\text{ms}^{-1} = 2.25\,\text{ms}^{-1}$$

25. (d) From the mirror formula,

$$\frac{1}{v} + \frac{1}{u} = \frac{1}{f} \quad (f = \text{constant})$$

$$-v^{-2}dv - u^{-2}du = 0$$

or $$|dv| = \left|\frac{v^2}{u^2}\right| |du|$$

Here, $|dv|$ = Size of image

$|du|$ = Size of object (short) lying along the axis

Further from Eq. (i), we can find

$$\frac{v^2}{u^2} = \left(\frac{f}{u-f}\right)^2$$

Substituting these values in Eq. (ii), we get

$$\text{Size of image} = b\left(\frac{f}{u-f}\right)^2$$

26. (b) Since, object and image move in opposite directions, the positioning should be as shown in the figure. Object lies between focus and centre of curvature $f < x < 2f$.

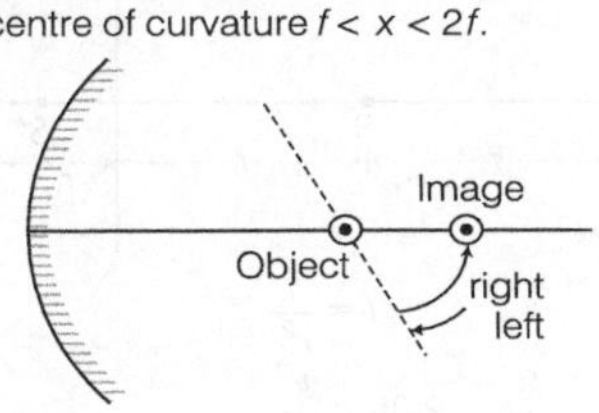

27. (a,b,c,d)

28. (a,d) (a) If AB = 5 cm

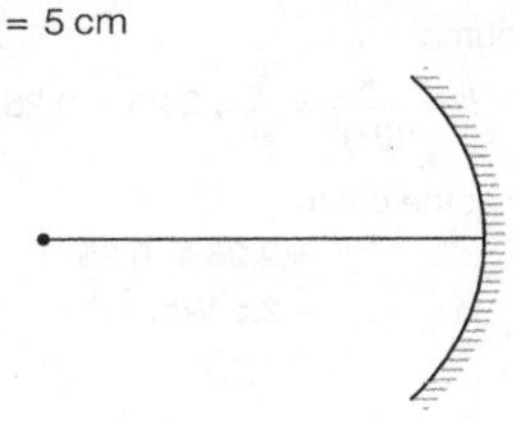

For point A, $\quad u = -30$ cm

$f = -20$ cm

$\because \quad \frac{1}{v} + \frac{1}{u} = \frac{1}{f}$

or $\quad \frac{1}{v} - \frac{1}{30} = -\frac{1}{20}$

$\therefore \quad v = -60$ cm

For point B,

$u' = -35$ cm, $f' = -20$ cm

$\because \quad \frac{1}{v'} + \frac{1}{u'} = \frac{1}{f'}$

or $\quad \frac{1}{v'} - \frac{1}{35} = -\frac{1}{20}$

or $$\frac{1}{v'} = \frac{1}{35} - \frac{1}{20} = \frac{4-7}{140} = \frac{-3}{140}$$

or $\quad u' = \frac{-140}{3}$ cm

The size of image = $A'B' = |v' - v|$

$$= 60 - \frac{140}{3} = \frac{180 - 140}{3} = \frac{40}{3}\text{ cm}$$

$= 13.33$ cm

(d) If AB = 5 mm, in this case, the length of object is very small with respect to object distance.

So, the formula of longitudinal magnification is applicable.

Here, $u = -30$ cm

$f = -20$ cm

$\therefore \quad \frac{1}{v} + \frac{1}{u} = \frac{1}{f}$

$v = -60$ cm

$\because$ Longitudinal magnification is

$$m_l = \frac{\Delta v}{\Delta u} = -\frac{v^2}{u^2}$$

or $$\frac{\Delta v}{5\text{ mm}} = -4$$

$\therefore \quad \Delta v = -20$ mm

Thus, size of image is 20 mm.

29. (a,c) (a) Here, $u = -30$ cm, $f = -20$ cm

$\because \quad \frac{1}{v} + \frac{1}{u} = \frac{1}{f}$

or $\quad \frac{1}{v} - \frac{1}{30} = -\frac{1}{20}$

or $\quad \frac{1}{v} = \frac{1}{30} - \frac{1}{20} = \frac{2-3}{60}$

$\therefore \quad v = -60$ cm

$\therefore$ Transverse magnification $= m = -\frac{v}{u}$

$$= -\frac{(-60)}{(-30)} = -2$$

$\therefore \quad \frac{l' \times l'}{l \times l} = m^2$

$\therefore$ Area of image $= m^2 l^2 = 4 \times 4 = 16\text{ mm}^2$

(c) Longitudinal magnification is found in side, which is parallel to optical axis. But transverse magnification is found in the side, which is perpendicular to optical axis.

$\therefore$ The side which is parallel to optical axis magnifies four times but the side which is perpendicular to optical axis magnifies two times.

$\therefore$ Area of image $= 8 \times 4 = 32\text{ mm}^2$

30. (a,b) (a) In this, the ray incident on one mirror does not incident on other mirror after reflection. So, successive reflections do not take place. Hence, maximum number of images is two.

(b) For convex mirror, $u = -20$ cm, $f = +20$ cm

$\because \quad \frac{1}{v} + \frac{1}{u} = \frac{1}{f}$ or $\frac{1}{v} - \frac{1}{20} = \frac{1}{20}$

or $\quad \frac{1}{v} = \frac{1}{20} + \frac{1}{20} = \frac{1}{10}$

$\therefore \quad v = +10$ cm

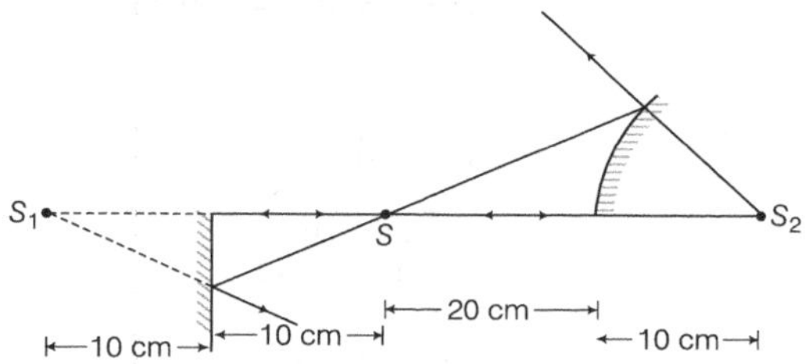

Thus, the separation between images

$= S_1S_2 = 10 + 10 + 20 + 10 = 50$ cm

31. (b,c) (b) Here, time period of vibration of object

$= 2\pi\sqrt{\frac{m}{k}} = 2\pi\sqrt{\frac{100 \times 10^{-3}}{20}}$

$= 2\pi\sqrt{5 \times 10^{-3}} = 0.44$s

(c) At mean position,

$mg = kx_0$

$\therefore \quad x_0 = \frac{mg}{k} = \frac{100 \times 10^{-3} \times 10}{20}$

$= 5 \times 10^{-2}$m $= 5$ cm

= Amplitude of vibration of object

For upper extreme position,

$u = -(20 + 5) = -25$ cm

$f = +10$ cm

$\because \quad \frac{1}{v} + \frac{1}{u} = \frac{1}{f}$

or $\quad \frac{1}{v} - \frac{1}{25} = \frac{1}{10}$

or $\quad \frac{1}{v} = \frac{1}{10} + \frac{1}{25} = \frac{5+2}{50}$

$\therefore \quad v = \frac{50}{7}$ cm

For lower extreme position,

$u' = -15$ cm

$f' = -10$ cm

$\because \quad \frac{1}{v'} + \frac{1}{u'} = \frac{1}{f'}$

or $\quad \frac{1}{v'} - \frac{1}{15} = -\frac{1}{10}$

or $\quad \frac{1}{v'} = \frac{1}{15} + \frac{1}{10} = \frac{2+3}{30}$

$\therefore \quad v' = 6$ cm

$\therefore$ The range of vibration of image

$= |v' - v| = \left|6 - \frac{50}{7}\right|$

$= \frac{50}{7} - 6$

$= \frac{50 - 42}{7} = \frac{8}{7}$ cm.

32. (a,b) Values of option (c) and (d) do not match with the mirror formula.

$\frac{1}{v} + \frac{1}{u} = \frac{1}{f}.$

DPP-4 **Spherical Mirror - II**

1. (c) For extreme position A, $u = -60$ cm, $f = -20$ cm

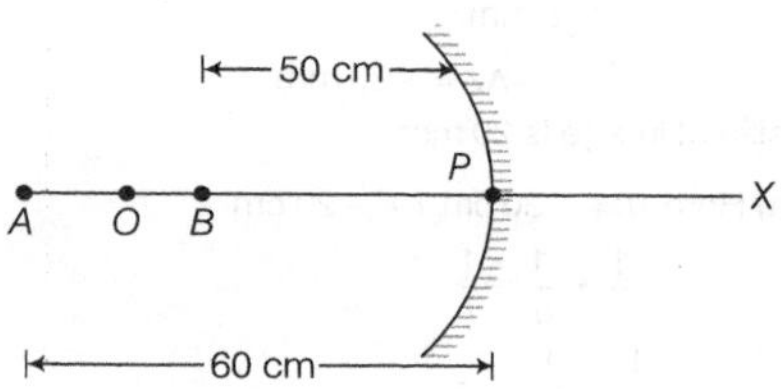

$\because \quad \frac{1}{v} + \frac{1}{u} = \frac{1}{f}$

or $\quad \frac{1}{v} - \frac{1}{60} = -\frac{1}{20}$

$\therefore \quad v = -30$ cm

For extreme position B, $u' = -50$ cm

$f' = -20$ cm

$\because \quad \frac{1}{v'} + \frac{1}{u'} = \frac{1}{f'}$ or $\frac{1}{v'} - \frac{1}{50} = \frac{1}{-20}$

or $\quad \frac{1}{v'} = \frac{1}{50} - \frac{1}{20} = \frac{2-5}{100}$

$\therefore \quad v' = -\frac{100}{3}$ cm $= -33.33$ cm

$\therefore$ The range of vibration of image is

$A'B' = |v' - v|$

$= (33.33 - 30)$ cm

$= 3.33$ cm.

2. (d) The maximum distance of image from mirror is at that time, when object is at extreme position B.

3. (b) The time period of vibration of image = time period of vibration of object

$= \frac{2\pi}{\omega} = \frac{2\pi}{10\pi} = 0.2$ s

4. (a) As, we know that intensity due to point source is

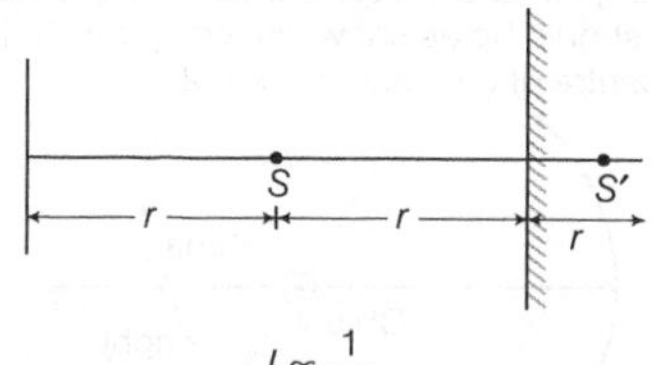

$I \propto \frac{1}{r^2}$

$\Rightarrow \quad 2.25 = \frac{k}{r^2}$

For reflected rays, image S' of point source S behaves as source

$\therefore \quad I' = \frac{k}{(3r)^2} = \frac{1}{9} \times 2.25 = 0.25$

$\therefore$ Total intensity at the centre,

$= 2.25 + 0.25$

$= 2.5$ Wm^{-2}

5. (b) In this case, source is at the focus of concave mirror. So, rays become parallel to optical axis after reflection from mirror. In parallel beam, intensity does not varry with distance. Thus, intensity at mirror is intensity at screen due to reflected rays.

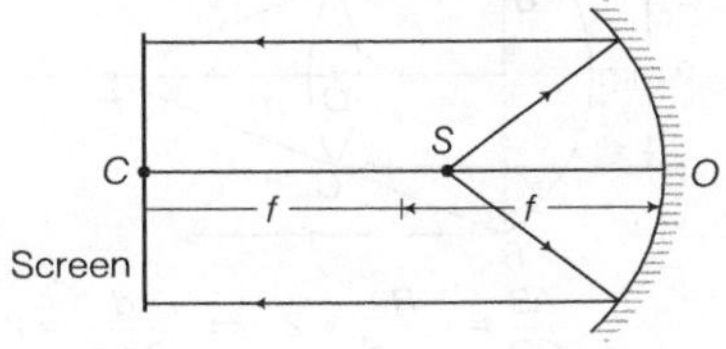

$\because \quad 2.25 = \frac{k}{f^2}$

$\therefore$ Total intensity at centre of screen

$$= 2.25 + 2.25$$
$$= 4.5 \text{ Wm}^{-2}$$

6. (c) Here, $u = -f$, $F = f$

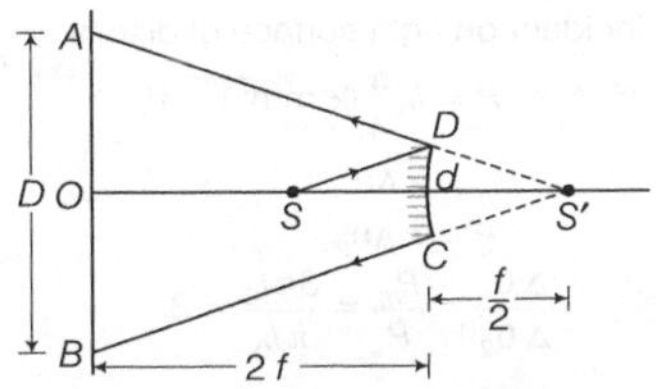

$$\therefore \quad \frac{1}{v} + \frac{1}{u} = \frac{1}{F}$$

$$\text{or} \quad \frac{1}{v} - \frac{1}{f} = \frac{1}{f}$$

$$\therefore \quad v = \frac{f}{2}$$

$\Delta AS'B$ and DCS' are similar triangles.

$$\therefore \quad \frac{D}{d} = \frac{2.5f}{0.5f} = 5$$

$$\Rightarrow \quad D = 5d = 5 \text{ mm}$$

Here, $\quad 2.25 = \frac{P_s}{4\pi f^2} \quad \ldots\text{(i)}$

The power incident on mirror is

$$P_m = \frac{P_s}{4\pi f^2} \frac{\pi d^2}{4}$$

The power received by screen due to reflected rays is $P = P_m$.

$\because$ Intensity due to reflected light on screen

$$I' = \frac{P}{\pi \frac{D^2}{4}} = \frac{P_s}{4\pi f^2} \frac{\frac{\pi d^2}{4}}{\frac{\pi D^2}{4}}$$

$$\text{or} \quad I' = \frac{2.25\,(4\pi f^2)}{(4\pi f^2)} \left(\frac{d}{D}\right)^2$$

$$= 2.25 \times \left(\frac{1}{5}\right)^2$$

$$= \frac{2.25}{25} = \frac{225}{25} \times 10^{-2} = 0.09$$

$\therefore$ Total intensity at centre of screen.

$$= 2.25 + 0.09$$
$$= 2.34 \text{ Wm}^{-2}$$

7. (b) If object is placed at centre of curvature of concave mirror, its image is at the centre of curvature. Thus, minimum distance between object and its real image is zero. It object is between pole and focus of concave mirror, its image is virtual. If object is beyond focus, its image is real. Thus, concave mirror forms real as will as virtual image.

8. (b) From figure, Statement I is correct.

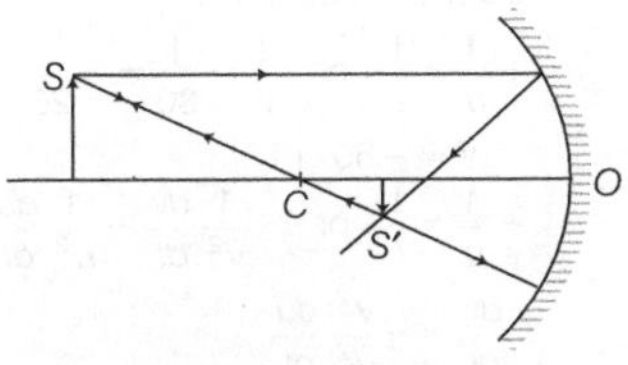

The transverse magnification is

$$m = \frac{\text{Image size}}{\text{Object size}} = \frac{-h_2}{h_1}$$

(i.e. negative)

Hence, Statement I and Statement II are correct.

9. (a) From figure, it is clear that object, centre of curvature and image are on a straight line. The ray which is passing through centre of curvature is normally incident on mirror.

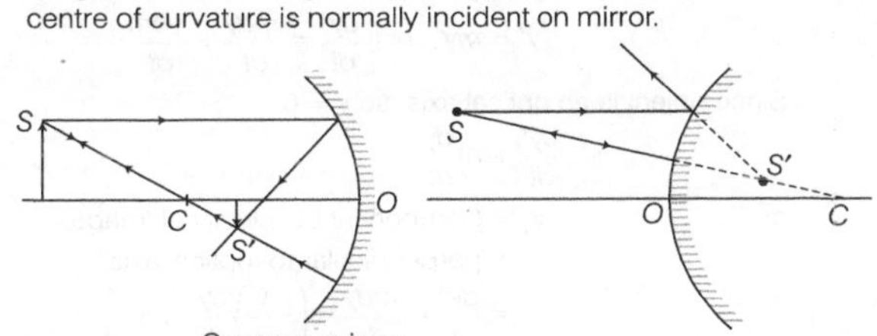

Concave mirror

So, it returns back on the same path.

10. (a) v-u graph for plane mirror

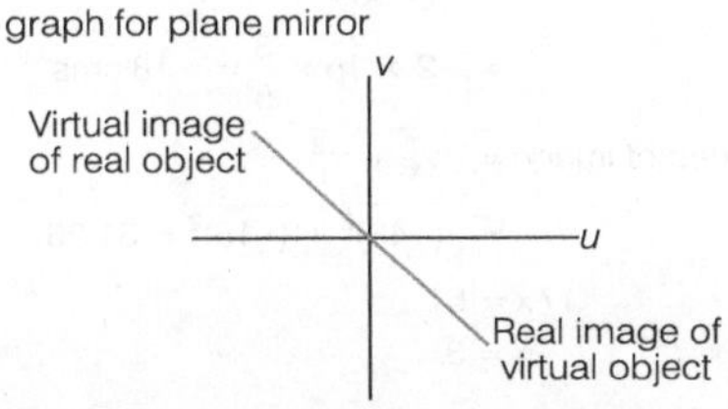

Thus, plane mirror forms virtual image of real object and real image of virtual object.

v-u graph for concave mirror

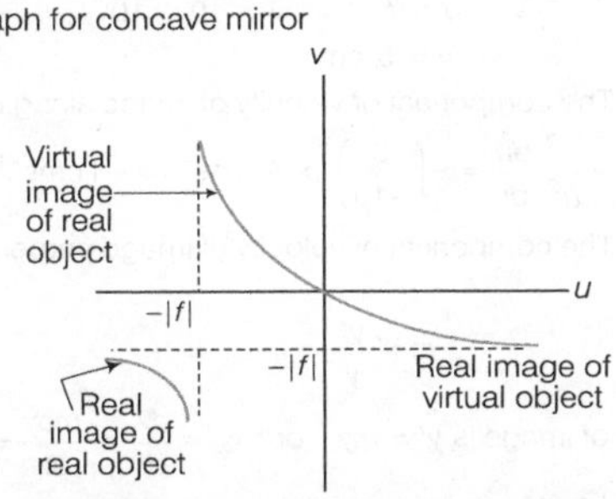

Thus, concave mirror never forms virtual image of virtual object.

v-u graph for convex mirror

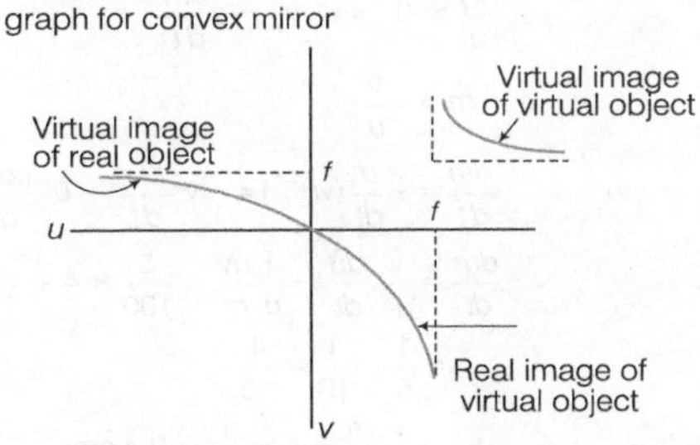

Thus, convex mirror never forms real image of real object.

Hence, (i) $\rightarrow$ (p,s), (ii) $\rightarrow$ (p,q,r), (iii) $\rightarrow$ (q), (iv) $\rightarrow$ (p,q,r).

11. (3) Here, $u = -30$ cm, $f = -20$ cm

$\because \quad \frac{1}{v} + \frac{1}{u} = \frac{1}{f}$ or $\frac{1}{v} - \frac{1}{30} = -\frac{1}{20}$

$\therefore \quad v = -60$ cm

$\because \quad \frac{1}{v} + \frac{1}{u} = \frac{1}{f}$ or $-\frac{1}{v^2}\frac{dv}{dt} - \frac{1}{u^2}\frac{du}{dt} = 0$

or $\quad \frac{dv}{dt} = -\frac{v^2}{u^2}\frac{du}{dt}$

or $\quad v_x$ = Velocity of image along optical axis

$= -4 \times 15\cos 37°$

$= -4 \times 15 \times \frac{4}{5} = -48 \text{ cms}^{-1}$

$\because \quad m = \frac{y'}{y}$

$\therefore$ Size of image $= y' = my$

$\because \quad y' = my$ or $\frac{dy'}{dt} = \frac{mdy}{dt} + \frac{ydm}{dt}$

Since object is an optical axis, so $y = 0$

$\therefore \quad \frac{dy'}{dt} = m\frac{dy}{dt}$

or $\quad v_y$ = Component of velocity of image perpendicular to optical axis.

$= \frac{dy'}{dt} = \frac{mdy}{dt} = \left(-\frac{v}{u}\right)\frac{dy}{dt}$

$= -\left(\frac{-60}{-30}\right) 15 \sin 37°$

$= -2 \times 15 \times \frac{3}{5} = -18 \text{ cms}^{-1}$

$\therefore$ Speed of image $= \sqrt{v_x^2 + v_y^2}$

$= \sqrt{(-48)^2 + (-18)^2} = 51.26$

$\therefore \quad 17x = 51$

$\therefore \quad x = 3.$

12. (4) Here, $u = -10$ cm, $f = +10$ cm

$\because \quad \frac{1}{v} + \frac{1}{u} = \frac{1}{f}$ or $\frac{1}{v} - \frac{1}{10} = \frac{1}{10}$

$\therefore \quad v = 5$ cm

Let v_x = The component of velocity of image along optical axis.

$= -\frac{v^2}{u^2}\frac{du}{dt} = -\left(\frac{5}{-10}\right)^2 \times 4 \text{ cms}^{-1} = -1 \text{ cms}^{-1}$

Let v_y = The component of velocity of image perpendicular to optical axis

$\because \quad m = \frac{y'}{y}$

The size of image is $y' = my$ or $v_y = \frac{dy'}{dt} = \frac{mdy}{dt} + \frac{ydm}{dt}$

But the object is moving parallel to optical axis.

$\therefore \quad \frac{dy}{dt} = 0 \Rightarrow v_y = y\frac{dm}{dt}$

$\because \quad m = -\frac{v}{u}$

or $\quad \frac{dm}{dt} = -\frac{d}{dt}(vu^{-1}) = -v\frac{du^{-1}}{dt} - u^{-1}\frac{dv}{dt}$

or $\quad \frac{dm}{dt} = \frac{v}{u^2}\frac{du}{dt} - \frac{1}{u}\frac{dv}{dt} = \frac{5}{100} \times 4 + \frac{1}{10}(-1)$

$= \frac{1}{5} - \frac{1}{10} = \frac{1}{10}$

$\therefore \quad v_y = \frac{ydm}{dt} = 4 \times \frac{1}{10} = 0.4 \text{ cms}^{-1}$

$\therefore \quad \frac{10|v_y|}{|v_x|} = \frac{4}{1} = 4$

13. (3) ΔABF and CDF are similar triangles.

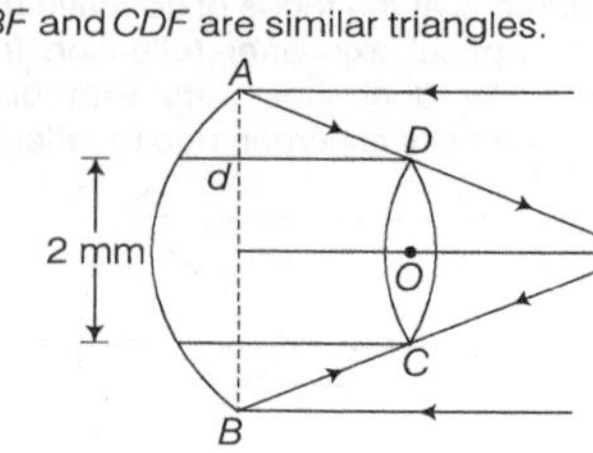

$\therefore \quad \frac{AB}{CD} = \frac{R}{2 \times \frac{R}{4}} = 2 \Rightarrow \frac{d}{2\text{mm}} = 2$

$\therefore \quad d = 4$ mm

The power incident on mirror is

$$P_m = I_0\left(\frac{\pi d^2}{4} - \frac{\pi 2^2}{4}\right) = I_0\frac{\pi}{4}(16-4) = 3\pi I_0$$

This power incident on left surface of disc.

The power incident on right surface of disc is

$$P = I_0\frac{\pi}{4}(2 \text{ mm}^2) = \pi I_0$$

$\therefore \quad P_m = k\Delta\theta_1$

and $\quad P = k\Delta\theta_2$

$\therefore \quad \frac{\Delta\theta_1}{\Delta\theta_2} = \frac{P_m}{P} = \frac{3\pi I_0}{\pi I_0} = 3.$

14. (2) For point A, $\quad u = -15$ cm, $f = +20$ cm

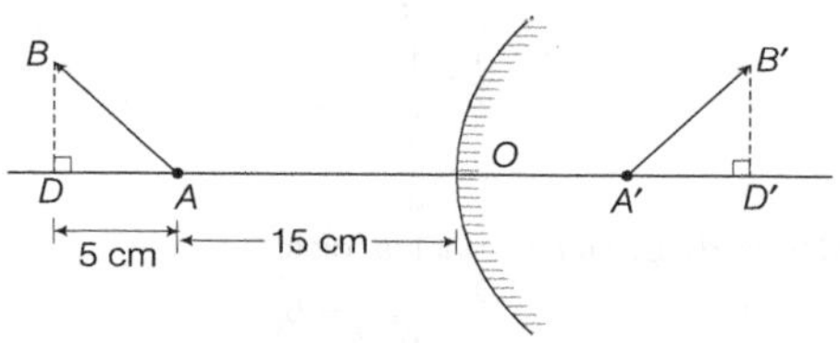

$\because \quad \frac{1}{v} + \frac{1}{u} = \frac{1}{f}$

or $\quad \frac{1}{v} - \frac{1}{15} = \frac{1}{20}$

or $\quad \frac{1}{v} = \frac{1}{20} + \frac{1}{15} = \frac{3+4}{60} = \frac{7}{60}$

$\therefore \quad v = \frac{60}{7}$ cm

For point D, $u' = -20$ cm, $f' = +20$ cm

$\because \quad \frac{1}{v'} + \frac{1}{u'} = \frac{1}{f'}$

or $\quad \frac{1}{v'} - \frac{1}{20} = \frac{1}{20}$

or $\quad \frac{1}{v'} = \frac{1}{20} + \frac{1}{20} = \frac{1}{10}$

$\therefore \quad v' = 10$ cm

$\because$ Transverse magnification is

$$m = \frac{B'D'}{BD} = \frac{-v'}{u'}$$

$\therefore \quad B'D' = \frac{-v'}{u'}(BD) = \frac{-10}{(-20)}(5) = -2.5$ cm

Also, $\quad A'D' = |v' - v| = \left|10 - \frac{60}{7}\right| = \frac{10}{7}$

$\therefore \quad A'B' = \sqrt{(A'D')^2 + (B'D')^2}$

$= \sqrt{\left(\frac{10}{7}\right)^2 + (-2.5)^2}$

$= \sqrt{\frac{100}{49} + 6.25} = 2.88$ cm

15. (7) For first reflection from mirror M_1

$u = -30$ cm, $f = -20$ cm

$\because \quad \frac{1}{v} + \frac{1}{u} = \frac{1}{f}$ or $\frac{1}{v} - \frac{1}{30} = -\frac{1}{20}$

or $\quad \frac{1}{v} = \frac{1}{30} - \frac{1}{20} = \frac{2-3}{60}$

$\therefore \quad v = -60$ cm

The image formed by first reflection through concave mirror M_1 behaves as object for first reflection through mirror M_2.

For mirror $M_2, u' = -10$ cm, $f' = -20$ cm

$\because \quad \frac{1}{v'} + \frac{1}{u'} = \frac{1}{f'}$

or $\quad \frac{1}{v'} - \frac{1}{10} = -\frac{1}{20}$

or $\quad \frac{1}{v'} = \frac{1}{10} - \frac{1}{20} = \frac{2-1}{20}$

$\therefore \quad v' = 20$ cm

The image formed by reflection through mirror M_2, behaves as object for mirror M_1.

For mirror M_1, $u'' = -90$ cm, $f'' = -20$ cm

$\because \quad \frac{1}{v''} + \frac{1}{u''} = \frac{1}{f''}$ or $\frac{1}{v''} - \frac{1}{90} = -\frac{1}{20}$

or $\quad \frac{1}{v''} = \frac{1}{90} - \frac{1}{20} = \frac{2-9}{180}$

$\therefore \quad v'' = \frac{-180}{7}$ cm

$x = 7.$

16. (4) $u = -30$ cm, $f = -20$ cm

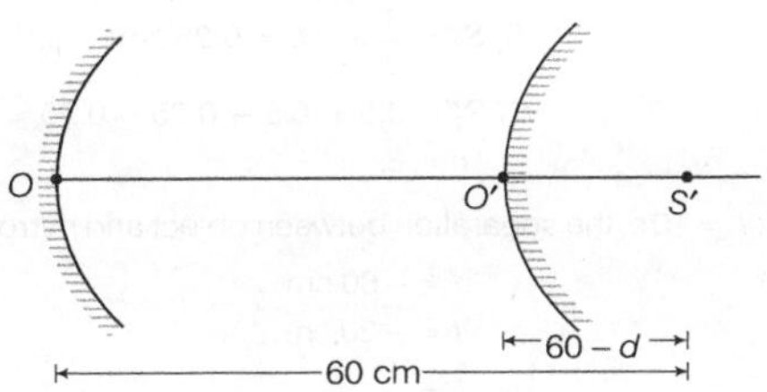

$\because \quad \frac{1}{v} + \frac{1}{u} = \frac{1}{f}$

or $\quad \frac{1}{v} - \frac{1}{30} = -\frac{1}{20}$

or $\quad \frac{1}{v} = \frac{1}{30} - \frac{1}{20} = \frac{2-3}{60} = \frac{-1}{60}$

$\therefore \quad v = -60$ cm

For convex mirror, S_1 behaves as virtual object,

$\therefore$ For convex mirror, $u' = 60 - d$

$f' = +10$ cm

$v' = -(d - 30)$

$\because \quad \frac{1}{v'} + \frac{1}{u'} = \frac{1}{f'}$

or $\quad \frac{1}{-(d-30)} + \frac{1}{60-d} = \frac{1}{10}$

or $\quad \frac{d - 60 + d - 30}{(60-d)(d-30)} = \frac{1}{10}$

or $\quad (2d - 90)\,10 = (60 - d)(d - 30)$

or $\quad (2d - 90)\,10 = 60d - d^2 - 1800 + 30d$

or $\quad 20d - 900 = 90d - d^2 - 1800$

or $\quad d^2 - 70d + 900 = 0$

$\therefore \quad d = \frac{70 \pm \sqrt{4900 - 3600}}{2}$

$= 53 \text{ cm} = 13x + 1$

$\therefore \quad x = 4.$

17. (1) $\because \tan\theta = \frac{AB}{5 \text{ cm}} = \frac{1 \text{ mm}}{5 \text{ cm}}$

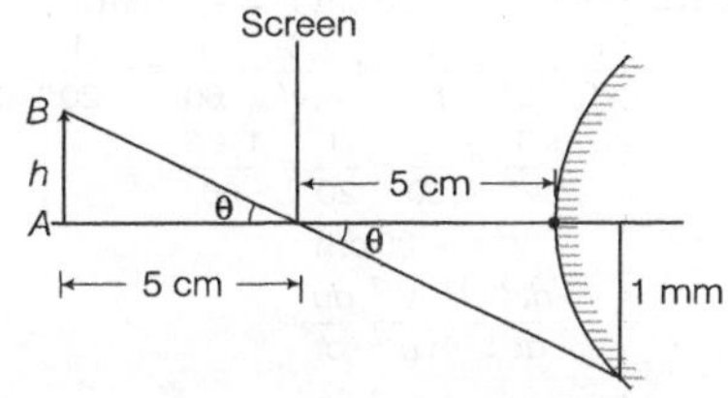

$\therefore \quad AB = \frac{1 \text{ mm}}{5 \text{ cm}} \times 5 \text{ cm} = 1 \text{ mm}$

18. (1) **First image** The first image is formed due to reflection through concave mirror.

Here, $u = -90$ cm, $f = -30$

$\because \quad \frac{1}{v} + \frac{1}{u} = \frac{1}{f}$

or $\quad \frac{1}{v} - \frac{1}{90} = -\frac{1}{30}$

or $\quad \frac{1}{v} = \frac{1}{90} - \frac{1}{30} = \frac{1-3}{90} = \frac{-2}{90}$

$\therefore \quad v = -45$ cm

$\because \quad m_1 = \frac{a'}{a} = -\frac{v}{u} = -\frac{45}{90} = -\frac{1}{2}$

$\therefore \quad a' = -\frac{a}{2}$

Second image The second image is formed by convex mirror for which first image behaves as object.

$\therefore$ For convex mirror, $u' = -45$ cm, $f' = +30$ cm

$\because \quad \frac{1}{v'} + \frac{1}{u'} = \frac{1}{f'}$ or $\frac{1}{v'} - \frac{1}{45} = \frac{1}{30}$

or $\quad \frac{1}{v'} = \frac{1}{45} + \frac{1}{30} = \frac{2+3}{90}$

$\therefore \quad v' = \frac{90}{5} = 18$ cm

$\therefore \quad m' = \frac{a''}{a'} = -\frac{v'}{u'}$

$\therefore \quad a'' = \frac{-18}{-(45)} \times \left(\frac{-a}{2}\right) = \frac{-a}{5}$

Thus, radius of second image is $a'' = \frac{5}{5} = 1$ mm.

19. (8) For convex mirror, $u = -10$ cm, $f = +10$ cm

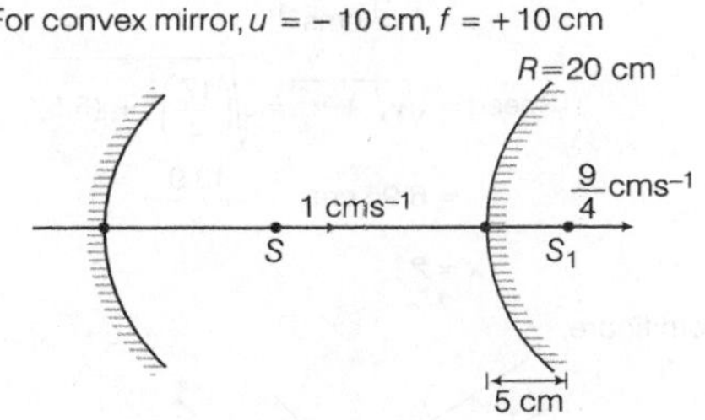

$\because \quad \frac{1}{v} + \frac{1}{u} = \frac{1}{f}$ or $\frac{1}{v} - \frac{1}{10} = \frac{1}{10}$

$\therefore \quad v = 5$ cm

$\because \quad \frac{dv}{dt} = \frac{-v^2}{u^2}\frac{du}{dt}$

or $\quad v_{\text{image mirror}} = -\frac{25}{100}(v_{\text{object + mirror}})$

or $\quad v_{\text{image}} - v_{\text{mirror}} = -\frac{1}{4}(1-2) = \frac{1}{4} \text{ cms}^{-1}$

$\therefore \quad v_{\text{image}} = \frac{1}{4} + 2 = \frac{9}{4} \text{ cms}^{-1}$

The image S_1 formed by convex mirror behaves as object for concave mirror.

For concave mirror, $u' = -60$ cm, $f' = -20$ cm

$\because \quad \frac{1}{v'} + \frac{1}{u'} = \frac{1}{f'}$ or $\frac{1}{v'} - \frac{1}{60} = -\frac{1}{20}$

or $\quad \frac{1}{v''} = \frac{1}{60} - \frac{1}{20} = \frac{1-3}{60}$

$\therefore \quad v' = -30$ cm

$\therefore \quad \frac{dv'}{dt} = -\frac{v'^2}{u'^2}\frac{du'}{dt}$

or $\quad v'_{\text{image mirror}} = -\frac{(-30)^2}{(-60)^2}(v'_{\text{object}} - v'_{\text{mirror}})$

or $\quad v'_{\text{image}} - v'_{\text{mirror}} = -\frac{1}{4}\left(\frac{-9}{4} - 1\right) = -\frac{1}{4}\left(\frac{-13}{4}\right) = \frac{13}{16}$

$\therefore \quad v'_{\text{image}} = \frac{13}{16} + v'_{\text{mirror}} = \frac{13}{16} + 1 = \frac{29}{16} = \frac{29}{2n}$

$\therefore \quad n = 8.$

20. (2) $u = -10$ cm, $f = +10$ cm

$\because \quad \frac{1}{v} + \frac{1}{u} = \frac{1}{f}$ or $\frac{1}{v} - \frac{1}{10} = \frac{1}{10}$

$\therefore \quad v = 5$ cm

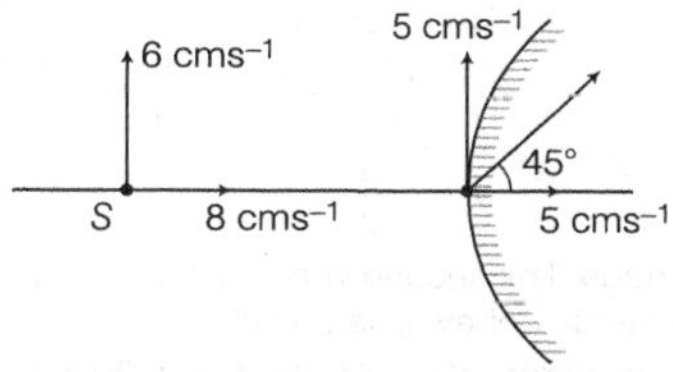

Let v_x = The component of velocity of image along optical axis.

$\therefore \quad v_x - 5 = -\left(\frac{v^2}{u^2}\right)\frac{du}{dt} = -\frac{1}{4}(8-5)$

$\therefore \quad v_x = 5 - \frac{3}{4} = \frac{17}{4}$ cms^{-1}

Let v_y = The component of velocity perpendicular to optical axis.

$\because \quad v_y - 5 = m\frac{dy}{dt} + y\frac{dm}{dt}$

or $\quad v_y - 5 = \left(-\frac{v}{u}\right)(6-5) + 0$

or $\quad v_y - 5 = -\left(\frac{5}{-10}\right) \times 1$ or $v_y - 5 = \frac{1}{2}$

$\therefore \quad v_y = 5 \cdot 5$ cms^{-1}

$\therefore \quad \text{Speed} = \sqrt{v_x^2 + v_y^2} = \sqrt{\left(\frac{17}{4}\right)^2 + (5.5)^2}$

$= 6.95$ cms$^{-1} = \frac{13.9}{x}$

$\therefore \quad x = 2.$

21. (6) From figure,

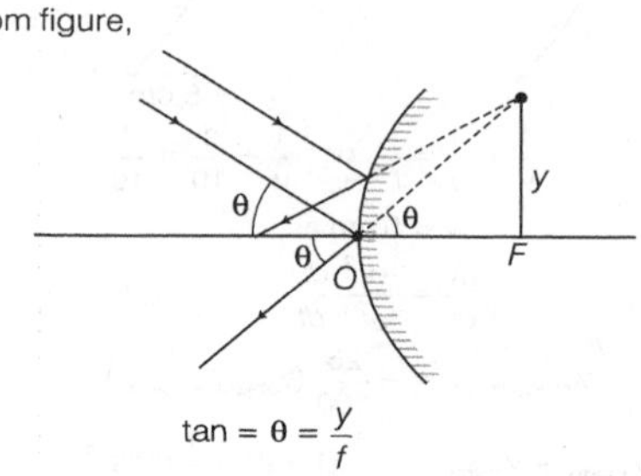

$\tan = \theta = \frac{y}{f}$

$\therefore \quad y = f\theta = 15 \times 2 \times \frac{\pi}{180°} = \frac{\pi}{6}$ cm.

22. (5) For mirror M_1, $u = -30$ cm, $f = -20$ cm

$\because \quad \frac{1}{v} + \frac{1}{u} = \frac{1}{f}$

or $\quad \frac{1}{v} - \frac{1}{30} = -\frac{1}{20}$

$\therefore \quad v = -60$ cm

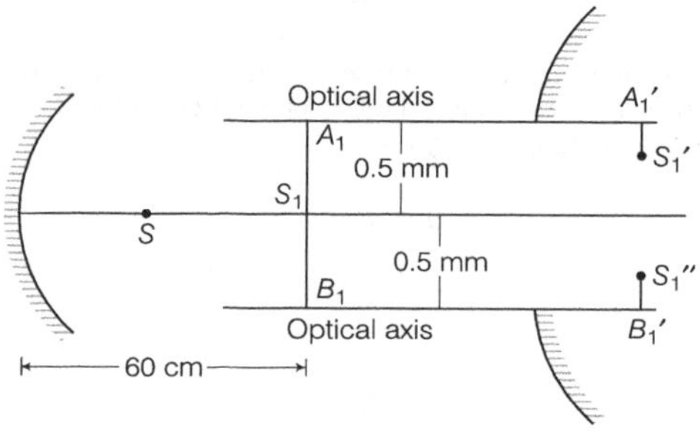

The image S_1 behaves as object for two halves.

For convex mirror, $u' = -10$ cm $\Rightarrow$ $f' = +10$ cm

$\frac{1}{v'} + \frac{1}{u'} = \frac{1}{f'}$

or $\quad \frac{1}{v'} - \frac{1}{10} = \frac{1}{10}$

$\therefore \quad v' = +5$ cm

$\therefore \quad m' = -\frac{v'}{u'} = \frac{-5}{-10} = \frac{1}{2}$

Thus, $\quad A_1' S_1' = \frac{1}{2} \times 0.5 = 0.25$ cm

and $\quad B_1' S_1'' = \frac{1}{2} \times 0.5 = 0.25$ cm

$\therefore \quad S_1' S_1'' = 0.5 + 0.5 - 0.25 - 0.25 = 0.5$ cm

$\therefore \quad n = 5.$

23. (9) At $t = 10$ s, the separation between object and mirror is 60 cm.

$\therefore \quad u = -60$ cm

$f = -20$ cm

$\therefore \quad \frac{1}{v} + \frac{1}{u} = \frac{1}{f}$

or $\quad \frac{1}{v} - \frac{1}{60} = -\frac{1}{20}$

or $\quad \frac{1}{v} = \frac{1}{60} - \frac{1}{20} = \frac{1-3}{60} = \frac{-2}{60}$

$\therefore \quad v = -30$cm

$\because \quad \frac{1}{v} + \frac{1}{u} = \frac{1}{f}$

or $\quad -\frac{1}{v^2}\frac{dv}{dt} - \frac{1}{u^2}\frac{du}{dt} = 0$

or $\quad \frac{dv}{dt} = -\frac{v^2}{u^2}\frac{du}{dt}$

or $\quad v_{\text{image mirror}} = \frac{-v^2}{u^2}(v_{\text{object + mirror}})$

or $\quad v_{\text{image}} - v_{\text{mirror}} = -\frac{1}{4}(v_{\text{object}} - v_{\text{mirror}})$

or $\quad v_{\text{image}} - 4 = -\frac{1}{4}(2-4)$

$\therefore \quad v_{\text{image}} = 4 + \frac{1}{2} = 4.5$ cms^{-1}

$= \frac{n}{2}$ cms^{-1}

$\therefore \quad n = 9.$

DPP-5 Refraction of Light-I

1. (i) (a)

(ii) (a,b,c) $1 \times \sin\theta = \mu \sin(90° - \alpha)$

or $\sin\theta = \mu\cos\alpha$

and for total internal reflection

$$\sin\alpha > \sin C$$

$$\therefore \quad \sin\alpha > \frac{1}{\mu}$$

(iii) (a) $\because \quad \sin\alpha > \frac{1}{\mu}$

or $\sin\alpha > \frac{1}{\sqrt{1.25}}$

or $\sin\alpha > \sqrt{\frac{100}{125}}$

or $\sin\alpha > \sqrt{4/5}$

(iv) (c) $\because \quad \sin\theta = \mu\cos\alpha$

$$\therefore \quad \sin\theta_0 = \mu\sqrt{1-\sin^2\alpha}$$

$$= \sqrt{1.25}\sqrt{1-\frac{4}{5}} = \sqrt{\frac{1.25}{5}}$$

$$= \sqrt{0.25} = 0.5$$

$$\therefore \quad \theta_0 = 30°$$

(v) (a)

(vi) (b) $1 \times \sin 45° = \sqrt{2}\sin(90° - \alpha)$

or $\sin(90° - \alpha) = \frac{1}{2} = \sin 30°$

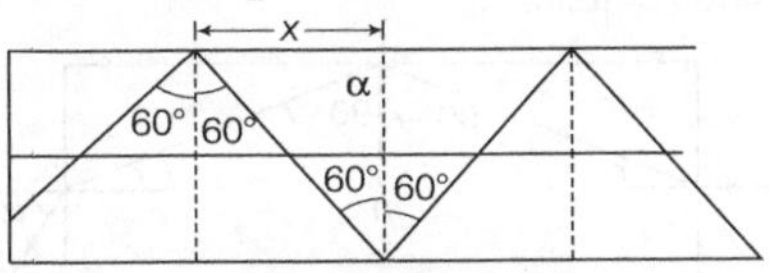

$$\therefore \quad \alpha = 60°$$

$$x = d\tan 60°$$

$$\therefore \quad N = \left[\frac{l}{d\tan 60°}\right] = \frac{10\sqrt{3}}{0.3\sqrt{3}}$$

$$= \left[\frac{100}{3}\right] = 33$$

(vii) (b) $\because \quad \sin\theta_0 = \mu\cos\alpha = \mu\sqrt{1-\sin^2\alpha} = \frac{1}{2}$

(viii) (b) Here, $\sin\alpha > \sin C$

or $\sin\alpha > \sqrt{2/3}$

$$\therefore \quad \cos\alpha_{max} = \sqrt{1-\sin^2\alpha_{min}} = \frac{1}{\sqrt{3}}$$

$$\therefore \quad \sin\theta_0 = \mu\cos\alpha_{max} = \sqrt{3} \times \frac{1}{\sqrt{3}} = 1$$

2. (i) (a) Slope $= \frac{dy}{dx} = 2x$

(ii) (a,c) Slope $= \tan(90° - \theta)$, where θ is angle with Y-axis

$= \cot\theta$

(iii) (a) According to Snell's law,

$$\mu\sin\theta = C$$

(iv) (a) For grazing incidence, $\theta = 90°, \mu = 1$

$$\therefore \quad 1 \times \sin 90° = C$$

$$\therefore \quad C = 1$$

(v) (b) $\because \mu\sin\theta = C = 1$

$$\therefore \quad \tan\theta = \frac{1}{\sqrt{\mu^2 - 1}}$$

(vi) (a,b,c,d) From above problem's solutions,

Slope $= \tan(90° - \theta) = \cot\theta = 2x$

$$\therefore \quad \sqrt{\mu^2 - 1} = 2x$$

$$\therefore \quad \mu = \sqrt{1+4x^2} = \sqrt{1+4y}$$

3. (i) (a,b)

(ii) (a) $\delta = i - r$

(iii) (c) $\delta = \pi - 2r = 180° - 2r$

(iv) (c) $\delta = 2(i - r) + 2(180° - 2r)$

(v) (a) $\because \delta = 2(i - r) + 2(180° - 2r)$

$$\therefore \quad \frac{d\delta}{di} = 2 - 2\frac{dr}{di} - 4\frac{dr}{di} = 2 - 6\frac{dr}{di}$$

(vi) (a,b) $\because \quad 1 \times \sin i = \mu\sin r$

or $\cos i\frac{di}{dr} = \mu\cos r$

$$\therefore \quad \frac{di}{dr} = \frac{\mu\cos r}{\cos i}$$

$\because$ For minimum deviation

$$\frac{d\delta}{di} = 0$$

$$\therefore \quad 2 - 6\frac{dr}{di} = 0$$

$$\therefore \quad \frac{dr}{di} = \frac{1}{3}$$

$$\therefore \quad \frac{di}{dr} = 3$$

(vii) (b) $\because \frac{di}{dr} = \frac{\mu\cos r}{\cos i}$ (from solutions of above problem)

(viii) (b) $\because \frac{di}{dr} = \frac{\mu\cos r}{\cos i} = 3$

or $\mu\cos r = 3\cos i$

or $\mu\sqrt{1-\sin^2 r} = 3\cos i$

or $\mu\sqrt{1-\frac{\sin^2 i}{\mu^2}} = 3\cos i$

or $\sqrt{\mu^2 - \sin^2 i} = 3\cos i$

or $\mu^2 - \sin^2 i = 9\cos^2 i$

or $\mu^2 - 1 = 8\cos^2 i$

$$\therefore \quad \cos i = \sqrt{\frac{\mu^2 - 1}{8}}$$

4. (i) (b) For grazing incidence, angle of incidence is 90°.

(ii) (b,c) According to Snell's law

$$\mu\sin\theta = C$$

or $(ky^{3/2} + 1)^{1/2}\sin\theta = C$

(iii) (a) $\because$ For initial point, $\mu = 1$, $\theta = 90°$

$$\therefore \quad 1 \times \sin 90° = C$$

$$\therefore \quad C = 1$$

(iv) (a) $\because \quad \mu\sin\theta = C = 1$

$$\therefore \quad \tan\theta = \frac{1}{\sqrt{\mu^2 - 1}} = \frac{1}{\sqrt{ky^{3/2} + 1 - 1}} = \frac{1}{\sqrt{y^{3/2}}}$$

$$\therefore \quad \tan\theta = y^{-3/4}$$

(v) (a) $\because \quad \frac{dx}{dy} = \tan\theta = y^{-3/4}$

(vi) (b) $\because \quad \frac{dy}{dx} = y^{-3/4}$

or $\int_0^x dx = \int_0^y y^{-3/4}dy$

or $x = \frac{y^{-\frac{3}{4}+1}}{\frac{-3}{4}+1} = \frac{y^{\frac{1}{4}}}{\frac{1}{4}}$

or $\frac{x}{4} = y^{\frac{1}{4}}$

$$\therefore \quad y = \frac{x^4}{256}.$$

(vii) (a)

(viii) (c) $\because y = \frac{x^4}{256}$ or $x^4 = 256$

$\therefore \quad x = 4$ m.

(ix) (a,b)

(x) (a)

(xi) (b)

5. (i) (b)

(ii) (a) $1 \times \sin i = \mu \sin\alpha$

$\therefore \quad \sin\alpha = \frac{\sin i}{\mu}$

(iii) (a) $\because \sin\alpha = \frac{\sin i}{\mu}$

$\therefore \sin\alpha_{max} = \frac{\sin 90^\circ}{\mu} = \frac{1}{\mu}$

$\therefore \quad \alpha_{max} = \sin^{-1}\left(\frac{1}{\mu}\right)$

(iv) (a) $\mu\sin(90^\circ - \alpha) = 1 \times \sin\theta$

$\mu\cos\alpha_{max} = \sin\theta$

or $\frac{1}{\sin\alpha_{max}}\cos\alpha_{max} = \sin\theta$

$\therefore \quad \cot\alpha_{max} = \sin\theta$

(v) (a) $90^\circ - \alpha_{max} \le C$

$90^\circ - \alpha_{max} \le \sin^{-1}\left(\frac{1}{\mu}\right)$

(vi) (b) $\because 90^\circ - \alpha_{max} \le \sin^{-1}\frac{1}{\mu}$

or $90^\circ - \alpha_{max} \le \alpha_{max}$ or $45^\circ \le \alpha_{max}$

$\therefore \quad \sin\alpha_{max} = \frac{1}{\mu}$

$\therefore \quad \mu \le \sqrt{2}$

6. (i) (a)

(ii) (b) $\because \quad \mu \propto \frac{1}{\lambda}$

$\therefore \quad \mu_v > \mu_r$

(iii) (a) Since, ray propagates from denser medium to rarer medium, so total internal reflection may take place.

(iv) (a) $\because \quad \sin C_r = \frac{1}{\mu_r}$

$\therefore \quad C_r = \sin^{-1}\left(\frac{1}{\mu_r}\right)$

(v) (b) $\because \mu \propto \frac{1}{\sin C}$ and $\mu \propto \frac{1}{\lambda}$

$\therefore \quad \sin C \propto \lambda$

$\therefore \quad C_r > C_v$

(vi) (b) $\because C_r > C_v$

$\therefore$ Hence, option (b) is correct.

(vii) (b) $\because \quad C_r > C_v$

(viii) (b) $\because \quad \theta > C_r$

$\therefore \quad \sin\theta > \frac{1}{\mu_r}$

$\therefore \quad \theta > \sin^{-1}\frac{1}{\mu_r}$

7. (i) (b) $C = \frac{c_0}{|\mu|} = \frac{3 \times 10^8}{3/2} = 2 \times 10^8 \text{ ms}^{-1}$

(ii) (c) $\lambda = \frac{\lambda_0}{|\mu|} = \frac{700}{3.5} = 200$ nm

(iii) (d)

(iv) (a) $1 \times \sin 45^\circ = |\mu|\sin r$

$\therefore \quad \sin r = \frac{1}{2}$

$\therefore \quad r = 30^\circ$

8. (d) From figure,

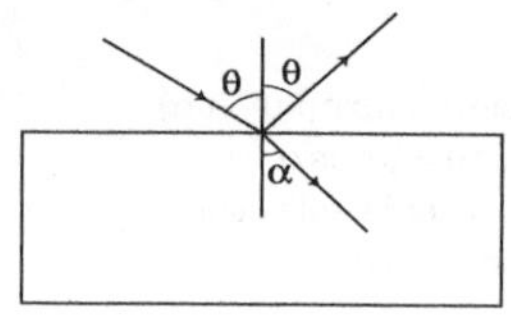

$\theta + \alpha = 90^\circ$

$\therefore \quad \alpha = 90^\circ - \theta$

According to Snell's law,

$1 \times \sin\theta = \mu\sin\alpha$

or $\sin\theta = \frac{3}{2}\sin(90^\circ - \theta)$

or $\tan\theta = \frac{3}{2}$

or $\cot\theta = 2/3$

$\therefore \quad \theta = \cot^{-1}(2/3)$.

9. (b) According to Snell's law,

$1 \times \sin i = \mu\sin r$

$\therefore \quad \mu$ = slope of the graph

$\because \quad C = \frac{c_0}{\mu}$

$\therefore \quad \mu = \frac{c_0}{C} = \frac{3 \times 10^8}{2 \times 10^8} = 3/2$

10. (b) $1 \times \sin 30^\circ = \mu\sin r$

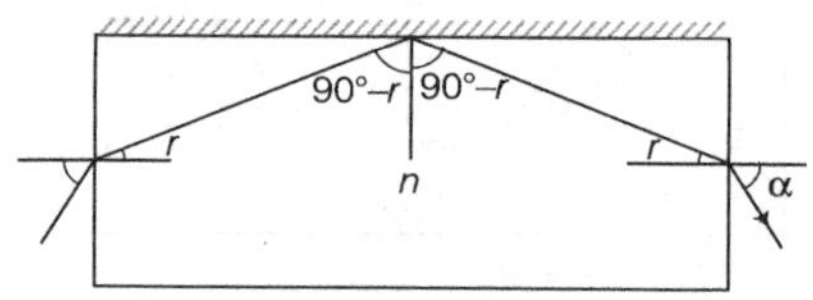

$\therefore \quad \sin r = \frac{\sin 30^\circ}{\mu}$

From figure, $\mu\sin r = 1 \times \sin\alpha$

$\therefore \quad \alpha = 30^\circ$

11. (c) Along the surface of refractive surface, momentum of photon remains conserved.

$\therefore \quad P_1\sin 45^\circ = P_2\sin 30^\circ$

or $\frac{P_1}{P_2} = \frac{\sin 30^\circ}{\sin 45^\circ} = \frac{1}{2 \times \frac{1}{\sqrt{2}}} = \frac{1}{\sqrt{2}}$

12. (a) Here, $\sin C = \frac{1}{\mu} = \frac{1}{2}$

$\Rightarrow \quad C = 30^\circ$

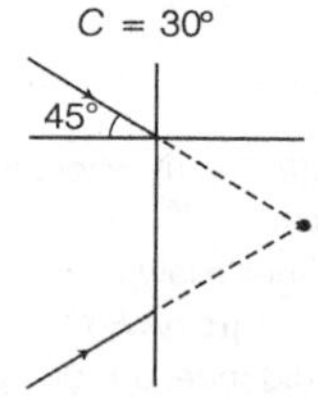

Since, angle of incident is greater than critical angle so, total internal reflection takes place. Thus, the surface behaves as plane mirror.

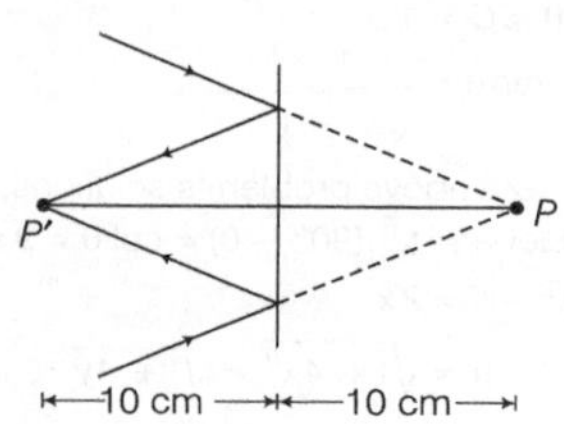

13. (a)

14. (a) Here, $\sin C = \dfrac{1}{\frac{4}{3}} = \dfrac{3}{4} = 0.75$

and $\sin\theta = \sin 53^\circ = \dfrac{4}{5} = 0.8$

Thus, $\theta > C$,

So, total internal reflection takes place,

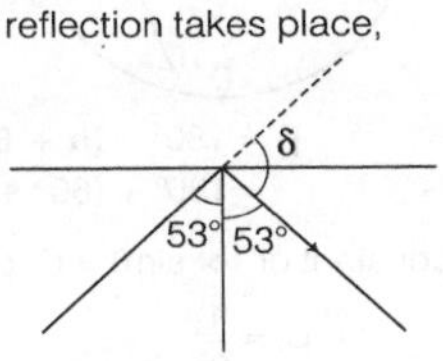

$\therefore$ $\delta = \pi - 2i = 180^\circ - 2 \times 53^\circ = 74^\circ$.

15. (c) According to Snell's law,

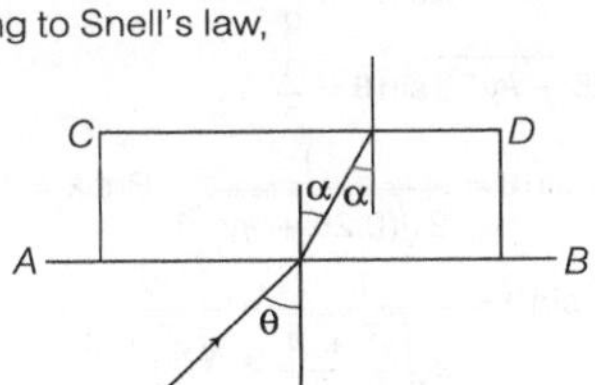

$$\frac{4}{3}\sin 53^\circ = \frac{3}{2}\sin\alpha$$

or $$\frac{4}{3} \times \frac{4}{5} = \frac{3}{2}\sin\alpha$$

or $$\sin\alpha = \frac{32}{45} = 0.711$$

At surface CD, $$\sin C = \frac{1}{3/2} = \frac{2}{3} = 0.67$$

$\therefore$ $\alpha > C$

Hence, total internal reflection takes place at the surface CD.

16. (b) According to Snell's law,

$$\mu \sin\theta = k$$

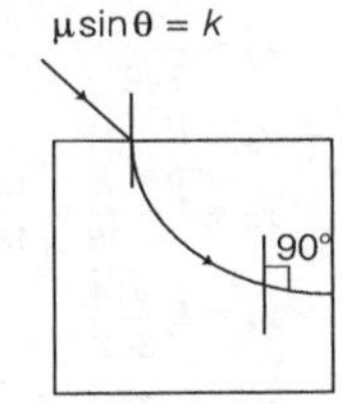

or $1 \times \sin 60^\circ = k$

$\therefore$ $k = \sqrt{3}/2$

$\Rightarrow$ $(\sqrt{3} - y_{max}) \sin 90^\circ = \dfrac{\sqrt{3}}{2}$

$\therefore$ $y_{max} = \dfrac{\sqrt{3}}{2}$

17. (c) According to Snell's law,

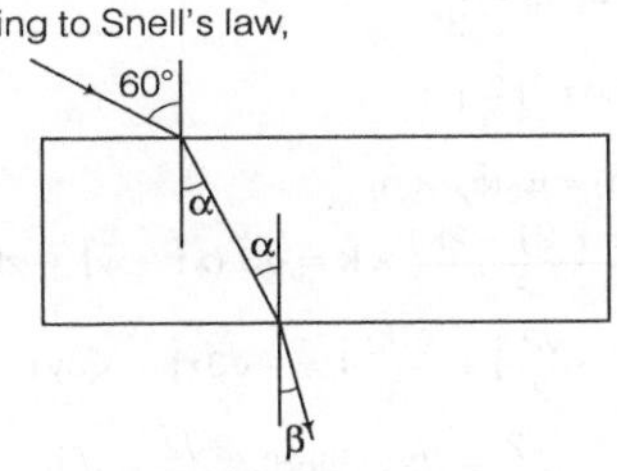

$$1 \times \sin 60^\circ = \frac{4}{3}\sin\alpha$$

or $\dfrac{\sqrt{3}}{2} = \dfrac{4}{3}\sin\alpha$ or $\sin\alpha = \dfrac{3\sqrt{3}}{8}$

and $\dfrac{4}{3}\sin\alpha = \dfrac{3}{2}\sin\beta$ or $\dfrac{4}{3} \times \dfrac{\sqrt{3}}{2} = \dfrac{3}{2}\sin\beta$

or $\sin\beta = \dfrac{4}{3} \times \dfrac{\sqrt{3}}{2} \times \dfrac{2}{3} = \dfrac{4\sqrt{3}}{9} = \dfrac{4}{3\sqrt{3}}$

$\therefore$ $\beta = \sin^{-1}\left(\dfrac{4}{3\sqrt{3}}\right)$.

18. (a) At upper boundary, $y = 1$

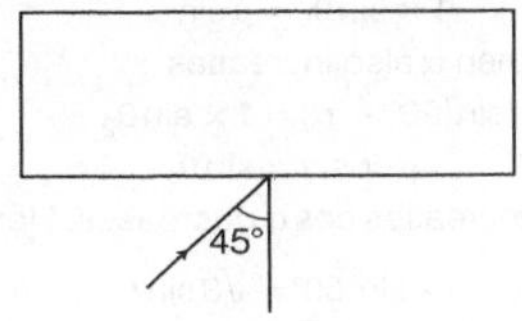

$\therefore$ $\mu = 2$

$\therefore$ $1 \times \sin 45^\circ = 2\sin\alpha$

$\therefore$ $\sin\alpha = \dfrac{1}{\sqrt{2} \times 2}$

$\therefore$ $\sin\alpha = \dfrac{1}{2\sqrt{2}}$.

19. (c) According to Snell's law,

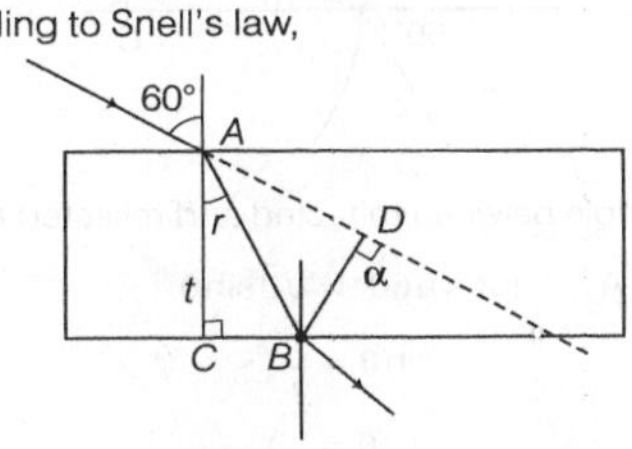

$$1 \times \sin 60^\circ = \mu \sin r$$

or $$\frac{\sqrt{3}}{2} = 2\sin r$$

$\therefore$ $$\sin r = \frac{\sqrt{3}}{4}$$

$\therefore$ $$\cos r = \sqrt{1 - \sin^2 r}$$

$$= \sqrt{1 - \frac{3}{16}} = \frac{\sqrt{13}}{16}$$

From triangle ABC, $$\sec r = \frac{AB}{t}$$

$\therefore$ $$AB = t\sec r$$

$$= 1 \times \sec 60^\circ = 2$$

The lateral shifting is

$$d = t\sec r \sin(i - r)$$
$$= 2\sin(i - r)$$
$$= 2\{\sin i \cos r - \cos i \sin r\}$$
$$= 2\{\sin 60^\circ \cos r - \cos 60^\circ \sin r\}$$
$$= 2\left\{\frac{\sqrt{3}}{2} \times \frac{\sqrt{13}}{4} - \frac{1}{2} \times \frac{\sqrt{3}}{4}\right\}$$
$$= \frac{\sqrt{39}}{4} - \frac{\sqrt{3}}{4} = \left(\frac{\sqrt{39} - \sqrt{3}}{4}\right)$$

20. (c) $$1 \times \sin i = n\sin\frac{i}{2}$$

$$\sin i = n\sin\frac{i}{2}$$

$$2\sin\frac{i}{2}\cos\frac{i}{2} = n\sin\frac{i}{2}$$

or $$2\cos\frac{i}{2} = n$$

or $$\cos\frac{i}{2} = \frac{n}{2}$$

$\therefore$ $$i = 2\cos^{-1}\left(\frac{n}{2}\right)$$

21. (a) Since, refractive index from top to bottom increases. So, angle made by ray with vertical decreases from top to bottom. Mathematically,

$$\mu \sin\theta = \text{constant}$$

$$\therefore \quad \sin\theta \propto \frac{1}{\mu}.$$

22. (c) According to Snell's law,

$$1 \times \sin\theta_1 = \mu \sin\alpha$$

If θ_1 increase, then α also increases

$$\mu \sin(90^\circ - \alpha) = 1 \times \sin\theta_2$$

or $$\mu \cos\alpha = \sin\theta_2$$

From this, if α increases $\cos\alpha$ decreases. Hence, θ_2 decreases.

23. (c) $$1 \times \sin 60^\circ = \sqrt{3} \sin r$$

or $$\frac{\sqrt{3}}{2} = \sqrt{3} \sin r \quad \text{or} \quad \sin r = \frac{1}{2}$$

$$r = 30^\circ$$

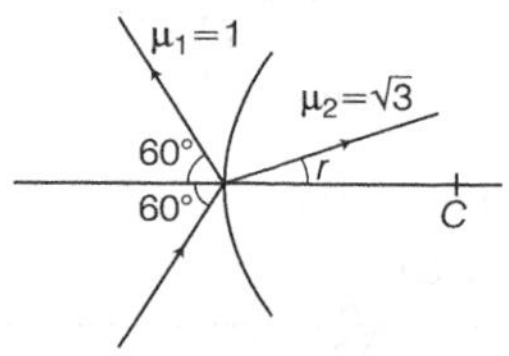

Thus, the angle between refracted and reflected ray is 90°.

24. (b) At point *A*, $$1 \times \sin 60^\circ = \sqrt{3} \sin\theta$$

$$\therefore \quad \sin\theta = \frac{1}{2}$$

$$\therefore \quad \theta = 30^\circ$$

At point *B*, $$\sqrt{3} \sin\theta = 1 \times \sin\alpha$$

or $$\sqrt{3} \sin 30^\circ = \sin\alpha$$

$$\therefore \quad \sin\alpha = \frac{\sqrt{3}}{2} \Rightarrow \alpha = 60^\circ$$

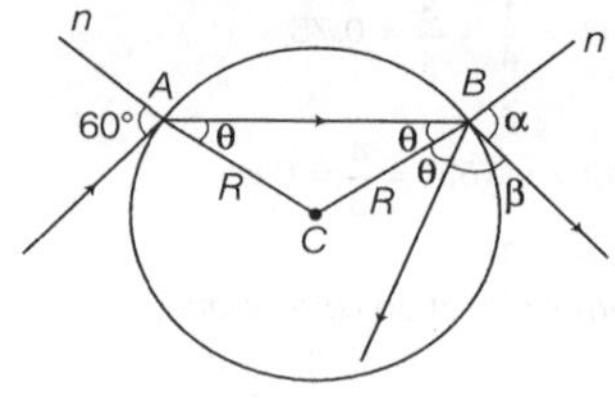

$$\therefore \quad \beta = 180^\circ - (\alpha + \theta)$$
$$= 180^\circ - (60^\circ + 30^\circ) = 90^\circ$$

25. (b) $\mu \sin\theta = C$ = constant or $1 \times \sin\theta = C$ or $\sin 30^\circ = C$

$$\therefore \quad C = \frac{1}{2}$$

$$\therefore \quad \mu \sin\theta = \frac{1}{2}$$

or $$\sqrt{0.25 + ky^{-2}} \sin\theta = \frac{1}{2}$$

or $$\sin\theta = \frac{1}{2\sqrt{(0.25 + ky^{-2})}} \quad \text{But } k = 1$$

$$\therefore \quad \sin\theta = \frac{1}{2\sqrt{\dfrac{y^2 + 4}{4y^2}}} = \frac{y}{\sqrt{y^2 + 4}}$$

or $$\tan\theta = \frac{y}{\sqrt{y^2 + 4 - y^2}} = \frac{y}{2}$$

or $$\frac{dx}{dy} = \frac{y}{2} \quad \text{or} \quad \int_0^x dx = \frac{1}{2}\int_0^y y\, dy$$

or $$x = \frac{y^2}{4}$$

$$\therefore \quad y^2 = 4x.$$

DPP-6 **Refraction of Light-II**

1. (a,b) (a) $\mu = \dfrac{c_0}{c} = \dfrac{3 \times 10^8}{\sqrt{3} \times 10^8} = \sqrt{3}$

(b) $\sin C = \dfrac{1}{\mu} = \dfrac{1}{\sqrt{3}}$

$$\therefore \quad C = \sin^{-1}\left(\frac{1}{\sqrt{3}}\right)$$

2. (a,b) The unit vector along incidence ray is

$$\hat{\mathbf{e}} = \frac{1}{\sqrt{2}}\hat{\mathbf{i}} - \frac{1}{\sqrt{2}}\hat{\mathbf{j}}$$

$$\because \quad \mu(\hat{\mathbf{e}} \times \hat{\mathbf{n}}) = \mu_1(\hat{\mathbf{e}}_1 \times \hat{\mathbf{n}}) = \mu_2(\hat{\mathbf{e}}_2 \times \hat{\mathbf{n}})$$

(a) $$1 \times \left(\frac{1}{\sqrt{2}}\hat{\mathbf{i}} - \frac{1}{\sqrt{2}}\hat{\mathbf{j}}\right) \times \hat{\mathbf{j}} = \frac{4}{3}(x_1\hat{\mathbf{i}} + y_1\hat{\mathbf{j}}) \times \hat{\mathbf{j}}$$

or $$\frac{1}{\sqrt{2}}\hat{\mathbf{k}} = \frac{4}{3}x_1\hat{\mathbf{k}} \quad \text{or} \quad x_1 = \frac{3}{4\sqrt{2}}$$

$$\because \quad x_1^2 + y_1^2 = 1 \quad \text{or} \quad \frac{9}{32} + y_1^2 = 1$$

or $$y_1^2 = 1 - \frac{9}{32} = \frac{23}{32}$$

$$\therefore \quad y_1 = \pm\sqrt{23/32}$$

$$\therefore \quad \hat{\mathbf{e}}_1 = \frac{3}{4\sqrt{2}}\hat{\mathbf{i}} - \frac{\sqrt{23}}{4\sqrt{2}}\hat{\mathbf{j}}$$

(b) $$\because \quad \mu(\hat{\mathbf{e}} \times \hat{\mathbf{n}}) = \mu_2(\hat{\mathbf{e}}_2 \times \hat{\mathbf{n}})$$

or $$\left(\frac{1}{\sqrt{2}}\hat{\mathbf{i}} - \frac{1}{\sqrt{2}}\hat{\mathbf{j}}\right) \times \hat{\mathbf{j}} = \frac{3}{2}(x_2\hat{\mathbf{i}} + y_2\hat{\mathbf{j}}) \times \hat{\mathbf{j}}$$

or $$\frac{1}{\sqrt{2}}\hat{\mathbf{k}} = \frac{3}{2}x_2\hat{\mathbf{k}}$$

$$\therefore \quad x_2 = \frac{2}{3\sqrt{2}}$$

$$\because \quad x_2^2 + y_2^2 = 1$$

$$\therefore \quad y_2^2 = 1 - \frac{4}{18} = \frac{14}{18}$$

$$\therefore \quad y_2 = \pm\frac{\sqrt{14}}{3\sqrt{2}}$$

$$\therefore \quad \hat{\mathbf{e}}_2 = x_2\hat{\mathbf{i}} + y_2\hat{\mathbf{j}} = \frac{2}{3\sqrt{2}}\hat{\mathbf{i}} - \frac{\sqrt{14}}{3\sqrt{2}}\hat{\mathbf{j}}$$

3. (a,b,c) Here, $\hat{\mathbf{e}}_1 = \dfrac{\hat{\mathbf{i}} + 2\hat{\mathbf{j}} - 2\hat{\mathbf{k}}}{\sqrt{1^2 + 2^2 + (-2)^2}} = \dfrac{1}{3}\hat{\mathbf{i}} + \dfrac{2}{3}\hat{\mathbf{j}} - \dfrac{2}{3}\hat{\mathbf{k}}$

$\hat{\mathbf{n}} = \hat{\mathbf{k}}$ and $\hat{\mathbf{e}}_2 = x\hat{\mathbf{i}} + y\hat{\mathbf{j}} + z\hat{\mathbf{k}}$

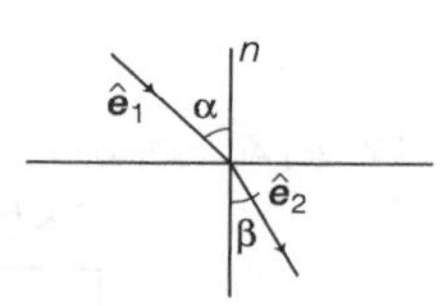

(a) $\cos\alpha = |\hat{\mathbf{e}}_1 \cdot \hat{\mathbf{n}}| = \dfrac{2}{3}$

$$\therefore \quad \alpha = \cos^{-1}\left(\frac{2}{3}\right)$$

(b) $\mu_1(\hat{\mathbf{e}}_1 \times \hat{\mathbf{n}}) = \mu_2(\hat{\mathbf{e}}_2 \times \hat{\mathbf{n}})$

or $$\sqrt{2}\frac{(\hat{\mathbf{i}} + 2\hat{\mathbf{j}} - 2\hat{\mathbf{k}})}{3} \times \hat{\mathbf{k}} = \sqrt{3}(x\hat{\mathbf{i}} + y\hat{\mathbf{j}} + z\hat{\mathbf{k}}) \times \hat{\mathbf{k}}$$

or $$-\frac{\sqrt{2}}{3}\hat{\mathbf{j}} + \frac{2\sqrt{2}}{3}\hat{\mathbf{i}} = -\sqrt{3}x\hat{\mathbf{j}} + \sqrt{3}y\hat{\mathbf{i}}$$

$$\therefore \quad \frac{\sqrt{2}}{3} = \sqrt{3}x \quad \text{and} \quad \frac{2\sqrt{2}}{3} = \sqrt{3}y$$

$$\therefore \quad x = \frac{\sqrt{2}}{3\sqrt{3}} \quad \text{and} \quad y = \frac{2\sqrt{2}}{3\sqrt{3}}$$

But $\quad x^2 + y^2 + z^2 = 1$ or $\frac{2}{27} + \frac{8}{27} + z^2 = 1$

or $\quad z^2 = 1 - \frac{10}{27} = \frac{17}{27}$

$\therefore \quad z = \pm \frac{\sqrt{17}}{3\sqrt{3}}$

$\therefore \quad \hat{\mathbf{e}}_2 = \frac{\sqrt{2}}{3\sqrt{3}}\hat{\mathbf{i}} + \frac{2\sqrt{2}}{3\sqrt{3}}\hat{\mathbf{j}} - \frac{\sqrt{17}}{3\sqrt{3}}\hat{\mathbf{k}}$

(c) $\cos\beta = |\hat{\mathbf{e}}_2 \cdot \hat{\mathbf{n}}| = \frac{\sqrt{17}}{3\sqrt{3}}\hat{\mathbf{k}}$

$\therefore \quad \beta = \cos^{-1}\left(\frac{\sqrt{17}}{3\sqrt{3}}\right)$.

4. (a,b) For $\mu_2 > \sqrt{2}$ (TIR will not take place)

$2\sin 45° = \mu_2 \sin r$

$\mu_2 \sin r = \sqrt{2}\sin i$

$i = 90°$

Hence, deviation is 45°

$\mu_2 < \sqrt{2}$ TIR will take place deviation is 90°.

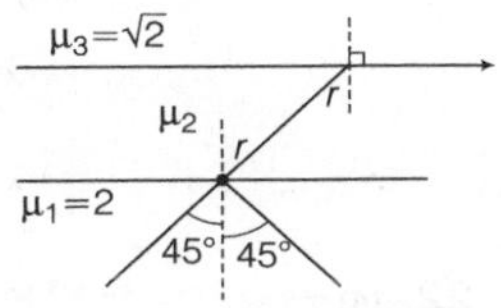

5. (a,b) From the surface, reflection as well as refraction take place. For refraction, option (a) is correct.
But for reflection option (b) is correct.

6. (b) The optical paths of ray (1) and ray (2) should be same.

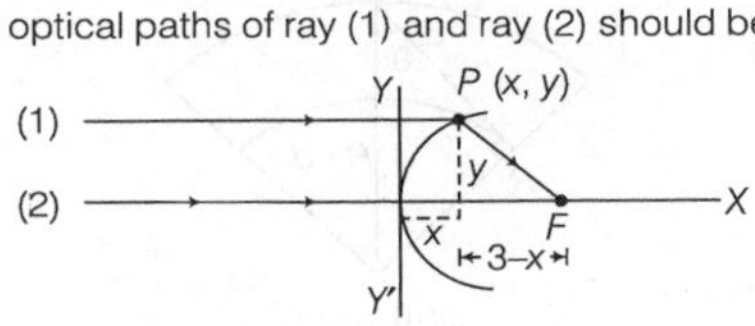

$\therefore \quad x + \mu\sqrt{(3-x)^2 + y^2} = 3\mu$

or $\quad x + 2\sqrt{(3-x)^2 + y^2} = 3 \times 2$

or $\quad x + 2\sqrt{9 + x^2 - 6x + y^2} = 6$

or $\quad 2\sqrt{x^2 + y^2 - 6x + 9} = 6 - x$

or $\quad 4(x^2 + y^2 - 6x + 9) = (6 - x)^2$

or $\quad 4x^2 + 4y^2 - 24x + 36 = 36 + x^2 - 12x$

or $\quad 3x^2 + 4y^2 - 12x = 0.$

7. (c) The maximum radius of beam is the maximum value of y.

$\because \quad 3x^2 - 12x + 4y^2 = 0$

$\therefore \quad D = (-12)^2 - 4 \times 3 \times 4y^2 = 144 - 48y^2$

For real value, $\quad D \geq 0$ or $144 - 48y^2 \geq 0$

or $\quad \frac{144}{48} \geq y^2$

or $\quad y^2 \leq 3$

$\therefore \quad y_{max} = \sqrt{3}$ cm

8. (b) Here, $\mu\sin\theta = k$, where θ is angle with axis.

or $\quad \frac{\sqrt{1 + (1+x)^2}}{1+x}\sin\theta = k$

For air, $\quad \mu = 1 \quad \theta = 90°$

$\therefore \quad 1 \times \sin 90° = k$

$\therefore \quad k = 1$

$\therefore \quad \frac{\sqrt{1 + (1+x)^2}}{1+x}\sin\theta = 1$

or For $x \to 0$

$\frac{\sqrt{1 + 1^2}}{1+0}\sin\theta = 1$ or $\sin\theta = \frac{1}{\sqrt{2}}$

$\theta = 45°$

9. (c) $\therefore \quad \frac{\sqrt{1 + (1+x)^2}}{1+x}\sin\theta = 1$

or $\quad \sin\theta = \frac{1+x}{\sqrt{1 + (1+x)^2}}$ or $\tan\theta = (1+x)$

or $\quad \frac{dy}{dx} = 1 + x \quad$ or $\int_0^y dy = \int_0^x (1+x)dx$

or $\quad y = x + \frac{x^2}{2}$

10. (a) Hence, (i) → (s), (ii) → (r), (iii) → (q), (iv) → (p).

11. (b) Laser beam follows exact path without deviation.

12. (b) Frequency does not depend on medium. But $\lambda \propto \frac{1}{\mu}$.

13. (b) When ray propagates from denser medium to rarer medium, refraction takes place if angle of incidence is critical angle. Because if angle of incidence is equal to critical angle Snell's law is applicable.

14. (c) According to particle in nature of light, interaction force between photon and the surface is along the normal to the surface. Hence, conservation principle of momentum is applicable along the surface. From water surface, refraction as well as reflection take place.

15. (c) $\sin C = \frac{1}{\mu}$ and $\mu \propto \frac{1}{\lambda}$

$\therefore \quad \sin C \propto \lambda$

$C_R > C_V \quad (\because \lambda_R > \lambda_V)$

Hence, Statement I is correct. But Statement II is incorrect.

16. (c) $\mu AB = u'CD \Rightarrow \quad \frac{AB}{CD} = \frac{\mu'}{\mu} \quad$ ...(i)

$t_1C_1 = AB, t_2C_2 = CD$

$\frac{t_1}{t_2} = \frac{AB}{CD} \times \frac{C_2}{C_1} \quad$ ...(ii)

$\mu \propto \frac{1}{C}$

$\frac{\mu}{\mu'} = \frac{C_2}{C_1} \quad$ ...(iii)

From Eqs. (i), (ii) and (iii), we get

$\frac{t_1}{t_2} = \frac{\mu'}{\mu} \times \frac{\mu}{\mu'}$

$\frac{t_1}{t_2} = 1 \Rightarrow t_1 = t_2$

Hence, Statement I is correct. Statement II is incorrect.

17. (4) According to Snell's law,

$1 \times \sin\theta = \mu\sin\alpha$

or $\quad \sin\theta = \sqrt{2}\sin\alpha$

Also, $\quad \mu\sin\beta = 1 \times \sin 90°$

or $\quad \sqrt{2}\sin\beta = 1$

or $\quad \sin\beta = \frac{1}{\sqrt{2}}$

$\therefore \quad \beta = \frac{\pi}{4}$

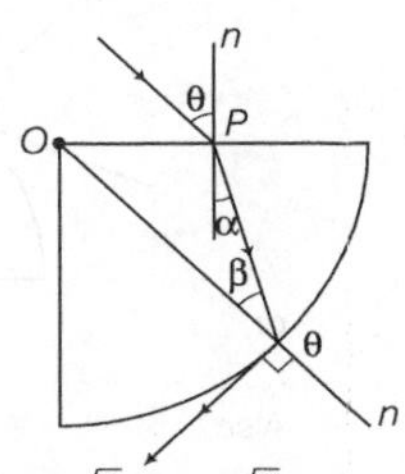

In ΔOPQ, $\quad \frac{OP}{\sin\beta} = \frac{OQ}{\sin\left(\frac{\pi}{2} + \alpha\right)}$ or $\frac{\sqrt{2}}{\sin\pi/4} = \frac{\sqrt{3}}{\cos\alpha}$

$\cos\alpha = \frac{\sqrt{3}}{2} \Rightarrow \quad \alpha = 30°$

$\therefore \quad \sin\theta = \sqrt{2}\sin 30° = \sqrt{2} \times \frac{1}{2} = \frac{1}{\sqrt{2}}$

$\sin\theta = \sin\frac{\pi}{4} \Rightarrow \theta = \frac{\pi}{4}$

Given, the minimum value of θ is $\frac{\pi}{n}$,

$\therefore \quad \frac{\pi}{n} = \frac{\pi}{4} \Rightarrow n = 4.$

18. (2) From figure, $\mu \sin\theta = 1 \sin 90°$

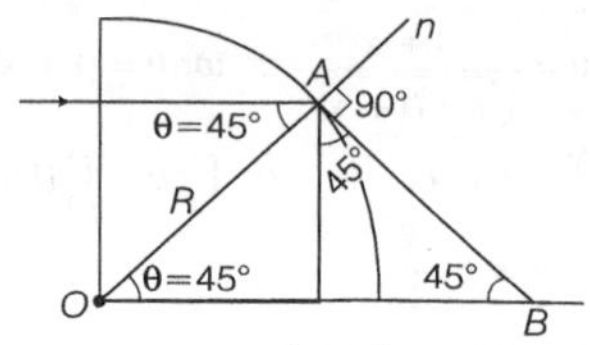

$\therefore \quad \sin\theta = \dfrac{1}{\mu} = \dfrac{1}{\sqrt{2}}$

$\therefore \quad \theta = 45°$

From ΔOAB, $\quad \dfrac{OB}{\sin 90°} = \dfrac{R}{\sin 45°}$

$\therefore \quad OB = \sqrt{2}R = \sqrt{2} \times \sqrt{2}\text{ cm} = 2\text{ cm}.$

19. (5) Here, $\sin i = \dfrac{h}{k} = \dfrac{10\sqrt{3}}{20}$

$\therefore \quad i = 60°$

According to Snell's law, $1 \times \sin i = \mu \sin r$

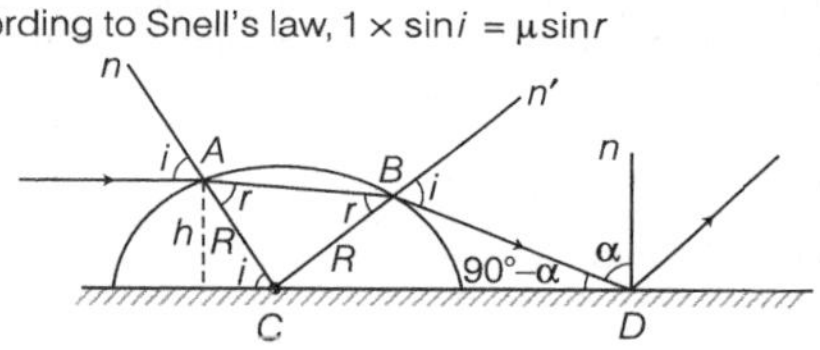

or $\quad \sin 60° = \sqrt{3/2}\,\sin r$

or $\quad \sin r = \dfrac{\sqrt{3} \times \sqrt{2}}{2 \times \sqrt{3}}$

$\therefore \quad r = 45°$

$\therefore \quad \angle BCD = 180° - i - (180° - 2r) = 2r - i$

$= 90° - 60° = 30°$

Also, $\quad \angle CBD = 180° - i = 180° - 45° = 135°$

In ΔBCD,

$90° - \alpha + 30° + 135° = 180°$

$\therefore \quad \alpha = 75°$

$\therefore$ The angle of incidence on the plane mirror is

$\alpha = 75° = 75 \times \dfrac{\pi}{180°} = \dfrac{5\pi}{12}$ radian

$\therefore \quad n = 5$

20. (3) From figure, $90° + i + i + 180° - 30° = 360°$

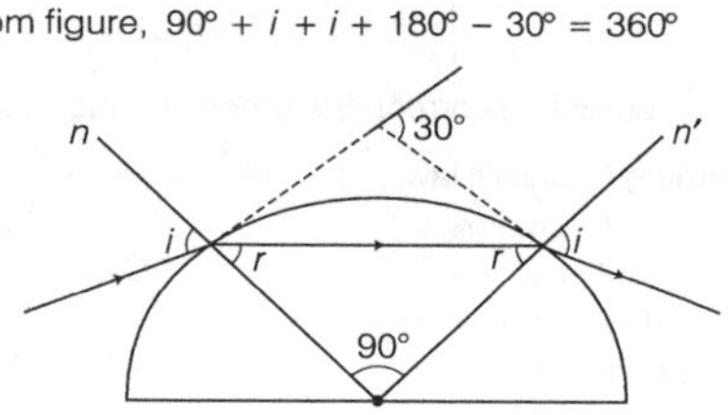

or $\quad 2i = 120°$

$\therefore \quad i = 60°$

Also, $\quad 2r + 90° = 180°$

$\therefore \quad r = 45°$

According to Snell's law,

$1 \times \sin i = \mu \sin r$

or $\quad \sin 60° = \mu \sin 45°$

$\therefore \quad \mu = \dfrac{\sqrt{3}}{2 \times \dfrac{1}{\sqrt{2}}} = \sqrt{\dfrac{3}{2}}$

$\therefore \quad n = 3.$

21. (2) From figure, $\mu \sin\theta = k$

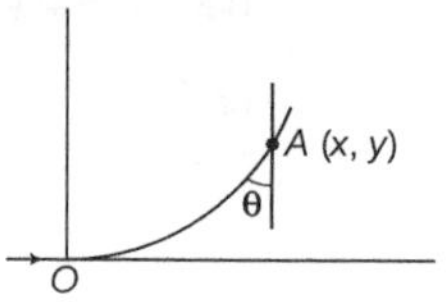

For point O, $\quad \mu = 1,\ \theta = 90°$

$\therefore \quad 1 \times \sin 90° = k$

$\therefore \quad k = 1$

$\therefore \quad \mu \sin\theta = 1$

or $\quad \sin\theta = \dfrac{1}{\mu}$

or $\quad \tan\theta = \dfrac{1}{\sqrt{\mu^2 - 1}}$

or $\quad \dfrac{dx}{dy} = \dfrac{1}{\sqrt{\mu^2 - 1}} = \dfrac{1}{\sqrt{ay}}$

or $\quad \int_0^y \dfrac{dy}{\sqrt{ay}} = \int_0^x dx$

$\therefore \quad x = 2\sqrt{\dfrac{y}{a}}$

$\therefore \quad x = 2\sqrt{\dfrac{2}{2 \times 10^{-6}}} = 2 \times 10^3\text{ m} = 2\text{ km}.$

22. (4) $\because \quad \sin\theta = \dfrac{R}{R + d}$

But $\quad \theta > C$

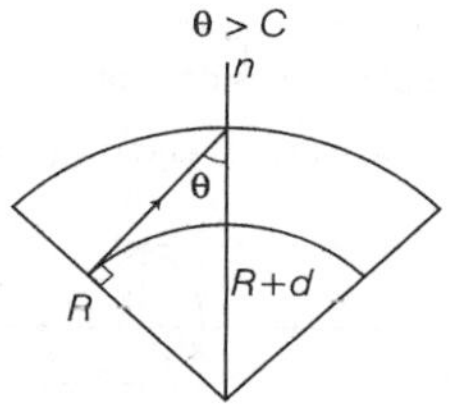

or $\quad \sin\theta > \sin C$

or $\quad \dfrac{R}{R + d} > \dfrac{1}{\mu}$

or $\quad \dfrac{d + R}{R} < \mu$ or $1 + \dfrac{d}{R} < \dfrac{3}{2}$

$\dfrac{d}{R} < \dfrac{1}{2}$ or $\dfrac{R}{d} > 2$

$\therefore \quad R_{min} = 4\text{ cm}.$

23. (2) In ΔABC, $\sin i = \dfrac{h}{R}$

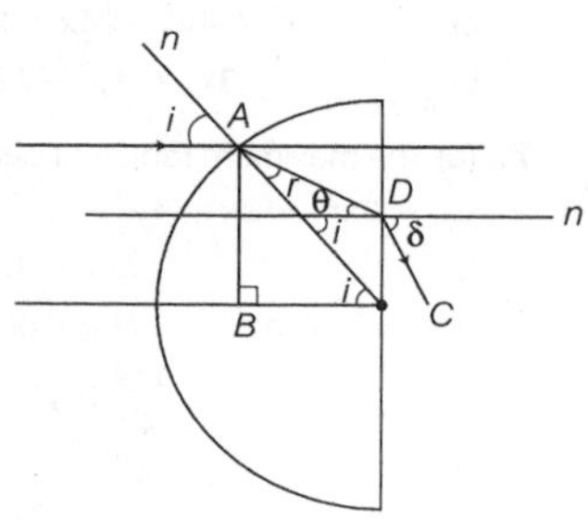

Applying Snell's law at curve surface,

$1 \times \sin i = \mu \sin r$

$\dfrac{h}{R} = \mu \sin r$

$\therefore \quad \dfrac{h}{R} = \mu \sin r$

But $\quad \mu \sin\theta = 1 \times \sin\delta$

$\therefore \quad \sin\delta = u \sin\theta$

$\therefore \quad i = r + \theta$

$\theta = i - r = \mu \sin(i - r)$

$= \mu \sin i \cos r - \mu \cos i \sin r$

$= \mu \dfrac{h}{R} \cos r - \mu \dfrac{\sqrt{R^2 - h^2}}{R} \dfrac{h}{\mu R}$

$= \dfrac{\mu h}{R} \dfrac{\sqrt{\mu^2 R^2 - h^2}}{\mu R} - \dfrac{\sqrt{R^2 - h^2}}{R^2} h$

$= \dfrac{h\sqrt{\mu^2 R^2 - h^2}}{R^2} - \dfrac{h\sqrt{R^2 - h^2}}{R^2}$

$\Rightarrow \quad n = 2$

DPP-7 **Image Formation due to Refraction through Spherical Surface - I**

1. (i) (c) $\mu_1 = \frac{4}{3}, \mu_2 = \frac{3}{2}, u = -40$ cm, $r = -10$ cm

$\therefore \quad \frac{\mu_2}{v} - \frac{\mu_1}{u} = \frac{\mu_2 - \mu_1}{r}$

or $\quad \frac{3}{2v} - \frac{4}{3(-40)} = \frac{\frac{3}{2} - \frac{4}{3}}{-10}$

or $\quad \frac{3}{2v} = -\frac{1}{60} - \frac{1}{30} = \frac{-1-2}{60} = -\frac{1}{20}$

$\therefore \quad v = -30$ cm

(ii) (b) $\therefore \frac{\mu_2}{v} - \frac{\mu_1}{u} = \frac{\mu_2 - \mu_1}{r}$

or $\quad -\frac{\mu_2}{v^2}\frac{dv}{dt} + \frac{\mu_1}{u^2}\frac{du}{dt} = 0$

$\therefore \quad \frac{dv}{dt} = \left(\frac{\mu_1}{\mu_2} - \frac{v^2}{v^2}\right)\frac{du}{dt}$

$= \frac{4}{3 \times 3/2} \times \frac{900}{1600} \times 10$

$= \frac{8}{9} \times \frac{900}{1600} \times 10 = 5 \text{ cms}^{-1}$

(iii) (b)

2. (i) (c) $\mu_1 = 1, \mu_2 = 3/2 \; r = +10$ cm, $u = -20$ cm

$\because \quad \frac{\mu_2}{v} - \frac{\mu_1}{u} = \frac{\mu_2 - \mu_1}{r}$

or $\quad \frac{3}{2v} - \frac{1}{-20} = \frac{\frac{3}{2} - 1}{10} = \frac{1}{20}$

or $\quad \frac{3}{2v} = \frac{1}{20} - \frac{1}{20} = 60$

$\therefore \quad v = \infty$

(ii) (a) For second surface,

$\mu_1' = \frac{3}{2}, \mu_2' = 1, u' = \infty, r' = -10$ cm

$\because \quad \frac{\mu_2'}{v'} - \frac{\mu_1'}{u'} = \frac{\mu_2' - \mu_1'}{r'} = \frac{1 - 3/2}{-10}$

or $\quad \frac{1}{v'} - \frac{3}{2 \times \infty} = \frac{1}{20}$

$\therefore \quad v' = 20$ cm

(iii) (b)

3. (i) (b) Here, $\mu_1 = 1, \mu_2 = 1.5, r = \infty, u = -12$ cm

$\because \quad \frac{\mu_2}{v} - \frac{\mu_1}{u} = \frac{\mu_2 - \mu_1}{\infty} = 0$

or $\quad \frac{3}{2v} - \frac{1}{-12} = 0$

or $\quad \frac{2v}{3} = -12$

$\therefore \quad v = -18$ cm

(ii) (a,c)

(iii) (b) Here, $\mu_1' = \frac{3}{2}, \mu_2' = 1,$

$u' = -21$ cm, $r' = +3$ cm

$\frac{\mu_2}{v'} - \frac{\mu_1'}{u'} = \frac{\mu_2' - \mu_1'}{r'}$

or $\quad \frac{1}{v'} - \frac{3}{2(-21)} = \frac{1 - 3/2}{3}$

or $\quad \frac{1}{v'} = -\frac{1}{6} - \frac{1}{14} = \frac{-7-3}{42}$

$\therefore \quad v' = -4.2$ cm

(iv) (a)

(v) (b) Here $\mu_1'' = 1, \mu_2'' = \frac{3}{2}, r'' = -3$ cm

$\mu'' = -10.2$ cm

$\frac{\mu_2''}{v''} - \frac{\mu_1''}{u''} = \frac{\mu_2'' - \mu_1''}{r''}$

or $\quad \frac{3}{2v''} + \frac{1}{10.2} = \frac{\frac{3}{2} - 1}{-3} = -\frac{1}{6}$

$\therefore \quad \frac{3}{2v''} = -\frac{1}{6} - \frac{10}{102}$

$= \frac{-1}{6} - \frac{10}{102}$

$= \frac{-17 - 10}{102}$

$\Rightarrow \quad \frac{3}{2v''} = \frac{-27}{102}$

$\Rightarrow \quad \frac{2v''}{3} = \frac{-102}{27}$

$\Rightarrow \quad v'' = \frac{102 \times 3}{27 \times 2}$

$\Rightarrow \quad v'' = \frac{-81}{9}$

$\Rightarrow \quad v'' = \frac{-17}{3}$ cm

(vi) (b)

(vii) (b)

4. (a) Here, $u = -x$, $r = -20$ cm,

$\mu_1 = 2, \; \mu_2 = \frac{3}{2}$

According to refractive surface formula,

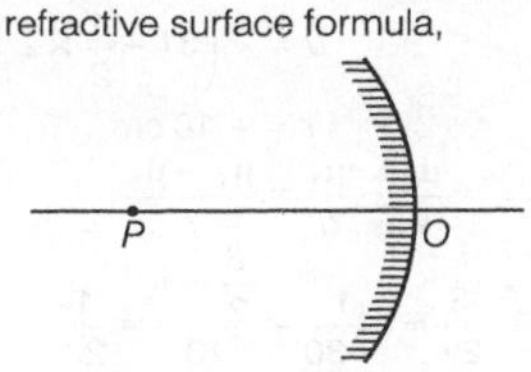

$\frac{\mu_2}{v} - \frac{\mu_1}{u} = \frac{\mu_2 - \mu_1}{r}$

or $\quad \frac{3}{2v} - \frac{2}{-x} = \frac{\frac{3}{2} - 2}{-20}$

or $\quad \frac{3}{2v} = \frac{1}{40} - \frac{2}{x}$

or $\quad \frac{3}{2v} = \frac{1}{40} - \frac{2}{x}$

or $\quad \frac{2v}{3} = \frac{40x}{x - 80}$

or $\quad v = \frac{120x}{2x - 160}$

For real image, $\quad v > 0$

$\therefore \quad \frac{120x}{2x - 160} > 0$

But x is positive, for positive value of $\frac{120x}{2x - 160}$,

$2x - 160 > 0$

or $\quad 2x > 160$

$\because \quad x > 80$ cm.

5. (b) Here, $\mu_1 = 2, \mu_2 = 3/2,\ u = -60$ cm, $r = -20$ cm

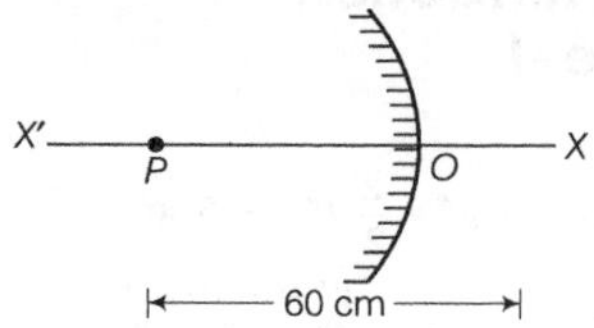

$$\frac{\mu_2}{v} - \frac{\mu_1}{u} = \frac{\mu_2 - \mu_1}{r}$$

or $$\frac{3}{2v} - \frac{2}{-60} = \frac{\frac{3}{2} - 2}{-20}$$

or $$\frac{3}{2v} + \frac{1}{30} = \frac{1}{40}$$

or $$\frac{3}{2v} = \frac{1}{40} - \frac{1}{30} = \frac{3-4}{120}$$

or $$\frac{3}{2v} = -\frac{1}{120}$$

or $$\frac{2v}{3} = -120$$

or $$v = -\frac{120 \times 3}{2} = -180 \text{ cm}$$

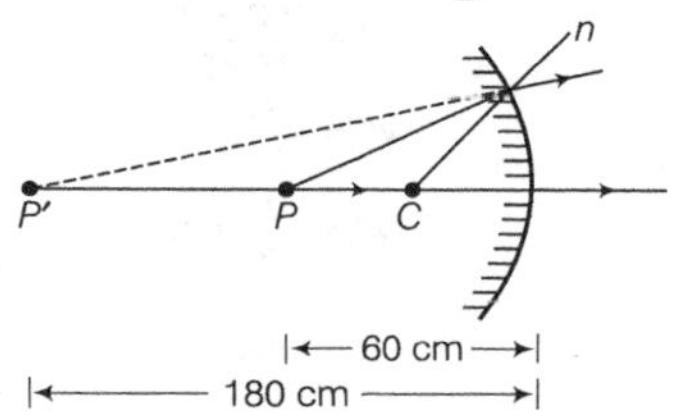

From ray diagram, it is clear that image is virtual.

6. (c) Here, $\mu_1 = 1, \mu_2 = \frac{3}{2}$

$$u = -\left(31 - \frac{1}{2} \times 2 \times 1^2\right) = -30 \text{ cm}$$

$$r = +10 \text{ cm}$$

$\because$ $$\frac{\mu_2}{v} - \frac{\mu_1}{u} = \frac{\mu_2 - \mu_1}{r}$$

or $$\frac{3}{2v} - \frac{1}{-30} = \frac{\frac{3}{2} - 1}{10} = \frac{1}{20}$$

or $$\frac{3}{2v} = \frac{1}{20} - \frac{1}{30} = \frac{3-2}{60}$$

or $$\frac{3}{2v} = \frac{1}{60}$$

or $$\frac{2v}{3} = 60$$

$\therefore$ $$v = \frac{60 \times 3}{2} = 90 \text{ cm}$$

$\because$ $$\frac{\mu_2}{v} - \frac{\mu_1}{u} = \frac{\mu_2 - \mu_1}{r}$$

or $$\frac{\mu_2}{v^2}\frac{dv}{dt} + \frac{\mu_1}{u^2}\frac{du}{dt} = 0$$

or $$\frac{dv}{dt} = \left(\frac{\mu_1}{\mu_2}\right)\left(\frac{v^2}{u^2}\right)\frac{du}{dt}$$

$$= \left(\frac{1}{\frac{3}{2}}\right)\left(\frac{90}{-30}\right)^2 (at)$$

$$= \frac{2}{3} \times 9 \times 2 \times 1 = 12 \text{ cms}^{-1}$$

7. (a) Here, $\mu_1 = \frac{3}{2}, \mu_2 = 1, u = -30$ cm, $r = -15$ cm

According to refractive surface formula,

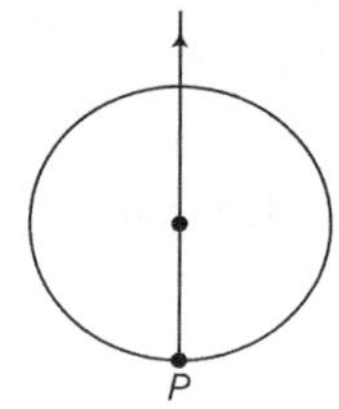

$$\frac{\mu_2}{v} - \frac{\mu_1}{u} = \frac{\mu_2 - \mu_1}{r}$$

or $$\frac{1}{v} - \frac{3}{2(-30)} = \frac{1 - \frac{3}{2}}{-15}$$

or $$\frac{1}{v} = \frac{1}{30} - \frac{1}{20} = \frac{2-3}{60}$$

$\therefore$ $$v = -60 \text{ cm}$$

8. (c) Here, $\mu_1 = 2, \mu_2 = 1,\ u = -30$ cm

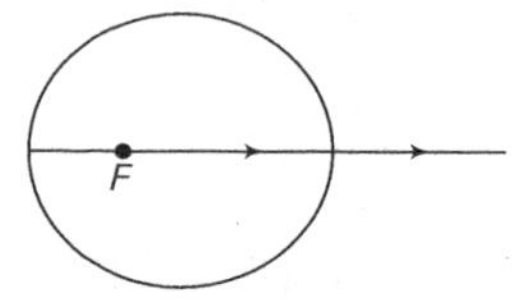

$$r = -20 \text{ cm}$$

$\because$ $$\frac{\mu_2}{v} - \frac{\mu_1}{v} = \frac{\mu_2 - \mu_1}{r}$$

or $$\frac{1}{v} - \frac{2}{-30} = \frac{1-2}{-20}$$

or $$\frac{1}{v} = \frac{1}{20} - \frac{1}{15} = \frac{3-4}{60} = -\frac{1}{60}$$

$\therefore$ $$v = -60 \text{ cm}$$

$\because$ $$\frac{\mu_2}{v} - \frac{\mu_1}{u} = \frac{\mu_2 - \mu_1}{r}$$

or $$-\frac{\mu_2}{v^2}\frac{dv}{dt} + \frac{\mu_1}{u^2}\frac{du}{dt} = 0$$

or $$\frac{dv}{dt} = \frac{\mu_1}{\mu_2}\left(\frac{v^2}{u^2}\right)\frac{dv}{dt} = \frac{2}{1}\left(\frac{-60}{-30}\right)^2 (+2)$$

$$= 2 \times 4 \times 2 = 16 \text{ cms}^{-1}$$

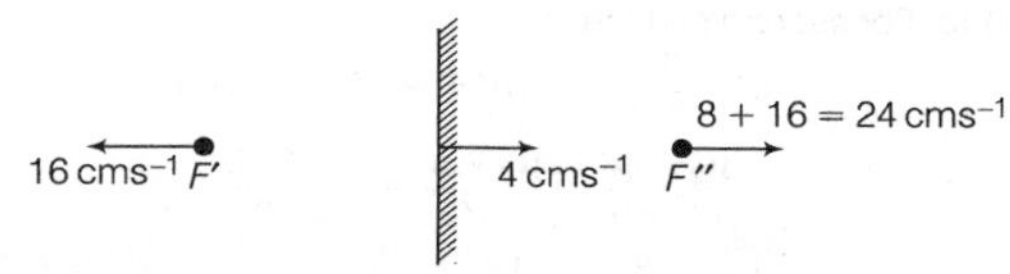

From figure, the speed of image by plane mirror is 24 cms^{-1}.

9. (b) For spherical surface,

$$\mu_1 = 1, \mu_2 = 2, u = -\infty, r = +10 \text{ cm}$$

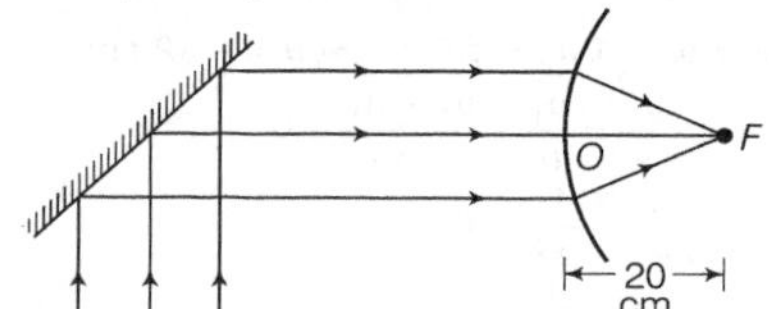

According to refractive surface formula

$$\frac{\mu_2}{v} - \frac{\mu_1}{u} = \frac{\mu_2 - \mu_1}{r}$$

or $$\frac{2}{v} - \frac{1}{\infty} = \frac{2-1}{10}$$

or $$\frac{2}{v} = \frac{1}{10}$$

$\because$ $$v = 20 \text{ cm}$$

Thus, focus is *at* distance 20 cm from optical centre of refractive surface.

10. (c) For refraction through cavity surface,

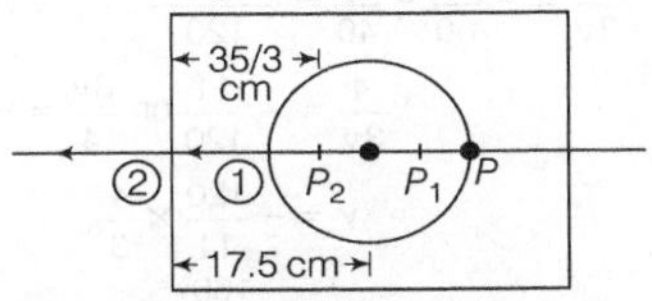

$\mu_1 = 1, \mu_2 = 3/2, u = -10$ cm, $r = -5$ cm

$\because \quad \dfrac{\mu_2}{v} - \dfrac{\mu_1}{u} = \dfrac{\mu_2 - \mu_1}{r}$

or $\quad \dfrac{3}{2v} - \dfrac{1}{-10} = \dfrac{\frac{3}{2} - 1}{-5}$

or $\quad \dfrac{3}{2v} = -\dfrac{1}{10} - \dfrac{1}{10} = -\dfrac{1}{5}$

or $\quad \dfrac{2v}{3} = -5$

$\therefore \quad v = -7.5$ cm

The image P_1 formed by refraction through cavity surface behaves as object for plane surface.

For plane surface,

$$\mu_1' = \frac{3}{2}, \mu_2' = 1, r' = \infty, u' = -17.5 \text{ cm}$$

$\because \quad \dfrac{\mu_2'}{v'} - \dfrac{\mu_1'}{u'} = \dfrac{\mu_2' - \mu_1'}{\infty} = 0$

or $\quad \dfrac{1}{v'} - \dfrac{3}{2(-17.5)} = 0$

$\therefore \quad v' = -\dfrac{35}{3}$ cm.

The ray diagram of problem is shown in the figure.

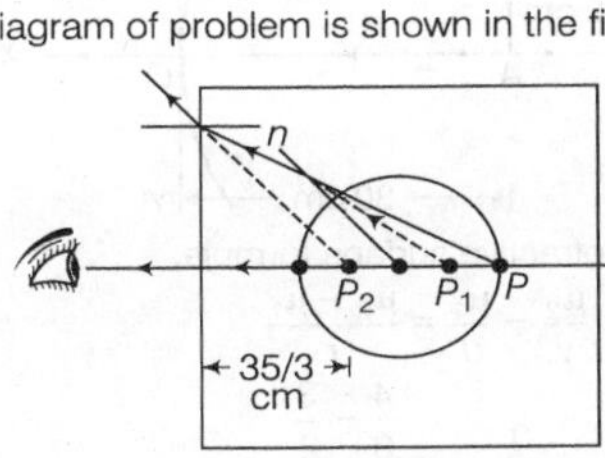

Thus, the distance through which point mark shifted is

$$l = PP_2 = 20 - \frac{35}{3}$$

$$= \frac{60 - 35}{3} = \frac{25}{3} \text{ cm}$$

11. (c) The nature of air bubble inside water is diverging. Thus, image is smaller, virtual and errect.

12. (d) For curve surface, $\mu_1 = 1$, $\mu_2 = \dfrac{3}{2}$, $u = -\infty$, $r = +20$ cm

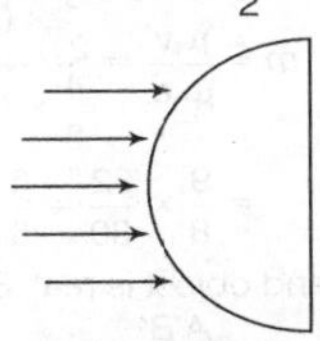

According to refractive surface formula,

$$\frac{\mu_2}{v} - \frac{\mu_1}{u} = \frac{\mu_2 - \mu_1}{r}$$

or $\quad \dfrac{3}{2v} - \dfrac{1}{-\infty} = \dfrac{3/2 - 1}{20}$

or $\quad \dfrac{3}{2v} = \dfrac{1}{40}$ or $\dfrac{2v}{3} = 40$

$\therefore \quad v = 60$ cm

For plane surface,

$$\mu_1' = 3/2 \;\; \mu_2' = 1, u' = v - 10 = 50 \text{ cm}, r' = \infty$$

$\therefore \quad \dfrac{\mu_2'}{v'} - \dfrac{\mu_1'}{u'} = \dfrac{\mu_2' - \mu_1'}{r'}$

or $\quad \dfrac{1}{v'} - \dfrac{3}{2 \times 50} = \dfrac{1 - 3/2}{\infty} = 0$ or $\dfrac{1}{v'} = \dfrac{3}{100}$

$\therefore \quad v' = \dfrac{100}{3}$ cm

13. (a) For coinciding image with object, ray should be incident on mirror normally. For curve surface

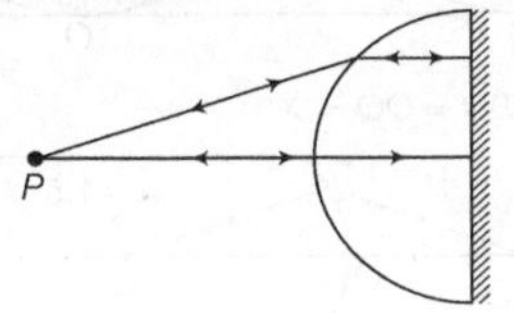

$$\mu_1 = 1, \mu_2 = \frac{3}{2}, v = \infty, r = +20 \text{ cm}$$

$\therefore \quad \dfrac{\mu_2}{v} - \dfrac{\mu_1}{u} = \dfrac{\mu_2 - \mu_1}{r}$

or $\quad \dfrac{3}{2 \times \infty} - \dfrac{1}{u} = \dfrac{3/2 - 1}{20} = \dfrac{1}{40}$

$$u = 40 \text{ cm.}$$

14. (d) For calculation of focal length, initial object distance should be infinity.

According to refractive surface formula,

$\because \quad \dfrac{\mu_2}{v} - \dfrac{\mu_1}{u} = \dfrac{\mu_2 - \mu_1}{r}$

or $\quad \dfrac{3}{2 \times v} - \dfrac{1}{-\infty} = \dfrac{\frac{3}{2} - 1}{-10}$

or $\quad \dfrac{3}{2f} = -\dfrac{1}{20}$

$$\frac{2f}{3} = -20$$

$$f = -30 \text{ cm}$$

15. (b) $\mu_1 = 1, \mu_2 = 2, u = -\infty, r = +10$ cm

$$\frac{\mu_2}{v} - \frac{\mu_1}{u} = \frac{\mu_2 - \mu_1}{r}$$

$$\frac{2}{v} - \frac{1}{-\infty} = \frac{2-1}{10}$$

$\Rightarrow \quad \dfrac{2}{v} = \dfrac{2-1}{10}$

$$v = 20 \text{ cm}$$

$\therefore$ The beam converges at point B.

Also theoretically, the net charge remains conserved and it divides itself in the ratio of capcitances.

16. (b) Here, $\mu_1 = 1, \mu_2 = \mu \; u = -\infty$

$$r = +10 \text{ cm}, v = +30 \text{ cm}$$

According to refractive surface formula,

$$\frac{\mu_2}{v} - \frac{\mu_1}{u} = \frac{\mu_2 - \mu_1}{r}$$

$$\frac{\mu}{30} - \frac{1}{-\infty} = \frac{\mu - 1}{10}$$

or $\quad \dfrac{\mu}{3} = \mu - 1$

or $\quad 1 = \dfrac{2\mu}{3}$

$\therefore \quad \mu = \dfrac{3}{2}$

But, $\quad \mu = 1 + \dfrac{t}{10}$

or $\quad \dfrac{3}{2} = 1 + \dfrac{t}{10}$ or $\dfrac{1}{2} = \dfrac{t}{10}$

$\therefore \quad t = 5$ s

17. (d) The ray diagram as shown in the figure. Therefore, the image will be real and between C and O.

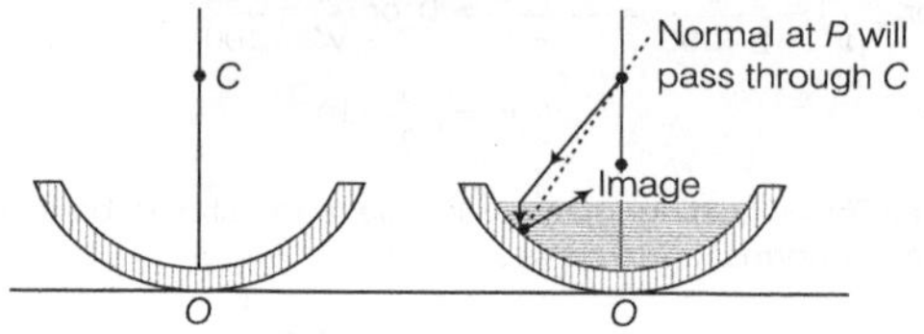

18. (a) Let us say $PO = OQ = X$

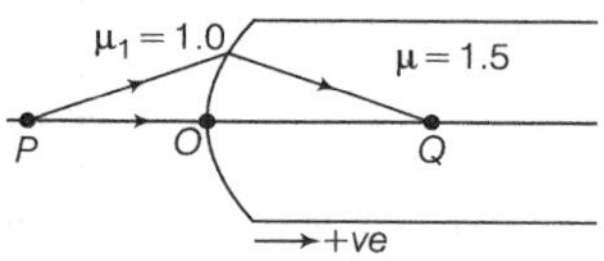

According to refractive surface formula,

$$\frac{\mu_2}{v} - \frac{\mu_1}{u} = \frac{\mu_2 - \mu_1}{R}$$

Substituting the value with sign $\frac{1.5}{+X} - \frac{1.0}{-X} = \frac{1.5 + 1.0}{+R}$

(Distances are measured from O and are taken as positive in the direction of ray of light)

$$\therefore \quad \frac{2.5}{X} = \frac{0.5}{R}$$

$$\therefore \quad X = 5R$$

19. (c) When the object is placed at the centre of the glass sphere, the rays from the object fall normally on the surface of the sphere and emerge undeviated.

20. (c) $v = \sqrt{2gh} = \sqrt{2 \times 10 \times 7.2} = 12\,\text{ms}^{-1}$

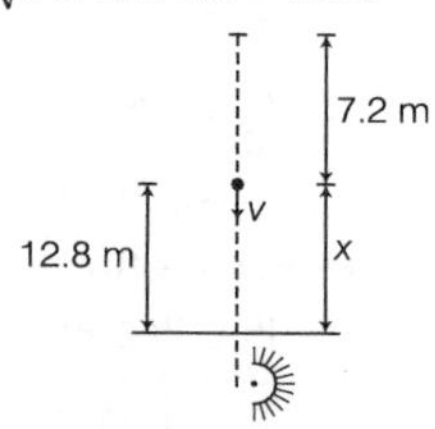

In this case when eye is inside water,

$$x_{\text{app.}} = \mu x$$

$$\therefore \quad \frac{dx_{\text{app.}}}{dt} = \mu \cdot \frac{dx}{dt}$$

or $\quad v_{\text{app.}} = \mu v = \frac{4}{3} \times 12 = 16\,\text{ms}^{-1}$

21. (a,c) Here, $\mu_1 = \frac{3}{2}, \mu_2 = \frac{4}{3}, u = -20$ cm, $r = +10$ cm

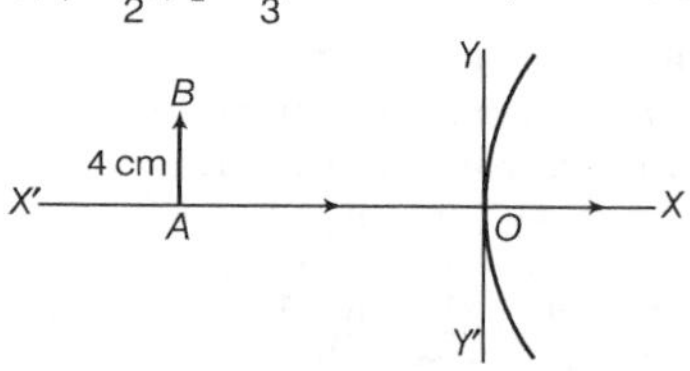

According to refractive surface formula,

$$\frac{\mu_2}{v} - \frac{\mu_1}{u} = \frac{\mu_2 - \mu_1}{r}$$

or $\quad \frac{4}{3v} - \frac{3}{2(-20)} = \frac{\frac{4}{3} - \frac{3}{2}}{+10}$

or $\quad \frac{4}{3v} = -\frac{1}{60} - \frac{3}{40} = \frac{-2-9}{120}$

or $\quad \frac{4}{3v} = -\frac{11}{120}$ or $\frac{3v}{4} = -\frac{-120}{11}$

$\therefore \quad v = -\frac{120}{11} \times \frac{4}{3}$

$\therefore \quad v = -\frac{160}{11}$ cm

$\because$ Transverse magnification $= m = \frac{A'B'}{AB} = \frac{\mu_1 v}{\mu_2 u}$

$\therefore \quad A'B' = \frac{\mu_1 v}{\mu_2 u} AB = \frac{3}{2 \times \frac{4}{3}} \times \frac{\left(-\frac{160}{11}\right)}{(-20)} \times 4$

$= 3.27$ cm

The ray diagram for formation of image.

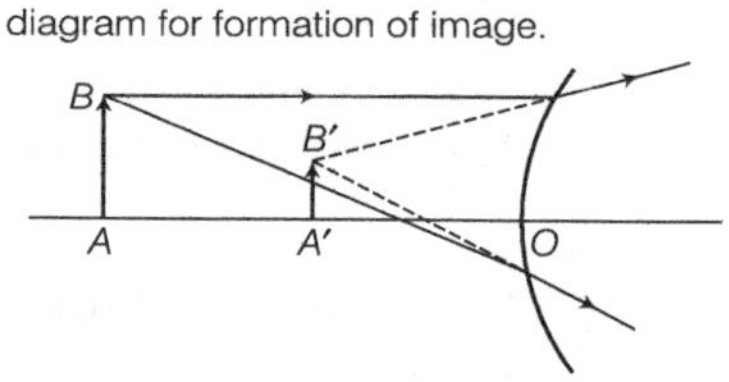

From ray diagram, image is virtual.

22. (a,b,c) Here, $\mu_1 = \frac{3}{2}, \mu_2 = \frac{4}{3}, u = -30$ cm, $r = -20$ cm

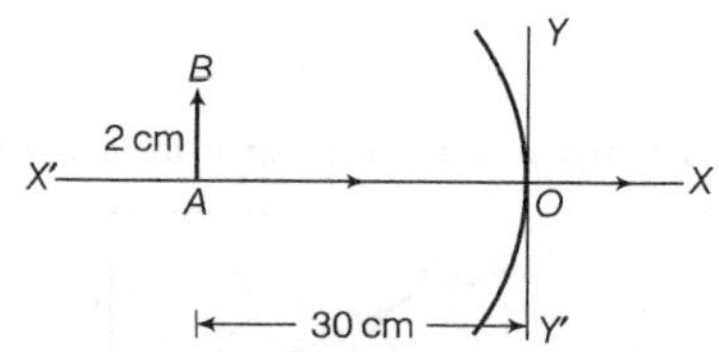

According to refractive surface formula,

$$\frac{\mu_2}{v} - \frac{\mu_1}{u} = \frac{\mu_2 - \mu_1}{r}$$

or $\quad \frac{4}{3v} - \frac{3}{2(-30)} = \frac{\frac{4}{3} - \frac{3}{2}}{-20}$

or $\quad \frac{4}{3v} = \frac{1}{120} - \frac{1}{20} = \frac{1-6}{120}$

or $\quad \frac{3v}{4} = -\frac{120}{5}$

$\therefore \quad v = -\frac{24 \times 4}{3} = -32$ cm

$\because$ Transverse magnification,

$$m = \frac{\mu_1 v}{\mu_2 u} = \frac{\frac{3}{2} \times (-32)}{\frac{4}{3} \times (-30)}$$

$$= \frac{9}{8} \times \frac{32}{30} = \frac{6}{5} = 1.2$$

Since, m is positive and object is real. So, image is virtual

$$m = \frac{A'B'}{AB}$$

$\because \quad A'B' = mAB$

$= 1.2 \times 2 = 2.4$ cm

Hence, image is virtual and its size is 2.4 cm.

23. (b,c) For refraction through curved surface,

$$\mu_1 = 1, \mu_2 = \frac{3}{2}, u = -10 \text{ cm}, r = +10 \text{ cm}$$

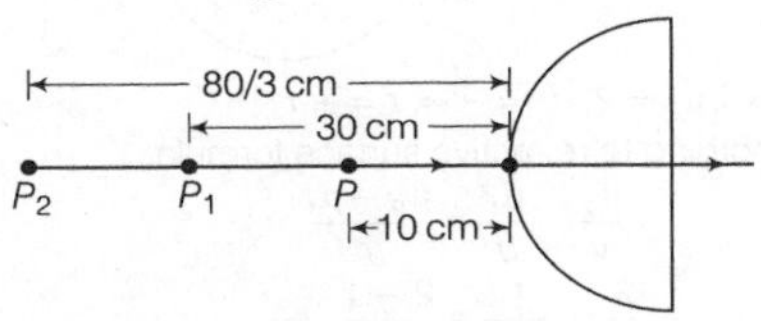

According to refractive surface formula,

$$\because \quad \frac{\mu_2}{v} - \frac{\mu_1}{u} = \frac{\mu_2 - \mu_1}{r} \quad \text{or} \quad \frac{3}{2v} - \frac{1}{-10} = \frac{\frac{3}{2} - 1}{10}$$

$$\text{or} \quad \frac{3}{2v} = \frac{1}{20} - \frac{1}{10} = -\frac{1}{20}$$

$$\text{or} \quad \frac{2v}{3} = -20$$

$$\therefore \quad v = -\frac{20 \times 3}{2} = -30 \text{ cm}$$

The image P_1, formed by refraction through curved surface behaves as object for plane surface.

For plane surface,

$$\mu_1' = \frac{3}{2}, \mu_2' = 1, u' = -40 \text{ cm}, r' = \infty$$

$$\therefore \quad \frac{\mu_2'}{v'} - \frac{\mu_1'}{u'} = \frac{\mu_2' - \mu_1'}{r'}$$

$$\frac{1}{v'} - \frac{3}{2(-40)} = \frac{1 - \frac{3}{2}}{\infty} = 0 \text{ or } \frac{1}{v'} = -\frac{3}{80}$$

$$\therefore \quad v' = -\frac{80}{3} \text{ cm}$$

DPP-8 Image Formation due to Refraction through Spherical Surface-II

1. (a,d) (a) For observer O_1,

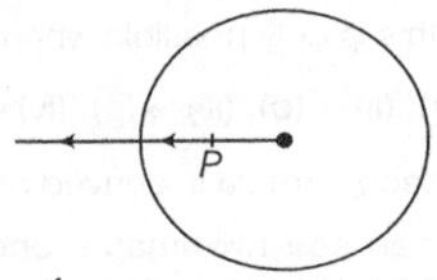

Here, $\mu_1 = \frac{3}{2}, \mu_2 = \frac{4}{3}, u = -5 \text{ cm}, r = -10 \text{ cm}$

According to refractive surface formula,

$$\frac{\mu_2}{v} - \frac{\mu_1}{u} = \frac{\mu_2 - \mu_1}{r}$$

$$\text{or} \quad \frac{4}{3v} - \frac{3}{2(-5)} = \frac{\frac{4}{3} - \frac{3}{2}}{-10}$$

$$\text{or} \quad \frac{4}{3v} = \frac{1}{60} - \frac{3}{10} = \frac{1 - 18}{60} = -\frac{17}{18}$$

$$\text{or} \quad \frac{3v}{4} = -\frac{18}{17}$$

$$\therefore \quad v = -\frac{18}{17} \times \frac{4}{3} = -\frac{24}{17}$$

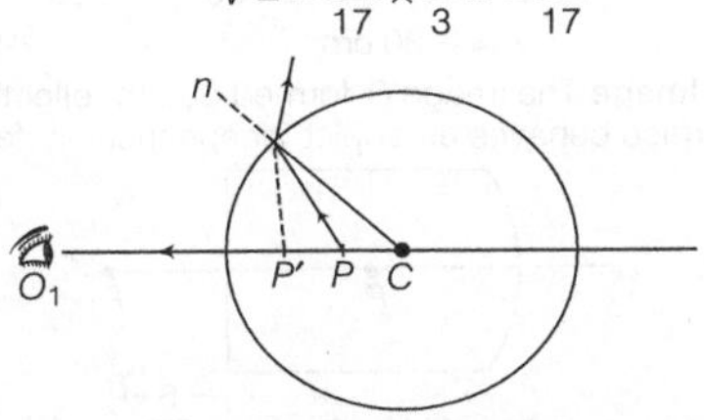

From ray diagram, the position of image from observer O_1 is

$$O_1P' = \frac{24}{17} + \frac{180}{17} - 10 = 2 \text{ cm}$$

(d)

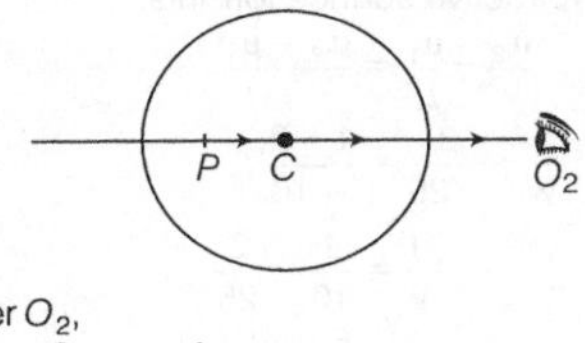

For observer O_2,

$$\mu_1' = \frac{3}{2}, \mu_2' = \frac{4}{3}, u' = -15 \text{ cm}, r' = -10 \text{ cm}$$

$$\because \quad \frac{\mu_2'}{v'} - \frac{\mu_1'}{u'} = \frac{\mu_2' - \mu_1'}{r'}$$

$$\text{or} \quad \frac{4}{3v'} - \frac{3}{2(-15)} = \frac{\frac{4}{3} - 3/2}{-10}$$

$$\text{or} \quad \frac{4}{3v'} = \frac{1}{60} - \frac{1}{10} = -\frac{1}{12}$$

$$\text{or} \quad \frac{3v'}{4} = -12$$

$$\therefore \quad v' = -\frac{12 \times 4}{3} = -16 \text{ cm}$$

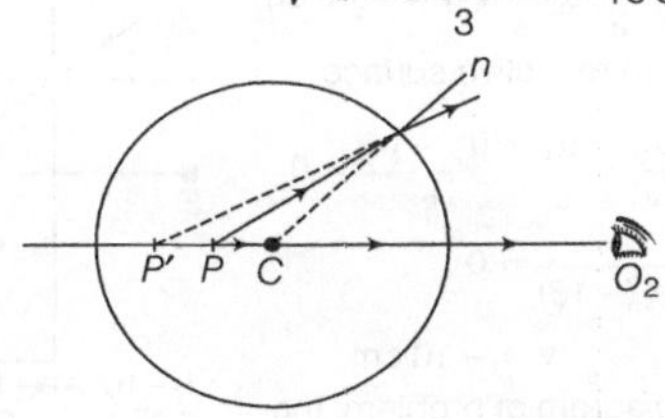

From ray diagram, the distance of image P'' from observer O_2 is

$$O_2P'' = O_2C + CP''$$
$$= (20 + 6) \text{ cm} = 26 \text{ cm.}$$

2. (a,c,d) For refraction through first surface (air-glass interface)

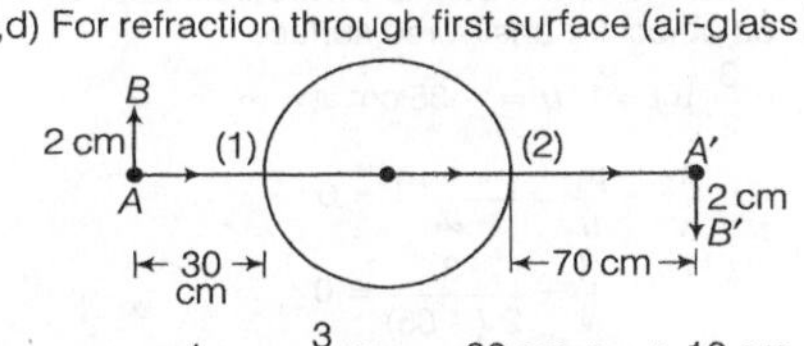

$$\mu_1 = 1, \mu_2 = \frac{3}{2}, u = -30 \text{ cm}, r = +10 \text{ cm}$$

According to refractive surface formula,

$$\frac{\mu_2}{v} - \frac{\mu_1}{u} = \frac{\mu_2 - \mu_1}{r}$$

$$\text{or} \quad \frac{3}{2v} - \frac{1}{-30} = \frac{\frac{3}{2} - 1}{10}$$

$$\text{or} \quad \frac{3}{2v} = \frac{1}{20} - \frac{1}{30} = \frac{3 - 2}{60}$$

$$\text{or} \quad \frac{2v}{3} = 60$$

$$\therefore \quad v = 90 \text{ cm}$$

$$\because \quad \frac{A'B'}{AB} = \frac{\mu_1 v}{\mu_2 u} = \frac{(1) \times 90}{\frac{3}{2} \times (-30)}$$

$$= -\frac{2 \times 90}{90} = -2 \text{ cm.}$$

Image $A'B'$ formed by refraction through first surface behaves as an object for second surface (glass-air interface).

For second surface,

$$\mu_1' = \frac{3}{2}, \mu_2' = 1, u' = +70 \text{ cm}$$

$$r' = -10 \text{ cm}$$

According to refractive surface formula,

$$\frac{\mu_2'}{v'} - \frac{\mu_1'}{u'} = \frac{\mu_2' - \mu_1'}{r'}$$

or $$\frac{1}{v'} - \frac{3}{2(70)} = \frac{1 - \frac{3}{2}}{-10}$$

or $$\frac{1}{v'} = \frac{1}{20} + \frac{3}{140} = \frac{7+3}{140}$$

$\because$ $$v' = 14 \text{ cm}$$

$\because$ $$\frac{A''B''}{A'B'} = \frac{\mu_1' v_1'}{\mu_2' u'} = \frac{3 \times 14}{2 \times 70} = 0.3$$

$\because$ $$A''B'' = 0.3 \times (-2) = -0.6 \text{ cm}$$

3. (a,b,c) From ray diagram, option (a), (b) and (c) are correct.

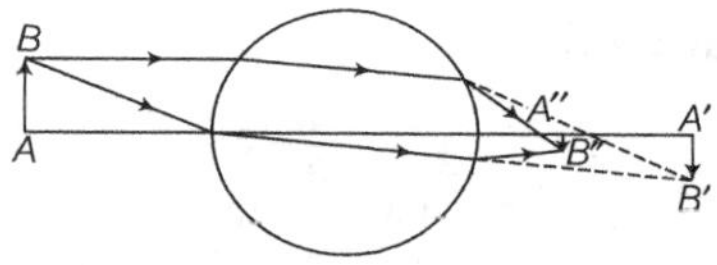

4. (a,c,d) Two images are seen by observer E.

First image is formed due to direct refraction through unsilvered surface.

Here, $\mu_1 = \frac{3}{2}, \mu_2 = 1, u = -15$ cm

According to refractive surface formula,

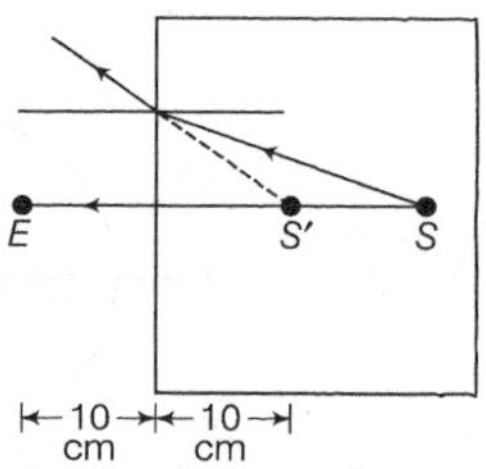

$\therefore$ $$\frac{\mu_2}{v} - \frac{\mu_1}{u} = \frac{\mu_2 - \mu_1}{\infty} = 0$$

or $$\frac{1}{v} - \frac{3}{2 \times (-15)} = 0$$

$\therefore$ $$v = -10 \text{ cm}$$

From ray diagram of problem, the position of image S' from observer E is $ES' = 20$ cm.

Second image The image $S_{\perp}$ formed due to reflection through silvered surface is 10 cm behind silvered surface. This image behaves as object for unsilvered surface,

Here, $\mu_1 = \frac{3}{2}, \mu_2 = 1, \; u = -35 \text{ cm}, r = \infty$

$\therefore$ $$\frac{\mu_2}{v} - \frac{\mu_1}{u} = \frac{\mu_2 - \mu_1}{\infty} = 0$$

or $$\frac{1}{v} - \frac{3}{2(-35)} = 0$$

$\therefore$ $$v = -\frac{70}{3} \text{ cm}$$

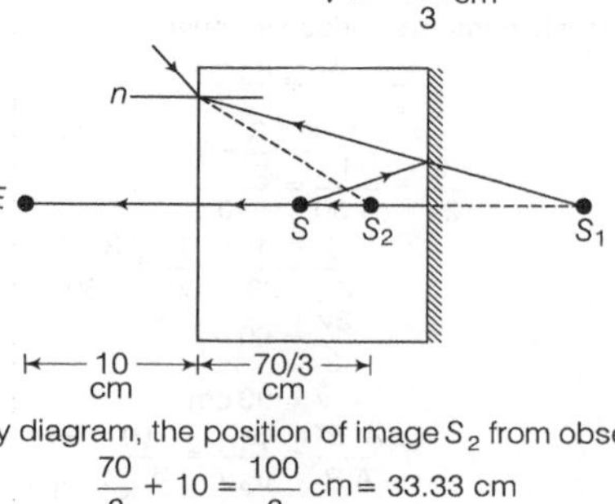

From ray diagram, the position of image S_2 from observer E is

$$\frac{70}{3} + 10 = \frac{100}{3} \text{ cm} = 33.33 \text{ cm}$$

5. (a,c)

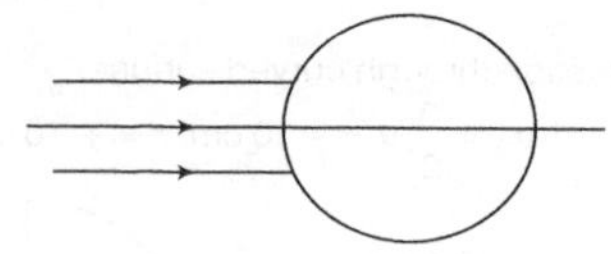

(a) $\mu_1 = 1, \mu_2 = 2, \; u = -\infty, r = +r$

According to refractive surface formula,

$\therefore$ $$\frac{\mu_2}{v} - \frac{\mu_1}{u} = \frac{\mu_2 - \mu_1}{r}$$

or $$\frac{2}{v} - \frac{1}{-\infty} = \frac{2-1}{R}$$

or $$\frac{2}{v} = \frac{1}{R} \text{ or } \frac{v}{2} = R$$

$\therefore v = 2R$, It means, beam converges at rear surface.

(c) $\mu_1 = 1, \mu_2 = m$

$$u = -\infty, r = +R, v = +R \text{ (for centre)}$$

According to refractive surface formula,

or $$\frac{\mu_2}{v} - \frac{\mu_1}{u} = \frac{\mu_2 - \mu_1}{r}$$

or $$\frac{\mu}{R} - \frac{1}{-\infty} = \frac{\mu - 1}{R}$$

or $$\frac{\mu}{R} = \frac{\mu - 1}{R}$$

or $\mu = \mu - 1$, this is only possible when, μ is so large.

6. (b) Hence, (i) $\rightarrow$(r), (ii)$\rightarrow$ (q), (iii)$\rightarrow$ (p), (iv) $\rightarrow$ (r)

7. (b) Refractive surface formula is derived only for paraxial rays.

8. (a) Observer E_1 can see two image, one image due to direct refraction through spherical surface and other image due to successive reflection through plane surface and then refraction through spherical surface.

9. (a) **First Image** By direct refraction through spherical surface,

$$\mu_1 = 2, \mu_2 = 1$$

$$u = -15 \text{ cm}, r = -10 \text{ cm}$$

According to refractive surface formula,

$\because$ $$\frac{\mu_2}{v} - \frac{\mu_1}{u} = \frac{\mu_2 - \mu_1}{r}$$

or $$\frac{1}{v} - \frac{2}{-15} = \frac{1-2}{-10} = \frac{1}{10}$$

or $$\frac{1}{v} = \frac{1}{10} - \frac{2}{15} = \frac{3-4}{30} = -\frac{1}{30}$$

$\therefore$ $$v = -30 \text{ cm}$$

Second Image The image P_1 formed due to reflection through plane surface behaves as object for spherical surface.

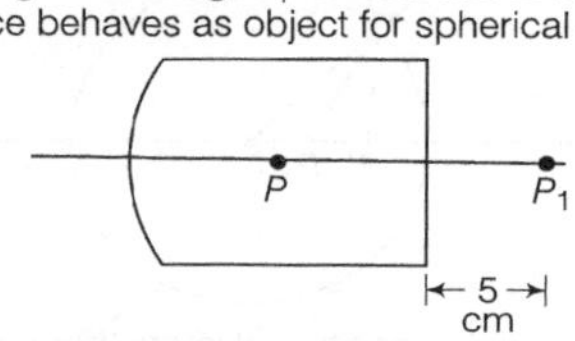

For spherical surface,

$$\mu_1 = 2, \mu_2 = 1, u = -25 \text{ cm}, \; r = -10 \text{ cm}$$

According to refractive surface formula,

$$\frac{\mu_2}{v} - \frac{\mu_1}{u} = \frac{\mu_2 - \mu_1}{r}$$

$$\frac{1}{v} - \frac{2}{-25} = \frac{1-2}{-10}$$

$$\frac{1}{v} = \frac{1}{10} - \frac{2}{25}$$

$$= \frac{5-4}{50} = \frac{1}{50}$$

$$v = 50 \text{ cm}$$

10. (b) **First Image** First image is due to direct refraction through plane surface.

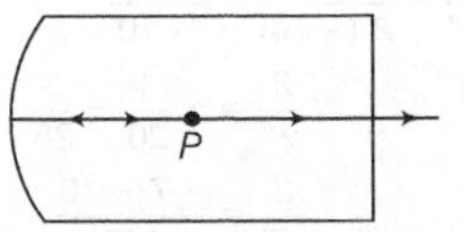

Here, $\mu_1 = 2, \mu_2 = 1, u = -5$ cm, $r = \infty$

According to refractive surface formula,

$$\because \quad \frac{\mu_2}{v} - \frac{\mu_1}{u} = \frac{\mu_2 - \mu_1}{\infty} = 0 \text{ or } \frac{1}{v} - \frac{2}{-5} = 0$$

$$\text{or} \quad \frac{1}{v} = -\frac{2}{5}$$

$$\therefore \quad v = -2.5 \text{ cm} = 2.5 \text{ cm left of } O_2$$

Second Image Image formed by concave mirror.

$$u = -15 \text{ cm}, f = -5 \text{ cm}$$

$$\frac{1}{v} + \frac{1}{u} = \frac{1}{f} \text{ or } \frac{1}{v} - \frac{1}{15} = \frac{1}{5}$$

$$\text{or} \quad \frac{1}{v} = -\frac{1}{5} + \frac{1}{15} = \frac{-3+1}{15} = -\frac{2}{15}$$

$$v = -7.5$$

This image behaves as object for plane surfaces.

Here, $\mu_1 = 2, \mu_2 = 1, u = -(20 - 7.5) = -12.5, r = \infty$

$$\because \quad \frac{\mu_2}{v} - \frac{\mu_1}{u} = \frac{\mu_2 - \mu_1}{r} = \frac{\mu_2 - \mu_1}{\infty} = 0$$

$$\text{or} \quad \frac{1}{v} - \frac{2}{-12.5} = 0$$

$$\therefore \quad v = -\frac{12.5}{2} = -6.25 \text{ cm} = 6.25 \text{ left of } O_2.$$

11. (6) The image formation, due to first refraction through unsilvered surface, $\mu_1 = 1, \mu_2 = 2, u = -20$ cm, $r = +10$ cm.

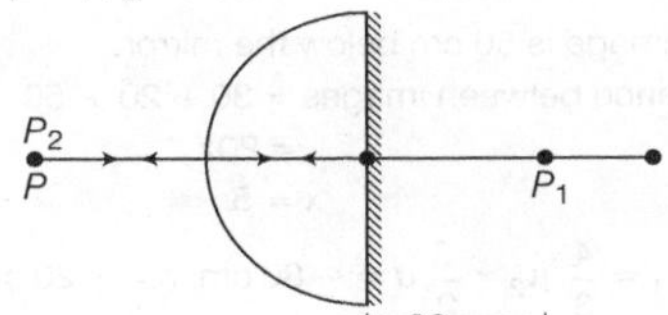

According to refractive surface formula,

$$\therefore \quad \frac{\mu_2}{v} - \frac{\mu_1}{u} = \frac{\mu_2 - \mu_1}{r}$$

$$\text{or} \quad \frac{2}{v} - \frac{1}{-20} = \frac{2-1}{10} \text{ or } \frac{2}{v} = \frac{1}{10} - \frac{1}{20} = \frac{1}{20}$$

$$\text{or} \quad \frac{v}{2} = 20$$

$$\therefore \quad v = 40 \text{ cm}$$

The image P_1 formed due to first refraction through unsilvered surface behaves as object for silvered surface behaves as plane mirror. The image P_2 formed by silvered surface behaves as object for second times refraction through unsilvered surface.

For second times refraction through unsilvered surface,

$\mu_1' = 2, \mu_2' = 1, u' = +30$ cm, $r' = -10$ cm

$$\frac{\mu_2'}{v'} - \frac{\mu_1'}{u'} = \frac{\mu_2' - \mu_1'}{r'} \text{ or } \frac{1}{v'} - \frac{2}{30} = \frac{1-2}{-10}$$

$$\text{or} \quad \frac{1}{v'} = \frac{1}{15} + \frac{1}{10} = \frac{2+3}{30} = \frac{5}{30}$$

$$\text{or} \quad v' = +6 \text{ cm}$$

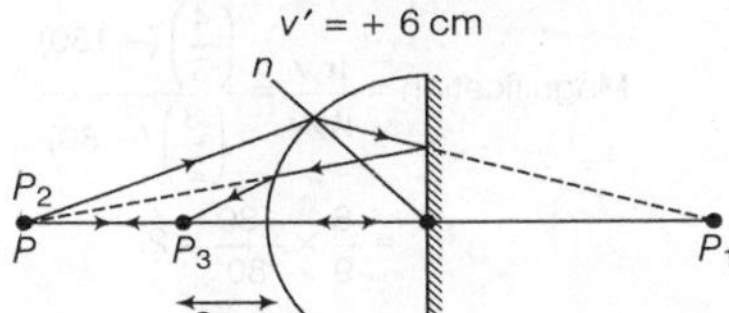

12. (3) For refraction through the first surface,

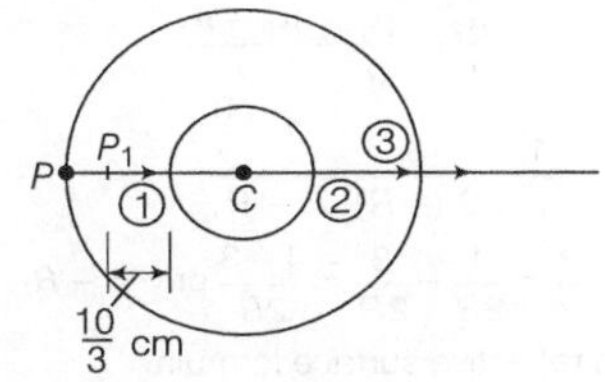

$\mu_1 = 2, \mu_2 = 1, u = -10$ cm, $r = +10$ cm

According to refractive surface formula,

$$\because \quad \frac{\mu_2}{v} - \frac{\mu_1}{u} = \frac{\mu_2 - \mu_1}{r}$$

$$\text{or} \quad \frac{1}{v} - \frac{2}{-10} = \frac{1-2}{10}$$

$$\text{or} \quad \frac{1}{v} = -\frac{1}{10} - \frac{1}{5}$$

$$= \frac{-1-2}{10} = -\frac{3}{10}$$

$$\because \quad v = -\frac{10}{3} \text{ cm}$$

The image P_1 formed by refraction through first surface behaves as object for second surface.

For second surface, $\mu_1' = 1, \mu_2' = 2$,

$$u' = -\left(20 + \frac{10}{3}\right) = -\left(\frac{70}{3}\right) \text{ cm} = \frac{-70}{3} \text{ cm}$$

$$r' = -10 \text{ cm}$$

$$\because \quad \frac{\mu_2'}{v'} - \frac{\mu_1'}{u'} = \frac{\mu_2' - \mu_1'}{r'}$$

$$\text{or} \quad \frac{2}{v'} - \frac{1}{-\frac{70}{3}} = \frac{2-1}{-10}$$

$$\text{or} \quad \frac{2}{v'} = -\frac{1}{10} - \frac{3}{70} = \frac{-7-3}{70}$$

$$\text{or} \quad \frac{v'}{2} = -7$$

$$\therefore \quad v' = -14 \text{ cm}$$

For refraction through third surface,

$$\mu_1' = 2, \mu_2'' = 1, u'' = -(14 + 10) = -24 \text{ cm}$$

$$r'' = -20 \text{ cm}$$

$$\because \quad \frac{\mu_2''}{v''} - \frac{\mu_1''}{u''} = \frac{\mu_2'' - \mu_1''}{r''}$$

$$\text{or} \quad \frac{1}{v''} - \frac{2}{-24} = \frac{1-2}{-20}$$

$$\text{or} \quad \frac{1}{v''} = \frac{1}{20} - \frac{2}{24} = \frac{1}{20} - \frac{20}{240}$$

$$\text{or} \quad v'' = -30 \text{ cm}$$

$$\because \quad v'' = -\frac{820}{39}$$

Thus, shifting in object = 40 − 30

$$10 \text{ cm} = \frac{10}{3}(n)$$

$$\therefore \quad n = 3.$$

13. (2) Here, $\mu_1 = \frac{3}{2}, \ \mu_2 = 1, r = -R, u = -R$

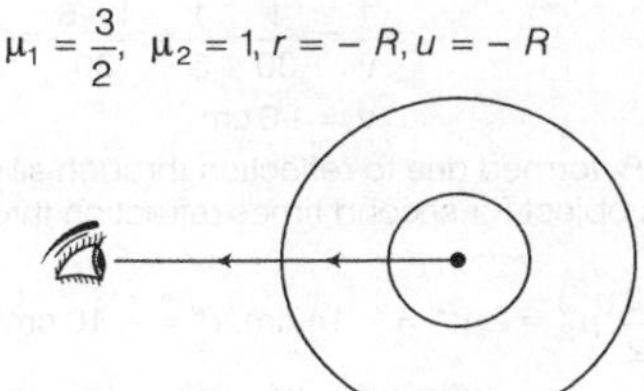

According to refractive surface formula,

$\because \quad \frac{\mu_2}{v} - \frac{\mu_1}{u} = \frac{\mu_2 - \mu_1}{r}$

or $\quad \frac{1}{v} - \frac{3}{2(-R)} = \frac{1 - \frac{3}{2}}{-R}$

or $\quad \frac{1}{v} = \frac{1}{2R} - \frac{3}{2R} = \frac{1-3}{2R}$ or $v = -R$

According to refractive surface formula,

$\because \quad \frac{\mu_2}{v} - \frac{\mu_1}{u} = \frac{\mu_2 - \mu_1}{r}$

or $\quad -\frac{\mu_2}{v^2}\frac{dv}{du} + \frac{\mu_1}{u^2} = 0$

or $\quad \frac{dv}{du} = \left(\frac{\mu_1}{\mu_2}\right)\left(\frac{v^2}{u^2}\right)$

or $\quad \frac{\Delta v}{\Delta u} = \left(\frac{3}{2}\right)\left(\frac{R^2}{R^2}\right)$

$\therefore \quad \Delta u = \frac{2}{3}\Delta v = \frac{2}{3} \times 3 \text{ mm} = 2 \text{ mm}.$

14. (5) Image formation due to first refraction through unsilvered surface,

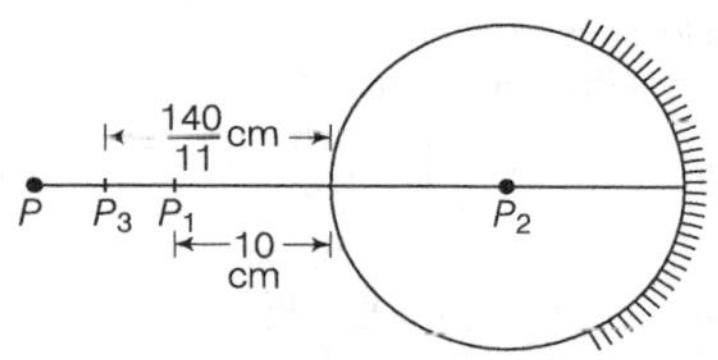

$\mu_1 = 2, \mu_2 = \frac{3}{2}, u = -20$ cm and $r = +10$ cm

According to refractive surface formula,

$\frac{\mu_2}{v} - \frac{\mu_1}{u} = \frac{\mu_2 - \mu_1}{r}$

or $\quad \frac{3}{2v} - \frac{2}{-20} = \frac{\frac{3}{2} - 2}{+10} = -\frac{1}{20}$

or $\quad \frac{3}{2v} = -\frac{1}{20} - \frac{1}{10} = \frac{-1-2}{20} = -\frac{3}{20}$

or $\quad \frac{2v}{3} = -\frac{20}{3}$

$\therefore \quad v = -10$ cm

The image P_1 formed due to refraction through unsilvered surface behaves as object for silvered surface. The silvered surface behaves as concave mirror.

For concave mirror,

$u' = -30$ cm

$f' = \frac{-10}{2} = -5$ cm

According to mirror formula,

$\frac{1}{v} + \frac{1}{u} = \frac{1}{f}$

or $\quad \frac{1}{v} - \frac{1}{30} = -\frac{1}{5}$

or $\quad \frac{1}{v} = \frac{1}{30} - \frac{1}{5} = \frac{1-6}{30} = \frac{-5}{30}$

$\therefore \quad v = -6$ cm

The image P_2 formed due to reflection through silivered surface behaves as object for second times refraction through unsilvered surface.

Here, $\mu_1'' = \frac{3}{2}, \mu_2'' = 2, \mu'' = -14$ cm, $r'' = -10$ cm

$\frac{\mu_2''}{v''} - \frac{\mu_1''}{u''} = \frac{\mu_2'' - \mu_1''}{r''}$

or $\quad \frac{2}{v''} - \frac{3}{2(-14)} = \frac{2 - \frac{3}{2}}{-10}$

or $\quad \frac{2}{v''} = -\frac{1}{20} - \frac{3}{28}$

or $\quad \frac{2}{v''} = \frac{-7-15}{140}$

$\therefore \quad v'' = -\frac{140}{22} \times 2 = -\frac{140}{11}$ cm

Thus the position of final image P_3 from centre of sphere

$= 10 + \frac{140}{11}$

$= \frac{110 + 140}{11} = \frac{250}{11}$ cm

$= \frac{50\,n}{11}$ cm

$\because \quad n = 5$

15. (5) **First image** First image is formed due to direct refraction through unsilvered surface.

For this, $\quad \mu_1 = 1, \mu_2 = 2, \mu = -15$ cm

$r = \infty$

$\because \quad \frac{\mu_2}{v} - \frac{\mu_1}{u} = \frac{\mu_2 - \mu_1}{\infty} = 0$

or $\quad \frac{2}{v} - \frac{1}{-15} = 0$

or $\quad \frac{v}{2} = -15$

$\therefore \quad v = -30$ cm

Second image The image formed by refractive surface behaves as object for plane mirror and final image is formed by plane mirror.

$\because$ Second image is 50 cm below the mirror.

Thus, distance between images = 30 + 20 + 50 = 100 cm

$= 20x$

$\because \quad x = 5.$

16. (2) Here, $\mu_1 = \frac{4}{3}, \mu_2 = \frac{3}{2}, u = -80$ cm, $r = +20$ cm

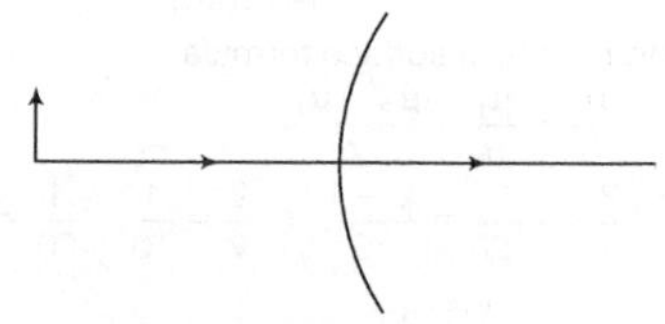

According to refractive surface formula,

$\frac{\mu_2}{v} - \frac{\mu_1}{u} = \frac{\mu_2 - \mu_1}{r}$

or $\quad \frac{3}{2v} - \frac{4}{3(-80)} = \frac{\frac{3}{2} - \frac{4}{3}}{+20}$

or $\quad \frac{3}{2v} = \frac{1}{120} - \frac{1}{60} = -\frac{1}{120}$

or $\quad \frac{2v}{3} = -120$

$\therefore \quad v = -120 \times \frac{3}{2} = -180$ cm

$\because \quad \text{Magnification} = \frac{\mu_1 v}{\mu_2 u} = \frac{\left(\frac{4}{3}\right)(-180)}{\left(\frac{3}{2}\right)(-80)}$

$= \frac{8}{9} \times \frac{180}{80} = 2.$

17. (4) Here, $\mu_1 = 1, \mu_2 = 3/2, u = -15, r = +10$ cm

According to refractive surface formula,

$$\frac{\mu_2}{v} - \frac{\mu_1}{u} = \frac{\mu_2 - \mu_1}{r}$$

or $$\frac{3}{2v} - \frac{1}{u} = \frac{1}{2 \times 10} = \frac{1}{20}$$

or $$\frac{3}{2v} = \frac{1}{20} - \frac{1}{15} = \frac{3-4}{60} = -\frac{1}{60}$$

or $$\frac{3}{2v} = -\frac{1}{60} \text{ or } \frac{2v}{3} = -60$$

or $$v = -\frac{180}{2} = -90 \text{ cm}$$

$\because$ $$\frac{\mu_2}{v} - \frac{\mu_1}{u} = \frac{\mu_2 - \mu_1}{r},$$

$$\frac{-\mu_2}{v^2}\frac{dv}{dt} + \frac{\mu_1}{u^2}\frac{du}{dt} = 0$$

$$\frac{dv}{dt} = \frac{\mu_1}{\mu_2}\frac{v^2}{u^2}\frac{du}{dt} = \frac{2}{3}\frac{90^2}{15^2}\frac{du}{dt} = \frac{2}{3} \times 36\frac{du}{dt}$$

$$= 24\frac{du}{dt} = 24(-2) = -48 \text{ cms}^{-1}$$

$$\left\{\because \frac{du}{dt} = v_{\text{object}} - v_{\text{surface}} = 2 - 4 = -2 \text{ cms}^{-1}\right\}$$

$\therefore$ $$\frac{dv}{dt} = v_{\text{image surface}} = v_{\text{image}} - v_{\text{surface}}$$

$\therefore$ $$v_{\text{image}} = \frac{dv}{dt} + v_{\text{mirror}} = -48 + 4 = -44 \text{ cms}^{-1}$$

$\therefore$ $$11x_0 = 44, \quad x_0 = 4$$

18. (2) Here, $\mu_1 = \frac{4}{3}, \mu_2 = 1, u = +\infty, r = +30$ cm

According to refractive surface formula,

$$\frac{\mu_2}{v} - \frac{\mu_1}{u} = \frac{\mu_2 - \mu_1}{r}$$

or $$\frac{1}{v} - 0 = \frac{1 - \frac{4}{3}}{30} \text{ or } \frac{1}{v} = -\frac{1}{90}$$

$\therefore$ $$v = -90 \text{ cm}$$

P = The power refracted $I_0\pi R^2$

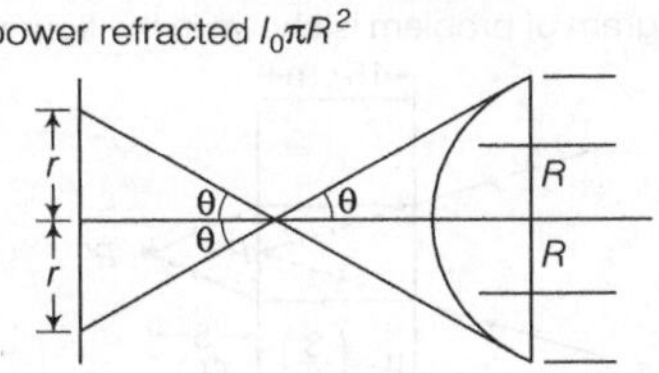

$\therefore$ Intensity on the surface of screen $= \frac{P}{\pi r^2} = \frac{I_0\pi R^2}{\pi r^2}$

Here $$\tan\theta = \frac{R}{90} = \frac{r}{9}$$

$\therefore$ $$R = r$$

$\therefore$ $$I = \frac{I_0\pi R^2}{\pi r^2}$$

$$I_0 = 10 \text{ Wm}^{-2} = 5n$$

$\therefore$ $$n = 2.$$

19. (2) Two refractions will take place, first from spherical surface and the other from the plane surface.

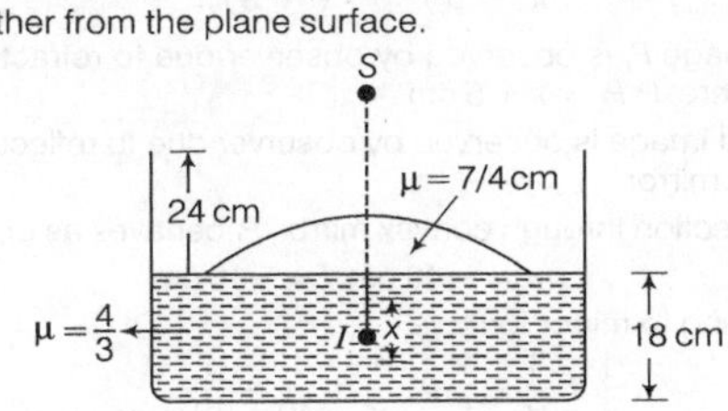

So, applying formula $\frac{\mu_2}{v} - \frac{\mu_1}{u} = \frac{\mu_2 - \mu_1}{R}$ two times with proper sign convention.

Ray of light is travelling downwards. Therefore, downward direction is taken as positive direction.

$$\frac{7/4}{v} - \frac{1.0}{-24} = \frac{7/4 - 1.0}{+6} \quad \ldots\text{(i)}$$

$$\frac{4/3}{(18-x)} - \frac{7/4}{v} = \frac{4/3 - 7/4}{\infty} \quad \ldots\text{(ii)}$$

Solving these equations, we get $x = 2$ cm.

20. $\left(\frac{4}{3}\right)$ Applying $\frac{\mu_2}{v} - \frac{\mu_1}{u} = \frac{\mu_2 - \mu_1}{R}$

First on plane surface

$$\frac{1.5}{AI_1} - \frac{1}{-mR} = \frac{1.5 - 1}{\infty} = 9$$

$\therefore$ $$AI_1 = -(1.5\, mR)$$

Then on curved surface

$$\frac{1}{\infty} - \frac{1.5}{-(1.5\, mR + R)} = \frac{1 - 1.5}{-R}$$

($v = \infty$ because final image is at infinity)

$\Rightarrow$ $$\frac{1.5}{(1.5\, m + 1)R} = \frac{0.5}{R}$$

$\Rightarrow$ $$3 = 1.5\, m + 1$$

$\Rightarrow$ $$\frac{3}{2}m = 2$$

$\therefore$ $$m = \frac{4}{3}$$

21. For refraction at first surface,

$$\frac{\mu_2}{v_1} = \frac{\mu_1}{-\infty} = \frac{\mu_2 - \mu_1}{+R} \quad \ldots\text{(i)}$$

For refraction at second surface

$$\frac{\mu_3}{v_2} - \frac{\mu_2}{v_1} = \frac{\mu_3 - \mu_2}{+R} \quad \ldots\text{(ii)}$$

Adding Eqs. (i) and (ii), we get

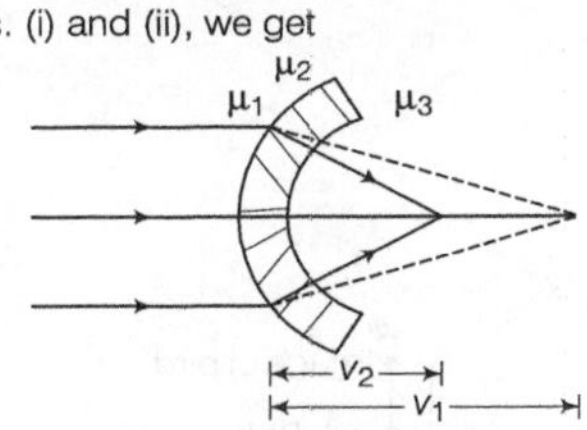

$$\frac{\mu_3}{v_2} = \frac{\mu_3 - \mu_1}{R}$$

or $$v_2 = \frac{\mu_3 R}{\mu_3 - \mu_1}$$

Therefore, focal length of the given lens system is

$$\frac{\mu_3 R}{\mu_3 - \mu_1}$$

22. Applying Snell's law on face AB,

$$(1)\sin 45^\circ = (\sqrt{2})\sin r$$

$\therefore$ $$\sin r = \frac{1}{2}$$

or $$r = 30^\circ$$

i.e. ray becomes parallel to AD inside the block.

Now, according to refractive surface formula.

$$\frac{\mu_2}{v} - \frac{\mu_1}{u} = \frac{\mu_2 - \mu_1}{R} \text{ on face } CD$$

$$\frac{1.514}{OE} - \frac{\sqrt{2}}{\infty} = \frac{1.514 - \sqrt{2}}{0.4}$$

Solving this equation, we get $OE = 6.06$ m.

DPP-9 Image Formation due to Refraction through Plane Surface-III

1. (i) (a,c) $\frac{\mu_2}{\mu_1} = \frac{v}{u}$

$$\frac{1}{4/3} = \frac{v}{-30}$$

$\therefore$ $v = \frac{3}{4}(-30) = -22.5$ cm

(ii) (a,c) Here, $\frac{\mu_2}{\mu_1} = \frac{v}{u}$ or $\frac{4}{3} = \frac{v}{-20}$

$\therefore$ $v = -80/3$ cm

(iii) (d) $\because$ $\frac{\mu_2}{\mu_1} = \frac{v}{u}$ or $\frac{1}{\mu} = \frac{v}{u}$

$\therefore$ $v = \frac{u}{\mu}$

$\therefore$ $\frac{dv}{dt} = \frac{1}{\mu}\frac{du}{dt} = \frac{3}{4} \times (-4) = -3\,\text{cms}^{-1}$

(iv) (b) $\because$ $\frac{\mu_2}{\mu_1} = \frac{v}{u}$ or $\frac{4}{3} = \frac{v}{u}$

$\therefore$ $v = \frac{4}{3}u$

$\therefore$ $\frac{dv}{dt} = \frac{4}{3}\frac{du}{dt} = \frac{4}{3} \times (-6) = -8$ cm

(v) (c) $\because$ $\frac{dv}{dt} = \frac{4}{3}\frac{du}{dt}$

or $v_{image} - (-1) = \frac{4}{3}(-6+1)$

$\therefore$ $v_{image} = \frac{-20}{3} - 1 = -\frac{23}{3}\,\text{cms}^{-1}$

(vi) (b) $\therefore$ $v_{image} - 4 = \frac{4}{3}(-6-4) = \frac{-40}{3} + 4 = \frac{-28}{3}\,\text{cms}^{-1}$

2. (i) (b)
(ii) (c)
(iii) (b)
(iv) (b)
(v) (b)
(vi) (b)

3. (c) Here, $\mu = \frac{y'}{y}$

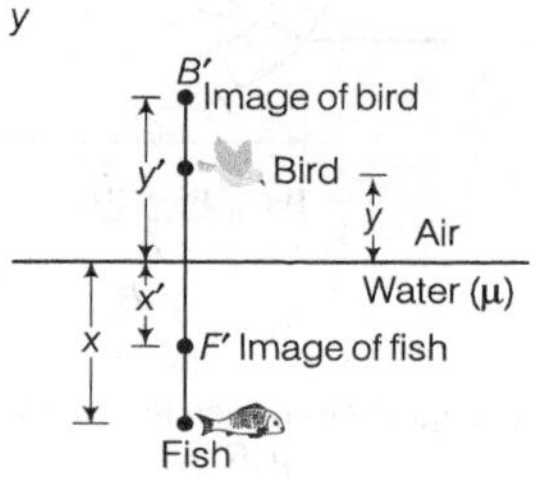

$\therefore$ $y' = \mu y$

and $\frac{1}{\mu} = \frac{x'}{x}$

$\therefore$ $x' = \frac{x}{\mu}$

According to the problem,

$\because$ $x' + y = 40$

$\frac{x}{\mu} + y = 40$ ...(i)

and $\because$ $y' + x = 50$

or $\mu y + x = 50$ ...(ii)

From Eqs. (i) and (ii), we get

$$\mu = \frac{50}{40} = \frac{5}{4} = 1.25$$

4. (d) dy' = apparent depth (let)

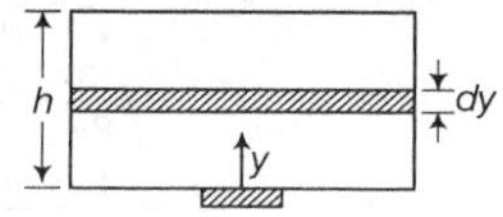

dy = real depth (let)

$dy' = \frac{dy}{\mu}$

$\therefore$ $\int_0^y dy' = \int_0^h \frac{dy}{\mu_0(1+y)}$

$$y = \left[\frac{1}{\mu_0}(\ln(1+y))\right]_0^h$$

$$y = \frac{1}{\mu_0}\ln(1+h)$$

5. (b) $x = t_1\left(1 - \frac{1}{\mu_1}\right) + t_2\left(1 - \frac{t}{\mu_2}\right) + \cdots$

or $x = 1\left(1 - \frac{2}{3}\right) + 2\left(1 - \frac{2}{3}\right)$

$= \frac{1}{3} + \frac{2}{3} = 1$ cm

$\Rightarrow$ No. of slabs $n = 2$

6. (c) For fish, $\mu_1 = \frac{4}{3}$, $\mu_2 = 1$

$\therefore$ $\frac{\mu_2}{v} - \frac{\mu_1}{u} = 0$ or $\frac{1}{v} = \frac{4}{3u}$

$\therefore$ $v = \frac{3u}{4}$ or $\frac{dv}{dt} = \frac{3}{4}\frac{du}{dt}$

or $\frac{dv}{dt} = \frac{3}{4} \times 4 = 3\,\text{cms}^{-1}$ towards the insect.

$\because$ $v_{ref} = v + v_o = 3 + 2t$ or $9 = 3 + 2t$

$\therefore$ $t = 3$ s.

7. (a) The object shiftness due to slab

$$= t\left(1 - \frac{1}{\mu}\right) = 15\left(1 - \frac{2}{3}\right) = 5\text{ cm.}$$

The ray diagram of problem is shown in the figure.

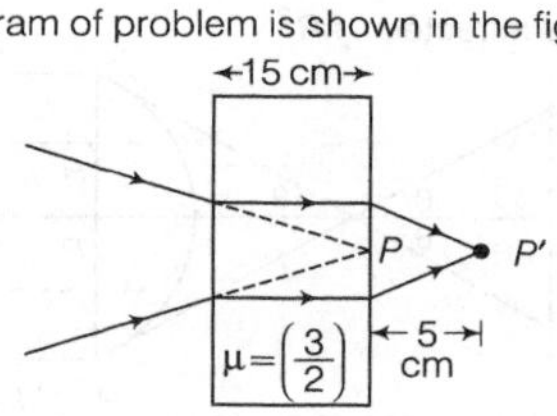

8. (b) The object shiftness due to slab is

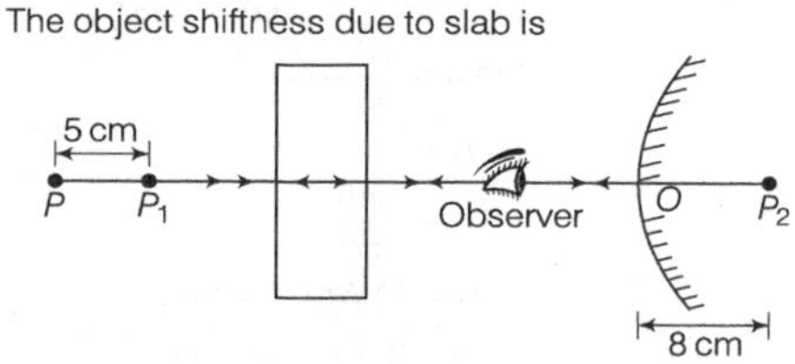

$$x = t\left(1 - \frac{1}{\mu}\right) = 15\left(1 - \frac{2}{3}\right) = 5\text{ cm}$$

First image P_1 is observed by observer due to refraction through slab. Here, $PP_1 = x = 5$ cm.

Second image is observed by observer due to reflection through convex mirror.

For reflection through convex mirror, P_1 behaves as object.

$\therefore$ $u = -40$ cm, $f = +10$ cm

According to mirror formula,

$\because$ $\frac{1}{v} + \frac{1}{u} = \frac{1}{f}$ or $\frac{1}{v} - \frac{1}{40} = \frac{1}{10}$

or $\qquad \frac{1}{v} = \frac{1}{40} + \frac{1}{10} = \frac{1+4}{40}$

$\therefore \qquad v = 8 \text{ cm}$

Thus, the distance between images is

$$P_1 P_2 = 40 + 8 = 48 \text{ cm.}$$

9. (b) Object shiftness $= x = 18\left(1 - \frac{4}{3 \times 3/2}\right)$

$$= 18\left(1 - \frac{8}{9}\right) = 18 \times \frac{1}{9}$$

$$= 2 \text{ cm rightward}$$

10. (a) Object shiftness $= t\left(1 - \frac{2}{\frac{3}{2}}\right) = t\left(1 - \frac{4}{3}\right)$

$$= 9 \times \left(\frac{-1}{3}\right) \text{cm} = -3 \text{ cm}$$

11. (b) The object shiftness due to portion of slab in air

$$= t\left(1 - \frac{1}{\mu}\right) = 18\left(1 - \frac{1}{\frac{3}{2}}\right)$$

$$= 18 \times \frac{1}{3} = 6 \text{ cm}$$

The object shiftness due to portion of slab in water.

$$= 18\left(1 - \frac{4}{3 \times \frac{3}{2}}\right)$$

$$= 18\left(1 - \frac{8}{9}\right)$$

$$= 18 \times \frac{1}{9} = 2 \text{ cm}$$

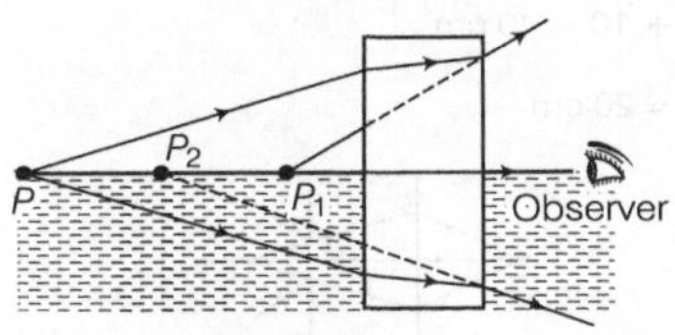

For ray diagram, the distance between images P_1 and P_2 is $P_1 P_2$

$$= PP_1 - PP_2$$

$$= 6 - 2 = 4 \text{ cm.}$$

12. (c) $\mu_{eq} = \frac{\Sigma t_i}{\Sigma \frac{t_i}{\mu_i}} = \frac{15 + 8 + 10}{\frac{15}{\frac{3}{2}} + \frac{8}{\frac{4}{3}} + \frac{10}{2}}$

$$= \frac{33}{10 + 6 + 5} = \frac{33}{21} = 1.57.$$

13. (d) The image formation due to refraction through first surface.

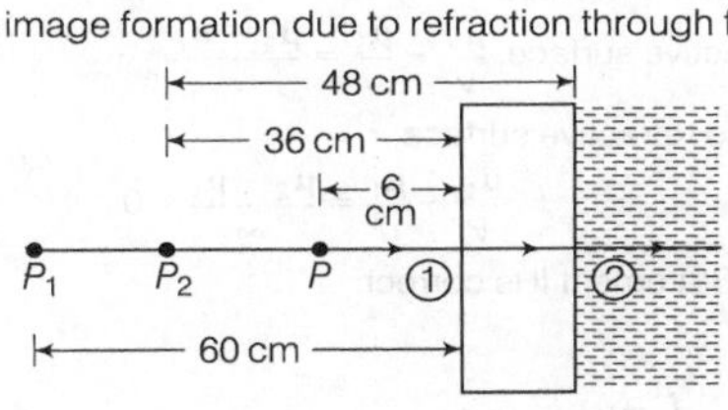

$$\mu_1 = 1, \mu_2 = 2, u = -30 \text{ cm}$$

$\because \qquad \frac{\mu_2}{\mu_1} = \frac{v}{u} \quad \text{or} \quad \frac{2}{1} = \frac{v}{-30}$

$\therefore \qquad v = -60 \text{ cm}$

The image P_1 formed due to refraction through first surface behaves as object for refraction through second surface.

For second surface, $\quad u_1' = 2, u_2' = \frac{4}{3}, u' = -72 \text{ cm}$

$\because \qquad \frac{u_2'}{u_1'} = \frac{v'}{u'}$

$\therefore \qquad v' = \frac{\mu_2'}{\mu_1'} u' = \frac{4}{3 \times 2}(-72)$

$\therefore \qquad v' = -48 \text{ cm}$

Thus, object shiftiness $= PP_2 = 36 - 30 = 6 \text{ cm.}$

14. (b) $\mu \sin \alpha = 1 \times \sin \beta$

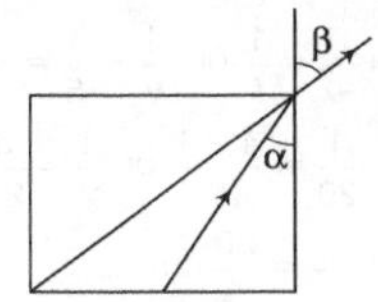

or $\qquad \mu \frac{5}{\sqrt{5^2 + (8.45)^2}} = \frac{10}{\sqrt{10^2 + (8.45)^2}}$

$\therefore \qquad \mu = 1.5$

15. (a) Object shiftness $= t\left(1 - \frac{1}{\mu}\right) = 15\left(1 - \frac{1}{3/2}\right) = 5 \text{ cm}$

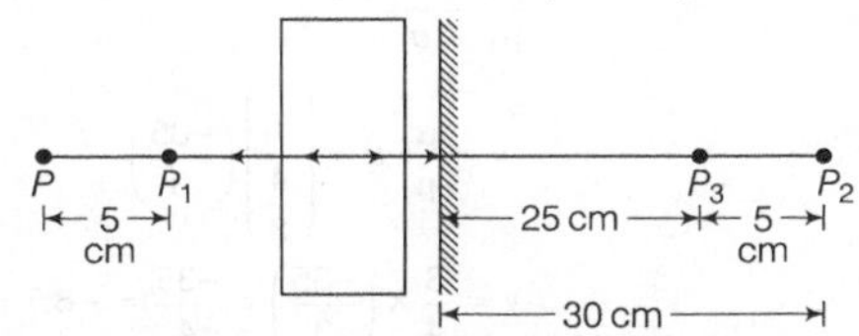

The image formation due to first refraction through slab is P_1. The image P_1 behaves as object for plane mirror. The image P_1 formed by plane mirror is P_2. The image P_2 behaves as object for second times refraction through the slab. The image due to second times refraction through slab is P_3.

16. (b)

17. (c)

18. (a,b,c,d) Here, $\mu_1 = 1, \mu_2 = \frac{4}{3}, u = -30 \text{ cm}, r = \infty$

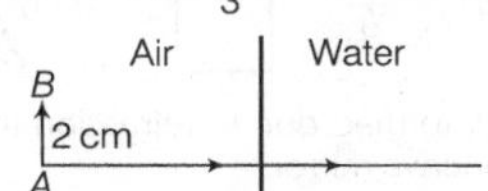

According to refractive surface formula,

$\because \qquad \frac{\mu_2}{v} - \frac{\mu_1}{u} = \frac{\mu_2 - \mu_1}{\infty} = 0$

or $\qquad \frac{4}{3v} - \frac{1}{-30} = 0$

or $\qquad \frac{4}{3v} = -\frac{1}{30}$

or $\qquad \frac{3v}{4} = -30$

$\therefore \qquad v = \frac{-30 \times 4}{3} = -40 \text{ cm}$

The ray diagram of problem is shown in the figure.

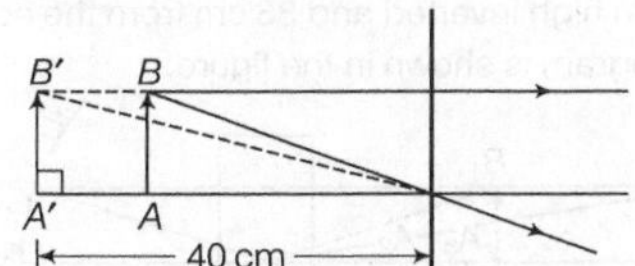

From figure, it is clear that image is virtual and of same size. Thus, options (a), (b), (c) and (d) are correct.

19. (a,c) Image formation due to reflection from concave mirror.
Here, $u = -5$ cm, $f = -20$ cm

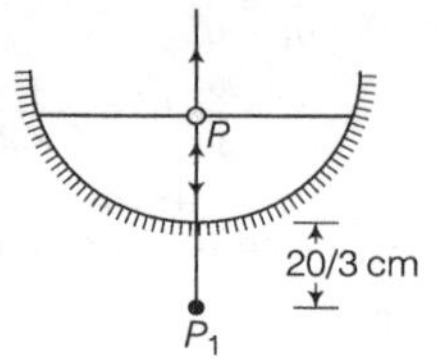

$\because \quad \frac{1}{v} + \frac{1}{u} = \frac{1}{f}$ or $\frac{1}{v} - \frac{1}{5} = -\frac{1}{20}$

or $\quad \frac{1}{v} = \frac{1}{5} - \frac{1}{20} = \frac{4-1}{20}$ or $\frac{1}{v} = \frac{3}{20}$

$\therefore \quad v = \frac{20}{3}$ cm

The image P_1 formed due to reflection through concave mirror behaves as object for refraction through water surface. For water surface,

$$\mu_1 = \frac{4}{3}, \mu_2 = 1, u = -\left(5 + \frac{20}{3}\right) = -\frac{35}{3} \text{ cm}$$

$\because \quad \frac{\mu_2}{\mu_1} = \frac{v}{u}$

$\therefore \quad v = \frac{\mu_2}{\mu_1} u = \left(\frac{1}{\frac{4}{3}}\right)\left(\frac{-35}{3}\right)$

$\therefore \quad v = \frac{3}{4} \times \left(\frac{-35}{3}\right) = \frac{-35}{4} = -8.75$ cm.

20. (a,c) The object shiftness due to slab is

$$x = t\left(1 - \frac{1}{\mu}\right) = 12\left(1 - \frac{3}{4}\right) = 3 \text{ cm}$$

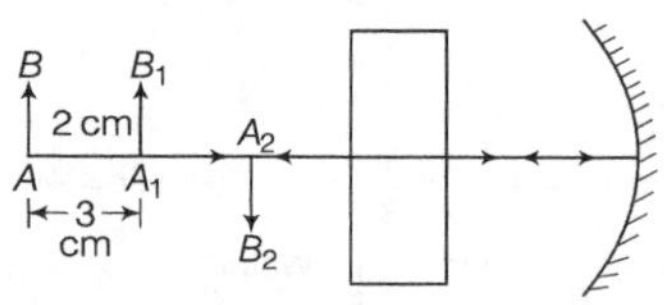

The image $A_1 B_1$ formed due to refraction through slab behaves as object for concave mirror.
For concave mirror,

$u = -60$ cm, $f = -20$ cm

According to mirror formula,

$\because \quad \frac{1}{v} + \frac{1}{u} = \frac{1}{f}$ or $\frac{1}{v} - \frac{1}{60} = -\frac{1}{20}$

or $\quad \frac{1}{v} = \frac{1}{60} - \frac{1}{20} = \frac{1-3}{60}$

$\therefore \quad v = -30$ cm

$$m = \frac{A_2B_2}{A_1B_1} = -\frac{v}{u} = -\frac{(-30)}{-60}$$

$\therefore \quad A_2B_2 = -1$ cm

$A_2 B_2$ behaves as virtual object for second times refraction through slab. The final image A_3B_3 is 3 cm leftward from A_2B_2 due to second times refraction through slab. Thus, final image is virtual, 1 cm high inverted and 33 cm from the concave mirror.
The ray diagram is shown in the figure.

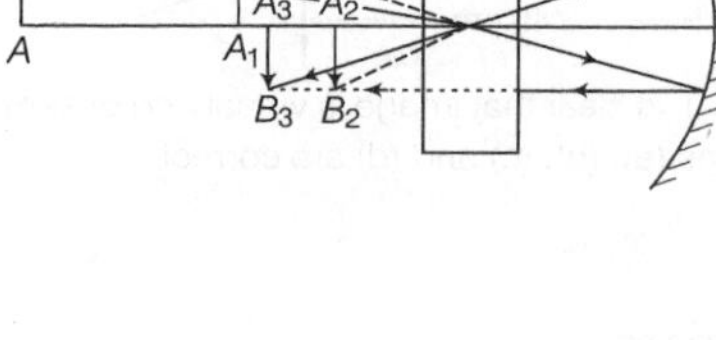

21. (a,c) Here, $u = -y$, $v = -y'$

$$\mu_1 = 1, \mu_2 = \frac{4}{3}$$

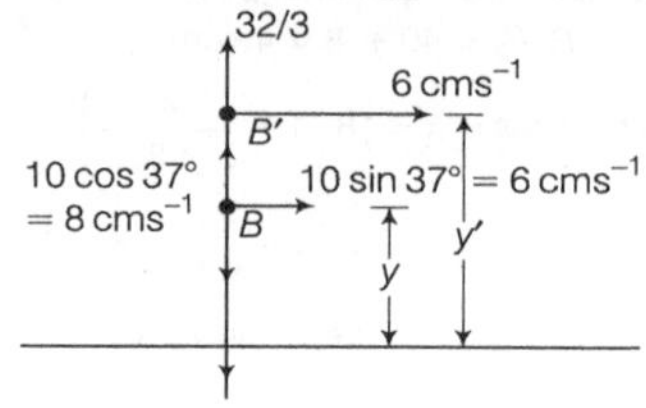

$\because \quad \frac{\mu_2}{\mu_1} = \frac{v}{u}$ or $\frac{4}{3} = \frac{-y'}{-y}$

$\therefore \quad y' = \frac{4}{3}y$

$\therefore \quad \frac{dy'}{dt} = \frac{4}{3}\frac{dy}{dt}$

or $\quad \frac{dy'}{dt} = \frac{4}{3} \times (-8) = \frac{-32}{3}$ cms^{-1}

$\therefore \quad$ Speed of image $= \sqrt{\left(\frac{32}{3}\right)^2 + 6^2} = 12.24$ cms^{-1}

$$\tan\theta = \frac{6}{32/3} = \frac{18}{32} = \frac{9}{16}$$

$\therefore \quad \theta = \tan^{-1}\left(\frac{9}{16}\right).$

22. (b)

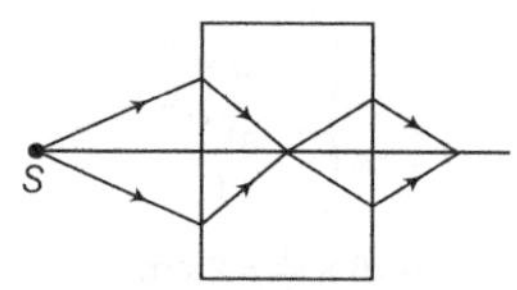

23. (a) 10 + 20 + 10 = 40 cm

24. (a) 10 + 10 = 20 cm

25. (a)

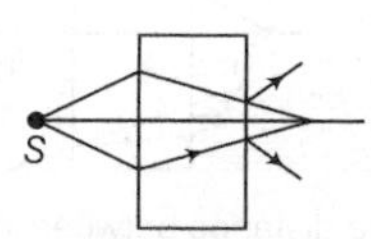

26. (b) Laser beam does not deviate.

27. (b) Paper partially lets through light. But owing to its fibrous structure and the great number of pores, the dissipation of light is very high all directions.

Due to this, it is impossible to read the words. When the sheet of paper is covered with water, water reduce the dissipation of light. Due to this, the light begins to pass through the paper without any appreciable deviation and words can easily read. Hence, Statement I is correct.

For refractive surface, $\frac{\mu_2}{v} - \frac{\mu_1}{u} = \frac{\mu_2 - \mu_1}{r}$

For plane refractive surface, $r = \infty$

$\therefore \quad \frac{\mu_2}{v} - \frac{\mu_1}{u} = \frac{\mu_2 - \mu_1}{\infty} = 0$

Hence, Statement II is correct.

28. (b)

(i) $\mu_{eq} = \frac{\int_0^1 dy}{\int_0^1 \frac{dy}{\mu}} = \frac{[y]_0^1}{\int_0^1 \frac{dy}{1 + \frac{y}{a}}} = \frac{1}{\ln 2} \qquad (\because a = 1 \text{ m}^{-1})$

$$= \frac{1}{0.693} = 1.44$$

(ii) Apparent depth $= \int \frac{dy}{\mu} = \int \frac{dy}{1+\frac{y}{a}} = \int_0^1 \frac{dy}{1+y}$ $(\because a = 1)$

$= [\ln(1+y)]_0^1 = \ln 2 = 0.693$ m.

(iii) Optical path $= \int \mu dy = \int_0^1 \left(1+\frac{y}{a}\right) dy = \int_0^1 (1+y)\, dy$ $(\because a = 1)$

$= \left[y + \frac{y^2}{2}\right]_0^1 = \frac{3}{2}$ m

(iv) Time $= \frac{\text{Optical path}}{3 \times 10^8} = 5 \times 10^{-9}$ s

Hecne, (i) → (s), (ii) → (q), (iii) → (r), (iv) → (p)

29. (5) Refraction through first surface,

Here, $\mu_1 = 1, \mu_2 = 2, u = -10$ cm

$\because \quad \frac{\mu_2}{\mu_1} = \frac{v}{u}$

$\therefore \quad v = \frac{\mu_2}{\mu_1} u = \frac{2}{1}(-10) = -20$ cm

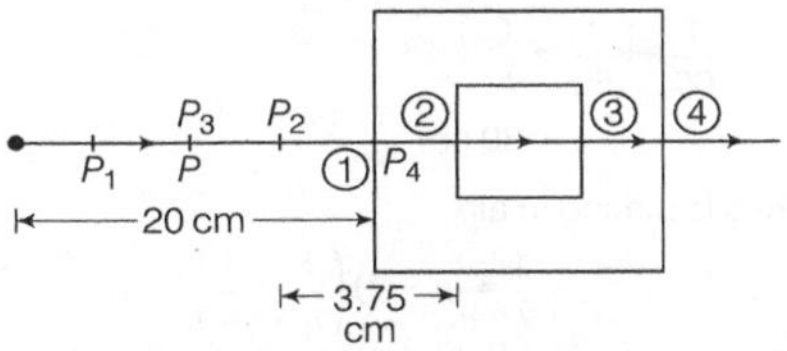

The image P_1 formed due to refraction through first surface, behaves as object for second surface.

For second surface,

$u' = -7.5$ cm, $\mu_1' = 2, \mu_2' = 1$

$\because \quad \frac{\mu_2'}{\mu_1'} = \frac{v'}{u'}$

$\therefore \quad v' = \frac{\mu_2'}{\mu_1'} u' = \frac{1}{2}(-7.5) = -3.75$ cm

The image P_2 formed due to refraction through second surface behaves as object for third surface.

For third surface, $u_1'' = 1, u_2'' = 2$

$u'' = -8.75 \text{ cm} = \frac{u_2''}{u_1''} = \frac{v''}{u''}$

$\therefore \quad v'' = \frac{2}{1}(-8.75) = -17.5$ cm

The image P_3 formed due to refraction through third surface behaves as object for fourth surface.

For fourth surface, $\mu_1''' = 2, \mu_2''' = 1, \mu_1''' = -20$ cm

$\because \quad \frac{\mu_2'''}{\mu_1'''} = \frac{v'''}{u'''}$

$\therefore \quad v''' = \left(\frac{1}{2}\right)(-2) = -10$ cm

30. (9) For refraction through water surface,

$\mu_1 = 1, \mu_2 = \frac{4}{3}, u = -x,$

$\because \quad \frac{\mu_2}{\mu_1} = \frac{v}{u}$

$\therefore \quad v = \frac{4}{3}(-x) = \frac{-4x}{3}$

Due to reflection through water surface, image P' is x cm below the water surface.

$\because \quad P_1 P' = 21$

or $\quad \frac{4x}{3} + x = 21$

or $\quad \frac{7x}{3} = 21$

$\therefore \quad x = \frac{21 \times 3}{7} = 9$ cm

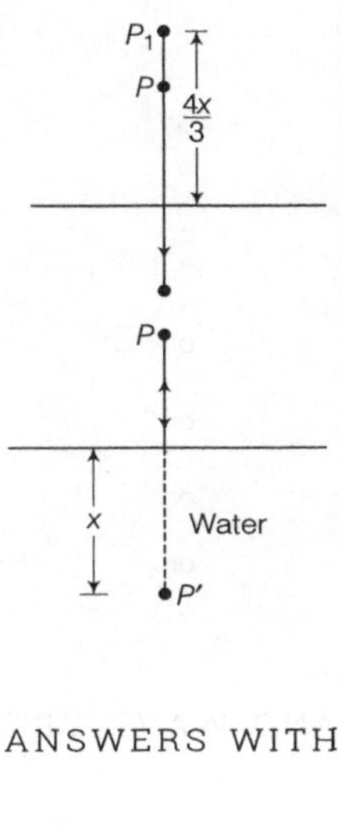

31. (5) The object shiftness due to slab $= t\left(1 - \frac{1}{\mu}\right)$

$= 15\left(1 - \frac{1}{3/2}\right) = 15\left(1 - \frac{2}{3}\right) = 5$ cm

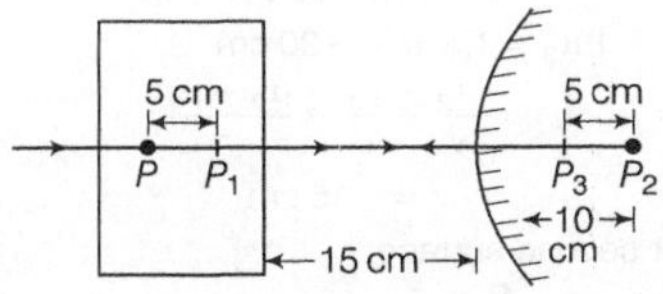

For convex mirror, image P_1 formed due to first refraction behaves as object.

For convex mirror, $u = -20$ cm, $f = +20$ cm

$\because \quad \frac{1}{v} + \frac{1}{u} = \frac{1}{f}$

or $\quad \frac{1}{v} - \frac{1}{20} = \frac{1}{20}$

$\therefore \quad v = 10$ cm

The image P_2 formed due to reflection through convex mirror behaves as object for second times refraction through slab. By second times refraction through slab, final image is P_3. Thus, final image is 5 cm from the pole of mirror.

32. (9) From figure, $\tan\theta = \frac{AC}{h}$

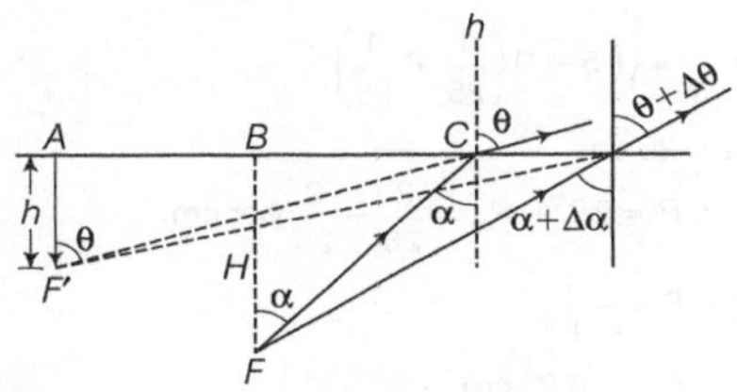

$\therefore \quad AC = h\tan\theta$

and $\quad \tan\alpha = \frac{BC}{H}$

$\therefore \quad BC = H\tan\alpha$

The distance between image plane and object plane remains constant.

$\therefore \quad AB = l = \text{constant}$

$\therefore \quad l = AC - BC = h\tan\theta - H\tan\alpha$

or $\quad \frac{dl}{d\alpha} = h\sec^2\theta \frac{d\theta}{d\alpha} - H\sec^2\alpha$

or $\quad O = h\sec^2\theta \frac{d\theta}{d\alpha} - H\sec^2\alpha$

$\therefore \quad \frac{d\theta}{d\alpha} = \frac{H \sec^2\alpha}{h \sec^2\theta}$ ...(i)

According to Snell's law,

$\mu\sin\alpha = 1 \times \sin\theta$

or $\quad \mu\cos\alpha = \cos\theta \frac{d\theta}{d\alpha}$

$\therefore \quad \frac{d\theta}{d\alpha} = \frac{\mu\cos\alpha}{\cos\theta}$ ...(ii)

From Eqs. (i) and (ii), we get

$\frac{H \sec^2\alpha}{h \sec^2\theta} = \frac{\mu\cos\alpha}{\cos\theta}$

or $\quad h = \frac{H\sec^3\alpha}{\mu\sec^3\theta} = \frac{3 \times \sec^3 37°}{\frac{4}{3}\sec^3 53°}$

$= \frac{9}{4}\left(\frac{5}{4 \times \frac{5}{3}}\right)^3 = \frac{9}{4} \times \frac{27}{64} = \frac{243}{256} = \frac{27n}{256}$

$\therefore \quad n = 9.$

DPP-10 Lens-I

1. (i) (a) $\frac{1}{-10} = (1.5 - 1)\left(-\frac{1}{R} - \frac{1}{R}\right)$

$\therefore \quad R = 10 \text{ cm}$

(ii) (c) $\mu_1 = 1, \mu_2 = 1.5, u = -20$ cm

$\because \quad \frac{\mu_2}{v} - \frac{\mu_1}{u} = \frac{\mu_2 - \mu_1}{r}$

$\therefore \quad v = -15 \text{ cm}$

(iii) (c) For second surface,

$\mu_1' = 1.5, \mu_2' = \frac{4}{3}, u' = -15 \text{ cm } r' = +10 \text{ cm},$

$\because \quad \frac{\mu_2'}{v'} - \frac{\mu_1'}{u'} = \frac{\mu_2' - \mu_1'}{r'}$

$$= \frac{4}{3v'} - \frac{3}{2(-15)} = \frac{\frac{4}{3} - \frac{3}{2}}{10}$$

$$\frac{4}{3v'} = \frac{1}{60} - \frac{1}{10} = \frac{1-6}{60}$$

$\therefore \quad v' = \frac{-12 \times 4}{3} = -16 \text{ cm}$

(iv) (b) Final image is in opposite direction of final refracted rays. Thus final image is virtual.

2. (i) (a)

(ii) (b) $\frac{1}{fl} = (1.5 - 1)\left(\frac{1}{25} + \frac{1}{25}\right)$

$\therefore fl = 25$ cm

(iii) (c) $\therefore P = 2P_l + P_m = \frac{2}{25} + \frac{2}{25}$ per cm

(iv) (b) $\because P = -\frac{1}{F}$

$\therefore \quad F = -6.25$ cm

(v) (a) For coinciding image with object rays should incident on silvered surface normally. This is possible only when the image formed by first refraction through unsilvered surface should be at the centre of curvature of silvered surface.

Here, $\quad \mu_1 = 1, \mu_2 = \frac{3}{2}$

$r = +25$ cm

$v = 25$ cm

$\therefore \quad \frac{\mu_2}{v} - \frac{\mu_1}{u} = \frac{\mu_2 - \mu_1}{r}$

$\therefore \quad u = 12.5$ cm

(vi) (a)

3. (i) (a) Here, object is placed between optical centre and focus. So, image is virtual errect and magnified.

(ii) (a) $v - u$ graph is shown in option (a).

(iii) (a) Here, object is placed between f and $2f$, the image is real, magnified and inverted.

(iv) (a) Here, $u = -40$ cm, $f = 20$ cm

$\because \quad \frac{1}{v} - \frac{1}{u} = \frac{1}{f}$

$\therefore \quad u = 40$ cm

$\therefore \quad m = \frac{v}{u} = -1$

Thus image is real, inverted and of same size.

(v) (b) Take, $\because u = -60$ cm, $f = 20$ cm

$$\frac{1}{v} - \frac{1}{u} = \frac{1}{f}$$

$\therefore \quad v = 30$ cm

$\therefore \quad m = \frac{v}{u} = -\frac{1}{2}$

Thus, image is real, inverted and diminished.

(vi) (a) The $v - u$ graph is shown in figure (A)

(vii) (b) P behaves as virtual object.

(viii) (a) $\because \quad \frac{1}{v} - \frac{1}{u} = \frac{1}{f}$

or $\quad \frac{1}{v} = \frac{1}{f} + \frac{1}{u}$

Here, u and f are positive so, v is positive. So image is real, diminished and errect.

(ix) (a) The $v - u$ graph is shown in option (a).

(x) (b) The general $v - u$ graph is shown in option (b).

4. (i) (a) (ii) (a)

(iii) (a) (iv) (a)

(v) (b)

5. (b) $u = -60$ cm, $v = +60$ cm

$$\frac{1}{v} - \frac{1}{u} = \frac{1}{f}$$

or $\quad \frac{1}{60} + \frac{1}{60} = \frac{1}{f}$

$f = 30$ cm

6. (b) If lens is placed in air

$$\frac{1}{f} = (\mu - 1)\left(\frac{1}{r_1} - \frac{1}{r_2}\right)$$

or $\quad \left(\frac{1}{30}\right) = (1.5 - 1)\left(\frac{1}{R} + \frac{1}{R}\right)$

$\therefore \quad R = 30$ cm

For refraction through liquid and lens surface,

$$\mu_1 = \mu, \ \mu_2 = \frac{3}{2}, u = -\infty, r = +30 \text{ cm}$$

$\because \quad \frac{\mu_2}{v} - \frac{\mu_1}{u} = \frac{\mu_2 - \mu_1}{r}$

or $\quad \frac{3}{2v} - \frac{\mu}{\infty} = \frac{\frac{3}{2} - \mu}{30}$

or $\quad \frac{3}{2v} = \frac{\frac{3}{2} - \mu}{30}$

For lens-air, surface, $\mu_1' = 1.5, \mu_2' = 1, \mu' = v$

$v' = +40$ cm, $r' = -30$ cm

$\because \quad \frac{\mu_2'}{v'} - \frac{\mu_1'}{u} = \frac{\mu_2' - \mu_1'}{r'}$

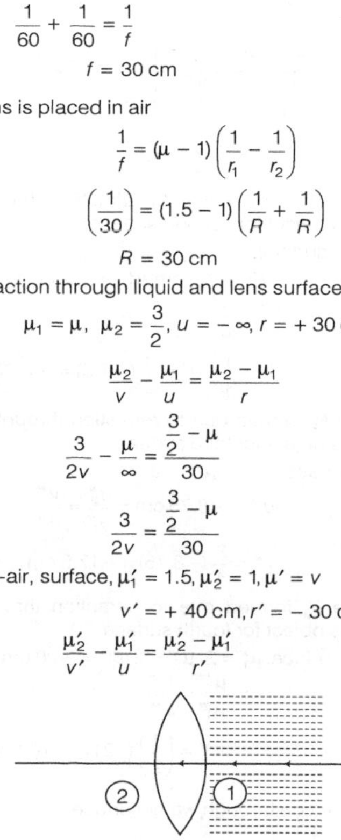

or $\quad \frac{1}{40} - \frac{1.5}{v} = \frac{1 - 1.5}{-30} = \frac{1}{60}$

or $\quad \frac{1}{40} - \frac{1}{60} = \frac{3}{2v} = \frac{\frac{3}{2} - \mu}{30}$

or $\quad \frac{3-2}{120} = \frac{\frac{3}{2} - \mu}{30}$

or $\quad \frac{3}{2} - \mu = \frac{1}{120} \times 30 = \frac{1}{4}$

or $\quad \frac{3}{2} - \frac{1}{4} = \mu$

or $\quad 1.5 - 0.25 = \mu$

$\therefore \quad \mu = 1.25$

7. (b) For L_1,

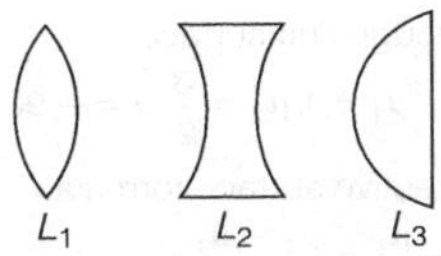

$$\frac{1}{f_1} = (\mu_1 - 1)\left(\frac{1}{r_1} - \frac{1}{r_2}\right)$$

$$= (2-1)\left(\frac{1}{\infty} + \frac{1}{10}\right) = \frac{1}{10}$$

$\therefore$ $f_1 = 10$ cm

For lens L_2, $\frac{1}{f_2} = (\mu_2 - 1)\left(\frac{1}{r_1} - \frac{1}{r_2}\right)$

$$= (1.5-1)\left(-\frac{1}{10} - \frac{1}{15}\right)$$

$$= \frac{1}{2}\left(\frac{-3-2}{30}\right) = \frac{-5}{60}$$

$\because$ $f_2 = -12$ cm

For lens L_3, $\frac{1}{f_3} = (\mu_3 - 1)\left(\frac{1}{r_1} - \frac{1}{r_2}\right)$

or $\frac{1}{f_3} = (2-1)\left(\frac{1}{15} - \frac{1}{\infty}\right) = \frac{1}{15}$

$\therefore$ $f_3 = 15$ cm

$\because$ $\frac{1}{f} = \frac{1}{f_1} + \frac{1}{f_2} + \frac{1}{f_3}$

$$= \frac{1}{10} - \frac{1}{12} + \frac{1}{15} = \frac{6-5+4}{60} = \frac{5}{60}$$

$\therefore$ $f = 12$ cm

8. (d) $u = -30$ cm

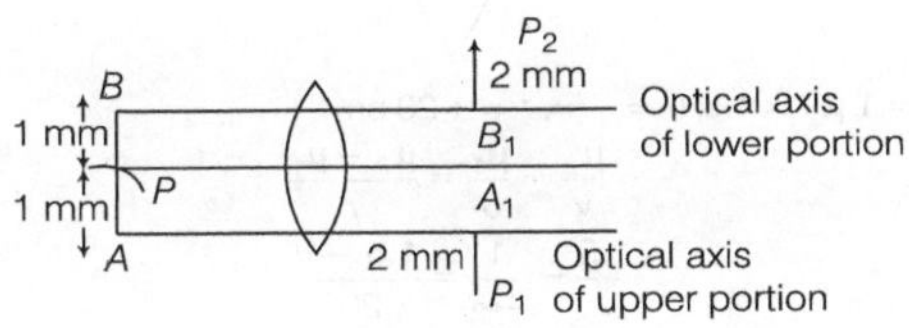

$f = 20$ cm

$\because$ $\frac{1}{v} - \frac{1}{u} = \frac{1}{f}$

$\therefore$ $v = 60$ cm

$\therefore$ $m = \frac{v}{u} = -2$

$\therefore$ $P_1 P_2 = 6$ cm.

9. (c) From figure, it is clear that, F_1 should be common focus,

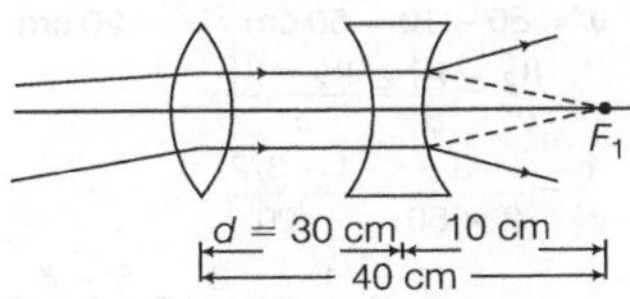

$\therefore$ Separation between lenses

$= d = 40 - 10 = 30$ cm.

10. (a) Here, $u = +10$ cm

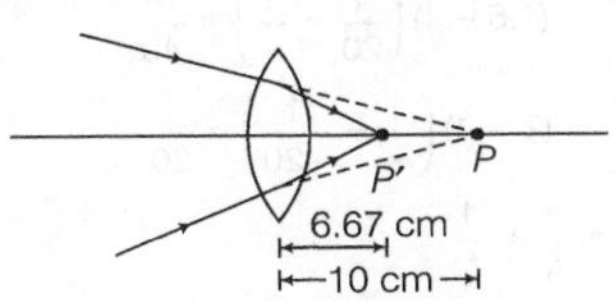

$f = +20$ cm

or, $\frac{1}{v} - \frac{1}{10} = \frac{1}{20}$

$$\frac{1}{v} = \frac{1}{10} + \frac{1}{20} = \frac{2+1}{20}$$

$\therefore$ $v = \frac{20}{3}$ cm $= 6.67$ cm

11. (c) $u = -20 \cos 60° = -10$ cm

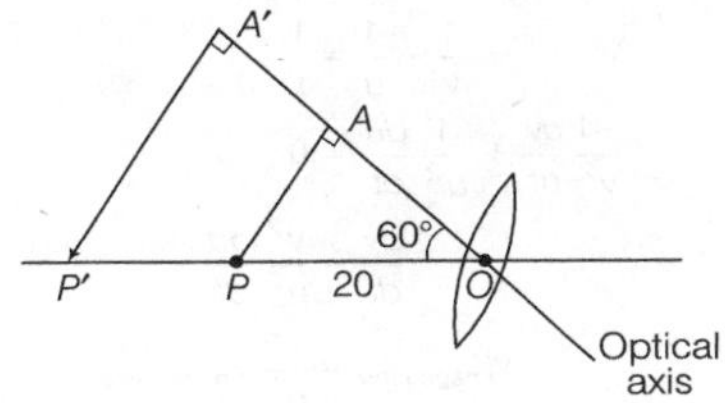

$f = +20$ cm

$\because$ $\frac{1}{v} - \frac{1}{u} = \frac{1}{f}$

or $\frac{1}{v} + \frac{1}{10} = \frac{1}{20}$

or $\frac{1}{v} = \frac{1}{20} - \frac{1}{10} = \frac{1-2}{20}$

$\therefore$ $v = -20$ cm

From triangle OAP',

$$\cos 60° = \frac{OA'}{OP'} = \frac{-20}{OP'}$$

$\therefore$ $OP' = \frac{-20}{\frac{1}{2}} = -40$ cm

12. (b) $\because$ $-\frac{1}{20} = (1.5-1)\left(-\frac{1}{R} - \frac{1}{R}\right)$

or $\frac{-1}{20} = -\frac{1}{R}$

$\therefore$ $R = 20$ cm

$\therefore$ The power of the silvered lens is

$$P = 2P_l + P_m$$

$$= \frac{-2}{20} - \frac{1}{10} = \frac{-2}{10} = -\frac{1}{5}$$

The focal length of equivalent mirror is

$$F = -\frac{1}{P} = 5 \text{ cm}$$

$\because$ $\frac{1}{v} + \frac{1}{u} = \frac{1}{f}$

or $\frac{1}{v} - \frac{1}{40} = \frac{1}{5}$

$\therefore$ $v = \frac{40}{9}$ cm

13. (c) For upper half,

$$\frac{1}{f_1} = \left(\frac{1.5}{1.2} - 1\right)\left(-\frac{1}{20} - \frac{1}{20}\right)$$

$$= \frac{0.3}{1.2} \times \left(\frac{-1}{10}\right) = \frac{-1}{40}$$

$\therefore$ $f_1 = -40$ cm

For lower part,

$$\frac{1}{f_2} = \left(\frac{1.5}{2.5} - 1\right)\left(-\frac{1}{20} - \frac{1}{20}\right)$$

$$= \frac{-1}{2.5} \times \left(\frac{-2}{20}\right) = \frac{1}{2.5 \times 10}$$

$f_2 = 25$ cm

Thus, separation between converging points is

$d = 25 + 40 = 65$ cm

14. (d) At, $t = 10$ s, the separation between object and lens is 60 cm.

Here, $u = -60$ cm, $f = 20$ cm

$\because \quad \frac{1}{v} - \frac{1}{u} = \frac{1}{f}$

or $\quad \frac{1}{v} + \frac{1}{60} = \frac{1}{20}$

$\therefore \quad v = 30$ cm

$\because \quad \frac{1}{v} - \frac{1}{u} = \frac{1}{f}$

or $\quad \frac{-1}{v^2}\frac{dv}{dt} + \frac{1}{u^2}\frac{du}{dt} = 0$

$\therefore \quad \frac{dv}{dt} = \frac{v^2}{u^2}\frac{du}{dt}$

or $\quad v_{\text{image lens}} = \frac{v^2}{u^2} v_{\text{object lens}}$

or $\quad v_{\text{image}} - v_{\text{lens}} = \left(\frac{30}{-60}\right)^2 (v_{\text{object}} - v_{\text{lens}})$

or $\quad v_{\text{image}} = 4 + \frac{1}{4}(2 - 4)$

$= 4 - 0.5 = 3.5 \text{ cms}^{-1}$

15. (a) $v_x = \frac{dv}{dt} = \frac{v^2}{u^2}\frac{du}{dt} = \left(\frac{60}{-30}\right)^2 \times 12 = 48 \text{ cms}^{-1}$

$\because \quad m = \frac{y'}{y}$

$\therefore \quad y' = my$

or $\quad \frac{dy'}{dt} = \frac{mdy}{dt} = \frac{ydm}{dt}$

or $\quad \frac{dy'}{dt} = \frac{mdy}{dt} = -2 \times 9 \qquad (\because y = 0)$

$= -18 \text{ cms}^{-1}$

16. (a) When lens is placed in air

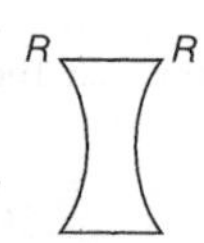

$\frac{1}{f_l} = (\mu - 1)\left(\frac{1}{r_1} - \frac{1}{r_2}\right)$

$\frac{-1}{30} = (1.5 - 1)\left(\frac{-1}{R} - \frac{1}{R}\right)$

$\therefore \quad R = 30$ cm

The power of liquid lens is $P_l' = (\mu - 1)\left(\frac{1}{r_1} - \frac{1}{r_2}\right)$

or $\quad P_l' = (2 - 1)\left(\frac{1}{\infty} - \frac{1}{-R}\right)$

$\therefore \quad P_l' = \frac{1}{30}$

$\therefore$ The power of combination

$\therefore \quad P = 2P_l' + 2P_l + 2P_l' + P_m$

$= \frac{1}{15} - \frac{1}{15} + \frac{1}{15} + 0 = \frac{1}{15}$

or $\quad \frac{-1}{F} = \frac{1}{15}$

$\therefore \quad F = -15$ cm

Thus, the combination behaves as a concave mirror of focal length 15 cm.

17. (b) If lens is placed in air,

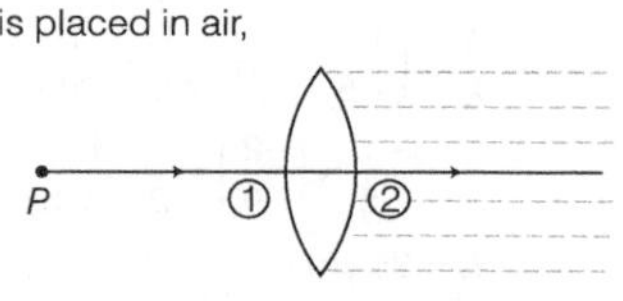

$\frac{1}{30} = (1.5 - 1)\left(\frac{1}{R} - \frac{1}{-R}\right)$

$\therefore \quad R = 30$ cm

For refraction through first surface,

$\mu_1 = 1, \mu_2 = \frac{3}{2}, u = -90$ cm, $r = +30$ cm

According to refractive surface formula,

$\frac{\mu_2}{v} - \frac{\mu_1}{u} = \frac{\mu_2 - \mu_1}{r}$

or $\quad \frac{3}{2v} - \frac{1}{-90} = \frac{\frac{3}{2} - 1}{30}$

or $\quad \frac{3}{2v} = \frac{1}{60} - \frac{1}{90} = \frac{3 - 2}{180}$

$\frac{2v}{3} = 180$

$\therefore \quad v = \frac{180 \times 3}{2} = +270$ cm

For refraction through second surface,

$\mu_1' = \frac{3}{2}, \mu_2' = \frac{4}{3}, u' = +270$ cm, $r' = -30$ cm

$\because \quad \frac{\mu_2'}{v'} - \frac{\mu_1'}{u'} = \frac{\mu_2' - \mu_1'}{r'}$

or $\quad \frac{4}{3v'} - \frac{3}{2 \times 270} = \frac{\frac{4}{3} - \frac{3}{2}}{-30}$

or $\quad \frac{4}{3v'} = \frac{1}{180} + \frac{3}{540}$

$\therefore \quad v' = +120$ cm

18. (a) For first surface,

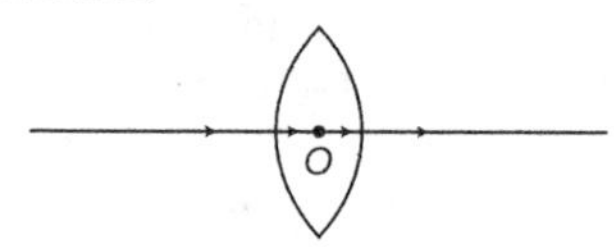

$\mu_1 = 1, \mu_2 = 1.5, u = -\infty, r = +20$ cm

$\because \quad \frac{\mu_2}{v} - \frac{\mu_1}{u} = \frac{\mu_2 - \mu_1}{r}$

or $\quad \frac{1.5}{v} - \frac{1}{-\infty} = \frac{1.5 - 1}{20}$

or $\quad \frac{3}{2v} = \frac{1}{40}$

or $\quad \frac{2v}{3} = 40$

$\therefore \quad v = \frac{120}{2} = 60$ cm

For second surface,

$\mu_1' = \frac{3}{2}, \mu_2' = 1$

$\mu' = 60 - 10 = 50$ cm, $r' = -20$ cm

$\because \quad \frac{\mu_2'}{v'} - \frac{\mu_1'}{u'} = \frac{\mu_2' - \mu_1'}{r'}$

or $\quad \frac{1}{v'} - \frac{3}{2 \times 50} = \frac{1 - 3/2}{-20}$

or $\quad \frac{1}{v'} = \frac{1}{40} + \frac{3}{100} = \frac{5 + 6}{200} = \frac{11}{200}$

$\therefore \quad v' = \frac{200}{11} = 18.18$ cm

19. (a) Here, $\frac{1}{f_1} = (1.5 - 1)\left(\frac{1}{20} - \frac{1}{\infty}\right) = \frac{1}{40}$

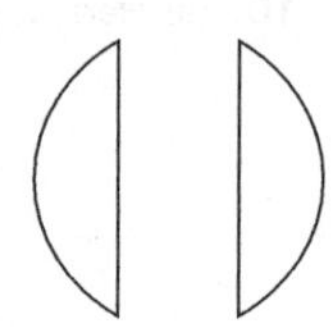

and $\quad \frac{1}{f_2} = (2 - 1)\left(\frac{1}{\infty} - \frac{1}{-20}\right) = \frac{1}{20}$

$\therefore \quad \frac{1}{f} = \frac{1}{f_1} + \frac{1}{f_2}$

or $\quad \frac{1}{f} = \frac{1}{40} + \frac{1}{20} = \frac{1 + 2}{40} = \frac{3}{40}$

$\because$ $$\frac{1}{v}-\frac{1}{u}=\frac{1}{f}$$

or $$\frac{1}{v}+\frac{1}{30}=\frac{3}{40}$$

or $$\frac{1}{v}=\frac{3}{40}-\frac{1}{30}=\frac{9-4}{120}=\frac{5}{120}$$

$\therefore$ $$v=\frac{120}{5}=24\text{ cm}$$

20. (b) For upper half,

$$\frac{1}{f_1}=(\mu-1)\frac{2}{R}$$

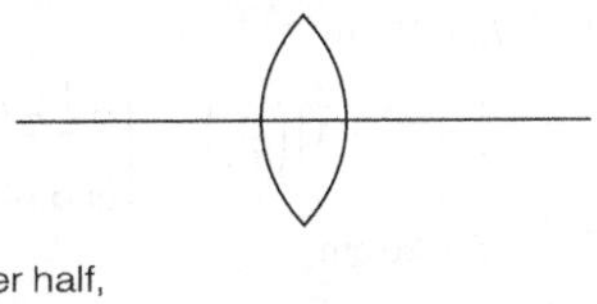

and for lower half,

$$\frac{1}{f_2}=(\mu+\Delta\mu+1)\frac{2}{R}$$

$\therefore$ $$\frac{1}{f_2}-\frac{1}{f_1}=(\mu+\Delta\mu-1)\frac{2}{R}-(\mu-1)\frac{2}{R}$$

$$=\frac{2\Delta\mu}{R}$$

or $$\frac{f_1-f_2}{f_1,f_2}=\frac{2\Delta\mu}{R}$$

or $$\frac{\Delta f}{f_1^2}=\frac{2\Delta\mu}{R}$$

or $$\Delta f=\frac{(\mu-1)^2 4}{R^2}=\frac{2\Delta\mu}{R}$$

$\therefore$ $$\Delta f=\frac{\Delta\mu R}{2(\mu-1)^2}$$

21. (b) $P=2P_l+P_m$

$$P_l=\frac{1}{f_l},\ P_m=\frac{-1}{f_m},\ f_l=30\text{ cm},\ f_m=-15\text{ cm}$$

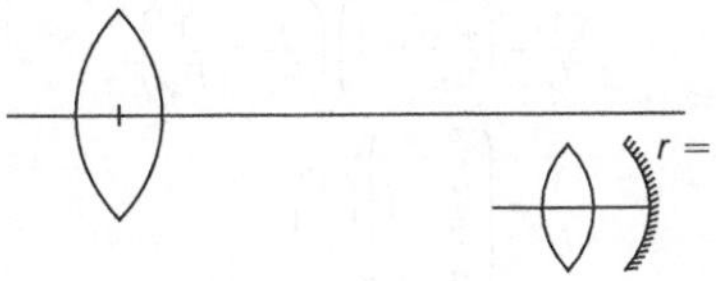

$$\frac{-1}{F}=\frac{2}{30}+\frac{1}{15}=\frac{1}{15}+\frac{1}{15}$$

$$-\frac{1}{F}=\frac{2}{15}$$

$F=-7.5$ cm, Equivalent mirror is concave mirror of focal 7.5 cm.

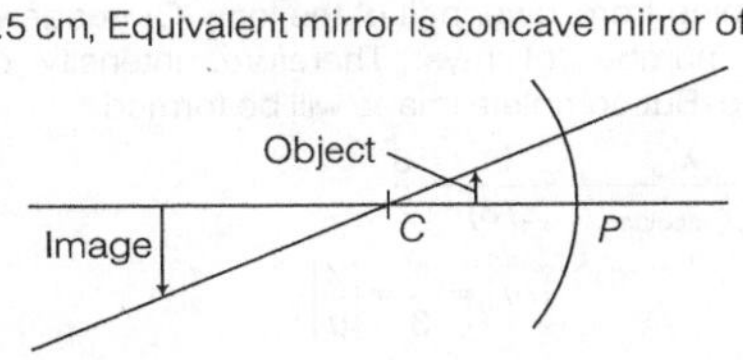

22. (c)

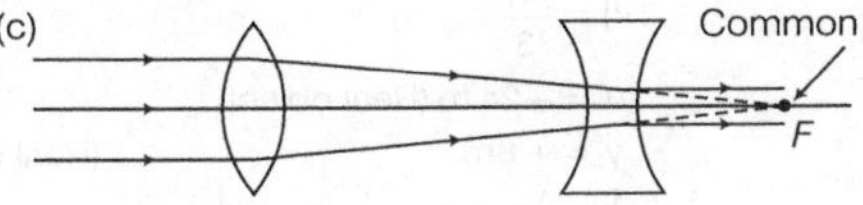

23. (c) $$\frac{1}{u}+\frac{1}{v}=\frac{1}{f}$$

$\therefore$ $$\frac{v}{u}+1=\frac{v}{f}$$

$\therefore$ $\frac{1}{y}+1=\frac{x}{f}$ graph cannot be a straight line.

$\therefore$ Option (a) is not correct. Graph II has increase in u increase v which is not correct.

Eq. $\frac{v}{u}+1=\frac{v}{f}$ gives $1+y=\frac{x}{f}$

$\therefore$ Linear graph (c) is not correct

$\therefore$ $\frac{1}{u}+\frac{1}{v}=\frac{1}{f}$ gives graph IV. Hence, option (d) is correct

24. (b) (a) Eyes receive the beam virtual image and real image in same manner. So, virtual and real images look the same to the eye. aberrations occurs due to large aperture of optical instrument and due to mixture of wavelength. So, aberrations occur for real image as weelas virtual image.

(b) The beam of light incidence on the diverging lens from real object, diverges after refraction through the lens. Hence, it form virtual image of a real object

(c) Only paraxial rays parallel to the axis of a spherical mirror are reflected through a single point.

(d) Only real image can be displayed on a screen.

25. (a) For $B, u=-30$ cm, $f=+20$ cm

Lens formula

$\therefore$ $$\frac{1}{v}-\frac{1}{u}=\frac{1}{f}$$

$\therefore$ $$v=60\text{ cm}$$

For A, $u'=-35$ cm, $f'=20$ cm,

Lens, $$\frac{1}{v'}-\frac{1}{u'}=\frac{1}{f'}$$

or $$\frac{1}{v'}+\frac{1}{35}=\frac{1}{20}$$

or $$\frac{1}{v'}=\frac{1}{20}-\frac{1}{35}=\frac{7-4}{140}=\frac{3}{140}$$

$\therefore$ $$v'=\frac{140}{3}$$

$\therefore$ $$A'B'=(v-v')>5$$

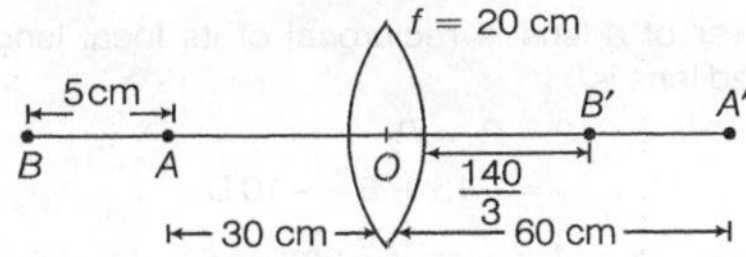

26. (c) The ray along PA does not deviate. So, image is always lying on line collinear PA.

27. (a) In case of ray diagrams we consider deviation at each curvature.

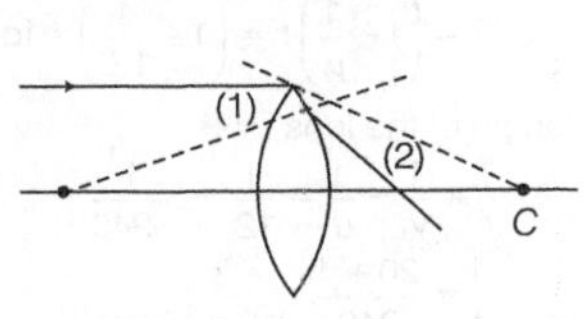

⇒ At point (1), it bends towards the normal as it enters a denser medium.

⇒ At point (2), it bends away from the normal as it enters a rares medium.

⇒ Also, its to be noted at that normals are lines joining the point of incidence to that curvature centre.

28. (a) A plano-convex lens behaves as a concave mirror, if its one surface (curved) is silvered. The ray refracted from plane surface are reflected from curved surface and again refract from plane surface.

Therefore, in this lens two refractions and 0 reflection occur.

Let the local length of silvered lens be F,

$$\frac{1}{F}=\frac{1}{f}+\frac{1}{f}+\frac{1}{f_m}=\frac{2}{f}+\frac{1}{f_m}$$

where, f = focal length of lens before silvering.

f_m = focal length of spherical mirror.

$$\frac{1}{F}=\frac{2}{f}+\frac{2}{R} \qquad [\because R = 2f_m]$$

Now, $$\frac{1}{f}=(\mu-1)\left(\frac{1}{R_1}-\frac{1}{R_2}\right)$$

Here, $R_1 = \infty$, $R_2 = 30$ cm

or $$\frac{1}{f}=(1.5-1)\left(\frac{1}{\infty}-\frac{1}{30}\right)$$

$\therefore$ $$\frac{1}{f}=-\frac{0.5}{30}=-\frac{1}{60}$$

or $$f = -60 \text{ cm}$$

Hence, from Eq. (i), we get

$$\frac{1}{F}=\frac{2}{60}+\frac{2}{30}=\frac{6}{60}$$

$$F = 10 \text{ cm}$$

Again given that,

Size of object = Size of image

i.e. $$O = I$$

$\therefore$ $$m=-\frac{v}{u}=\frac{I}{O}$$

$\Rightarrow$ $$\frac{v}{u}=-1$$

or $$v = -u$$

Thus, from lens formula

$$\frac{1}{F}=\frac{1}{v}-\frac{1}{u}$$

$\Rightarrow$ $$\frac{1}{10}=\frac{1}{-u}-\frac{1}{u}$$

$$\frac{1}{10}=-\frac{2}{u}$$

$\therefore$ $$u = -20 \text{ cm}$$

Hence, to get a real image, object must be placed at a distance 20 cm on the left side of lens.

29. (b) Power of a lens is reciprocal of its focal length. Power of combined lens is

$$P = P_1 = P_2$$
$$= -15 + 5 = -10 \text{ D}$$
$$f=\frac{1}{P}=\frac{100}{-10}\text{ cm}$$

or $$f = -10 \text{ cm}$$

30. (d) Shift image position due to glass plate,

$$S=\left(1-\frac{1}{\mu}\right)t=\left(1-\frac{1}{1.5}\right)\times 1\text{cm}=\frac{1}{3}\text{cm}$$

For the focal length of the lens

$$\frac{1}{f}=\frac{1}{v}-\frac{1}{u}=\frac{1}{12}-\frac{1}{-240}$$

$$\frac{1}{f}=\frac{20+1}{240}$$

$$f=\frac{240}{21}\text{ cm}$$

Now, get back image on the film, lens has to form image at $\left(12-\frac{1}{3}\right)$ cm $=\frac{35}{3}$ cm such that the glass plate will shift the image on the film.

As, $$\frac{1}{f}=\frac{1}{v}-\frac{1}{u}$$

$$\frac{1}{u}=\frac{1}{v}-\frac{1}{f}=\frac{3}{35}-\frac{21}{340}$$

$$=\frac{48\times 3-7\times 21}{1680}=-\frac{1}{560}$$

$$u = -5.6 \text{ m}$$

31. (c) $n=\frac{3}{2}$ $\qquad \left[\because n=\frac{c}{v}\right]$

$$3^2 + (R - 3\text{ mm})^2 = R^2$$

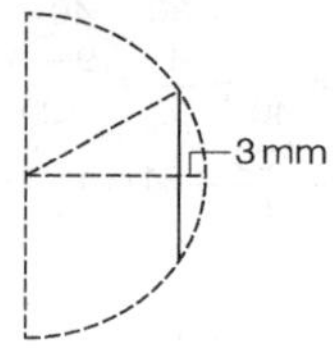

$\Rightarrow$ $$3^2 + R^2 - 2R(3\text{ mm}) + (3\text{ mm})^2 = R^2$$

$\Rightarrow$ $$R \approx 15 \text{ cm}$$

$\therefore$ $$\frac{1}{f}=\left(\frac{3}{2}-1\right)\left(\frac{1}{15}\right) \qquad \left[\because \frac{1}{f}=(n-1)\left(\frac{1}{R_1}-\frac{1}{R_2}\right) \text{ and } R_1=\infty,\ R_2=-15\right]$$

$\Rightarrow$ $$f = 30 \text{ cm}$$

32. (b) According to lens maker's formula, when the lens in the air

$$\frac{1}{f}=\left(\frac{3}{2}-1\right)\left(\frac{1}{R_1}-\frac{1}{R_2}\right) \Rightarrow \frac{1}{f}=\frac{1}{2x}\Rightarrow f=2x$$

Here, $$\frac{1}{x}=\frac{1}{R_1}-\frac{1}{R_2}$$

In case of liquid, where refractive indices are $\frac{4}{3}$ and $\frac{5}{3}$, we get

Focal length in first liquid

$$\frac{1}{f_1}=\left(\frac{\mu_s}{\mu_{l_1}}-1\right)\left(\frac{1}{R_1}-\frac{1}{R_2}\right)$$

$$\frac{1}{f_1}=\left(\frac{3/2}{4/3}-1\right)\frac{1}{x}$$

$\Rightarrow$ F_1 is positive.

Nature of lens is not change

$$\frac{1}{f_1}=\frac{1}{8x}=\frac{1}{4(2x)}=\frac{1}{4f}\Rightarrow f_1=4f$$

Focal length in second liquid,

$$\frac{1}{f_2}=\left(\frac{\mu_s}{\mu_{l_2}}-1\right)\left(\frac{1}{R_1}-\frac{1}{R_2}\right)$$

$$\frac{1}{f_2}=\left(\frac{\frac{3}{2}}{\frac{5}{3}}-1\right)\left(\frac{1}{x}\right)$$

$\Rightarrow$ f_2 is negative.

Nature of lens change i.e convex behave as concave.

33. (b) When upper half of the lens is covered, image is formed by the rays coming from lower half of the lens. Or image will be formed by less number of rays. Therefore, intensity of image will decrease. But complete image will be formed.

34. (c) $$\mu=\frac{\lambda_{\text{air}}}{\lambda_{\text{medium}}}=\frac{1}{(2/3)}=\frac{3}{2}$$

Further, $$|m|=\frac{1}{3}=\left|\frac{v}{u}\right|$$

$\therefore$ $$|v|=\frac{|u|}{3}$$

$\Rightarrow$ $u = -24$ m (Real object)

$\therefore$ $v = +8$m (Real image)

Now, $$\frac{1}{v}-\frac{1}{u}=\frac{1}{f}=(\mu-1)\left(\frac{1}{+R}-\frac{1}{\infty}\right)$$

$\therefore$ $$\frac{1}{8}+\frac{1}{24}=\left(\frac{3}{2}-1\right)\left(\frac{1}{R}\right)$$

$\therefore$ $$R = 3 \text{ m}$$

35. (b) Using the lens formula,

$$\frac{1}{v}-\frac{1}{u}=\frac{1}{F}=\frac{1}{f_1}+\frac{1}{f_2}$$

or
$$\frac{1}{v}=\frac{1}{u}+\frac{1}{f_1}+\frac{1}{f_2}$$

$$=\frac{1}{u}+(n_1-1)\left(\frac{1}{R_1}-\frac{1}{R_2}\right)+(n_2-1)\left(\frac{1}{R_1'}-\frac{1}{R_2}\right)$$

Substituting the values, we get

$$\frac{1}{v}=\frac{1}{-40}+(1.5-1)\left(\frac{1}{14}-\frac{1}{\infty}\right)+(1.2-1)\left(\frac{1}{\infty}-\frac{1}{-14}\right)$$

Solving this equations, we get

$$v=+40\text{ cm}$$

36. (b) Object is placed at distance $2f$ from the lens. So first image I_2 will be formed at distance $2f$ on other side. This image will behave like a virtual object for mirror. The second image I_2 will be formed at distance 20 cm in front of the mirror or, at distance 10 cm to the left hand side of the lens.

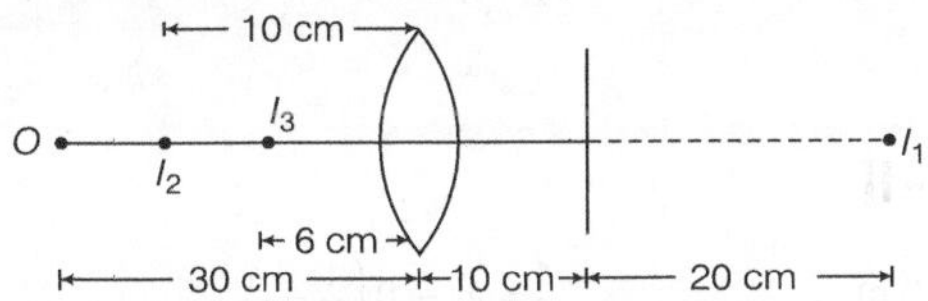

Now, applying lens formula

$$\frac{1}{v}-\frac{1}{u}=\frac{1}{f}$$

$\therefore$
$$\frac{1}{v}-\frac{1}{+10}=\frac{1}{+15}$$

or
$$v=6\text{ cm}$$

Therefore, the final image is at distance 16 cm from the mirror. But this image will be real.

This is because ray of light is travelling from right to left.

37. (d) Let focal length of convex lens is $+f$, then focal length of concave lens would be $-\frac{3}{2}f$.

From the given condition, $\frac{1}{30}=\frac{1}{f}-\frac{2}{3f}=\frac{1}{3f}$

$\therefore$
$$f=10\text{ cm}$$

Therefore, focal length of convex lens $=+10$ cm

and that of concave lens $=-15$ cm

38. (c) Refraction from lens $\frac{1}{v_1}-\frac{1}{-20}=\frac{1}{15}$

$\therefore$ $v=60$ cm $\overrightarrow{+\text{ve-direction}}$

i.e. first image is formed at 60 cm to the right of lens system.

Reflection from mirror

After reflection from the mirror, the second image will be formed at a distance of 60 cm to the left of lens system.

Refraction from lens

$$\frac{1}{v_3}-\frac{1}{60}=\frac{1}{15}\qquad\longleftarrow +\text{ ve-direction}$$

or
$$v_3=12\text{ cm.}$$

Therefore, the final image is formed at 12 cm to the left of the lens system.

39. (b) From the lens formula

$$\frac{1}{f}=\frac{1}{v}-\frac{1}{u}$$

we have,
$$\frac{1}{f}=\frac{1}{10}-\frac{1}{-10}$$

or $f=+5$

Further, $\Delta u=0.1$

and $\Delta v=0.1$ (from the graph)

Now, differentiating the lens formula, we have

$$\frac{\Delta f}{f^2}=\frac{\Delta v}{v^2}+\frac{\Delta u}{u^2}\text{ or }\Delta f=\left(\frac{\Delta v}{v^2}+\frac{\Delta u}{u^2}\right)f^2$$

Substituting the values, we have

$$\Delta f=\left(\frac{0.1}{10^2}+\frac{0.1}{10^2}\right)(5)^2=0.05$$

$\therefore$
$$f\pm\Delta f=5\pm0.05$$

40. (b) $r=f\tan\theta$

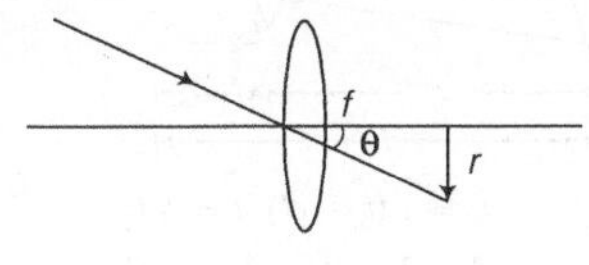

or
$$r\propto f$$

$\therefore$
$$\pi r^2\propto f^2$$

41. (b) Image formed by convex lens at I_1 will act as a virtual object for concave lens. For concave lens.

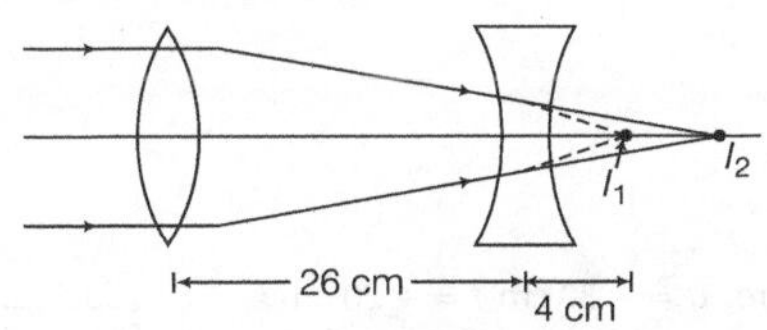

$$\frac{1}{v}-\frac{1}{u}=\frac{1}{f}$$

or
$$\frac{1}{v}-\frac{1}{4}=\frac{1}{-20}\quad\text{or}\quad v=5\text{ cm}$$

Magnification for concave lens

$$m=\frac{v}{u}=\frac{5}{4}=1.25$$

As size of the image at I_1 is 2 cm. Therefore, size of images I_2 will be $2\times1.25=2.5$ cm.

42. (d) The lens makers' formula

$$\frac{1}{f}=\left(\frac{n_L}{n_m}-1\right)\left(\frac{1}{R_1}-\frac{1}{R_2}\right)$$

where n_L = Refractive index of lens and

n_m = Refractive index of medium

In case of double concave lens, R_1 is negative and R_2 is positive. Therefore, $\left(\frac{1}{R_1}-\frac{1}{R_2}\right)$ will be negative.

For the lens to be diverging in nature, focal length f should be negative or $\left(\frac{n_L}{n_m}-1\right)$ should be positive or $n_L>n_m$ but since $n_2>n_1$ (given), therefore the lens should be filled with L_2 and immersed in L_1.

43. (a) $R_1=-R$, $R_2=+R$, $\mu_g=1.5$ and $\mu_m=1.75$

$\therefore$
$$\frac{1}{f}=\left(\frac{\mu_g}{\mu_m}-1\right)\left(\frac{1}{R_1}-\frac{1}{R_2}\right)$$

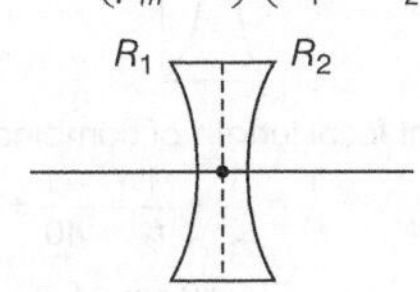

Substituting the values, we have

$$\frac{1}{f}=\left(\frac{1.5}{1.75}-1\right)\left(\frac{1}{-R}-\frac{1}{R}\right)=\frac{1}{3.5R}$$

$$f=+3.5R$$

44. (c) From the first lens parallel beam of light is focused at its focus i.e. at a distance f_1 from it. This image I_1 acts as virtual object for second lens L_2. Therefore for L_2.

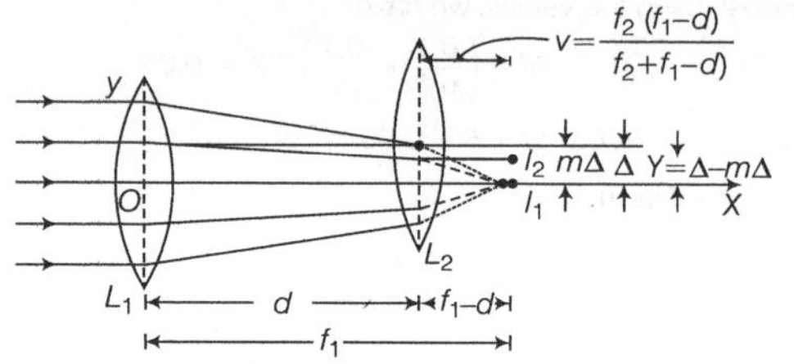

$$u = +(f_1 - d),\ f = +f_2$$

$$\therefore \quad \frac{1}{v} = \frac{1}{f} + \frac{1}{u} = \frac{1}{f_2} + \frac{1}{f_1 - d}$$

Hence, $$v = \frac{f_2(f_1 - d)}{f_2 + f_1 + d}$$

Therefore, x-coordinate of its focal point will be

$$x = d + v = d + \frac{f_2(f_1 - d)}{f_2 + f_1 - d}$$

$$= \frac{f_1 f_2 + d(f_1 - d)}{f_1 + f_2 - d}$$

Linear magnification for L_2

$$m = \frac{v}{u} = \frac{f_2(f_1 - d)}{f_2 + f_1 - d}\,\frac{1}{f_1 - d} = \frac{f_2}{f_2 + f_1 - d}$$

Therefore, second image will be formed at a distance of n

or $\left(\dfrac{f_2}{f_2 + f_1 - d}\right)\Delta$ below its optic axis.

Therefore, y-coordinate of the focus of system will be

$$y = \Delta - \left(\frac{f_2\,\Delta}{f_2 + f_1 - d}\right)$$

or $$y = \frac{(f_1 - d)\,\Delta}{f_2 + f_1 - d}$$

DPP-11 Lens-II

1. (a,b,c)

(a) Here, $u = -30$ cm, $f = +20$ cm

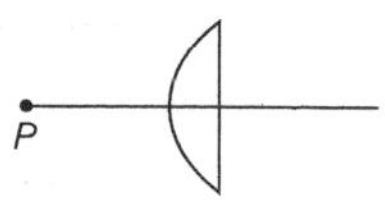

$$\because \quad \frac{1}{v} - \frac{1}{u} = \frac{1}{f}$$

or $$\frac{1}{v} + \frac{1}{30} = \frac{1}{20}$$

or $$\frac{1}{v} = \frac{1}{20} - \frac{1}{30} = \frac{3-2}{60} = \frac{1}{60}$$

$$\therefore \quad v = 60 \text{ cm}$$

(b) $$\because \quad \frac{1}{20} = (1.5 - 1)\left(\frac{1}{R} - \frac{1}{\infty}\right)$$

or $$\frac{1}{20} = \frac{1}{2R}$$

$$\therefore \quad R = 10 \text{ cm}$$

$$\because \quad \frac{1}{f'} = \left(\frac{3}{2 \times \frac{4}{3}} - 1\right)\left(\frac{1}{R} - \frac{1}{\infty}\right) = \frac{1}{8} \times \frac{1}{10} = \frac{1}{80}$$

$$\therefore \quad f' = 80 \text{ cm}$$

$$\because \quad \frac{1}{v} - \frac{1}{u} = \frac{1}{f'}$$

or $$\frac{1}{v} + \frac{1}{30} = \frac{1}{80}$$

or $$\frac{1}{v} = \frac{1}{80} - \frac{1}{30} = \frac{3-8}{240}$$

$$\therefore \quad v = -\frac{240}{5} = -48 \text{ cm}$$

(c)

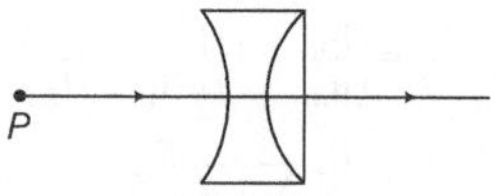

The equivalent focal length of combination is

$$\frac{1}{F} = \frac{1}{f_1} + \frac{1}{f_2} = \frac{-1}{40} + \frac{1}{20} = \frac{-1+2}{40} + \frac{1}{40}$$

$$\therefore \quad F = 40 \text{ cm}$$

$$\because \quad \frac{1}{v} - \frac{1}{u} = \frac{1}{F} \quad \text{or} \quad \frac{1}{v} + \frac{1}{30} = \frac{1}{40}$$

or $$\frac{1}{v} = \frac{1}{40} - \frac{1}{30} = \frac{3-4}{120} = \frac{-1}{120}$$

$$\therefore \quad v = -120 \text{ cm}$$

2. (a,b,c) $$\because \quad \frac{1}{f} = (\mu - 1)\left(\frac{1}{r_1} - \frac{1}{r_2}\right)$$

or $$\frac{1}{20} = \left(\frac{3}{2} - 1\right)\left(\frac{1}{r_1} - \frac{1}{r_2}\right)$$

or $$\frac{1}{20} = \frac{1}{2}\left(\frac{1}{r_1} - \frac{1}{r_2}\right)$$

$$\therefore \quad \frac{1}{r_1} - \frac{1}{r_2} = \frac{1}{10}$$

$\because$ When lens is placed in the medium of refractive index 2.

$$\because \quad \frac{1}{F} = \left(\frac{3}{2 \times 2}\right)\left(\frac{1}{r_1} - \frac{1}{r_2}\right)$$

or $$\frac{1}{F} = \left(\frac{3}{4}\right) \times \frac{1}{10}$$

$$\therefore \quad F = \frac{40}{3} \text{ cm}$$

$$\because \quad u = -40,\ F = +\frac{40}{3} \text{ cm}$$

$$\because \quad \frac{1}{v} - \frac{1}{u} = \frac{1}{F} \quad \text{or} \quad \frac{1}{v} + \frac{1}{40} = \frac{3}{40}$$

or $$\frac{1}{v} = \frac{3}{40} - \frac{1}{40} = \frac{2}{40} = \frac{1}{20}$$

$$\therefore \quad v = 20 \text{ cm}$$

$$m = \frac{A'B'}{AB} = \frac{v}{u} = \frac{20}{-40} = \frac{-1}{2}$$

$$\therefore \quad A'B' = -1 \text{ cm}$$

3. (a,b)

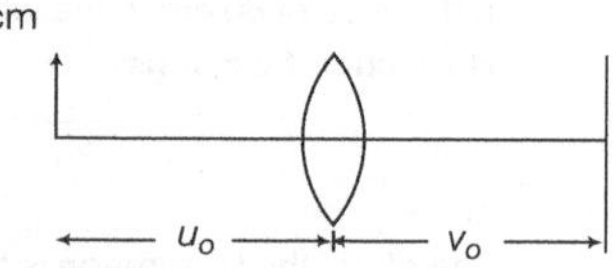

(a) Here, $u_o + v_o = 180$

and $$\frac{v_o}{u_o} = 2$$

$$\therefore \quad v_o = 2\mu_o$$

$$\because \quad u_o + v_o = 180 \quad \text{or} \quad u_o + 2u_o = 180$$

$$\therefore \quad u_o = 60 \text{ cm} \quad \text{and} \quad v_o = +120 \text{ cm}$$

$$\because \quad \frac{1}{v} - \frac{1}{u} = \frac{1}{f} \quad \text{or} \quad \frac{1}{120} + \frac{1}{60} = \frac{1}{f}$$

or $$\frac{1+2}{120} = \frac{1}{f}$$

$$\therefore \quad f = \frac{120}{3} = 40 \text{ cm}$$

(b) $f = \dfrac{D^2 - x^2}{4D} = \dfrac{180^2 - x^2}{4 \times 180}$

or $\quad 40 = \dfrac{180^2 - x^2}{720}$

or $\quad x^2 = 180^2 - 40 \times 720$

$\therefore \quad x = 60$ cm.

4. (a,b,c) Object shiftness $x = t\left(1 - \dfrac{1}{\mu}\right) = 15\left(1 - \dfrac{2}{3}\right) = 5$ cm

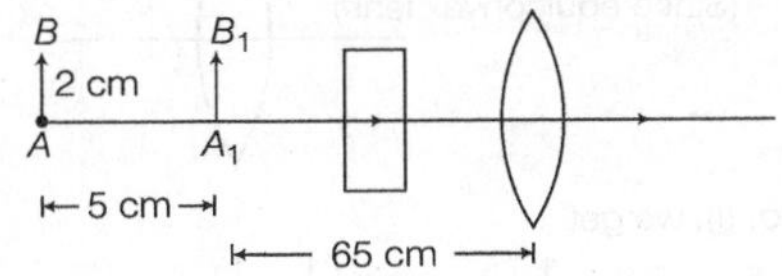

The image of AB is A_1B_1 due to refraction through slab, which is shifted 5 cm from AB. The size of A_1B_1 is same as that of AB.

$A_1 B_1$ behaves as object for the lens.

For convex lens, $A_1 B_1$ behaves as object.

For lens, $\quad u = -60$ cm, $f = 20$ cm

$\because \quad \dfrac{1}{v} - \dfrac{1}{u} = \dfrac{1}{f}$ or $\dfrac{1}{v} + \dfrac{1}{60} = \dfrac{1}{20}$

or $\quad \dfrac{1}{v} = \dfrac{1}{20} - \dfrac{1}{60} = \dfrac{3-1}{60}$

$\therefore \quad v = +30$ cm

$\because \quad \dfrac{A_2 B_2}{A_1 B_1} = \dfrac{v}{u}$

or $\quad A_2 B_2 = \dfrac{v}{u} A_1 B_1 = \left(\dfrac{30}{-60}\right) 2 \text{ cm} = -1 \text{ cm}$

Thus, final image $A_2 B_2$ is real, inverted, 1 cm in length and is 30 cm from the lens.

5. (a,d) **First possibility** If image due to first refraction through lens is at the pole of mirror, then image concides with object.

For this, $\quad v = +20, \quad f = +10$ cm

$u = -x$

$\because \quad \dfrac{1}{v} - \dfrac{1}{u} = \dfrac{1}{f}$ or $\dfrac{1}{20} - \dfrac{1}{u} = \dfrac{1}{10}$

or $\quad \dfrac{1}{u} = \dfrac{1}{20} - \dfrac{1}{10} = \dfrac{1-2}{20}$

$\therefore \quad u = -20$ cm

or $\quad -x = -20$

$\therefore \quad x = 20$ cm

Second possibility If image due to first refraction through lens is at the centre of curvature of mirror, then image coincides with object.

For this, $\quad v = -10$ cm, $f = +10$ cm, $u = -x$

$\because \quad \dfrac{1}{v} - \dfrac{1}{u} = \dfrac{1}{f}$ or $\dfrac{1}{-10} + \dfrac{1}{x} = \dfrac{1}{10}$

or $\quad \dfrac{1}{x} = \dfrac{2}{10} = \dfrac{1}{5}$

$\therefore \quad x = 5$ cm

6. (a,b) For concave lens, $u = -\infty, f = -10$ cm

$\because \quad \dfrac{1}{v} - \dfrac{1}{u} = \dfrac{1}{f}$

$\therefore \quad v = -10$ cm

For lens L_2, $u = -20$ cm, $f = 20$ cm

$\therefore \quad \dfrac{1}{v} - \dfrac{1}{u} = \dfrac{1}{f}$

$\therefore \quad v = \infty$

Thus, for any value of d, beam converges at the focus of lens L_3.

7. (a,b) **First possibility** If image formed due to first refraction through the thin convex lens is at the pole mirror. Thin image coincides with object.

For this, $u = -30$ cm, $f = +20$ cm $v = +d$

$\dfrac{1}{v} - \dfrac{1}{u} = \dfrac{1}{f}$ or $\dfrac{1}{d} + \dfrac{1}{30} = \dfrac{1}{20}$

or $\quad \dfrac{1}{d} = \dfrac{1}{20} - \dfrac{1}{30} = \dfrac{3-2}{60}$

$\therefore \quad d = +60$ cm

Second possibility If image formed due to first refraction through lens is at the centre of curvature of the mirror.

Then, image coincides with the object.

For this, $\quad v = d + R$

$60 = d + 10$

$\therefore \quad d = 50$ cm

8. (a,b,c) Image formation due to first refraction through convex lens.

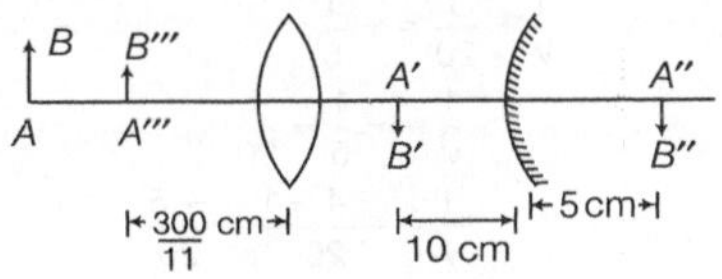

$u = -30, f = +20$

$\because \quad \dfrac{1}{v} - \dfrac{1}{u} = \dfrac{1}{f}$ or $\dfrac{1}{v} + \dfrac{1}{30} = \dfrac{1}{20}$

$\therefore \quad v = +60$ cm

$\because \quad \dfrac{A'B'}{AB} = \dfrac{v}{u} = \dfrac{60}{-30} = -2$

$\therefore \quad A'B' = -2 \times 2 = -4$ cm

The image formed by first refraction through lens behaves as object for convex mirror.

For convex mirror, $u' = -10$ cm, $f' = +10$ cm

$\dfrac{1}{v'} + \dfrac{1}{u'} = \dfrac{1}{f'}$

$\therefore \quad v' = +5$ cm

$\therefore \quad m' = \dfrac{A'' B''}{A'B'} = \dfrac{-v'}{u'}$

$\therefore \quad A'' B'' = -\dfrac{5}{-10} \times (-4) = -2$ cm

$A'' B''$ behaves as object for second times refraction through lens.

$\therefore \quad u'' = -75$ cm

$f'' = +20$ cm

$\therefore \quad \dfrac{1}{v''} - \dfrac{1}{u''} = \dfrac{1}{f''}$

or $\quad \dfrac{1}{v''} = \dfrac{1}{20} - \dfrac{1}{75} = \dfrac{15-4}{300}$

$\therefore \quad v'' = \dfrac{300}{11}$ cm

$\therefore \quad \dfrac{A''' B'''}{A'' B''} = \dfrac{v''}{u''} = \dfrac{300}{11 \times (-75)} = \dfrac{-4}{11}$

$\therefore \quad A''' B''' = \dfrac{-4}{11} \times (-2) = \dfrac{8}{11}$ cm

9. (a,b,c)

(a) Laser beam passing through centre is unaffected because they pass through pole so no deviation takes place in lens as well as in tank.

(b) Because parallel beam through lens cut at focus point.

(c) All the focus in Fig (A).

DPP-12 Lens-III

1. (b) For plane concave lens

$$u = -20 \text{ cm}, f = -5 \text{ cm}$$

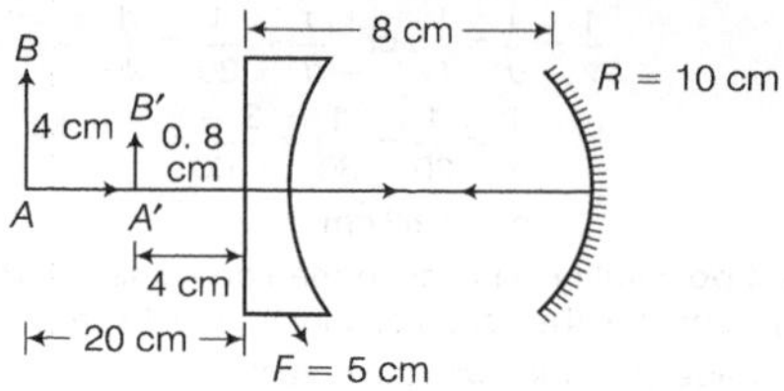

Lens formula,

$$\frac{1}{v} - \frac{1}{u} = \frac{1}{f}$$

$$\frac{1}{v} - \frac{1}{-20} = -\frac{1}{5}$$

$$\frac{1}{v} + \frac{1}{20} = -\frac{1}{5}$$

$$\frac{1}{v} = -\frac{1}{5} - \frac{1}{20}$$

$$\frac{1}{v} = \frac{-4-1}{20} = \frac{-5}{20}$$

$$v = -4 \text{ cm}$$

Magnification, $\dfrac{A'B'}{AB} = \dfrac{v}{u}$

$\therefore$ $A'B' = \dfrac{v}{u} \times AB$

$$= \frac{-4}{-20} \times 4 = 0.8 \text{ cm}$$

2. (a) For mirror, the image formed by refraction though lens behaves as object for mirror.

$$u' = -12 \text{ cm}$$

$$f' = -R/2$$

$$f' = \frac{-10}{2} = -5 \text{ cm}$$

Mirror formula,

$$\frac{1}{v'} + \frac{1}{u'} = \frac{1}{f'}$$

$$\frac{1}{v'} - \frac{1}{12} = -\frac{1}{5}$$

$$\frac{1}{v'} = -\frac{1}{5} + \frac{1}{12}$$

$$= \frac{-12+5}{60} = \frac{-7}{60}$$

$$v' = \frac{-60}{7}$$

The image formed by reflection through concave mirror behave as object for second time refraction and lens formed the final image.

For second time refraction through lens

$$u'' = \frac{-60}{7} + 8, f'' = +5$$

$$\frac{1}{v''} - \frac{1}{u''} = \frac{1}{f''}$$

$\Rightarrow$ $v'' = -0.645$ cm

Hence, final image is $(0.645 + 8) = 8.645$ cm from mirror.

3. (b) He can see two images, one due to first refraction through lens and other due to reflection through the mirror.

4. (b) A paraxial parallel beam of light incident on it. If the final image is formed after two internal reflections.

$$\frac{1}{f} (\mu - 1)\left(\frac{1}{r_1} - \frac{1}{r_2}\right) \quad \text{...(i) (in air)}$$

$\mu = 1.5, r_1 = r, r_2 = -r$ (Since equiconvex lens)

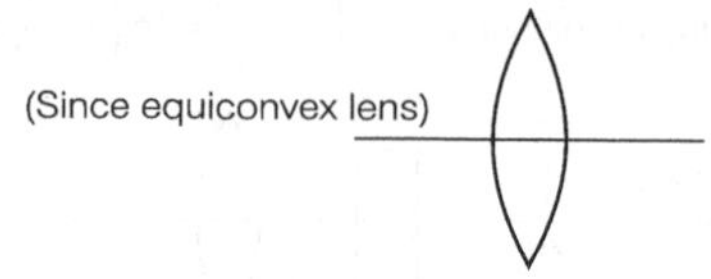

From Eq. (i), we get

$$\frac{1}{f} = (1.5 - 1)\left(\frac{1}{r} + \frac{1}{r}\right)$$

$$\frac{1}{f} = \frac{1}{2} \cdot \left(\frac{2}{r}\right) \Rightarrow f = r$$

where, $f = 30$ cm

So that, $r = 30$ cm

5. (a) First reflection from first surface,

$$\mu_1 = 1, \mu_2 = \frac{3}{2}, r = +30 \text{ cm}, u = -\infty$$

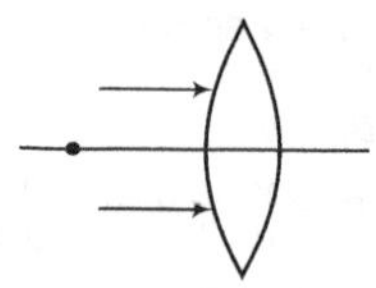

$$\frac{\mu_2}{v} - \frac{\mu_1}{u} = \frac{\mu_2 - \mu_1}{r}$$

$$\frac{3}{2v} - \frac{1}{-\infty} = \frac{\frac{3}{2} - 1}{30} \Rightarrow \frac{3}{2v} = \frac{1}{60}$$

$$2v = 180 \Rightarrow v = 90 \text{ cm}$$

6. (b) First reflection from second surface.

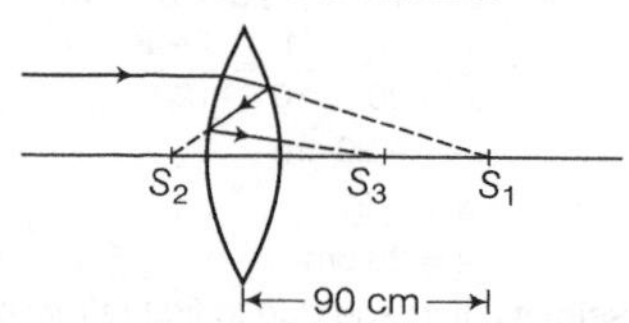

$\mu' = +90$ cm, $f' = -15$ cm

$$\frac{1}{v'} + \frac{1}{u'} = \frac{1}{f'}$$

$$\frac{1}{v'} + \frac{1}{90} = -\frac{1}{15}$$

or $$\frac{1}{v'} = \frac{-1}{15} - \frac{1}{90} = \frac{-6-1}{90} = \frac{-7}{90} \text{ cm}$$

Now, second reflection from 1st surface

$$u'' = v', f'' = +15 \text{ cm}$$

$$\frac{1}{v''} + \frac{1}{u''} = \frac{1}{f''}$$

$$\frac{1}{v''} - \frac{7}{90} = \frac{1}{15}$$

$$\frac{1}{v''} = \frac{7}{90} + \frac{1}{15} = \frac{7+6}{90} = \frac{13}{90}$$

For final refraction from second surface

$$\mu_1''' = \frac{3}{2}, \mu_2''' = 1, \mu''' = v''$$

$$r''' = -30 \text{ cm}$$

$$\frac{\mu_2'''}{v'''} - \frac{\mu_1'''}{u'''} = \frac{\mu_2''' - \mu_1'''}{r'''}$$

$$\frac{1}{v'''} - \frac{3 \times 13}{2 \times 90} = \frac{1 - \frac{3}{2}}{30}$$

$$\therefore \quad \frac{1}{v'''} = \frac{39}{180} + \frac{1}{60}$$

$$= \frac{39 + 3}{180} = \frac{7}{30}$$

$$\therefore \quad v''' = \frac{30}{7} = 4.2$$

7. (d) $\frac{1}{f} = (\mu - 1)\left(\frac{1}{r_1} - \frac{1}{r_2}\right)$

$$\frac{1}{f} = (1.5 - 1)\left(\frac{1}{20} - \frac{1}{60}\right) = \frac{1}{2}\left(\frac{3-1}{60}\right) = \frac{2}{120}$$

$\Rightarrow f = 60$ cm

8. (b) For lens, $u = -80$ cm, $f = 60$ cm

$$\frac{1}{v} - \frac{1}{u} = \frac{1}{f}$$

or $$\frac{1}{v} + \frac{1}{80} = \frac{1}{60}$$

or $$\frac{1}{v} = \frac{1}{60} - \frac{1}{80} = \frac{4-3}{240} = \frac{1}{240}$$

$\therefore \quad v = 240$ cm

$\therefore$ Distance to image from mirror is $= 240 - 160 = 80$ cm

9. (a) For plane mirror image formed by the lens behaves as object, which is situated by the plane mirror. This image behaves virtual object for the plane mirror. Hence, image must be real.

10. (b) This image formed by plane mirror is 80 cm below.
This image behave as object for water surface.
For water surface

$$\mu = 1, \mu_2 = \mu, v = -20 \text{ cm}, r = \infty, u = -15 \text{ cm}$$

$$\therefore \quad \frac{\mu_2}{v} - \frac{\mu_1}{u} = \frac{\mu_2 - \mu_2}{\infty} = 0$$

or $$\frac{\mu}{-20} - \frac{1}{-15} = 0$$

or $$\frac{\mu}{20} = \frac{1}{15}$$

$$\therefore \quad \mu = \frac{4}{3}$$

11. (b) For virtual object

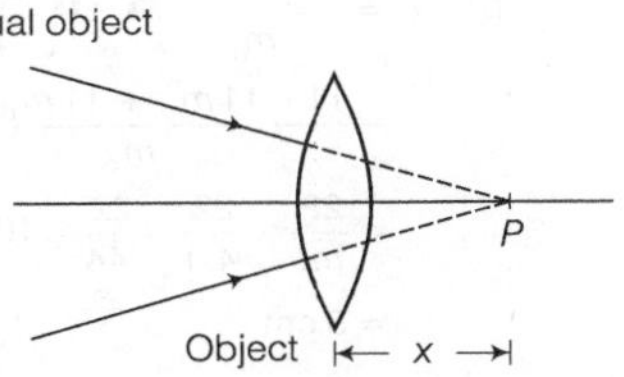

$$u = +x, F = +f$$

$\because$ $$\frac{1}{v} - \frac{1}{u} = \frac{1}{F}$$

or $$\frac{1}{v} - \frac{1}{x} = \frac{1}{F}$$

or $$\frac{1}{v} = \frac{1}{F} + \frac{1}{x}$$

$\therefore \quad v > 0$, it means image is real.

12. (a)

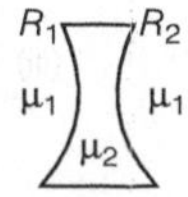

$$\frac{1}{f} = \left(\frac{\mu_2}{\mu_1} - 1\right)\left(\frac{1}{r_1} - \frac{1}{r_2}\right)$$

$$= \left(\frac{\mu_2}{\mu_1} - 1\right)\left(-\frac{1}{R_1} - \frac{1}{R_2}\right)$$

Here, $\mu_2 < \mu_1$, the value of f is positive. Hence, lens behaves as converging lens.

13. (a) If a portion of lens is covered by black paper, radio of surfaces does not change. Hence, focal length does not change.

14. (a) $\because$ $$\frac{1}{f_1} = \left(\frac{3}{2} - 1\right)\left(\frac{1}{r_1} - \frac{1}{r_2}\right)$$

and $$\frac{1}{f_2} = \left(\frac{3}{2 \times \frac{4}{3}} - 1\right)\left(\frac{1}{r_1} - \frac{1}{r_2}\right)$$

$$= \frac{1}{8}\left(\frac{1}{r_1} - \frac{1}{r_2}\right)$$

Hence, $f_2 > f_1$.

15. (d) Here, $u = -10$ cm, $f = -10$ cm

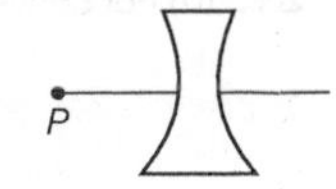

$$\therefore \quad \frac{1}{v} - \frac{1}{u} = \frac{1}{f}$$

or $$\frac{1}{v} + \frac{1}{10} = -\frac{1}{10}$$

$\therefore \quad v = -5$ cm

So, Statement I is incorrect. Hence, option (d) is correct

16. (b) Statement I

$\because$ $$\frac{1}{v} - \frac{1}{u} = \frac{1}{f}$$

or $$\frac{1}{v} = \frac{1}{u} + \frac{1}{f}$$

Here, $u > 0$ and $f < 0$

If $|u| > |f|$, $v < 0$, image is virtual.

If $|u| < |f|$, $v > 0$, image is real. Hence, Statement I is correct.

17. (a)

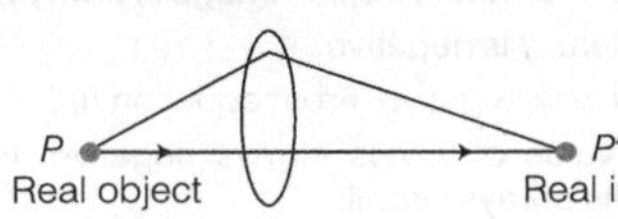

18. (c) When mirror is rotated, second refraction through lens does not take place. So, statement I is correct. Law of reflection does not depend on orientation of mirror. So statement II is incorrect.

19. (a)

(i) $\because$ $$P = P_1 + P_2$$

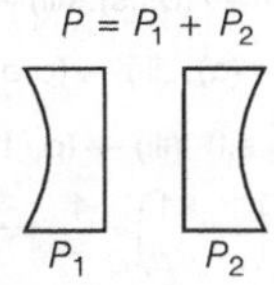

or $$-\frac{1}{20} = 2P_1$$

$\because$ $$P_1 = -\frac{1}{40}$$

(ii) $\because \frac{1}{f} = (\mu_2 - \mu_1)\left(\frac{1}{r_1} - \frac{1}{r_2}\right) = -\frac{1}{2.0}$

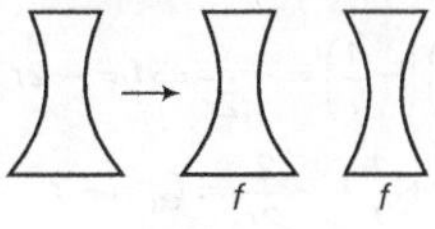

(iii) $\because \frac{1}{F} = \left(\frac{\mu_2}{\mu_1} - 1\right)\left(\frac{1}{r_1} - \frac{1}{r_2}\right) = 0$ $\quad (\because \mu_2 = \mu_1)$

$\therefore F = \infty$, So it acts as slab.

(iv) $\frac{1}{F} = \left(\frac{\mu_2}{\mu_1} - 1\right)\left(\frac{1}{r_1} - \frac{1}{r_2}\right) > 0$

$\therefore$ It behaves as converging lens.

Hence, (i) → (q), (ii) → (r), (iii) → (p), (iv) → (s)

20. (d)

(a) Observer A will get refracted beam through the lens only. Also, it will get incidence beam from the object.

(b) Observer B will get two refracted beam one due to convex lens and other due to concave lens.

(c) Observer C will get refracted rays from concave lens and other reflected rays from the mirror.

(d) Observer D will get non of beam.

Hence, (i) → (q,s) (ii) → (p), (iii) → (p) (iv) → (r)

21. (d) For converging lens, focal length is positive.

$$\frac{1}{v} - \frac{1}{u} = \frac{1}{f}$$

$$\frac{1}{v} = \frac{1}{f} + \frac{1}{u} \quad \text{...(i)}$$

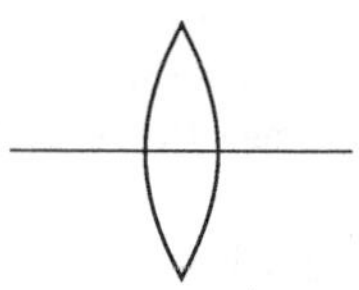

For real object, u is negative.

From equation (i) for negative value of u, v may be positive or may be negative depends upon value of f.

Hence, for real object image is either virtual or real.

For virtual object, u is positive. For positive value of u, v is always positive. Hence, for virtual object, image is always real.

For diverging lens, f is negative.

For real object, u is negative. From equation (i).

For negative value of u, v is always negative. Hence, for real object, image is always virtual.

But for virtual object, u is positive

For positive value of u, v is either positive or negative depends upon the value of f.

Hence, for virtual object, image is either real or virtual.

Hence, (i) → (p,q,s), (ii) → (p,r,s), (iii) → (p,q,s), (iv) → (p,r,s)

22. (a) (i) → (p,q,r,s), (ii) → (q), (iii) → (p,q,r,s), (iv) → (p,q,r,s)

23. (a) (i) → (p,r), (ii) → (q,s,t), (iii) → (p,r,t), (iv) → (q,s)

24. (c) (p) 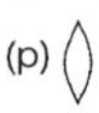 $\frac{1}{f} = \left(\frac{3}{2} - 1\right)\left(\frac{1}{r} + \frac{1}{r}\right) = \frac{1}{r} \Rightarrow f = r$

$\Rightarrow$  $\frac{1}{f_{eq}} = \frac{1}{f} + \frac{1}{f} = \frac{2}{r} \Rightarrow f_{eq} = \frac{r}{2}$

(q) $\frac{1}{f} = \left(\frac{3}{2} - 1\right)\left(\frac{1}{r}\right) \Rightarrow f = 2r$

$\Rightarrow$  $\frac{1}{\phi} + \frac{1}{\phi} = \frac{2}{\phi} = \frac{1}{\rho} \Rightarrow f_{eq} = r$

(r)  $\frac{1}{f} = \left(\frac{3}{2} - 1\right)\left(-\frac{1}{r}\right) = -\frac{1}{2r} \Rightarrow f = -2r$

$\Rightarrow$ 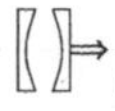 $\Rightarrow \frac{1}{f_{eq}} = \frac{1}{f} + \frac{1}{f} = -\frac{2}{2r} \Rightarrow f_{eq} = -r$

(s) 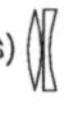 $\Rightarrow \frac{1}{f_{eq}} = \frac{1}{r} + \frac{1}{-2r} = \frac{1}{2r} \Rightarrow f_{eq} = 2r$

25. (4) From figure,

$BA = \frac{d}{2}$, where d = aperture of the lens.

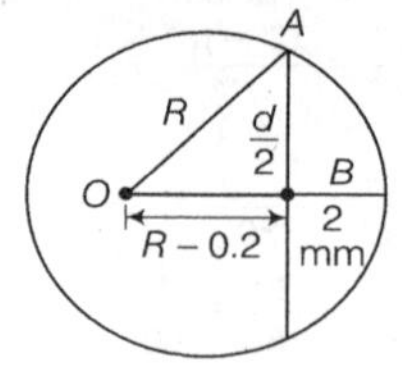

In ΔOAB, $\quad R^2 = (R - 0.2)^2 + \frac{d^2}{4}$

or $\quad 10^2 = R^2\left(1 - \frac{0.2}{10}\right)^2 + \frac{d^2}{4}$

or $\quad 100 = 100\left(1 - \frac{0.4}{10}\right) + \frac{d^2}{4}$

or $\quad \frac{d^2}{4} = 4$

$\because \quad d = 4$ cm

26. (5) If image is real, $\quad m = -m_o = \frac{v}{u}$

$\because \quad v = -m_o u$

$\because \quad \frac{1}{v} - \frac{1}{u} = \frac{1}{11}$ or $-\frac{1}{m_o u} - \frac{1}{u} = \frac{1}{11}$

or $\quad -\frac{1}{u}\left(\frac{1}{m_o} + 1\right) = \frac{1}{11}$

or $\quad \frac{m_o u}{m_o + 1} = -11$

$\because \quad u = \frac{(m_o + 1)(-11)}{m_o}$

If image is virtual, $m + m_o$

$\because \quad m = m_o = \frac{v'}{u'}$

$\therefore \quad v' = m_o u'$

$\therefore \quad \frac{1}{v'} - \frac{1}{u'} = \frac{1}{11}$

or $\quad \frac{1}{m_o u'} - \frac{1}{u'} = \frac{1}{11}$

or $\quad \frac{1}{u'}\left(\frac{1 - m_o}{m_o}\right) = \frac{1}{11}$

or $\quad u' = \frac{11(1 - m_o)}{m_o}$

$\therefore \quad u' - u = \frac{11(1 - m_o)}{m_o} + 11\left(\frac{m_o + 1}{m_o}\right)$

$$= \frac{11 - 11m_o + 11m_o + 11}{m_o}$$

$$= \frac{22}{m_o} = \frac{22}{4.4} = \frac{22}{44} \times 10$$

$$= 5 \text{ cm}$$

27. (2) $\because$ Intensity $I \propto d^2$

$\therefore \quad 10 = Kd^2$

and $\quad I' = K\left(\frac{d}{2}\right)^2 = 2.5$

$\therefore \quad$ Intensity of image $= 10 - 2.5$

$= 7.5$ Wm^{-2}

$= 3.75\ x$

$\therefore \quad x = \frac{7.5}{3.75} = 2$

28. (9) From figure,

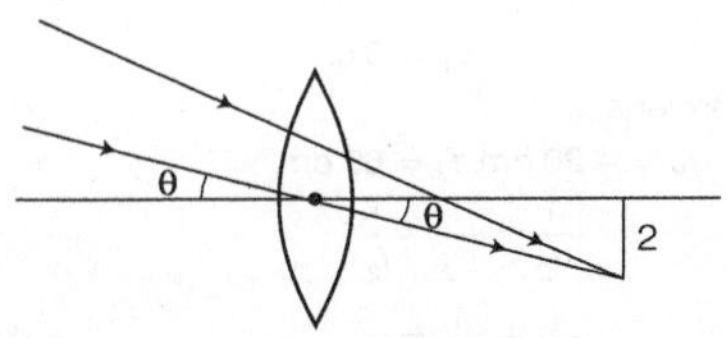

$$\tan\theta = \theta = \frac{2}{f}$$

or $\quad 2 \times \frac{\pi}{180} = \frac{\pi}{f}$

$\therefore \quad f = \frac{180}{2} = 90 \text{ cm}$

29. (5) Here, $\quad P = \frac{1}{f_1} + \frac{1}{f_1} = \frac{2}{f_1}$

$$= 2(1.5 - 1)\left(\frac{1}{R} + \frac{1}{R}\right)$$

$$= 2 \times \frac{1}{2} \times \frac{2}{R} = \frac{2}{R}$$

$\therefore \quad R = \frac{2}{P} \quad \ldots\text{(i)}$

When gap is filled with liquid,

$$\frac{P}{3} = P_l + P_l' + P_l = 2P_l + P_l'$$

$$= 2P_l + (\mu - 1)\left(-\frac{1}{R} - \frac{1}{R}\right)$$

$$= 2P_l - \frac{2(\mu - 1)}{R}$$

$$\frac{P}{3} = P - \frac{2}{R}(\mu - 1)$$

or $\quad \frac{2}{R}(\mu - 1) = \frac{2P}{3}$

or $\quad \mu - 1 = \frac{PR}{3} = \frac{P}{3} \times \frac{2}{P} = \frac{2}{3}$

$\because \quad \mu = 1 + \frac{2}{3} = \frac{5}{3}$

30. (5) For point A,

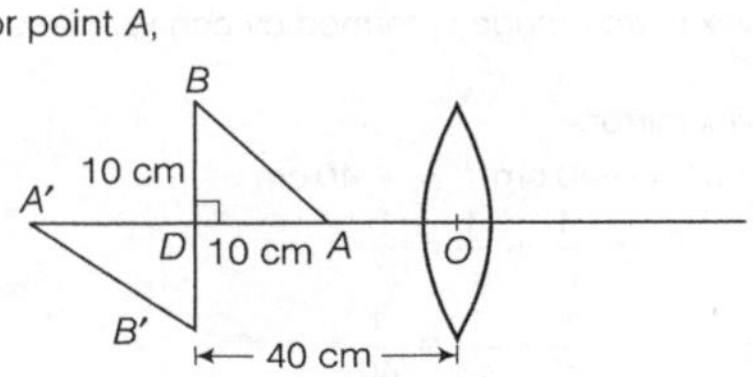

$$u = -30 \text{ cm}$$

$$f = 20 \text{ cm}$$

$$\frac{1}{v} - \frac{1}{u} = \frac{1}{f}$$

or $\quad \frac{1}{v} + \frac{1}{30} = \frac{1}{20}$

or $\quad \frac{1}{v} = \frac{1}{20} - \frac{1}{30} = \frac{3 - 2}{60}$

$\therefore \quad v = +60 \text{ cm}$

For point D

$$u' = -40 \text{ cm}, f' = +20 \text{ cm}$$

$\therefore \quad \frac{1}{v'} - \frac{1}{u'} = \frac{1}{f}$

or $\quad \frac{1}{v'} + \frac{1}{40} = \frac{1}{20}$

or $\quad \frac{1}{v'} = \frac{1}{20} - \frac{1}{40} = \frac{2 - 1}{40}$

$\because \quad v' = 40 \text{ cm}$

$\because \quad \frac{B'D'}{BD} = \frac{v'}{u'} = \frac{40}{-40}$

$\because \quad B'D' = -1 \times 10 \text{ cm} = -10 \text{ cm}$

$\because \quad A'B' = \sqrt{(A'D)^2 + (B'D')^2}$

$$= \sqrt{(20)^2 + 10^2} = \sqrt{500} = 10\sqrt{5} \text{ cm}$$

$$= 10\sqrt{n} \text{ cm}$$

$\therefore \quad n = 5.$

31. (9) The beam converges in focal plane of the lens.

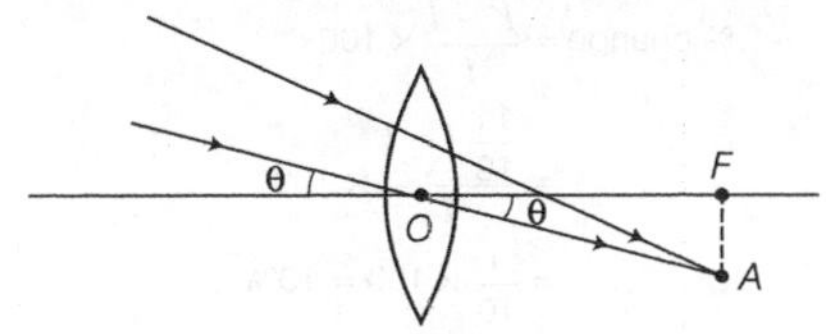

$\therefore \quad \tan\theta = \theta = \frac{AF}{OF}$

or $\quad 2 \times \frac{\pi}{180} = \frac{AF}{20}$

$\therefore \quad AF = \frac{\pi}{90} \times 20 = \frac{2\pi}{9} \text{ cm}$

$\therefore \quad n = 9.$

32. (4) If the focal length of the unsilvered lens is f.

$\because \quad \frac{1}{f} = (\mu - 1)\left(\frac{1}{r_1} - \frac{1}{r_2}\right)$

or $\quad \frac{1}{f} = (1.5 - 1)\left(\frac{1}{40} + \frac{1}{40}\right)$

$$= \frac{1}{40}$$

$\because \quad f = 40 \text{ cm}$

For image formation due to unsilvered part,

$u = -30 \text{ cm}, f = +40 \text{ cm}$

$\because \quad \frac{1}{v} - \frac{1}{u} = \frac{1}{f}$

or $\quad \frac{1}{v} + \frac{1}{30} = \frac{1}{40}$

or $\quad \frac{1}{v} = \frac{1}{40} - \frac{1}{30} = \frac{3 - 4}{120}$

$\because \quad v = -120 \text{ cm.}$

For silvered part, the equivalent power is

$$P = 2P_l + P_m = \frac{2}{f} - \frac{1}{f_m}$$

$$-\frac{1}{F} = -\frac{2}{40} + \frac{1}{20} = \frac{1}{20} + \frac{1}{20}$$

$\because \quad F = -10 \text{ cm}$

Thus, the silvered part behaves as a concave mirror.

$$u' = -30 \text{ cm}, F = -10 \text{ cm}$$

$$\frac{1}{v'} + \frac{1}{u'} = \frac{1}{F}$$

or $\quad \frac{1}{v'} - \frac{1}{30} = -\frac{1}{10}$

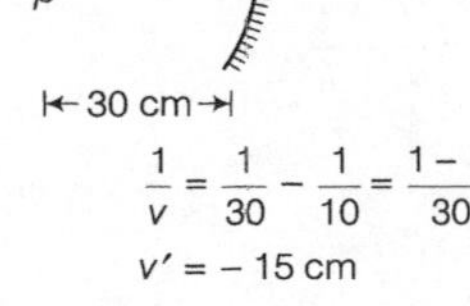

or $\quad \frac{1}{v} = \frac{1}{30} - \frac{1}{10} = \frac{1 - 3}{30}$

$\therefore \quad v' = -15 \text{ cm}$

Thus, the distance between images is $|v' - v|$

$$= |15 + 120| = 105 \text{ cm}$$

$$\Rightarrow \quad \frac{420}{n}$$

$$\therefore \quad n = \frac{420}{105} = 4$$

33. (5) $$m = \frac{l}{10} = \frac{l'}{11}$$

$$\because \quad \frac{l'}{l} = \frac{11}{10}$$

$$\% \text{ change} = \frac{l' - l}{l} \times 100$$

$$= \frac{\frac{11}{10} l - l}{l} \times 100$$

$$= \frac{1}{10} \times 100 = 10\%$$

$$= 2n$$

$$\therefore \quad n = 5$$

34. (9) From figure, $d = f_1 + f_2 = 5 + 4$

$$= 9 \text{ cm}$$

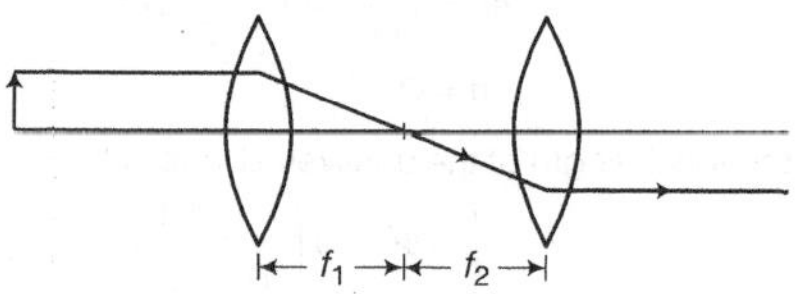

35. (2) Here, $u = -30$ cm, $f = +15$ cm

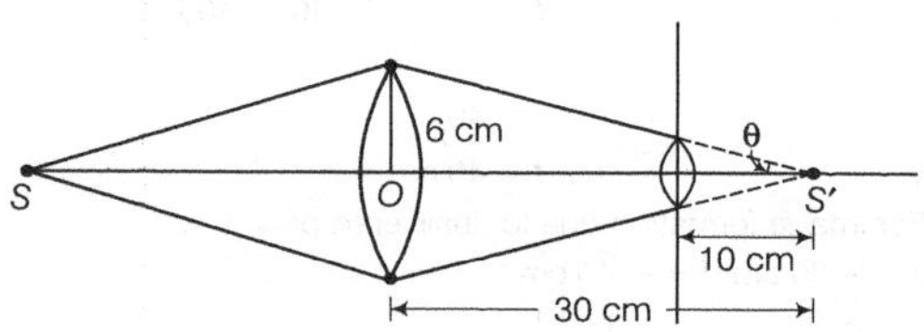

$$\because \quad \frac{1}{v} - \frac{1}{u} = \frac{1}{f}$$

$$\text{or} \quad \frac{1}{v} + \frac{1}{30} = \frac{1}{15}$$

$$\therefore \quad v = 30 \text{ cm}$$

$$\tan\theta = \frac{r}{10 \text{ cm}} = \frac{6 \text{ mm}}{30 \text{ cm}}$$

$$\therefore \quad r = \frac{6 \text{ mm}}{30 \text{ cm}} \times 10 \text{ cm} = 2 \text{ mm}$$

36. (8) For concave lens,

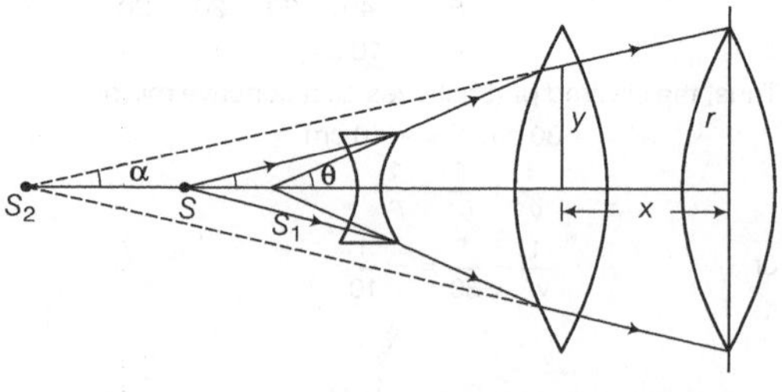

$$u_1 = -10 \text{ cm}, f_1 = -10 \text{ cm}$$

$$\because \quad \frac{1}{v_1} - \frac{1}{u_1} = \frac{1}{f_1}$$

$$\text{or} \quad \frac{1}{v_1} + \frac{1}{10} = -\frac{1}{10}$$

$$\therefore \quad v_1 = 5 \text{ cm}$$

For convex lens,

$$u_2 = -20 \text{ cm}, f_2 = 60 \text{ cm}$$

$$\because \quad \frac{1}{v_2} - \frac{1}{u_2} = \frac{1}{f_2}$$

$$\text{or} \quad \frac{1}{v_2} + \frac{1}{20} = \frac{1}{60}$$

$$\text{or} \quad \frac{1}{v_2} = \frac{1}{60} - \frac{1}{20} = \frac{1-3}{60}$$

$$\therefore \quad v_2 = -30 \text{ cm}$$

From figure, $\tan\theta = \frac{y}{20} = \frac{1 \text{ mm}}{5}$

$$\therefore \quad y = 4 \text{ mm}$$

And $\tan\alpha = \frac{y}{30} = \frac{r}{x + 30}$

$$\therefore \quad r = \frac{4 \text{ mm}}{30}(x + 30)$$

$$= \frac{4 \text{ mm}}{30} \times 60$$

$$= 8 \text{ mm}$$

37. (4) For concave lens,

$$u = -60 \text{ cm}, f = -60 \text{ cm}$$

$$\frac{1}{v} - \frac{1}{u} = \frac{1}{f}$$

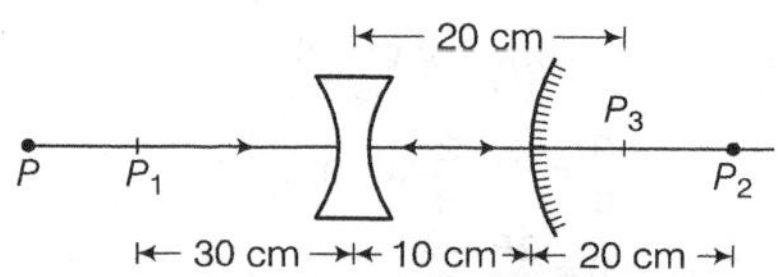

$$\text{or} \quad \frac{1}{v} + \frac{1}{60} = -\frac{1}{60}$$

$$\text{or} \quad \frac{1}{v} = -\frac{1}{30}$$

$$\therefore \quad v = -30 \text{ cm}$$

For convex mirror image P_1 formed by concave lens behaves as object.

For convex mirror,

$$u' = -40 \text{ cm}, f' = +40 \text{ cm}$$

$$\because \quad \frac{1}{v'} + \frac{1}{u'} = \frac{1}{f'}$$

$$\text{or} \quad \frac{1}{v'} - \frac{1}{40} = \frac{1}{40}$$

$$\therefore \quad v' = +20 \text{ cm}$$

Image P_2 formed by convex mirror behaves as object for second times refraction through lens.

Here, $u'' = -30$ cm, $f'' = -60$ cm

$$\therefore \quad \frac{1}{v''} - \frac{1}{u''} = \frac{1}{f''}$$

$$\text{or} \quad \frac{1}{v''} + \frac{1}{30} = -\frac{1}{60}$$

$$\text{or} \quad \frac{1}{v''} = -\frac{1}{30} - \frac{1}{60} = \frac{-2-1}{60}$$

$$\therefore \quad v'' = -\frac{60}{3} = -20 \text{ cm}$$

$$\therefore \quad 5x_0 = 20$$

$$\therefore \quad x_0 = 4$$

38. (8) Here, $\tan\delta = \dfrac{\Delta\mu l}{d}$

(Here, $\Delta\mu l$ is difference in optical path in lower and upper face ray).

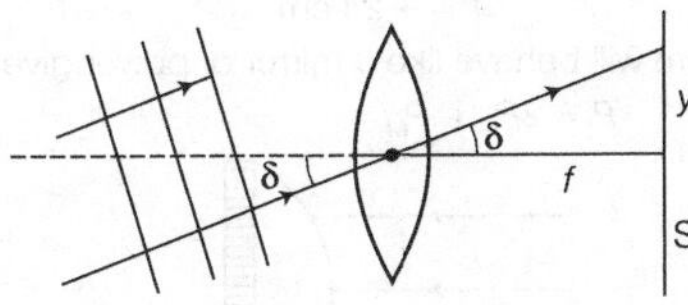

or $\dfrac{y_0}{f} = \dfrac{\Delta\mu l}{d}$

$\because$ $y_0 = \dfrac{f\Delta\mu l}{d} = \dfrac{40 \times 0.1 \times 10}{5} = 8$ cm

39. (4) If lens is placed in air

$$\frac{1}{f} = (\mu - 1)\left(\frac{1}{r_1} - \frac{1}{r_2}\right)$$

or $\dfrac{1}{f} = (1.5 - 1)\left(\dfrac{1}{R} + \dfrac{1}{R}\right)$

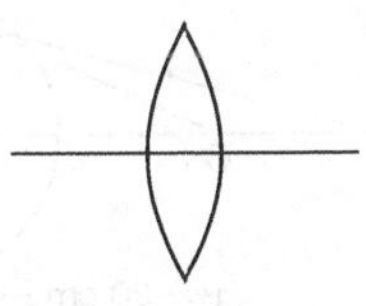

or $\dfrac{1}{20} = \dfrac{1}{R}$

$\therefore$ $R = 20$ cm

For refraction through first surface of lens,

$$\mu_1 = 1, \mu_2 = 1.5$$

$$u = -\infty, r = +20 \text{ cm}$$

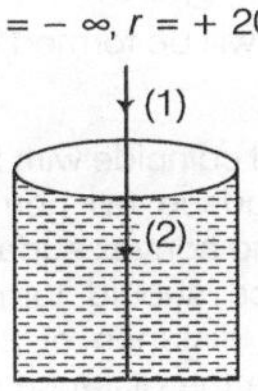

$\because$ $\dfrac{\mu_2}{v} - \dfrac{\mu_1}{u} = \dfrac{\mu_2 - \mu_1}{r}$

or $\dfrac{1.5}{v} - \dfrac{1}{-\infty} = \dfrac{1.5 - 1}{20}$

or $\dfrac{1.5}{v} = \dfrac{1}{40}$

$\therefore$ $v = 60$ cm

For second surface of lens,

$$u' = v = 60 \text{ cm}, r' = -20 \text{ cm}$$

$$v' = +h = \text{height of vessel}$$

$$\mu_1' = \frac{3}{2}, \mu_2' = \frac{4}{3}$$

$\therefore$ $\dfrac{\mu_2'}{v'} - \dfrac{\mu_1'}{u'} = \dfrac{\mu_2' - \mu_1'}{r'}$

$\therefore$ $h = 40 \text{ cm} = 10\,n$

$\therefore$ $n = 4$

40. (5) For refraction through first surface,

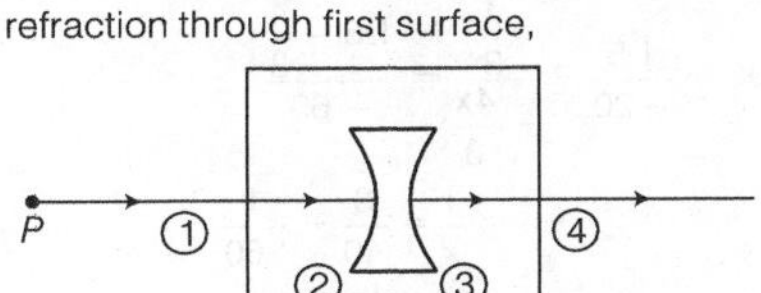

$$\mu_1 = 1, \mu_2 = \frac{3}{2}, u = -10 \text{ cm}$$

$$r = \infty$$

$\therefore$ $\dfrac{\mu_2}{v} - \dfrac{\mu_1}{u} = \dfrac{\mu_2 - \mu_1}{\infty} = 0$

or $\dfrac{3}{2v} - \dfrac{1}{-10} = 0$

$\therefore$ $v = -15$ cm

For second surface

$$\mu_1' = \frac{3}{2}, \mu_2' = 1, u' = -30 \text{ cm}$$

$$r' = -20 \text{ cm}$$

$\because$ $\dfrac{\mu_2'}{v'} - \dfrac{\mu_1'}{u'} = \dfrac{\mu_2' - \mu_1'}{r'}$

or $\dfrac{1}{v'} - \dfrac{3}{2(-30)} = \dfrac{1 - \frac{3}{2}}{-20} = \dfrac{1}{40}$

or $\dfrac{1}{v'} = \dfrac{1}{40} - \dfrac{1}{20}$

$= \dfrac{1 - 2}{40} = -\dfrac{1}{40}$

$\therefore$ $v' = -40$ cm

For third surface,

$$\mu_1'' = 3/2\ \mu_2'' = 1, r'' = +20 \text{ cm}, u'' = -40 \text{ cm}$$

$\therefore$ $\dfrac{\mu_2''}{v''} - \dfrac{\mu_1''}{u''} = \dfrac{\mu_2'' - \mu_1''}{r''}$

or $\dfrac{1}{v''} - \dfrac{3}{2(-40)} = \dfrac{\frac{3}{2} - 1}{20} = \dfrac{1}{40}$

or $\dfrac{1}{v''} = \dfrac{1}{40} - \dfrac{3}{80} = \dfrac{2 - 3}{80} = -\dfrac{1}{80}$

$\therefore$ $v'' = -80$ cm

For fourth surface, $\mu_1''' = \dfrac{3}{2}, \mu_2''' = 1, r''' = \infty$

$u''' = 90$ cm $\dfrac{\mu_2'''}{v'''} - \dfrac{\mu_1'''}{u'''} = 0$

$\therefore$ $v''' = -60$ cm

$$5x = 60 - 35 = 25$$

$\therefore$ $x = 5$

41. (1) For first part L_1 of lens 1. Here, $u = -30$ cm, $f = 20$

$\because$ $\dfrac{1}{v} - \dfrac{1}{u} = \dfrac{1}{f}$

$\because$ $v = +60$ cm

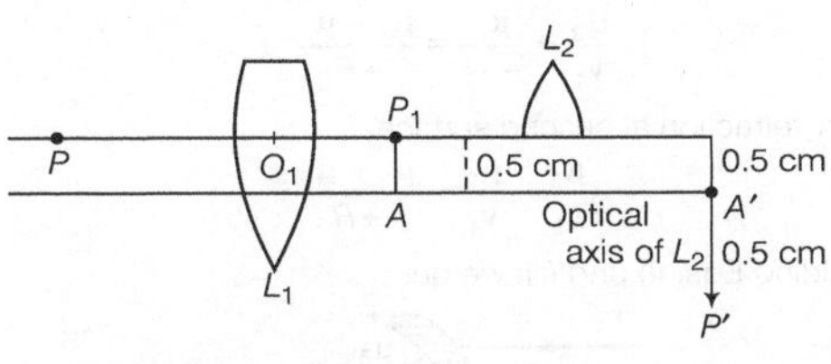

For second part L_2,

$$u' = -40 \text{ cm}, f' = 20 \text{ cm}$$

$\because$ $\dfrac{1}{v'} - \dfrac{1}{u'} = \dfrac{1}{f'}$

or $\dfrac{1}{v'} + \dfrac{1}{40} = \dfrac{1}{20}$

or $\dfrac{1}{v'} = \dfrac{1}{20} - \dfrac{1}{40} = \dfrac{2 - 1}{40}$

$\therefore$ $v' = +40$ cm

$\therefore$ $\dfrac{A'P'}{AP} = \dfrac{v'}{u'}$

$\therefore$ $A'P' = \dfrac{40}{-40}(0.5)$

$= -0.5$ cm

42. (6) $\frac{1}{v} - \frac{1}{u} = \frac{1}{f}$

or $$\frac{u}{v} - 1 = \frac{u}{f}$$

or $$\frac{u}{v} = \left(\frac{u+f}{f}\right)$$

$$m = \frac{v}{u} = \left(\frac{f}{u+f}\right)$$

$$\frac{m_{25}}{m_{50}} = \frac{\left(\frac{20}{-25+20}\right)}{\left(\frac{20}{-50+20}\right)} = 6.$$

43. (11) The given system behaves like a mirror of power given by

$$P = 2P_L + P_M$$

or $$-\frac{1}{F} = 2\left(\frac{1}{0.2}\right) + \left(\frac{-2}{0.22}\right)$$

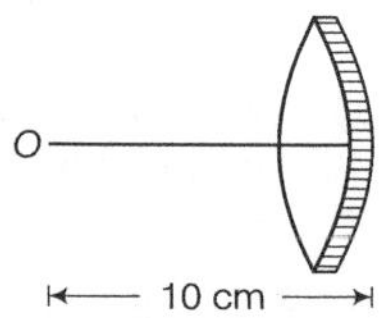

As $$P_L = \frac{1}{f(m)}$$

and $$P_M = \frac{-1}{f(m)} = \frac{-2}{R(m)}$$

Solving this equation, we get

$$F = -1.1\,\text{m} = -110\,\text{cm}$$

i.e. the system behaves as a concave mirror of focal length 110 cm.

Using the mirror formula $\frac{1}{v} + \frac{1}{u} = \frac{1}{f}$ we have

$$\frac{1}{v} - \frac{1}{10} = -\frac{1}{110}$$

or $$v = 11\,\text{cm}$$

i.e. virtual image will be formed at a distance of 11 cm.

44. For refraction at first surface,

$$\frac{\mu_2}{v_1} - \frac{\mu_1}{-\infty} = \frac{\mu_2 - \mu_1}{+R} \quad \text{...(i)}$$

For refraction at second surface.

$$\frac{\mu_3}{v_2} - \frac{\mu_2}{v_1} = \frac{\mu_3 - \mu_2}{+R} \quad \text{...(ii)}$$

Adding Eqs. (i) and (ii), we get

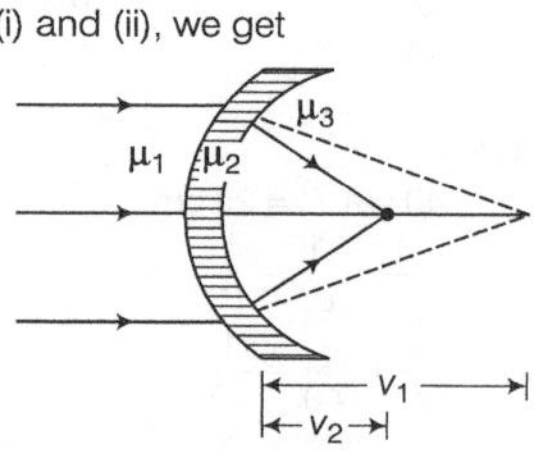

$$\frac{\mu_3}{v_2} = \frac{\mu_3 - \mu_1}{R}$$

or $$v_2 = \frac{\mu_3 R}{\mu_3 - \mu_1}$$

Therefore, focal length of the given lens system is

$$\frac{\mu_3 R}{\mu_3 - \mu_1}.$$

45. (i) $\frac{1}{f} = (\mu - 1)\left(\frac{1}{R_1} - \frac{1}{R_2}\right) = (1.5 - 1)\left(\frac{1}{12} - \frac{1}{\infty}\right) = \frac{1}{24}$

$\therefore$ $$f = +24\,\text{cm}$$

(ii) The system will behave like a mirror of power given by

$$P = 2P_L + P_M$$

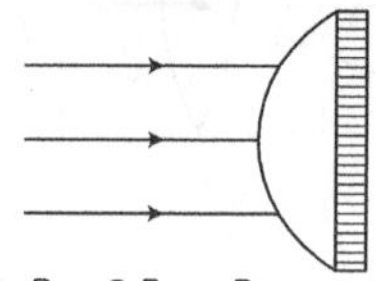

$$P = 2P_L + P_M$$

$\therefore$ $$-\frac{1}{F(m)} = 2\left(\frac{1}{0.24}\right) + 0$$

$\therefore$ $$F = = -0.12\,\text{m} = -12\,\text{cm}$$

Hence, the system will behaves like a concave mirror of focal length 12 cm. Therefore parallel rays will converge at a distance of 12 cm (to the left) of the system.

(iii)

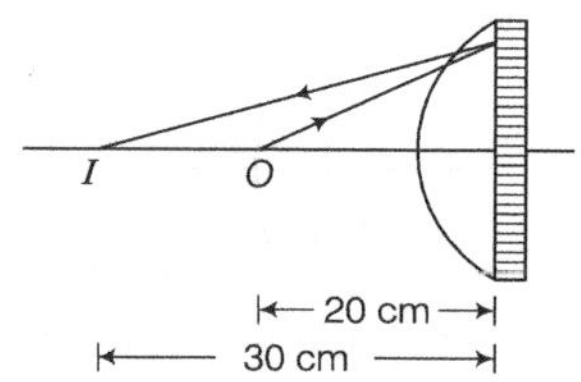

(iv) Using the mirror formula

$$\frac{1}{v} - \frac{1}{20} = \frac{-1}{12}$$

On solving, we get $v = -30$ cm.

Therefore, the image will be formed at a distance of 30 cm to the left of system.

46. (i) Image of object will coincide with it if ray of light after refraction from the concave surface fall normally on concave mirror so formed by silvering the convex surface. Or image after refraction from concave surface should form at centre of curvature of concave mirror or a distance of 20 cm on same side of the combination. Let x be the distance of pin from the given optical system.

Applying, $$\frac{\mu_2}{v} - \frac{\mu_1}{u} = \frac{\mu_2 - \mu_1}{R}$$

With proper signs

$$\frac{1.5}{-20} - \frac{1}{-x} = \frac{1.5 - 1}{-60}$$

or $$\frac{1}{x} = \frac{3}{40} - \frac{1}{120} = \frac{8}{120}$$

$\therefore$ $$x = \frac{120}{8} = 15\,\text{cm}$$

(ii) Now, before striking with the concave surface, the ray is first refracted from a plane surface. So, let x be the distance of pin, then the plane surface will from its image at a distance $\frac{4}{3}x$ ($h_{app} = \mu h$) from it.

Now, using $\frac{\mu_2}{v} - \frac{\mu_1}{u} = \frac{\mu_2 - \mu_1}{R}$ with proper signs,

we have $$\frac{1.5}{-20} - \frac{\frac{4}{3}}{-\frac{4x}{3}} = \frac{1.5 - \frac{4}{3}}{-60}$$

or $$\frac{1}{x} = \frac{3}{40} - \frac{1}{360}$$

or $$x = 13.84\,\text{cm}$$

$\therefore$ $$\Delta x = x_1 - x_2 = 15\,\text{cm} - 13.84\,\text{cm}$$
$$= 1.16\,\text{cm}$$ (downwards)

47. **Refer figure (a)**

In this case refraction of the rays starting from O takes place from a plane surface. So, we can use

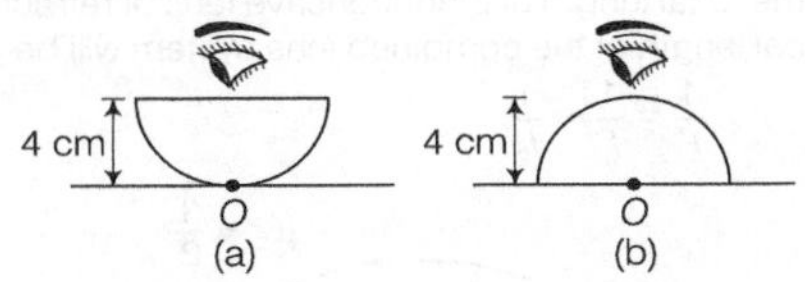

$$d_{app} = \frac{d_{actual}}{\mu} \quad \text{or} \quad 3 = \frac{4}{\mu} \quad \text{or} \quad \mu = \frac{4}{3}$$

Refer figure (b) In this case refraction takes place from a spherical surface. Hence, applying

$$\frac{\mu_2}{v} - \frac{\mu_1}{u} = \frac{\mu_2 - \mu_1}{R}$$

We have, $$\frac{1}{(-25/8)} = \frac{4/3}{-4} = \frac{1-4/3}{-R}$$

or $$\frac{1}{3R} = \frac{1}{3} - \frac{8}{25} = \frac{1}{75}$$

$\therefore$ $$R = 25 \text{ cm}$$

Now, to find the focal length we will use the lens Maker's formula

$$\frac{1}{f} = (\mu - 1)\left(\frac{1}{R_1} - \frac{1}{R_2}\right)$$

$$= \left(\frac{4}{3} - 1\right)\left(\frac{1}{\infty} - \frac{1}{-25}\right) = \frac{1}{75}$$

$\therefore$ $$f = 75 \text{ cm}$$

48. For both the halves, position of object and image is same. Only difference is of magnification. Magnification for one of the halves is given as 2 (> 1). This can be for the first one because for this, $|v| > |u|$. Therefore, magnifications $|m| = |v/u| > 1$.

So, for the first half

$$|v/u| = 2 \quad \text{or} \quad |v| = 2|u|$$

Let $u = -x$ then $v = +2x$ and $|u| + |v| = 1.8$ m

i.e $\quad 3x = 1.8$ m or $x = 0.6$ m

Hence, $\quad u = -0.6$ m and $v = +1.2$ m

Using $$\frac{1}{f} = \frac{1}{v} - \frac{1}{u} = \frac{1}{1.2} - \frac{1}{-0.6} = \frac{1}{0.4}$$

$\therefore$ $$f = 0.4 \text{ m}$$

For the second half

$$\frac{1}{f} = \frac{1}{1.2-d} - \frac{1}{-(0.6+d)}$$

or $$\frac{1}{0.4} = \frac{1}{1.2-d} + \frac{1}{0.6+d}$$

Solving this, we get $d = 0.6$ m

Magnification for the second half will be

$$m_2 = \frac{v}{u} = \frac{0.6}{-(1.2)} = -\frac{1}{2}$$

and magnification for the first half is

$$m_1 = \frac{v}{u} = \frac{1.2}{-(0.6)} = -2$$

The ray diagram is as follows:

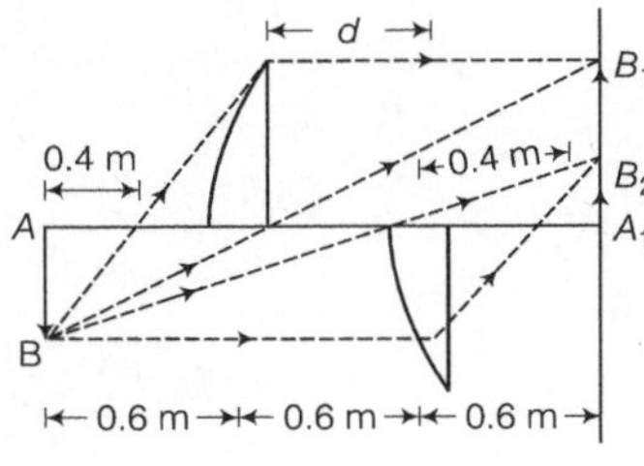

49. From lens maker's formula

$$\frac{1}{f} = (\mu - 1)\left(\frac{1}{R_1} - \frac{1}{R_2}\right)$$

We have $$\frac{1}{0.3} = \left(\frac{3}{2} - 1\right)\left(\frac{1}{R} - \frac{1}{-R}\right)$$

(Here, $R_1 = R$ and $R_2 = -R$)

$$R = 0.3$$

Now applying $$\frac{\mu_2}{v} - \frac{\mu_1}{u} = \frac{\mu_2 - \mu_1}{R}$$

at air glass surface, we get

$$\frac{3/2}{v_1} - \frac{1}{-(0.9)} = \frac{3/2 - 1}{0.3}$$

$\therefore$ $$v_1 = 2.7 \text{ m}$$

i.e. first image I_1 will be formed at 2.7 m from the lens. This will act as the virtual object for glass water surface. Therefore, applying $\frac{\mu_2}{v} - \frac{\mu_1}{u} = \frac{\mu_2 - \mu_1}{R}$ at glass water surface, we have

$$\frac{4/3}{v_2} - \frac{3/2}{2.7} = \frac{4/3 - 3/2}{-0.3}$$

$\therefore$ $$v_2 = 1.2 \text{ m}$$

i.e. second image I_2 is formed at 1.2 m from the lens or 0.4 m from the plane mirror. This will act as a virtual object for mirror. Therefore, third real image I_3 will be formed at a distance of 0.4 m in front of the mirror after reflection from it. Now this image will work as a real object for water-glass interface. Hence, applying

$$\frac{\mu_2}{v} - \frac{\mu_1}{u} = \frac{\mu_2 - \mu_1}{R}$$

We get, $$\frac{3/2}{v_4} - \frac{4/3}{-(0.8-0.4)} = \frac{3/2 - 4/3}{0.3}$$

$\therefore$ $$v_4 = -0.54 \text{ m}$$

i.e. fourth image is formed to the right of the lens at a distance of 0.54 m front it. Now finally applying the same formula for glass-air surface.

$$\frac{1}{v_5} - \frac{3/2}{-0.54} = \frac{1 - 3/2}{-0.3}$$

$\therefore$ $$v_5 = -0.9 \text{ m}$$

i.e. position of final image is 0.9 m relative to the lens (rightwards) or the image is formed 0.1 m behind the mirror.

50. Rays combing from object AB first refract from the lens and then reflect from the mirror.

Reflection from the lens

$$u = -20 \text{ cm}, f = +15 \text{ cm}$$

Using lens formula $\frac{1}{v} - \frac{1}{u} = \frac{1}{f}$

$\Rightarrow$ $$\frac{1}{v} - \frac{1}{-20} = \frac{1}{15}$$

$\therefore$ $$v = +60 \text{ cm}$$

and linear magnification, $m_1 = \frac{v}{u} = \frac{+60}{-20} = -3$

i.e. First image formed by the lens will be at 60 cm from it (or 30 cm from mirror) towards left and 3 times magnified but inverted. Length of first image A_1, B_1 would be $1.2 \times 3 = 3.6$ cm (inverted).

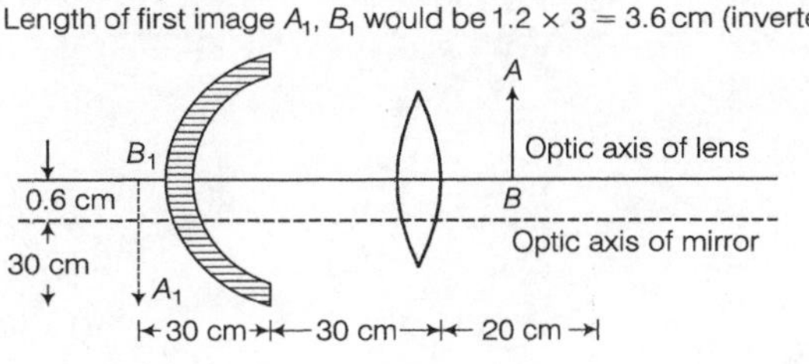

Reflection from mirror Image formed by lens ($A_1 B_1$) will behave like a virtual object for mirror at a distance of 30 cm from it as shown. There fore, $u = +\ 30$ cm, $f = -\ 30$ cm.

Using mirror formula

$$\frac{1}{v} + \frac{1}{u} = \frac{1}{f}$$

or $$\frac{1}{v} + \frac{1}{30} = -\frac{1}{30}$$

$\therefore$ $$v = -\ 15 \text{ cm}$$

and linear magnification

$$m_2 = -\frac{v}{u} = -\frac{-15}{+30} = +\frac{1}{2}$$

i.e. final image $A'B'$ will be located at a distance of 15 cm from the mirror (towards right) and since magnification is $+\frac{1}{2}$, length of final image would be

$$3.6 \times \frac{1}{2} = 1.8 \text{ cm}$$

$\therefore$ $$A'B' = 1.8 \text{ cm}$$

Point B_1 is 0.6 cm above the optic axis of mirror, therefore, its image B' would be $(0.6)\left(\frac{1}{2}\right) = 0.3$ cm above optic axis. Similarly, point A_1 is 3 cm below the optic axis, therefore, its image A' will be $3 \times \frac{1}{2} = 1.5$ cm below the optic axis as shown below.

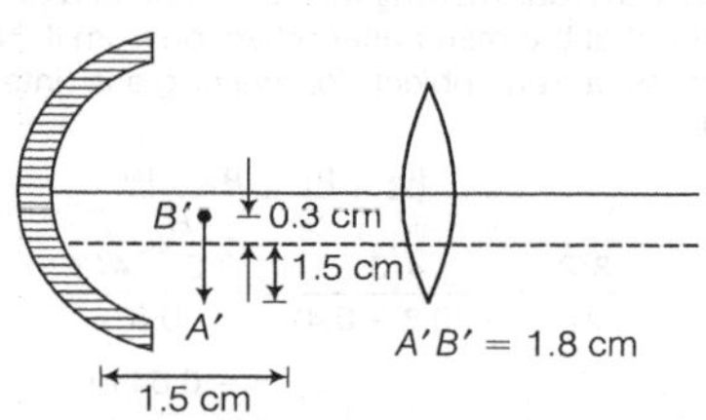

Total magnification of the image,

$$m = m_1 \times m_2 = (-\ 3)\left(+\frac{1}{2}\right) = -\frac{3}{2}$$

$$A'B' = (m)(AB) = \left(-\frac{3}{2}\right)(1.2) = -\ 1.8 \text{ cm.}$$

Note that, there is no need of drawing the ray diagram, if not asked in the question.

Note *With reference to the pole of an optical instrument (whether lens it is a lens or a mirror) the coordinates of the object (X_o, Y_o) are generally known to us. The corresponding coordinates of image $(X_i + Y_i)$ are found as follow. X_1 is obtained using*

$$\frac{1}{v} \pm \frac{1}{u} = \frac{1}{f}$$

Here v, is actually X_i and u is X_o the above formula can be written as $\frac{1}{X_i} \pm \frac{1}{X_o} = \frac{1}{f}$.

Similarly, Y_i is obtained from $m = \frac{I}{O}$

Here, I is Y_i and O is Y_o i.e. the above formula can be written as $m = Y_i / Y_o$ or $Y_r = mY_o$.

51. Let R be the radius of curvature of both the surfaces of the equiconvex lens. In the first case.

Let f_1 be the focal length of equiconvex lens of refractive index μ_1 and f_2 the focal length of planoconcave lens of refractive index μ_2. The focal length of the combined lens system will be given by

$$\frac{1}{F} = \frac{1}{f_1} + \frac{1}{f_2}$$

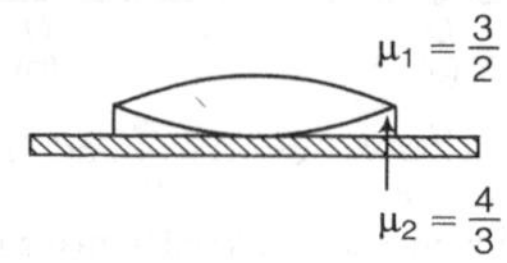

$$= (\mu_1 - 1)\left(\frac{1}{R} - \frac{1}{-R}\right) + (\mu_2 - 1)\left(\frac{1}{-R} - \frac{1}{\infty}\right)$$

$$= \left(\frac{3}{2} - 1\right)\left(\frac{2}{R}\right) + \left(\frac{4}{3} - 1\right)\left(-\frac{1}{R}\right)$$

$$= \frac{1}{R} - \frac{1}{3R} = \frac{2}{3R}$$

or $$F = \frac{3R}{2}.$$

Now, image coincides with the object, when ray of light retraces its path or it falls normally on the plane mirror. This is possible only when object is at focus of the lens system.

Hence, $F = 15$ cm (distance of object = 15 cm)

or $$\frac{3R}{2} = 15 \text{ cm}$$

or $$R = 10 \text{ cm.}$$

In the second case, let μ be the refractive index of the liquid filled between lens the mirror and let F' be the focal of new lens system. Then,

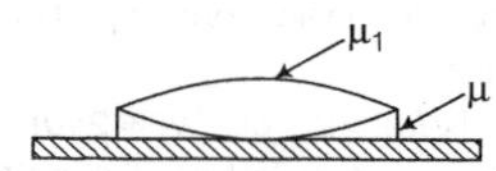

$$\frac{1}{F'} = (\mu_1 - 1)\left(\frac{1}{R} - \frac{1}{-R}\right) + (\mu - 1)\left(\frac{1}{-R} - \frac{1}{\infty}\right)$$

or $$\frac{1}{F'} = \left(\frac{3}{2} - 1\right)\left(\frac{2}{R}\right) - \frac{(\mu - 1)}{R}$$

or $$= \frac{1}{R} - \frac{\mu - 1}{R} = \frac{(2 - \mu)}{R}$$

$\therefore$ $$F' = \frac{R}{2 - \mu} = \frac{10}{2 - \mu} \qquad (R = 10 \text{ cm})$$

Now, the image coincides with object, when it is placed at 25 cm distance.

Hence, $$F' = 25$$

or $$\frac{10}{2 - \mu} = 25$$

or $$50 - 25\mu = 10$$

or $$25\mu = 40$$

$\therefore$ $$\mu = \frac{40}{25} = 1.6$$

or $$\mu = 1.6.$$

DPP-13 **Prism**

1. (i) (a) Deviation = δ = angle between emergent ray and incident ray = 30°.

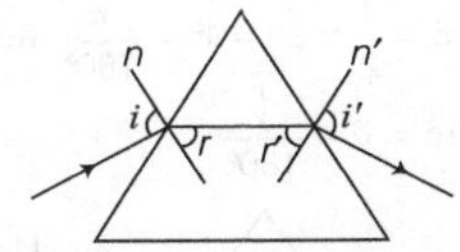

(ii) (c) $\because$ $\delta = i + i' - A$

or $30° = 60° + i' - 30°$

$\therefore$ $i' = 0$

(iii) (a) Since, angle of emergence is zero. So, emergent ray emerges out perpendicular to the face through which it emerges.

(iv) (b) $\because \mu = \dfrac{\sin\dfrac{\delta_{min} + A}{2}}{\sin A/2} = \dfrac{\sin\left(\dfrac{60° + 60°}{2}\right)}{\sin 30°} = \sqrt{3}$

(v) For not trans mitted the ray, $r_{min} > C$

or $A - r_{max} > C$

or $A - C > C$

$A > 2C$

or $\dfrac{A}{2} > C$

or $\sin\dfrac{A}{2} > \sin C$

or $\sin\dfrac{A}{2} > \dfrac{1}{\mu}$

or $\text{cosec}\,\dfrac{A}{2} < \mu$

$\therefore$ $\mu > \text{cosec}\,\dfrac{A}{2}$

$\therefore$ $\mu > 2$

$\therefore$ Option (c) is correct.

(vi) (a)

(vii) (a) $\because$ $r + r' = A$

or $r + 0 = 30°$

$\therefore$ $r = 30°$

(viii) (c) $\because$ $1 \times \sin 45° = \mu \sin 30°$

$\therefore$ $\mu = \sqrt{2}$

2. (i) (a) $\tan\delta = \delta = \dfrac{h}{5}$

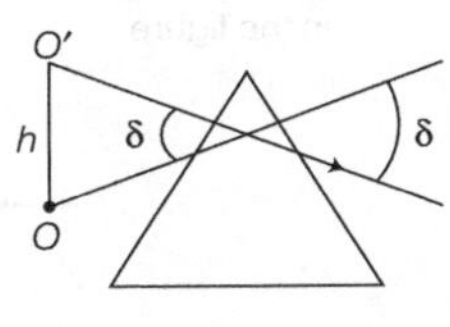

$\therefore$ $h = 5\delta = 5(\mu - 1)A$

$= 5\left(\dfrac{3}{2} - 1\right)2 \times \dfrac{\pi}{180°}$

$= \dfrac{\pi}{36}$ cm

(ii) (a,c)

(iii) (b) Here, $u = -25$ cm, $f = -30$ cm

$\dfrac{1}{v} + \dfrac{1}{u} = \dfrac{1}{f}$ or $\dfrac{1}{v} - \dfrac{1}{25} = \dfrac{1}{-30}$

or $\dfrac{1}{v} = \dfrac{1}{25} - \dfrac{1}{30} = \dfrac{6-5}{150}$

$\therefore$ $v = 150$ cm

$\therefore$ $n = 150 + 25 = 175$ cm

(iv) (c) $\because$ $\dfrac{h'}{h} = \dfrac{-v}{u}$

$\therefore$ $h' = -\dfrac{(150)h}{-25} = 6 \times \dfrac{\pi}{36} = \dfrac{\pi}{6}$ cm

3. (b) $\mu = \dfrac{\sin\dfrac{\delta_{min} + A}{2}}{\sin\dfrac{A}{2}}$

$= \dfrac{\sin\dfrac{50° + 30°}{2}}{\sin 25°}$

$= \dfrac{\sin 40°}{\sin 25°} = \dfrac{\cos 50°}{\sin 25°}$.

4. (b) The deviation produced by prism $= \delta = (\mu - 1))A$

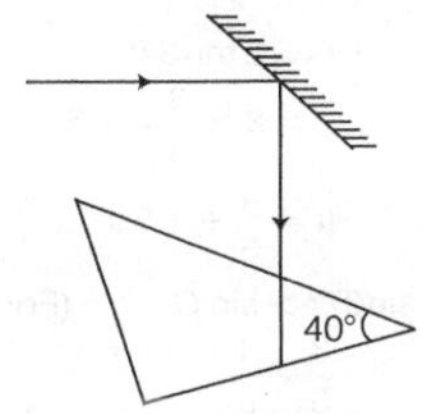

$= (1.5 - 1)4° = 2°$

For 90° deviation in ray, mirror is rotated through 1° in anticlockwise.

5. (c) Here, $i - i' = 10°$ ...(i)

and $\delta = i + i' - A$

or $40° = i + i' - 60°$

$\therefore$ $i + i' = 100°$ ...(ii)

After solving Eqs. (i) and (ii), we get

$i = 55°$ and $i = 45°$

6. (c) For this, $\mu > \text{cosec}\,A/2$

or $\mu > \text{cosec}\,45°$ or $\mu > \sqrt{2}$

7. (b) For total internal reflection,

$90° - \theta > c$

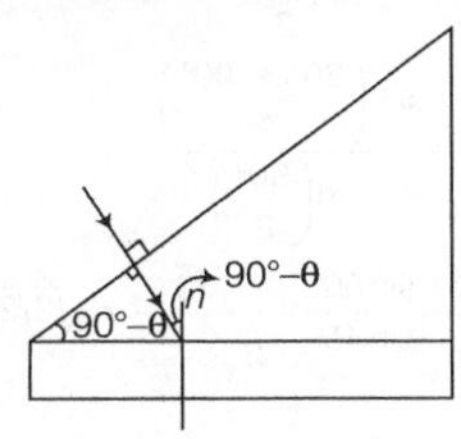

or $\sin(90° - \theta) > \sin c$

or $\cos\theta > \dfrac{6}{5 \times 3/2}$

or $\cos\theta > \dfrac{12}{15}$

or $\cos\theta > \dfrac{4}{5}$

or $\theta = < 37°$

8. (c) According to Snell's law,

$1 \times \sin i = \mu \sin 30°$

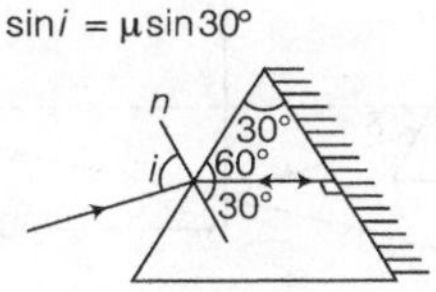

or $\sin i = \sqrt{2} \times \dfrac{1}{2}$

$\therefore$ $i = 45°$

9. (a) From figure, incident and emergent ray are anti parallel to each other.

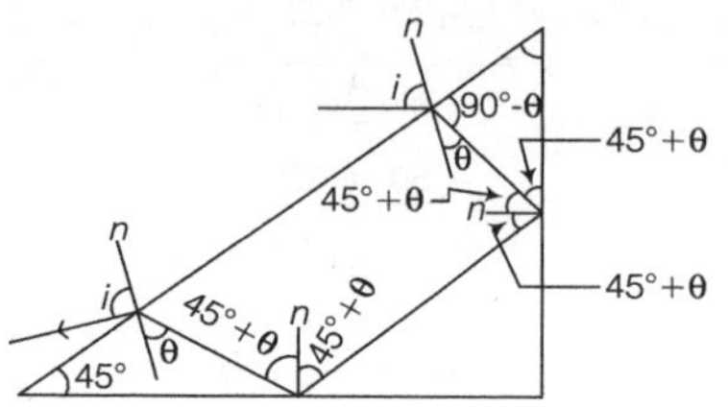

10. (b) Here, $\tan 37° = \frac{DE}{2}$

$\therefore \quad DE = 2\tan 37°$

$= 2 \times \frac{3}{4} = 1.5$

$\therefore \quad \mu = \frac{4}{3} + 1.5b$

$\because \quad \sin 37° > \sin C$ (For total internal reflection)

or $\quad \frac{3}{5} > \frac{1}{\mu}$

or $\quad \frac{3}{5} > \frac{1}{\frac{4}{3} + 1.5b}$

or $\quad 4 + 4.5b > 5$

or $\quad 4.5b > 1$

$\therefore \quad b > \frac{2}{9}.$

11. (a) $\because \delta = i + i' - A$ or $\delta = 50° + 40° - 60°$

$\therefore \quad \delta = 30°$

But $\quad \delta_{min} < \delta$

or $\quad \delta_{min} < 30°$

12. (c) $\mu = \frac{\sin\left(\frac{\delta_{min} + A}{2}\right)}{\sin\frac{A}{2}}$

$= \frac{\sin\left(\frac{30° + 90°}{2}\right)}{\sin\left(\frac{90°}{2}\right)}$

$= \frac{\sin 60°}{\sin 45°} = \frac{\sqrt{3}}{2 \times \frac{1}{\sqrt{2}}} = \sqrt{3/2}$

13. (c) $\because \delta = i + i' - A$

or $\quad 44° = 42° + 62° - A$

or $\quad A = 104° - 44° = 60°$

$\because \quad \delta_{min} = 2i - A \quad (\because i = i')$

or $\quad 38° = 2i - 60°$

$\therefore \quad i = \frac{60° + 38°}{2} = \frac{98°}{2} = 49°.$

14. (a)

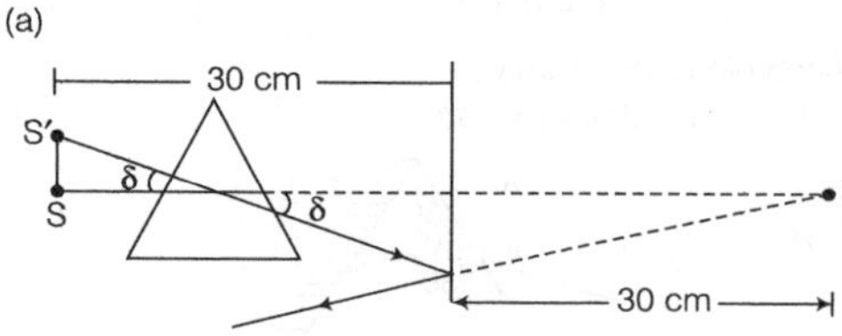

$\therefore 30 + 30 = 60$ cm

15. (b) $\delta = (\mu - 1)A$

$\delta = \left(\frac{3}{2} - 1\right)A$

$\delta = \frac{1}{2} \times 2° = 1° = \frac{\pi}{180°}$ rad

$\tan\delta = \delta = \frac{y}{180°}$ (from figure)

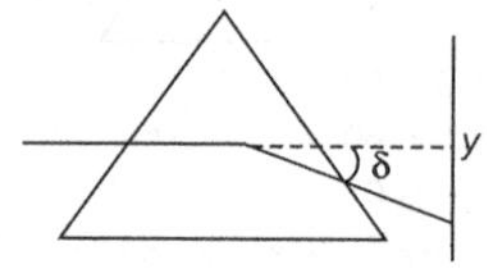

$\therefore \quad y = 180°\ \delta = 180° \times \frac{\pi}{180°}$

$\therefore \quad \pi = 3.14$ cm

16. (b) From graph, $i = 15°, i' = 60°$

$\delta = 30°$

$\delta = i + i' - A$

$30° = 75° - A$

$\therefore \quad A = 75° - 30° = 45°.$

17. (a) $\because$ Angle of prism $= A = 3.6°, \mu = 1.5 f = 20$ cm,

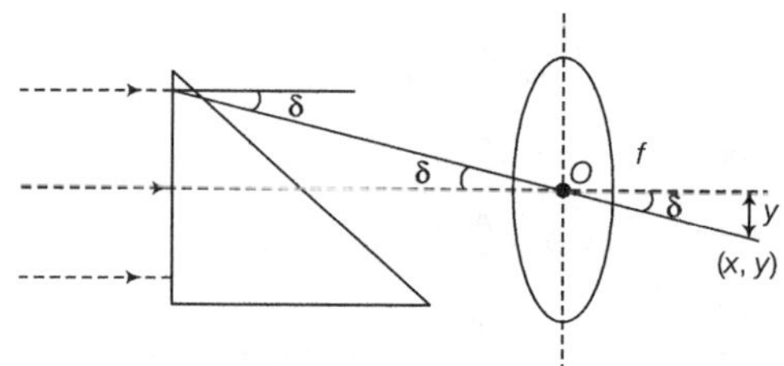

$\because \quad \delta = (\mu - 1)\ A = (1.5 - 1) \times 3.6 = 0.5 \times 3.6$

$\because \quad \tan\delta = \frac{y}{f} \Rightarrow y = f\tan\delta = f\delta$

$= 20 \times 1.8 \times \frac{\pi}{180°} = \frac{\pi}{5}$

$\because \quad$ Coordinate $= \left(20, -\frac{\pi}{5}\right)$

18. (c) The diagramatic representation of the given problem it shown in the figure.

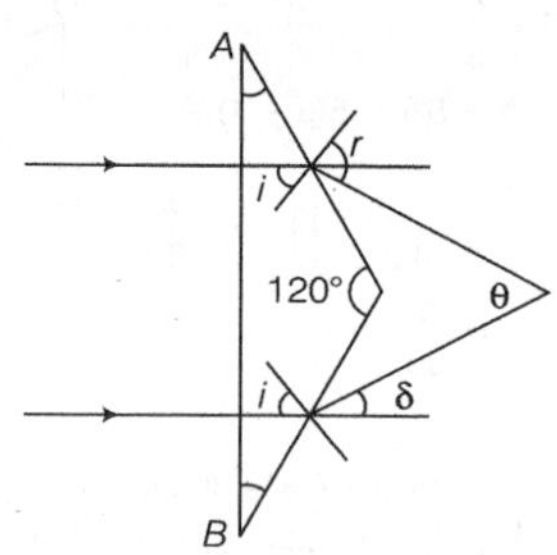

From figure it follows that $\angle i = \angle A = 30°$

From Snell's law, $n_1 \sin i = n_2 \sin r$

or $\quad \sin r = \frac{1.44 \sin 30°}{1} = 0.72$

Now, $\quad \angle\delta = \angle r - \angle i = \sin^{-1}(0.72) - 30°$

$\therefore \quad \theta = 2(\angle\delta) = 2\{\sin^{-1}(0.72) - 30°\}.$

19. (c) Figure (a) is part of an equilateral prism of figure (b) as shown in the figure, which is a magnified image of figure (c). Therefore, the ray will suffer the same deviation in figure (a) and (c).

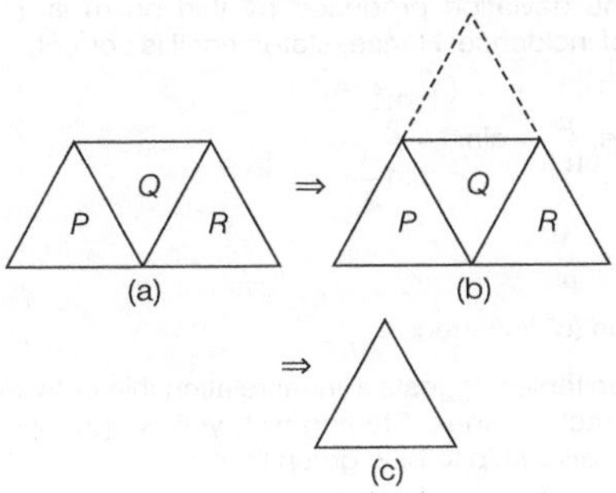

Note *Questions are often asked based on part of a prism. For example, section as shown in the figure (a) is part of a prism shown in the figure (c).*

20. (b) During minimum deviation the ray inside the prism is parallel to the base of the prism in case of an equilateral prism.

21. (a,c,d) Here, $2\alpha = \theta$

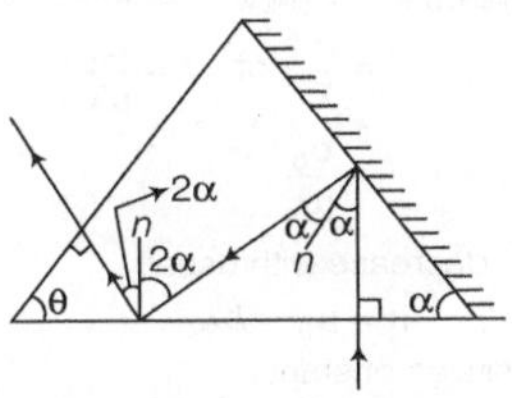

and $\quad 2\theta + \alpha = 180^\circ$

or $\quad 4\alpha + \alpha = 180^\circ$

or $\quad \alpha = \dfrac{180^\circ}{5} = 36^\circ$

$\therefore \quad \theta = 2\alpha = 72^\circ$

For total internal reflection,

$\sin 2\alpha > \sin c$

or $\quad \sin 72^\circ > \dfrac{1}{\mu}$

or $\quad \mu > \text{cosec } 72^\circ.$

22. (a,c) From figure, $\theta + 30^\circ = 90^\circ$

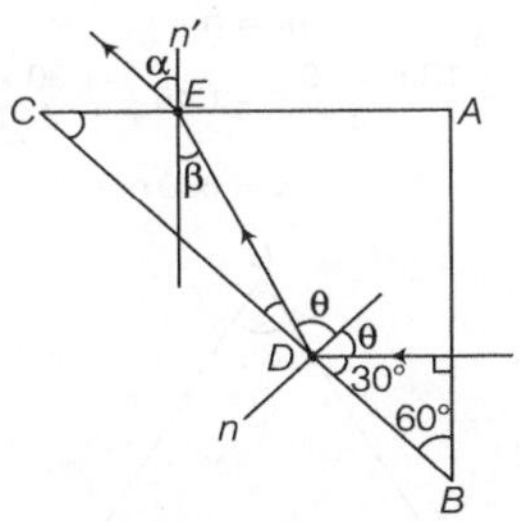

$\theta = 60^\circ$

$\therefore \quad \angle CDE = 90^\circ - \theta = 90^\circ - 60^\circ = 30^\circ$

$\therefore \quad \angle CED = 180^\circ - (30^\circ + 30^\circ)$

$\therefore \quad \angle CED = 120^\circ$

$\therefore \quad \beta = 120^\circ - 90^\circ = 30^\circ$

According to Snell's law.

$$\frac{5}{3}\sin\beta = \frac{4}{3} \times \sin\alpha$$

$\therefore \quad \sin\alpha = \dfrac{5}{4} \times \dfrac{1}{2} = \dfrac{5}{8}.$

(c) Here, $\sin c = \dfrac{5}{2\sqrt{3} \times \dfrac{5}{3}} = \dfrac{\sqrt{3}}{2}$

$\therefore \quad c = 60^\circ$

But $\quad \theta = 60^\circ$

Hence, total internal reflection does not take place.

23. (a,b,c)

(a) For this prism, $\delta = (\mu - 1)A$

So, it does not depend on angle of incidence.

(b) $\because \delta = (\mu - 1)A$

$\therefore$ When μ increases, δ increases

(c) $\because \delta = (\mu - 1)A$

$\therefore \delta \propto A$

24. (b,c,d)

(b) Form minimum deviation,

$r = r'$ and $i = i'$

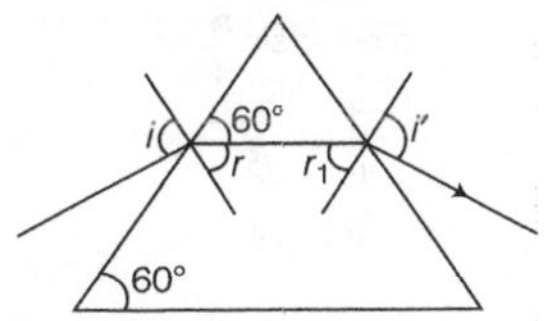

$\because \quad r + r' = A = 60^\circ$

or $\quad 2r = 60^\circ$

$\therefore \quad r = 30^\circ$

and $\quad r' = 30^\circ$

(c) $\because \quad \delta = i + i' - A$

For maximum deviation either $i = 90^\circ$ or $i' = 90^\circ$

Hence, two angle of incidence are possible for maximum deviation.

(d) $\because \mu = \dfrac{\sin\left(\dfrac{\delta_{min} + A}{2}\right)}{\sin\dfrac{A}{2}}$

From this expression, it is clear that when μ increases, δ_{min} increases.

25. (a,b) For $\delta_{min}, r = r'$

But $\quad r + r' = A$

$2r = A$

$r = A/2 = 30^\circ \qquad (\because A = 60^\circ)$

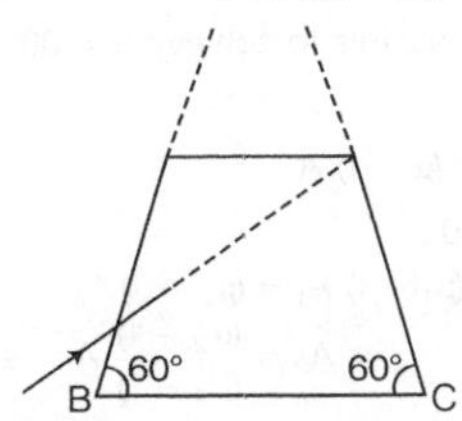

26. (a,b) The value of θ is

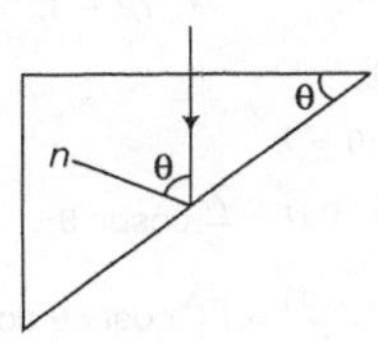

$\sin\theta > \sin c$

$\sin\theta > \sqrt{3}/2$

$\theta > 60^\circ.$

27. (a,c) Here, $\theta > c$

and $\alpha > c$

Case I If $\theta = \alpha$

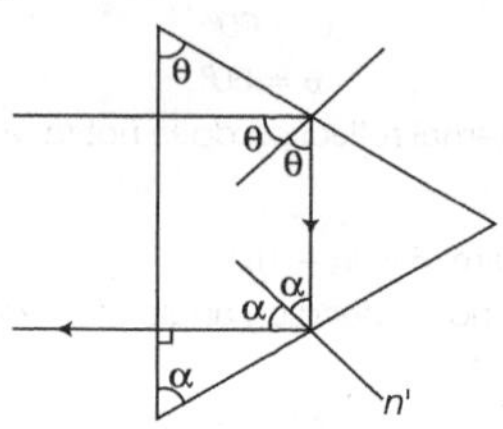

Then $\theta = \alpha = \frac{\pi}{4}$

$\therefore$ $\frac{\pi}{4} > c$

$\therefore$ $c < \frac{\pi}{4}$

and $\sin c < \sin\frac{\pi}{4}$

or $\frac{1}{\mu} < \frac{1}{\sqrt{2}}$

or $\mu > \sqrt{2}$

Case II If $\alpha < \theta$,

then $\alpha < \frac{\pi}{4}$ and $\theta > \frac{\pi}{4}$

But, $\alpha > c$

or $\sin\alpha > \sin c$

or $\sin\alpha > \frac{1}{\mu}$

or $\frac{1}{\sin\alpha} < \mu$

or $\frac{1}{\mu} < \sin\alpha < \sin\frac{\pi}{4}$

or $\frac{1}{\mu} < \frac{1}{\sqrt{2}}$

or $\mu > \sqrt{2}$

Thus, $\mu_{min} = \sqrt{2}$

(c) $\because$ $\alpha > c$

or $30° > c$

But $\sin c = \frac{1}{\mu} = \frac{3}{5}$

$\therefore$ $c = 37°$

Hence, it is not possible to achieve $\alpha = 30°$.

For $\mu = 5/3$.

28. (c) Deviation $\delta = (\mu - 1)A$

Given, $\delta_{net} = 0$

$$(\mu_1 - 1)A_1 = (\mu_2 - 1)A_2$$

or $A_2 = \frac{(\mu_2 - 1)}{(\mu_2 - 1)} A_1$

$$= \frac{(1.54 - 1)}{(1.72 - 1)}(4°) = 3°$$

Correct option is (c).

29. (a,b,c) Here, $2d\sin\theta = \lambda$

$\therefore$ $d = \frac{\lambda}{2}\operatorname{cosec}\theta$

or $\frac{d(d)}{d\theta} = \frac{-\lambda}{2}\operatorname{cosec}\theta\cot\theta$

or $\Delta d = \frac{-\lambda\cos\theta}{2\sin^2\theta}$

as θ increases, Δd decreases.

30. (b) $\because$ $\delta = \left(\frac{\mu}{\mu_s} - 1\right)A$, Thus the deviation decreases. So, statement I is correct. For thin prism the deviation is $\delta = (\mu - 1)A$. Thus the deviation produced by thin prism is independent of angle of incidence. Hence, statement II is correct.

31. (b) Here, $\frac{\mu}{\mu_s} = \frac{\sin\frac{(\delta_{min} + A)}{2}}{\sin\frac{A}{2}}$

32. (b) $\because c \propto \frac{1}{\mu}$

$\therefore$ Option (b) is correct.

33. (c) Given table suggests a linear relationship between wavelength and refractive index. Starting with yellow light, we can find this linear relationship to blue green light.

As, compared to yellow light waves, blue green light waves are 60 nm shorter. In the given table, reading from bottom to top, we see that each 20 nm decrease in wavelength corresponds to a 0.002 increase in refractive index μ. Since yellow and green light differ in wavelength by 60 nm, the corresponding μ's should differ by 0.006.

$\therefore$ $\mu_{\text{green blue}} = \mu_{\text{yellow}} + 0.006 = 1.506$

34. (a) $\because$ $f = \frac{c}{\lambda}$ or $f = \frac{c_0}{\mu\lambda}$

or $f = \frac{c_0}{\mu\lambda}$

Since, μ linearly decrease with usually,

$\therefore$ $\mu = \mu_0 - b\lambda$

where, b is positive constant

$\therefore$ $\lambda = \frac{\mu_0 - \mu}{b}$

$\therefore$ $f = \frac{c_0 b}{\mu(\mu_0 - \mu)} = \frac{c_0 b}{\mu\mu_0 - \mu^2}$

Here, $\mu < \mu_0$

$\therefore$ $\mu\mu_0 - \mu^2 > 0$

Hence, option (a) is correct.

35. (b) The deviation produced by thin prism depends on refractive index of prism. For different deviation, refractive index should be different. Hence, option (b) is correct.

36. (a) Since, ray goes without deviation,

$\therefore$ $n_1 = n_2$

or $1.2 + \frac{10.8 \times 10^4}{\lambda^2} = 1.45 + \frac{1.80 \times 10^4}{\lambda^2}$

$\Rightarrow$ $\lambda = 600$ nm

37. (b)

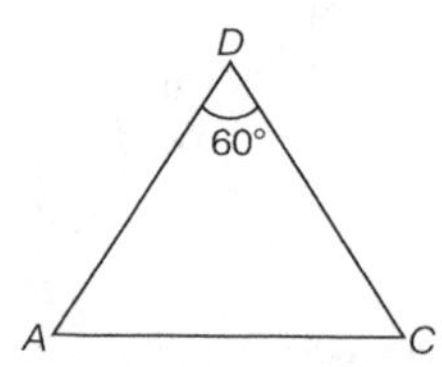

38. (c) Here, $\mu = 1.2 + \frac{10.8 \times 10^4}{600 \times 600} = 1.2 + 0.3 = 1.5$

For minimum deviation, $i = i'$, $r = r'$

But $r + r' = A = 60°$

$\therefore$ $r = 30°$

$\therefore$ $1 \times \sin i = \mu\sin r$

$\sin i = 1.5 \times \frac{1}{2} = \left(\frac{3}{4}\right)$

$\therefore$ $i = \sin^{-1}\left(\frac{3}{4}\right)$

39. (b)

(i) Without changing medium, refraction does not take place.

(ii) Ray goes without deviated

(iii) Ray come back

(iv) Ray goes goes without deviation

40. (a)

(i) When light incident on a surface, a part of light may be reflected and a part of light may be refracted. For total internal refraction, ray must propagate from denser medium to rarer medium.

(ii) Monochromatic light only consists of single wavelength (colour). Hence, dispersion does not take place.

(iii) Simple microscope is based upon refraction through converging lens.

(iv) In the lens refraction takes place. Also, lens is combination of thin prisms.

41. (a) (i) $\rightarrow$ (p), (ii) $\rightarrow$ (p), (iii) $\rightarrow$ (s), iv $\rightarrow$ (r).

42. (d) For $e \rightarrow i$

$$45° > \theta_c$$

$$\Rightarrow \quad \sin 45° > \sin\theta$$

$$\frac{1}{\sqrt{2}} > \frac{\mu_2}{\mu_1}$$

$$\mu_1 > \sqrt{2}\mu_2$$

For $e \rightarrow f$

Angle of refraction is lesser than angle of incidence, so $\mu_2 > \mu_1$ and then $\mu_2 > \mu_3$.

For $e \rightarrow g$,

$$\mu_1 = \mu_2$$

For $e \rightarrow h$,

$$\mu_2 < \mu_1 < \sqrt{2}\mu \text{ and } \mu_2 > \mu_3.$$

Hence, (i) $\rightarrow$ (q), (ii) $\rightarrow$ (r), (iii) $\rightarrow$ (s), (iv) $\rightarrow$ (p).

43. (5) The deviation produced by combination of lenses is

$$\delta = \delta_1 - \delta_2$$

$$= (\mu_1 - 1)A_1 - (\mu_2 - 1)A_2$$

$$= (1.5 - 1)4 - \left(\frac{4}{3} - 1\right)6 = 2 - 2 = 0°$$

For lens L_1,

$$u = -30 \text{ cm}, f = +20 \text{ cm}$$

$$\because \quad \frac{1}{v} - \frac{1}{u} = \frac{1}{f}$$

$$\text{or} \quad \frac{1}{v} + \frac{1}{30} = \frac{1}{20}$$

$$\text{or} \quad \frac{1}{v} = \frac{1}{20} - \frac{1}{30} = \frac{3-2}{60}$$

$$\therefore \quad v = 60 \text{ cm}$$

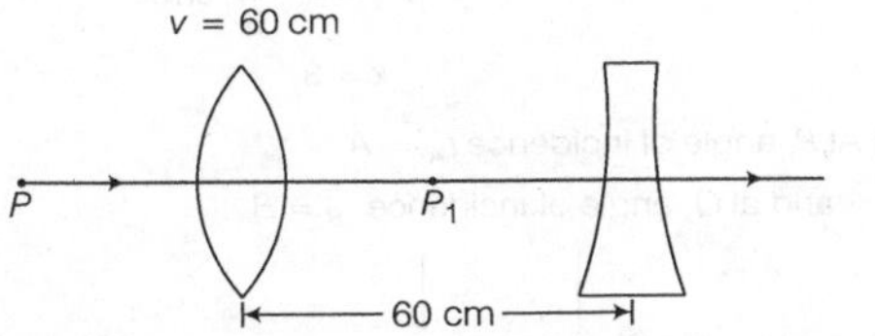

The image P_1 formed by lens L_1 behaves as object for concave lens L_2.

For Lens L_2,

$$u' = -10 \text{ cm}, f' = -10 \text{ cm}$$

$$\because \quad \frac{1}{v'} - \frac{1}{u'} = \frac{1}{f'}$$

$$\text{or} \quad \frac{1}{v'} + \frac{1}{10} = \frac{-1}{10}$$

$$\therefore \quad v' = -5 \text{ cm}$$

44. (4) Now, $1 \times \sin 30° = \mu \sin\alpha$

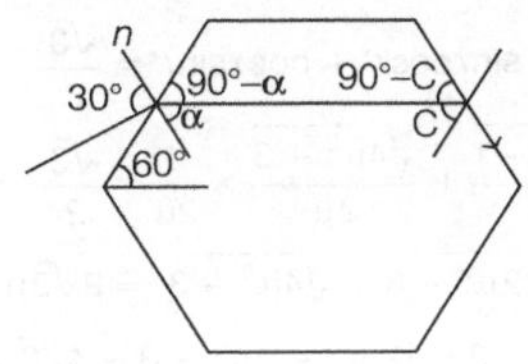

$$\therefore \quad \sin\alpha = \frac{1}{2\mu}$$

$$\text{Also} \quad \sin c = \frac{1}{\mu}$$

Here, $90° - \alpha + 120° + 120° + 90° - c$

$$= 360°$$

$$\text{or} \quad \alpha + c = 240° + 180° - 360°$$

$$= 240° - 180° = 60°$$

$$\text{or} \quad \sin(\alpha + c) = \sin 60°$$

$$\text{or} \quad \sin\alpha\cos c + \cos\alpha\sin c = \frac{\sqrt{3}}{2}$$

$$\text{or} \quad \frac{1}{2\mu}\frac{\sqrt{\mu^2 - 1}}{\mu} + \frac{\sqrt{4\mu^2 - 1}}{2\mu} \times \frac{1}{\mu} = \frac{\sqrt{3}}{2}$$

$$\therefore \quad \mu = \left(\frac{13}{14}\right)^{1/2}$$

$$\therefore \quad n = 4$$

45. (1) Here, $\delta = (\mu - 1)\,\theta = \left(\frac{3}{2} - 1\right)2 = 1°$

$$\delta = \frac{h}{f}$$

$$\therefore \quad h = f\delta = \frac{180°}{\pi} \times 1 \times \frac{\pi}{180°} = 1 \text{ cm}$$

46. (3) $\because$ $\quad i - i' = 30°$ ...(i)

and $\quad \delta = i + i' - A$

$$30° = i + i' - 60°$$

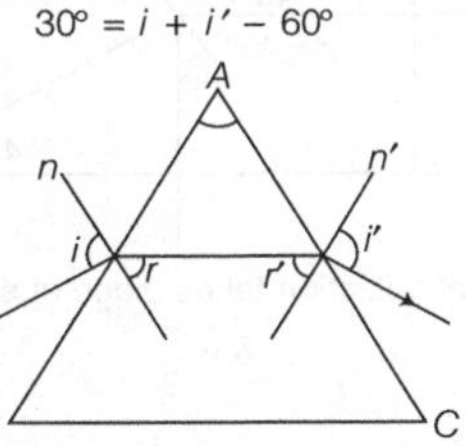

$i + i' = 90°$ ...(ii)

From Eqs. (i) and (ii), we get

$$i = 60° \text{ and } i' = 30°$$

$$\because \quad r + r' = A = 60°$$

$$\text{and} \quad \mu = \frac{\sin i}{\sin r} = \frac{\sin i'}{\sin r'}$$

$$\text{or} \quad \sin i = \mu \sin r$$

$$\text{and} \quad \sin i' = \mu \sin r'$$

$\therefore \quad \sin r = \frac{\sin 60^\circ}{\mu} = \frac{\sqrt{3}}{2\mu}$

and $\quad \sin r' = \frac{\sin 30^\circ}{\mu} = \frac{1}{2\,\mu}$

$\because \quad r + r' = 60^\circ$

or $\quad \sin(r + r') = \sin 60^\circ = \frac{\sqrt{3}}{2}$

or $\quad \sin r \cos r' + \cos r \sin r' = \frac{\sqrt{3}}{2}$

or $\quad \frac{\sqrt{3}}{2\mu}\frac{\sqrt{4\mu^2 - 1}}{2\mu} + \frac{\sqrt{4\mu^2 - 3}}{2\mu} \times \frac{1}{2\mu} = \frac{\sqrt{3}}{2}$

or $\quad \sqrt{12\mu^2 - 3} + \sqrt{4\mu^2 - 3} = 2\sqrt{3}\mu$

or $\quad 12\,\mu^2 - 3 = 12\,\mu^2 + 4\mu^2 - 3 - 4\sqrt{3}\sqrt{4\mu^3 - 3}$

or $\quad 4\sqrt{3}\sqrt{4\mu^2 - 3} = 4\mu^2$

or $\quad 48(4\,\mu^2 - 3) = 16\,\mu^4$

or $\quad 12\,\mu^2 - 9 = \mu^4$

or $\quad \mu^4 - 12\,\mu^2 + 9 = 0$

or $\quad \mu^2 = \frac{12 \pm \sqrt{144 - 36}}{2}$

$= \frac{12 \pm \sqrt{108}}{2}$

$= 6 \pm \sqrt{27} = 6 \pm 5.2$

$= 11.2 \text{ or } 0.8$

$\therefore \quad \mu = \sqrt{11.2} = 3.3$

47. (0) Zero, whenever there exists minimum deviation, the refracted ray and the prism will be parallel.

48. (2) Here, $r' = 45^\circ$

But $\quad \sin c = \frac{1}{\mu}$

$\sin c = \frac{2}{3} = 0.67$

$\sin r' = \sin 45^\circ = \frac{1}{\sqrt{2}} = 0.707$

$\therefore \quad r' > c$

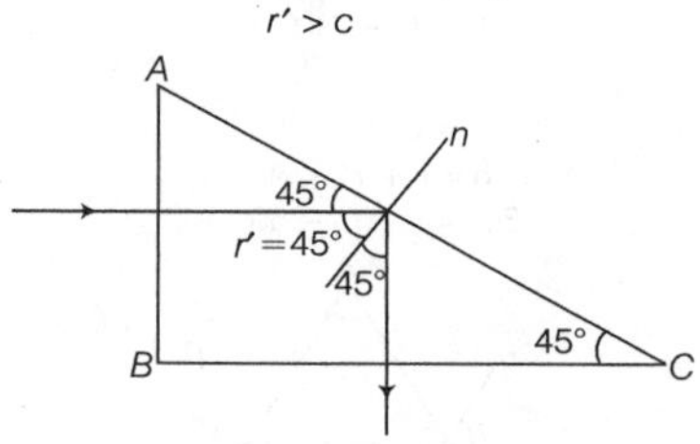

So, total internal reflection takes place at surface AC.

$\therefore \quad \delta = \frac{\pi}{2}.$

49. (3) For first surface,

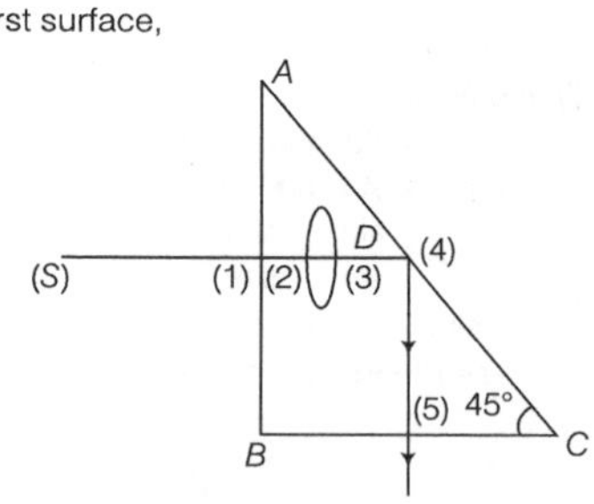

$\mu_1 = 1, \quad \mu_2 = 1.5,$

$r = \infty, \quad u = -20 \text{ cm}$

$\because \quad \frac{\mu_2}{v} - \frac{\mu_1}{u} = \frac{\mu_2 - \mu_1}{\infty} = 0$

or $\quad \frac{v}{1.5} - \frac{1}{-20} = 0$

or $\quad \frac{v}{1.5} = -20$

$\therefore \quad v = -30 \text{ cm}$

For second surface, $\mu_1' = 1.5, \mu_2' = 1, u' = -(30 + 30) = -60$ cm, $r' = +20$ cm

$\because \quad \frac{\mu_2'}{v'} - \frac{\mu_1'}{u'} = \frac{\mu_2' - \mu_1'}{r'}$

or $\quad \frac{1}{v'} - \frac{3}{2(-60)} = \frac{1 - 312}{20} = -\frac{1}{40}$

or $\quad \frac{1}{v'} = -\frac{1}{40} - \frac{1}{40} = -\frac{1}{20}$

$\because \quad v' = -20 \text{ cm}$

For refraction through third surface,

$\mu_1'' = 1, \mu_2'' = 1{\cdot}5,$

$r'' = -20 \text{ cm}, u'' = -20 \text{ cm}$

$\because \quad \frac{\mu_2''}{v''} - \frac{\mu_1''}{u''} = \frac{\mu_2'' - \mu_1''}{r''}$

or $\quad \frac{3}{2v''} - \frac{1}{-20} = \frac{1{\cdot}5 - 1}{-20}$

or $\quad \frac{3}{2v''} = -\frac{1}{40} - \frac{1}{20}$

$= \frac{-1 - 2}{40} = \frac{-3}{40}$

or $\quad v'' = \frac{-40}{3} \times \frac{3}{2} = -20 \text{ cm}$

Total internal reflection takes place at surface AC.

So, it behaves as plane mirror.

It forms image 500 m above point D, which behaves as object for surface BC.

For BC,

$\mu_1''' = \frac{3}{2}, \mu_2''' = 1,$

$u''' = -80 \text{ cm}, r''' = 0,$

$\because \quad \frac{\mu_2'''}{v'''} - \frac{\mu_1'''}{u'''} = 0$

or $\quad \frac{1}{v'''} - \frac{3}{2(-80)} = 0$

or $\quad \frac{1}{v'''} = \frac{-3}{160}$

$\therefore \quad v''' = -\frac{160}{3} \text{ cm}$

$\therefore \quad x = 3$

50. (i) At P, angle of incidence $i_A = A$

and at Q, angle of incidence $i_B = B$

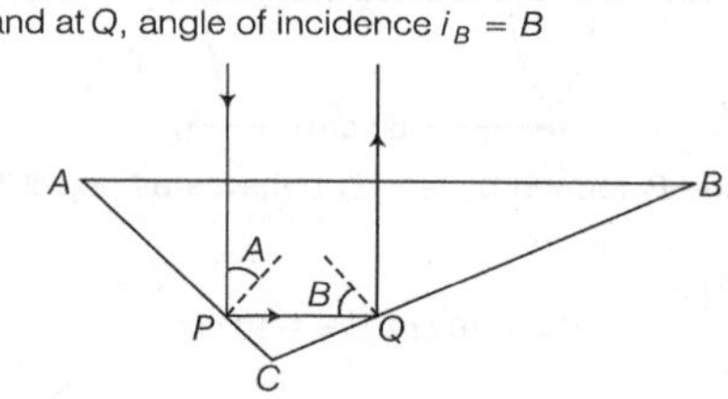

If TIR satisfies for the smaller angle of incidence than for larger angle of incidence is automatically satisfied.

$$B \le A$$

$$\therefore \quad i_B \le i_A$$

Maximum value of B can be 45°. Therefore, if condition of TIR is satisfied, then condition of TIR will be satisfied for all value of i_A and i_B.

Thus, $\quad 45° \ge \theta_c$

or $\quad \sin 45° \ge \sin\theta_c$ or $\frac{1}{\sqrt{2}} \ge \frac{1}{\mu}$

or $\quad \mu \ge \sqrt{2}$

$\therefore$ Maximum value of μ or n is $\sqrt{2}$.

(ii) For $n = \frac{5}{3}$,

$$\sin\theta_c = \frac{1}{n} = \sin^{-1}\left(\frac{3}{5}\right) = 37°$$

If $B = 30°$, then

$$i_B = 30°$$

then $\quad A = 60°$ or $i_A = 60°$

$$i_A > \theta_c \text{ but } i_B < \theta_c$$

i.e. TIR will take place at A but not at B,

or we write : $\quad \sin i_B < \sin\theta_c < \sin i_A$

or $\quad \sin 30° < \frac{3}{5} < \sin 60°$

or $\quad 0.5 < 0.6 < 0.86$

51. (i) Critical angle θ_c at face AC will be given by

$$\theta_c = \sin^{-1}\left(\frac{n_1}{n}\right)$$

or $\quad \sin\theta_c = \frac{n_1}{n}$

Now, it is given that $r_2 = \theta_c$

$$\therefore \quad r_1 = A - r_2 = (45° - \theta_c)$$

Applying Snell's law at face AB, we have

$$n = \frac{\sin i_1}{\sin r_1}$$

or $\quad \sin i_1 = n\sin r_1$

$$\therefore \quad i_1 = \sin^{-1}(n\sin r_1)$$

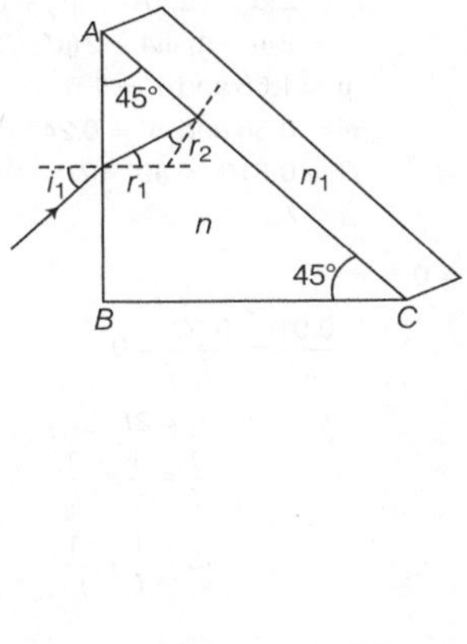

Substituting value of r_1, we get

$$i_1 = \sin^{-1}\{n\sin(45° - \theta_c)\}$$

$$= \sin^{-1}\{n(\sin 45°\cos\theta_c - \cos 45°\sin\theta_c)\}$$

$$= \sin^{-1}\left\{\frac{n}{\sqrt{2}}(\sqrt{1-\sin^2\theta_c} - \sin\theta_c\right\}$$

$$= \sin^{-1}\left\{\frac{n}{\sqrt{2}}\left(\sqrt{1-\frac{n_1^2}{n^2}} - \frac{n_1}{n}\right)\right\}$$

$$i_1 = \sin^{-1}\left\{\frac{1}{\sqrt{2}}(\sqrt{n^2-n_1^2} - n_1\right\}$$

Therefore, required angle of incidence (i_1) at face AB for which the ray strikes at AC at critical angle is

$$i_1 = \sin^{-1}\left\{\frac{1}{\sqrt{2}}(\sqrt{n^2-n_1^2}) - n_1\right\}$$

(ii) The ray will pass undeviated through face AC when either $n_1 = n$ or $r_2 = 0°$ i.e. ray falls normally on face AC.

Hence, $n_1 \ne n$ (because $n_1 < n$ is given)

$$\therefore \quad r_2 = 0°$$

or $\quad r_1 = A - r_2 = 45° - 0° = 45°$

Now applying Snell's law at face AB, we have

$$n = \frac{\sin i_1}{\sin r_1}$$

or $\quad 1.352 = \frac{\sin i_1}{\sin 45°}$

$$\therefore \quad \sin i_1 = (1.352)\left(\frac{1}{\sqrt{2}}\right)$$

$$\sin i_1 = 0.956$$

$$\therefore \quad i_1 = \sin^{-1}(0.956) = 73°$$

Therefore, required angle of incidence is

$$i = 73°$$

52. (i) At minimum deviation, $r_1 = r_2 = 30°$

$\therefore$ From Snell's law

$$\mu = \frac{\sin i_1}{\sin r_1}$$

or $\quad \sqrt{3} = \frac{\sin i_1}{\sin 30°}$

$$\sin i_1 = \frac{\sqrt{3}}{2} \text{ or } i_1 = 60°$$

(ii) In the position shown net deviation suffered by the ray of light should be minimum. Therefore, the second prism should be rotated by 60° (anticlockwise).

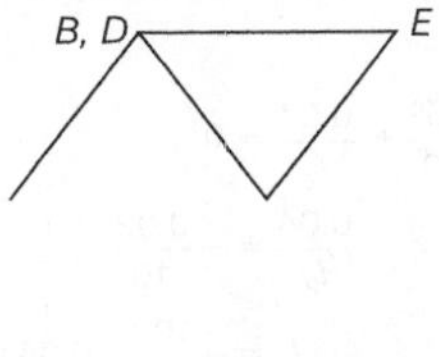

DPP-14 Dispersion and Spectrum

1. (b) $\sin\theta = \mu \sin\theta$

or $\sin\theta' = \dfrac{\sin\theta}{\mu}$

or $\cos\theta' \dfrac{d\theta'}{d\lambda} = \dfrac{-\sin\theta}{\mu^2}\dfrac{d\mu}{d\lambda}$

or $\delta\theta' = \dfrac{\sin\theta'}{\mu\cos\theta'}(\delta\lambda)\dfrac{d\mu}{d\lambda}$

$= \left(\dfrac{\tan\theta'}{\mu}\right)(\delta\lambda)\left(\dfrac{d\mu}{d\lambda}\right)$

2. (b) Dispersive power, $\omega = \dfrac{\text{Angular dispersion}}{\text{Deviation in mean ray}} = \dfrac{d\theta}{\theta}$

3. (c) For a chromatic combination

$$\frac{\omega_1}{f_1} + \frac{\omega_2}{f_2} = 0$$

This is possible only, when sign of f_1 and f_2 are opposite.

4. (b) Axial chromatic aberration is $f_r - f_v = \omega f_y$

$= 0.04 \times 10$ cm

$= 0.4$ cm

5. (b,c) A parallel beam of hight incident on different part of mirror converge at different point.

6. (a,b,c,d)

7. (a,b,c,d)

(a) $\mu_y = \dfrac{\mu_v + \mu_r}{2} = \dfrac{1.623 + 1.642}{2} = 1.632$

(b) $\omega = \dfrac{\mu_v - \mu_r}{\mu_y - 1} = \dfrac{1.642 - 1.623}{1.632 - 1} = 0.03$

(c) $D = (\mu_v - \mu_r)A = (1.642 - 1.623)4 = 0.076°$

(d) $\delta = (\mu_y - 1)A = (1.632 - 1)A = 2.528°$

8. (c)

(i) $\dfrac{\omega_1}{f_1} + \dfrac{\omega_2}{f_2} = 0$

or $\dfrac{0.04}{20} + \dfrac{0.02}{f_2} = 0$

$\therefore$ $f_2 = -5$ cm

$\dfrac{1}{f} = \dfrac{1}{f_1} + \dfrac{1}{f_2} = \dfrac{1}{10} - \dfrac{1}{5} = \dfrac{1-2}{10}$

$\therefore$ $f = -10$ cm

(ii) $\dfrac{\omega_1}{f_1} + \dfrac{\omega_2}{f_2} = 0$

or $\dfrac{0.02}{10} + \dfrac{0.04}{f_2} = 0$

or $\dfrac{0.04}{f_2} = -\dfrac{0.02}{10}$

$\therefore$ $f_2 = \dfrac{-10}{0.02} \times 0.04 = -20$ cm

$\dfrac{1}{f} = \dfrac{1}{f_1} + \dfrac{1}{f_2}$

or $\dfrac{1}{f} = \dfrac{1}{10} - \dfrac{1}{20} = \dfrac{2-1}{10}$

$\therefore$ $f = 20$ cm

(iii) $\dfrac{\omega_1}{f_1} + \dfrac{\omega_2}{f_2} = 0$

or $\dfrac{1}{f_1} + \dfrac{1}{f_2} = 0$

or $\dfrac{1}{f_1} = -\dfrac{1}{f_2}$

$f_2 = -f_1 = -10$ cm

$\because$ $\dfrac{1}{f} = \dfrac{1}{f_1} + \dfrac{1}{f_2}$

or $\dfrac{1}{f} = \dfrac{1}{10} - \dfrac{1}{10} = 0$

$f = \infty$.

Thus combination behaves as thin slab.

(iv) $f_2 = +20$ cm, is not possible because second lens behaves as diverging lens.

Hence, (i) → (q,s), (ii) → (p), (iii) → (t), (iv) → (r).

9. (b)

10. (a) Dispersive power of a material is $\omega = \dfrac{\mu_v - \mu_r}{\mu_y - 1}$.

Since, $\mu_v > \mu_r$, so ω is positive.

11. (a) Dispersive power only depends on material.

12. (c)

13. (d) The focal length of mirror is independent of refractive index.

14. (a)

15. (a) At high temperature, atoms emitting light are moving at tremendous speeds. Due to this motion of source (Doppler's effect of light), the wavelength of light emitted in spread a little on lower and higher sides.

16. (7) In this case, net deviation should be zero.

$\therefore$ $2(\mu - 1)A - 2(\mu' - 1)A' = 0$

D = Net dispersion

$= 2(\mu_v' - \mu_r)A - 2(\mu_v' - \mu_r')A'$

$= 2(\mu - 1)\omega A - 2(\mu' - 1)A'\omega'$

Here, $\mu = 1.68$ and $u' = 1.5$

$\omega = 0.36$ and $\omega' = 0.24$

$\therefore$ $D = 0.816° = 49' = 7n$

$\therefore$ $n = 7$

17. (5) $\dfrac{\omega_1}{f_1} + \dfrac{\omega_2}{f_2} = 0$

or $\dfrac{0.01}{f_1} + \dfrac{0.02}{f_2} = 0$

$\therefore$ $f_2 = 2f_1$

$\because$ $\dfrac{1}{f} = \dfrac{1}{f_1} + \dfrac{1}{f_2}$

or $5 = \dfrac{1}{f} + \dfrac{1}{f^2}$.

DPP-15 Optical Instrument

1. (b) According to mirror formula,

$$\frac{1}{v}-\frac{1}{-60}=\frac{1}{10}$$

or $\frac{1}{v}=\frac{1}{10}-\frac{1}{60}=\frac{6-1}{60}$

$\therefore$ $v = 12$ cm

The range of movement necessary between lens and film $= 12 - 10 = 2$ cm

2. (d) f-number $=\frac{f}{D}=\frac{50}{10}=5$

3. (a) $t \propto \frac{1}{D^2}$ and f-number $\propto \frac{1}{D}$

$\therefore$ $t \propto (f-\text{number})^2$

$\therefore$ $\frac{t}{\frac{1}{400}}=\left(\frac{6}{5}\right)^2$

$\therefore$ $t=\frac{36}{25}\times\frac{1}{400}=\frac{9}{2500}$ s.

4. (d) In the case of pinhole camera,

$\therefore$ $\frac{h'}{h}=\frac{v}{u}$

or $\frac{10}{200}=\frac{20}{u}$

$\therefore$ $u=\frac{200\times 20}{10}$

$= 400$ cm $= 4$ m.

5. (c) From figure, $\alpha=\frac{d}{f}=\frac{d}{75}$ and $\theta=\frac{d}{D}=\frac{d}{25}$

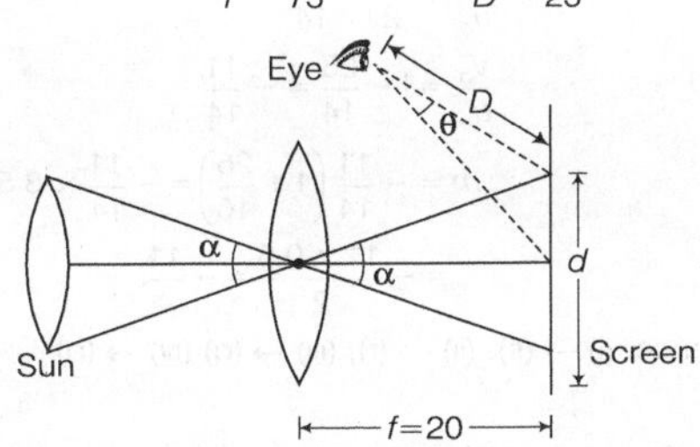

$\therefore$ Angular magnification $= m = \frac{-|\theta|}{|\alpha|}=-\frac{d}{25}\times\frac{75}{d}=-3.$

6. (c) Resolving power $=\frac{D}{1.22\lambda}=\frac{1.22}{1.22\times 500\times 10^{-9}}$

$= 0.2\times 10^7 = 2\times 10^6$.

7. (d) $\because \Delta\theta=\frac{d}{r}$

$\therefore$ $d = r\Delta\theta = r\times\frac{1}{\text{Resolving power}}=\frac{4\times 10^8}{2\times 10^6}=2\times 10^2$ m

$= 200$ m.

8. (c) Here, $u = -25$ cm, $v = -30$ cm

$\because$ $\frac{1}{v}-\frac{1}{u}=\frac{1}{f}$ or $\frac{1}{-30}+\frac{1}{25}=\frac{1}{f}$

or $\frac{-5+6}{150}=\frac{1}{f}$

$\therefore$ $f = 150$ cm $=\frac{150}{100}$ m

$\therefore$ Power of lens $=\frac{1}{f}=\frac{100}{150}=\frac{2}{3}D$

9. (b) According to mirror formula,

$\because$ $\frac{1}{v}-\frac{1}{u}=\frac{1}{f}$

or $-\frac{1}{180}+\frac{1}{d}=\frac{1}{150}$

or $\frac{1}{d}=\frac{1}{180}+\frac{1}{150}=\frac{5+6}{900}$

$\therefore$ $d=\frac{900}{11}$ cm $= 81.82$ cm

10. (c) Objective of compound microscope is a convex lens. Convex lens form real and enlarged image, when an object is placed between its focus and lens.

11. (b) Resolving power of telescope, $RP=\frac{d}{1.22\lambda}$

12. (a) Optical fibres work on the principle of total internal reflection.

13. (a,b)

(a) $m=\frac{D}{f}$ (for normal adjustment) $=\frac{25}{10}=2.5$

(b) $m=1+\frac{D}{f}=1+\frac{25}{10}=3.5.$

14. (a,c,d)

(a) Here, $u = -4$ cm, $f = 5$ cm

According to mirror formula,

$\because$ $\frac{1}{v}-\frac{1}{u}=\frac{1}{f}$

or $\frac{1}{v}+\frac{1}{4}=\frac{1}{5}$

or $\frac{1}{v}=\frac{1}{5}-\frac{1}{4}$

$=\frac{4-5}{20}$

$\therefore$ $v = -20$ cm

Thus, a virtual image is formed at distance 20 cm from the lens.

(c) The linear magnification $=\frac{v}{u}=\frac{-20}{-4}=5$

(d) $m=\frac{\beta}{\alpha}=\frac{D}{u}=\frac{25}{4}=6.25$

15. (a,c)

(a) $l = f_o + f_e = 100 + 2 = 102$ cm

(c) The angular magnification is

$$m=-\frac{f_o}{f_e}=\frac{100}{2}=-50$$

16. (b,d) The image formed by objective is $A'B'$, which is in the focal plane of objective. This image behaves as object for eyepiece, which forms image $A''B''$ at least distance of distinct vision.

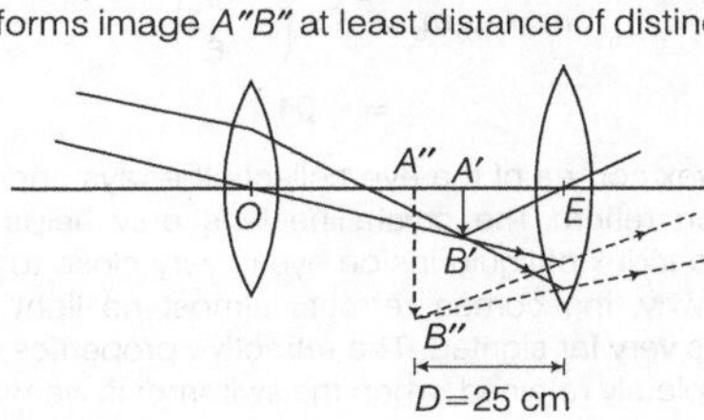

For eyepiece, $v = -25$ cm, $f = 2$ cm

$\because$ $\frac{1}{v}-\frac{1}{u}=\frac{1}{f}$

or $-\frac{1}{25}-\frac{1}{u}=\frac{1}{2}$

or $\quad \frac{1}{u} = -\frac{1}{25} - \frac{1}{2} = \frac{-2-25}{50} = -\frac{27}{50}$

$\therefore \quad u = -\frac{50}{27}$

(b) Thus length of tube is

$$l = f_o + |u|$$
$$= 100 + \frac{50}{27} = \frac{2750}{27}$$
$$= 101.85 \text{ cm}$$

(d) The angular magnification is

$$m = -\frac{f_o}{|u|} = \frac{-100}{\frac{50}{57}} = -54$$

17. (a,d) (a) The image is formed in focal plane (i.e. retina)

$\therefore$ The distance between retina and eye lens $= f = 2.5$ cm

(d) $\because \frac{1}{v} - \frac{1}{u} = \frac{1}{f}$

or $\quad \frac{1}{2.5} - \frac{3}{5} = \frac{1}{u}$

or $\quad \frac{2}{3} - \frac{3}{5} = \frac{1}{u}$

$\therefore \quad u = -5$ cm

Thus, near point is at 5 cm.

18. (a)

For eyepiece $= \frac{1}{-25} - \frac{1}{u_e} = \frac{1}{5}$

$\therefore \quad u_e = -\frac{25}{6}$

$\therefore \quad m_e = \frac{v_e}{u_e} = -\frac{25}{-25/6} = 6$

For objective, $\quad v = L - |u_e|$

$\therefore \quad v = \frac{125}{6} - \frac{25}{6} = \frac{100}{6}$

It object is at distacne u from objective.

$\therefore \quad \frac{1}{v} - \frac{1}{u} = \frac{1}{1}$

or $\quad \frac{6}{100} - \frac{1}{u} = 1$

$\therefore \quad -\frac{1}{u} = 1 - \frac{6}{100} = \frac{94}{100}$

$\therefore \quad u = -\frac{100}{94}$

$\therefore \quad m_0 = \frac{v}{u} = \frac{\frac{100}{6}}{-\frac{100}{94}} = -\frac{94}{6}$

19. (c) $\therefore$ Total magnification,

$$m = m_o m_e = 6 \times \left(-\frac{94}{6}\right)$$
$$= -94$$

20. (a) Convex cornea of the eye collects the rays and produces an image on retina. The crystalline lens only helps in this. The refractive index of liquid inside eye is very close to that of water. That is why, the cornea refracts almost no light and the eye becomes very far sighted. The refractive properties of the cornea are completely retained, when the swimmer is wearing a mask.

21. (a) It is difficult to estimate the distance of the object with one eye with both eyes open, the distance can be correctly estimated and threading the needle becomes easier.

22. (a) Scattering $\propto \frac{1}{\lambda^4}$

23. (b) (i) The focal length of objective is lesser than that of eyepiece.

$\therefore$ The focal length of objective = 2 cm

(ii) Image due to objective should be at the focal plane of eyepiece.

$\therefore \quad v_o = 15 - 10 = 5$ cm, $f_o = 2$ cm

$$\frac{1}{v_o} - \frac{1}{u_o} = \frac{1}{f_o} \text{ or } \frac{1}{5} - \frac{1}{u_o} = \frac{1}{2}$$

or $\quad \frac{1}{u_o} = \frac{1}{5} - \frac{1}{2} = \frac{2-5}{10}$

$\therefore \quad u_o = -\frac{10}{3}$ cm

(iii) $m = \frac{v_o}{u_o}\frac{D}{f_e} = \frac{5}{\frac{-10}{3}}\frac{25}{10} = -\frac{375}{100} = -3.75$

(iv) For clear vision, $m = \frac{v_o}{u_o}\left(1 + \frac{D}{f_e}\right)$

Here, for eyepiece, $f_e = 10$ cm, $v_e = -25$ cm

$\therefore \quad \frac{1}{-25} - \frac{1}{u_e} = \frac{1}{10}$

or $\quad -\frac{1}{10} - \frac{1}{25} = \frac{1}{u_e}$

or $\quad \frac{-5-2}{50} = \frac{1}{u_e}$

$\therefore \quad u_e = \frac{-50}{7}.$

For objective, $\quad v_o = 15 - \frac{50}{7} = \frac{25}{7}$

$\therefore \quad \frac{1}{v_o} - \frac{1}{u_o} = \frac{1}{2}$

or $\quad 1 - \frac{v_o}{u_o} = \frac{v_o}{2} = \frac{25}{14}$

$\therefore \quad \frac{v_o}{u_o} = 1 - \frac{25}{14} = -\frac{11}{14}$

$\therefore \quad m = -\frac{11}{14}\left(1 + \frac{25}{10}\right) = -\frac{11}{14} \times 3.5$

$$= -\frac{11 \times 0.5}{2} = -\frac{11}{4}.$$

Hence, (i) $\rightarrow$ (s), (ii) $\rightarrow$ (r), (iii) $\rightarrow$ (q) (iv) $\rightarrow$ (p)

24. (b)

(i) $L = f_o - |f_e| = 100 - 10 = 90$ cm

(ii) $m = -\frac{f_o}{f_e} = \frac{-100}{-10} = 10$

(iii) For objective,

According to mirror formula,

$$\frac{1}{v} - \frac{1}{u} = \frac{1}{f}$$

or $\quad \frac{1}{v} = \frac{1}{100} - \frac{1}{400} = \frac{4-1}{400}$

$\therefore \quad v = \frac{400}{3}$ cm

$\therefore \quad L = v - |f_e| = \frac{400}{3} - 10 = \frac{370}{3}$

$$= 123.7$$

(iv) $m = -\frac{v}{f_e} = \frac{400}{3 \times 10} = \frac{40}{3} = 13.33$

Hence, (i) $\rightarrow$ (r), (ii) $\rightarrow$ (q), (iii) $\rightarrow$ (s), (iv) $\rightarrow$ (p)

25. (a) Hence, (i) $\rightarrow$ (p,s), (ii) $\rightarrow$ (q,r), (iii) $\rightarrow$ (r,s), (iv) $\rightarrow$ (q)

26. (7) $m = \frac{v}{u}\frac{D}{f_e}$

$$-15 = \frac{v}{u}\frac{25}{5}$$

$\therefore \quad \frac{v}{u} = \frac{-15 \times 5}{25} = -3$

$\because \quad \frac{1}{v} - \frac{1}{u} = \frac{1}{f_o}$

or $\quad \frac{1}{-3u} - \frac{1}{u} = \frac{1}{0.5}$

or $\quad -\frac{1}{u} \times \frac{4}{3} = \frac{10}{5}$ or $-\frac{1}{u} = \frac{30}{20}$

$\therefore \quad u = -2/3$ cm

$\therefore \quad u = -3u = 2$ cm

$\therefore \quad$ Separation $= v + f_e = 2 + 5 = 7$ cm

27. (9) For initial condition,

$$\frac{1}{v} - \frac{1}{-3} = \frac{1}{2}$$

$\therefore \quad v = 6$ cm

If one of lens is removed.

$$\frac{1}{f'} = \frac{1}{f} - \frac{1}{f_1} = \frac{1}{2} - \frac{1}{10}$$

$\therefore \quad f' = 2.5$ cm

$\therefore \quad \frac{1}{v'} - \frac{1}{-3} = \frac{1}{2.5}$

$\therefore \quad v' = 15$ cm

$\therefore \quad$ Shifting $= 15 - 6 = 9$ cm.

28. (3) Here, $-\frac{1}{\infty} - \frac{1}{u_e} = \frac{1}{f_e}$

$\therefore \quad u_e = -3$ cm

$\because \quad 5n = v_o + |u_e| = 12 + 3 = 15$ cm

$\therefore \quad 5n = 15$

$\therefore \quad n = 3.$

29. (9) The condenser should be produce a real image of the source on the lens having the size of the lens.

Here, $\quad \frac{1}{v} - \frac{1}{u} = \frac{1}{f}$ and $-\frac{0.6}{2} = \frac{v}{u}$

or $\quad -\frac{1}{0.3u} - \frac{1}{u} = \frac{1}{f}$

or $\quad -\frac{1}{u}\left(\frac{10}{3} + 1\right) = \frac{1}{f}$

or $\quad -\frac{1}{u}\left(\frac{13}{3}\right) = \frac{1}{f}$

$\therefore \quad u = -\frac{13}{3}f$

$|v| + |u| = 39$

or $\quad |0.3u| + |u| = 39$

or $\quad 1.3|u| = 39$

$\therefore \quad |u| = \frac{39}{1.3} = 30$

or $\quad \frac{13}{3}f = 30$

or $\quad f = \frac{30 \times 3}{13} = \frac{90}{13}$

30. (3) When the child looks at objects at large distance through his spectacles, he sees them as he would see objects at a distance $x_2 = 60$ cm.

Let f = maximum focal of eye,

f_o = focal length of spectacles

y = depth of the eye

When the child is wearing spectacles,

$$\frac{1}{x} + \frac{1}{y} = \frac{1}{f} + \frac{1}{f_o}$$

where, $x = \infty$

$\therefore \quad \frac{1}{y} = \frac{1}{f} + \frac{1}{f_o} \quad \ldots$(i)

When the child is without spectacles,

$$\frac{1}{x_2} + \frac{1}{y} = \frac{1}{f}$$

or $\quad \frac{1}{60} + \frac{1}{f} + \frac{1}{f_o} = \frac{1}{f}$

$\therefore \quad f_o = -60$ cm

For nearest point of accomodation of the eye with spectacles,

$$\frac{1}{x_1} + \frac{1}{y_1} = \frac{1}{f_1}$$

and $\quad \frac{1}{5n} + \frac{1}{y_1} = \frac{1}{f_1} + \frac{1}{f_o}$

$\therefore \quad \frac{1}{5n} + \frac{1}{f_1} - \frac{1}{x_1} = \frac{1}{f_1} + \frac{1}{f_o}$

or $\quad \frac{1}{5n} - \frac{1}{12} = \frac{1}{-60}$

or $\quad \frac{1}{5n} = \frac{1}{12} - \frac{1}{60} = \frac{5-1}{60}$

or $\quad 5n = 15$

$\therefore \quad n = 3.$

31. (2) When Ram wears Shyam's spectacles, he can see only far objects. So, the distance y_2 of best vision of the eye of Ram can be determined by the equation, $\frac{1}{\infty} - \frac{1}{y_2} = \frac{1}{f_1} \quad \ldots$(i)

where, f_1 = focal length of spectalces of Shyam.

If the focal length of spectacles of Ram is f_2.

Then, $\quad \frac{1}{25} - \frac{1}{y_2} = \frac{1}{f_2} \quad \ldots$(ii)

If y_3 is distance of best vision of Shyam's eye.

Then, $\quad \frac{1}{25} - \frac{1}{y_3} = \frac{1}{f_1} \quad \ldots$(iii)

If Shyam wears spectacles of Ram, the minimum distance y at which Shyam can easily read a small word, can be determined from the relation,

$$\frac{1}{y} - \frac{1}{y_3} = \frac{1}{f_2}$$

or $\quad \frac{1}{y} + \frac{1}{f_1} - \frac{1}{25} = \frac{1}{25} - \frac{1}{y_2} \quad$ (from Eq. (iii))

or $\quad \frac{1}{y} + \frac{1}{f_1} + \frac{1}{y_2} = \frac{2}{25} \quad$ (from Eq. (i))

or $\quad \frac{1}{y} + 0 = \frac{2}{25}$

$\therefore \quad y = \frac{25}{2} = \frac{25}{n}$

$\therefore \quad n = 2 \quad \ldots$(ii)

32. (5) $m = \frac{f_o}{f_e}$

and $\quad l = f_o = f_e \quad \ldots$(i)

$\therefore \quad \frac{d}{D} = \frac{b}{l} = \frac{b}{f_o + f_e}$

and $\quad \frac{1}{b} - \frac{1}{f_o + f_e} = \frac{1}{f_e} \quad \ldots$(iii)

or $\quad \frac{D}{d(f_o + f_e)} + \frac{1}{f_o + f_e} = \frac{1}{f_e}$

$\therefore \quad m = \frac{f_o}{f_e} = \frac{D}{d} = 25$

$\therefore \quad n = 5$

Revisal Probelms (JEE Main)

1. (b) Limit is resolution

$$= 1.22\frac{\lambda}{D} = \theta = \frac{100}{r}$$

$$\therefore \qquad r = \frac{100 \times 0.25}{1.22 \times 500 \times 10^{-7}} \text{ cm} = 4.1 \text{ km}$$

2. (c)

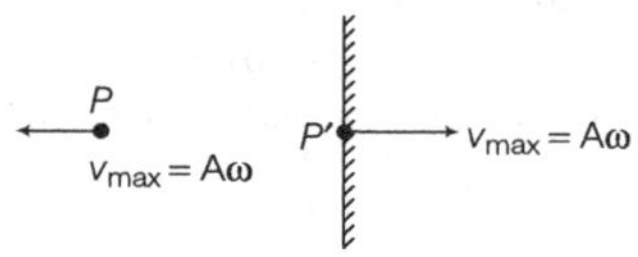

From figure,

$$v_{image} = v_{object} + 2v_{mirror}$$

$$= A\omega + 2A\omega = 5 \times 4 + 40 = 60 \text{ cms}^{-1}$$

3. (d) If converging point of incident beam is at focus, reflected beam will be parallel beam.

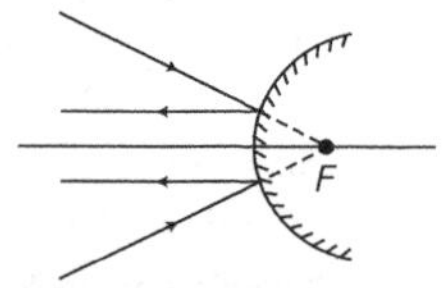

If converging point of incident beam is between pole and focus, the reflected beam converges left of mirror. So, the reflected beam is converging in nature.

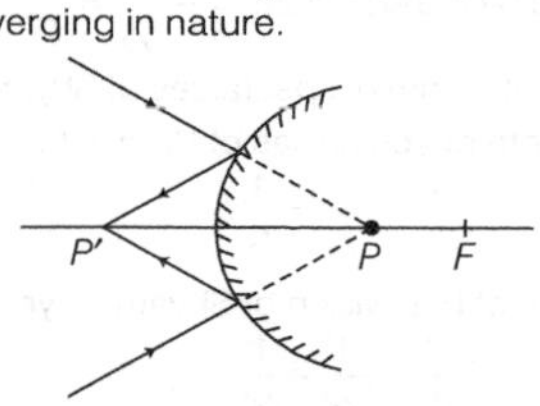

If the converging point of incident beam is beyond the focus. The coverging point of reflected beam is right of mirror. So, the reflected beam is diverging in nature.

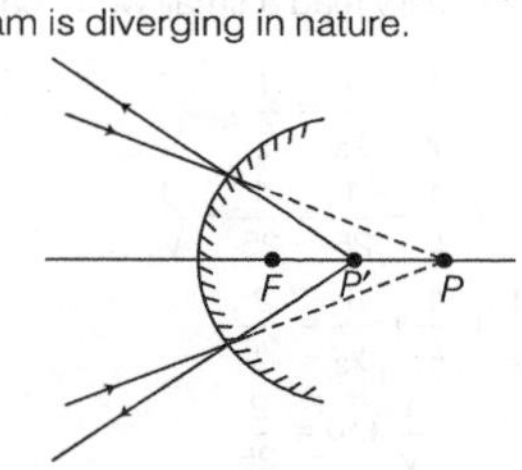

4. (d) When object P will reach at mirror, image P' will also reach at mirror. So, the distance travelled by image P' is less that of point object P. Hence, the speed of mass is less than that of object.

5. (b)

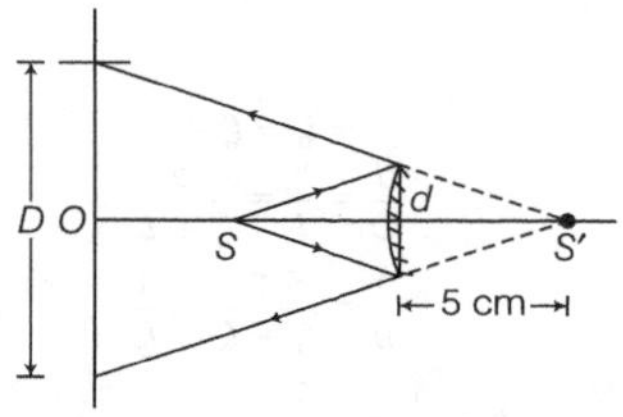

Here, $I_0 = \frac{P_0}{4\pi (0.2)^2}$, where P_0 is power of source.

The intensity at the side of mirror is $I = \frac{P_0}{4\pi (0.1)^2} = \frac{I_0 4\pi (0.2)^2}{4\pi (0.1)^2}$

$$= 4I_0$$

$\therefore$ The power of incident on mirror is

$$P_m = \frac{\pi d^2}{4} \times 4I_0 = \pi d^2 I_0$$

Here, d is aparture of mirror.

$$\because \qquad u = -10 \text{ cm}, f = +10 \text{ cm}$$

$$\because \qquad \frac{1}{v} + \frac{1}{u} = \frac{1}{f}$$

$$\therefore \qquad v = 5 \text{ cm}$$

From similar triangles,

$$\frac{D}{d} = \frac{35}{5}$$

$$\therefore \qquad D = 7d$$

Intensity due to reflected light on the screen is

$$I_1 = \frac{4P_m}{\pi (7d)^2} = \frac{4\pi d^2 I_0}{\pi 49 d^2} = \frac{4I_0}{49}$$

$$\therefore \qquad \text{Total intensity} = I_0 + \frac{4I_0}{49} = \frac{53 I_0}{49}$$

$$= \frac{53 \times 49}{49} = 53 \text{ Wm}^{-2}$$

6. (c) Here, $u = -10$ cm, $f = +F$, $m = +m_0$

$$\because \qquad \frac{v}{u} = +m_0 \quad \text{or} \quad v = -10\, m_0$$

$$\text{or} \qquad \frac{1}{v} - \frac{1}{u} = \frac{1}{f} \quad \text{or} \quad -\frac{1}{10\, m_0} + \frac{1}{10} = \frac{1}{F}$$

$$\text{or} \qquad \frac{-1 + m_0}{10\, m_0} = \frac{1}{F} \qquad \ldots\text{(i)}$$

If $u' = -30$ cm, $f' = F$

$$\therefore \qquad m' = -m_0 = \frac{v'}{u'}$$

$$\therefore \qquad v' = -m_0 u'$$

$$= -m_0(-30) = 30\, m_0$$

$$\therefore \qquad \frac{1}{30\, m_0} + \frac{1}{30} = \frac{1}{F} \qquad \ldots\text{(ii)}$$

From Eqs. (i) and (ii) we get,

$$\frac{-1 + m_0}{10\, m_0} = \frac{1}{30\, m_0} + \frac{1}{30}$$

$$\text{or} \qquad -\frac{1 + m_0}{10\, m_0} = \frac{1 + m_0}{30\, m_0}$$

$$\text{or} \quad -30\, m_0 + 30 m_0^2 = 10 m_0 + 10 m_0^2$$

$$\text{or} \qquad 20 m_0^2 = 40 m_0$$

$$\therefore \qquad m_0 = 2$$

$$\therefore \qquad \frac{1}{F} = \frac{1}{60} + \frac{1}{30} = \frac{1 + 2}{60} = \frac{1}{20}$$

$$\therefore \qquad P = \frac{1}{F} = \frac{100}{20} = 5\text{D}$$

7. (a) For refraction through spherical surface,

$$\mu_1 = 1.2, \mu_2 = 1.5$$

$$r = +30 \text{ cm}, u = -40 \text{ cm}$$

$$\because \qquad \frac{\mu_2}{v} - \frac{\mu_1}{u} = \frac{\mu_2 - \mu_1}{r}$$

$$\text{or} \qquad \frac{1.5}{v} - \frac{1.2}{-40} = \frac{1.5 - 1.2}{30} \quad \text{or} \quad \frac{1.5}{v} + \frac{1.2}{400} = \frac{1}{300}$$

$$\text{or} \qquad \frac{1.5}{v} = \frac{1}{300} - \frac{12}{400} = \frac{4 - 36}{1200}$$

$$\text{or} \qquad \frac{1.5}{v} = \frac{-32}{1200} = \frac{-8}{300}$$

$$\text{or} \qquad v = \frac{-300}{8} \times 1.5 = -56.25 \text{ cm}$$

For, refraction through plane surface,

$$\mu_1' = 1.5, \mu_2' = 1, u' = -56.25 \text{ cm}, r' = \infty$$

$\therefore \quad \dfrac{\mu_2'}{v'} - \dfrac{\mu_1'}{u'} = \dfrac{\mu_2' - \mu_1'}{\infty} = 0$

or $\quad \dfrac{1}{v'} + \dfrac{1.5}{56.25} = 0$

$\therefore \quad v' = -\dfrac{56.25}{1.5} = -37.5 \text{ cm}$

8. (b) For first refraction through lens,

$u = -30 \text{ cm}, f = +20 \text{ cm}$

$\because \quad \dfrac{1}{v} - \dfrac{1}{u} = \dfrac{1}{f}$

$\therefore \quad v = +60 \text{ cm}$

$$v_{\text{image lens}} = \frac{v^2}{u^2}\frac{du}{dt} = \frac{60^2}{(-30)^2}(6-2)$$

or $\quad v_{\text{image}} - v_{\text{lens}} = 16 \text{ cms}^{-1}$

$\therefore \quad v_{\text{image}} = 16 + 2 = 18 \text{ cms}^{-1}$

For reflectioon through mirror,

$u' = -20 \text{ cm}, f' = -10 \text{ cm}$

$\therefore \quad v' = -20 \text{ cm}$

$\therefore \quad v_{\text{image mirror}} = -\dfrac{(-20)^2}{(-20)^2}(18-4) = -14 \text{ cms}^{-1}$

$\therefore \quad v_{\text{image}} = -14 + 4 = -10 \text{ cms}^{-1}$

9. (b) Image formation due to concave lens,

$u = -20 \text{ cm}, f = -20 \text{ cm}$

$\because \quad \dfrac{1}{v} - \dfrac{1}{u} = \dfrac{1}{f}$

or $\quad \dfrac{1}{v} + \dfrac{1}{20} = -\dfrac{1}{20}$

$\therefore \quad v = -10 \text{ cm}$

$\because \quad \dfrac{1}{v} - \dfrac{1}{u} = \dfrac{1}{f}$

or $\quad -\dfrac{1}{v^2}\dfrac{dv}{dt} + \dfrac{1}{u^2}\dfrac{du}{dt} = 0$

or $\quad \dfrac{dv}{dt} = \dfrac{v^2}{u^2}\dfrac{du}{dt}$

or $\quad v_x =$ component of velocity along optical axis

$$= \frac{100}{400} \times 10 \cos 37°$$

$$= \frac{1}{4} \times 8 = 2 \text{ cms}^{-1}$$

$\because \quad \dfrac{y'}{y} = m$

or $\quad y' = my$

$\therefore \quad \dfrac{dy'}{dt} = m\dfrac{dy}{dt} + y\dfrac{dm}{dt}$

Here, $y = 0$ $\quad$ ($\because$ object is on optical axis)

$\therefore \quad \dfrac{dy'}{dt} = v_y = \dfrac{v}{u} \times 6 = \dfrac{-10}{-20} \times 6$

$$= -3 \text{ cms}^{-1}$$

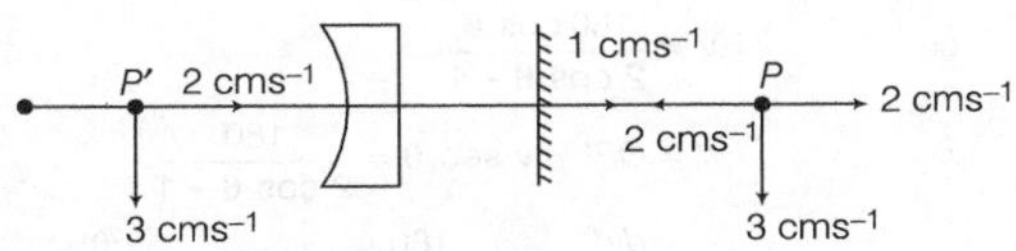

From figure, $\quad \mathbf{v}_{\text{image}} = (-3\hat{\mathbf{j}}) \text{cms}^{-1}$

10. (a) After refraction from concave lens, the beam becomes parallel to optical axis. So, beam converges at focus of convex lens after refraction through convex lens.

11. (a) $\quad \delta = \delta_1 - \delta_2 + \delta_3 - \delta_4$

But $\delta_1 = \delta_2 = \delta_3 = \delta_4 = \delta_0$

$\therefore \quad \delta = 0$

12. (b) $\delta = (\mu - 1) A = 0.5 \times \dfrac{5\pi}{180} = \dfrac{\pi}{72}$ radian

$\therefore \quad \delta = \dfrac{h}{b}$

$\therefore \quad h = b\delta = 50.4 \times \dfrac{\pi}{72} = \dfrac{50.4}{72} \times \dfrac{22}{7} = 2.2 \text{ cm}$

The separation between images = 4.4 cm

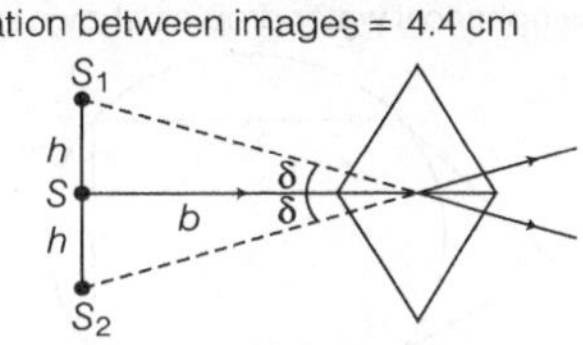

13. (a) For first surface of lens, $u = -20 \text{ cm}, \mu_1 = 1, \mu_2 = 1.5$, $r = +20 \text{ cm}$

$\therefore \quad \dfrac{\mu_2}{v} - \dfrac{\mu_1}{u} = \dfrac{\mu_2 - \mu_1}{r}$

or $\quad \dfrac{1.5}{v} - \dfrac{1}{-20} = \dfrac{1.5-1}{20} - \dfrac{1}{40}$

or $\quad \dfrac{3}{2v} = \dfrac{1}{40} - \dfrac{1}{20} = \dfrac{1-2}{40}$

$\therefore \quad v = \dfrac{-40 \times 3}{2} = -60 \text{ cm}$

For second surface of lens,

$u' = -60 \text{ cm}, r' = -20 \text{ cm}, \mu_1' = 1.5, \mu_2' = 2$

$\because \quad \dfrac{\mu_2'}{v'} - \dfrac{\mu_1'}{u'} = \dfrac{\mu_2' - \mu_1'}{r'}$

$\Rightarrow \quad \dfrac{2}{v'} - \dfrac{1.5}{-60} = \dfrac{2-1.5}{-20}$

$\Rightarrow \quad \dfrac{2}{v'} = -\dfrac{1}{40} - \dfrac{3}{180} = \dfrac{-9-6}{360}$

$\Rightarrow \quad \dfrac{v'}{2} = -\dfrac{360}{15}$

$\therefore \quad v' = \dfrac{-360}{15} \times 2 = -\dfrac{720}{15} = -48 \text{ cm}$

For bottom surface, $u'' = -60 \text{ cm}, r'' = +20 \text{cm}, \mu_1'' = 2, \mu_2'' = 1$

$\because \quad \dfrac{\mu_2''}{v''} - \dfrac{\mu_1''}{u''} = \dfrac{\mu_2'' - \mu_1''}{r''}$

or $\quad \dfrac{1}{v''} - \dfrac{2}{-60} = \dfrac{1-2}{20}$

or $\quad \dfrac{1}{v''} = -\dfrac{1}{20} - \dfrac{1}{30} = \dfrac{-3-2}{60}$

$\therefore \quad v'' = \dfrac{-60}{5} = -12 \text{ cm}$

14. (b)

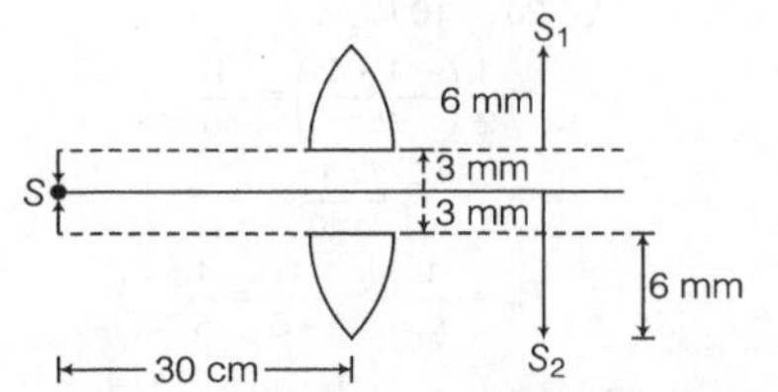

Here, $u = -30 \text{ cm}, f = 20 \text{ cm}$

$\because \quad \dfrac{1}{v} - \dfrac{1}{u} = \dfrac{1}{f}$

$\therefore \quad v = +60 \text{ cm}$

$\therefore \quad m = \dfrac{v}{u} = -2$

The separation between images = S_1S_2 = 18 mm

15. (c) $\because \sin c = \dfrac{\text{Refractive index of rarer medium}}{\text{Refractive index of denser medium}}$

$$\sin c_{ag} = \frac{1}{3/2} = 0.67$$

$$\sin c_{wg} = \frac{4}{3 \times \frac{3}{2}} = \frac{8}{9} = 0.89$$

16. (c) Secondary rainbow is formed, when the incident sunlight under goes two internal reflections under goes two internal reflections in a spherical water drop as shown in figure.

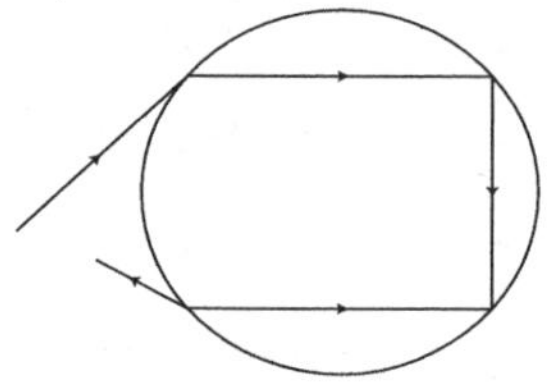

17. (b) $\because \delta = (\mu - 1) A = (1.5 - 1) \dfrac{14}{11} = \dfrac{1}{2} \times \dfrac{14}{11} = \dfrac{7}{11}$

$\therefore \quad \tan \delta = \delta = \dfrac{h}{15}$

$\Rightarrow \quad \dfrac{14}{11} \times \dfrac{\pi}{180} = \dfrac{h}{15}$

$\therefore \quad h = \dfrac{14}{11} \times 15 \times \dfrac{22}{7 \times 180} = \dfrac{1}{3}$ cm

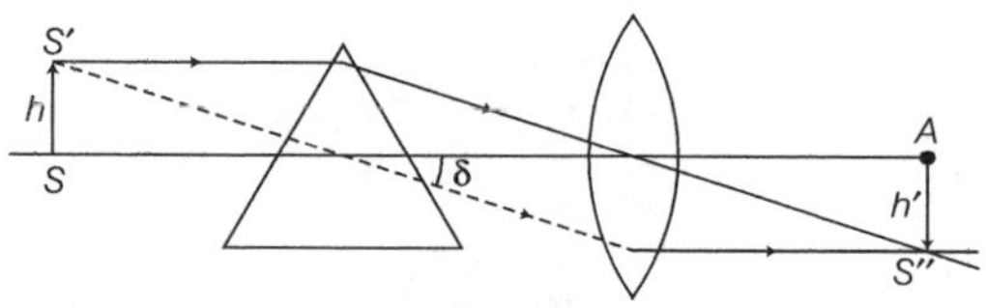

S' behaves as object for the thin convex lens

For the lens,

$$u = -30 \text{ cm}, f = +20 \text{ cm}$$

$\therefore \quad \dfrac{1}{v} - \dfrac{1}{u} = \dfrac{1}{f}$

$\because \quad \dfrac{1}{v} - \dfrac{1}{-30} = \dfrac{1}{20}$

or $\quad \dfrac{1}{v} = \dfrac{1}{20} - \dfrac{1}{30} = \dfrac{3-2}{60}$

$\therefore \quad v = 60$ cm

$\therefore \quad \dfrac{h'}{h} = \dfrac{v}{u} = \dfrac{60}{-30}$

$\therefore \quad h' = -2 \times \dfrac{1}{3} = \dfrac{-2}{3}$ cm

18. (b) $\dfrac{1}{f_l} = (1.5 - 1)\left(-\dfrac{1}{20} + \dfrac{1}{10}\right)$

$$= \frac{1}{2}\left(\frac{-1+2}{20}\right) = \frac{1}{40}$$

$\therefore \quad P_l = \dfrac{1}{40}$

and $\quad P_m = \dfrac{1}{f_m} = -\dfrac{1}{-5} = \dfrac{1}{5}$

$\therefore \quad P = 2P_l + P_m = \dfrac{1}{20} + \dfrac{1}{5} = \dfrac{1+4}{20}$

$\therefore \quad P = \dfrac{1}{4}$

So, equivalent mirror is concave mirror of focal length 4 cm.

So, object should be at the centre of curvature of equivalent concave mirror.

19. (a) $\quad \dfrac{1}{f} = (\mu - 1)\left(\dfrac{1}{R} + \dfrac{1}{R}\right)$

or $\quad \dfrac{1}{20} = (1.5 - 1)\dfrac{2}{R}$

$\therefore \quad R = 20$ cm

$\therefore$ Power of equivalent mirror is

$$P = 2P_l + P_m = \frac{2}{20} - \frac{1}{-10}$$

$$= \frac{1}{10} + \frac{1}{10} = \frac{1}{5}$$

$\therefore \quad \dfrac{-1}{F} = \dfrac{1}{5}$

$\therefore \quad F = -5$ cm

$\because \quad \dfrac{1}{v} + \dfrac{1}{u} = \dfrac{1}{F}$ or $\dfrac{1}{v} - \dfrac{1}{15} = -\dfrac{1}{5}$

or $\quad \dfrac{1}{v} = \dfrac{1}{15} - \dfrac{1}{5} = \dfrac{1-3}{15} = \dfrac{-2}{15}$

$\therefore \quad v = \dfrac{-15}{2}$ cm

$\therefore \quad v_{image} = \dfrac{-v^2}{u^2}\dfrac{du}{dt} = -\dfrac{\left(-\frac{15}{2}\right)^2}{(-15)^2} \times 4 = -1 \text{ cms}^{-1}$

20. (c)

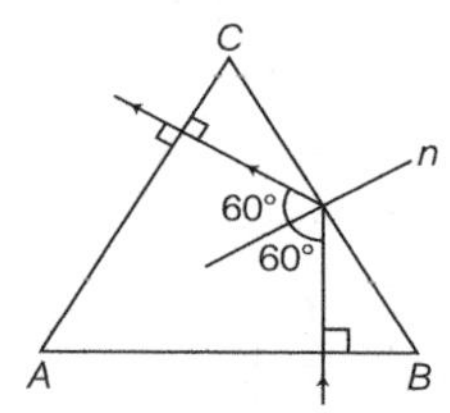

Here, $\sin c = \dfrac{1}{\mu} = \dfrac{2}{3} = 0.67$

and $\sin \theta = \sin 60° = \dfrac{\sqrt{3}}{2} = 0.866$

$\therefore \quad \theta > c$

So, total internal reflection takes place on side BC.

Hence, beam will emerge from face AC normally.

21. (b)

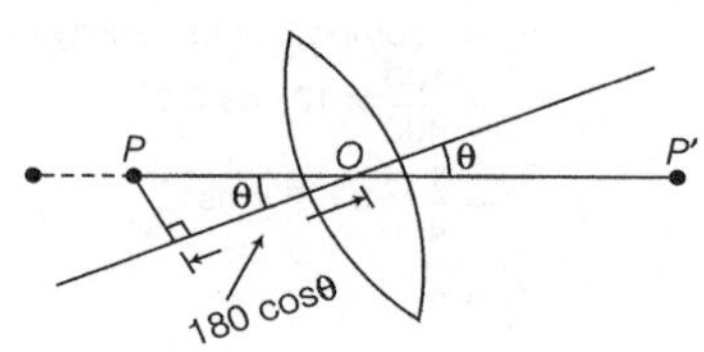

Here, $u = -180 \cos \theta$, $f = +90$ cm

$\because \quad \dfrac{1}{v} - \dfrac{1}{u} = \dfrac{1}{f}$

or $\quad \dfrac{1}{v} + \dfrac{1}{180 \cos \theta} = \dfrac{1}{90}$

or $\quad \dfrac{1}{v} = \dfrac{1}{90} - \dfrac{1}{180 \cos \theta} = \dfrac{2 \cos - 1}{180 \cos \theta}$

or $\quad v = \dfrac{180 \cos \theta}{2 \cos \theta - 1}$

$\therefore \quad v' = OP' = v \sec \theta = \dfrac{180}{2 \cos \theta - 1}$

$\therefore \quad v_{image} = \dfrac{dv'}{dt} = \dfrac{180}{(2 \cos \theta - 1)^2} = 2 \sin \theta \dfrac{d\theta}{dt}$

or $\quad \dfrac{dv'}{dt} = \dfrac{3600w}{(2-1)^2}$ ($\because \theta$ is very small)

$$= \frac{360 \times \pi}{180} \times \frac{1}{\pi} = 2 \text{ cms}^{-1}$$

22. (d)

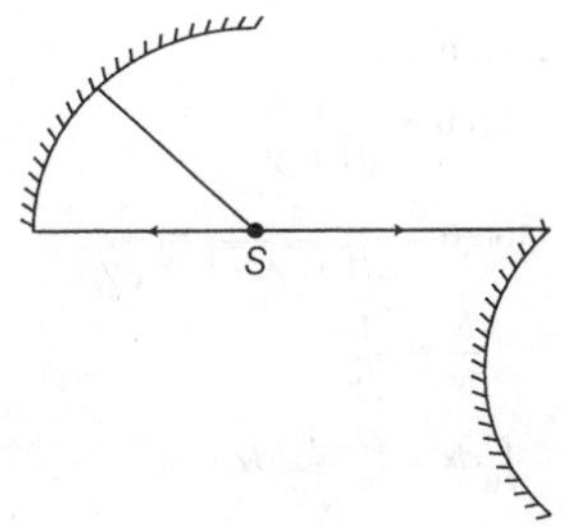

23. (b) As, we know,

$$\mu \propto \frac{1}{\lambda} \text{ and } \sin c = \frac{1}{\mu}$$

$$\therefore \quad \sin c \propto \lambda$$

Thus, the critical angle for blue is least. So, total internal reflection of blue light takes place first. Hence, option (b) is correct.

24. (a) $\sin\theta = \sin 45^\circ$

$$\frac{1}{\sqrt{2}} = \frac{1}{1.414}$$

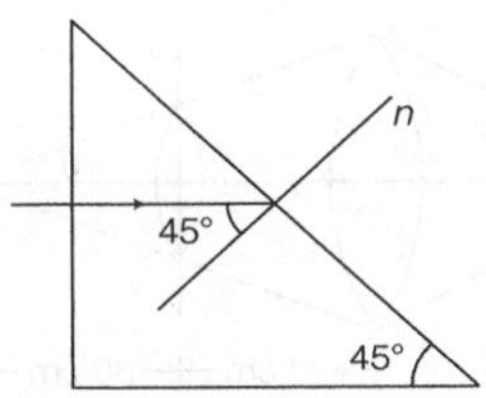

$$\sin C_{red} = \frac{1}{1.39}$$

$$\sin C_{green} = \frac{1}{1.44}$$

$$\sin C_{blue} = \frac{1}{1.47}$$

From this critical angle for red is larger than incident angle. But critical angles for green and blue lesser than that of incident angle.

25. (a) $u = -2.0$ m, $f = +1.5$ m

Image due to lens is formed at

$$v = \frac{uf}{f+u} = 6.0 \text{ m}$$

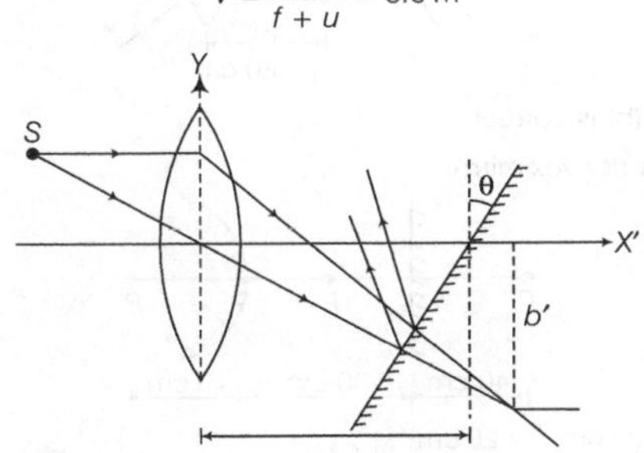

Its y-coordinate is $b' = -\frac{6}{2} \times 0.1 = -0.3$ m

$$\tan\theta = \frac{|b|}{v-d}$$

$$\therefore \quad d = v - |b'| \cos\theta = 5.0 \text{ m}$$

Hence, position of the mirror is 5.0 from the lens. Coordinates of the final image is (4.0 m, 0.3 m)

26. (b) The focal length of converging lens in water is f_1.

$$\therefore \quad \frac{1}{f_1} = \left(\frac{\mu_1}{\mu_w} - 1\right)\left(\frac{1}{r_1} - \frac{1}{r_2}\right)$$

$$= \left(\frac{2}{4/3} - 1\right)\left(\frac{2}{10}\right)$$

$$= \left(\frac{6-4}{4}\right) \times \frac{2}{10} = \frac{1}{2} \times \frac{2}{10}$$

$$\therefore \quad f_1 = 10 \text{ cm}$$

and
$$\frac{1}{f_2} = \left(\frac{1.5}{4/3} - 1\right)\left(\frac{-2}{10}\right)$$

or
$$\frac{1}{f_2} = \frac{0.5}{4}\left(\frac{-2}{10}\right) = \frac{-1}{40}$$

$$\therefore \quad f_2 = -40 \text{ cm}$$

$$\therefore \quad \frac{1}{f} = \frac{1}{f_1} + \frac{1}{f_2} = \frac{1}{10} - \frac{1}{40}$$

$$= \frac{4-1}{40}$$

$$\therefore \quad f = \frac{40}{3} \text{ cm}$$

27. (b) Let incident ray is along unit vector $\hat{\mathbf{e}} = e_x\hat{\mathbf{i}} + e_y\hat{\mathbf{j}} + e_z\hat{\mathbf{k}}$. If first plane mirror is YZ-plane unit vector along reflected ray is $\hat{\mathbf{e}}_1 = e_x\hat{\mathbf{i}} + e_y\hat{\mathbf{j}} + e_z\hat{\mathbf{k}}$. This reflected ray behaves as incident for second mirror which is in XZ-plane. Thus, the unit vector along reflected ray from second mirror is $\hat{\mathbf{e}}_q = -e_x\hat{\mathbf{i}} - e_y\hat{\mathbf{j}} + e_z\hat{\mathbf{k}}$. Thus, final emergent ray is $\hat{\mathbf{e}}_3 = -e_x\hat{\mathbf{i}} - e_y\hat{\mathbf{j}} - e_z\hat{\mathbf{k}}$. Which is anti parallel to unit vector along incident ray.

28. (d) The speed of sound wave in air is lesser than that in water. So, for sound wave air is denser than water. When it is passing from air to water, it bends away from normal.

29. (a) Galaxies are at distance of billions of light years from Earth. So, light takes billions of years to the earth from galaxy. If astronomer focus his telescope on a galaxy today, his telescope receive light emitted by galaxy billions of years ago. That is why astronomers talk about looking back in time.

30. (a) A person inside well lighted room receives two types of light. One coming from object out side the room and other due to reflection from inner surface of window. The intensity of light reflected from inner surface of window is more than that of coming from outside.

When the light of room is switched OFF, the person only receives the light coming from object outside the room.

31. (a) $\because \quad \frac{1}{f} = (\mu - 1)\left(\frac{1}{r_1} - \frac{1}{r_2}\right)$

and $\quad \mu \propto \frac{1}{\lambda}$

32. (b) For lens, $D \geq 4f$ for real image. In the case of concave mirror if objeet is at centre of curvature, image is also at the centre of curvature.

33. (b) $\therefore \frac{1}{f} = \left(\frac{\mu_l}{\mu_s} - 1\right)\left(\frac{1}{r_1} - \frac{1}{r_2}\right)$.

34. (a) $\lambda = \frac{h}{mv}$ (for electron)

35. (d) Since, no atmosphere is present on moon. So, no irregularity is present in the light coming from star. So, light from star coming continuously without twinkling.

36. (a) Hence, (i) → (p), (ii) → (p), (iii) → (r), (iv) → (q), (v) → (q)

Revisal Problems (JEE Advanced)

1. (d) Here, $\tan\theta = \frac{0.5}{1} = \frac{1+b}{2}$

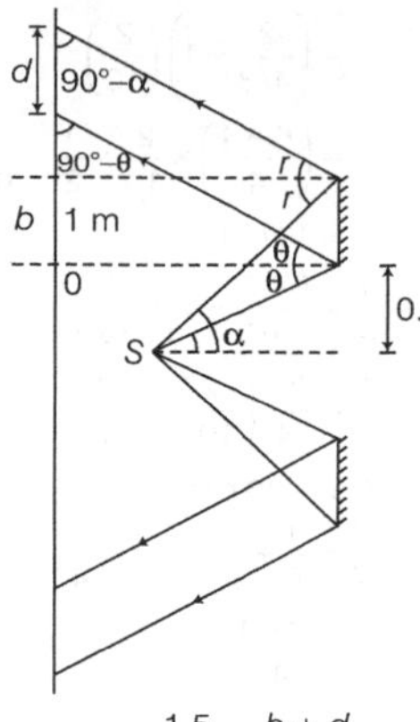

and $\qquad \tan\alpha = \frac{1.5}{1} = \frac{b+d}{2}$

$\because \qquad 1 + b = 1$

$\therefore \qquad b = 0$ and $b + d = 3$

$\therefore \qquad d = 3$

Thus, length of reflected path = $2d$ = $6m$

2. (b) From figure, the time period is

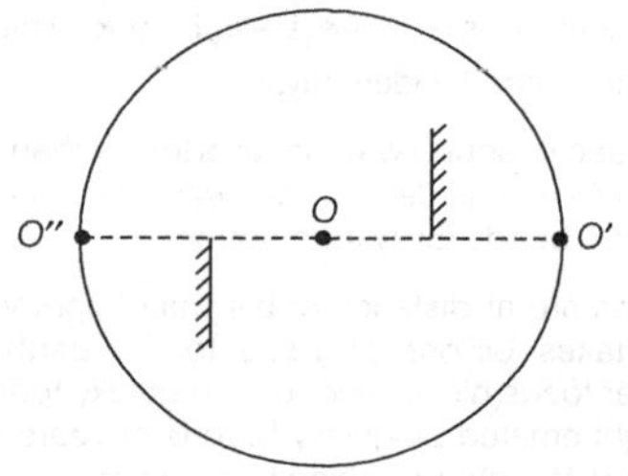

$$T = \frac{\pi r}{v}$$

$$\therefore \qquad f = \frac{1}{T} = \frac{v}{\pi r} = \frac{\pi}{\pi \times 20}\text{Hz}$$

$$= \frac{1}{20}\text{Hz} = 0.05\text{ Hz}$$

3. (a)

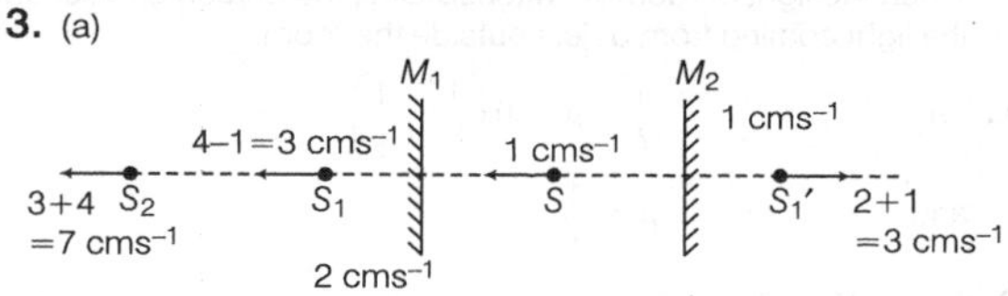

4. (b)

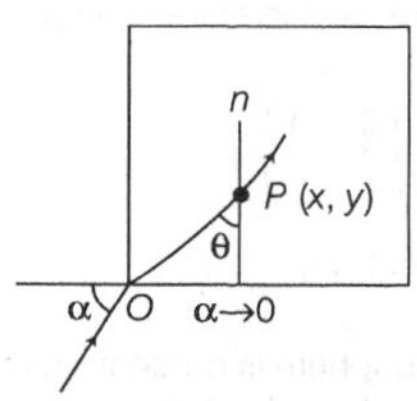

Here, $\mu \sin\theta = k$

or $\qquad \sqrt{1+y}\sin\theta = k$

For air $\qquad 1 \times \sin 90° = k$

$\therefore \qquad k = 1$

$$\sqrt{1+y}\sin\theta = 1$$

$$\therefore \qquad \sin\theta = \frac{1}{\sqrt{1+y}}$$

$$\Rightarrow \qquad \tan\theta = \frac{1}{\sqrt{1+y-1}} = \frac{1}{\sqrt{y}}$$

$$\Rightarrow \qquad \frac{dx}{dy} = \frac{1}{\sqrt{y}}$$

$$\Rightarrow \qquad \int_0^x dx = \int_0^y \frac{1}{\sqrt{y}}dy$$

$$\Rightarrow \qquad x = \left[\frac{y^{-\frac{1}{2}+1}}{-\frac{1}{2}+1}\right]_0^y$$

$$\Rightarrow \qquad x = [2\sqrt{y}]_0^y$$

$$\Rightarrow \qquad x = 2\sqrt{y}$$

$$\Rightarrow \qquad x^2 = 4y$$

This is equation of parabola.

5. (a) For lens, $u = -30$ cm, $f = 20$ cm

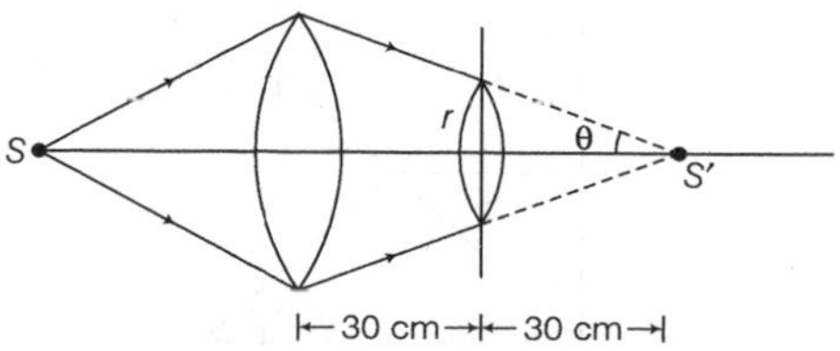

$\because \qquad \frac{1}{v} - \frac{1}{u} = \frac{1}{f}$ or $\frac{1}{v} + \frac{1}{30} = \frac{1}{20}$

$\therefore \qquad v = +60$ cm

From figure, $\qquad \tan\theta = \frac{1\text{ mm}}{30\text{ cm}} = \frac{r}{30\text{ cm}}$

$\therefore \qquad r = 1$ mm.

6. (b)

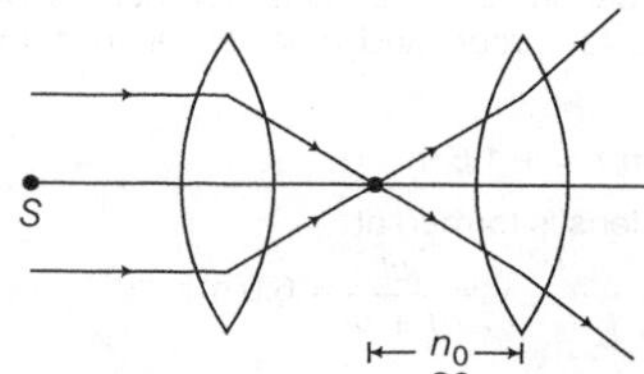

Here, (b) is correct

7. (c) For convex mirror,

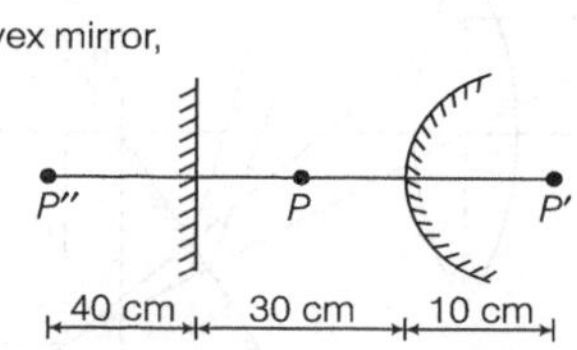

$u = -20$ cm, $f = 20$ cm

$\because \qquad \frac{1}{v} + \frac{1}{u} = \frac{1}{f}$

or $\qquad \frac{1}{v} - \frac{1}{20} = \frac{1}{20}$

$\therefore \qquad v = 10$ cm

Image P' formed by convex mirror behaves as object for plane mirror. Thus, the image formed by plane mirror is P''. When plane mirror is rotating with angular velocity = ω, image P'' is rotating with $\omega' = 2\omega = 4$ rad/s.

$\therefore$ The speed of image is $v_{image} = r\omega' = 40 \times 4 = 160\text{ cms}^{-1}$

8. (b) The image of point object P is on line PA.

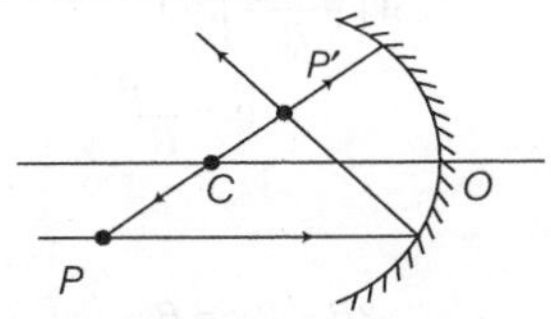

9. (b) $\mu = \frac{\text{Real depth}}{\text{Apparent depths}} = \frac{x}{y}$

$$\Rightarrow x = \mu y$$
$$\Rightarrow \frac{dx}{dt} = \mu \frac{dy}{dt}$$
$$\Rightarrow v_{\text{fish surface}} = \frac{4}{3} v_{\text{image surface}}$$
$$\Rightarrow v_{\text{fish}} - v_{\text{surface}} = \frac{4}{3}(v_{\text{image}} - v_{\text{surface}})$$
$$\Rightarrow 4 - 1 = \frac{4}{3}(v_{\text{image}} - 1)$$
$$\Rightarrow 3 \times \frac{3}{4} + 1 = v_{\text{image}}$$
$$\Rightarrow \frac{13}{4} = v_{\text{image}}$$
$$v_{\text{rel}} = v_{\text{image}} - v_{\text{bird}}$$
$$\Rightarrow 0 = v_{\text{image}} - v_{\text{bird}}$$
$$\therefore v_{\text{bird}} = v_{\text{image}} = \frac{13}{4} = 3.25 \text{ cms}^{-1}$$

10. (a) $\sin c = \frac{1}{\mu} - \frac{1}{5/3} = \frac{3}{5}$

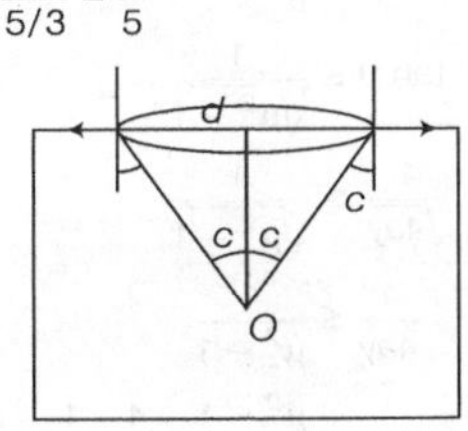

$$\therefore c = 37°$$
$$\therefore \tan c = \frac{d}{2 \times 5}$$
$$\therefore d = 10 \tan 37° = 10 \times \frac{3}{4} = 7.5 \text{ cm}$$

11. (b) When the plano convex lens is viewed from plane surface

$$2 = \frac{t}{\mu}$$
$$\therefore \mu = \frac{t}{2}$$

when it is viewed from curved surface, $\mu_1 = \mu$, $\mu_2 = 1$, $\mu = -t$

$r = -10$ cm, $v = -2.22$ cm

$$\therefore \frac{\mu_2}{v} - \frac{\mu_1}{u} = \frac{\mu_2 - \mu_1}{r}$$
$$\Rightarrow \frac{1}{v} - \frac{\mu}{-t} = \frac{1-\mu}{-10} = \frac{\mu - 1}{10}$$
$$\Rightarrow \frac{1}{-2.22} + \frac{1}{2} = \frac{\mu - 1}{10} = \frac{-1 + 1.11}{2.22} = \frac{\mu - 1}{10}$$
$$\frac{0.11}{2.22} = \frac{\mu - 1}{10}$$
$$\Rightarrow \mu - 1 = \frac{1.11}{2.22} = \frac{11}{22.2}$$
$$\Rightarrow \mu = 1 + \frac{11}{22.2} = \frac{33.2}{22.2} = 1.5$$

12. (b) For maximum deviation,

$$i = 90° \text{ or } i' = 90°$$

Taking, $i = 90°$

$$\because \mu = \frac{\sin i}{\sin r} = \frac{1}{\sin r}$$
$$\therefore \sin r = \frac{1}{\mu}$$
$$\therefore \cos r = \frac{\sqrt{\mu^2 - 1}}{\mu}$$

Also, $\mu = \frac{\sin i'}{\sin r} = \frac{\sin i'}{\sin (A - r)}$

$$\therefore \sin i' = \mu \sin (90° - r) = \mu \cos r = \sqrt{\mu^2 - 1}$$
$$\delta_{\max} = 90 + i' - A = i'$$
$$\therefore \sin \delta_{\max} = \sin i' = \sqrt{\mu^2 - 1}$$

Here $\mu = \frac{\mu_{\text{prism}}}{\mu_{\text{surrounding}}} = \frac{\sqrt{656}}{15 \times 4} \times 3 = \frac{\sqrt{41}}{5}$

$$\therefore \sin \delta_{\max} = \sqrt{\frac{41}{25} - 1} = \sqrt{\frac{41 - 25}{5}} = \frac{4}{5}$$
$$\therefore \delta_{\max} = 53°$$

13. (b) $\sin \theta = \frac{4}{5}$

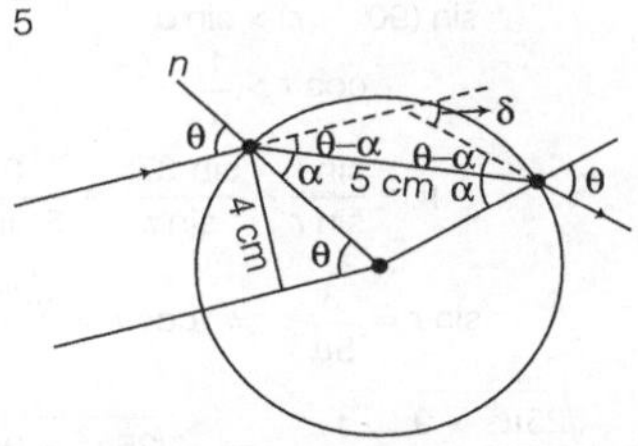

$$\therefore \theta = 53°$$
$$1 \times \sin \theta = \mu \sin \alpha$$
$$\frac{4}{5} = \mu \sin \alpha$$
$$\therefore \sin \alpha = \frac{4}{5\mu}$$
$$\therefore \alpha = \frac{4}{5 \times \frac{8}{5}} = \frac{1}{2}$$
$$\therefore \alpha = 30°$$
$$\therefore \delta = 2(\theta - \alpha) = 2(53° - 30°) = 46°$$

14. (a) According to snell's law,

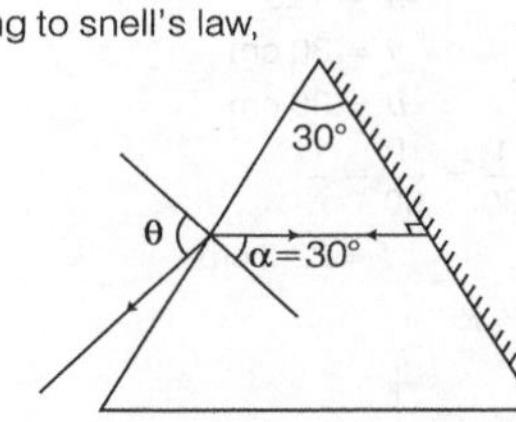

$$1 \times \sin \theta = \mu \sin \alpha$$

or $\sin 45° = \sqrt{2} \sin \alpha$

or $\sin \alpha = \frac{1}{2}$

$\therefore \alpha = 30°$, then, the ray incident on mirror normally. So, it retraces its path.

15. (b) For convex lens, $u = -30$ cm, $f = 20$ cm

$\because \quad \frac{1}{v} - \frac{1}{u} = \frac{1}{f}$

$\therefore \quad v = 60$ cm

The object shiftness by slab $= x = t\left(1 - \frac{1}{\mu}\right) = 5$ cm

$\therefore$ Object distance for convex mirror is

$u' = -(85 - 60 - 5) = -20$ cm

$\because \quad \frac{1}{v'} + \frac{1}{u'} = \frac{1}{f'}$

or $\quad \frac{1}{v'} - \frac{1}{20} = \frac{1}{20}$

$\therefore \quad v' = 10$ cm

16. (a) Here $r' > c$

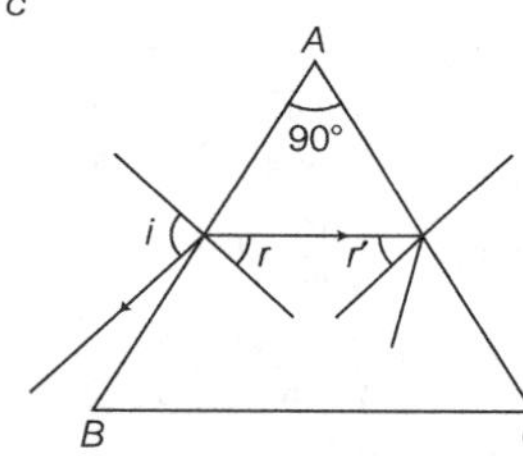

$\Rightarrow \quad 90° - r > c$

$\Rightarrow \quad \sin(90° - r) > \sin c$

$\Rightarrow \quad \cos r > \frac{1}{\mu}$

But, $\quad \mu = \frac{\sin i}{\sin r} = \frac{\sin 37°}{\sin r} = \frac{3}{5 \sin r}$

$\Rightarrow \quad \sin r = \frac{3}{5\mu} \quad \Rightarrow \cos r = \frac{\sqrt{25\mu^2 - 9}}{5\mu}$

$\Rightarrow \quad \frac{\sqrt{25\mu^2 - 9}}{5\mu} > \frac{1}{\mu} \quad \Rightarrow \sqrt{25\mu^2 - 9} > 5$

$\Rightarrow \quad 25\mu^2 - 9 > 25 \quad \Rightarrow \quad 25\mu^2 > 34$

$\Rightarrow \quad \mu > \sqrt{\frac{34}{25}}$

$\therefore \quad \mu > \frac{\sqrt{34}}{5}$

17. (a,b) Here, $\frac{v^2}{u^2} = \frac{1}{9} \Rightarrow \frac{v}{u} = \frac{1}{3}$

$\therefore \quad v = \frac{1}{3}u$

$\because \quad |v| + |u| = 120$

$\Rightarrow \quad 4v = 120$

$\therefore \quad v = 30$ cm

and $\quad u = 90$ cm

$\frac{1}{30} + \frac{1}{90} = \frac{1}{f}$

$\therefore \quad f = 22.5$ cm

18. (a,b) (a) $\mu \sin \theta = k$

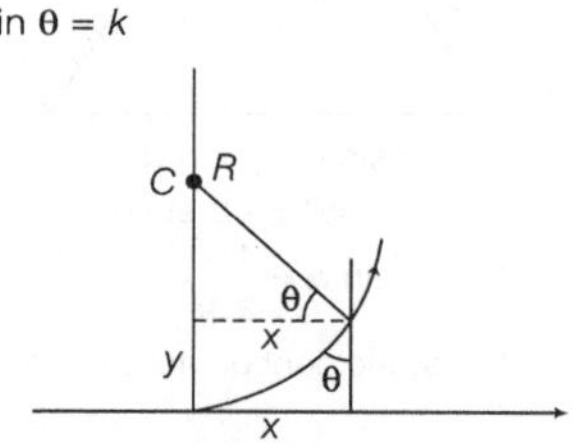

$\Rightarrow \quad 1 \times \sin 90° = k$

$\therefore \quad k = 1$

$\mu \sin\theta = 1$

$\Rightarrow \quad \mu\left(\frac{R - y}{R}\right) = 1$

$\therefore \quad \mu = \frac{R}{R - y}$

(b) $2 = \frac{R}{R - y}$

$2R - 2y = R$

$\therefore \quad y = \frac{R}{2}$

19. (a,b) Primary rainbow is formed due to only one internal reflection while secondary rainbow is formed due to two internal reflections.

20. (a,d) $\because y = ax^2$

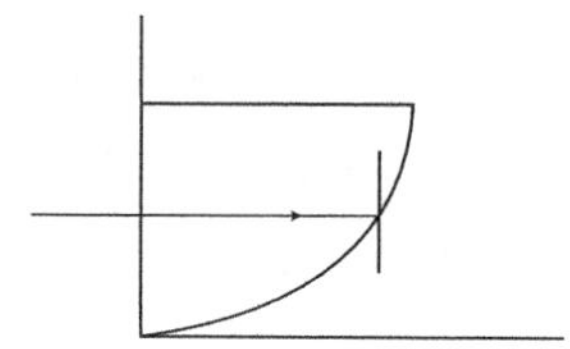

$\Rightarrow \quad \frac{dy}{dx} = 2ax$

$\Rightarrow \quad \frac{dy}{dx} = 2a\sqrt{\frac{y}{a}} = \sqrt{4ay}$

$\Rightarrow \quad \cot \theta = \sqrt{4ay}$

For refraction, $\quad \sin \theta \le \sin c$

$\Rightarrow \quad \sin \theta \le \frac{1}{\mu}$

$\therefore \quad \tan \theta \le \frac{1}{\sqrt{\mu^2 - 1}}$

$\Rightarrow \quad \frac{1}{\sqrt{4ay}} \le \frac{1}{\sqrt{\mu^2 - 1}}$

$\Rightarrow \quad \frac{1}{4ay} \le \frac{1}{\mu^2 - 1}$

$\Rightarrow \quad y_{min} = \frac{\mu^2 - 1}{4a} = \frac{4 - 1}{4 \times 1} = \frac{3}{4}$ m

21. (a,b,c) (a) $\tan \theta = \frac{1}{30}$

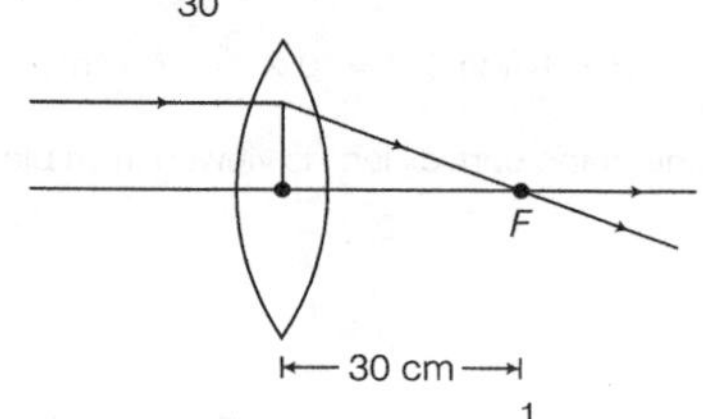

$\because \quad y = mx + c \quad$ or $y = -\frac{1}{30}x + 1$

(b) $u = -39$ cm, $f = 30$ cm

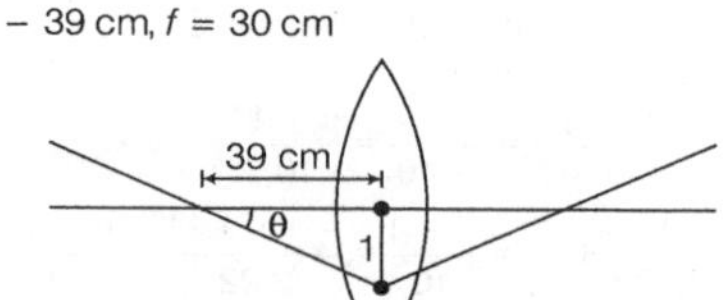

$\because \quad \frac{1}{v} - \frac{1}{u} = \frac{1}{f}$

or $\quad \frac{1}{v} + \frac{1}{39} = \frac{1}{30}$

or $\quad \frac{1}{v} = \frac{1}{30} - \frac{1}{39} = \frac{13 - 10}{390} = \frac{3}{390} = \frac{1}{130}$

$\because \quad v = 130$ cm

Slope of refracted ray = $\frac{1}{130}$

$\therefore \quad y = \frac{1}{130}x - 1$

$\therefore \quad \frac{x}{130} - y - 1 = 0$

(c)

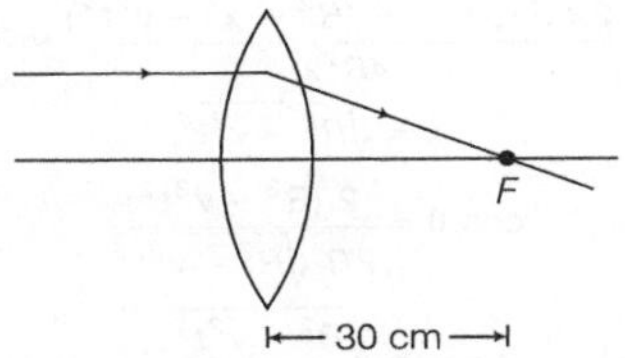

The ray whose equation is $y = 1$, will incident on the lens parallel to optical axis. So, it will pass through focus after refraction.

22. (b) $u_x = 20\,\text{ms}^{-1}, u_y = 20\,\text{ms}^{-1}$

$a_x = 0, a_y = -10\,\text{ms}^{-2}$

For first particle,

$$v_1 = \mu + a_1 t = 20\hat{i} + 20\hat{j} - 10\hat{j} = 20\hat{i} + 10\hat{j}$$

For second particle,

$$v_2 = 20\hat{j} - 10 \times 1\hat{j} = 10\hat{j}$$

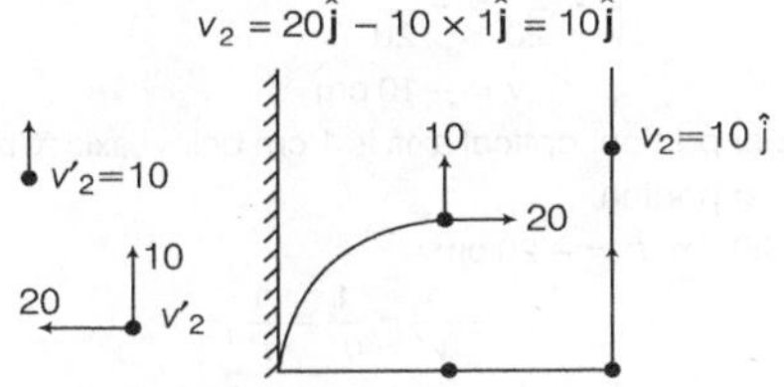

$\therefore \quad v_{rel} = 20\,\text{ms}^{-1}$

23. (d) Collision takes place at the maximum height.

$\therefore \quad m \times 20 = 2m\, v_x$

$\therefore \quad v_x = 10\,\text{ms}^{-1}$

24. (c) $\because x + \sqrt{3}y - 10 = 0$

or $\quad y = -\frac{1}{\sqrt{3}}x + \frac{10}{\sqrt{3}}$

$\therefore \quad \tan\theta = -\frac{1}{\sqrt{3}}$

$\therefore \quad \theta = -30°$

Thus, the ray incident on face OA normally.

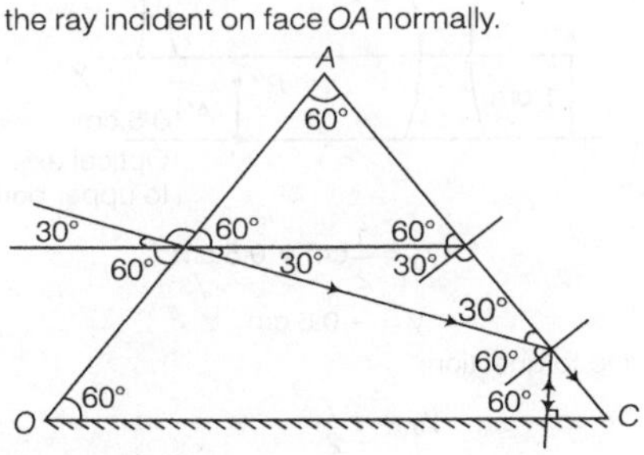

For face AC, $\mu \sin 60° = 1 \times \sin 90°$

$\therefore \quad \mu = \frac{1}{\sin 60°} = \frac{2}{\sqrt{3}}$

25. (b) Here, $\sin c = \frac{1}{\mu} = \frac{2}{3} = 0.67$

But $\sin\theta = \sin 60° = 0.866$

$\because \theta > c$, so the ray is totally reflected from face AC.

26. (a) In this case, the ray incident on silvered surface normally. Thus, it retraces its path. So, equationof ray is $x + \sqrt{3}y - 10 = 0$.

27. (b) $\because \quad \mu \sin\theta = k$

$\Rightarrow \quad 1 \times \sin 90° = k$

$\therefore \quad k = 1$

$\mu \sin\theta = 1$

$\therefore \quad \sin\theta = \frac{1}{\mu} \qquad$ (Here, $\mu = \sqrt{1 + e^{\frac{x}{d}}}$)

$$\tan\theta = \frac{1}{\sqrt{\mu^2 - 1}} = e^{-\frac{x}{2d}}$$

$\Rightarrow \quad \frac{dy}{dx} = e^{-\frac{x}{2d}}$

$\Rightarrow \quad \int_0^y dy = \int_0^x e^{-\frac{x}{2d}}\, dx$

$\Rightarrow \quad y = 2d\,(1 - e^{-\frac{x}{2d}})$

28. (a) $\because \quad y = 2d\,(1 - e^{-\frac{x}{2d}})$

put, $y = d$

$$\frac{1}{2} = 1 - e^{-\frac{x}{2d}}$$

$\Rightarrow \quad e^{-\frac{x}{2d}} = \frac{1}{2}$

$\Rightarrow \quad -\frac{x}{2d} = -1n^2$

$\Rightarrow \quad x = 2d\,1n^2$

29. (d) $\mu = \sqrt{1 + e^{\frac{2d\ln^2}{d}}} = \sqrt{1 + 4} = \sqrt{5}$

30. (a) $\frac{1}{v_1} + \frac{1}{30} = \frac{1}{10}$

$\therefore \quad v_1 = 15\,\text{cm}$

For mirror, $\frac{1}{v_2} - \frac{1}{x - v_1} = -\frac{1}{30}$

$$v_2 = \left(\frac{x - 15}{45 - x}\right)30$$

$$\frac{dv_2}{dt} = \frac{\{(45 - x) + (x - 15)\}30}{(45 - x)^2}\frac{dx}{dt}$$

$$= \left\{\frac{45 \times 30 - 30x + 30x - 15 \times 30}{(45 - x)^2}\right\}\frac{dx}{dt}$$

or $\quad \frac{dv_2}{dt} = \frac{900}{(45 - x)^2}v_m$

$$= \frac{900}{(45 - x)^2} \times \frac{(45 - x)^2}{300} = 3\,\text{cms}^{-1}$$

$\because \quad \frac{dv_2}{dt} = v_{\text{image mirror}} = v_{\text{image}} - v_{\text{mirror}}$

$\therefore \quad v_{\text{image}} = \frac{dv_2}{dt} + v_m = 3 + \frac{(45 - x)^2}{300}$

The net magnification $m = (m_1)(m_2)$

$$= \frac{1}{2} \times \left|\frac{30}{45 - x}\right| = \frac{15}{45 - x}$$

For $x = 30$ cm, $m = 1$

For $x = 40$ cm, $m = 3$

31. (4) $u = -15$ cm, $f = +10$ cm

$$\because \quad \frac{1}{v} - \frac{1}{u} = \frac{1}{f}$$

$$\Rightarrow \quad \frac{1}{v} + \frac{1}{15} = \frac{1}{10}$$

$$\Rightarrow \quad \frac{1}{v} = \frac{1}{10} - \frac{1}{15}$$

$$= \frac{3-2}{30} = \frac{1}{30}$$

$$v = 30 \text{ cm}$$

According to formula

$$\frac{1}{v} - \frac{1}{u} = \frac{1}{f}$$

$$\Rightarrow \quad -\frac{1}{v^2}\frac{dv}{dt} + \frac{1}{u^2}\frac{du}{dt} = 0$$

$$\frac{dv}{dt} = \frac{v^2}{u^2}\frac{du}{dt}$$

Along X-axis,

$$v_{\text{image lens}} = \left(\frac{30}{-15}\right)^2 v_{\text{objectlens}}$$

$$= 4(8-4)$$

$$\Rightarrow \quad v_{\text{image}} - v_{\text{lens}} = 16$$

$$\Rightarrow \quad v_{\text{image}} = 16 + 4 = 20 \text{ cms}^{-1} \text{ along optical axis.}$$

$$\therefore \quad v_x = 20 \text{ cms}^{-1}$$

$$\frac{y'}{y} = m$$

$$y' = my$$

$$\Rightarrow \quad \frac{dy'}{dt} = m\frac{dy}{dt} + y\frac{dm}{dt}$$

$$= m\frac{dy}{dt} + 0$$

$$\frac{dy'}{dt} = m\frac{dy}{dt} = \left(\frac{30}{-15}\right)(11-2)$$

$$v_{\text{image lens}} = -2 \times 9 = -18$$

$$v_{\text{image}} - v_{\text{lens}} = -18$$

$$\therefore \quad v_y = -18 + 2 = -16 \text{ cms}^{-1}$$

$$v_{\text{image}} = (20\hat{\mathbf{i}} - 16\hat{\mathbf{j}}) \text{ cms}^{-1}$$

$$|v| = \sqrt{400 + 256} = \sqrt{656}$$

$$= 4\sqrt{41} \text{ cms}^{-1}$$

$$\therefore \quad n = 4.$$

32. (5) Thin dark ring will be visible, if ray from source get total internal reflection from the spherical shell. For this,

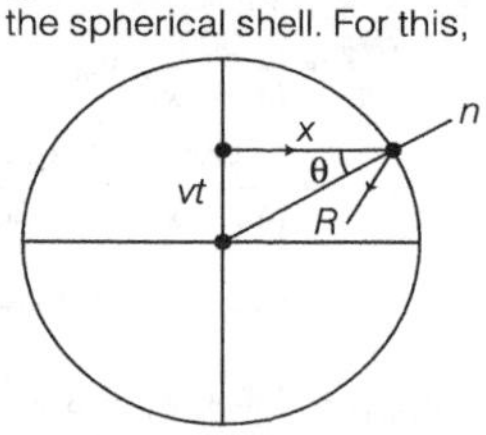

$$\theta > c$$

$$\Rightarrow \quad \sin\theta > \sin c \Rightarrow \sin\theta > \frac{1}{\mu}$$

$$\Rightarrow \quad \sin\theta > \frac{1}{\sqrt{2}}$$

$$\therefore \quad \theta > \frac{\pi}{4}$$

From figure,

$$\cos\theta = \frac{R^2 + x^2 - v^2t^2}{2Rx}$$

For minimum value of θ,

$$\frac{d(\cos\theta)}{dx} = 0$$

$$\Rightarrow \quad \frac{2x(2Rx) - 2R(R^2 + x^2 - v^2t^2)}{4R^2x^2} = 0$$

$$x = \sqrt{R^2 - v^2t^2}$$

$$\therefore \quad \cos\theta = \frac{2(R^2 - v^2t^2)}{2R\sqrt{R^2 - v^2t^2}}$$

$$= \frac{\sqrt{R^2 - v^2t^2}}{R}$$

For minimum time, $\theta = 45°$

$$\Rightarrow \quad \cos 45° = \frac{\sqrt{R^2 - v^2t^2}}{R}$$

$$\therefore \quad t = \frac{R}{\sqrt{2}v} = \frac{10}{\sqrt{2} \times \sqrt{2}} = 5 \text{ s}$$

33. (1) For lower portion, $u = -20$ cm, $f = -20$ cm

$$\because \quad \frac{1}{v} - \frac{1}{u} = \frac{1}{f}$$

$$\Rightarrow \quad \frac{1}{v} + \frac{1}{20} = -\frac{1}{20}$$

$$\Rightarrow \quad v = -10 \text{ cm}$$

For upper portion, optical axis is 1 cm below axis X-axis.

For upper portion,

$u' = -20$ cm, $f' = -20$ cm

$$= \frac{1}{v'} - \frac{1}{u'} = \frac{1}{f'}$$

$$\Rightarrow \quad \frac{1}{v'} + \frac{1}{20} = -\frac{1}{20}$$

$$\Rightarrow \quad \frac{1}{v'} = -\frac{1}{20} - \frac{1}{20}$$

$$= -\frac{2}{20} = -\frac{1}{10}$$

$$\therefore \quad v' = -10 \text{ cm}$$

$$m = \frac{v'}{u'} = \frac{y_0}{1}$$

$$\text{or} \quad \frac{-10}{-20} = \frac{y_0}{1}$$

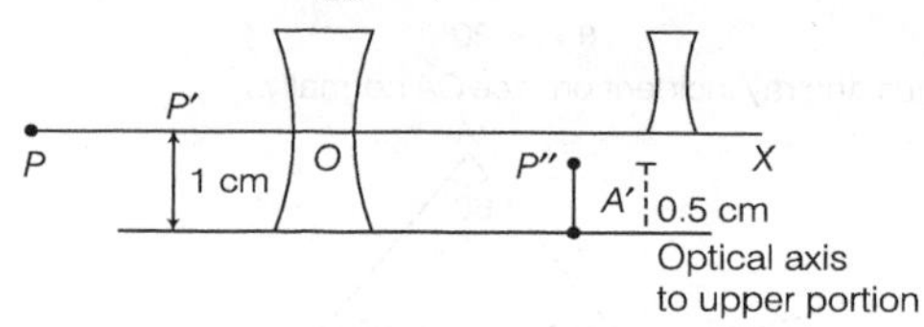

$$\therefore \quad y_0 = \frac{1}{2} \text{cm} = 0.5 \text{ cm}$$

$$\therefore \quad y = -0.5 \text{ cm}$$

According to question,

$$y_0 = -\frac{y}{2}$$

$$\Rightarrow \quad y = 2y_0$$

$$= 2 \times 0.5$$

$$= 1$$

$$\text{and} \quad y^2 = (1)^2 = 1$$

34. (5) For refraction through top surface,

$$\mu_1 = 1, \mu_2 = \mu, u = -\infty, r = +10 \text{ cm}$$

$\because \quad \dfrac{\mu_2}{v} - \dfrac{\mu_1}{u} = \dfrac{\mu_2 - \mu_1}{r}$

$\Rightarrow \quad \dfrac{\mu}{v} - \dfrac{1}{-\infty} = \dfrac{\mu - 1}{+10}$

$\Rightarrow \quad \dfrac{\mu}{v} = \dfrac{\mu - 1}{10} \quad \ldots(1)$

For cavity surface, $\mu_1 = \mu, \mu_2 = 1, u' = v - 10$

$r' = +5 \text{ cm}, \quad v^1 = +10 \text{ cm}$

$\because \quad \dfrac{\mu'_2}{v'} - \dfrac{\mu'_1}{u'} = \dfrac{\mu'_2 - \mu'_1}{r'}$

$\Rightarrow \quad \dfrac{1}{10} - \dfrac{\mu}{v - 10} = \dfrac{1 - \mu}{5}$

$\Rightarrow \quad \dfrac{\mu}{v - 10} = \dfrac{1}{10} - \dfrac{1 - \mu}{5}$

$\Rightarrow \quad \dfrac{\mu}{v - 10} = \dfrac{1 - (2 - 2\mu)}{10}$

$\Rightarrow \quad \dfrac{\mu}{v - 10} = \dfrac{2\mu - 1}{10}$

$\Rightarrow \quad 10\mu = (2\mu - 1)(v - 10)$

$\therefore \quad v - 10 = \dfrac{10\mu}{2\mu - 1}$

$\Rightarrow \quad v = \dfrac{10\mu}{2\mu - 1} + 10$

$\Rightarrow \quad \dfrac{10\mu}{\mu - 1} = \dfrac{10\mu}{2\mu - 1} + 10$

After solving, $\quad \mu = \dfrac{3 \pm \sqrt{5}}{2} = \dfrac{3 + \sqrt{5}}{2}$

$= \dfrac{3 + 2.2}{2} = \dfrac{5.2}{2} = 2.6$

$\Rightarrow \quad \dfrac{13}{n} = 2.6$

$n = \dfrac{13}{2.6} = \dfrac{130}{26}$

$n = 5$

35. (4) For coinciding image with object, the ray should incident on silvered surface normally.

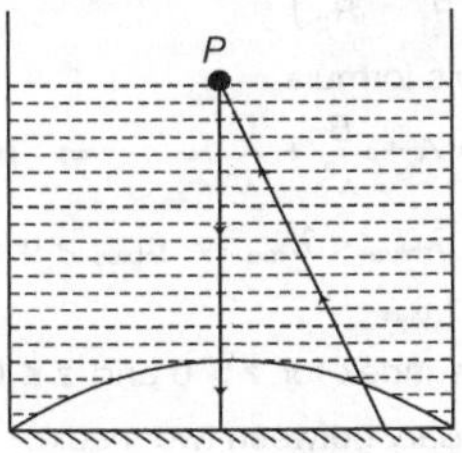

For curve surface of lens,

$\mu_1 = \dfrac{4}{3}$

$\mu_2 = \dfrac{3}{2}$

$v = \infty$

$\dfrac{\mu_2}{v} - \dfrac{\mu_1}{u} = \dfrac{\mu_2 - \mu_1}{20}$

$\Rightarrow \quad \dfrac{3}{2 \times \infty} - \dfrac{4}{3u} = \dfrac{\frac{3}{2} - \frac{4}{3}}{20}$

$= \dfrac{1}{120}$

$\Rightarrow \quad -\dfrac{3u}{4} = 120$

$\therefore \quad u = -\dfrac{120 \times 4}{3}$

$= -160 \text{ cm}$

Thus, the height of water inside vessel in 160 cm.

$\therefore \quad x = \dfrac{160}{40} = 4$

36. (3) For figure, $\sin\theta = \dfrac{R}{2R}$

$\therefore \quad \theta = 30°$

Here, $\quad \sqrt{3} \sin 30° = 1 \times \sin\beta$

$\therefore \quad \beta = 60°$

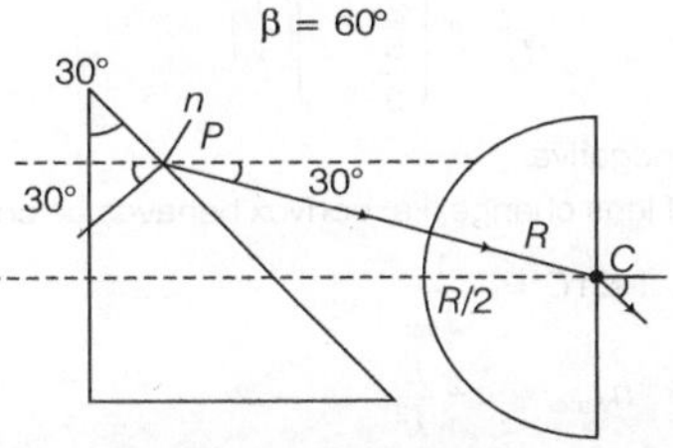

From figure, the ray strikes on curved surface normally

$\because \quad \sqrt{3} \sin\theta = 1 \times \sin\delta$

$\Rightarrow \quad \dfrac{\sqrt{3}}{2} = \sin\delta$

$\delta = \dfrac{\pi}{n}$

$\sin 60° = \sin\delta$

$\delta = 60°$

$\dfrac{\pi}{n} = 60°$

$n = \dfrac{180°}{60°}$

$n = 3$

JEE Main & AIEEE Archive

1. (b) According to lens maker's formula, when the lens in the air

$$\frac{1}{f}=\left(\frac{3}{2}-1\right)\left(\frac{1}{R_1}-\frac{1}{R_2}\right) \Rightarrow \frac{1}{f}=\frac{1}{2x} \Rightarrow f=2x$$

Here, $\frac{1}{x}=\frac{1}{R_1}-\frac{1}{R_2}$

In case of liquid,where refractive indices are $\frac{4}{3}$ and $\frac{5}{3}$, we get

Focal length in first liquid,

$$\frac{1}{f_1}=\left(\frac{\mu_s}{\mu_{l_1}}-1\right)\left(\frac{1}{R_1}-\frac{1}{R_2}\right)$$

$$\frac{1}{f_1}=\left(\frac{3/2}{4/3}-1\right)\frac{1}{x}$$

$\Rightarrow F_1$ is positive.

Nature of lens is not change.

$$\frac{1}{f_1}=\frac{1}{8x}=\frac{1}{4(2x)}=\frac{1}{4f} \Rightarrow f_1=4f$$

Focal length in second liquid,

$$\frac{1}{f_2}=\left(\frac{\mu_s}{\mu_{l_2}}-1\right)\left(\frac{1}{R_1}-\frac{1}{R_2}\right)$$

$$\frac{1}{f_2}=\left(\frac{\frac{3}{2}}{\frac{5}{3}}-1\right)\left(\frac{1}{x}\right)$$

$\Rightarrow$ f_2 is negative.

Nature of lens change, i.e. convex behaves as concave.

2. (b) Here, $\sin C=\frac{1}{n_{\text{water}}}$

and $n_{\text{water}}=a+\frac{b}{\lambda^2}$

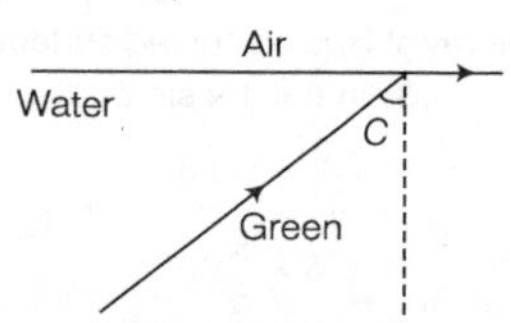

If frequency is less, then λ is greater and hence, RI $n_{(\text{water})}$ is less and therefore critical angle increases. So, they do not suffer reflection and come out at angle less than 90° of the air medium.

3. (c)

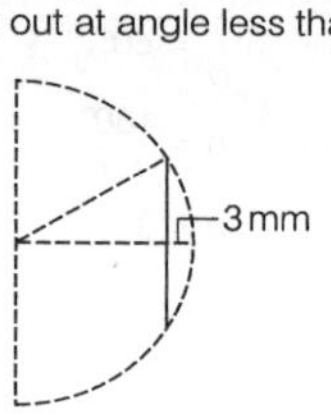

$$n=\frac{3}{2} \qquad \left[\because n=\frac{c}{v}\right]$$

$$3^2+(R-3\text{ mm})^2=R^2$$

$\Rightarrow 3^2+R^2-2R(3\text{ mm})+(3\text{ mm})^2=R^2$

$\Rightarrow$ $R \approx 15$ cm

$\therefore$ $\frac{1}{f}=\left(\frac{3}{2}-1\right)\left(\frac{1}{15}\right)$ $\left[\because \frac{1}{f}=(n-1)\left(\frac{1}{R_1}-\frac{1}{R_2}\right) \text{ and } R_1=\infty, R_2=-15\right]$

$\Rightarrow$ $f=30$ cm

4. (c) We know that, the angle of deviation depends upon the angle of incidence. If we determine experimentally, the angles of deviation corresponding to different angles of incidence and then plot i (on x-axis) and δ (on y-axis), we get a curve as shown in the figure.

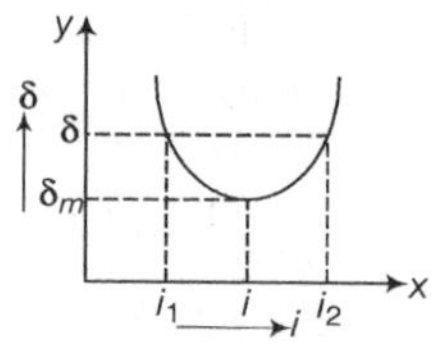

Clearly, if angle of incidence is gradually increased from a small value, the angle of deviation first decreases, becomes minimum for a particular angle of incidence and then begins to increase.

5. (d) Shift in image position due to glass plate,

$$S=\left(1-\frac{1}{\mu}\right)t=\left(1-\frac{1}{1.5}\right)\times 1\text{ cm}=\frac{1}{3}\text{ cm}$$

For focal length of the lens,

$$\frac{1}{f}=\frac{1}{v}-\frac{1}{u}=\frac{1}{12}-\frac{1}{-240}$$

or $$\frac{1}{f}=\frac{20+1}{240}$$

$\Rightarrow$ $$f=\frac{240}{21}\text{ cm}$$

Now, to get back image on the film, lens has to form image at $\left(12-\frac{1}{3}\right)$ cm $=\frac{35}{3}$ cm such that the glass plate will shift the image on the film.

As, $$\frac{1}{f}=\frac{1}{v}-\frac{1}{u}$$

$\Rightarrow$ $$\frac{1}{u}=\frac{1}{v}-\frac{1}{f}=\frac{3}{35}-\frac{21}{240}$$

$$=\frac{48\times 3-7\times 21}{1680}=-\frac{1}{560}$$

$\Rightarrow$ $u=-5.6$ m

6. (b) $\frac{1}{f}=(\mu-1)\left(\frac{1}{R_1}-\frac{1}{R_2}\right)$

Also, by Cauchy's formula,

$$\mu=A+\frac{B}{\lambda^2}+\frac{C}{\lambda^4}+\ldots \Rightarrow \mu\propto\frac{1}{\lambda}$$

As, $\lambda_{\text{blue}}<\lambda_{\text{red}} \Rightarrow \mu_{\text{blue}}>\mu_{\text{red}}$

Hence, $f_{\text{red}}>f_{\text{blue}}$

7. (a) As refractive index for $z>0$ and $z\le 0$ is different, xy-plane should be boundary between two media.

Angle of incidence

$$\cos i=\left|\frac{A_z}{\sqrt{A_x^2+A_y^2+A_z^2}}\right|=\frac{1}{2}$$

$\therefore$ $i=60°$

From Snell's law, $\frac{\sin i}{\sin r}=\frac{\mu_2}{\mu_1}=\frac{\sqrt{3}}{\sqrt{2}}$

$$\sin r=\frac{\sqrt{2}}{\sqrt{3}}\times\sin 60°=\frac{\sqrt{2}}{\sqrt{3}}\times\frac{\sqrt{3}}{2}=\frac{1}{\sqrt{2}}=45°$$

$\Rightarrow$ $r=45°$

8. (a) For the mirror, $\frac{1}{u}+\frac{1}{v}=\frac{1}{f}$

Differentiate this equation w.r.t. t, we get

$$-\frac{1}{u^2}\frac{du}{dt}-\frac{1}{v^2}\frac{dv}{dt}=0$$

$$\Rightarrow \quad \frac{dv}{dt}=-\frac{v^2}{u^2}\left(\frac{du}{dt}\right)$$

But, $$\frac{v}{u}=\frac{f}{u-f}$$

$$\therefore \quad \frac{dv}{dt}=-\left(\frac{f}{u-f}\right)^2\left(\frac{du}{dt}\right)$$

$$=\left(\frac{0.2}{-2.8-02}\right)^2\times 15=\frac{1}{15}\,\text{ms}^{-1}$$

9. (b) As intensity is maximum at axis.

Therefore, μ will be maximum and speed will be minimum on the axis of the beam.
Hence, beam will converge.

10. (d) For a parallel cylindrical beam, wavefront will be planar.

11. (c) The speed of light in the medium is directly proportional to the intensity I.

12. (a) It is possible when object kept at centre of curvature.

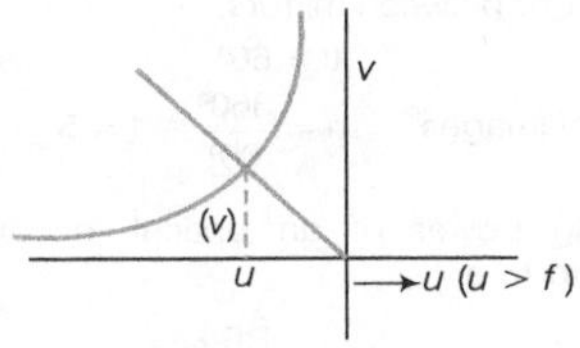

For $$u=v$$

$$u=2f,\ v=2f$$

13. (d) $$\sin C=\frac{\sqrt{3}}{2} \quad \ldots\text{(i)}$$

$$\sin r=\sin(90^\circ-C)=\cos C=\frac{1}{2}$$

$$\frac{\sin\theta}{\sin r}=\frac{\mu_2}{\mu_1}$$

$$\sin\theta=\frac{2}{\sqrt{3}}\times\frac{1}{2}\Rightarrow \theta=\sin^{-1}\left(\frac{1}{\sqrt{3}}\right)$$

14. (c) Experimental observation at $v=\infty, u=-f$

At, $u=\infty, v=f$

$$\frac{1}{v}-\frac{1}{u}=\frac{1}{f}=\text{constant}$$

15. (b) Power of a lens is reciprocal of its focal length. Power of combined lens is

$$P=P_1+P_2$$
$$=-15+5$$
$$=-10\text{ D}$$

$$\therefore \quad f=\frac{1}{P}=\frac{100}{-10}\text{ cm}$$

$$\Rightarrow \quad f=-10\text{ cm}$$

16. (a) $D=(\mu-1)A$

For blue light, μ is greater than that for red light, so $D_2>D_1$.

Alternate Method

$$D_1=(1.520-1)A$$
$$D_2=(1.525-1)A$$

$$\Rightarrow \quad \frac{D_1}{D_2}<1$$
$$D_2>D_1$$

17. (b) The situation is shown in figure.

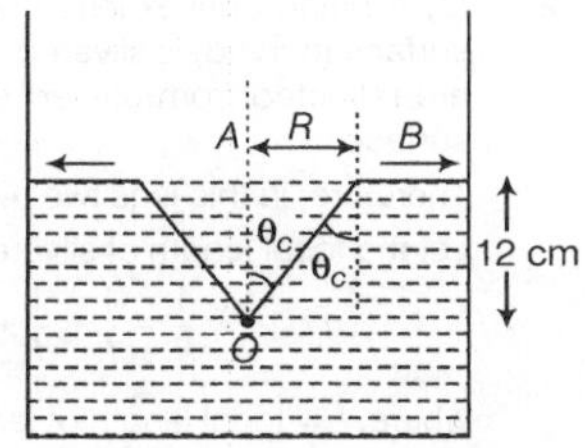

$$\sin\theta_C=\frac{1}{\mu}$$

$$\tan\theta_C=\frac{AB}{OA}$$

$$\therefore \quad AB=OA\tan\theta_C$$

$$=OA\frac{\sin\theta_C}{\cos\theta_C}$$

$$=\frac{\sin\theta_C}{\sqrt{1-\sin^2\theta_C}}\times OA$$

$$=\frac{1/\mu}{\sqrt{1-1/\mu^2}}\times OA=\frac{OA}{\sqrt{\mu^2-1}}$$

or $$AB=\frac{OA}{\sqrt{\mu^2-1}}=\frac{12}{\sqrt{\left(\frac{4}{3}\right)^2-1}}=\frac{36}{\sqrt{7}}$$

18. (a) We know, $\frac{y}{D}\geq 1.22\frac{\lambda}{d} \Rightarrow D\leq\frac{yd}{1.22\lambda}$

$$\therefore \quad D=\frac{10^{-3}\times 3\times 10^{-3}}{1.22\times 5\times 10^{-7}}=5\text{ m} \Rightarrow D_{\text{max}}=5\text{ m}$$

19. (a) $$\frac{1}{f_a}=(\mu-1)\left(\frac{1}{R_1}-\frac{1}{R_2}\right) \quad \ldots\text{(i)}$$

$$=(1.5-1)\left(\frac{1}{R_1}-\frac{1}{R_2}\right)$$

and $$\frac{1}{f_m}=\left(\frac{\mu_g-\mu_m}{\mu_m}\right)\left(\frac{1}{R_1}-\frac{1}{R_2}\right) \quad \ldots\text{(ii)}$$

$$\frac{1}{f_m}=\left(\frac{1.5}{1.6}-1\right)\left(\frac{1}{R_1}-\frac{1}{R_2}\right)$$

Thus, $$\frac{f_m}{f_a}=\frac{(1.5-1)}{\left(\frac{1.5}{1.6}-1\right)}=-8$$

or $$f_m=-8\times f_a=-8\times\frac{-1}{5} \quad \left[\because f_a=\frac{1}{P}=-\frac{1}{5}\text{ m}\right]$$

$$=1.6\text{ m}$$

$$\therefore \quad P_m=\frac{\mu}{f_m}=\frac{1.6}{1.6}=1\text{ D}$$

20. (b) For total internal reflection from glass-air interface, critical angle C must be less than angle of incidence.

i.e. $$C<i$$

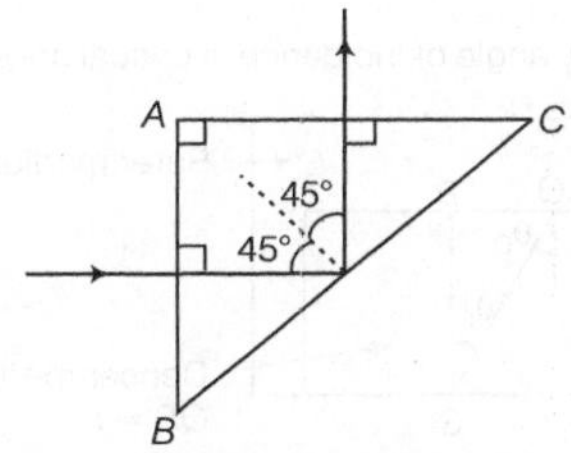

or $$C<45^\circ \quad [\because \angle i=45^\circ]$$

But, $n=\frac{1}{\sin C} \Rightarrow C=\sin^{-1}\left(\frac{1}{n}\right)$

$$\Rightarrow \quad \sin^{-1}\left(\frac{1}{n}\right)<45^\circ \quad \text{or} \quad \frac{1}{n}<\sin 45^\circ$$

$$\Rightarrow \quad n>\frac{1}{\sin 45^\circ} \Rightarrow n>\frac{1}{(1/\sqrt{2})}$$

$$\Rightarrow \quad n>\sqrt{2}$$

21. (a) A plano-convex lens behaves as a concave mirror, if its one surface (curved) is silvered. The rays refracted from plane surface are reflected from curved surface and again refract from plane surface.

Therefore, in this lens two refractions and one reflection occur.

Let the focal length of silvered lens be F.

$$\frac{1}{F}=\frac{1}{f}+\frac{1}{f}+\frac{1}{f_m}=\frac{2}{f}+\frac{1}{f_m}$$

where, f = focal length of lens before silvering

f_m = focal length of spherical mirror

$$\frac{1}{F}=\frac{2}{f}+\frac{2}{R} \qquad [\because R=2f_m] \ldots \text{(i)}$$

Now, $$\frac{1}{f}=(\mu-1)\left(\frac{1}{R_1}-\frac{1}{R_2}\right) \qquad \ldots \text{(ii)}$$

Here, $R_1=\infty, R_2=30$ cm

$\therefore$ $$\frac{1}{f}=(1.5-1)\left(\frac{1}{\infty}-\frac{1}{30}\right)$$

or $$\frac{1}{f}=-\frac{0.5}{30}=-\frac{1}{60}$$

or $$f=-60 \text{ cm}$$

Hence, from Eq. (i), we get

$$\frac{1}{F}=\frac{2}{60}+\frac{2}{30}=\frac{6}{60}$$

$$F=10 \text{ cm}$$

Again given that,

Size of object = Size of image

i.e. $$O=I$$

$\therefore$ $$m=-\frac{v}{u}=\frac{I}{O}$$

$\Rightarrow$ $$\frac{v}{u}=-1$$

or $$v=-u$$

Thus, from lens formula,

$$\frac{1}{F}=\frac{1}{v}-\frac{1}{u}$$

$\Rightarrow$ $$\frac{1}{10}=\frac{1}{-u}-\frac{1}{u}$$

$$\frac{1}{10}=-\frac{2}{u}$$

$\therefore$ $$u=-20 \text{ cm}$$

Hence, to get a real image, object must be placed at a distance 20 cm on the left side of lens.

22. (d) The particular angle of incidence for which reflected light is totally polarised for reflection from air to glass, is called the angle of polarisation (i_P) or Brewster's law.

Accordingly, $$n=\tan i_p$$ $$i_p=\tan^{-1}(n)$$

where, n is refractive index of glass.

23. (c) Objective of compound microscope is a convex lens. Convex lens form real and enlarged image when an object is placed between its focus and lens.

24. (a) The spectrum is correctly given by Rayleigh Jeans law.

25. (b) Number of images, $n=\dfrac{360^\circ}{\theta}-1$

where, θ is angle between mirrors.

$\therefore$ $$3=\frac{360^\circ}{\theta}-1 \text{ or } \theta=90^\circ$$

26. (a) Number of images, $n=\dfrac{360^\circ}{\theta}-1$

where, θ is angle between mirrors.

Thus, $\theta=60^\circ$ [given]

So, number of images $n=\dfrac{360^\circ}{60^\circ}-1=5$

27. (d) Resolving power of an optical instrument is inversely proportional to λ i.e.

$$\text{RP} \propto \frac{1}{\lambda}$$

$\therefore$ $$\frac{\text{Resolving power at } \lambda_1}{\text{Resolving power at } \lambda_2}=\frac{\lambda_2}{\lambda_1}=\frac{5000}{4000}=5:4$$

28. (b) Resolving power of telescope, $\text{RP}=\dfrac{d}{1.22\lambda}$

29. (a) Optical fibres work on the principle of total internal reflection.

JEE Advanced & IIT JEE Archive

1. (c) At point Q, angle of incidence is critical angle θ_C, where

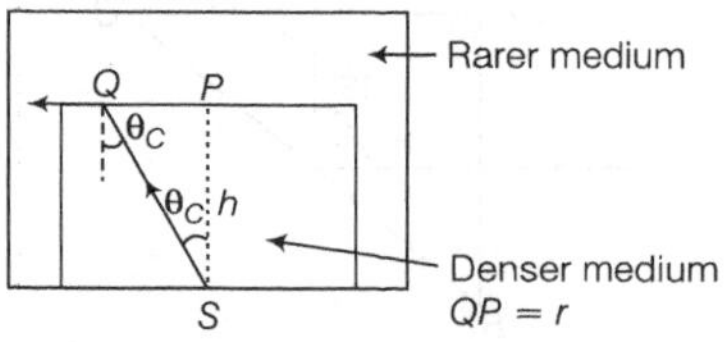

$$\sin \theta_C=\frac{\mu_l}{\mu_{\text{block}}}$$

In ΔPQS, $$\sin \theta_C=\frac{r}{\sqrt{r^2+h^2}}$$

$\therefore$ $$\frac{\mu_l}{\mu_{\text{block}}}=\frac{r}{\sqrt{r^2+h^2}}$$

$\Rightarrow$ $$\mu_l=\frac{r}{\sqrt{r^2+h^2}} \times 2.72$$

$$=\frac{5.77}{11.54} \times 2.72=1.36$$

2. (a)

(i) $\dfrac{1}{f}=\left(\dfrac{3}{2}-1\right)\left(\dfrac{1}{r}+\dfrac{1}{r}\right)=\dfrac{1}{r} \Rightarrow f=r$

$\Rightarrow$ $\dfrac{1}{f_{\text{eq}}}=\dfrac{1}{f}+\dfrac{1}{f}=\dfrac{2}{r}$, $f_{\text{eq}}=\dfrac{r}{2}$

(ii) $\dfrac{1}{f}=\left(\dfrac{3}{2}-1\right)\left(\dfrac{1}{r}\right) \Rightarrow f=2r$

$\Rightarrow$ $$\frac{1}{f}+\frac{1}{f}=\frac{2}{f}=\frac{1}{r}$$

$\Rightarrow$ $$f_{\text{eq}}=r$$

(iii) $\dfrac{1}{f}=\left(\dfrac{3}{2}-1\right)\left(-\dfrac{1}{r}\right)=-\dfrac{1}{2r} \Rightarrow f=-2r$

$\Rightarrow$ $$\frac{1}{f_{\text{eq}}}=\frac{1}{f}+\frac{1}{f}=-\frac{2}{2r}$$

$\Rightarrow$ $$f_{\text{eq}}=-r$$

(iv) $\dfrac{1}{f_{\text{eq}}}=\dfrac{1}{r}+\dfrac{1}{-2r}=\dfrac{1}{2r} \Rightarrow f_{\text{eq}}=2r$

Hence, (i) $\rightarrow$ (q), (ii) $\rightarrow$ (s), (iii) $\rightarrow$ (r), (iv) $\rightarrow$ (p)

3. (a,c) $\frac{1}{f_{\text{film}}} = (n_1 - 1)\left(\frac{1}{R} - \frac{1}{R}\right) \Rightarrow f_{\text{film}} = \infty$ (infinite)

$\therefore$ There is no effect of presence of film.

From Air to Glass

Using the equation $\frac{n_2}{v} - \frac{1}{u} = \frac{n_2 - 1}{R}$

$$\frac{1.5}{v} - \frac{1}{\infty} = \frac{1.5 - 1}{R} \Rightarrow v = 3R$$

$$\therefore \quad f_1 = 3R$$

From Glass to Air Again using the same equation,

$$\frac{1}{v} - \frac{n_2}{u} = \frac{1 - n_2}{-R}$$

$$\Rightarrow \quad \frac{1}{v} - \frac{1.5}{\infty} = \frac{1 - 1.5}{-R}$$

$$\Rightarrow \quad v = 2R$$

$$\therefore \quad f_2 = 2R$$

4. (c) $\mu = \frac{\lambda_{\text{air}}}{\lambda_{\text{medium}}} = \frac{1}{(2/3)} = \frac{3}{2}$

Further, $|m| = \frac{1}{3} = \left|\frac{v}{u}\right|$

$\therefore \quad |v| = \frac{|u|}{3} \Rightarrow u = -24 \text{ m}$ (Real object)

$\therefore \quad v = +8 \text{ m}$ (Real image)

Now, $\frac{1}{v} - \frac{1}{u} = \frac{1}{f} = (\mu - 1)\left(\frac{1}{+R} - \frac{1}{\infty}\right)$

$$\therefore \quad \frac{1}{8} + \frac{1}{24} = \left(\frac{3}{2} - 1\right)\left(\frac{1}{R}\right)$$

$$\therefore \quad R = 3 \text{ m}$$

5. (a) Let angle between the directions of incident ray and reflected ray be θ

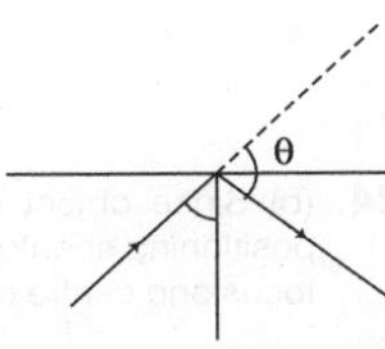

$$\cos\theta = \frac{1}{2}(\hat{i} + \sqrt{3}\hat{j}) \cdot \frac{1}{2}(\hat{i} - \sqrt{3}\hat{j})$$

$$\cos\theta = \frac{-1}{2}$$

$$\theta = 120°$$

6. (d) For $e \rightarrow i$, $45° > \theta_C \Rightarrow \sin 45° > \sin\theta_C$

$$\frac{1}{\sqrt{2}} > \frac{\mu_2}{\mu_1}$$

$$\mu_1 > \sqrt{2}\mu_2$$

For $e \rightarrow f$

angle of refraction is lesser than angle of incidence, so $\mu_2 > \mu_1$ and then $\mu_2 > \mu_3$

For $e \rightarrow g$, $\mu_1 = \mu_2$

For $e \rightarrow h$, $\mu_2 < \mu_1 < \sqrt{2}\mu_2$ and $\mu_2 > \mu_3$

Hence, (i) $\rightarrow$ (q), (ii) $\rightarrow$ (r), (iii) $\rightarrow$ (s), (iv) $\rightarrow$ (p)

7. (b) Using the lens formula,

$$\frac{1}{v} - \frac{1}{u} = \frac{1}{F} = \frac{1}{f_1} + \frac{1}{f_2}$$

or $$\frac{1}{v} = \frac{1}{u} + \frac{1}{f_1} + \frac{1}{f_2}$$

$$= \frac{1}{u} + (n_1 - 1)\left(\frac{1}{R_1} - \frac{1}{R_2}\right) + (n_2 - 1)\left(\frac{1}{R_1'} - \frac{1}{R_2'}\right)$$

Substituting the values, we get

$$\frac{1}{v} = \frac{1}{-40} + (1.5 - 1)\left(\frac{1}{14} - \frac{1}{\infty}\right) + (1.2 - 1)\left(\frac{1}{\infty} - \frac{1}{-14}\right)$$

Solving this equations, we get

$$v = +40 \text{ cm}$$

8. (b) Since value of n in meta-material is negative.

$$\therefore \quad v = \frac{c}{|n|}$$

9. (c) According to the paragraph, refracted ray in meta-material should be on same side of normal.

10. (c) After critical angle reflection will be 100% and transmission is 0%. Options (b) and (c) satisfy this condition. But option (c) is the correct option. Because in option (b) transmission is given 100% at $\theta = 0°$, which is not true.

11. (2)

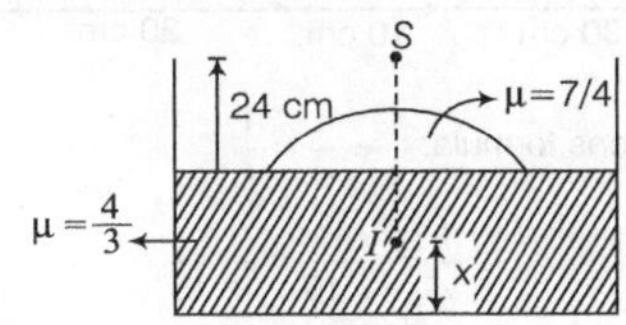

Two refractions will take place, first from spherical surface and the other from the plane surface.

So, applying $\frac{\mu_2}{v} - \frac{\mu_1}{u} = \frac{\mu_2 - \mu_1}{R}$

two times with proper sign convention.

Ray of light is travelling downwards. Therefore, downward direction is taken as positive direction.

$$\frac{7/4}{v} - \frac{1.0}{-24} = \frac{7/4 - 1.0}{+6} \quad \text{...(i)}$$

$$\frac{4/3}{(18 - x)} - \frac{7/4}{v} = \frac{4/3 - 7/4}{\infty} \quad \text{...(ii)}$$

Solving these equations, we get

$$x = 2 \text{ cm}$$

12. (6) $\frac{1}{v} - \frac{1}{u} = \frac{1}{f} \Rightarrow \frac{u}{v} - 1 = \frac{u}{f}$

$$\Rightarrow \quad \frac{u}{v} = \left(\frac{u + f}{f}\right)$$

$$\therefore \quad m = \frac{v}{u} = \left(\frac{f}{u + f}\right)$$

$$\frac{m_{25}}{m_{50}} = \frac{\left(\frac{20}{-25 + 20}\right)}{\left(\frac{20}{-50 + 20}\right)} = 6$$

13. (3) Using mirror formula twice,

$$\frac{1}{+25/3} + \frac{1}{-u_1} = \frac{1}{+10}$$

or $$\frac{1}{u_1} = \frac{3}{25} - \frac{1}{10} \text{ or } u_1 = 50 \text{ m}$$

and $$\frac{1}{(+50/7)} + \frac{1}{-u_2} = \frac{1}{+10}$$

$$\therefore \quad \frac{1}{u_2} = \frac{7}{50} - \frac{1}{10} \text{ or } u_2 = 25 \text{ m}$$

Speed of object $= \frac{u_1 - u_2}{\text{Time}} = \frac{25}{30} \text{ ms}^{-1} = 3 \text{ kmh}^{-1}$

14. (6) $\frac{R}{t} = \tan\theta_C$

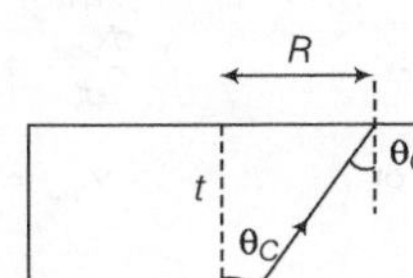

$\Rightarrow \quad R = t\,(\tan\theta_C)$

But, $\sin\theta_C = \frac{1}{\mu} = \frac{3}{5}$

$\therefore \quad \tan\theta_C = \frac{3}{4}$

$\therefore \quad R = \frac{3}{4}t = \frac{3}{4}(8\text{cm}) = 6 \text{ cm}$

15. (b) Object is placed at distance $2f$ from the lens. So, first image I_1 will be formed at distance $2f$ on other side. This image I_1 will behave like a virtual object for mirror. The second image I_2 will be formed at distance 20 cm infront of the mirror, or at distance 10 cm to the left hand side of the lens.

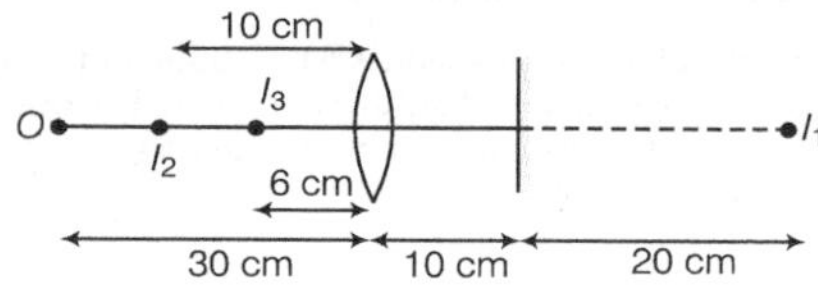

Now, applying lens formula, $\frac{1}{v} - \frac{1}{u} = \frac{1}{f}$

$\therefore$ $$\frac{1}{v} - \frac{1}{+10} = \frac{1}{+15}$$

$\Rightarrow$ $$v = 6 \text{ cm}$$

Therefore, the final image is at distance 16 cm from the mirror. But, this image will be real.

This is because ray of light is travelling from right to left.

16. (a)

(i) since $\mu_1 < \mu_2$, the ray of light will bend towards normal after first refraction.

(ii) $\mu_1 > \mu_2$, the ray of light will bend away from the normal after first refraction.

(iii) since $\mu_2 = \mu_3$ means in second refraction there will be no change in the path of ray of light.

(iv) Since $\mu_2 > \mu_3$, ray of light will bend away from the normal after second refraction.

Hence, (i) → (p,r), (ii) → (q,s,t), (iii) → (p,r,t), (iv) → (q,s)

17. (a,b,c) $\sqrt{3} = \frac{\sin 60°}{\sin r}$

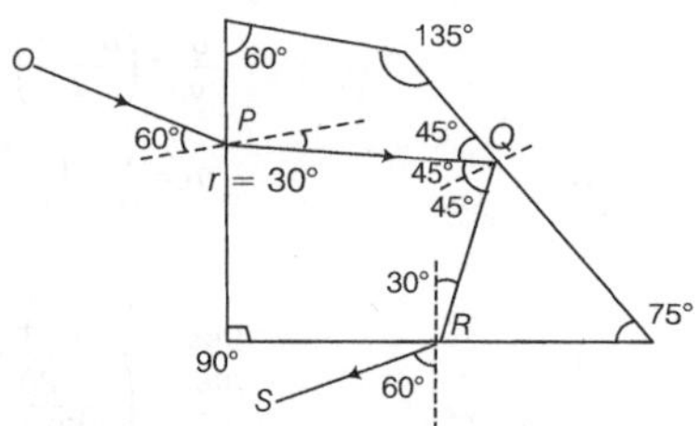

$\therefore$ $r = 30°$

$$\theta_C = \sin^{-1}\left(\frac{1}{\sqrt{3}}\right)$$

or $\sin\theta_C = \frac{1}{\sqrt{3}}$

$= 0.577$

At point Q, angle of incidence inside the prism is $i = 45°$.

Since, $\sin i = \frac{1}{\sqrt{2}}$ is greater than $\sin\theta_C = \frac{1}{\sqrt{2}}$, ray gets totally internally reflected at face CD. Path of ray of light after point Q is shown in the figure.

From the figure, we can see that angle between incident ray OP and emergent ray RS is 90°.

Therefore, correct options are (a), (b) and (c).

18. (c) $v = \sqrt{2gh} = \sqrt{2 \times 10 \times 7}$

$= 12 \text{ ms}^{-1}$

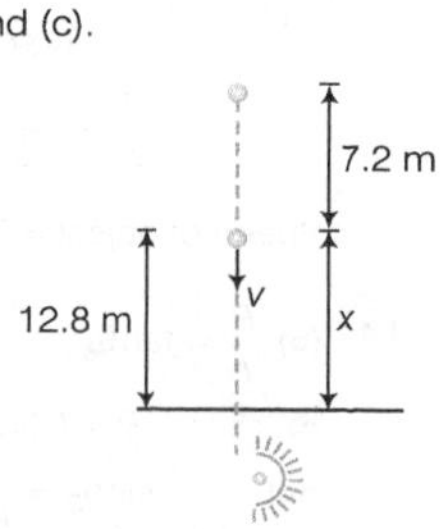

In this case when eye is inside water,

$$x_{\text{app.}} = \mu x$$

$\therefore$ $$\frac{dx_{\text{app.}}}{dt} = \mu \cdot \frac{dx}{dt}$$

or $$v_{\text{app.}} = \mu v = \frac{4}{3} \times 12$$

$$= 16 \text{ ms}^{-1}$$

19. (c,d) Values of options (c) and (d) don't match with the mirror formula.

$$\frac{1}{v} + \frac{1}{u} = \frac{1}{f}$$

20. (a) (i), (iii) and (iv) In case of concave mirror or convex lens image can be real, virtual, diminished magnified or of same size.

(ii) In case of convex mirror image is always virtual (for real object).

Hence, (i) → (p,q,r), (ii) → (q), (iii) → (p,q,r), (iv) → (p,q,r)

21. (a) At minimum deviation ($\delta = \delta_m$):

$$r_1 = r_2 = \frac{A}{2} = \frac{60°}{2} = 30° \quad \text{(For both colours)}$$

22. (b) Critical angle from Region III to Region IV,

$$\sin\theta_C = \frac{n_0/8}{n_0/6} = \frac{3}{4}$$

Now, applying Snell's law in Region I and Region III

$$n_0 \sin\theta = \frac{n_0}{6}\sin\theta_C$$

or $$\sin\theta = \frac{1}{6}\sin\theta_C = \frac{1}{6}\left(\frac{3}{4}\right) = \frac{1}{8}$$

$\therefore$ $$\theta = \sin^{-1}\left(\frac{1}{8}\right)$$

23. (c) Since $\theta < \theta_C$, both reflection and refraction will take place. From the figure, we can see that angle between reflected and refracted rays α is less than $180° - 2\theta$.

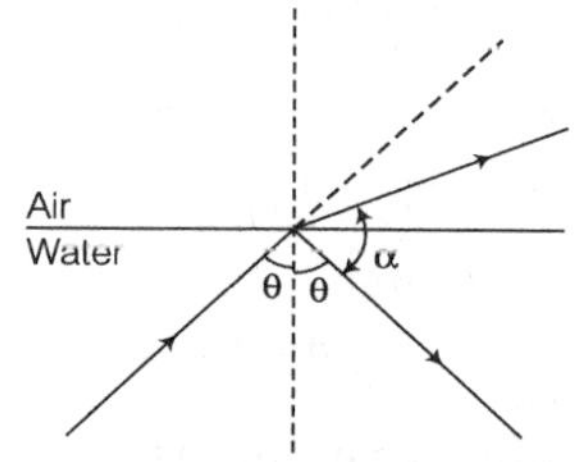

24. (b) Since object and image move in opposite directions, the positioning should be as shown in the figure. Object lies between focus and centre of curvature $f < x < 2f$.

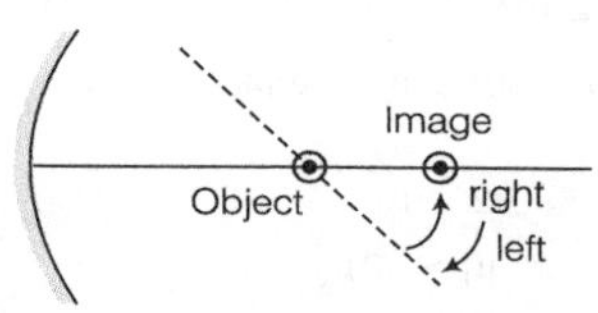

25. (a) Wavefronts are parallel in both media. Therefore, light which is perpendicular to wavefront travels as a parallel beam in each medium.

26. (c) All points on a wavefront are at the same phase.

$\therefore$ $$\phi_d = \phi_c \text{ and } \phi_f = \phi_e$$

$\therefore$ $$\phi_d - \phi_f = \phi_c - \phi_e$$

27. (b) In medium-2, wavefront bends away from the normal after refraction. Therefore, ray of light which is perpendicular to wavefront bends towards the normal in medium-2 during refraction. So, medium-2 is denser or its speed in medium-1 is more.

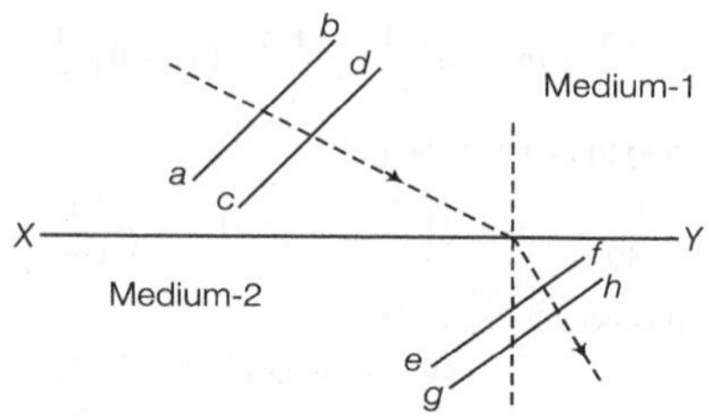

28. (c) Laws of reflection can be applied to any type of surface.

29. (c) Refraction from lens : $\frac{1}{v_1} - \frac{1}{-20} = \frac{1}{15}$

$\therefore \quad v = 60 \text{ cm} + \overrightarrow{\text{ve direction}}$

i.e. first image is formed at 60 cm to the right of lens system.

Reflection from mirror

After reflection from the mirror, the second image will be formed at a distance of 60 cm to the left of lens system.

Refraction from lens

$$\frac{1}{v_3} - \frac{1}{60} = \frac{1}{15} \quad \longleftarrow + \text{ve direction}$$

or $\quad v_3 = 12$ cm

Therefore, the final image is formed at 12 cm to the left of the lens system.

30. (b) From the lens formula, $\frac{1}{f} = \frac{1}{v} - \frac{1}{u}$

we have, $\quad \frac{1}{f} = \frac{1}{10} - \frac{1}{-10}$ or $f = +5$

Further, $\quad \Delta u = 0.1$

and $\quad \Delta v = 0.1 \quad$ (from the graph)

Now, differentiating the lens formula, we have

$$\frac{\Delta f}{f^2} = \frac{\Delta v}{v^2} + \frac{\Delta u}{u^2}$$

or $\quad \Delta f = \left(\frac{\Delta v}{v^2} + \frac{\Delta u}{u^2}\right) f^2$

Substituting the values, we have

$$\Delta f = \left(\frac{0.1}{10^2} + \frac{0.1}{10^2}\right)(5)^2 = 0.05$$

$\therefore \quad f \pm \Delta f = 5 \pm 0.05$

31. (b)

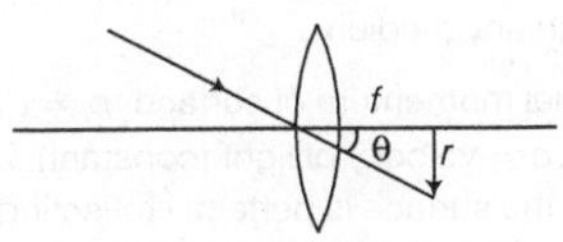

$$r = f \tan\theta \text{ or } r \propto f$$

$\therefore \quad \pi r^2 \propto f^2$

32. (a) Hence, (i) → (p), (ii) → (r), (iii) → (r), (iv) → (p,q,r)

33. (c) Distance of object from mirror

$$= 15 + \frac{33.25}{1.33} = 40 \text{ cm}$$

Distance of image from mirror

$$= 15 + \frac{25}{1.33} = 33.8 \text{ cm}$$

For the mirror, $\quad \frac{1}{v} + \frac{1}{u} = \frac{1}{f}$

$\therefore \quad \frac{1}{-33.8} + \frac{1}{-40} = \frac{1}{f}$

$\therefore \quad f = -18.3$ cm

34. (d) Let focal length of convex lens is $+f$, then focal length of concave lens would be $-\frac{3}{2}f$.

From the given condition,

$$\frac{1}{30} = \frac{1}{f} - \frac{2}{3f} = \frac{1}{3f}$$

$\therefore \quad f = 10$ cm

Therefore, focal length of convex lens = + 10 cm and that of concave lens = – 15 cm

35. Critical angles at 1 and 2

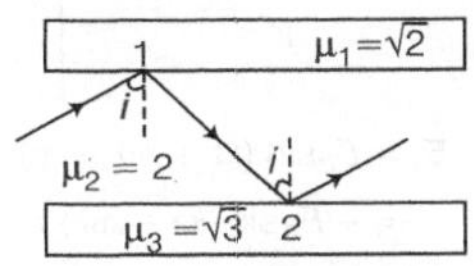

$$\theta_{C_1} = \sin^{-1}\left(\frac{\mu_1}{\mu_2}\right)$$

$$= \sin^{-1}\left(\frac{1}{\sqrt{2}}\right) = 45^\circ$$

$$\theta_{C_2} = \sin^{-1}\left(\frac{\mu_3}{\mu_2}\right)$$

$$= \sin^{-1}\left(\frac{\sqrt{3}}{2}\right) = 60^\circ$$

Therefore, minimum angle of incidence for total internal reflection to take place on both slabs should be 60°.

$$i_{min} = 60^\circ$$

36. (i) At minimum deviation, $r_1 = r_2 = 30^\circ$

$\therefore$ From Snell's law,

$$\mu = \frac{\sin i_1}{\sin r_1}$$

$\Rightarrow \quad \sqrt{3} = \frac{\sin i_1}{\sin 30^\circ}$

$\therefore \quad \sin i_1 = \frac{\sqrt{3}}{2}$

$\Rightarrow \quad i_1 = 60^\circ$

(ii) In the position shown net deviation suffered by the ray of light should be minimum. Therefore, the second prism should be rotated by 60° (anti-clockwise).

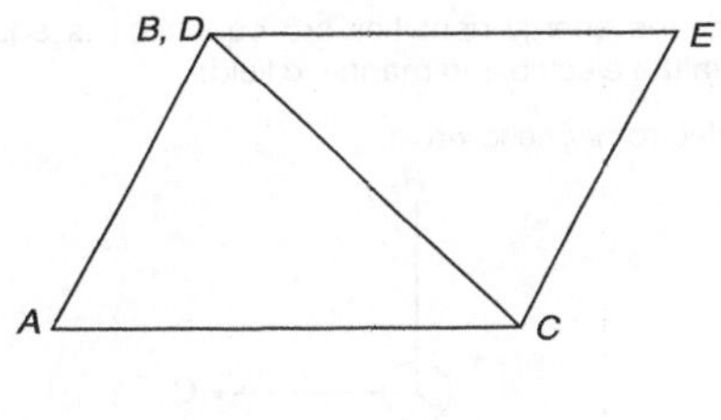

2. Wave Optics

DPP-1 Electromagnetic Waves

1. (a) When wave reflects from denser medium, phase changes through π.

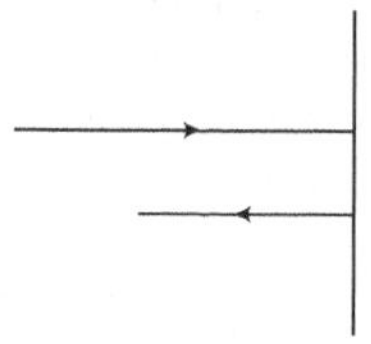

$$\therefore \quad \mathbf{E}_r = E_0 \sin(kz + \omega t + \pi)\,\hat{\mathbf{i}}$$
$$= -E_0 \sin(kz + \omega t)\,\hat{\mathbf{i}}$$

2. (b) $P = \frac{E}{c} = \frac{60}{3 \times 10^8} = 20 \times 10^{-8} = 2 \times 10^{-7}\,\text{kg ms}^{-1}$

3. (c) Intensity, $I = \frac{\text{Power}}{4\pi r^2} = \frac{100}{4\pi r^2}$

or $\quad \frac{1}{2}\varepsilon_0 E_0^2 c \propto \frac{1}{r^2}$

or $\quad E_0 \propto \frac{1}{r}$ or $\frac{E'}{E} = \frac{1}{2}$

$\therefore \quad E' = \frac{E}{2}$

4. (b) $\hat{\mathbf{c}}$, $\hat{\mathbf{E}}$ and $\hat{\mathbf{B}}$ should be mutually perpendicular.

5. (a) $\because \quad I = \frac{1}{2}\varepsilon_0 E_0^2 c$

$$\therefore \quad E_0 = \sqrt{\frac{2I}{\varepsilon_0 c}} = \sqrt{\frac{2 \times 1400}{8.85 \times 10^{-12} \times 3 \times 10^8}}$$
$$= 10.27 \times 10^2$$
$$= 1.03 \times 10^3\,\text{N/C}$$

$\because \quad c = \frac{E_0}{B_0}$

$$\therefore \quad B_0 = \frac{E_0}{c} = \frac{1.03 \times 10^3}{3 \times 10^8}$$
$$= 0.34 \times 10^{-5} = 3.4\,\mu\text{T}$$

6. (c) Due to conductivity, conduction current is present but due to variation of electric field, displacement current is present.

7. (c) Both the energy densities are equal i.e., is equally divided between the electric and magnetic field.

8. (b) In electromagnetic wave,

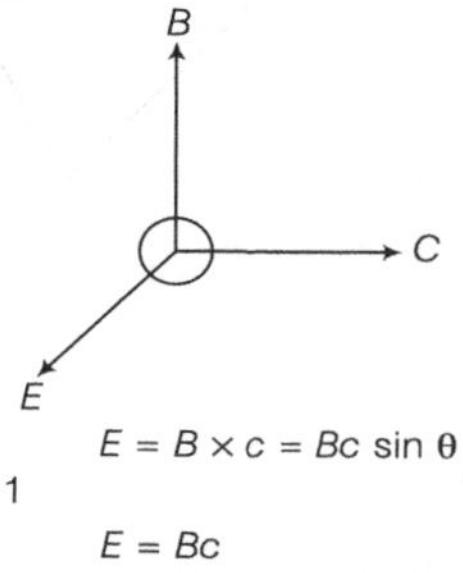

$$E = B \times c = Bc \sin\theta$$

$\theta = 90°, \sin\theta = 1$

$\therefore \quad E = Bc$

Given. $\quad B = 20\,n\text{T} = 20 \times 10^{-9}\,\text{T}$

$c = 3 \times 10^8\,\text{ms}^{-1}$

$E = 20 \times 10^{-9} \times 3 \times 10^8$

$\Rightarrow \quad E = 6\,\text{V m}^{-1}$

9. (b) In electromagnetic wave, the direction of propagation of wave, electric field and magnetic field are mutually perpendicular, i.e., wave propagates perpendicular to **E** and **B** or along **E** × **B**. While polarisation of wave takes place parallel to electric field vector.

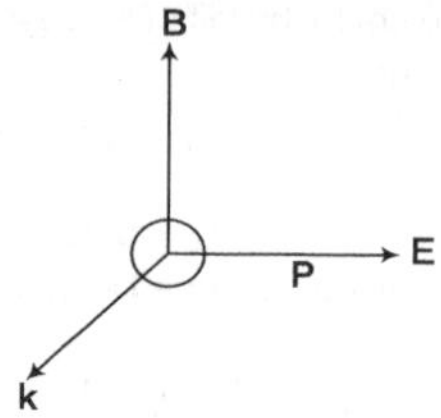

10. (a) Total average energy $= \varepsilon_0 E_{rms}^2$
$$= 8.85 \times 10^{-12} \times (720)^2$$
$$= 4.58 \times 10^{-6}\,\text{Jm}^{-3}$$

11. (c) In vacuum, $\varepsilon_0 = 1$

In medium, $\quad \varepsilon = 4$

$\therefore$ Refractive index, $\quad \mu = \sqrt{\frac{\varepsilon}{\varepsilon_0}} = \sqrt{\frac{4}{1}} = 2$

Wavelength, $\quad \lambda' = \frac{\lambda}{\mu} = \frac{\lambda}{2}$

and wave velocity, $\quad v = \frac{c}{\mu} = \frac{c}{2} \qquad [\because \mu = \frac{c}{v}]$

Hence, it is clear that wavelength and velocity will become half but frequency remains unchanged, when the wave is passing through any medium.

12. (b) Initial momentum of surface, $p_i = E/c$

where, c = velocity of light (constant)

Since, the surface is perfectly reflecting, so the same momentum will be reflected completely. Find momentum,

$$pf = -E/c \qquad \text{[negative value]}$$

$\therefore$ Change in momentum,

$$\Delta p = p_i - p_i = -\frac{E}{c} - \frac{E}{c} = -\frac{2E}{c}$$

Thus, momentum transferred to the surface is

$$\Delta p' = |\Delta p| = \frac{2E}{c}$$

13. (a) The spectrum of electromagnetic waves stretches, over an infinite range of wavelengths from 10^{-12}m to 10^6m least wavelength being of gamma rays from 10^{-10}m to 10^{-14}m (γ has highest frequency)

14. (c) Cosmic rays, γ-rays and X-rays are part of electromagnetic spectrum, while β-rays are emitted by radioactive elements. Hence, β-rays is not electromagnetic waves.

15. (a) Polarisation, using polariser and analyser, we get change in intensity of light at different rotations, hence we conclude that light waves are transverse in nature.

16. (b) $\quad c = \frac{1}{\sqrt{\mu_0 \mu_r \varepsilon_0 \varepsilon_r}}$ and $c_0 = \frac{1}{\sqrt{\mu_0 \varepsilon_0}}$

$\therefore$ Refractive index $= \frac{c_0}{c} = \sqrt{\mu_r \varepsilon_r}$

17. (a,c) $\quad E = \frac{hc}{\lambda}$ or $11\,\text{eV} = \frac{hc}{\lambda}$

$$\lambda = \frac{hc}{11\,\text{eV}} = \frac{1242\,\text{nm eV}}{11\,\text{eV}} = 112.91\,\text{nm}.$$

This wavelength lying in ultraviolet region.

18. (b,c) $\because$ $c = \frac{E_0}{B_0}$

$\therefore$ $B_0 = \frac{E_0}{c}$

In electromagnetic wave, electric field and magnetic field are mutually perpendicular and are in same phase.

19. (a,c) E, B and propagation of wave are mutually perpendicular.

20. (a,b)

(a) $\lambda = \frac{c}{f} = \frac{3 \times 10^8}{10^9} = 3 \times 10^{-1} = 0.3 \text{ m} = 30 \text{ cm}$

(b) This wave is lying in radio wave region.

21. (a,c,d) If a charged particle is accelerated, electromagnetic wave is produced.

22. (a) As, electromagnetic wave is plane polarised, so the receiving antenna should be parallel to electric/magnetic part of the electromagnetic wave.

23. (d)

24. (d) Here, conduction current

$$I_0 = \frac{\varepsilon_0}{X_C} = \frac{\varepsilon_0}{\frac{1}{\omega C}} = \varepsilon_0 \omega C$$

But current = displacement current.

Hence, when frequency increase displacement current increases.

25. (b) Electromagnetic radiation exerts pressure.

26. (b) Hence, (i) $\to$ (r), (ii) $\to$ (p), (iii) $\to$ (s), (iv) $\to$ (q)

27. (a) Hence, (i) $\to$ (r), (ii) $\to$ (q), (iii) $\to$ (p), (iv) $\to$ (s)

28. (a) Hence, (i) $\to$ (s), (ii) $\to$ (q), (iii) $\to$ (r), (iv) $\to$ (p)

29. (c) Hence, (i) $\to$ (r), (ii) $\to$ (s), (iii) $\to$ (q), (iv) $\to$ (p)

30. (d) (a) Infrared rays are used to treat muscular strain.

(b) Radio waves are used for broadcasting purposes.

(c) X-rays are used to detect fracture of bones.

(d) Ultraviolet rays are absorbed by ozone.

Hence, (i) $\to$ p, (ii) $\to$ q, (iii) $\to$ r, (iv) $\to$ r,

31. (2) $E = \frac{\varepsilon}{\varepsilon_0} = \frac{q}{A\varepsilon_0} = \frac{I_0 t}{A\varepsilon_0}$

$\therefore$ $I_D = \varepsilon_0 \frac{d}{dt}\left(E\frac{A}{2}\right) = \frac{\varepsilon_0 A}{2}\frac{dE}{dt}$

$= \frac{\varepsilon_0 A}{2} \times \frac{I_0}{A\varepsilon_0} = \frac{I_0}{2}$

$\therefore$ $n = 2$

32. (6) Here, $q = q_0 e^{-\frac{t}{RC}}$

$I = -\frac{dq}{dt} = \frac{q_0}{RC} e^{-\frac{t}{RC}}$

At $t = 0$, $I_0 = \frac{q_0}{RC}$

$= \frac{42 \times 10^{-11}}{2 \times 6 \times 10^{-12}}$

$= \frac{7}{2} \times 10 = 35 \ A$

$\therefore$ $I = I_d = \varepsilon_0 A \frac{dE}{dt}$

$\frac{dE}{dt} = \frac{I_d}{\varepsilon_0 A} = \frac{35}{8.85 \times 10^{-12} \times 10^{-2}}$

$= 395 \times 10^{12} \text{ Vm}^{-1}\text{s}^{-1}$

$\therefore$ $n = 6$

33. (3) $\frac{J_C}{J_D} = \frac{\sigma}{\omega\varepsilon} = \frac{35 \times 10^6}{4 \times 10^{10} \times 10^{-9}} \times 36\pi = 9.9 \times 10^7$

$\therefore$ $J_D = \frac{J_C}{9.9 \times 10^7} = \frac{5.5}{\pi (1.5 \times 10^{-3})^2 \times 9.9 \times 10^7}$

$= 7.86 \times 10^{-3} \mu\text{Am}^{-2}$

34. (9) $\because$ $J_C = J_D$

But $\frac{J_C}{J_D} = \frac{\sigma}{\omega\varepsilon}$

$\therefore$ $\frac{\sigma}{\omega\varepsilon} = 1$

$\therefore$ $\omega = \frac{\sigma}{\varepsilon_0 \varepsilon_r} = \frac{5}{8.854 \times 10^{-12}}$

$= 565 \times 10^9 \text{rads}^{-1}$

$\therefore$ $n = 9$

35. (5) $I_d = \varepsilon_0 \varepsilon_r A \frac{dE}{dt}$

$= \varepsilon_0 \varepsilon_r 2\pi r l \frac{d}{dt}\left(\frac{q}{2\pi\varepsilon_0 \varepsilon_r l r}\right)$

$= \frac{\varepsilon_0 \varepsilon_r 2\pi r l}{2\pi \varepsilon_0 \varepsilon_r l r} \times c \frac{dv}{dt}$

$= c\frac{dv}{dt}$

$= \frac{2\pi \varepsilon_0 \varepsilon_r l}{\ln\left(\frac{b}{a}\right)} \times 250 \times 377 \cos(377t)$

$\therefore$ $I_{d\,max} = 9.63 \times 10^{-5} A$

$\therefore$ $n = 5$

DPP-2 **Interference of Light - I**

1. (i) (a) Here, screen is parallel to line joining point sources. So, shape of fringes is hyperbolic but central bright fringe is straight line.

(ii) (d) The line joining point sources S_1 and S_2 is parallel to screen, so, shape of fringes is hyperbolic but central bright fringe is straight line.

(iii) (d) Since, sources are broad, so the shapes of fringes are straight line parallel to broadness.

(iv) (a) Here, $\Delta x = d \sin\theta \approx d \tan\theta$

$= \frac{yd}{D}$

(v) (b) $\frac{yd}{D} = m\lambda$

$\therefore$ $y = \frac{m\lambda D}{d}$

$= \frac{3 \times 650 \times 10^{-9} \times 1.2}{2 \times 10^{-3}}$

$\therefore$ $y = 1.17 \times 10^{-3} \text{m}$

(vi) (c,b) $\Delta x = \frac{yd}{D} = m_1\lambda_1 = m_2\lambda_2$

$\therefore$ $\frac{m_1}{m_2} = \frac{\lambda_2}{\lambda_1} = \frac{520}{650} = \frac{4}{5} = \frac{8}{10}$

(vii) (b) $\because \frac{yd}{D} = m_1\lambda_1 = 4\lambda_1$

$\therefore \quad y = \frac{4\lambda_1 D}{d}$

$= \frac{4 \times 650 \times 10^{-9} \times 1.2}{2 \times 10^{-3}}$

$= 1.56 \times 10^{-3}$ m

2. (i) (a) Here, point source S and its image S' behave as coherent sources. SS' is parallel to screen, so shape will be hyperbolic.

(ii) (a)

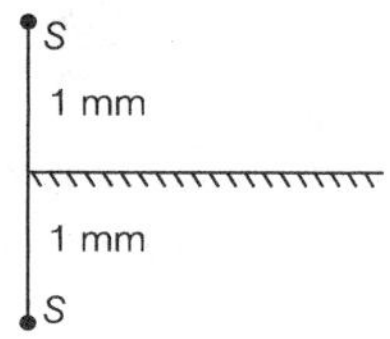

(iii) (a) $\beta = \frac{D}{d}\lambda$

$= \frac{2}{2 \times 10^{-3}} \times 500 \times 10^{-9}$

$= 0.5 \times 10^{-3}\text{m} = 0.5\text{ mm}$

(iv) (b)

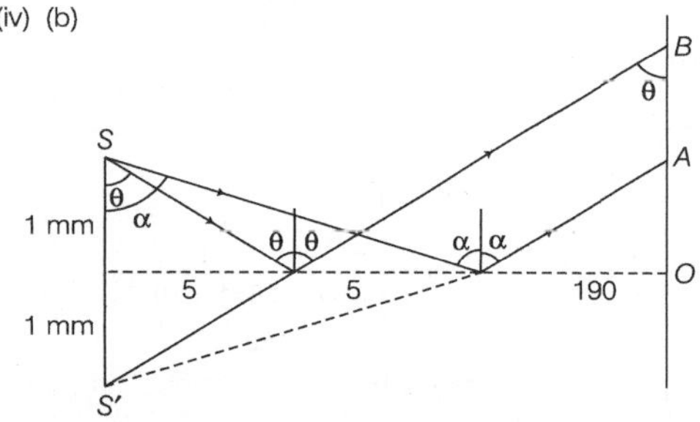

Length of fringe pattern is

$= AB = OB - OA$

$= 2$ cm

(v) (c) $N = \left[\frac{AB}{\beta}\right] = \frac{2 \times 10^{-2}}{0.5 \times 10^{-3}}$

$= 40$

(vi) (b) $\Delta x = S'O + \frac{\lambda}{2} - SO = \frac{\lambda}{2}$

(vii) (b) Since, path difference at point O is $\frac{\lambda}{2}$.

It is a point of dark fringe.

3. (b) Here, path difference is

$$\Delta x = \frac{yd}{D} = \frac{d^2}{2D}$$

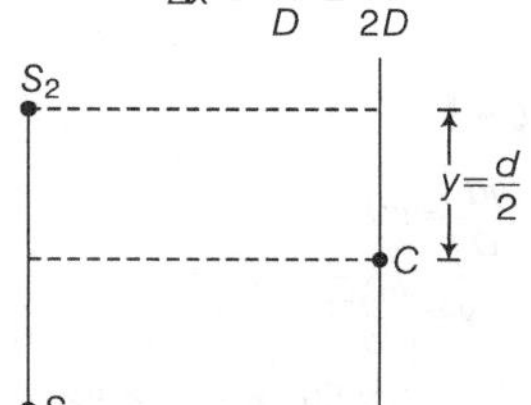

For first order minima,

$$\Delta x = \frac{\lambda}{2}$$

or $\quad \frac{d^2}{2D} = \frac{\lambda}{2}$

$\therefore \quad \lambda = \frac{d^2}{D}$

4. (c) Here, $I_1 = I_0$ and $I_2 = 4I_0$

$\because \quad I = I_1 + I_2 + 2\sqrt{I_1 I_2}\cos\phi$

$= I_0 + 4I_0 + 2 \times 2I_0\left(-\frac{1}{2}\right)$

$= 5I_0 - 2I_0 = 3I_0$

5. (a)

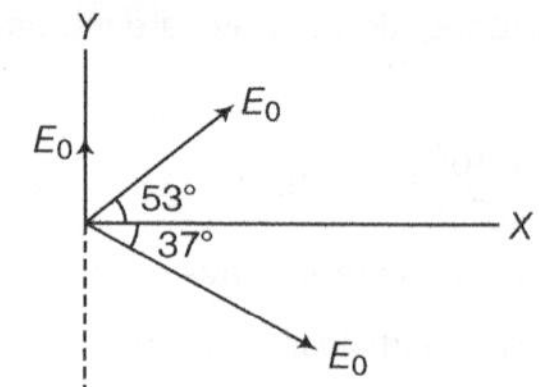

Here, $\quad E_x = E_0\cos 37° + E_0\cos 53°$

$= \frac{4}{5}E_0 + \frac{3}{5}E_0 = \frac{7}{5}E_0$

and $\quad E_y = E_0 + E_0\sin 53° - E_0\sin 37°$

$= E_0 + \frac{4}{5}E_0 - \frac{3}{5}E_0 = E_0 + \frac{1}{5}E_0 = \frac{6E_0}{5}$

$\therefore \quad E_0' = \sqrt{E_x^2 + E_y^2} = \sqrt{\frac{49}{25} + \frac{36}{25}}\,E_0$

$= \sqrt{\frac{85}{25}}\,E_0 = \frac{\sqrt{85}}{5}E_0$

Here, $\quad \tan\phi = \frac{E_y}{E_x} = \frac{\frac{6}{5}E_0}{\frac{7}{5}E_0} = \frac{6}{7}$

$\therefore \quad \phi = \tan^{-1}\left(\frac{6}{7}\right)$

Thus, $\quad E = E_0'\sin(\omega t + \phi)$

$= \frac{\sqrt{85}}{5}\sin\left[\omega t + \tan^{-1}\left(\frac{6}{7}\right)\right]$

6. (a) $\because x^2 + y^2 - z = 6$

Here, $\quad f = x^2 + y^2 - z$

Here, gradient is $\frac{\partial f}{\partial x}\hat{\mathbf{i}} + \frac{\partial f}{\partial y}\hat{\mathbf{j}} + \frac{\partial f}{\partial z}\hat{\mathbf{k}}$

$= 2x\hat{\mathbf{i}} + 2y\hat{\mathbf{j}} - \hat{\mathbf{k}}$

$= 4\hat{\mathbf{i}} + 6\hat{\mathbf{j}} - \hat{\mathbf{k}}$

Thus, unit vector along ray is

$$\hat{\mathbf{e}} = \frac{4\hat{\mathbf{i}} + 6\hat{\mathbf{j}} - \hat{\mathbf{k}}}{\sqrt{16 + 36 + 1}} = \frac{4\hat{\mathbf{i}} + 6\hat{\mathbf{j}} - \hat{\mathbf{k}}}{\sqrt{53}}$$

7. (b)

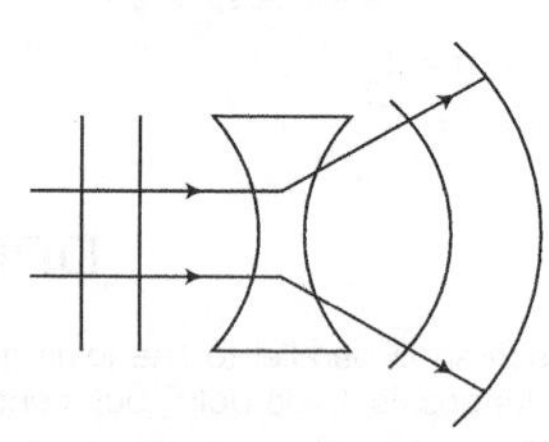

8. (a) Ray is perpendicular to wavefront.

Thus, the vector along ray is same as a vector perpendicular to wavefront $2x + y + 2z$.

Thus, the vector along the ray is $2\hat{\mathbf{i}} + \hat{\mathbf{j}} + 2\hat{\mathbf{k}}$.

Let ray makes α angle with X-axis.

$\therefore \quad \cos\alpha = \frac{2}{\sqrt{2^2 + 1^2 + 2}} = \frac{2}{3}$

$\therefore \quad \alpha = \cos^{-1}(2/3)$

9. (a) For interference, sources should be coherent.

10. (c)

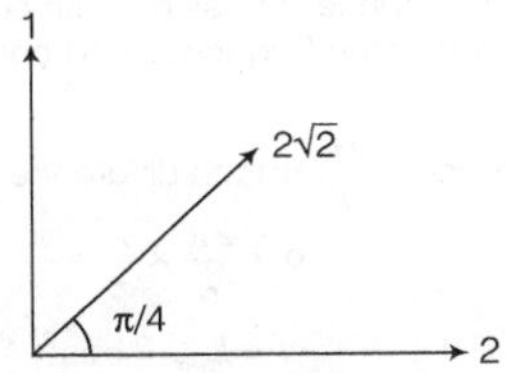

Here, $a_x = 2 + 2\sqrt{2}\cos\frac{\pi}{4} = 4$ unit

$$a_y = 2\sqrt{2}\sin 45° + 1 = 3 \text{ unit}$$

$$\therefore \quad A = \sqrt{a_x^2 + a_y^2} = \sqrt{4^2 + 3^2} = 5 \text{ unit}$$

11. (b) Here, $\Delta x = d\sin\alpha + d\sin\theta$

For centre of screen, $\theta = 0$

$$\therefore \quad \Delta x = d\sin\alpha = (2m-1)\frac{\lambda}{2}$$

$$\therefore \quad \sin\alpha = (2m-1)\frac{\lambda}{2d}$$

$$\therefore \quad \alpha_{min} = \sin^{-1}\left(\frac{\lambda}{2d}\right)$$

12. (b) Here, path difference is

$$\Delta x = d\sin\theta$$

$$\therefore \text{ Phase difference} = \phi = \frac{2\pi}{\lambda}\Delta x = \frac{2\pi}{\lambda}(d\sin\theta) = \frac{2\pi}{\frac{c}{f}}(d\sin\theta)$$

$$\therefore \quad \phi = \frac{2\pi}{3\times10^8}\times10^6\times150\sin30° = \frac{\pi}{2}$$

$$\because \quad I = 4I_0\cos^2\frac{\theta}{2}$$

$$= 4\times1\cos^2\frac{\pi}{4}$$

$$= 4\times\frac{1}{2} = 2\ \text{Wm}^{-2}$$

13. (a) The path difference at point P is

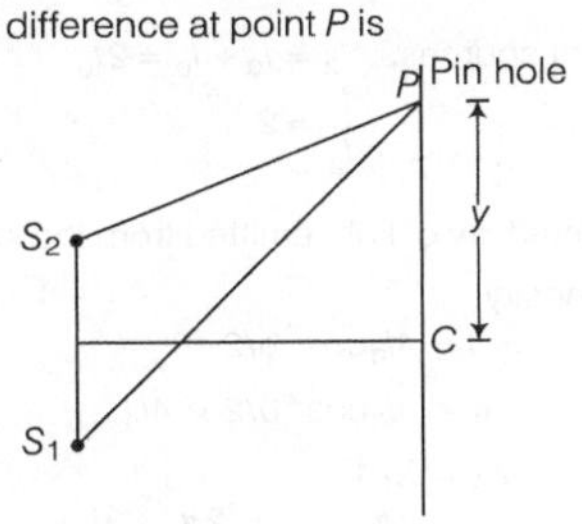

$$\Delta x = \frac{yd}{D}$$

$$= \frac{2\times10^{-3}\times1\times10^{-3}}{1}$$

$$= 2\times10^{-6}$$

$$= 2000\times10^{-9}\text{ m} = 2000 \text{ nm.}$$

For absent wavelength, minima is formed.

$$\therefore \quad \Delta x = (2m-1)\frac{\lambda}{2}$$

or $\quad 2000 = (2m-1)\frac{\lambda}{2}$

$$\therefore \quad \lambda = \frac{4000}{2m-1}$$

Here, $\quad m = 1, 2, 3, \cdots$

For $m = 1$, $\quad \lambda_1 = 4000$ nm

For $m = 2$, $\quad \lambda_2 = \frac{4000}{3}$ nm

For $m = 3$, $\quad \lambda_3 = 800$ nm

For $m = 4$, $\quad \lambda_4 = 571.4$ nm

For $m = 5$, $\quad \lambda_5 = 444.4$ nm

For $m = 6$, $\quad \lambda_6 = 363.6$ nm

14. (c) The path difference at the centre of screen is

$$\Delta x = (\mu - 1)t$$

For maxima, $\quad \Delta x = m\lambda$

or $\quad (\mu - 1)t = m\lambda$

$$\therefore \quad t = \frac{m\lambda}{\mu - 1}$$

$$\therefore \quad t_{min} = \frac{\lambda}{\mu - 1} = \frac{500}{1.4 - 1}$$

$$= \frac{500}{0.4} = 1250 \text{ nm}$$

15. (a) $N = \frac{(\mu - 1)t}{\lambda}$

or $\quad 20\mu - 20 = \frac{(\mu - 1)t}{\lambda}$

or $\quad 20\times2 - 20 = \frac{(2-1)t}{\lambda}$

$$\therefore \quad t = 24\times10^{-6}\text{m}$$

16. (a) For path difference, $\Delta x = m\lambda$, $m = 0, 1, 2, \cdots$, maxima is formed.

17. (c) By Snell's law, $\frac{\sin i}{\sin r} = \frac{\mu_2}{\mu_1}$

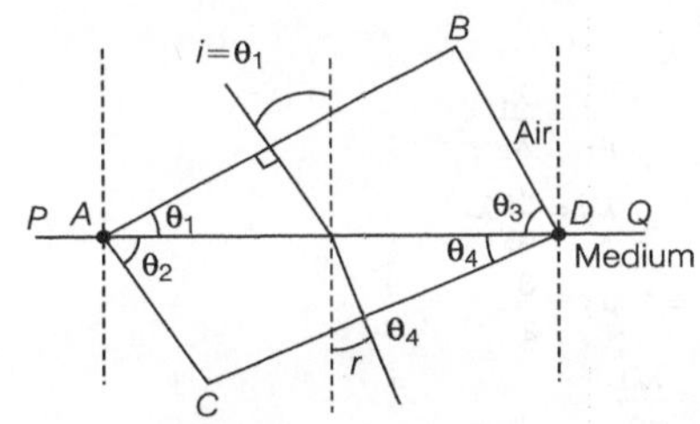

For $i = \theta_1$, $r = \theta_4$ and $\mu_1 = 1$

$$\mu_2 = \frac{\sin\theta_1}{\sin\theta_4}$$

18. (c) $\frac{I_{max}}{I_{min}} = \frac{(\sqrt{I_1} + \sqrt{I_2})^2}{(\sqrt{I_1} - I_2)^2}$

$$\frac{I_{max}}{I_{min}} = \left(\frac{1 + \sqrt{\frac{I_2}{I_1}}}{1 - \sqrt{\frac{I_2}{I_1}}}\right)^2$$

For given information, $\quad I_{max} = \frac{105}{100}I$

and $\quad I_{min} = \left(\frac{95}{100}\right)I$

where, $I = (I_1 + I_2)$ is average intensity.

$$\frac{I_{max}}{I_{min}} = \frac{105}{95}\left(\frac{1 + \sqrt{\frac{I_2}{I_1}}}{1 - \sqrt{\frac{I_2}{I_1}}}\right)^2$$

$$\Rightarrow \quad \sqrt{\frac{I_2}{I_1}} = 0.0244$$

$$\Rightarrow \quad \frac{I_1}{I_2} = 1681$$

19. (c) Since wavelength of violet ray is minimum. So, its deviation will be maximum. Hence, it will be closer to central band.

20. (b)

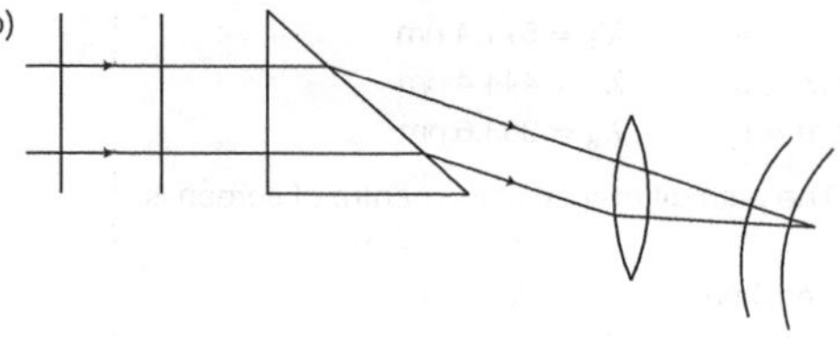

21. (b) $\beta = \dfrac{\lambda D}{d}$

$$\beta_1 - \beta_2 = \frac{\lambda(D_1 - D_2)}{d} \text{ or } \lambda = \frac{d(\beta_1 - \beta_2)}{D_1 - D_2}$$

or $$\lambda = \frac{3 \times 10^{-5} \times 10^{-3}}{5 \times 10^2} = 6000 Å$$

22. (a) At any point P, path difference = 0

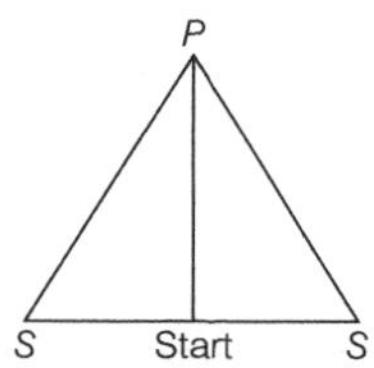

For maxima path difference must be zero.
So, at every point of path, maxima exists.

23. (b) $\mu \propto \dfrac{1}{\lambda}$

$$\frac{\mu_2}{\mu_1} = \frac{\lambda_1}{\lambda_2}$$

$$\lambda_2 = \frac{\mu_1}{\mu_2}\lambda_1$$

$$\mu_1 = 1, \ \mu_2 = \frac{3}{2}$$

24. (d) $\beta_w = \dfrac{\lambda D}{\mu d}$

25. (a)

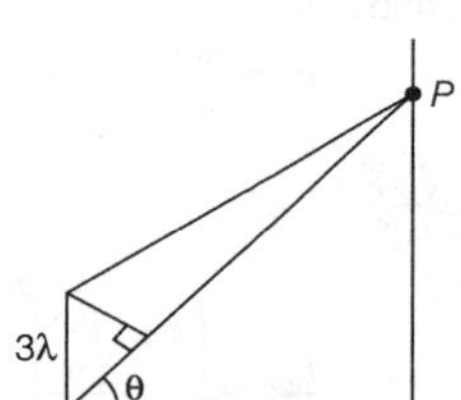

Here, path difference on Y-axis is $\Delta x = 3\lambda \sin\theta$

But for minima, $3\lambda \sin\theta = (2m+1)\dfrac{\lambda}{2}$

But $-1 < \sin\theta < 1$, or $-1 < \dfrac{(2m+1)}{6} < 1$

$\therefore$ $$m = -3, -2, -1, 0, 1, 2, 3$$

Thus, number of minima on Y-axis is 6.
Hence, option (a) is correct.

26. (b) If slits are of unequal widths intensities on the screen due to both slits on screen are not same.

$\therefore$ $$I_{min} = (\sqrt{I_1} - \sqrt{I_2})^2 \neq 0$$

Hence, option (b) is correct.

27. (c) $I_{min} = (\sqrt{I + \delta I} - \sqrt{I})^2 = I\left(\left[1 + \dfrac{\delta I}{I}\right]^{1/2} - 1\right)^2$

$$= I\left(\frac{\delta I}{2I}\right)^2 = \frac{(\delta I)^2}{4I}$$

28. (d) Sustained interference is possible with coherent sources only i.e. sources having same frequency and constant or zero phase difference.

29. (b) Phase difference $= \dfrac{2\pi}{\lambda} \times$ Path difference

i.e. $$\phi = \frac{2\pi}{\lambda} \times \frac{\lambda}{6} = \frac{\pi}{3}$$

or $$I = I_{max} = \cos^2\left(\frac{\phi}{2}\right)$$

or $$\frac{I}{I_{max}} = \cos^2\left(\frac{\phi}{2}\right)$$

$$\frac{I}{I_0} = \cos^2\left(\frac{\pi}{6}\right) = \frac{3}{4} \qquad [\because I_{max} = I_0]$$

30. (a) The shape of the interference fringes formed on a screen is hyperbola.

31. (b) For possible interference maxima on the screen the condition is

$$d \sin\theta = n\lambda \qquad \ldots(i)$$

Given, $$d = \text{slit width} = 2\lambda$$

$\therefore$ $$2\lambda \sin\theta = n\lambda$$

or $$2 \sin\theta = n$$

The maximum value of $\sin\theta$ is 1, hence

$$n = 2 \times 1 = 2$$

Thus, Eq. (i) must be satisfied by 5 integer values i.e., $-2, -1, 0, 1, 2$. Hence, the maximum number of possible interference maxima is 5.

32. (c) $3\lambda_1 = 4\lambda_2$ $\qquad \left[\because \dfrac{3D\lambda_1}{d} = 4\dfrac{D\lambda_2}{d}\right]$

$\Rightarrow$ $$\lambda_2 = \frac{3}{4}\lambda_1 = \frac{3}{4} \times 590$$

$$= \frac{1770}{4} = 442.5 \text{ nm}$$

33. (d) For a parallel cylindrical beam, wavefront will be planar.

34. (b) For coherent sources, $I_1 = 4I_0 \cos^2 \dfrac{\phi}{2} = 4I_0$

For incoherent sources, $I_2 = I_0 + I_0 = 2I_0$

$\therefore$ $$\frac{I_1}{I_2} = 2$$

35. (b) Let I_0 be intensity of light emitted from the source, then

Resultant intensity

$$I = 4I_0 \cos^2 \phi/2$$

$\therefore$ $$I_1 = 4I_0 \cos^2 0/2 = 4I_0 \qquad [\because \phi = 0]$$

Now, $$\Delta x = \lambda/4$$

$\therefore$ $$\phi = \frac{2\pi}{\lambda} \times \Delta x = \frac{2\pi}{\lambda} \times \frac{\lambda}{4} \Rightarrow \phi = \frac{\pi}{2}$$

and $$I_2 = 4I_0 \cos^2 \frac{\pi}{4} = 2I_0 \qquad \left[\because \phi = \frac{2\pi}{\lambda}\Delta x\right]$$

$\therefore$ $$I_1 : I_2 = 2 : 1$$

36. (d) Given, $a_1 = 2a_2$

$\Rightarrow$ $$I_1 = 4I_2 = 4I_0 \qquad [\because I \propto a^2]$$

$\therefore$ $$I_{max} = (\sqrt{I_1} + \sqrt{I_2})^2 = (3\sqrt{I_2})^2 = 9I_2 = 9I_0$$

Now, $$I = I_1 + I_2 + 2\sqrt{I_1 I_2}\cos\phi$$

$$= 4I_0 + I_0 + 2\sqrt{4I_0 \cdot I_0}\cos\phi$$

$$= 5I_0 + 4I_0 \cos\phi$$

$$= \frac{I_m}{9}(5 + 4\cos\phi) \qquad \left[\because I_0 = \frac{I_m}{9}\right]$$

$$= \frac{I_m}{9}[1 + 4(1 + \cos\phi)]$$

$$= \frac{I_m}{9}(1 + 8\cos^2 \phi/2) \qquad \left[\because \cos\phi = 2\cos^2\frac{\phi}{2} - 1\right]$$

37. (c) $$I = I_{max}\cos^2\left(\frac{\phi}{2}\right)$$

∴ $$\frac{I_{max}}{4} = I_{max}\cos^2\frac{\phi}{2}$$

$$\cos\frac{\phi}{2} = \frac{1}{2}$$

or $$\frac{\phi}{2} = \frac{\pi}{3}$$

∴ $$\phi = \frac{2\pi}{3} = \left(\frac{2\pi}{\lambda}\right)\Delta x \quad \text{...(i)}$$

where, $$\Delta x = d\sin\theta$$

Substituting in Eq. (i), we get

$$\sin\theta = \frac{\lambda}{3d}$$

or $$\theta = \sin^{-1}\left(\frac{\lambda}{3d}\right)$$

38. (d) Fringe width, $\beta = \frac{\lambda D}{d}$

or $$\beta \propto \lambda$$

Now, $$\lambda_R > \lambda_G > \lambda_B$$

∴ $$\beta_R > \beta_G > \beta_B$$

39. (a) Path difference due to slab should be integral multiple of λ or

$$\Delta x = n\lambda$$

or $$(\mu - 1)t = n\lambda, \; n = 1, 2, 3, \ldots$$

or $$t = \frac{n\lambda}{\mu - 1}$$

For minimum value of t, $n = 1$

∴ $$t = \frac{\lambda}{\mu - 1} = \frac{\lambda}{1.5 - 1} = 2\lambda$$

40. (b) Fringe width, $\omega = \frac{\lambda D}{d} \propto \lambda$

When the wavelength is decreased from 600 nm to 400 nm, fringe width will also decrease by a factor of $\frac{4}{6}$ or $\frac{2}{3}$ or number of fringes in the same segment will increase by a factor of 3/2.

Therefore, number of fringes observed in the same segment

$$= 12 \times \frac{3}{2} = 18$$

41. (b) $I(\phi) = I_1 + I_2 + 2\sqrt{I_1 I_2}\cos\phi$ ...(i)

Here, $I_1 = I$ and $I_2 = 4I$

At point A, $$\phi = \frac{\pi}{2}$$

∴ $$I_A = I + 4I = 5I$$

At point B, $$\phi = \pi$$

∴ $$I_B = I + 4I - 4I = I$$

∴ $$I_A - I_B = 4I$$

42. (d) In interference we know that

$$I_1 = (\sqrt{I_1} + \sqrt{I_2})^2$$

and $$I_{min} = (\sqrt{I_1} - \sqrt{I_2})^2$$

Under normal conditions (when the widths of both the slits are equal)

$$I_1 = I_2 = I \text{ (say)}$$

∴ $$I_{max} = 4I \text{ and } I_{min} = 0$$

When the width of one of the slits is increased. Intensity due to that slit would increase, while that of the other will remains same. So, let

$$I_1 = I \text{ and } I_2 = \eta I \qquad [\because \eta > 1]$$

Then, $$I_{max} = I(1 + \sqrt{\eta})^2 > 4I$$

and $$I_{max} = I(\sqrt{\eta} - 1)^2 > 0$$

Intensity of both maxima and minima is increased.

DPP-3 **Interference of Light - II**

1. (a,b,c,d) (a) In light wave, $E = E_0\sin(\omega t - kx)$

and $B = B_0\sin(\omega t - kx)$

Thus, electric field E and magnetic field B are in same phase.

(b)

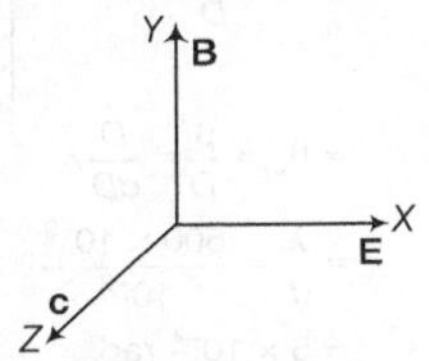

(c) Light is an electromagnetic wave.

(d) The speed of light in vacuum does not depend on reference frame.

2. (a,d) For interference between two waves, sources should be coherent.

3. (b,d) $9 = \frac{(\sqrt{I_1} + \sqrt{I_2})^2}{(\sqrt{I_1} - \sqrt{I_2})^2}$

⇒ $$9(\sqrt{I_1} - \sqrt{I_2})^2 = (\sqrt{I_1} + \sqrt{I_2})^2$$

⇒ $$3(\sqrt{I_1} - \sqrt{I_2}) = (\sqrt{I_1} + \sqrt{I_2})$$

⇒ $$3\sqrt{I_1} - \sqrt{I_1} = 4\sqrt{I_2}$$

⇒ $$2\sqrt{I_1} = 4\sqrt{I_2} \quad \Rightarrow \quad 4I_1 = 16I_2$$

∴ $$\frac{I_1}{I_2} = \frac{16}{4} = 4$$

∴ $$\frac{a_1}{a_2} = \sqrt{4} = 2$$

4. (a,b,c)

(a) For central bright fringe,

$$m = 0$$

$$\Delta x = 0$$

Thus, zero order bright fringe is at the same place for all wavelengths. Here, central bright fringe is white.

(b) $$\lambda_v < \lambda_r$$

and $$\beta = \frac{D}{d}\lambda$$

(c) ∵ $$\beta = \frac{D}{d}\lambda$$

∴ $$\beta \propto \lambda$$

Hence, first order bright fringe is coloured.

5. (a,c,d) If sources are coherent, vector method finding resultant amplitudes is applicable. According to polygon law amplitude is zero.

Thus, minimum intensity is zero.

Hence, option (a) is correct.

Also, maximum amplitude is na_0

∴ $$I_{max} = kn^2a_0^2 = n^2 I_0$$

Hence, option (c) is correct.

For incoherent sources,

$$I = nI_0$$

Hence, option (d) is correct.

6. (a,c,d) $\because I = 4I_0 \cos^2 \frac{\phi}{2}$

$$\Rightarrow \quad 2I_0 = 4I_0 \cos^2 \frac{\phi}{2}$$

$$\Rightarrow \quad \cos^2 \frac{\phi}{2} = \pm \frac{1}{2}$$

$$\Rightarrow \quad \cos \frac{\phi}{2} = \pm \frac{1}{\sqrt{2}}$$

$$\Rightarrow \quad \phi = \frac{\pi}{2}, \frac{3\pi}{2}, \frac{5\pi}{2}, \ldots$$

$$\Rightarrow \quad \frac{2\pi}{\lambda}(\mu - 1)t = \frac{\pi}{2}, \frac{3\pi}{2}, \frac{5\pi}{2}, \ldots$$

$$\Rightarrow \quad \frac{\pi}{\lambda} t = \frac{\pi}{2}, \frac{3\pi}{2}, \frac{5\pi}{2}, \ldots$$

$$\Rightarrow \quad t = \frac{\lambda}{2}, \frac{3\lambda}{2}, \frac{5\lambda}{2}, \ldots$$

$$\therefore \quad t = 300 \text{ nm}, 900 \text{ nm}, 1500 \text{ nm}$$

7. (a,b,c,d)

(a) The path difference at the centre of screen is $\Delta x = (\mu - 1)t$
It may be equal to integral multiple of λ. Hence, maxima may be formed at the centre of screen.

(b) For central maxima, $\Delta x = 0$

or $\quad (\mu - 1)t = 0$ or $\mu = 1$

But $\mu \neq 1$. Hence, central maxima is not formed on screen.

(c) For minima, $\quad \Delta x = (2m - 1)\frac{\lambda}{2}$

or $\quad (\mu - 1)t = (2m - 1)\frac{\lambda}{2}$

where, $m = 1, 2, 3, \cdots$

(d) For being intensity zero at centre of screen, minima will be formed at the centre of screen.

$$\therefore \quad \Delta x = (2m - 1)\frac{\lambda}{2}$$

or $\quad (\mu - 1)t = (2m - 1)\frac{\lambda}{2}$

or $\quad t = \frac{(2m - 1)\lambda}{2(\mu - 1)}$

For $m = 3$, $\quad t = \frac{5\lambda}{2(\mu - 1)}$

8. (a,b,c) Here, $\Delta x = \frac{yd}{D}$

$$= \frac{1 \times 10^{-3} \times 2 \times 10^{-3}}{1}$$

$$= 2 \times 10^{-6} = m\lambda$$

or $\quad \lambda = \frac{2000 \text{ nm}}{m}$

$$= 2000 \text{ nm}, 1000 \text{ nm}, 666.67 \text{ nm}, 500 \text{ nm}, 400 \text{ nm}$$

9. (a,b,c,d)

(a) Here, $I_1 = I_0$ and $I_2 = \frac{I_0}{4}$

$$\because \quad I_1 = ka_1^2 \text{ and } I_2 = ka_2^2$$

$$\therefore \quad \frac{a_{max}}{a_{min}} = \frac{a_1 + a_2}{a_1 - a_2} = \frac{\sqrt{\frac{I_0}{k}} + \sqrt{\frac{I_0}{4k}}}{\sqrt{\frac{I_0}{k}} - \sqrt{\frac{I_0}{4k}}}$$

$$= \frac{\sqrt{\frac{I_0}{k}}\left(1 + \frac{1}{2}\right)}{\sqrt{\frac{I_0}{k}}\left(1 - \frac{1}{2}\right)} = \frac{3}{2 \times \frac{1}{2}} = 3$$

(b) Here, $\Delta x = \frac{yd}{D} - (\mu - 1)t = m\lambda$

For central maxima, $m = 0$

$$\therefore \quad \frac{yd}{D} = (\mu - 1)t = 0.45 \times 0.02 \times 10^{-3}$$

$$= 90 \times 10^{-7}$$

$$\therefore \quad y = \frac{90 \times 10^{-7} \times 1}{1 \times 10^{-3}} = 90 \times 10^{-4} \text{ m}$$

$$= 9 \text{ mm}$$

(c) Number of fringe shifting,

$$= \frac{(\mu - 1)t}{\lambda} = \frac{0.45 \times 2 \times 10^{-5}}{600 \times 10^{-9}}$$

$$= \frac{0.9}{600} \times 10^4$$

$$= \frac{9}{6} \times 10^1 = 15$$

Thus, when an identical transparent sheet is pasted, the number of fringes crosses through centre of screen = 15

(d) $\frac{I_{max}}{I_{min}} = \frac{(\sqrt{I_1} + \sqrt{I_2})^2}{(\sqrt{I_1} + \sqrt{I_2})^2}$

$$= \frac{(\sqrt{I_0} + \sqrt{I_0/4})^2}{\left(\sqrt{I_0} - \sqrt{\frac{I_0}{4}}\right)^2} = \frac{I_0\left(1 + \frac{1}{2}\right)^2}{I_0\left(1 - \frac{1}{2}\right)^2}$$

$$= \frac{9}{4 \times \frac{1}{4}} = 9$$

10. (a,b,c,d)

(a) Here, $\Delta x = \frac{yd}{D} = \frac{0.5 \times 10^{-3} \times 10^{-3}}{1} = 0.5 \times 10^{-6}$ m

$$\therefore \quad \phi = \frac{2\pi}{\lambda} \Delta x = \frac{2\pi}{500 \times 10^{-9}} \times 0.5 \times 10^{-6} = 2\pi$$

$$\therefore \quad I = 4I_0 \cos^2 \frac{\phi}{2} = 4 \times 2 \cos^2 \pi = 8 \text{ Wm}^{-2}$$

(b) Angular fringe width,

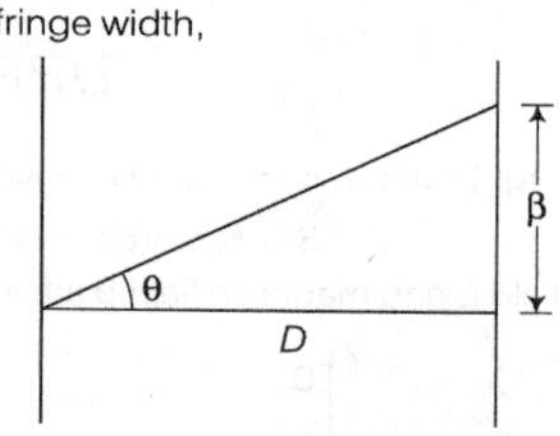

$$= \theta_0 = \frac{\beta}{D} = \frac{D}{dD}\lambda$$

$$= \frac{\lambda}{d} = \frac{500 \times 10^{-9}}{10^{-3}}$$

$$= 5 \times 10^{-4} \text{ rad}$$

(c) For minima,

$$\Delta x = \frac{yd}{D} = (2m - 1)\frac{\lambda}{2}$$

or $\quad \frac{yd}{D} = (2 \times 1 - 1)\frac{\lambda}{2}$

or $\quad \frac{yd}{D} = \frac{\lambda}{2}$

or $\quad y = \frac{D\lambda}{2d} = \frac{1 \times 500 \times 10^{-9}}{2 \times 10^{-3}}$

$$= 2.5 \times 10^{-4} \text{ m}$$

$$\Rightarrow \quad = 0.25 \text{ mm}$$

(d) $\because \beta = \frac{D}{d}\lambda \Rightarrow \beta \propto \lambda$

when, experiment is performed in water, wavelength decreases. Hence, fringe width decreases.

11. (a,c) One key idea here is that the phase difference of two light waves can change, if they travel through different media, with different indices of refraction. The reason is that their wavelengths are different in the different media. We can calculate the change in phase difference by counting the number of wavelengths that fits into each medium and then subtracting those numbers. When the path lengths of the waves in the two media are identical. Here, we have $n_1 = 1.000$ (for the air), $n_2 = 1.600$, $L = 2.600\ \mu m$ and $\lambda = 550.0$ nm. Thus,

$$N_2 - N_1 = \frac{1}{\lambda}(n_2 - n_1)$$

$$= \frac{2.600 \times 10^{-6}\ m}{5.500 \times 10\ m}(1.600 - 1.000)$$

$$= 2.84$$

A second key idea is that the effective difference is the decimal part of the actual phase difference expressed in wavelengths. Thus, we have effective phase difference = 0.84 wavelength.

12. (a,c,d)

13. (b) $\Delta x_1 = 0$

$$\Delta x_2 = \mu_w \frac{yd}{D}$$

$$\Delta x_3 = (\mu - \mu_w)t$$

$$\Delta x = \mu_w \frac{yd}{D} + (\mu - \mu_w)t$$

For bright fringes,

$$\Delta x = m\lambda$$

$$\frac{\mu_w yd}{D} + (\mu + \mu_w)t = m\lambda$$

For central bright fringe, $m = 0$

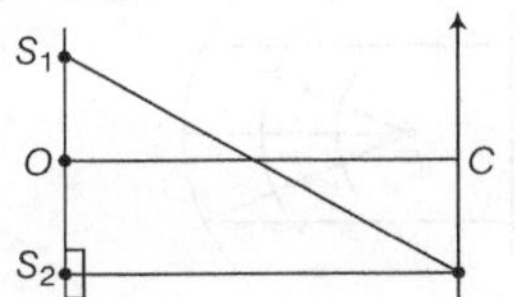

$$\frac{\mu_w yd}{D} + (\mu - \mu_w)t = 0$$

$$\frac{\mu_w yd}{D} = -(\mu - \mu_w)t$$

$$y = \frac{(\mu - \mu_w)tD}{\mu_w d}$$

$$y = -\frac{\left[\frac{3}{2} - \frac{4}{3}\right]10.4 \times 10^3 \times 3 \times 1.5}{4 \times 45 \times 10^{-5}}$$

$$y = \frac{10.4 \times 3 \times 1.5 \times 10^{-3}}{6 \times 4 \times 45 \times 10^{-5}}$$

$$\Rightarrow \quad y = -4.33\ m$$

14. (b) At point O, $\quad y = 0$

$$\Delta x = (\mu - 1)t$$

$$\phi = \frac{2\pi}{\lambda} \times (\mu - 1)t$$

$$\phi = \frac{2\pi}{600 \times 10^{-9}} \times (0.5)10.4 \times 10^{-3}$$

$$= \frac{2\pi}{6} \times 5 \times 104 \times 10^{-5} \times 10^7$$

$$= 2\pi \times 86.66 \times 10^2$$

$$\phi = 2\pi \times 866.6$$

$$I = 4I_0 \cos^2 \frac{\phi}{2}$$

$$\Rightarrow \quad I = 4I_0$$

15. (a) $\Delta x = (\mu - 1)t$

$$\Delta x = m\lambda$$

$$m\lambda = (\mu - 1)t$$

$$\lambda = \frac{(\mu - 1)t}{m}$$

where $m = 0, 1, 2, 3, \cdots$

$$\lambda = \frac{(1.5 - 1)\ 10.4 \times 10^3}{m}$$

$$\lambda = 650\ \text{nm},\ 433.3\ \text{nm}$$

16. (c) $\theta_0 = \frac{\beta}{D} = \frac{D}{dD}\lambda = \frac{\lambda}{d}$

$$\therefore \quad d = \frac{\lambda}{\theta_0} = \frac{6000 \times 10^{-10}}{0.1 \times \pi} \times 180$$

$$\Delta x = (\mu_A - 1)t \text{ at the point } O$$

$$I = I_0 \cos^2 \frac{\phi}{2}$$

$$\text{or} \quad \frac{I_0}{2} = I_0 \cos^2 \frac{\phi^2}{2}$$

$$\text{or} \quad \cos \frac{\phi}{2} = \frac{1}{\sqrt{2}}$$

$$\therefore \quad \theta = 90° = \frac{\pi}{2}$$

$$\therefore \quad \phi = \frac{2\pi}{\lambda}\Delta x$$

$$= \frac{2\pi}{\lambda}(\mu_A - 1)t$$

$$\text{or} \quad \frac{\pi}{2} = \frac{2\pi}{6000 \times 10^{-10}} \times (\mu_A - 1)\ 250 \times 10^{-9}$$

$$\therefore \quad \mu_A = 1.6$$

17. (a) $\quad (\mu_B - 1)t = \frac{\lambda}{2},$

$$\therefore \quad \mu_B = 2$$

18. (a) $\because \frac{yd}{D} - (\mu_A - 1)t = m\lambda = 0$

$$\therefore \quad \frac{d}{D} = \frac{(\mu_A - 1)t}{y}$$

Putting the value of d, we get

$$D = 1.1\ m$$

19. (c) $\quad \phi = \frac{2\pi}{\lambda}\Delta x = \frac{2\pi}{5000 \times 10^{-10}} \times 11250 \times 10^{-10}$

$$= 4.5\pi = \frac{9}{2}\pi$$

$$\therefore \quad I = I_0 \cos^2 \frac{\phi}{2}$$

$$= I_0 \cos^2 \frac{9}{4} \times 180°$$

$$= I_0 \cos^2 405°$$

$$= \frac{I_0}{2}$$

20. (c) $\because \quad \Delta x = \frac{yd}{D} = 11250 \times 10^{-10}$ (at point P)

$$\text{or} \quad 10^{-3}\frac{d}{D} = 11250 \times 10^{-10}$$

$$\therefore \quad \frac{d}{D} = 11250 \times 10^{-7}$$

$$\therefore \quad \beta = \frac{D}{d}\lambda$$

$$= \frac{1}{11250 \times 10^{-7}} \times 5000 \times 10^{-10}$$

$$= 0.44\ \text{mm}$$

21. (a) Angular fringe width $= 0.1 \times \pi/180 = \dfrac{\lambda}{d}$

$\therefore \quad d = \dfrac{5000 \times 10^{-10} \times 180}{0.1 \times \pi} = 0.29$ mm

22. (c) Here, path difference is

$$\Delta x = d \sin 30° + (\mu - 1)t_1 + \frac{yd}{D} - (\mu - 1)t_2$$
$$= \frac{0.1}{2} \times 10^{-3} + \frac{1}{2} \times 10 \times 10^{-6} + \frac{yd}{D} - \frac{1}{2} \times 5 \times 10^{-6}$$
$$= 50 \times 10^{-6} + 5 \times 10^{-6} + \frac{yd}{D} - 2.5 \times 10^{-6}$$

At point O, $y = 0$

$\therefore \quad \Delta x = 52.5 \times 10^{-6}$

$\therefore \quad \phi = \dfrac{2\pi}{\lambda} \Delta x$

$$= \frac{2\pi}{525 \times 10^{-9}} \times 52.5 \times 10^{-6}$$
$$= \frac{2\pi}{10^{-8}} \times 10^{-6} = 200\pi$$

$\therefore \quad I = I_1 + I_2 + 2\sqrt{I_1 I_2} \cos \phi$
$= 1 + 9 + 2 \times 3$
$= 16 \text{ Wm}^{-2}$

23. (a) From the question,

$$\Delta x = 52.5 \times 10^{-6} + \frac{yd}{D}$$

$\because \quad \Delta x = m\lambda$ (for maxima)

$\because$ For central maxima, $m = 0$

or $\quad 52.5 \times 10^{-6} + \dfrac{yd}{D} = 0$

$\therefore \quad y = -\dfrac{D}{d} \times 52.5 \times 10^{-6}$

$$= -\frac{1}{0.1 \times 10^{-3}} \times 52.5 \times 10^{-6}$$
$$= -52.5 \times 10^{-2} \text{m}$$
$$= -52.5 \text{ cm}$$

24. (a) $\beta = \dfrac{D}{d}\lambda$

$$= \frac{1}{0.1 \times 10^{-3}} \times 500 \times 10^{-9}$$
$$= 500 \times 10^{-5} \text{m}$$
$$= 5 \times 10^{-3} \text{cm}$$

25. (a)

(i) $I = I_1 + I_2 + 2\sqrt{I_1 I_2} \cos \phi$
$= 4I_0 + 6I_0 - 2\sqrt{4I_0 \cdot 6I_0}$
$= (10 - 4\sqrt{6}) I_0 = 0.2 I_0$

(ii) $I = I_1 + I_2 + 2\sqrt{I_1 I_2} \cos \phi$
$= 9I_0 + 25I_0 + 2\sqrt{9I_0 \times 25I_0} \cos \pi$
$= 34I_0 - 30I_0 = 4I_0$

(iii) $I = 81 I_0 + 49 I_0 - 2\sqrt{81I_0 \times 49I_0}$
$= 130 I_0 - 56 I_0 = 74 I_0$

(iv) $I = 4I_0 + 4I_0 + 2\sqrt{4I_0 4I_0}$
$= 8 I_0 + 8I_0 = 16 I_0$

Hence, (i) → (q), (ii) → (s), (iii) → (p), (iv) → (r)

26. (b)

(i) Before introducing transparent sheet, intensity of maxima is $4I_0$. After introducing transparent sheet, intensity of maxima is

$$\left(\sqrt{I_0} + \sqrt{\frac{I_0}{4}}\right)^2 = I_0\left(1 + \frac{1}{2}\right)^2 = \frac{9}{4} I_0$$

Thus, intensity of maxima will decrease.

(ii) Before introducing transparent sheet, intensity of minima is zero. After introducing transparent sheet, intensity of minima is $(\sqrt{I_0} - \sqrt{I_0/4})^2 = \dfrac{I_0}{4}$

(iii) Fringe width, $\beta = \dfrac{D}{d}\lambda$. Hence, it remains unchanged.

(iv) Since, transparent sheet is used infront S_1, so fringe pattern will shift downward.

Hence, (i) → (q), (ii) → (p), (iii) → (r), (iv) → (s).

27. (c)

(a)

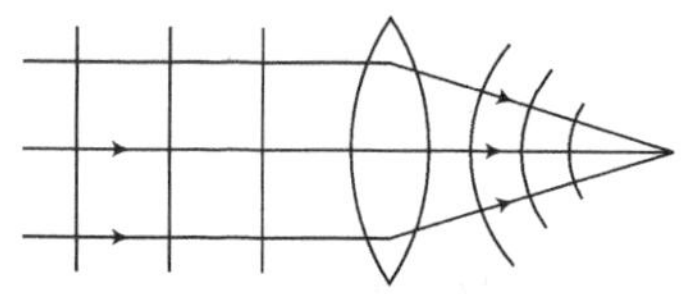

(b)

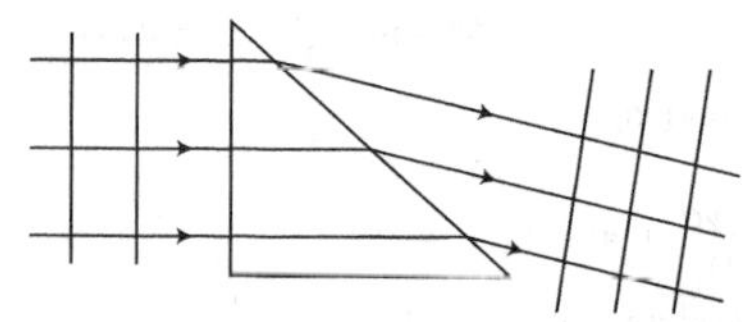

(c)

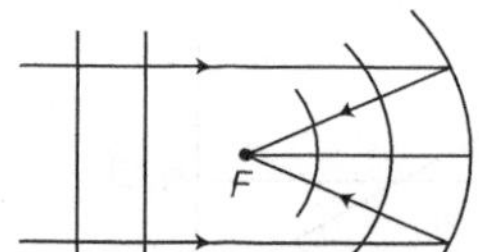

(d)

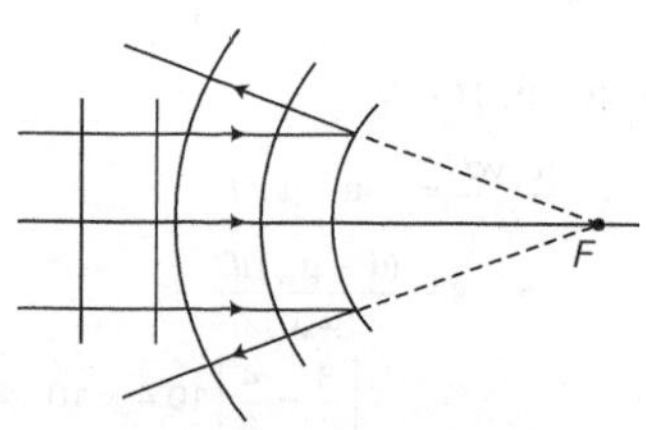

Hence, (i) → (r), (ii) → (p), (iii) → (q), (iv) → (r)

28. (b) Hence, (i) → (p), (ii) → (r), (iii) → (t), (iv) → (r)

29. (a)

(i) Interference takes place in coherent sources.

(ii) Newton's corpuscular theory is based on particle theory. It does not explain the phenomenon of wave.

(iii) Photoelectric effect is based on particle nature of light. So, it is not explain by Huygen's wave theory of light.

(iv) When amplitudes of both waves are same, intensity of dark fringe is zero. For maximum contrast in YDSE, amplitudes of waves should be same.

Hence, (i) → (r), (ii) → (q), (iii) → (s), (iv) → (p,r).

30. (a) Hence, (i) → (q,r,s) (ii) → (p,r), (iii) → (q,r,s) (iv) → (p,r).

DPP-4 Interference of light - III

1. (d) The frequency of light does not change due to change of medium. So, colour of light does not change due to change of medium. But $\lambda = \frac{\lambda_0}{\mu}$. Thus, wavelength of light in different medium are different.

2. (d) If light wave superposed, light energy is not converted into the form of energy.

The presence of minima on the interference pattern means that no quantity of light enters the given section of space.

3. (a)

4. (b) Here, $I_1 \neq I_2$

$\therefore \quad I_{max} = (\sqrt{I_1} + \sqrt{I_2})^2$

and $\quad I_{min} = (\sqrt{I_1} - \sqrt{I_2})^2$

Hence, both statements are correct.

5. (c) $\quad \mu AB = \mu' CD$

$$\Rightarrow \quad \frac{AB}{CD} = \frac{\mu'}{\mu} \quad \dots(i)$$

$$t_1 C_1 = AB$$

$$t_2 C_2 = CD$$

$$\frac{t_1}{t_2} = \frac{AB}{CD} \times \frac{C_2}{C_1} \quad \dots(ii)$$

$$\mu \propto \frac{1}{C}$$

$$\frac{\mu}{\mu'} = \frac{C_2}{C_1} \quad \dots(iii)$$

From Eqs. (i), (ii) and (iii), we get

$$\frac{t_1}{t_2} = \frac{\mu'}{\mu} \times \frac{\mu}{\mu'}$$

$$\frac{t_1}{t_2} = 1$$

$$\Rightarrow \quad t_1 = t_2$$

Hence, Statement I is correct, Statement II is incorrect.

6. (1) $\Delta x = (S_1P - S_2P) + (\mu - 1)t = m\lambda$

For central maxima, $m = 0$

$$\therefore \quad \frac{yd}{D} + (\mu - 1)t = \lambda$$

$$\text{or} \quad y = \frac{-(\mu - 1)t\,D}{d}$$

$$\therefore \quad y_1 - y_2 = \frac{(\mu_1 - \mu_2)t\,D}{d}$$

$$= \frac{(1.5 - 1.4)\,0.04 \times 10^{-3} \times 1}{4 \times 10^{-3}}$$

$$= \frac{0.1 \times 0.04 \times 10^{-3}}{4 \times 10^{-3}}$$

$$= 0.1 \times 0.01$$

$$= 0.1 \times 10^{-3}\text{m} = 1\,\text{mm}$$

7. (5) Here, $\Delta x = d\sin\theta - (SS_2 - SS_1)$

$$= d\sin\theta - 40\,\mu\text{m} = m\lambda$$

$$\text{or} \quad d\sin\theta = m\lambda + 40\,\mu\text{m}$$

$$\text{or} \quad \sin\theta = \frac{m\lambda}{d} + 40 \times 10^{-6}\text{m}$$

$$\because \quad -1 < \sin\theta < 1$$

$$\text{or} \quad -1 < \frac{m\lambda}{d} + 40 \times 10^{-6} < 1$$

$$\text{or} \quad -1 < \frac{m \times 400 \times 10^{-9}}{10 \times 10^{-6}} + 40 \times 10^{-6} < 1$$

$$\text{or} \quad -1 < 4m \times 10^{-2} + 40 \times 10^{-6} < 1$$

$$\text{or} \quad -1 < (1000\,m + 1)\,40 \times 10^{-6} < 1$$

$$\text{or} \quad -2500 < 1000\,m + 1 < 2500$$

$$\text{or} \quad -2.5 < m < 2.5$$

$$\text{or} \quad \therefore \quad m = -2, -1, 0, 1, 2$$

Thus, possible number of maxima is 5.

8. (4) Here, path difference at point P is

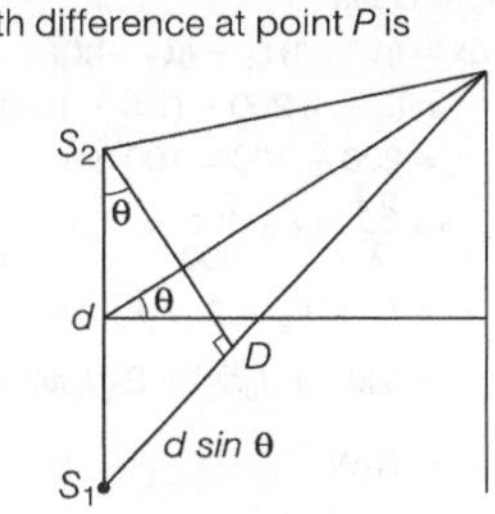

$$\Delta x = d\sin\theta$$

$$\text{For minima,} \quad d\sin\theta = (2m - 1)\frac{\lambda}{2}$$

$$\text{or} \quad 2\lambda\sin\theta = (2m - 1)\frac{\lambda}{2}$$

$$\text{or} \quad \sin\theta = \frac{(2m - 1)}{4}$$

$$-1 < \sin\theta < 1$$

$$\text{or} \quad -1 < \frac{(2m - 1)}{4} < 1$$

$$\text{or} \quad -4 < (2m - 1) < 4$$

$$\text{or} \quad -3 < 2m < 5$$

$$\text{or} \quad -1.5 < m < 2.5$$

$$\therefore \quad m = -1, 0, 1, 2$$

Thus, number of minima is 4.

9. (8) The path difference is $\Delta x = \sqrt{D^2 + d^2} - D$

$$+ (\mu_2 - \mu_3)t_2 + \mu_3 S_2C - (\mu_1 - \mu_3)t_1 - \mu_3 S_1C$$

$$= D\left(1 + \frac{d^2}{D^2}\right)^{\frac{1}{2}} - D + (\mu_2 - \mu_3)t_2 - (\mu_1 - \mu_3)t_1$$

$$= D\left(1 + \frac{d^2}{2D^2}\right) - D + (\mu_2 - \mu_3)t_2 - (\mu_1 - \mu_3)t_1$$

$$= \frac{d^2}{2D} + (\mu_2 - \mu_3)t_2 - (\mu_1 - \mu_3)t_1$$

$$= \frac{4 \times 10^{-6}}{2} + (1.4 - 1.8)12 \times 10^{-6} - (1.2 - 1.8)\,6 \times 10^{-6}$$

$$= 2 \times 10^{-6} - 4.8 \times 10^{-6} + 3.6 \times 10^{-6}$$

$$= 0.8 \times 10^{-6}\text{ m}$$

$$\text{For first maxima,} \quad \Delta x = \lambda$$

$$\text{or} \quad 0.8 \times 10^{-6} = \lambda$$

$$\therefore \quad \lambda = 8 \times 10^{-7} = 800 \times 10^{-9}\text{m} = 800\text{ nm}$$

10. (5) Here, path difference

$$\Delta x = d\sin\theta$$

$$\text{For maxima,} \quad \Delta x = m\lambda$$

$$\text{or} \quad d\sin\theta = m\lambda \quad \text{or} \quad 3\lambda\sin\theta = m\lambda$$

$$\text{or} \quad \sin\theta = \frac{m}{3}$$

$$\text{or} \quad -1 < \sin\theta < 1$$

$$\text{or} \quad -3 < m < 3$$

$$\therefore \quad m = -2, -1, 0, 1, 2$$

Thus, number of maxima on screen is 5.

11. (2) Here, $\Delta x = \frac{yd}{D} = (2m_1 - 1)\frac{\lambda_1}{2}$

$$= n\lambda_2$$

or $(2 \times 4 - 1)\frac{400}{2} = n \times 700$

$\therefore$ $n = \frac{7}{2} \times \frac{4}{7}$

$\therefore$ $n = 2$

12. (7) Here, path difference is

$$\Delta x = (\mu_1 - 1)t_1 - (\mu_2 - 1)t_2$$
$$= (2 - 1)\,200 - (1.5 - 1)\,200$$
$$= 200 - 100 = 100 \text{ nm}$$

$\therefore$ $\phi = \frac{2\pi}{\lambda}\Delta x = \frac{2\pi}{600} \times 100 = \frac{\pi}{3}$

$\therefore$ $P = P_1 + P_2 + 2\sqrt{P_1P_2}\cos\phi$

$$= I_0\pi r_1^2 + I_0\pi r_2^2 + 2\sqrt{I_0\pi r_1^2 \times I_0\pi r_2^2}\cos\frac{\pi}{3}$$

$$= 7\mu\text{W}$$

13. (4) The path difference at the centre of screen is

$$\Delta x = \int_0^x (\mu - 1)\,dx = \int_0^x \sqrt{x}\,dx$$
$$= \frac{x}{3/2} = \frac{2}{3}x = \frac{2}{3} \times 3 \times 10^{-6}\text{m}$$
$$= 2 \times 10^{-6}\text{m}$$

$\therefore$ $\phi = \frac{2\pi}{\lambda}\Delta x$

$$= \frac{2\pi}{400 \times 10^{-9}} \times 2 \times 10^{-6}$$
$$= \pi \times 10 = 10\pi$$

$\because$ $I = 4I_0\cos^2\frac{\phi}{2}$

$$= 4 \times 10^{-15}\cos^2 5\pi$$
$$= 4 \times 10^{-15}\,\text{Wm}^{-2}$$
$$= 4 \text{ femtowatt/m}^2$$

14. (4) Here, $\Delta x = \frac{yd}{D} = (2m + 1)\frac{\lambda_1}{2} = m\lambda_2$

Here, $(2m + 1)\frac{\lambda_1}{2} = m\lambda_2$

or $(2m + 1)\,200 = m \times 400.2$

or $\frac{2m + 1}{m} = \frac{400.2}{200}$

or $\frac{1}{m} = \frac{400.2}{200} - 2$

or $\frac{1}{m} = \frac{0.2}{200} = \frac{2}{2000}$

$\therefore$ $m = 1000$

$\therefore$ $y = \frac{1000 \times 4002 \times 10^{-10}}{10 \times 10^{-3}}$

$$= 4002 \times 10^{-5} = 0.04 \text{ m} = 4 \text{ cm}$$

15. Let n_1 bright fringe corresponding to wavelength $\lambda_1 = 500$ nm coincides with n_2 bright fringe corresponding to wavelength $\lambda_2 = 700$ nm

$\therefore$ $n_1\frac{\lambda_1 D}{d} = n_2\frac{\lambda_2 D}{d}$ or $\frac{n_1}{n_2} = \frac{\lambda_2}{\lambda_1} = \frac{7}{5}$

This implies that 7th maxima of λ_1 coincides with 5th maxima of λ_2. Similarly, 14th maxima of λ_1 will coincides with 10th maxima of λ_2 and so on.

$\therefore$ Minimum distance $= \frac{n_1\lambda_1 D}{d} = 7 \times 5 \times 10^{-7} \times 10^3$

$$= 3.5 \times 10^{-3}\text{ m} = 3.5 \text{ mm}$$

16. Given, $y_1 = 40$ cm, $D_1 = 2\text{m} = 200$ cm, $D_2 = 10$ cm

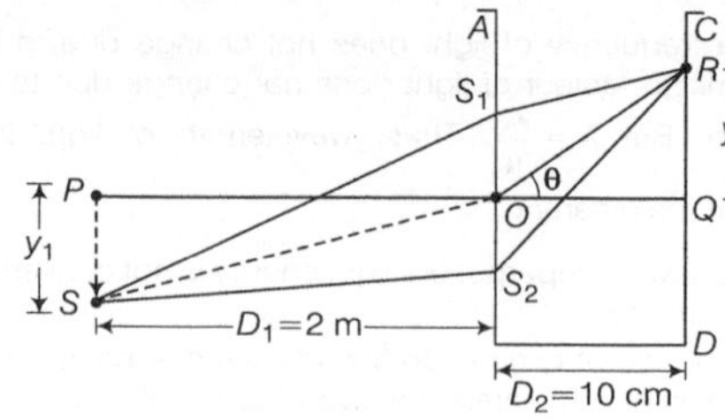

$$\tan\alpha = \frac{y_1}{D_1} = \frac{40}{200} = \frac{1}{5}$$

$\Rightarrow$ $\therefore \alpha = \tan^{-1}(1/5)$

$$\sin\alpha = \frac{1}{\sqrt{26}} = \frac{1}{5} = \tan\alpha$$

Path difference between SS_1 and SS_2 is

$$\Delta x_1 = SS_1 - SS_2 \quad \text{or} \quad \Delta x_1 = d\sin\alpha = (0.8 \text{ mm})\left(\frac{1}{5}\right)$$

$\Rightarrow$ $\Delta x_1 = 0.16$ mm ...(i)

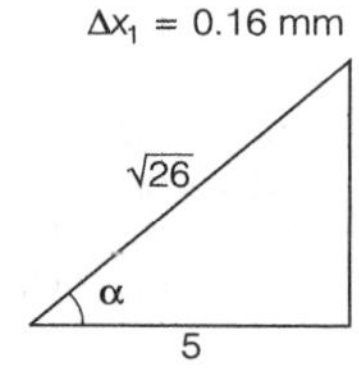

Now, let at point R on the screen, central bright fringe is observed (i.e. net path difference = 0).

Path difference between S_2R and S_1R would be

$$\Delta x_2 = S_2R - S_1R$$

$\Rightarrow$ $\Delta x_2 = d\sin\theta$ ...(ii)

Central bright fringe will be observed when net path difference is zero.

$\Rightarrow$ $\Delta x_2 - \Delta x_1 = 0$

$$\Delta x_2 = \Delta x_1$$

$\Rightarrow$ $d\sin\theta = 0.16$

or $(0.8)\sin\theta = 0.16$

$\Rightarrow$ $\sin\theta = \frac{0.16}{0.8} = \frac{1}{5}$

$$\tan\theta = \frac{1}{\sqrt{24}} = \sin\theta = \frac{1}{5}$$

Hence, $\tan\theta = \frac{y_2}{D_2} = \frac{1}{5}$

$\therefore$ $y_2 = \frac{D_2}{5} = \frac{10}{5} = 2$ cm

Therefore, central bright fringe is observed at 2 cm above point Q on side CD.

Alternate Method

ΔX at B will be zero if $\Delta x_1 = \Delta x_2$

$\Rightarrow$ $d\sin\alpha = d\sin\theta \Rightarrow \alpha = \theta$

$\Rightarrow$ $\tan\alpha = \tan\theta \Rightarrow \frac{y_1}{D_1} = \frac{y_2}{D_2}$

$\Rightarrow$ $y_2 = \frac{D_2}{D_1}y_1 = \left(\frac{10}{200}\right)(40 \text{ cm})$

$\Rightarrow$ $y_2 = 2$ cm

(ii) The central bright fringe will be observed at point Q. If the path difference created by the liquid slab of thickness $t = 10$ cm or 100 mm is equal to Δx_1, so that the net path difference at Q becomes zero.

So, $(\mu - 1)t = \Delta x_1 \Rightarrow (\mu - 1)(100) = 0.16$

$\Rightarrow$ $\mu - 1 = 0.0016$

$\Rightarrow$ $\mu = 1.0016$

17. Given, $\lambda = 600 \text{ nm} = 6 \times 10^{-7} \text{m}$

$$d = 0.45 \text{ mm} = 0.45 \times 10^{-3} \text{m}$$

$$D = 1.5 \text{ m}$$

Thickness of glass sheet, $t = 10.4 \, \mu\text{m} = 10.4 \times 10^{-6} \text{m}$

Refractive index of medium, $\mu_m = 4/3$

and refractive index of glass sheet, $\mu_g = 1.5$

(i) Let central maximum is obtained at a distance y below point O. Then,

$$\Delta x_1 = S_1P - S_2P = \frac{yd}{D}$$

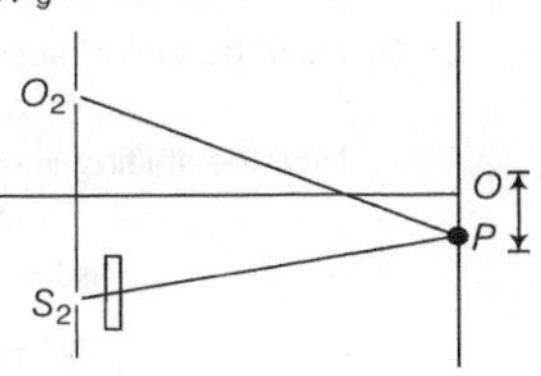

Path difference due to glass sheet

$$\Delta x_2 = \left(\frac{\mu_g}{\mu_w} - 1\right)t$$

Net path difference will be zero when

$$\Delta x_1 = \Delta x_2$$

or $$\frac{yd}{D} = \left(\frac{\mu_g}{\mu_m} - 1\right)t$$

$\therefore$ $$y = \left(\frac{\mu_g}{\mu_m} - 1\right)t\frac{D}{d}$$

Substituting the values, we have

$$y = \left(\frac{1.5}{4/3} - 1\right)\frac{10.4 \times 10^{-6}(1.5)}{0.45 \times 10^{-3}}$$

$$y = 4.33 \times 10^{-3} \text{m}$$

or we can say, $y = 4.33$ mm

(ii) At O, $\Delta x_1 = 0$ and $\Delta x_2 = \left(\frac{\mu_g}{\mu_m} - 1\right)t$

$\therefore$ Net path difference, $\Delta x = \Delta x_2$

Corresponding phase difference, $\Delta\phi$ or simple $\phi = \frac{2\pi}{\lambda} \cdot \Delta x$

Substituting the values, we have

$$\phi = \frac{2\pi}{6 \times 10^{-7}}\left(\frac{1.5}{4/3} - 1\right)(10.4 \times 10^{-6})$$

$$\phi = \left(\frac{13}{3}\right)\pi$$

Now, $$I(\phi) = I_{\text{max}} \cos^2\left(\frac{\phi}{2}\right)$$

$\therefore$ $$I = I_{\text{max}} \cos^2\left(\frac{13\pi}{6}\right)$$

$$I = \frac{3}{4} I_{\text{max}}$$

(iii) At O, path difference $\Delta x = \Delta x_2 = \left(\frac{\mu_g}{\mu_m} - 1\right)t$

For maximum intensity at O

$\Delta x = n\lambda$ (Here, $n = 1, 2, 3, \cdots$)

$\therefore$ $$\lambda = \frac{\Delta x}{1}, \frac{\Delta x}{2}, \frac{\Delta x}{3} \cdots \text{ and so on}$$

$$\Delta x = \left(\frac{1.5}{4/3} - 1\right)(10.4 \times 10^{-6} \text{ m})$$

$$= \left(\frac{1.5}{4/3} - 1\right)(10.4 \times 10^{3} \text{nm})$$

$$\Delta x = 1300 \text{ nm}$$

$\therefore$ Maximum intensity will be corresponding to

$$\lambda = 1300 \text{ nm}, \frac{1300}{2} \text{ nm}, \frac{1300}{3} \text{ nm}, \frac{1300}{4} \text{ nm} \ldots$$

$$= 1300 \text{ nm}, 650 \text{ nm}, 433.33 \text{ nm}, 325 \text{ nm} \ldots$$

The wavelengths in the range 400 to 700 nm are 650 nm and 433.33 nm.

18. Given, $\lambda = 0.5 \text{ mm}, d = 1.0 \text{ mm}, D = 1 \text{ m}$

(i) When the incident beam falls normally,

Path difference between the two rays S_2P and S_1P is

$$\Delta x = S_2P - S_1P = d \sin\theta$$

For minimum intensity,

$$d \sin\theta = (2n - 1)\frac{\lambda}{2}, \quad \text{where } n = 1, 2, 3, \cdots$$

or $$\sin\theta = \frac{(2n-1)\lambda}{2d} = \frac{(2n-1)\, 0.5}{2 \times 1.0} = \frac{2n-1}{4}$$

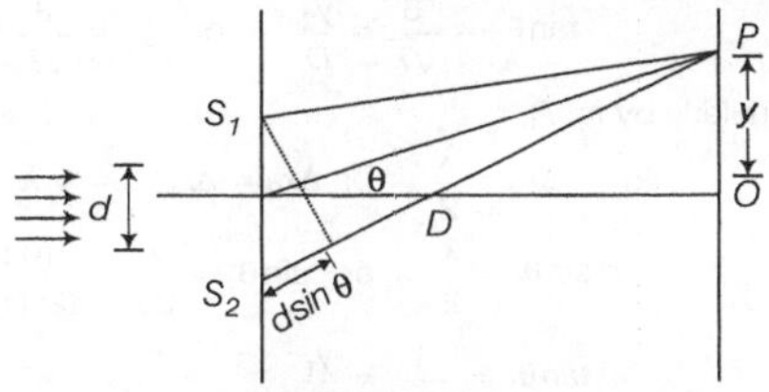

As $\sin\theta \le 1$, therefore $\frac{(2n-1)}{4} \le 1$ or $n \le 2.5$

So, n can be either 1 or 2.

When $n = 1$, $$\sin\theta_1 = \frac{1}{4}$$

$\Rightarrow$ $$\tan\theta_1 = \frac{1}{\sqrt{15}}$$

When $n = 2$, $$\sin\theta_2 = \frac{3}{4}$$

$\Rightarrow$ $$\tan\theta_2 = \frac{3}{\sqrt{7}}$$

$\therefore$ $$y = D\tan\theta = \tan\theta \quad [\because D = 1 \text{m}]$$

So, the position of minima will be

$$y_1 = \tan\theta_1 = \frac{1}{\sqrt{15}} \text{ m} = 0.26 \text{ m}$$

$$y_2 = \tan\theta_2 = \frac{3}{\sqrt{7}} \text{ m} = 1.13 \text{ m}$$

And as minima can be on either side of centre O.

Therefore, there will be four minimas at positions $\pm$ 0.26 m and $\pm$ 1.13 m on the screen.

(ii) When $\alpha = 30°$, path difference between the rays before reaching S_1 and S_2 is

$$\Delta x_1 = d \sin\alpha = (1.0)\sin 30° = 0.5 \text{ mm} = \lambda$$

So, there is already a path difference of λ between the rays.

Position of central maximum Central maximum is defined as a point where net path difference is zero. So,

$$\Delta x_1 = \Delta x_2 \quad \text{or} \quad d \sin\alpha = d\sin\theta$$

or $$\theta = \alpha = 30°$$

or $$\tan\theta = \frac{1}{\sqrt{3}} = \frac{y_0}{D}$$

$\therefore$ $$y_0 = \frac{1}{\sqrt{3}} \text{ m} \quad [\because D = 1 \text{m}]$$

$$y_0 = 0.58 \text{ m}$$

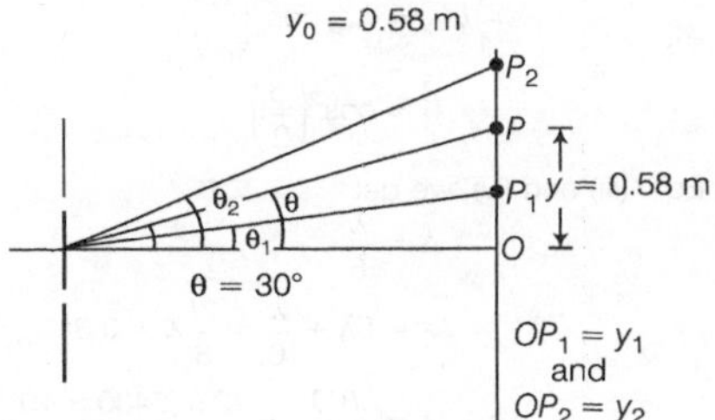

At point P, $\Delta x_1 = \Delta x_2$

Above point P, $\Delta x_2 > \Delta x_1$ and

Below point P, $\Delta x_1 > \Delta x_2$

Now, let P_1 and P_2 be the minima on either side of central maxima. Then for P_2,

$$\Delta x_2 - \Delta x_1 = \frac{\lambda}{2}$$

or $$\Delta x_2 = \Delta x_1 + \frac{\lambda}{2} = \lambda + \frac{\lambda}{2} = \frac{3\lambda}{2}$$

or $$d \sin\theta_2 = \frac{3\lambda}{2}$$

or $$\sin\theta_2 = \frac{3\lambda}{2d} = \frac{(3)(0.5)}{(2)(1.0)} = \frac{3}{4}$$

$\therefore$ $$\tan\theta_2 = \frac{3}{\sqrt{7}} = \frac{y_2}{D} \quad \text{or} \quad y_2 = \frac{3}{\sqrt{7}} = 1.33 \text{ m}$$

Similarly by for P_1,

$$\Delta x_1 - \Delta x_2 = \frac{\lambda}{2} \quad \text{or} \quad \Delta x_2 = \Delta x_1 - \frac{\lambda}{2} = \lambda - \frac{\lambda}{2} = \frac{\lambda}{2}$$

or $$d \sin\theta_1 = \frac{\lambda}{2} \quad \text{or} \quad \sin\theta_1 = \frac{\lambda}{2d} = \frac{(0.5)}{(2)(1.0)} = \frac{1}{4}$$

$\therefore$ $$\tan\theta_1 = \frac{1}{\sqrt{15}} = \frac{y_1}{D}$$

or $$y_1 = \frac{1}{\sqrt{15}} \text{ m} = 0.26 \text{ m}$$

Therefore, y-coordinates of the first minima on either side of the central maximum are $y_1 = 0.26$ m and $y_2 = 1.33$ m.

19. Given, $\mu_1 = 1.4$ and $\mu_2 = 1.7$ and let t be the thickness of each glass plates. Path difference at O, due to insertion of glass plates will be

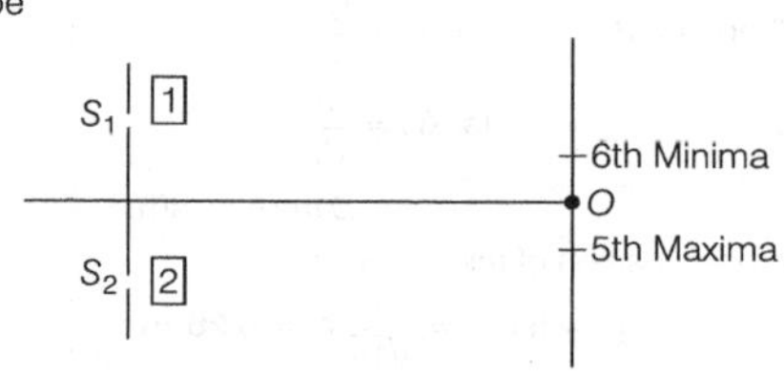

$$\Delta x = (\mu_2 - \mu_1)t = (1.7 - 1.4)t = 0.3t \quad \text{...(i)}$$

Now, since 5th maxima (earlier) lies below O and 6th minima lies above O.

This path difference should lie between 5λ and $5\lambda + \frac{\lambda}{2}$.

So, let $$\Delta x = 5\lambda + \Delta \quad \text{...(ii)}$$

where, $$\Delta < \frac{\lambda}{2}$$

Due to the path difference Δx, the phase difference at O will be

$$\phi = \frac{2\lambda}{\lambda}\Delta x = \frac{2\pi}{\lambda}(5\lambda + \Delta)$$

$$= \left(10\pi + \frac{2\pi}{\lambda}\Delta\right) \quad \text{...(iii)}$$

Intensity at O is given $\frac{3}{4}I_{max}$ and since

$$I(\phi) = I_{max}\cos^2\left(\frac{\phi}{2}\right)$$

$$\frac{3}{4}I_{max} = I_{max}\cos^2\left(\frac{\phi}{2}\right)$$

or $$\frac{3}{4} = \cos^2\left(\frac{\phi}{2}\right) \quad \text{...(iv)}$$

From Eqs. (iii) and (iv), we get

$$\Delta = \frac{\lambda}{6}$$

i.e. $$\Delta x = 5\lambda + \frac{\lambda}{6} = \frac{3l}{6}\lambda = 0.3t$$

$$t = \frac{31\lambda}{6(0.3)} = \frac{(31)(5400 \times 10^{-10})}{1.8}$$

or $$t = 9.3 \times 10^{-6} \text{m} = 9.3 \, \mu\text{m}$$

20. (i) Path difference due to the glass slab,

$$\Delta x = (\mu - 1)t = (1.5 - 1)t = 0.5t$$

Due to this slab, 5 red fringes have been shifted upwards. Therefore,

$$\Delta x = 5\lambda_{red}$$

or $$0.5t = (5)(7 \times 10^{-2}\text{m})$$

$\therefore t$ = thickness of glass slab $= 7 \times 10^{-6}$ m

(ii) Let u' be the refractive index for green light, then

$$\Delta x' = (\mu' - 1)t$$

Now the shifting is of 6 fringes of red light. Therefore,

$$\Delta x' = 6\lambda_{red}$$

$\therefore$ $$(\mu' - 1)t = 6\lambda_{red}$$

$\therefore$ $$(\mu' - 1) = \frac{(6)(7 \times 10^{-7})}{7 \times 10^{-6}} = 0.6$$

$$\mu' = 1.6$$

In part (i), shifting of 5 bright fringes was equal to 10^{-3} m. Which implies that,

$$5\omega \text{ red} = 10^{-3} \text{ m} \qquad \text{[Here, fringe width]}$$

$$\omega_{red} = \frac{10^{-3}}{5} \text{ m} = 0.2 \times 10^{-3} \text{ m}$$

Now since, $$\omega = \frac{\lambda D}{d}$$

or $$\omega \propto \lambda$$

$\therefore$ $$\frac{\omega_{green}}{\omega_{red}} = \frac{\lambda_{green}}{\lambda_{red}}$$

$\therefore$ $$\omega_{green} = \frac{\omega_{red}\lambda_{green}}{\lambda_{red}} = (0.2 \times 10^{-3})\left(\frac{5 \times 10^{-7}}{7 \times 10^{-7}}\right)$$

$$\omega_{green} = 0.143 \times 10^{-3} \text{m}$$

$\therefore$ $$\Delta\omega = \omega_{green} - \omega_{red} = (0.143 - 0.2) \times 10^{-3} \text{m}$$

$$\Delta\omega = -5.71 \times 10^{-5} \text{m}$$

21. Given, $\mu = 1.33$, $d = 1$ mm, $D = 1.33$ m, $\lambda = 6300$ Å

(i) Wavelength of light in the given liquid,

$$\lambda' = \frac{\lambda}{\mu} = \frac{6300}{1.33} \text{ Å} \approx 4737 \text{ Å} = 4737 \times 10^{-10} \text{ m}$$

$\therefore$ Fringe width, $\omega = \frac{\lambda'}{d}D$

$$\omega = \frac{(4737 \times 10^{-10} \text{ m})(1.33 \text{ m})}{(1 \times 10^{-3} \text{ m})} = 6.3 \times 10^{-4} \text{ m}$$

$$\omega = 0.63 \text{ mm}$$

(ii) Let t be the thickness of the glass slab.

Path difference due to glass slab at centre O,

$$\Delta x = \left(\frac{\mu_{glass}}{\mu_{liquid}} - 1\right)t = \left(\frac{1.53}{1.33} - 1\right)t$$

or $$\Delta x = 0.15t$$

Now, for the intensity to be minimum at O, this path difference should be equal to $\frac{\lambda'}{2}$.

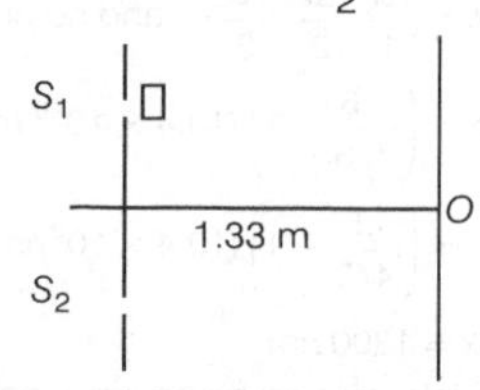

$\therefore$ $$\Delta x = \frac{\lambda'}{2} \quad \text{or} \quad 0.15t = \frac{4737}{2} \text{ Å}$$

$\therefore$ $$t = 15790 \text{ Å}$$

or $$t = 1.5790 \, \mu\text{m}$$

22. (i) The desired distance will be

$$y_1 = 3\omega_1 = 3\left(\frac{\lambda_1 D}{d}\right) = \frac{(3)(6500 \times 10^{-10})(1.2)}{(2 \times 10^{-3})}$$

$$= 11.7 \times 10^{-4} \text{m} = 1.17 \text{ mm}$$

(ii) Let n_1 bright fringe of λ_1 coincides with n_2 bright fringe of λ_2. Then,

$$\frac{n_1\lambda_1 D}{d} = \frac{n_2\lambda_2 D}{d}$$

or $$\frac{n_1}{n_2} = \frac{\lambda_2}{\lambda_1} = \frac{5200}{6500} = \frac{4}{5}$$

Therefore, 4th bright of λ_1 coincides with 5th bright of λ_2. Similarly, 8th bright of λ_1 will coincide with 10th bright of λ_2 and so on. The least distance from the central maximum will therefore corresponding to 4th bright of λ_1 (or 5th bright of λ_2). Hence,

$$Y_{min} = \frac{4\lambda_1 D}{d} = \frac{4(6500 \times 10^{-10})(1.2)}{(2 \times 10^{-3})}$$

$$= 15.6 \times 10^{-4} \text{m} = 1.56 \text{ mm}$$

23. Shifting of fringes due to introduction of slab in the path of one of the slits, comes out to be,

$$\Delta y = \frac{(\mu - 1)t\, D}{d} \quad \text{...(i)}$$

Now, the distance between the screen and slits is doubled. Hence, the new fringe width will become.

$$\omega' = \frac{\lambda(2D)}{d} \quad \text{...(ii)}$$

Given, $$\Delta y = \omega'$$

or $$\frac{(\mu - 1)tD}{d} = \frac{2\lambda D}{d}$$

$\therefore$ $$\lambda = \frac{(\mu - 1)t}{2}$$

$$= \frac{(1.6 - 1)(1.946 \times 10^{-6})}{2}$$

$$= 0.5892 \times 10^{-6} \text{ m}$$

$$= 5892 \text{ Å}$$

DPP-5 Interference of Light - IV

1. (i) (b) $\Delta x = 2\mu t - \frac{\lambda}{2}$

(ii) (a) $\Delta x = 2\mu t - \frac{\lambda}{2} = (2m - 1)\frac{\lambda}{2}$ or $2\mu t = m\lambda$

$\therefore$ $\lambda = \frac{2\mu t}{m}$, $m = 1, 2, 3 \cdots$

$\therefore$ $\lambda = 400$ nm, 466.7 nm, 560 nm, 700 nm

(iii) (b) For strong reflection

$$2\mu t - \frac{\lambda}{2} = m\lambda,\ m = 0, 1, 2, 3,$$

or $$2\mu t = (2m + 1)\frac{\lambda}{2}$$

$\therefore$ $$\lambda = \frac{4\mu t}{2m + 1}$$

Putting the values

$\lambda = 622.2$ nm, 509.1 nm, 430.8 nm.

Hence, option (b) is correct.

(iv) (a) $\Delta x = 2\mu t = m\lambda$

$\therefore$ $\lambda = \frac{2\mu t}{m}$

Here, $m = 0, 1, 2, 3, 4, \cdots$

$\therefore$ $\lambda = 400$ nm, 466.7 nm 560 nm and 700 nm

(v) (b) $\Delta x = 2\mu t - \frac{\lambda}{2} = (2m - 1)\frac{\lambda}{2}$

$\therefore$ $\lambda = \frac{2\mu t}{m}$

$\therefore$ $\lambda = 430.8$ nm, 509.1 nm, 622.2 nm

(vi) (b, c) $\delta l' < \frac{\beta}{2}$

$\therefore$ $$\delta l_{max} = \frac{\lambda}{4\alpha}\left(1 + \frac{r}{b}\right) = 43 \times 10^{-6} \text{ m}$$

2. (i) (b) $\alpha = \frac{d}{2r}$

$\therefore$ $d = 2r\alpha$

(ii) (a) $D = b + r$

(iii) (a, b) $\beta = \frac{D}{d}\lambda = \left(\frac{b + r}{2r\alpha}\right)\lambda$

(iv) (a) $\tan \alpha = \frac{l}{2b}$

$\therefore$ $l = 2b\,\alpha$ $\quad$ ($\because \alpha$ is very small)

(v) (d) $2\left[\frac{l}{2\beta}\right] + 1 = 2\left[\frac{b\alpha}{\beta}\right] + 1 = 9$

(vi) (b, d) $\delta\alpha = \frac{\delta l}{r}$

$\therefore$ $\tan \delta\alpha = \frac{\delta l'}{b}$

$\therefore$ $\delta l' = \left(\frac{\delta l}{r}\right)b = 13$ mm

3. (i) (a) $\cos \theta = \frac{SA}{SO}$

$\therefore$ $SA = a \cos \theta = a$

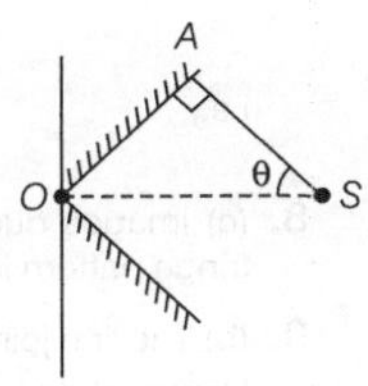

(ii) (a) $SS_1 = SS_2 = 2SA$

$$= 2a \cos \theta = 2a$$

(iii) (c) $\sin \theta = \theta = \frac{S_1S_2}{2 \times 2a}$

$\therefore$ $S_1S_2 = 4a\theta$

(iv) (d) $\beta = \frac{D}{d}\lambda = \left(\frac{b + 2a}{4a\theta}\right)\lambda$

(v) (a) The path difference at c is zero. So, it is a point of central bright fringe.

(vi) (a)

4. (c) The path difference at point P is

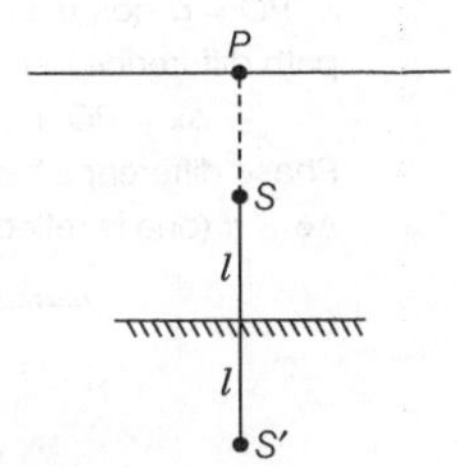

$$\Delta x = S'P + \frac{\lambda}{2} - SP$$

$$= 2l + \frac{\lambda}{2} = m\lambda$$

or $$2l = m\lambda - \frac{\lambda}{2}$$

$\therefore$ $$l = \frac{m\lambda}{2} - \frac{\lambda}{4} \quad \text{...(i)}$$

When the mirror is shifted downward a distance, y_0

$$\Delta x' = 2(l + l_0) + \frac{\lambda}{2} = (m + 1)\lambda$$

or $$l + l_0 = (m + 1)\frac{\lambda}{2} - \frac{\lambda}{4}$$

or $$m\frac{\lambda}{2} - \frac{\lambda}{4} + l_0 = m\frac{\lambda}{2} + \frac{\lambda}{2} - \frac{\lambda}{4}$$

or $$l_0 = \frac{\lambda}{2} = \frac{600}{2} = 300 \text{ nm}$$

5. (c) Here, lens is combination of two media of refractive indices and $(\mu - \delta\mu)$.

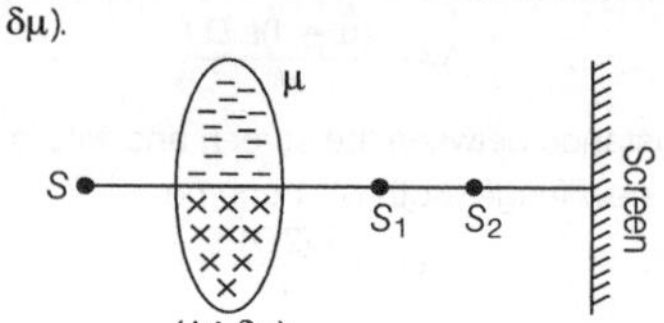

$(14\ \delta\mu)$

The line joining images S_1 and S_2 are perpendicular to screen. So, shape of fringe on the sceen is circular.

6. (b) Here, $50\lambda = l\,\alpha\,\Delta\theta$

$$\therefore \qquad \alpha = \frac{50 \times 500 \times 10^{-9}}{10 \times 10^{-2} \times 25} = 10^{-5}/^{\circ}\text{C}$$

7. (b) The situation can be taken as, if there are two sources S_1 and S_2 as shown in figure.

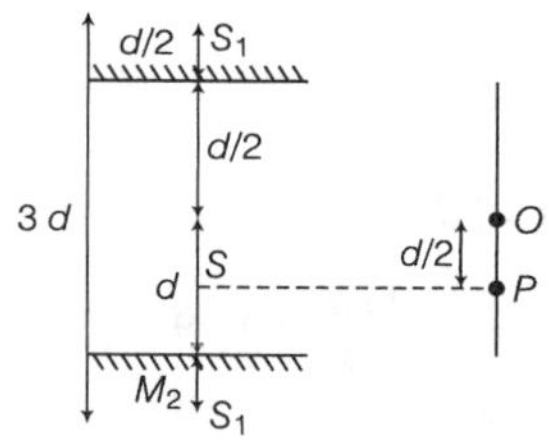

Due to these S_1 and S_2 the central maxima will be at P at a distance $d/2$ from O.

$$y = \frac{m\lambda D}{3d} = \frac{d}{2}$$

$$\frac{d}{2} = \frac{m\lambda D}{3d} \lambda = \frac{3d^2}{2mD}$$

i.e.,
$$\lambda = \frac{3d^2}{2d}, \frac{3d^2}{4d}$$

8. (c) Images due to both parts of lens are at same point. So, no fringe pattern is formed on screen.

9. (b) The line joining sources is parallel to screen.
Hence, shape of fringes on screen is circular.

10. (d) A fringe is a locus of points having constant path difference from the two coherent sources S_1 and S_2. It will be concentric circle.

11. (b) $PR = d$

$\therefore$ $PO = d\cos\theta$ and $CO = PO\cos 2\theta = d\sec\theta\cos 2\theta$

path difference between two rays is

$$\Delta x = PO + OC = (d\sec\theta + d\sec\theta\cos 2\theta)$$

Phase difference between two rays is

$\Delta\phi = \pi$ (one is reflected, while another is direct)

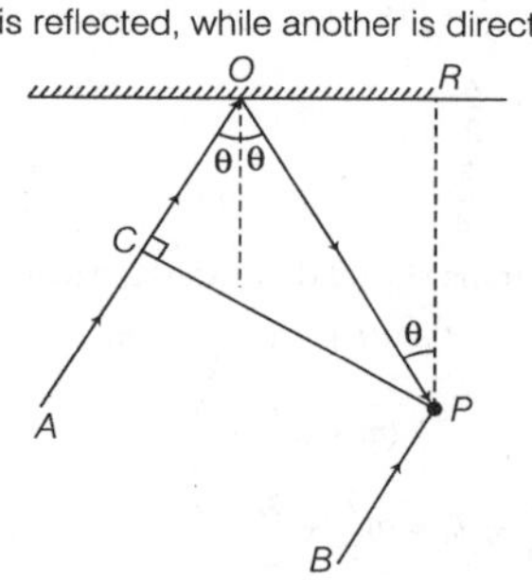

Therefore, condition for constructive interference should be

$$\Delta x = \frac{\lambda}{2}, \frac{3\lambda}{2} \ldots \quad \text{or} \quad d\sec\theta\,(1 + \cos 2\theta) = \frac{\lambda}{2}$$

or
$$\left(\frac{d}{\cos\theta}\right)(2\cos^2\theta) = \frac{\lambda}{2}$$

or
$$\cos\theta = \frac{\lambda}{4d}$$

12. (a) Locus of equal path difference are the lines running parallel to the axis of the cylinder. Hence, straight fringes are obtained.

13. (a,b,c) (a) Here, $\Delta x = \left| 2\mu t - \frac{\lambda}{2} \right|$

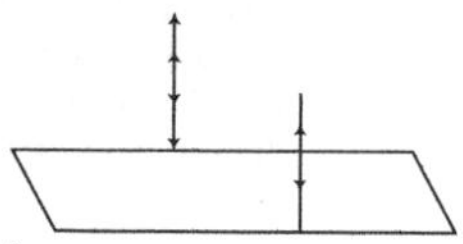

But $t << \lambda$

$\therefore$ $\Delta x = \frac{\lambda}{2}$, Hence, minima is formed.

(b) $\Delta x = 2\mu t - \frac{\lambda}{2} = 2 \times \frac{4}{3} \times \frac{3\lambda}{8} - \frac{\lambda}{2} = \frac{\lambda}{2}$

Hence, it appears dark.

(c) $\Delta x = 2\mu t = 2 \times \frac{4}{3} \times 3\lambda = 8\lambda$

Hence, it appears bright.

14. (a,c,d) For $n = \infty$, $\Delta x = 0 \Rightarrow m = 0$

For $x = 0$, $\Delta x = 10\lambda \Rightarrow m = 10$

Thus, number of minima between central maxima and 10th maxima is 10.

15. (b,c) (b) From figure, resultant amplitude is zero. Hence, intensity is zero.

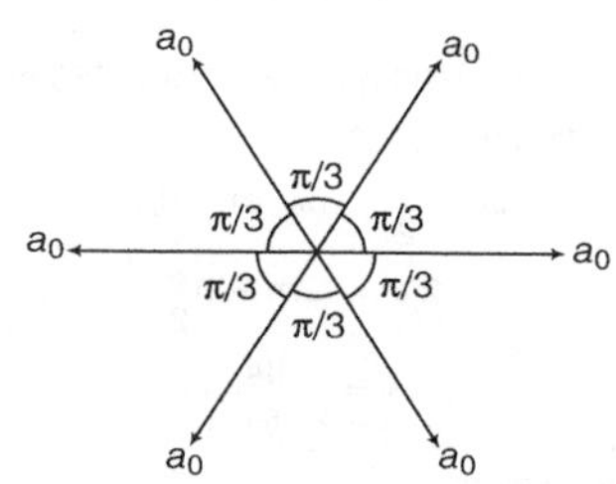

(c) If sources are incoherent, $I = 6I_0$.

16. (a,b,c) The path difference at point P is

$$\Delta x = S_1E = 2\lambda\cos\theta$$

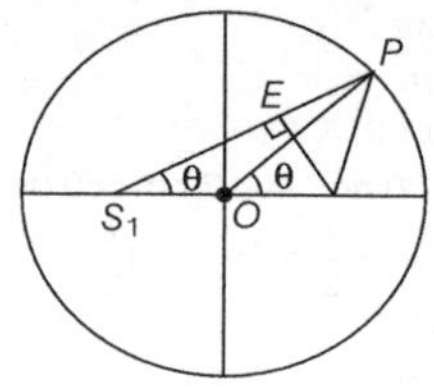

For maxima, $\qquad 2\lambda\cos\theta = m\lambda$

$$\therefore \qquad \cos\theta = \frac{m}{2}$$

The possible value of m in first quadrant is 0, 1, 2.

Thus number of maxima in all four quadrants is eight.

For minima, $\qquad \Delta x = (2m - 1)\frac{\lambda}{2}$

or $\qquad 2\lambda\cos\theta = (2m - 1)\lambda/2$

or $\qquad \cos\theta = \frac{(2m - 1)}{4}$

The possible value of m in first quadrant is 1, 2.

Thus, total number of minima in all four qudrants is eight.

17. (a,b,c,d) (a) The line joining point source S_1 and S_2 is perpendicular to screen. Hence, shape of fringe is circular.

(b) The parth difference at point P is

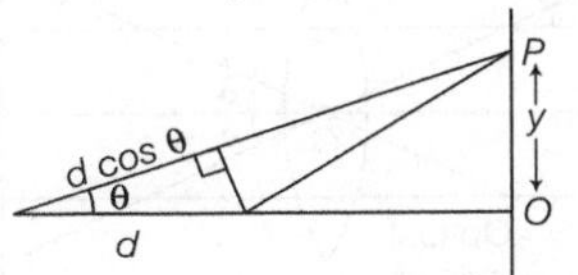

$$\Delta x = d \cos \theta$$

For first minima, $\Delta x = \frac{\lambda}{2}$

$\therefore \quad d \cos \theta = \frac{\lambda}{2}$ or $\cos \theta = \frac{\lambda}{2d}$

or $\frac{y}{\sqrt{D^2 + y^2}} = \frac{\lambda}{2d}$ or $\frac{y^2}{D^2 + y^2} = \frac{\lambda^2}{4d^2} = \frac{1}{4 \times 4.8^2}$

or $92.16y^2 = 1 + y^2$ or $91.16y^2 = 1$

$\therefore \quad y = 0.1\,\text{m} = 10\,\text{cm}.$

(c) $\Delta x = d \cos\theta = (2m-1)\frac{\lambda}{2}$

or $\cos \theta = (2m-1)\frac{\lambda}{2d} = \frac{(2m-1)}{9.6}$

or $-1 < \cos \theta < 1$ or $-9.6 < 2m - 1 < 9.6$

or $-8.6 < 2m < 10.6$ or $-4.3 < m < 5.3$

$\therefore \quad m = -4, -3, -2, -1, 0, 1, 2, 3, 4, 5$

Thus, total number of minima is 10.

(d) The path difference at point O is $\Delta x = d = 4.8\lambda$. Hence, it is neither a point of maxima nor a point of minima.

18. (a,b) When ray is reflected from denser medium, phase changes through it.

(a) For (1), reflections take place from rarer medium at both surfaces. Thus, phase difference between them is zero.

For (4), reflections at both surfaces take place from denser medium. So, phase difference between them is zero. So, option (a) is correct.

(b) In (2) and (3), reflections at upper surfaces are from rarer medium but at lower surfaces, reflections take place from denser medium. Hence, phase difference between them is π.

19. (a) Since, ray goes without deviation,

$\therefore \quad n_1 = n_2$

or $1.2 + \frac{10.8 \times 10^4}{\lambda^2} = 1.45 + \frac{1.80 \times 10^4}{\lambda^2}$

$\Rightarrow \quad l = 600\,\text{nm}$

20. (b)

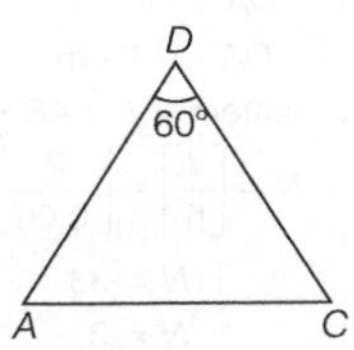

21. (c) Here, $m = 1.2 + \frac{10.8 \times 10^4}{600 \times 600} = 1.2 + 0.3 = 1.5$

For minimum deviation, $i = 1, r = r'$

But $r + r' = A = 60°$

$\therefore \quad r = 30°$

$\therefore \quad 1 \times \sin i = m \sin r$

$\sin i = 1.5 \times \frac{1}{2} = \left(\frac{3}{4}\right)$

$\therefore \quad i = \sin^{-1}\left(\frac{3}{4}\right)$

22. (a) $\Delta x = \Delta x_2 - x_1 = 2t_2 + nt_1 + \frac{\lambda}{2} - \left(\frac{\lambda}{2}\right)$

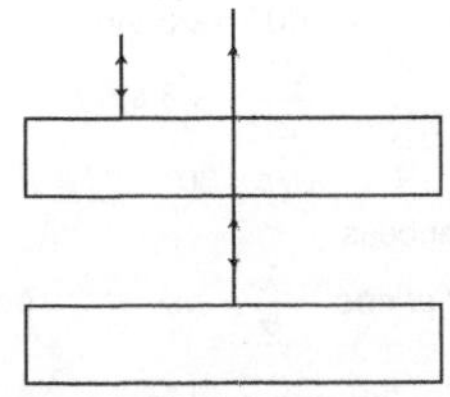

$$\Delta x = (2t_2 + 2nt_1)$$
$$\Delta x = 2(t_2 + nt_1)$$

23. (a) $2(t_2 + nt_1) = m\lambda$

or $2(127 + 1.5 \times 64)\,\text{nm} = m\lambda$

or $446\,\text{nm} = m\lambda$

$\therefore \quad \lambda = \frac{446}{m}\,\text{nm}$

For $m = 1, \lambda = 446$ nm

For $m = 2, \lambda = 223$ nm (not lying in visible region)

24. (c) Δx = path difference

$$\Delta x = 2d - d'$$

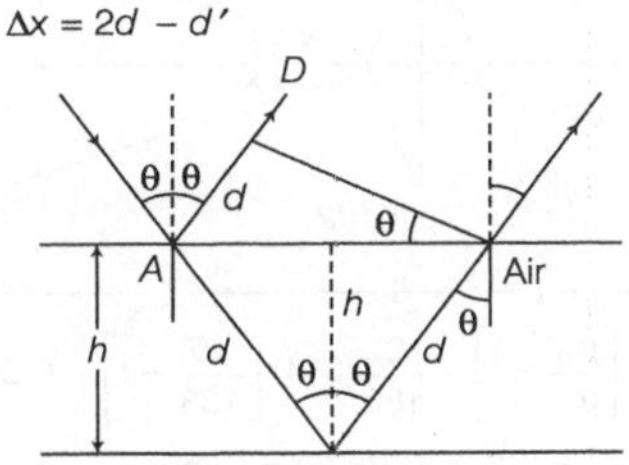

$\sec \theta = \frac{d}{h}$

$\therefore \quad d = h\sec \theta$

Also, $\tan \theta = \frac{AC}{2h}$

$\therefore \quad AC = 2h \tan \theta$

$\therefore \quad \sin \theta = \frac{AD}{AC} = \frac{d'}{2h \tan \theta}$

$\therefore \quad d' = 2h \tan \theta \sin \theta$

$\therefore \quad \Delta x = 2d - d'$

$\Delta x = 2h\sec \theta - 2h \tan \theta \sin \theta$

$\Delta x = \frac{2h}{\cos \theta} - \frac{2h\sin^2 \theta}{\cos \theta} = \frac{2h \cos^2 \theta}{\cos \theta}$

$\Delta x = 2h \cos \theta = m\lambda$ for construction interference

$\therefore \quad 2 \times 223 \cos 20° = m\lambda$

$446 \cos 20° = m\lambda$

$\therefore \quad \lambda = \frac{446 \cos 20°}{m} = \frac{446 \times 0.9397}{m}$

$= \frac{419}{m}\,\text{nm},$

For $m = 1, \lambda = 419$ nm

$m = 2, \lambda = 209.5$ nm (not visible.)

25. (d) Depending on the value of refractive index, the path travelled by light would be decided, which is turn decides the shapes of wavefronts. For more refractive index, speed of light is less and hence wavefront location would be somewhat located back side.

26. (a) For $\mu_1 = \mu_2$, the two waves in the reflected system differ by optical phase difference of π and and hence due to destructive interference, the film appears to the dark. In transmitted system, the two waves are in phase and hence film appears to be shiny. Similarly, we can go for the other conditions.

27. (2)

Here, $1 \times \sin i = \mu \sin r$

or $\sin 60° = \sqrt{3} \sin r$

or $\frac{\sqrt{3}}{2} = \sqrt{3} \sin r$

$r = 30°$

Here, path difference is

$$2\,\mu t \cos - \frac{\lambda}{2} = m\lambda \quad \text{(for maximum reflection)}$$

or $$2\mu t \cos r = \left(m\lambda + \frac{\lambda}{2}\right)$$

For minimum thickness, $m = 0$,

$\therefore$ $$t = \frac{\lambda}{2\mu \cos r} = \frac{600}{2\sqrt{3} \times \frac{\sqrt{3}}{2}}$$

$= 200$ nm

$\therefore$ $x = 2$

28. (71) Angle of deviation of the prism with water on one side is

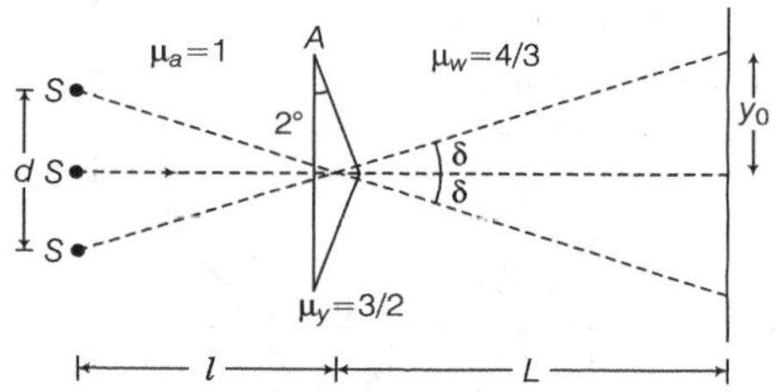

$$\delta = A\left[\frac{\mu_g}{\mu_w} - 1\right] = \frac{2}{180}(314)\left[\frac{15}{133} - 1\right] = 0.0043 \text{ rad}$$

$d = 2\delta l = 2(0.0043)(575) = 5$ mm

$D = l + L = 150$ cm

$\therefore$ Fringe width, $\omega = \frac{\lambda D}{\mu_w d} = \frac{(0.5 \times 10^{-6})\,(1.5)}{(133)(5 \times 10^{-3})} = 0.1125$ mm

$\tan \delta = \delta = \frac{y_0}{L}$

$\therefore$ $y_0 = L\delta = 3.975$

$\therefore$ No of bright fringe $= 2\left[\frac{y_0}{\omega}\right] + 1 = 2[35.35] + 1$

$= 2 \times 35 + 1 = 71.$

29. (7) Here, path difference is $\Delta x = 2\lambda \cos \theta + \frac{\lambda}{2}$

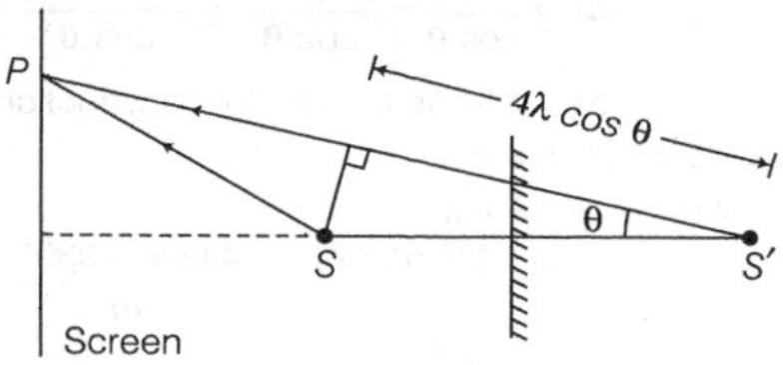

For minima, $\Delta x = (2m + 1)\frac{\lambda}{2}$

$$4\lambda \cos \theta + \frac{\lambda}{2} = (2m + 1)\frac{\lambda}{2}$$

or $$4\lambda \cos \theta = (2m + 1)\frac{\lambda}{2} - \frac{\lambda}{2}$$

or $4\lambda \cos \theta = m\lambda$ or $\cos \theta = \frac{m}{4}$

$$-1 < \cos \theta < 1$$

$$-1 < \frac{m}{4} < 1$$

or $-4 < m < 4$ or $-3, -2, -1, 0, 1, 2, 3.$

Thus, the number of minima is 7.

30. (9)

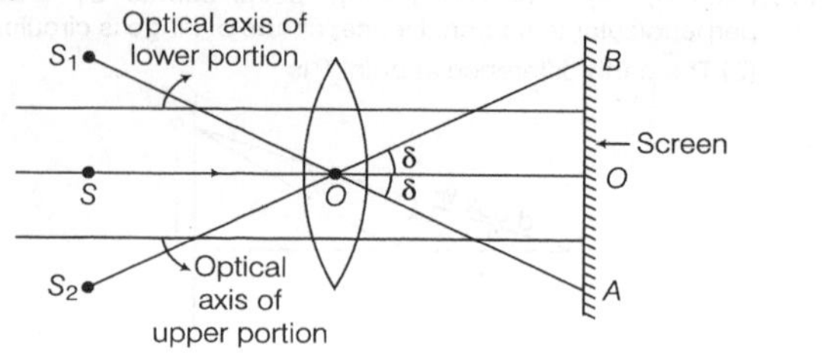

Here, $\tan \delta = \delta = \frac{h}{2f} = \frac{d}{2v_0}$ (Here, $S_1S_2 = d$)

$\therefore$ $d = \frac{hv_0}{f}$

$\therefore$ $\beta = \frac{D}{d}\lambda = \frac{(b + v_0)}{hv_0} f\lambda = \frac{f\lambda}{h}\left(1 + \frac{b}{v_0}\right)$

Since point source is in focal plane, so images are at infinity.

$\therefore$ $v_0 = \infty$

$\therefore$ $\beta = \frac{f\lambda}{h} = \frac{1 \times 500 \times 10^{-9}}{2 \times 10^{-3}} = 2.5 \times 10^{-4}$ m

From figure, $\tan \delta = \delta = \frac{OB}{b}$

$\therefore$ $l = OB = b$, $\delta = \frac{bh}{2f} = \frac{1 \times 10^{-3} \times 2}{2} = 10^{-3}$m

$\therefore$ No. of maxima $= 2\left[\frac{l}{\beta}\right] + 1 = 2 \times 4 + 1 = 9$

31. (3) Here, $d = 2 \times 10^{-3}$ m

$D = (190 + 5 + 5)$ cm $= 2$m

$\therefore$ $\beta = \frac{D}{d}\lambda = \frac{2}{2 \times 10^{-3}} \times 600 \times 10^{-9}$

$= 6 \times 10^{-4}$m $= 0.6$ mm

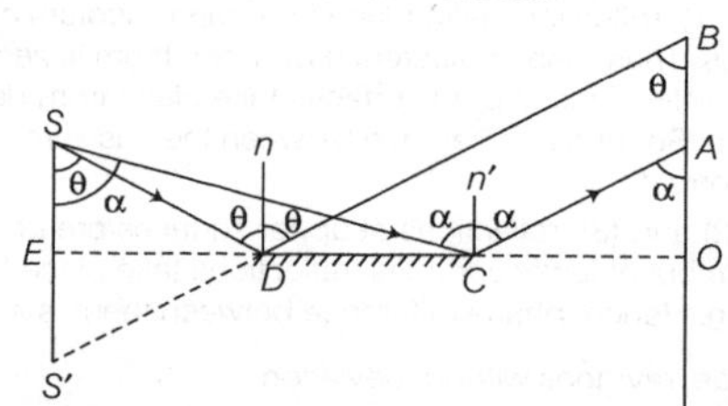

Here, $\tan\theta = \frac{DE}{SE} = \frac{195}{OB}$ or $OB = \frac{195 \times 0.1}{5} = 3.9$ cm

Similarly, $\tan \alpha = \frac{10}{0.1} = \frac{190}{OA}$

$\therefore$ $OA = 1.9$ cm

$\therefore$ The length of fringe pattern $= l = AB = 3.91.9 = 2$cm

$\therefore$ The no. of fringes $= N = \left[\frac{l}{\beta}\right] = \left[\frac{2}{0.06}\right] = 33$

$\therefore$ $11N = 33$

$\therefore$ $N = 3$

32. (8) Here,

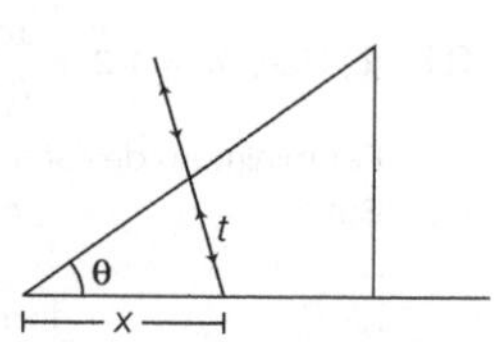

$\tan \theta = \theta = \frac{48 \times 10^{-3}}{120} = \frac{t}{x}$

$\therefore$ $t = 4 \times 10^{-4} x$

The path difference is $\Delta x = 2t + \frac{\lambda}{2}$

For maxima, $\Delta x = m\lambda$ or $2t + \frac{\lambda}{2} = m\lambda$

or $2t = (2m - 1)\frac{\lambda}{2}$ or $t = (2m - 1)\frac{\lambda}{4}$

or $\quad 4 \times 10^{-4} n = \dfrac{(2m-1)\, 6000 \times 10^{-10}}{4}$

or $\quad n = \dfrac{(2m-1)\, 6000 \times 10^{-10}}{4 \times 10^{-4}} = (2m-1)\, 1500 \times 10^{-6}$

$= (2m-1) 15 \times 10^{-4}\ m = 1.5(2m-1)$ mm

$\therefore \quad n_1 = 1.5$ mm

$n_2 = 4.5$ mm

$\therefore \quad \beta = 4.5 - 1.5 = 3$ mm

$\therefore$ No. of maxima $= \dfrac{l}{\beta} = \dfrac{120}{3} = 40$

33. Incident ray AB is partly reflected as ray 1 from the upper surface and partly reflected as ray 2 from the lower surface and the layer of thickness and refractive index $\mu_1 = 1.8$ as shown in the figure. Path difference between the two rays would by

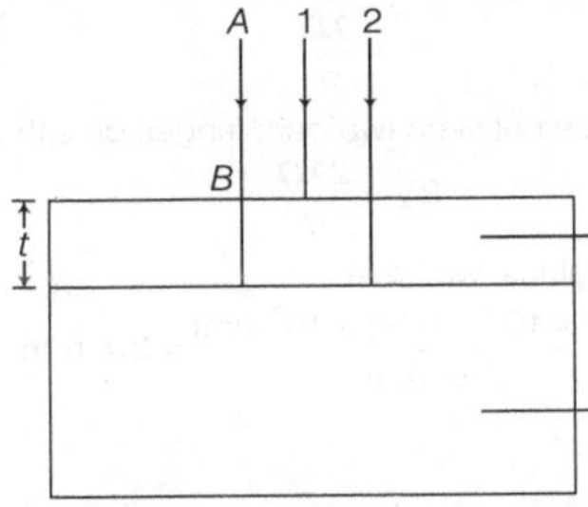

Ray 1 is reflected from a denser medium, therefore, it undergoes a phase change of π, whereas they ray 2 gets reflected from a rarer medium, therefore, there is no change in phase of ray 2. Hence, phase difference between rays 1 and 2 would be $\Delta\phi = \pi$. Therefore, condition of constructive interference will be

$$\Delta x = \left(n - \frac{1}{2}\right)\lambda \text{ where, } n = 1, 2, 3$$

or $\quad 3.6\, t = \left(n - \dfrac{1}{2}\right)\lambda$

Least value of t is corresponding to $n = 1$ or

$$t_{max} = \frac{\lambda}{2 \times 3.6} \quad \text{or} \quad t_{max} = \frac{648}{7.2} \text{ nm}$$

or $\quad t_{max} = 90$ nm

34. (i) Shape of interference fringes will be circular.

(ii) Intensity of light reaching on the screen directly from the source $I_1 = I_0$ (say) and intensity of light reaching on the screen after reflecting from the mirror is $I_2 = 36\%$ of $I_0 = 0.36 I_0$

$$\therefore \quad \frac{I_1}{I_2} = \frac{I_0}{0.36 I_0} = \frac{1}{0.36} \text{ or } \sqrt{\frac{I_1}{I_2}} = \frac{1}{0.6}$$

$$\frac{I_{min}}{I_{max}} = \frac{\left(\sqrt{\frac{I_1}{I_2}} - 1\right)^2}{\left(\sqrt{\frac{I_1}{I_2}} + 1\right)^2} = \frac{\left(\frac{1}{0.6} - 1\right)^2}{\left(\frac{1}{0.6} + 1\right)^2} = \frac{1}{16}.$$

(iii) Initially path difference at P between two waves reaching from S and S' is $2h$.

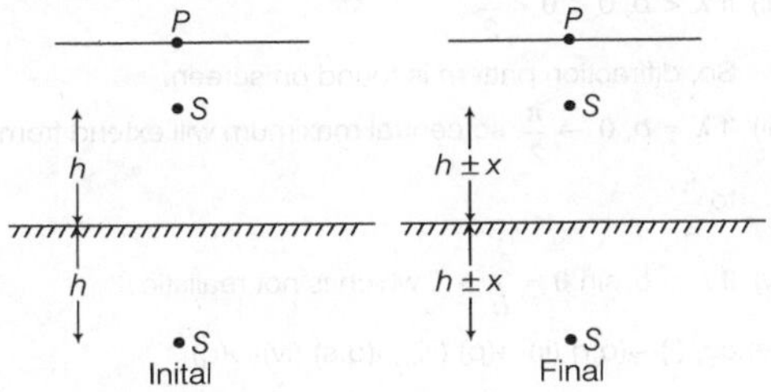

Therefore, for maximum intensity at P

$$2h = \left(n - \frac{1}{2}\right)\lambda \qquad \text{...(i)}$$

Now, let the source S is path difference will be $2h + 2x$ or $2h - 2x$. So, for displaced by x (away or towards mirror) then, maximum intensity at P.

$$2h + 2x = \left[x + 1 - \frac{1}{2}\right]\lambda \qquad \text{...(ii)}$$

$$\text{or } 2h - 2x = \left[x - 1 - \frac{1}{2}\right]\lambda \qquad \text{...(iii)}$$

Solving Eqs. (i) and (ii) or Eqs. (i) and (iii), we get

$$x = \frac{\lambda}{2} = \frac{600}{2} = 300 \text{ nm}$$

35. Each plate reflects 25% and transmits 75%

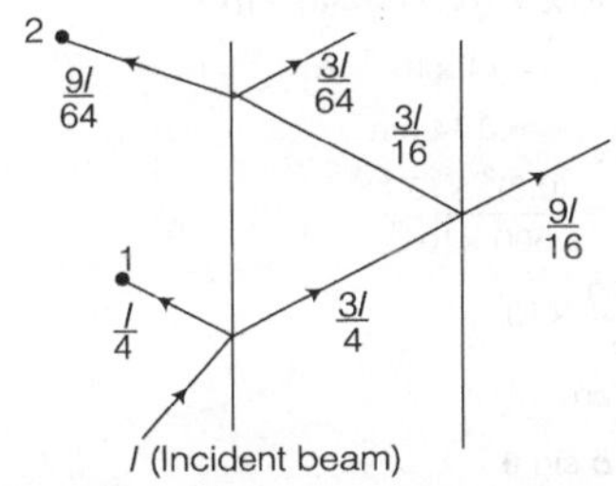

Incident beam has an intensity I. Thus beam undergoes multiple reflections and refraction. The corresponding intensity after each reflection and refraction (transmission) are shown in the figure. Interference pattern is to take place between rays 1 and 2.

$$I_1 = \frac{I}{4} \text{ and } I_2 = 9I/64$$

$$\frac{I_{min}}{I_{max}} = \left(\frac{\sqrt{I_1} - \sqrt{I_2}}{\sqrt{I_1} + \sqrt{I_2}}\right)^2 = \frac{1}{49}$$

36. Power received be aperture A,

$$= P_A = I(\pi r_A^2)$$

$$= \frac{10}{\pi}(\pi)(0.001)^2 = 10^{-5}$$

Power received by aperture B,

$$P_B = I(\pi r_B^2)$$

$$= \frac{10}{\pi}(\pi)(0.002)^2$$

$$= 4 \times 10^5 \text{ W}$$

Only 10% of P_A and P_B goes to the original direction.

Hence $\quad$ 10% of $P_A = 10^{-6} = P_1$ (say)

and $\quad$ 10% of $P_B = 4 \times 10^{-6} = P_2$ (say)

Path difference created by slab

$$\Delta x = (\mu - 1)t = (1.5 - 1)(2000) = 1000 \text{ Å}$$

Corresponding phase difference,

$$\phi = \frac{2\pi}{\lambda}\Delta x = \frac{2\pi}{6000} \times 1000 = \frac{\pi}{3}$$

Now, resultant power at the focal point

$$P = P_1 + P_2 = 2\sqrt{P_1 P_2} \cos\phi$$

$$= 10^{-6} + 4 \times 10^{-6}$$

$$= 2\sqrt{(10^{-6})(4 \times 10^{-6})} \cos\frac{\pi}{3}$$

$$= 7 \times 10^{-6} \text{ W}$$

DPP-6 Diffraction of Light

1. (d)

2. (a) Angular width of central maxima is proportional to wavelength.

$\because$ $$\lambda = \frac{h}{mv}$$

when v decreases, λ increases.

3. (c) Number of blocked half-period zone increases, when d decreases. Hence, intensity decreases.

4. (b) Fringe width $\propto \lambda$

5. (a) For diffraction, the size of obstacle must be order of wavelength.

6. (b) $r_4 \approx \sqrt{4P\lambda} = \sqrt{4 \times 1 \times 490 \times 10^{-9}}$

$= 14 \times 10^{-4}$ m

$= 0.14$ cm

7. (a) $f_1 = \frac{r_n^2}{n\lambda} = \frac{(0.3)^2 \times 10^{-6}}{500 \times 10^{-9}}$

$= \frac{0.09}{5} \times 10^3$

$= 18$ cm

8. (b) $\because$ $b \sin \theta = \lambda$

$\therefore$ $\lambda = 0.02 \times 10^{-3} \times \frac{90}{60} \times \frac{\pi}{180}$

$= 0.02 \times 10^{-3} \times 1.5 \times \frac{3.14}{180}$

$= 523.3$ nm

9. (d) Diffraction is obtained when the slit width is of the order wavelength of light (or any electromagnetic wave) used. Here, wavelength of X-rays (1-100 Å) << slit width (0.6 mm). Therefore, no diffraction pattern will be observed.

10. (d) At first minima, $b \sin \theta = \lambda$

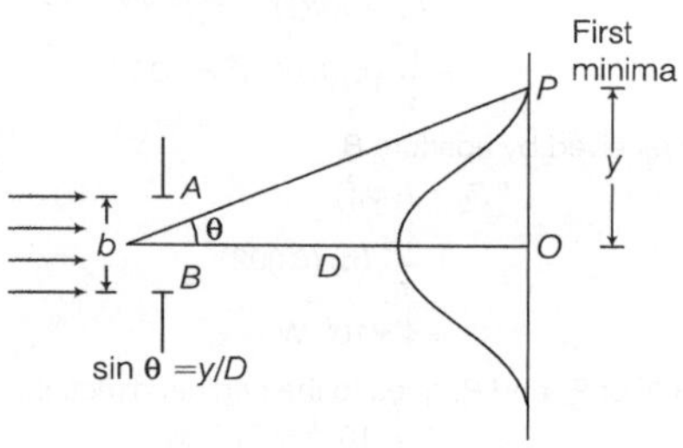

$\Rightarrow$ $b\theta = \lambda$

$\Rightarrow$ $b\left(\frac{y}{D}\right) = \lambda$

$\Rightarrow$ $y = \frac{\lambda D}{b}$

$\Rightarrow$ $\frac{yb}{D} = \lambda$ ...(i)

Now, at P (first minima) path difference between the rays reaching from two edges (A and B) will be

$\Delta x = \frac{yb}{D}$ (Compare with $\Delta x = \frac{yd}{D}$ in YDSE)

$\Rightarrow$ $\Delta x = \lambda$ [From Eq. (i)]

Corresponding phase difference (ϕ) will be

$$\phi = \left(\frac{2\pi}{\lambda}\right) \cdot \Delta x, \ \phi = \frac{2\pi}{\lambda} \cdot \lambda = 2\pi$$

11. (d) For first dark fringe on either side, $d \sin \theta = \lambda$

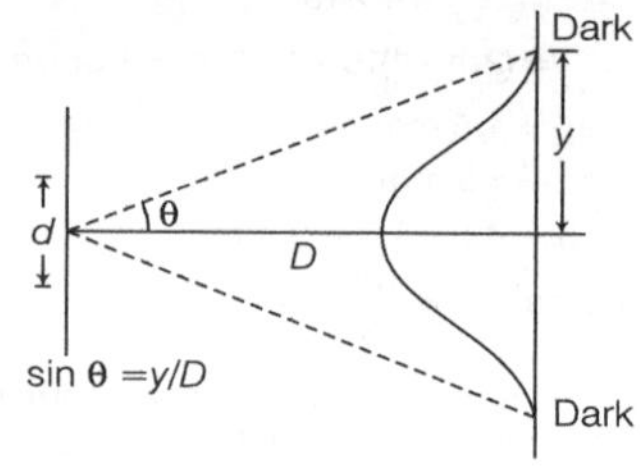

or $\frac{dy}{D} = \lambda$

$\therefore$ $y = \frac{\lambda D}{d}$

Therefore, distance between two dark fringes on either side

$$= 2y = \frac{2\lambda D}{d}$$

Substituting the values, we have

$$\text{Distance} = \frac{2(600 \times 10^{-6}\text{mm})(2 \times 10^3\text{mm})}{(1.0 \text{ mm})} = 2.4 \text{ mm}$$

12. (a,b,c)

13. (a,b,c) If we increase the slit width, the envelope of the fringe pattern changes so that its central peak is sharper. The fringe spacing which depends on slit separation does not change. Hence, less interference maxima fall within the central diffraction maximum.

14. (a) $b \sin \theta = \pm \lambda$ or $\frac{b}{2} = \lambda$

$\therefore$ $b = 2\lambda = 2 \times 500 \times 10^{-9} = 10^{-6}$ m $= 10^{-3}$ mm

15. (c) $\because$ $b \sin \theta = \lambda$

$\therefore$ $b = \frac{\lambda}{\sin 90°} \Rightarrow \lambda = 500$ nm

16. (a) (i) If $\lambda << b$, $\sin \theta = \theta \to 0$

So, spreading of light will take place. Hence, no diffraction pattern is observed on screen.

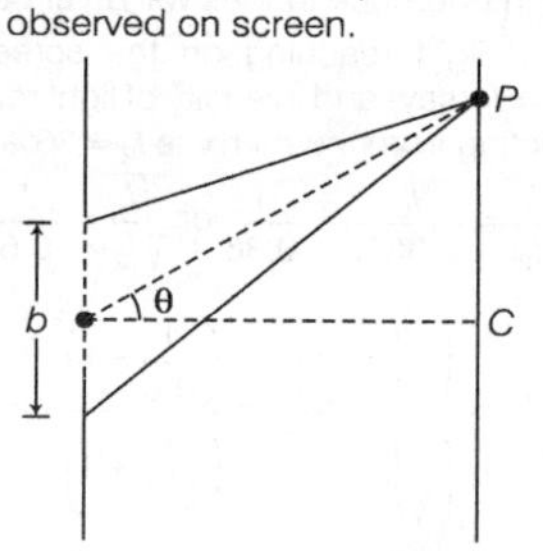

But a sharp image of slit is found on screen.

(ii) If $\lambda < b$, $0 < \theta < \frac{\pi}{2}$.

So, diffraction pattern is found on screen.

(iii) If $\lambda = b$, $\theta \to \frac{\pi}{2}$, so central maximum will extend from $-\frac{\pi}{2}$ to $\frac{\pi}{2}$.

(iv) If $\lambda > b$, $\sin \theta = \frac{\lambda}{b} > 1$, which is not realistic.

Hence, (i)→(q,r) (ii)→(p) (iii)→(q,s) (iv)→(q).

17. (b) The angular limit of resolution is equal to the angular separation between the centre of central maximum and first minimum.

18. (b) $\because b \sin = \pm \lambda$ (for central maxima), when the b decreases, θ increases. So, central maxima becomes wider.

19. (5)
$$b \sin \theta = \lambda$$
$$\therefore \quad \theta = \frac{\lambda}{b}$$
$$\therefore \quad \lambda = b\theta = b \times \frac{5 \times 10^{-3}}{2}$$
$$= \frac{0.2 \times 10^{-3} \times 5 \times 10^{-3}}{2}$$
$$= 5 \times 10^{-7} \text{ m}$$
$$\therefore \quad n = 5$$

20. (5) Here,
$$b \sin \theta = \lambda$$
$$\text{or} \quad \sin \theta = \frac{\lambda}{b}$$
$$\text{or} \quad \theta = \frac{\lambda}{b}$$
$$\text{or} \quad \frac{Y}{f} = \frac{\lambda}{b}$$
$$\therefore \quad Y = \frac{\lambda}{b} f = \frac{500 \times 10^{-9} \times 60 \times 10^{-2}}{0.2 \times 10^{-3}}$$
$$= 15 \times 10^{-4} \text{ m} = 15 \times 10^{-2} \text{ cm}$$
$$\therefore \quad x = 5$$

21. (1)
$$f = \frac{r_n^2}{n\lambda}$$
$$= \frac{\left(\frac{\sqrt{2}}{2} \times 10^{-3}\right)^2}{500 \times 10^{-9}}$$
$$= \frac{2}{4 \times 500} \times 10^3$$
$$= 1 \text{ m}$$

22. (5) $\frac{1}{v} + \frac{1}{u} = \frac{n\lambda}{r_n^2} = \frac{1}{f}$
$$\text{or} \quad \frac{1}{20} + \frac{1}{20} = \frac{1}{f}$$
$$\therefore \quad f = 10 \text{ cm}$$
$$\therefore \quad n = \frac{r_n^2}{f\lambda} = 2500$$
$$\therefore \quad x = 5$$

23. (i) Given, $\lambda = 6000$ Å

Let b be the width of slit and D the distance between screen and slit.

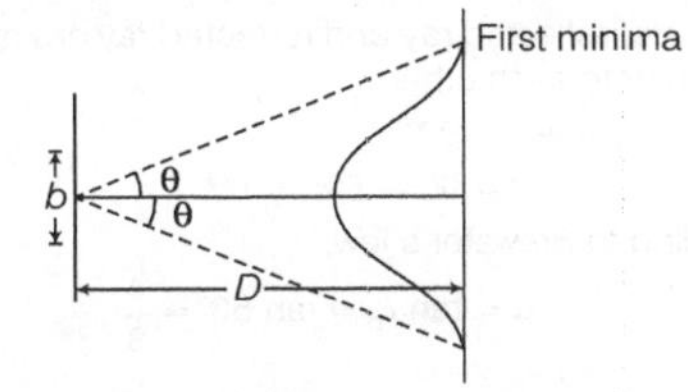

First minima is obtained at $b \sin \theta = \lambda$

or $\quad b\theta = \lambda$ as $\sin \theta = \theta$ or $\theta = \frac{\lambda}{b}$

Angular width of first maxima $= 2\theta = \frac{2\lambda}{b} = \lambda$

Angular width will decrease by 30%, when λ is decreased by 30%.

Therefore, new wavelength
$$\lambda' = \left\{(6000) - \left(\frac{30}{100}\right)6000\right\} \text{ Å}$$
$$\lambda' = 4200 \text{ Å}$$

(ii) When the apparatus is immersed in a liquid of refractive index μ, the wavelength is decreased μ times. Therefore,
$$4200 \text{ Å} = \frac{6000 \text{ Å}}{\mu}$$
$$\mu = \frac{6000}{4200} \Rightarrow \mu = 1.429 \approx 1.43$$

DPP-7 **Polarisation of Light**

1. (d) Polarisation takes place in transverse wave, but not in sound wave.

2. (b)
$$I = I_0 \cos^2 45°$$
$$kA'^2 = kA^2 \cos^2 45°$$
$$A'^2 = \frac{A^2}{2}$$
$$\therefore \quad A' = \frac{A}{\sqrt{2}}$$

3. (b) $B_0 = E_0 / c$, E_0 and B_0 should be mutually perpendicular.

4. (b) Here, $\tan \theta = \frac{E_2}{E_1}$ = constant. Thus, wave is plane polarised.

5. (a) All the vibrations of an unpolarised light at a given instant can be resolved in two mutually perpendicular direction.

6. (a) If incident light is unpolarised, then as vibrations are equally probable in all directions.
$$\therefore \quad \langle I \rangle = \langle I_0 \cos^2 \theta \rangle$$
$$= I_0 \langle \cos^2 \theta \rangle = \frac{I_0}{2}$$

7. (d) By law of Malus,
$$I = I_0 \cos^2 \theta$$

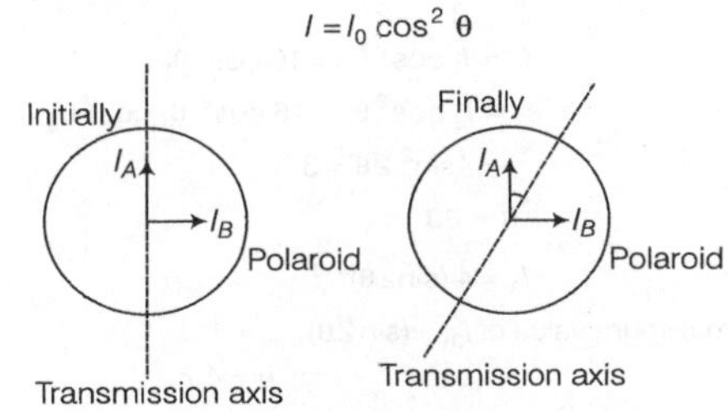

Now,
$$I_{A'} = I_A \cos^2 30°$$
$$I_B' = I_B \cos^2 60°$$
As,
$$I_A' = I_B'$$
$$I_A \cos^2 30° = I_B \cos^2 60°$$
$$\Rightarrow \quad I_A\left(\frac{3}{4}\right) = I_B\left(\frac{1}{4}\right)$$
$$\Rightarrow \quad \frac{I_A}{I_B} = \frac{1}{3}$$

8. (c) Relation between intensities is

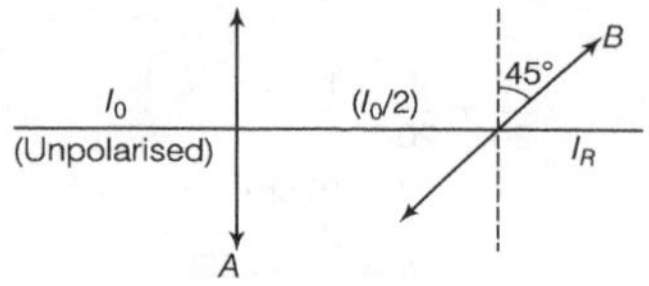

$$I_R = \left(\frac{I_0}{2}\right)\cos^2(45°) = \frac{I_0}{2} \times \frac{1}{2} = \frac{I_0}{4}$$

9. (d)

10. (d) Longitudinal wave never be polarised.

11. (a,d) (a) Here, reflected ray and refracted ray are mutually perpendicular to each other.

$\therefore \quad i + r = 90°$

$\therefore \quad r = 90° - 53° = 37°$

(d) According to Brewster's law,

$$\mu = \tan i_p = \tan 53° = \frac{4}{3}$$

12. (b) If at any instant, the axis of polariser makes angle θ with the direction of vibration of incident light.

From Law of Malus,

$$I = I_0 \cos^2 \theta$$

As polariser is rotating, all values of θ are possible.

$$\therefore \quad \langle I \rangle = \langle I_0 \cos^2 \theta \rangle = \frac{I_0}{2} = \frac{\text{Power}}{2 \times \text{area}}$$

$$= \frac{10^{-3}}{3 \times 10^{-4} \times 2} = \frac{5}{3}\ \text{Wm}^{-2}$$

13. (a) Energy of light passing through the polariser per revolution

$$= \langle I \rangle \text{ area} \times \frac{2\pi}{\omega}$$

$$= \frac{5}{3} \times 3 \times 10^{-4} \times \frac{1}{5} = 10^{-4}\ \text{J/rev}$$

14. (a) If θ_1 is the angle between the transmission axes of first polaroid and second polaroid while θ_2 between the transmission axes of second polaroid and third polaroid.

$\because \quad \theta_1 + \theta_2 = 90°$

$\therefore \quad \theta_2 = 90° - \theta_1$

Now, if I_0 is intensity of unpolarised light incident on first polaroid, the intensity of transmission light from it is

$$I_1 = \frac{I_0}{2} = \frac{32}{2} = 16\ \text{Wm}^{-2}$$

But $\quad I_2 = I_1 \cos^2 \theta_1 = 16 \cos^2 \theta_1$

and $\quad I_3 = I_2 \cos^2 \theta_2 = 16 \cos^2 \theta_1 \cos^2 \theta_2$

$= 4\sin^2 2\theta = 3$

$\Rightarrow \quad \theta = 30°$

15. (b) $\because \quad I_3 = 4(\sin 2\theta)^2$

For maximum value of I_3, $(\sin 2\theta)_{max} = 1$

$\therefore \quad \theta = 45°$

16. (c) The transmitted intensity the unpolarised light will be constant for all the orientations, of the polarised sheet. But intensity of polarised light will be found by law of Malus,

$$I_P' = I_P \cos^2 \theta$$

For $\theta = \frac{\pi}{2}$, $\quad I_P' = 0$

For $\theta = 0$, $\quad I_P' = I_P$

But for all orientation, $\quad I_0' = \frac{I_0}{2}$

$\therefore \quad I_{max} = I_P + \frac{I_0}{2}$ and $I_{min} = \frac{I_0}{2}$

$$\frac{I_{max}}{I_{min}} = 4 \quad \text{or} \quad \frac{I_P + \frac{I_0}{2}}{\frac{I_0}{2}} = 4$$

or $\quad I_P = 2I_0 - \frac{I_0}{2} = \frac{3I_0}{2}$

$\therefore \quad \frac{I_P}{I_0} = \frac{3}{2}$

17. (d) $I = I_P \cos^2 45° + \frac{I_0}{2} = \frac{3I_0}{2} \times \frac{1}{2} + \frac{I_0}{2} = \frac{5}{4} I_0$

18. (a) (i) $\tan \theta = \frac{E_z}{E_y} = \frac{E_2}{E_1}$ = constant and electric field is always along a fixed direction.

Then, light is linearly or plane polarised.

(ii) same as (i)

(iii) $\tan \theta = \frac{E_z}{E_y} = \frac{E_2 \sin(\omega t - kx)}{E_1 \sin\left(\omega t - kx + \frac{\pi}{2}\right)} = \tan(\omega t - kx)$

$\Rightarrow \quad \theta = (\omega t - kx)$

Thus, electric field rotates with constant angular velocity. Thus, light is circularly polarised.

(iv) If E_1, E_2, then light is elliptically polarised.

Hence, (i)→(r,s) (ii)→(r,s) (iii)→(p) (iv)→(q).

19. (b) Coloured or dark glasses reduce glare by absorbing incident light. This is turn affects the visibility of object adversely. However, the glare results due to horizontally polarised light produced by reflection from near by smooth surface such as water or table top is avoided by polariods without appreciable decrease in intensity of light scattered from object.

20. (a) According to Brewster's law,

$$\tan \theta_0 = \frac{\mu_2}{\mu_1}, \mu_2 < \mu_1$$

When light travels in such a medium,

$$\sin C = \frac{\mu_2}{\mu_1}, \text{ where } \mu_2 < \mu_1$$

As $|\tan \theta_0| > \sin C$ for large angles $\theta_0 < C$.

Thus, polarisation by reflection will definitely occur.

21. (b)

22. (3) According to Brewster's law,

$$\tan \theta = \mu = \sqrt{3}$$

$$\theta = \frac{\pi}{3}$$

$\therefore \quad n = 3$

23. (2) $\because$ Path difference $= \Delta x = \mu_o t - \mu_e t$

$\therefore \quad \phi = \frac{2\pi}{\lambda} \Delta x$

or $\quad \frac{6\pi}{10^n} = \frac{2\pi}{600 \times 10^{-9}} (\mu_o - \mu_e) t$

Putting the values, we get

$n = 2$

24. (7) Here, $\frac{\lambda}{4} = (\mu_o - \mu_e)t$

$\therefore \quad t = \frac{\lambda}{4(\mu_o - \mu_e)t}$

$$= \frac{400 \times 10^{-9}}{4(1.5632 - 1.5541)}$$

$$= \frac{0.1099}{10^7}\ \text{m}$$

$\therefore \quad n = 7$

Revisal Problems (JEE Main)

1. (c) $\beta = \frac{D}{d}\lambda = \frac{1}{0.2 \times 10^{-3}} \times 500 \times 10^{-9} = 2.5 \times 10^{-3}$ m $= 2.5$ mm

$\therefore$ The distance between 12th minima and 8th maxima.

$= 11.5\,\beta - 8\,\beta = 3.5\,\beta = 3.5 \times 2.5$ mm $= 8.75$ mm

2. (b) $\phi = \frac{2\pi}{\lambda}\Delta x = \frac{2\pi}{\lambda} \times \frac{\lambda}{4} = \frac{\pi}{2}$

$\therefore \quad I = 4I_0 \cos^2 \frac{\theta}{2} = 4I_0 \cos^2 \frac{\pi}{4} = 4I_0 \times \frac{1}{2} = 2I_0$

3. (c) $I_1 = 81\,k,\ I_2 = k$

$\therefore \quad \frac{I_{max}}{I_{min}} = \frac{(\sqrt{I_1} + \sqrt{I_2})^2}{(\sqrt{I_1} - \sqrt{I_2})^2} = \frac{(\sqrt{81k} + \sqrt{k})^2}{(\sqrt{81k} - \sqrt{k})^2} = \frac{100}{64} = \frac{25}{16}$

4. (a) $I = I_0 \cos^2 \frac{\theta}{2} \Rightarrow \frac{I_0}{4} = I_0 \cos^2 \frac{\theta}{2}$ or $\cos \frac{\theta}{2} = \frac{1}{2}$

$\therefore \quad \phi = \frac{2\pi}{3} = \frac{2\pi}{3}\Delta x$ or $\Delta x = \frac{\lambda}{3}$ or $Y\frac{d}{D} = \lambda/3$

$\therefore \quad Y = \frac{D\lambda}{3d}$

5. (b) $I = I_1 + I_2 + 2\sqrt{I_1 I_2}\cos\phi = I_1 + I_2 + 2\sqrt{I_1 I_2}\cos\left(\frac{2\pi}{\lambda}\right)\frac{xd}{D}$

$= (\sqrt{I_1} - \sqrt{I_2})^2 + 4\sqrt{II_2}\cos^2\left(\frac{2\pi}{\lambda}\right)\frac{xd}{D} = A + B\cos^2\frac{kx}{2}$

6. (b) If interferring waves are polarised, they must be in the same state of polarisation, otherwise instead of interference, unpolarised light will result. Hence, option (b) is correct.

7. (a) If the angle of biprism increased, the distance between virtual sources would be increase because $d = 2a(\mu - 1)A$. Due to this, fringe width decreases. So, the fringes will not be separately visible and may be disappear ultimately.

8. (c) $\Delta x = \frac{yd}{D} = m_1\lambda_1 = m_2\lambda_2$

$\therefore \quad \frac{m_1}{m_2} = \frac{\lambda_1}{\lambda_2} = \frac{480}{600} = \frac{48}{60} \Rightarrow \frac{12}{m_2} = \frac{48}{60}$

$\therefore \quad m_2 = \frac{12 \times 60}{48} = 15$ th

9. (a) Here, path difference is $\Delta x = \left(2\mu t + \frac{\lambda}{2} - \frac{\lambda}{2}\right) = 2\mu t$

For minimum intensity, $\Delta x = (2m - 1)\frac{\lambda}{2}$

or $\quad 2\mu t = (2m - 1)\frac{\lambda}{2}$

$\therefore \quad t_{min} = \frac{\lambda}{4\mu} = \frac{480 \times 10^{-9}}{4 \times 1.2} = 10^{-7}$m

10. (a) For constructive interference,

$2\mu t = (2m + 1)\frac{\lambda}{2},\ m = 0, 1, 2, \cdots$

$\therefore \quad t_{min} = \frac{\lambda}{4\mu} = \frac{400 \times 10^{-9}}{4 \times 1.4} = 714.3$ Å

11. (c) $D_n^2 = 4n\lambda R$ or $(0.5 \times 10^{-2})^2 = 4 \times 10 \times 5.9 \times 10^{-7}\,R$

$\therefore \quad R = 1.06$ m

But $\quad 2t = \frac{D^2}{4R}$

$\therefore \quad t = \frac{(0.5 \times 10^{-2})}{8 \times 1.06} = 3 \times 10^{-6}$ m

12. (c) According to Brewster's law.

$\frac{\mu_g}{\mu_w} = \tan\theta = \sqrt{3}$

$\therefore \quad \mu_g = \sqrt{3}.\sqrt{2} = \sqrt{6}$

13. (c) Shifting $= \frac{(\mu - 1)t}{\lambda}\beta$

$7\beta = \frac{(\mu - 1)t\beta}{\lambda}$

$\therefore \quad \lambda = \frac{(\mu - 1)t}{7} = \frac{(1.6 - 1) \times 7 \times 10^{-6}}{7}$

$= 600 \times 10^{-9}$ m $= 600$ nm.

14. (b) $\beta = \frac{D}{d}\lambda$

and $\quad d = \sqrt{d_1 \cdot d_2} = \sqrt{4.5 \times 2.9} = 3.6$ mm

$\therefore \quad \lambda = \frac{\beta d}{D} = \frac{0.36 \times 10^{-2} \times 0.0326 \times 10^{-2}}{200 \times 10^{-2}}$

$= 5.87 \times 10^{-7}$ m $= 5870$ Å

15. (d) Let N fringes are observed in field of view of length l.

$\therefore \quad \beta = \frac{l}{N} = \frac{D}{d}\lambda$

$\therefore \quad N\lambda =$ constant

$N_1\lambda_1 = N_2\lambda_2$

$\therefore \quad N_2 = \frac{N_1\lambda_1}{\lambda_2} = \frac{62 \times 5893}{5461} = 67$

16. (b) The path difference at line of contact is $\Delta x = \frac{\lambda}{2}$. Hence, it appears as dark.

17. (a) If white light is used, the central bright fringe is white, but other bright fringes are coloured.

18. (a) $c = \frac{E_0}{B_0}$

$\therefore \quad B_0 = \frac{18}{3 \times 10^8} = 6 \times 10^{-8}$ T

19. (b) $u = \frac{1}{2}\varepsilon_0 E_0^2 = \frac{1}{2} \times 8.85 \times 10^{-12} \times 50^2 = 22.13 \times 10^{-9}$ Jm^{-3}

20. (b) $E_0 = nhf$ and angular momentum, $L = \frac{nh}{2\pi} = \frac{E_0}{2\pi f}$

21. (a) They behave as incoherent sources.

22. (b) Huygens' principle is applicable for any type of wave.

23. (a) Sound wave wavelengths are from 15 m to 15 mm. Diffraction effect are seen, if $b \sim \lambda$.

24. (c) **Statement I** Here, $D_n = \sqrt{2(2n - 1)\lambda R}$

$\therefore \quad D_n \propto \sqrt{2n - 1}$

Statement II At contact point, path difference is $\frac{\lambda}{2}$. Hence, the contact point is not a point of maxima.

25. (b) In the case of interference, energy remains conserved. In this case, energy from the region of dark fringe is accumulated in the region of bright fringe.

26. (b) In this case, light beams are not interfered at central point of screen.

27. (b) In the biprism experiment, the wavefront emerging from narrow slit S_0 is divided in width by the biprism. After refraction, one part of the wavefront appears to diverge from S_1 and other from S_2. The two coherent sources S_1 and S_2 have a definite relative position and the interference fringes can be obtained anywhere in the region which permits both the coherent sources to be seen. These non-localised fringes have a good contrast. An extended source would be equivalent to a number of adjacent narrow slits, each producing its own set of fringes. This results in a poor contrast due to overlapping.

Revisal Problems (JEE Advanced)

1. (a) Intensity of reflected light from upper surface is

$$I_1 = I_0 \times \frac{25}{100} = \frac{I_0}{4}$$

and intensity of transmitted light from upper surface is

$$I' = I_0 - \frac{I_0}{4} = \frac{3I_0}{4}$$

The intensity of reflected light from lower surface is

$$I_2 = \frac{3I_0}{4} \times \frac{50}{100} = \frac{3I_0}{8}$$

$$\therefore \quad \frac{I_{min}}{I_{max}} = \frac{(\sqrt{I_1} - \sqrt{I_2})^2}{(\sqrt{I_1} + \sqrt{I_2})^2} = \frac{\left(\sqrt{\frac{I_0}{4}} - \sqrt{\frac{3I_0}{8}}\right)^2}{\left(\sqrt{\frac{I_0}{4}} + \sqrt{\frac{3I_0}{8}}\right)^2} = \frac{\left(\frac{1}{2} - \sqrt{\frac{3}{8}}\right)^2}{\left(\frac{1}{2} + \sqrt{\frac{3}{8}}\right)^2}$$

2. (a) $\quad 30\,\beta = \frac{(\mu - 1)t\,\beta}{\lambda}$

$\therefore \quad \mu - 1 = \frac{30\lambda}{t} = 1.5$, for $\lambda = 600$ nm

3. (d) The required angular width is $\phi = \frac{\lambda}{d}$ or $\frac{b}{D} = \frac{\lambda}{d}$

$$\therefore \quad b = \frac{600 \times 10^{-9}}{0.1 \times 10^{-3}} \times 2 = 12 \times 10^{-3} \text{ m} = 12 \text{ mm}$$

4. (b) The condition for spatial coherence is that the angular width of the slit at either point source be less than angular fringe width.

$$\phi = \frac{\lambda}{d} \quad \text{or} \quad \frac{b}{D} = \frac{\lambda}{d} \quad \text{or} \quad \frac{D}{b} = \frac{d}{\lambda}$$

or $\quad D = \frac{bd}{\lambda} = \frac{1.2 \times 10^{-3} \times 0.1 \times 10^{-3}}{600 \times 10^{-9}} = 0.2 \text{ m}$

5. (b) Here, $n = \frac{dD}{\frac{1.22\,\lambda}{\mu}} = \frac{1.5 \times 7 \times 10^{-3} \times 1.36}{1.22 \times 550 \times 10^{-9}}$

$= 21.28 \times 10^3 \text{ m} = 21.28$ km

6. (d) Angular position in first order is

$$\theta = \sin^{-1}\left(\frac{\lambda}{d}\right) = \sin^{-1}(5500 \times 10^2 \lambda)$$

The position of this wavelength on screen is $y = L\tan\theta$

For $\lambda = 630$ nm, $y = 0.37$ m

For $\lambda = 490$ nm, $y = 0.29$ m

The separation of colours in first order

$= 0.37 - 0.29 = 0.08 \text{ m} = 8$ cm

7. (a) $\Delta x = 2t + \frac{\lambda}{2} = m\lambda$ or $2t = m\lambda - \frac{\lambda}{2} = (2m - 1)\frac{\lambda}{2}$

or $\quad t = (2m - 1)\frac{\lambda}{4}$ or $\theta = \frac{t}{x} = \frac{d}{L}$

$\therefore \quad t = \frac{xd}{L}$

or $\quad \frac{xd}{L} = (2m - 1)\frac{\lambda}{4}$ or $x = (2m - 1)\frac{L\lambda}{4d}$

$\therefore \quad x_1 = \frac{L\lambda}{4d}$ and $x_2 = \frac{3L\lambda}{4d}$

$\therefore \quad \beta = x_2 - x_1 = \frac{L\lambda}{2d}$

$\therefore \quad N = \left[\frac{L}{\beta}\right]$ or $125 = \frac{L}{\frac{L\lambda}{2d}} = \frac{2d}{\lambda}$

$\therefore \quad d = \frac{125\lambda}{2} = \frac{125}{2} \times 500 \times 10^{-9} = 31.25 \times 10^{-6}$ m

$\therefore \quad R = \frac{d}{2} = 15.625 \times 10^{-6}$ m

8. (b) Here, $\Delta x = d\sin\theta = (2m + 1)\frac{\lambda}{2}$

But $\quad 1 < \sin\theta < -1$ or $1 > (2m + 1)\frac{\lambda}{2d} > -1$

or $\quad \frac{2d}{\lambda} > 2m + 1 > -\frac{2d}{\lambda}$

or $\quad \frac{2 \times 0.1 \times 10^{-3}}{400 \times 10^{-9}} > 2m + 1 > \frac{-2 \times 0.1 \times 10^{-3}}{400 \times 10^{-9}}$

or $\quad 500 > 2m + 1 > -500$ or $499 > 2m > -501$

or $\quad 249.5 > m > -250.5$

$\therefore$ Number of minima = 500

9. (a,b,c) (a) At point 'O' path difference is $\frac{\lambda}{2}$.

So, dark fringe is formed.

(b) In this case, sources are broad.

So, the shapes of fringes are straight lines.

(c) Here, path difference, $\Delta x = 2\mu t - \frac{\lambda}{2}$

$\therefore \quad 2\mu t - \frac{\lambda}{2} = m\lambda$

or $\quad 2\mu t = (2m + 1)\frac{\lambda}{2}$ or $2\mu x\theta = (2m + 1)\frac{\lambda}{2}$

or $\quad x - (2m + 1)\frac{\lambda}{4\mu\theta}$

or $\quad x_1 = \frac{\lambda}{4\mu\theta}, m = 0$ or $x_2 = \frac{3\lambda}{4\mu\theta}, m = 1$

$\therefore$ Spacing between fringes $= x_2 - x_1 = \frac{\lambda}{2\mu 0}$

10. (a,b,c) (a) At central point O, path difference is zero. Hence, it is a bright spot.

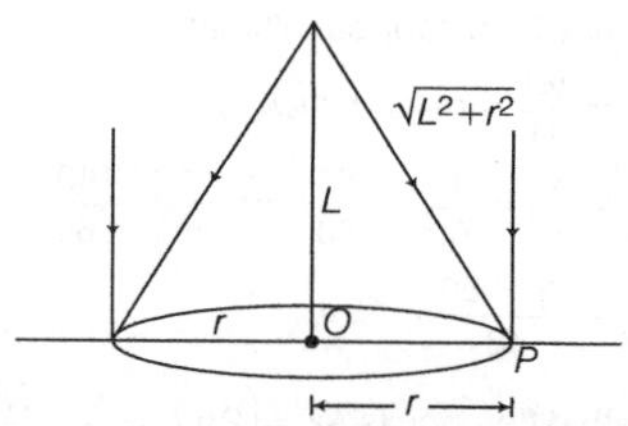

(b) Here, path difference on a circle of radius r is constant. Hence, shape of fringes are circular.

(c) $\Delta x = \sqrt{L^2 + r^2} - L = m\lambda \Rightarrow \sqrt{L^2 + r^2} = m\lambda + L$

$\Rightarrow \quad L^2 + r^2 = m^2\lambda^2 + L^2 + 2m\lambda L$

$\Rightarrow \quad r = \sqrt{2m\lambda L}$ [$\because m^2\lambda^2$ may be neglected]

$\Rightarrow \quad r_n = \sqrt{2n\lambda L}$

$\Rightarrow \quad r_{n-1} = \sqrt{2(n-1)\lambda L}$

$\therefore \quad$ Spacing $= \sqrt{2\lambda L}(\sqrt{n} - \sqrt{n-1})$

11. (a,c) (a) Here, $\Delta x = \frac{yd}{D}$ and $\Delta x' = \frac{y'd}{D}$

$y' - y = a$

$\therefore \quad \delta(\Delta x) = \Delta x' - \Delta x = \frac{(y' - y)d}{D} = \frac{ad}{D}$

$\therefore \quad \delta(\Delta\phi) = \frac{2\pi}{\lambda} \cdot \frac{ad}{D}$

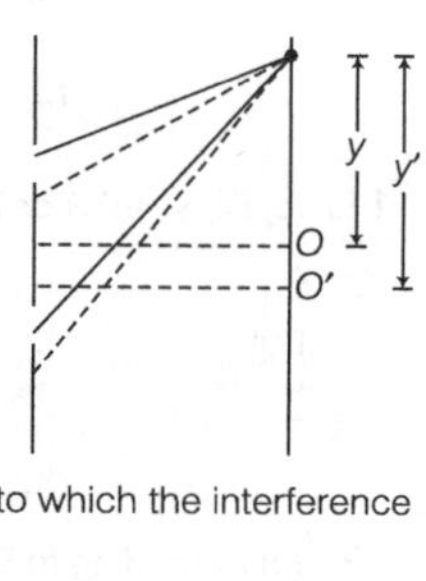

(c) If $\delta(\Delta x) = \frac{\lambda}{2}$, the maximum from top edges will be superimposed on the minimum from the bottom edges, owing to which the interference pattern will disappear completely,

$$\frac{\lambda}{2} = \frac{ad}{D}$$

$$\therefore \quad a = \frac{\lambda D}{2d}$$

12. (a,b,d)

(a) In the direction, $\theta = \pi$

$$\phi = \frac{2\pi}{\lambda}d - \phi_0$$

For maxima, $\frac{2\pi}{\lambda}d - \phi_0 = 2m\pi$

$$\therefore \quad \frac{2\pi}{\lambda}d = 2m\pi + \phi_0 \qquad \ldots\text{(i)}$$

(b) For $\theta = 0°$, $\phi = \frac{2\pi}{\lambda}d + \phi_0$

For minima, $\frac{2\pi}{\lambda} + \phi_0 = (2m + 1)\pi \qquad \ldots\text{(ii)}$

(d) From Eqs. (i) and (ii), we get

$$\phi_0 = \frac{\pi}{2} \text{ and } \frac{d}{\lambda} = \left(m + \frac{1}{4}\right)$$

where, $m = 0, 1, 2, 3, \ldots$

13. (b,d) $\frac{I_{max}}{I_{min}} = \frac{(\sqrt{I_1} + \sqrt{I_2})^2}{(\sqrt{I_1} - \sqrt{I_2})^2} = \left(\frac{\sqrt{I_1/I_2} + 1}{\sqrt{I_1/I_2} - 1}\right)^2 = 9$ (given)

Solving this, we have $\frac{I_1}{I_2} = 4$

but $I \propto A^2$

$$\therefore \quad \frac{A_1}{A_2} = 2$$

Hence, option (b) and (d) are correct.

14. (a,b,c) Fringe width, $\beta = \frac{\lambda D}{d}$ or $\beta \propto \lambda$

$\therefore \quad \lambda_2 > \lambda_1$

So, $\beta_2 > \beta_1$

Number of fringes in a given width,

$$m = \frac{y}{\beta} \text{ or } m \propto \frac{1}{\beta}$$

$\Rightarrow \quad m_2 < m_1$ and $\beta_2 > \beta_1$

Distance of 3rd maximum of λ_2 from central maximum

$$= \frac{3\lambda_2 D}{d} = \frac{1800D}{d}$$

Distance of 5th minimum of λ_1 from central maximum

$$= \frac{9\lambda_1 D}{2d} = \frac{1800D}{d}$$

So, 3rd maximum of λ_2 will overlap with 5th minimum of λ_1.

Angular separation (or angular fringe width)

$$= \frac{\lambda}{d} \propto \lambda$$

$\Rightarrow$ Angular separation for λ_1 will be lesser.

15. (a,b,c) The intensity of light is $I(\theta) = I_0\cos^2\left(\frac{\delta}{2}\right)$

where, $\delta = \frac{2\pi}{\lambda}(\Delta x) = \left(\frac{2\pi}{\lambda}\right)(d\sin\theta)$

(a) For $\theta = 30°$, $\lambda = \frac{c}{v} = \frac{3 \times 10^8}{10^6} = 300$ m and $d = 150$ m

$$\delta = \left(\frac{2\pi}{300}\right)(150)\left(\frac{1}{2}\right) = \frac{\pi}{2}$$

$$\therefore \quad \frac{\delta}{2} = \frac{\pi}{4}$$

$$\therefore \quad I(\theta) = I_0\cos^2\left(\frac{\pi}{4}\right) = \frac{I_0}{2}$$

(b) For $\theta = 90°$, $\delta = \left(\frac{2\pi}{300}\right)(150)(1) = \pi$

or $\frac{\delta}{2} = \frac{\pi}{2}$ and $I(\theta) = 0$

(c) For $\theta = 0°$, $\delta = 0$ or $\frac{\delta}{2} = 0$

$$\therefore \quad I(\theta) = I_0$$

16. (a,b) For $d = \lambda$, there will be only one, central maxima.

For $\lambda < d < 2\lambda$, there will be three maximas on the screen corresponding to path difference, $\Delta x = 0$ and $\Delta x = \pm\lambda$.

Hence, correct options are (a) and (b).

17. (a) According to Fermat's principle, if light reflects from plane surface, it follows a path of least time.

18. (a) According to Fermat's principle, if light refracts from plane surface, it follows a path of least time.

19. (b) According to Fermat's principle, if light reflects from curved path it follows a path of maximum time.

20. (b)

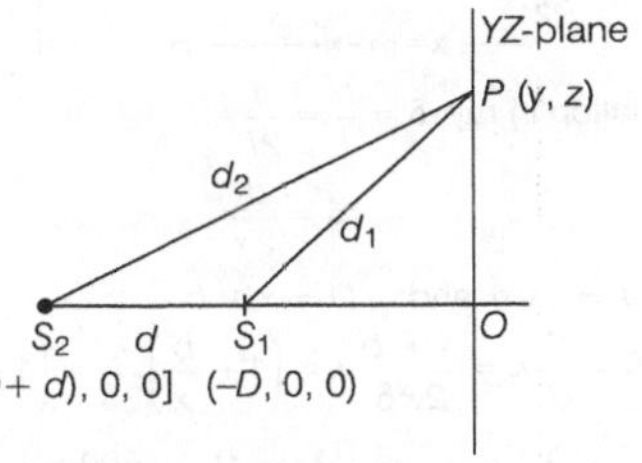

Here, path difference at point P is

$$\Delta x = d_2 - d_1 = i\lambda$$

or $\sqrt{(D + d)^2 + y^2 + z^2} - \sqrt{D^2 + y^2 + z^2} = i\lambda$

or $(D + d)^2 + y^2 + z^2 = i^2\lambda^2 + D^2 + y^2 + z^2 + 2i\lambda\sqrt{D^2 + y^2 + z^2}$

or $(D + d)^2 - i^2\lambda^2 - D^2 = 2i\lambda\sqrt{D^2 + y^2 + z^2}$

$$\frac{(2D + d)d - i^2\lambda^2}{2i\lambda} = \sqrt{D^2 + y^2 + z^2}$$

$$\therefore \quad y^2 + z^2 = \frac{(2Dd + d^2 - i^2\lambda^2)^2}{2i^2\lambda^2} - D^2$$

Thus, the locus is a circle whose centre is at O.

21. (a) For nearest bright band,

$\Delta x = (n - 1)\lambda$

$\Delta x = d_2 - d_1 = (n - 1)\lambda$

or $\sqrt{(D + d)^2 + b^2} - \sqrt{D^2 + b^2} = (n - 1)\lambda$

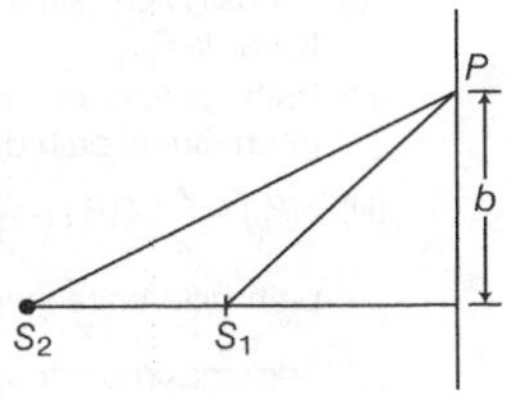

But $d = n\lambda >> \lambda$

and $D >> d$

$$b \approx \sqrt{\frac{2D(D + n\lambda)}{n}}$$

22. (b) For ith ring

$$\Delta x = d_2 - d_1$$

$$= \sqrt{(2n\lambda)^2 + r_i^2} - \sqrt{(n\lambda)^2 + r_i^2} = i\lambda$$

$$\therefore \quad r_i = \frac{\lambda}{2i}\sqrt{(9n^2 - i^2)(n^2 - i^2)}$$

23. (a) Wavefronts are parallel in both media. Therefore, light which is perpendicular to wavefront travels as a parallel beam in each medium.

24. (c) All points on a wavefront are at the same phase.

$\therefore \quad \phi_d = \phi_c$

and $\phi_f = \phi_e$

$\therefore \quad \phi_d - \phi_f = \phi_c - \phi_e$

25. (b) In medium-2 wavefront bends away from the normal after refraction. Therefore, ray of light which is perpendicular to wavefront bends towards the normal in medium-2 during refraction. So, medium-2 is denser or its speed in medium-1 is more.

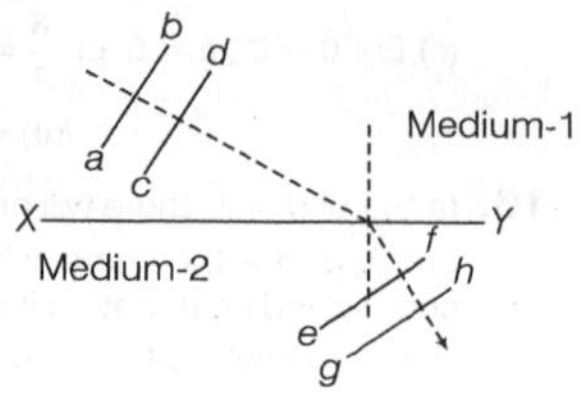

26. (a)

(i) Source is at focus, So, images will be at infinity.

(ii)

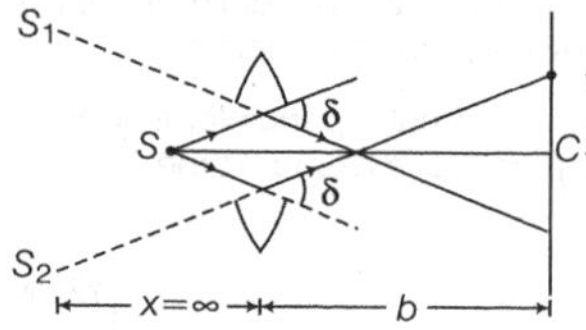

Deviation in ray, $\delta = \frac{h}{f} = \frac{a}{2f}$

$\therefore \quad \delta = \frac{d}{2x}$

$\therefore \quad d = 2x\delta$ and $D = x + b$

$\therefore \quad \beta = \frac{D}{d}\lambda = \frac{x+b}{2x\delta}\lambda = \left(1 + \frac{b}{x}\right)\frac{\lambda}{2\delta} = \left(1 + \frac{b}{\delta}\right)\frac{\lambda}{2\delta}$

$\therefore \quad \beta = \frac{\lambda}{2\delta}$ or $\beta = \frac{\lambda}{\frac{2a}{2f}} = \frac{f\lambda}{a} = \frac{600 \times 10^{-9} \times 0.25}{0.001}$

$= 1.5 \times 10^{-4}$ m $= 0.15$ mm

(iii) Here, $\frac{2l}{d} = \frac{b}{x}$

$\therefore \quad 2l = \frac{ab}{f} = \frac{0.001 \times 0.5}{0.25} = 2 \times 10^{-3}$ m $= 2$ mm

$\therefore$ Length of fringe pattern = 2 mm.

(iv) $\therefore N = 2\left[\frac{l}{\beta}\right] + 1 = 2\left[\frac{10^{-3}}{0.15 \times 10^{-3}}\right] + 1 = 2 \times 6 + 1 = 13$

27. (b)

(i) Intensity at P_0 is maximum. It will continuously decrease from P_0 towards P_2.

(ii) Path difference due to slap will be compensated by geometrical path difference. Hence, $\delta(P_1) = 0$.

(iii) $\delta(P_0) = \frac{\lambda}{2}$, $\delta(P_1) = \frac{\lambda}{2} - \frac{\lambda}{4} = \frac{\lambda}{4}$ and $\delta(P_2) = \frac{\lambda}{2} - \frac{\lambda}{3} = \frac{\lambda}{6}$. When path difference increases from 0 to $\frac{\lambda}{2}$, intensity will decrease from maximum to zero. Hence, in this case,

$$I(P_2) > I(P_1) > I(P_0)$$

(iv) $\delta(P_0) = \frac{3\lambda}{4}$, $\delta(P_1) = \frac{3\lambda}{4} - \frac{\lambda}{4} = \frac{\lambda}{2}$ and $\delta(P_2) = \frac{3\lambda}{4} - \frac{\lambda}{3} = \frac{5\lambda}{12}$.

In this case $I(P_1) = 0$.

Hence, (i) → (p,s), (ii) → (q), (iii) → (t), (iv) → (r,s,t)

28. (5)

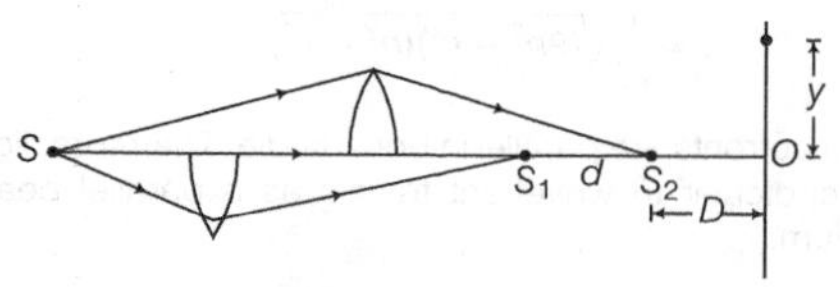

Here, $u = -40$ cm, $f = 20$ cm

$\because \quad \frac{1}{v_1} - \frac{1}{u} = \frac{1}{f}$

$\therefore \quad v_1 = 40$ cm

and $\quad \frac{1}{v_2} - \frac{1}{-41} = \frac{1}{20}$

$\therefore \quad v_2 = 39.0476$ cm

$\therefore \quad d = S_1S_2 = 39.0476 + 1 - 40 = 0.0476$ cm $= 0.476$ mm

Optical path difference at point P is

$= \Delta x = SS_2 + S_2P - (SS_1 + S_1P) = S_1S_2 + S_2P - S_1P$

$= d - d\left(1 - \frac{y^2}{2D^2}\right) = \frac{y^2 d}{2D^2} = m\lambda$

$\therefore \quad y = D\sqrt{\frac{2m\lambda}{d}}$

For first maxima, $m = 1$

$\therefore \quad y = D\sqrt{\frac{2\lambda}{d}} = 44.72$ mm $= \frac{2236}{5} \times 10^{-4}$ m

$\therefore \quad n = 5$

29. (5)

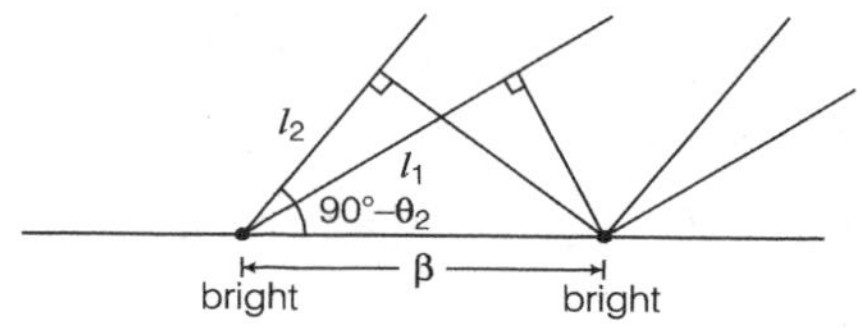

Here, $l_2 = \beta \sin\theta_2$ and $l_1 = \beta \sin\theta_1$

$\therefore \quad \Delta x = l_1 - l_2 = \lambda = \beta\sin\theta_1 - \beta\sin\theta_2 = \lambda$

$\therefore \quad \beta = \frac{\lambda}{\sin\theta_1 - \sin\theta_2} = \frac{\lambda}{\sin 53^\circ - \sin 37^\circ} = \frac{\lambda}{\frac{4}{5} - \frac{3}{5}} = 5\lambda$

30. (5) Here, $\delta = (\mu - 1)A = 0.001$

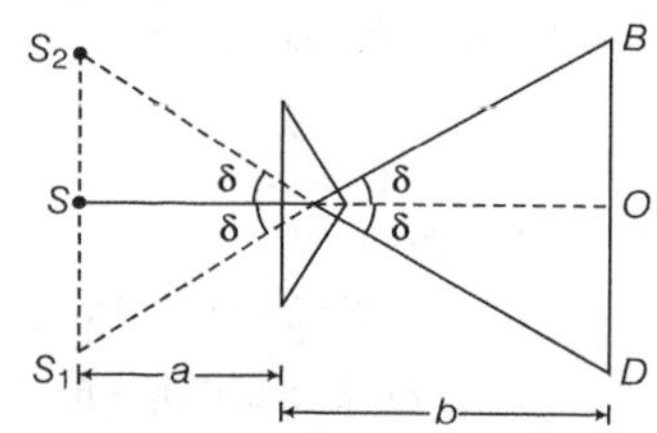

$\because \quad \tan\delta = \delta = \frac{d}{2a}$

$\therefore \quad d = 2a\delta = 2 \times 1 \times 0.001 = 0.002$ m

$\therefore \quad \beta = \frac{D}{d}\lambda = \left(\frac{a+b}{a}\right)\lambda = 15 \times 10^{-4}$ m

Here, $\quad \tan\delta = \delta = \frac{OB}{b}$

$\therefore \quad l = OB = b\delta = 4 \times 0.001 = 0.004$

$\therefore$ Number of bright fringe

$= 2\left[\frac{l}{\beta}\right] + 1 = 2\left[\frac{4 \times 10^{-3}}{15 \times 10^{-4}}\right] + 1 = 2\left[\frac{40}{15}\right] + 1 = 2\left[\frac{8}{3}\right] + 1$

$= 2 \times 2 + 1 = 5$

31. (4) Here, $\quad \Delta x = 2\mu t - \frac{\lambda}{2} = m\lambda$

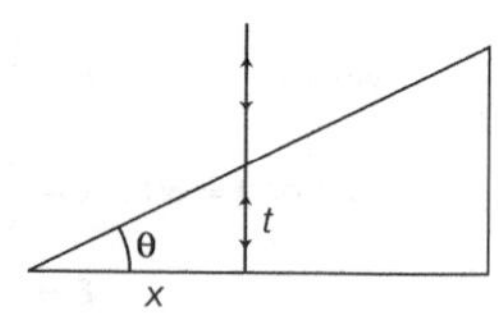

$\therefore \quad 2\mu t = (2m + 1)\frac{\lambda}{2}$

or $\quad 2\mu x\theta = (2m + 1)\frac{\lambda}{2}$

or $\quad x = (2m + 1)\frac{\lambda}{4\mu\theta}$

$\therefore \quad x_1 = \frac{\lambda}{4\mu\theta}$ and $x_2 = \frac{3\lambda}{4\mu\theta}$

$\therefore \quad \beta = x_2 - x_1 = \frac{\lambda}{2\mu\theta} = \frac{450 \times 10^{-9}}{2 \times 1.5 \times \frac{1}{20} \times \frac{\pi}{180^\circ}}$

$= 1.72 \times 10^{-6}$

$\therefore \quad n = 4$

32. (7)

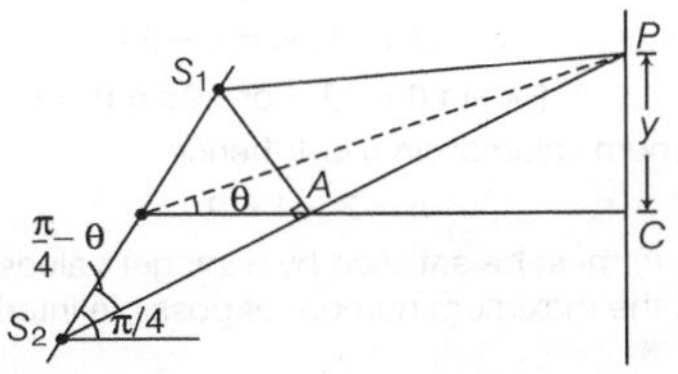

$\therefore \quad \Delta x_1 = S_2A = d\cos\left(\frac{\pi}{4} - \theta\right) = \frac{d}{\sqrt{2}}(\sin\theta + \cos\theta)$

$= \frac{d}{\sqrt{2}}(\theta + 1)$ $\quad$ ($\because \theta$ is very small)

$$\Delta x_2 = SS_1 - SS_2 = \text{constant} = b$$

$$\therefore \quad \Delta x = \Delta x_1 - \Delta x_2 = \frac{d}{\sqrt{2}}(1+\theta) - b$$

$\therefore$ For maxima, $\Delta x = m\lambda$

$\therefore \quad \frac{d}{\sqrt{2}}(1+\theta) - b = m\lambda$

But $\theta \simeq \frac{y}{D}$

$\therefore \quad \frac{y}{D} = (m\lambda + b)\frac{\sqrt{2}}{d} - 1$

$\therefore$ For first maxima, $\frac{y_1}{D} = (\lambda + b)\frac{\sqrt{2}}{d} - 1$

For second maxima,

$$\frac{y_2}{D} = (2\lambda + b)\frac{\sqrt{2}}{d} - 1$$

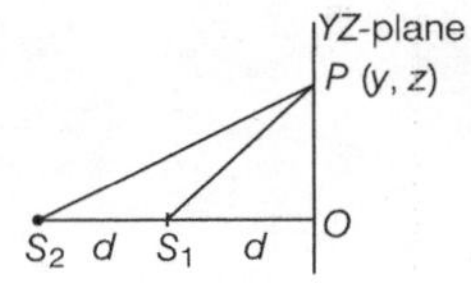

$\therefore \quad \beta = y_2 - y_1 = \frac{D}{d}(\sqrt{2}\lambda)$

$= 0.7 \times 10^{-3}$ m

$= 7 \times 10^{-4}$ m

JEE Main & AIEEE Archive

1. (d) By law of Malus, i.e. $I = I_0 \cos^2\theta$

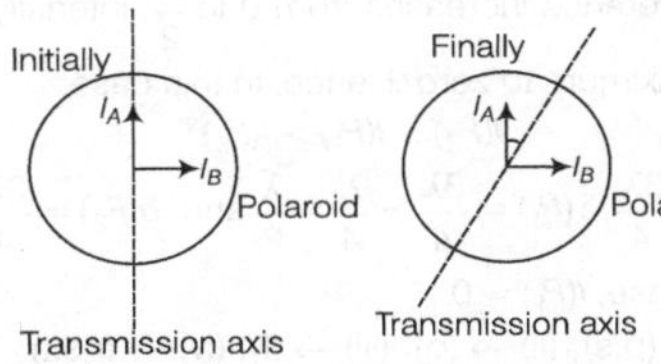

Now, $I_{A'} = I_A \cos^2 30°, I_{B'} = I_B \cos^2 60°$

As, $I'_A = I'_B$

$$I_A \cos^2 30° = I_B \cos^2 60°$$

$\Rightarrow \quad I_A \frac{3}{4} = I_B \frac{1}{4} \Rightarrow \frac{I_A}{I_B} = \frac{1}{3}$

2. (c) Relation between intensities is

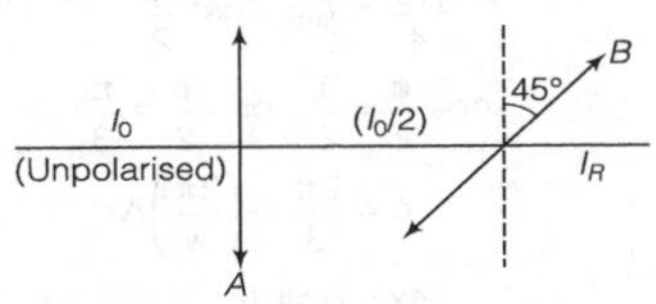

$$I_R = \left(\frac{I_0}{2}\right)\cos^2(45°) = \frac{I_0}{2} \times \frac{1}{2} = \frac{I_0}{4}$$

3. (d)

For bright fringe, $S_1P - S_2P = n\lambda$

So, fringes are concentric circles (centre of origin).

[$\because$ due to point source, wavefronts will be spherical]

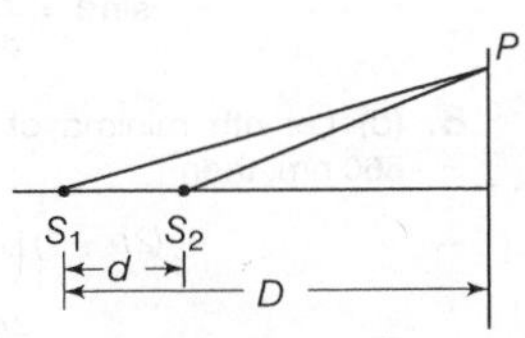

4. (d) Given, $a_1 = 2a_2 \Rightarrow I_1 = 4I_2 = 4I_0$ $\quad [\because I \propto a^2]$

$\therefore \quad I_{max} = (\sqrt{I_1} + \sqrt{I_2})^2 = (3\sqrt{I_2})^2 = 9I_2 = 9I_0$

Now, $\quad I = I_1 + I_2 + 2\sqrt{I_1 I_2}\cos\phi$

$= 4I_0 + I_0 + 2\sqrt{4I_0 \cdot I_0}\cos\phi = 5I_0 + 4I_0\cos\phi$

$= \frac{I_m}{9}(5 + 4\cos\phi)$ $\quad \left[\because I_0 = \frac{I_m}{9}\right]$

$= \frac{I_m}{9}[1 + 4(1 + \cos\phi)]$

$= \frac{I_m}{9}(1 + 8\cos^2\phi/2)$ $\quad \left[\because \cos\phi = 2\cos^2\frac{\phi}{2} - 1\right]$

5. (b) Let I_0 be intensity of light emitted from the source, then resultant intensity, $I = 4I_0\cos^2\phi/2$

$\therefore \quad I_1 = 4I_0\cos^2 0/2 = 4I_0$ $\quad [\because \phi = 0]$

Now, $\quad \Delta x = \lambda/4$

$\therefore \quad \phi = \frac{2\pi}{\lambda} \times \Delta x = \frac{2\pi}{\lambda} \times \frac{\lambda}{4} \Rightarrow \phi = \frac{\pi}{2}$

and $\quad I_2 = 4I_0\cos^2\frac{\pi}{4} = 2I_0$ $\quad \left[\because \phi = \frac{2\pi}{\lambda}\Delta x\right]$

$\therefore \quad I_1 : I_2 = 2 : 1$

6. (c) We know, apparent shift, $\Delta h = \left(1 - \frac{1}{\mu}\right)h$

$\therefore$ Apparent shift produced by water, $\Delta h_1 = \left(1 - \frac{1}{\mu_1}\right)h_1$

and apparent shift produced by kerosene, $\Delta h_2 = \left(1 - \frac{1}{\mu_2}\right)h_2$

$\therefore \quad \Delta h = \Delta h_1 + \Delta h_2 = \left(1 - \frac{1}{\mu_1}\right)h_1 + \left(1 - \frac{1}{\mu_2}\right)h_2$

7. (c) Scattering for blue light is largest and it is polarised by scattering.

Also, for polarised light, $I = I_0\cos^2\theta$

8. (b) For coherent sources, $I_1 = 4I_0\cos^2\frac{\phi}{2} = 4I_0$

For incoherent sources, $I_2 = I_0 + I_0 = 2I_0$

$\therefore \quad \frac{I_1}{I_2} = 2$

9. (b) Both statements are correct but Statement II does not explain Statement I.

10. (c) $\quad 3\lambda_1 = 4\lambda_2$ $\quad \left[\because \frac{3D\lambda_1}{d} = 4\frac{D\lambda_2}{d}\right]$

$\Rightarrow \quad \lambda_2 = \frac{3}{4}\lambda_1 = \frac{3}{4} \times 590 = \frac{1770}{4} = 442.5$ nm

11. (d) Phase difference $= \frac{2\pi}{\lambda} \times$ Path difference

i.e. $\quad \phi = \frac{2\pi}{\lambda} \times \frac{\lambda}{6} = \frac{\pi}{3}$

As, $\quad I = I_{max}\cos^2\left(\frac{\phi}{2}\right) \Rightarrow \frac{I}{I_{max}} = \cos^2\left(\frac{\phi}{2}\right)$

$\Rightarrow \quad \frac{I}{I_0} = \cos^2\left(\frac{\pi}{6}\right) = \frac{3}{4}$ $\quad [\because I_{max} = I_0]$

12. (d) The shape of the interference fringes formed on a screen is parabola.

13. (a) $I = I_0 \cos^2 \theta$

Intensity of polarised light $= I_0/2$

$\therefore$ Intensity of untransmitted light $= I_0 - \frac{I_0}{2} = \frac{I_0}{2}$

14. (b) For possible interference maxima on the screen, the condition is

$$d \sin \theta = n\lambda \quad \text{...(i)}$$

Given, d = slit-width $= 2\lambda$

$\therefore$ $2\lambda \sin \theta = n\lambda$ or $2 \sin \theta = n$

The maximum value of $\sin \theta$ is 1, hence

$$n = 2 \times 1 = 2$$

Thus, Eq. (i) must be satisfied by 5 integer values i.e. $-2, -1, 0, 1, 2$. Hence, the maximum number of possible interference maxima is 5.

15. (d) Sustained interference is possible with coherent sources only i.e. sources having same frequency and constant or zero phase difference.

JEE Advanced & IIT JEE Archive

1. (a,b,c) Fringe width, $\beta = \frac{\lambda D}{d}$ or $\beta \propto \lambda$

$\therefore$ $\lambda_2 > \lambda_1$

So $\beta_2 > \beta_1$

Number of fringes in a given width, $m = \frac{y}{\beta}$ or $m \propto \frac{1}{\beta}$

$\Rightarrow$ $m_2 < m_1$ as $\beta_2 > \beta_1$

Distance of 3rd maximum of λ_2 from central maximum

$$= \frac{3\lambda_2 D}{d} = \frac{1800D}{d}$$

Distance of 5th minimum of λ_1 from central maximum

$$= \frac{9\lambda_1 D}{2d} = \frac{1800D}{d}$$

So, 3rd maximum of λ_2 will overlap with 5th minimum of λ_1

Angular separation (or angular fringe width) $= \frac{\lambda}{d} \propto \lambda$

$\Rightarrow$ Angular separation for λ_1 will be lesser.

2. (d) $2d \sin\theta = \lambda \Rightarrow d = \frac{\lambda}{2\sin\theta}$

Differentiate, $\partial(d) = \frac{\lambda}{2} \partial(\text{cosec}\theta)$

$$\partial(d) = \frac{\lambda}{2}(-\text{cosec } \theta \cot \theta)\, \partial\theta$$

$$\partial(d) = \frac{-\lambda\cos\theta}{2\sin^2\theta}$$

As θ = increases, $\frac{\lambda\cos\theta}{2\sin^2\theta}$ decreases.

3. (b) $I = I_{max} \cos^2 \frac{\phi}{2}$ ...(i)

Given, $I = \frac{I_{max}}{2}$ ...(ii)

From Eqs. (i) and (ii), we have

$$\phi = \frac{\pi}{2}, \frac{3\pi}{2}, \frac{5\pi}{2}$$

Or path difference, $\Delta x = \left(\frac{\lambda}{2\pi}\right) \cdot \phi$

$\therefore$ $\Delta x = \frac{\lambda}{4}, \frac{3\lambda}{4}, \frac{5\lambda}{4} \ldots \left(\frac{2n+1}{4}\right)\lambda$

4. (b) $\Delta d = \Delta l = \frac{0.5}{100}$ mm $= 0.005$ mm

$$Y = \frac{4MLg}{\pi l d^2}$$

$\therefore$ $\left(\frac{\Delta Y}{Y}\right)_{max} = \left(\frac{\Delta l}{l}\right) + 2\left(\frac{\Delta d}{d}\right)$

$$\left(\frac{\Delta l}{l}\right) = \frac{0.5/100}{0.25} = 0.02$$

and $\frac{2\Delta d}{d} = \frac{(2)(0.5/100)}{0.5} = 0.02$ or $\frac{\Delta l}{l} = 2 \cdot \frac{\Delta d}{d}$

5. (a) (i) Intensity at P_0 is maximum. It will continuously decrease from P_0 towards P_2.

(ii) Path difference due to slap will be compensated by geometrical path difference. Hence, $\delta(P_1) = 0$.

(iii) $\delta(P_0) = \frac{\lambda}{2}$, $\delta(P_1) = \frac{\lambda}{2} - \frac{\lambda}{4} = \frac{\lambda}{4}$ and $\delta(P_2) = \frac{\lambda}{2} - \frac{\lambda}{3} = \frac{\lambda}{6}$. When path difference increases from 0 to $\frac{\lambda}{2}$, intensity will decrease from maximum to zero. Hence, in this case,

$$I(P_2) > I(P_1) > I(P_0)$$

(iv) $\delta(P_0) = \frac{3\lambda}{4}$, $\delta(P_1) = \frac{3\lambda}{4} - \frac{\lambda}{4} = \frac{\lambda}{2}$ and $\delta(P_2) = \frac{3\lambda}{4} - \frac{\lambda}{3} = \frac{5\lambda}{12}$.

In this case, $I(P_1) = 0$.

Hence, (i)$\rightarrow$ (p,s), (ii) $\rightarrow$ (q), (iii) $\rightarrow$ (t), (iv) $\rightarrow$ (r,s,t)

6. (d) Fringe width, $\beta = \frac{\lambda D}{d}$ or $\beta \propto \lambda$

Now, $\lambda_R > \lambda_G > \lambda_B$

$\therefore$ $\beta_R > \beta_G > \beta_B$

7. (c) $I = I_{max} \cos^2\left(\frac{\phi}{2}\right)$

$\therefore$ $\frac{I_{max}}{4} = I_{max} \cos^2 \frac{\phi}{2}$

$$\cos\frac{\phi}{2} = \frac{1}{2} \quad \text{or} \quad \frac{\phi}{2} = \frac{\pi}{3}$$

$\therefore$ $\phi = \frac{2\pi}{3} = \left(\frac{2\pi}{\lambda}\right)\Delta x$ ...(i)

where, $\Delta x = d\sin\theta$

On substituting in Eq. (i), we get

$$\sin\theta = \frac{\lambda}{3d} \quad \text{or} \quad \theta = \sin^{-1}\left(\frac{\lambda}{3d}\right)$$

8. (d) Let nth minima of 400 nm coincides with mth minima of 560 nm, then

$$(2n-1)\left(\frac{400}{2}\right) = (2m-1)\left(\frac{560}{2}\right)$$

or $\frac{2n-1}{2m-1} = \frac{7}{5} = \frac{14}{10} = \ldots$

i.e. 4th minima of 400 nm coincides with 3rd minima of 560 nm. Location of this minima is,

$$Y_1 = \frac{(2 \times 4 - 1)(1000)(400 \times 10^{-6})}{2 \times 0.1} = 14 \text{ mm}$$

Next 11th minima of 400 nm will coincide with 8th minima of 560 nm. Location of this minima is,

$$Y_2 = \frac{(2 \times 11 - 1)(1000)(400 \times 10^{-6})}{2 \times 0.1} = 42 \text{ mm}$$

$\therefore$ Required distance $= Y_2 - Y_1 = 28$ mm

Last 3 Years' Questions
JEE Main & Advanced
(2016-2018)

JEE Main

2016

1. An observer looks at a distant tree of height 10 m with a telescope of magnifying power of 20. To observer the tree appears **[Geometrical Optics]**

(a) 10 times taller (b) 10 times nearer
(c) 20 times taller (d) 20 times nearer

2. The box of a pin hole camera of length L, has a hole of radius a. It is assumed that when the hole is illuminated by a parallel beam of light of wavelength λ the spread of the spot (obtained on the opposite wall of the camera) is the sum of its geometrical spread and the spread due to diffraction. The spot would then have its minimum size (say b_{min}) when **[Geometrical Optics]**

(a) $a = \frac{\lambda^2}{L}$ and $b_{min} = \left(\frac{2\lambda^2}{L}\right)$ (b) $a = \sqrt{\lambda L}$ and $b_{min} = \left(\frac{2\lambda^2}{L}\right)$

(c) $a = \sqrt{\lambda L}$ and $b_{min} = \sqrt{4\lambda L}$ (d) $a = \frac{\lambda^2}{L}$ and $b_{min} = \sqrt{4\lambda L}$

3. In an experiment for determination of refractive index of glass of a prism by i-δ plot, it was found that a ray incident at an angle 35° suffers a deviation of 40° and that it emerges at an angle 79°. In that case, which of the following is closest to the maximum possible value of the refractive index? **[Geometrical Optics]**

(a) 1.5 (b) 1.6 (c) 1.7 (d) 1.8

2017

4. In a Young's double slit experiment, slits are separated by 0.5 mm and the screen is placed 150 cm away. A beam of light consisting of two wavelengths, 650 nm and 520 nm, is used to obtain interference fringes on the screen. The least distance from the common central maximum to the point where the bright fringes due to both the wavelengths coincide, is **[Wave Optics]**

(a) 7.8 mm (b) 9.75 mm
(c) 15.6 mm (d) 1.56 mm

5. A diverging lens with magnitude of focal length 25 cm is placed at a distance of 15 cm from a converging lens of magnitude of focal length 20 cm. A beam of parallel light falls on the diverging lens. The final image formed is **[Geometrical Optics]**

(a) virtual and at a distance of 40 cm from convergent lens
(b) real and at a distance of 40 cm from the divergent lens
(c) real and at a distance of 6 cm from the convergent lens
(d) real and at a distance of 40 cm from convergent lens

2018

6. Unpolarised light of intensity I passes through an ideal polariser A. Another identical polariser B is placed behind A. The intensity of light beyond B is found to be $\frac{I}{2}$. Now, another identical polariser C is placed between A and B. The intensity beyond B is now found to be $\frac{1}{8}$. The angle between polariser A and C is **[Wave Optics]**

(a) 0° (b) 30° (c) 45° (d) 60°

7. The angular width of the central maximum in a single slit diffraction pattern is 60°. The width of the slit is 1 μm. The slit is illuminated by monochromatic plane waves. If another slit of same width is made near it, Young's fringes can be observed on a screen placed at a distance 50 cm from the slits. If the observed fringe width is 1 cm, what is slit separation distance? (i.e. distance between the centres of each slit.) **[Wave Optics]**

(a) 25 μm (b) 50 μm
(c) 75 μm (d) 100 μm

JEE Advanced

2016

1. A parallel beam of light is incident from air at an angle α on the side PQ of a right angled triangular prism of refractive index $n = \sqrt{2}$. Light undergoes total internal reflection in the prism at the face PR when α has a minimum value of 45°. The angle θ of the prism is **[Geometrical Optics, Type 1]**

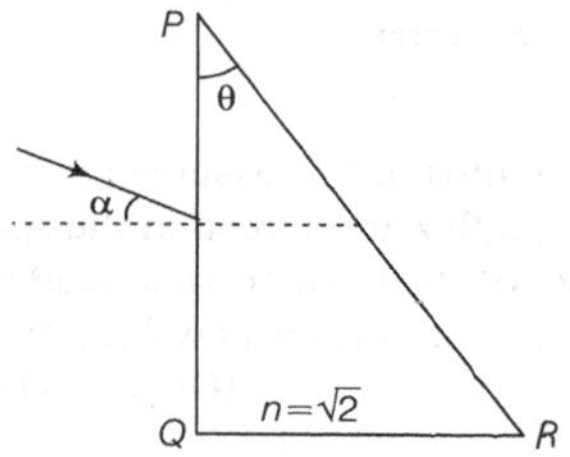

(a) 15° (b) 22.5°
(c) 30° (d) 45°

2. A plano-convex lens is made of material of refractive index n. When a small object is placed 30 cm away in front of the curved surface of the lens, an image of double the size of the object is produced. Due to reflection from the convex surface of the lens, another faint image is observed at a distance of 10 cm away from the lens. Which of the following statement(s) is (are) true? **[Geometrical Optics, Type 2]**

(a) The refractive index of the lens is 2.5
(b) The radius of curvature of the convex surface is 45 cm
(c) The faint image is erect and real
(d) The focal length of the lens is 20 cm

3. A transparent slab of thickness d has a refractive index $n(z)$ that increases with z. Here, z is the vertical distance inside the slab, measured from the top. The slab is placed between two media with uniform refractive indices n_1 and $n_2 (> n_1)$, as shown in the figure. A ray of light is incident with angle θ_i from medium 1 and emerges in medium 2 with refraction angle θ_f with a lateral displacement l.

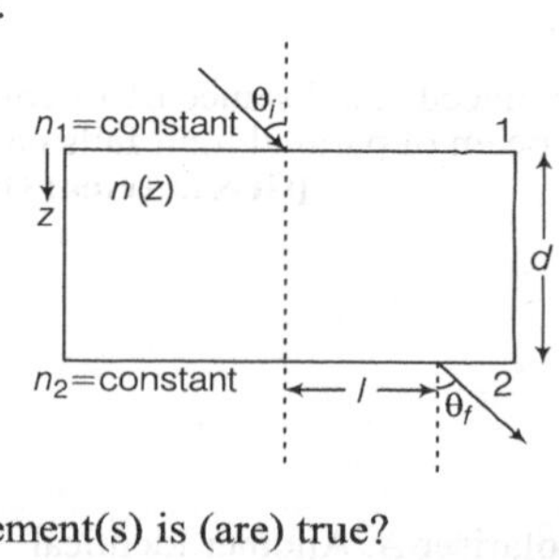

Which of the following statement(s) is (are) true? **[Geometrical Optics, Type 2]**

(a) l is independent on $n(z)$ (b) $n_1 \sin\theta_i = (n_2 - n_1)\sin\theta_f$
(c) $n_1 \sin\theta_i = n_2 \sin\theta_f$ (d) l is independent of n_2

Note *Type-1* *Only one correct option,* *Type-2* *More then one correct option,* *Type-3* *Single digit Answer,* *Type-4* *Numerical value based*

4. A small object is placed 50 cm to the left of a thin convex lens of focal length 30 cm. A convex spherical mirror of radius of curvature 100 cm is placed to the right of the lens at a distance of 50 cm. The mirror is tilted such that the axis of the mirror is at an angle $\theta = 30°$ to the axis of the lens, as shown in the figure.

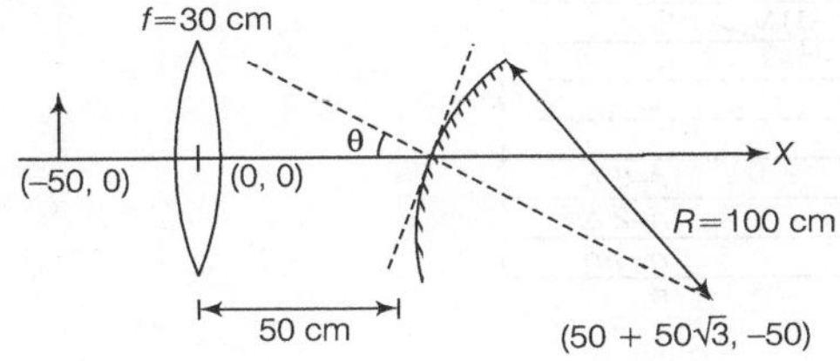

If the origin of the coordinate system is taken to be at the centre of the lens, the coordinates (in cm) of the point (x, y) at which the image is formed are **[Geometrical Optics, Type 1]**

(a) $(125/3, 25/\sqrt{3})$ (b) $(50 - 25\sqrt{3}, 25)$

(c) (0, 0) (d) $(25, 25\sqrt{3})$

5. While conducting the Young's double slit experiment, a student replaced the two slits with a large opaque plate in the *XY*-plane containing two small holes that act as two coherent point sources (S_1, S_2) emitting light of wavelength 600 mm. The student mistakenly placed the screen parallel to the *XZ*-plane (for $z > 0$) at a distance $D = 3$ m from the mid-point of S_1S_2, as shown schematically in the figure. The distance between the source $d = 0.6003$ mm. The origin O is at the intersection of the screen and the line joining S_1S_2.

[Wave Optics, Type 2]

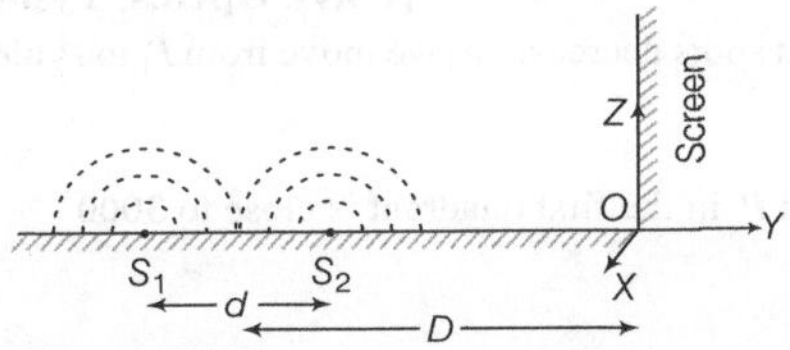

Which of the following is (are) true of the intensity pattern on the screen?

(a) Semi circular bright and dark bands centered at point O
(b) The region very close to the point O will be dark
(c) Straight bright and dark bands parallel to the X-axis
(d) Hyperbolic bright and dark bands with foci symmetrically placed about O in the x-direction

2017

6. For an isosceles prism of angle A and refractive index μ, it is found that the angle of minimum deviation $\delta_m = A$. Which of the following options is/are correct?

[Geometrical Optics, Type 2]

(a) For the angle of incidence $i_1 = A$, the ray inside the prism is parallel to the base of the prism
(b) At minimum deviation, the incident angle i_1 and the refracting angle r_1 at the first refracting surface are related by $r_1 = \left(\frac{i_i}{2}\right)$
(c) For this prism, the emergent ray at the second surface will be tangential to the surface when the angle of incidence at the first surface is $i_1 = \sin^{-1}\left[\sin A\sqrt{4\cos^2\frac{A}{2} - 1} - \cos A\right]$
(d) For this prism, the refractive index μ and the angle prism A are related as $A = \frac{1}{2}\cos^{-1}\left(\frac{\mu}{2}\right)$

7. A monochromatic light is travelling in a medium of refractive index $n = 1.6$. It enters a stack of glass layers from the bottom side at an angle $\theta = 30°$. The interfaces of the glass layers are parallel to each other. The refractive indices of different glass layers are monotonically decreasing as $n_m = n - m\Delta n$, where n_m is the refractive index of the mth slab and

$\Delta n = 0.1$ (see the figure). The ray is refracted out parallel to the interface between the $(m-1)$th and mth slabs from the right side of the stack. What is the value of m?

[Geometrical Optics, Type 3]

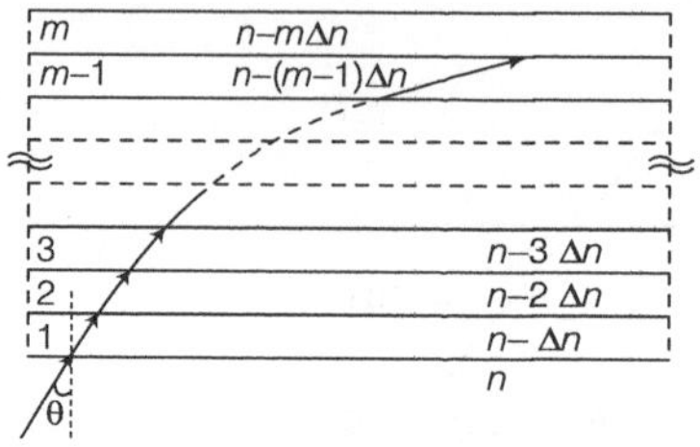

8. Two coherent monochromatic point sources S_1 and S_2 of wavelength $\lambda = 600\,\text{nm}$ are placed symmetrically on either side of the centre of the circle as shown. The sources are separated by a distance $d = 1.8\,\text{mm}$. This arrangement produces interference fringes visible as alternate bright and dark spots on the circumference of the circle. The angular separation between two consecutive bright spots is $\Delta\theta$. Which of the following options is/are correct?

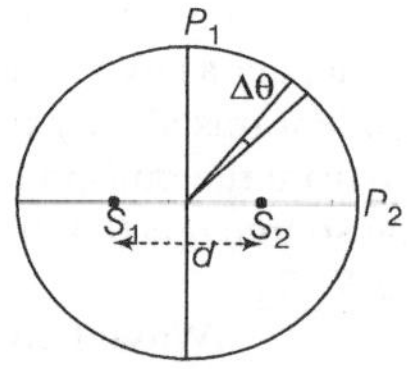

[Wave Optics, Type 2]

(a) The angular separation between two consecutive bright spots decreases as we move from P_1 to P_2 along the first quadrant

(b) A dark spot will be formed at the point P_2

(c) The total number of fringes produced between P_1 and P_2 in the first quadrant is close to 3000

(d) At P_2 the order of the fringe will be maximum

2018

9. Sunlight of intensity $1.3\,\text{kWm}^{-2}$ is incident normally on a thin convex lens of focal length 20 cm. Ignore the energy loss of light due to the lens and assume that the lens aperture size is much smaller than its focal length. The average intensity of light, in kW m^{-2}, at a distance 22 cm from the lens on the other side is .... **[Geometrical Optics, Type 4]**

10. A wire is bent in the shape of a right angled triangle and is placed in front of a concave mirror of focal length f as shown in the figure. Which of the figures shown in the four options qualitatively represent(s) the shape of the image of the bent wire? (These figures are not to scale.) **[Wave Optics, Type 2]**

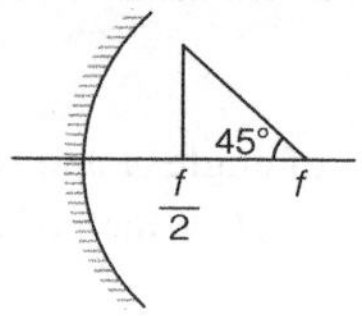

(a) 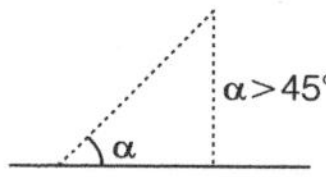

(b) 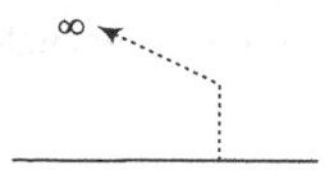

(c) 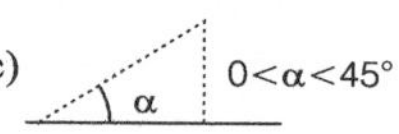

(d) 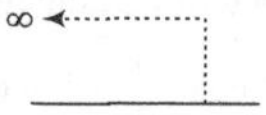

Answer *with* Explanations

JEE Main

1. *(d)* Telescope resolves and brings objects closer. Hence, telescope with magnifying power of 20, the tree appears 20 times nearer.

2. *(c)*

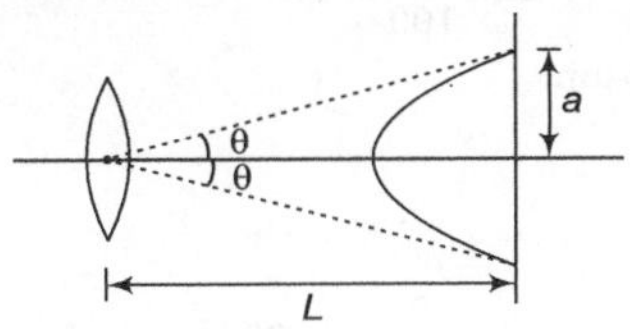

$$\because \quad a\sin\theta \approx \lambda$$

$$\Rightarrow \quad a\left(\frac{a}{L}\right) \approx \lambda$$

$$\Rightarrow \quad a = \sqrt{\lambda L}$$

$$\therefore \quad \text{Spread} = 2a = \sqrt{4\lambda L}$$

3. *(a)* $\delta = (i_1 + i_2) - A$

$$\Rightarrow \quad 40^\circ = (35^\circ + 79^\circ) - A$$

$$\Rightarrow \quad A = 74^\circ$$

Now, we know that $\mu = \dfrac{\sin\left(\dfrac{A+\delta_m}{2}\right)}{\sin\left(\dfrac{A}{2}\right)}$

It we take the given deviation as the minimum deviation then,

$$\mu = \frac{\sin\left(\dfrac{74^\circ + 40^\circ}{2}\right)}{\sin\left(\dfrac{74^\circ}{2}\right)} = 1.51$$

The given deviation may or may not be the minimum deviation. Rather it will be less than this value. Therefore, μ will be less than 1.51.

Hence, maximum possible value of refractive index is 1.51.

4. *(a)* Let n_1th fringe formed due to first wavelength and n_2th fringe formed due to second wavelength coincide i.e. their distance from common central maxima will be same

i.e. $\quad y_{n_1} = y_{n_2}$

$$\Rightarrow \quad \frac{n_1\lambda_1 D}{d} = \frac{n_2\lambda_2 D}{d}$$

$$\Rightarrow \quad \frac{n_1}{n_2} = \frac{\lambda_1}{\lambda_2} = \frac{520}{650} = \frac{4}{5}$$

Hence, distance of the point of coincidence from the central maxima is

$$y = \frac{n_1\lambda_1 D}{d} = \frac{n_2\lambda_2 D}{d}$$

$$= \frac{4 \times 650 \times 10^{-9} \times 1.5}{0.5 \times 10^{-3}}$$

$$= 7.8\text{ mm}$$

5. *(d)* Focal length of diverging lens is 25 cm.

As the rays are coming parallel, so the image (I_1) will be formed at the focus of diverging lens i.e. at 25 cm towards left of diverging lens.

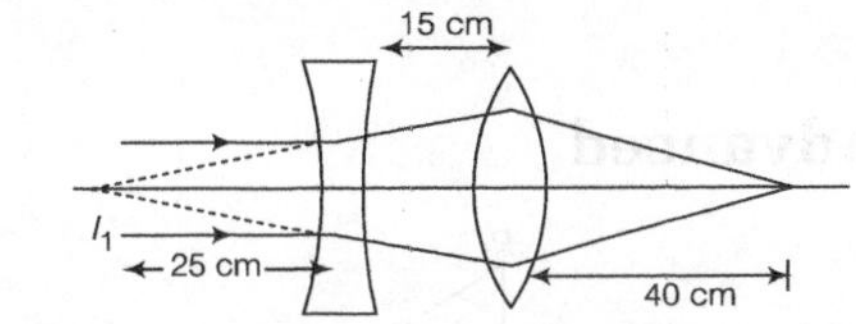

Now, the image (I_1) will work as object for converging lens.

For convergying lens, distance of object u

(i.e. distance of I_1) $= -(25 + 15) = -40$ cm

$$f = 20\text{ cm}$$

$\therefore$ From len's formula $\dfrac{1}{f} = \dfrac{1}{v} - \dfrac{1}{u}$

$$\frac{1}{20} = \frac{1}{v} - \frac{1}{-40}$$

$$\Rightarrow \quad \frac{1}{v} = \frac{1}{20} - \frac{1}{40}$$

$$\frac{1}{v} = \frac{1}{90}$$

$$\Rightarrow \quad v = 40\text{ cm}$$

v is positive so image will be real and will form at right side of converging lens at 40 cm.

6. (c)

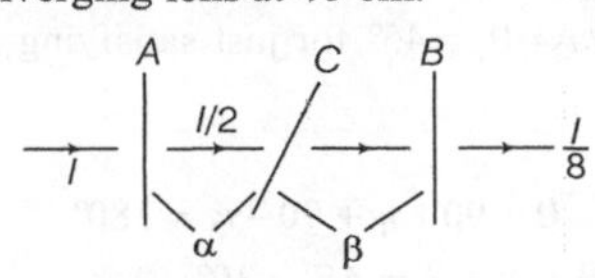

Using Malus law, intensity available after $C = \dfrac{I}{2} \times \cos^2\alpha$

and intensity available after

$$B = \frac{I}{2}\cos^2\alpha \times \cos^2\beta = \frac{I}{8} \qquad \text{(given)}$$

$$\text{So,} \quad \frac{I}{2} \times \cos^2\alpha \cdot \cos^2\beta = \frac{I}{8}$$

$$\Rightarrow \quad \cos^2\alpha \cdot \cos^2\beta = \frac{1}{4}$$

This is satisfied with $\alpha = 45^\circ$ and $\beta = 45^\circ$

So, angle between A and C is 45°.

7. (a) Angular width of diffraction pattern $= 60^\circ$

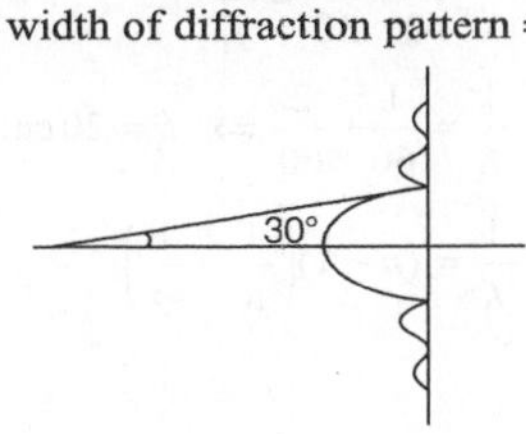

For first minima,

$$\frac{a}{2}\sin\theta = \frac{\lambda}{2}, \quad [\text{here, } a = 10^{-6}\text{ m}, \theta = 30°]$$

$$\Rightarrow \quad \lambda = 10^{-6} \times \sin 30°$$

$$\Rightarrow \quad \lambda = \frac{10^{-6}}{2}\text{m}$$

Now, in case of interference caused by bringing second slit,

$\therefore$ Fringe width, $\beta = \frac{\lambda D}{d}$

$$\left[\text{here, } \lambda = \frac{10^{-6}}{2}\text{ m}, \beta = 1\text{ cm} = \frac{1}{100}\text{ m}, \right.$$

$$\left. d = ? \text{ and } D = 50\text{ cm} = \frac{50}{100}\text{ m}\right]$$

$$\text{So, } d = \frac{\lambda D}{\beta} = \frac{10^{-6} \times 50}{2 \times \frac{1}{100} \times 100} = 25 \times 10^{-6}\text{ m}$$

or $\quad d = 25\ \mu\text{m}$

JEE Advanced

1. *(a)*

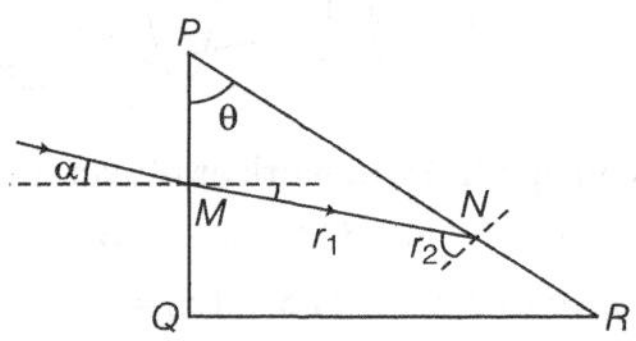

Applying Snell's law at M,

$$n = \frac{\sin\alpha}{\sin r_1}$$

$$\Rightarrow \quad \sqrt{2} = \frac{\sin 45°}{\sin r_1}$$

$$\Rightarrow \quad \sin r_1 = \frac{\sin 45°}{\sqrt{2}} = \frac{1/\sqrt{2}}{\sqrt{2}} = \frac{1}{2}$$

$$r_1 = 30°$$

$$\sin\theta_c = \frac{1}{n} = \frac{1}{\sqrt{2}}$$

$$\Rightarrow \quad \theta_c = 45°$$

Let us take $r_2 = \theta_c = 45°$ for just satisfying the condition of TIR.

In ΔPNM,

$$\theta + 90 + r_1 + 90 - r_2 = 180°$$

$$\text{or} \quad \theta = r_2 - r_1 = 45° - 30° = 15°$$

Note *If $\alpha > 45°$ (the given value). Then, $r_1 > 30°$ (the obtained value)*

$$r_2 > \theta_c \ (\text{as } r_2 - r_1 = \theta \text{ or } r_2 = \theta + r_1)$$

or TIR will take place. So, for taking TIR under all conditions α should be greater than 45° or this is the minimum value of α.

2. *(a,d)* **Case 1**

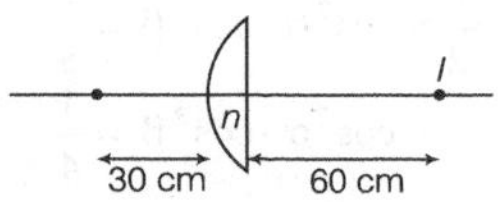

Using lens formula,

$$\frac{1}{60} + \frac{1}{30} = \frac{1}{f_1}$$

$$\Rightarrow \quad \frac{1}{f_1} = \frac{1}{60} + \frac{2}{60} \Rightarrow f_1 = 20\text{ cm}$$

$$\text{Further,} \quad \frac{1}{f_1} = (n-1)\left(\frac{1}{R} - \frac{1}{\infty}\right)$$

$$\Rightarrow \quad f_1 = \frac{R}{n-1} = +20\text{ cm}$$

Case 2

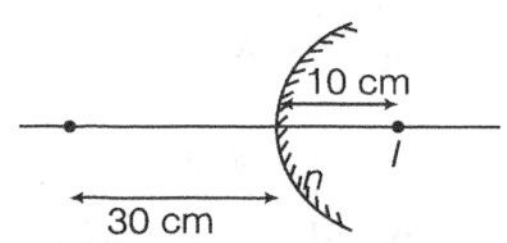

Using mirror formula, $\frac{1}{10} - \frac{1}{30} = \frac{1}{f_2}$

$$\frac{3}{30} - \frac{1}{30} = \frac{1}{f_2} = \frac{2}{30}$$

$$f_2 = 15 = \frac{R}{2}$$

$$\Rightarrow \quad R = 30$$

$$R = 30\text{ cm}$$

$$\frac{R}{n-1} = +20\text{ cm} = \frac{30}{n-1}$$

$$\Rightarrow \quad 2n - 2 = 3 \quad \Rightarrow \quad f_1 = +20\text{ cm}$$

Refractive index of lens is 2.5. Radius of curvature of convex surface is 30 cm. Faint image is erect and virtual. Focal length of lens is 20 cm.

3. *(a,c,d)* From Snell's law,

$$n \sin\theta = \text{constant}$$

$$\therefore \quad n_1 \sin\theta_i = n_2 \sin\theta_f$$

Further, l will depend on n_1 and $n(z)$. But it will be independent of n_2.

4. *(d)*

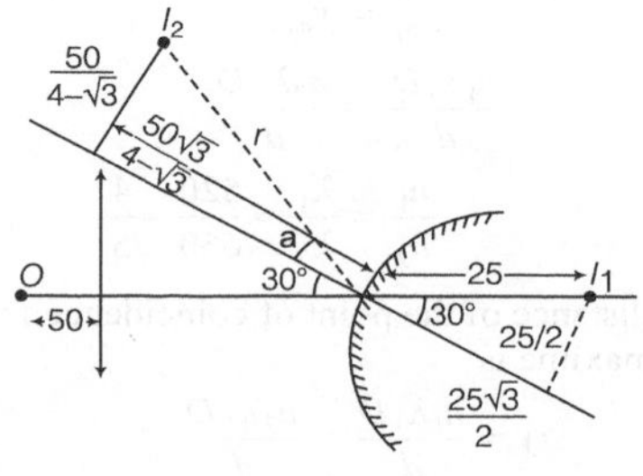

For Lens

$$\frac{1}{v} - \frac{1}{u} = \frac{1}{f}$$

$$\Rightarrow \qquad v = \frac{uf}{u+f}$$

$$\Rightarrow \qquad v = \frac{(-50)(30)}{-50+30} = 75 \text{ cm}$$

For Mirror

$$\frac{1}{v} + \frac{1}{u} = \frac{1}{f}$$

$$\Rightarrow \qquad v = \frac{uf}{u-f}$$

$$\Rightarrow \qquad v = \frac{\left(\frac{25\sqrt{3}}{2}\right)(50)}{\frac{25\sqrt{3}}{2} - 50} = \frac{-50\sqrt{3}}{4-\sqrt{3}} \text{ cm}$$

$$m = -\frac{v}{u} = \frac{h_2}{h_1}$$

$$\Rightarrow \qquad h_2 = -\left(\frac{\frac{-50\sqrt{3}}{4-\sqrt{3}}}{\frac{25\sqrt{3}}{2}}\right) \cdot \frac{25}{2}$$

$$\Rightarrow \qquad h_2 = \frac{+50}{4-\sqrt{3}}$$

The x-coordinate of the images

$$= 50 - v\cos 30 + h_2 \cos 60 \approx 25$$

The y-coordinate of the images

$$= v\sin 30 + h_2 \sin 60 \approx 25\sqrt{3}$$

5. *(a,b)*

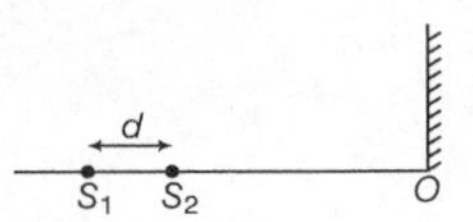

Path difference at point $O = d = 0.6003$ mm

$$= 600300 \text{ nm}.$$

This path difference is equal to $\left(1000\lambda + \frac{\lambda}{2}\right)$.

$\Rightarrow$ Minima is formed at point O.

Line S_1S_2 and screen are perpendicular to each other, so fringe pattern is circular (semi-circular because only half of screen is available).

6. *(a,b,c)* The minimum deviation produced by a prism

$$\delta_m = 2i - A = A$$

$\therefore \qquad i_1 = i_2 = A$ and $r_1 = r_2 = A/2$

$\therefore \qquad r_1 = i_1/2$

Now, using Snell's law

$$\sin A = \mu \sin A/2$$

$$\Rightarrow \qquad \mu = 2\cos(A/2)$$

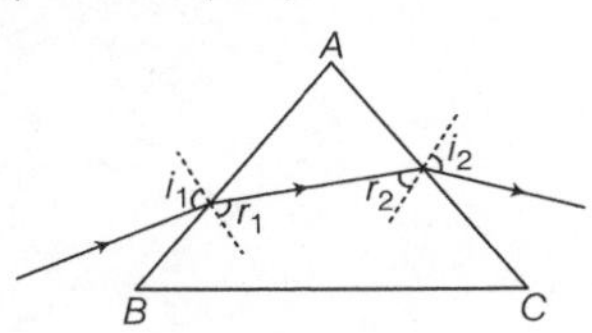

For this prism when the emergent ray at the second surface is tangential to the surface

$$i_2 = \pi/2$$

$\therefore \qquad r_2 = \theta_c$

$\therefore \qquad r_1 = A - \theta_c$

so, $\sin i_1 = \mu \sin(A - \theta_c)$

so, $\quad i_1 = \sin^{-1}\left[\sin A\sqrt{4\cos^2\frac{A}{2} - 1} - \cos A\right]$

For minimum deviation through isosceles prism, the ray inside the prism is parallel to the base of the prism if $\angle B = \angle C$.

But it is not necessarily parallel to the base if, $\angle A = \angle B$ or $\angle A = \angle C$

7. *(8)* But this value of refractive index is not possible.

$$1.6\sin\theta = (n - m\Delta n)\sin 90°$$

$$1.6\sin\theta = n - m\Delta n$$

$$1.6 \times \frac{1}{2} = 1.6 - m(0.1)$$

$$0.8 = 1.6 - m(0.1)$$

$$m \times 0.1 = 0.8$$

$$m = 8$$

8. *(c,d)*

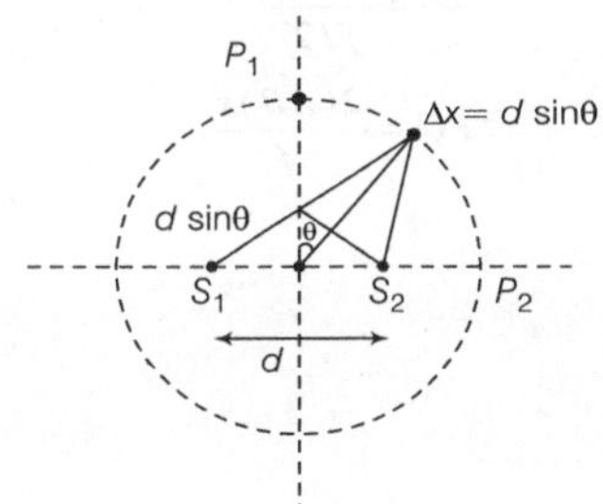

$$\lambda = 600 \text{ nm}$$

at $P_1 \qquad \Delta x = 0$

at $P_2 \qquad \Delta x = 1.8 \text{ mm} = n\lambda$

Number of maximas will be

$$n = \frac{\Delta x}{\lambda}$$

$$= \frac{1.8 \text{ mm}}{600 \text{ nm}} = 3000$$

at $P_2 \qquad \Delta x = 3000\lambda$

Hence, bright fringe will be formed.

At P_2, 3000th maxima is formed.

For (a) option

$$\Delta x = d\sin\theta$$

$$d\Delta x = d\cos d\theta$$

$$R\lambda = d\cos\theta R d\theta$$

$$R d\theta = \frac{R\lambda}{d\cos\theta}$$

As we move from P_1 to P_2

$$\theta \uparrow \cos\theta \downarrow Rd\theta \uparrow$$

9. (130.0 $\mathbf{kW/m^2}$)

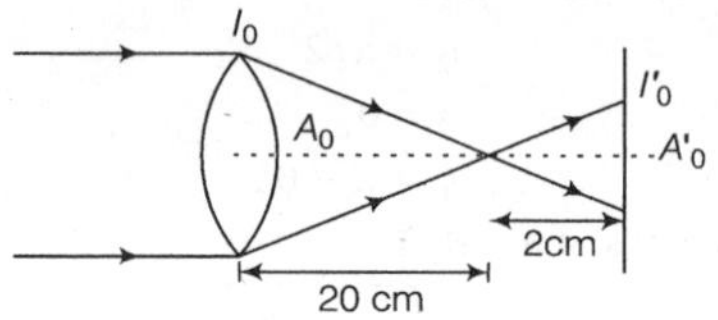

$$\frac{A_0'}{A_0}=\left(\frac{2}{20}\right)^2=\frac{1}{100}$$

$$\Rightarrow \quad A_0'=\frac{A_0}{100}$$

$$P=I_0A_0=I_0'A_0'$$

$$\Rightarrow \quad I_0'=\frac{I_0A_0}{\frac{A_0}{100}}=100I_0=130\,\text{kW/m}^2$$

10. (d) Image of point A

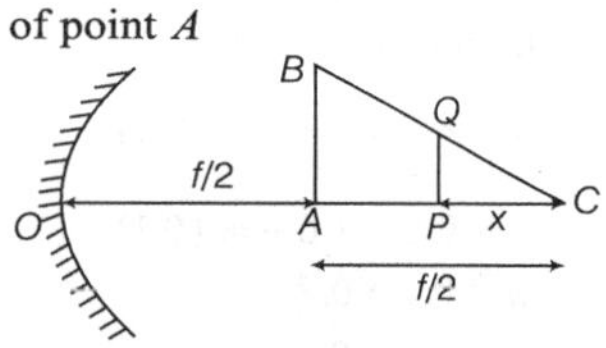

$$\frac{PQ}{x}=\frac{AB}{f/2}$$

$$\Rightarrow \quad PQ=\frac{2(AB)x}{f}$$

For A **:**

$$\frac{1}{v}+\frac{1}{[-(f/2)]}=\frac{1}{-f}$$

$$\Rightarrow \quad v=f$$

$$\Rightarrow \quad \frac{I_{AB}}{AB}=-\frac{v}{u}=-\frac{f}{\left(-\frac{f}{2}\right)}$$

$$\Rightarrow \quad I_{AB}=2\,AB$$

For height of PQ,

$$\frac{1}{v}+\frac{1}{-[(f-x)]}=\frac{1}{-f}$$

$$\Rightarrow \quad \frac{1}{v}=\frac{1}{(f-x)}-\frac{1}{f}$$

$$\Rightarrow \quad v=\frac{f(f-x)}{x}$$

$$\Rightarrow \quad \frac{I_{PQ}}{PQ}=-\frac{v}{u}=\frac{f(f-x)}{x[(f-x)]}=\left(\frac{f}{x}\right)$$

$$\Rightarrow \quad I_{PQ}=\frac{f}{x}PQ$$

$$=\left(\frac{f}{x}\right)\left(\frac{2\,(AB)x}{f}\right)\quad\left[\because PQ=\frac{2(AB)x}{f}\right]$$

$$I_{PQ}=2AB$$

(Size of image is independent of x. So, final image will be of same height terminating at infinity)